Mathematics for Electronics and Computers

Mathematics
for Electronics and
Computers

Nigel P. Cook

Prentice
Hall

Upper Saddle River, New Jersey
Columbus, Ohio

Library of Congress Cataloging-in-Publication Data

Cook, Nigel P.,
    Mathematics for electronics and computers / Nigel P. Cook.
        p. cm.
    Includes index.
    ISBN 0-13-081162-9
    1. Mathematics.   2. Electronics--Mathematics   3. Computer science--Mathematics   I.
Title.

QA39.3.C68 2003
510′.886213--dc21   2002074972

Editor in Chief: Stephen Helba
Executive Editor: Frank I. Mortimer, Jr.
Media Development Editor: Michelle Churma
Production Editor: Louise N. Sette
Production Supervision: Clarinda Publication Services
Design Coordinator: Diane Ernsberger
Cover Designer: Jason Moore
Cover art: Photonica
Production Manager: Brian Fox
Marketing Manager: Tim Peyton

This book was set by The Clarinda Company. It was printed and bound by Courier Kendallville, Inc.
The cover was printed by The Lehigh Press, Inc.

Pearson Education Ltd.
Pearson Education Australia Pty. Limited
Pearson Education Singapore Pte. Ltd.
Pearson Education North Asia Ltd.
Pearson Education Canada, Ltd.
Pearson Educación de Mexico, S.A. de C.V.
Pearson Education—Japan
Pearson Education Malaysia Pte. Ltd.
Pearson Education, *Upper Saddle River, New Jersey*

10 9 8 7 6 5 4 3 2 1
ISBN: 0-13-081162-9

To Dawn, Candy, and Jon

Books by Nigel P. Cook

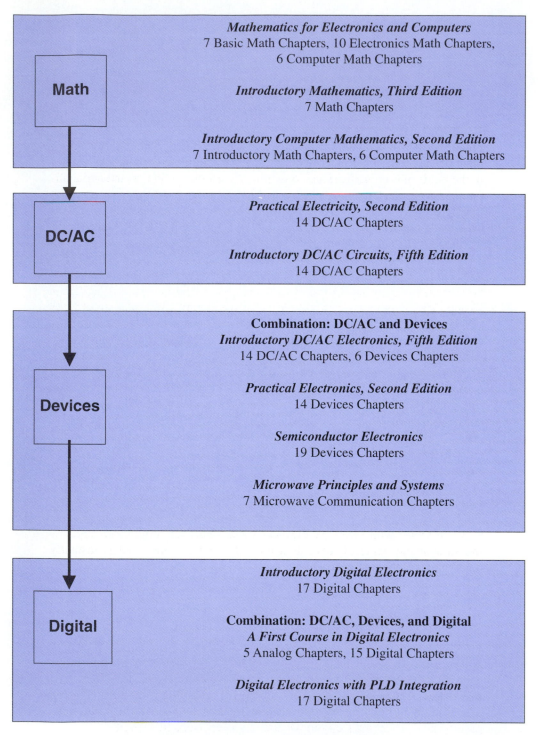

Math

Mathematics for Electronics and Computers
7 Basic Math Chapters, 10 Electronics Math Chapters,
6 Computer Math Chapters

Introductory Mathematics, Third Edition
7 Math Chapters

Introductory Computer Mathematics, Second Edition
7 Introductory Math Chapters, 6 Computer Math Chapters

DC/AC

Practical Electricity, Second Edition
14 DC/AC Chapters

Introductory DC/AC Circuits, Fifth Edition
14 DC/AC Chapters

Devices

Combination: DC/AC and Devices
Introductory DC/AC Electronics, Fifth Edition
14 DC/AC Chapters, 6 Devices Chapters

Practical Electronics, Second Edition
14 Devices Chapters

Semiconductor Electronics
19 Devices Chapters

Microwave Principles and Systems
7 Microwave Communication Chapters

Digital

Introductory Digital Electronics
17 Digital Chapters

Combination: DC/AC, Devices, and Digital
A First Course in Digital Electronics
5 Analog Chapters, 15 Digital Chapters

Digital Electronics with PLD Integration
17 Digital Chapters

For more information on any of the other textbooks by Nigel Cook, see his web page at
www.prenhall.com/cook or ask your local Prentice Hall representative.

Preface

INTRODUCTION

Since World War II no branch of science has contributed more to the development of the modern world as electronics. It has stimulated dramatic advances in the fields of communication, computing, consumer products, industrial automation, test and measurement, and health care. It has now become the largest single industry in the world, exceeding the automobile and oil industries, with annual sales of electronic systems greater than $2 trillion. One of the most important trends in this huge industry has been a gradual shift from *analog electronics* to *digital electronics*. This movement began in the 1960s and is almost complete today. In fact, a recent statistic stated that, on average, 90% of the circuitry within electronic systems is now digital, and only 10% is analog. This digitalization of the electronics industry is merging sectors that were once separate. For example, two of the largest sectors or branches of electronics are *computing* and *communications*. Being able to communicate with each other using the common language of digital has enabled computers and communications to interlink, so that computers can now function within communication-based networks, and communications networks can now function through computer-based systems. Industry experts call this merging *convergence* and predict that digital electronics will continue to unite the industry and stimulate progress in practically every field of human endeavor.

Mathematics is interwoven into the very core of science, and therefore an understanding of mathematics is imperative for anyone pursuing a career in technology. As with any topic to be learned, the method of presentation can make a big difference between clear comprehension and complete confusion. Employing an "integrated math applications" approach, this text reinforces all math topics with extensive electronic and computer applications to show the student the value of math as a tool; therefore, if the need is instantly demonstrated, the tool is retained.

OUTLINE

After 12 textbooks, 21 editions, and 19 years of front-line education experience, best-selling author Nigel Cook has written *Mathematics for Electronics and Computers* as a complete math course for technology students. To assist educators in curriculum chapter selection, the text has been divided into three parts to give it the versatility to adapt to a variety of course lengths.

Part A Basic Math

Chapter 1   Fractions
Chapter 2   Decimal Numbers
Chapter 3   Positive and Negative Numbers
Chapter 4   Exponents and the Metric System
Chapter 5   Algebra, Equations, and Formulas
Chapter 6   Geometry and Trigonometry
Chapter 7   Logarithms and Graphs

Part B Electronics Math

Chapter 8   Current and Voltage
Chapter 9   Resistance and Power
Chapter 10   Series DC Circuits
Chapter 11   Parallel DC Circuits
Chapter 12   Series–Parallel DC Circuits and Theorems
Chapter 13   Alternating Current (AC)
Chapter 14   Capacitors
Chapter 15   Inductors and Transformers
Chapter 16   *RLC* Circuits and Complex Numbers
Chapter 17   Diodes and Transistors

Part C Computer Math

Chapter 18   Analog to Digital
Chapter 19   Number Systems and Codes
Chapter 20   Logic Gates
Chapter 21   Boolean Expressions and Algebra
Chapter 22   Binary Arithmetic
Chapter 23   Introduction to Computer Programming

SUPPLEMENTS

- Instructor's Solutions Manual (ISBN 0-13-048778-3)
- Test Item File (ISBN 0-13-049025-3)
- PH Test Manager (ISBN 0-13-048779-1)
- Companion Website: http://www.prenhall.com/cook

■ **INTEGRATED MATH APPLICATION:**

In Figure 6-2(c), a 200-foot piece of string is stretched from the top of a 125-foot cliff to a point on the beach. What is the distance from this point to the cliff?

■ *Solution:*

In this example, B is unknown, and therefore

$$B = \sqrt{C^2 - A^2}$$
$$= \sqrt{200^2 - 125^2}$$
$$= \sqrt{40,000 - 15,625}$$
$$= \sqrt{24,375}$$
$$= 156 \text{ ft}$$

6-2-3   *Vectors and Vector Diagrams*

Vector or Phasor
A quantity that has magnitude and direction and that is commonly represented by a directed line segment whose length represents the magnitude and whose orientation in space represents the direction.

A **vector** or **phasor** is an arrow used to represent the magnitude and direction of a quantity. Vectors are generally used to represent a physical quantity that has two properties. For example, Figure 6-8(a) shows a motorboat heading north at 12 miles per hour. Figure 6-8(b) shows how vector **A** could be used to represent the vessel's direction and speed. The size of the vector represents the speed of 12 mph by being 12 cm long, and because we have made the top

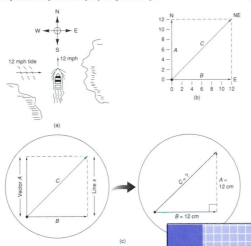

FIGURE 6-8   Vectors and Vector Diagrams.

146   CHAPTER 6 / GEOMETRY AND TRIGONOMETRY

Once into the pages of this book, the reader will quickly be put at ease by the **student-friendly writing style** and the author's ability to simplify traditionally difficult subjects.

Each chapter opens with a **humanistic vignette,** a brief story that describes the history of a mathematical discovery, written in terms of the people who made these discoveries.

PART A   **BASIC MATH**

Fractions

There's No Sleeping When He's Around!

Carl Friedrich Gauss was born April 30, 1777, to poor, uneducated parents in Brunswick, Germany. He was a child of precocious abilities, particularly in mental computation. In elementary school he soon impressed his teachers, who said that mathematical ability came easier to Gauss than speech.

In secondary school he rapidly distinguished himself in ancient languages and mathematics. At 14, Gauss was presented to the court of the duke of Brunswick, where he displayed his computing skill. Until his death in 1806, the duke generously supported Gauss and his family, encouraging the boy with textbooks and a laboratory.

In the early years of the nineteenth century, Gauss's interest was in astronomy, and his accumulated work on celestial mechanics was published in 1809. In 1828, at a conference in Berlin, Gauss met physicist Wilhelm Weber, who would eventually become famous for his work on electricity. They worked together for many years and became close friends, investigating electromagnetism and the use of a magnetic needle for the measurement of current. In 1833 they constructed an electric telegraph system that could communicate across Göttingen from Gauss's observatory to Weber's physics laboratory. (This telegraph system of communication was later developed independently by U.S. inventor Samuel Morse.)

Gauss conceived almost all of his fundamental mathematical discoveries between the ages of 14 and 17. There are many stories of his genius in his early years, one of which involved a sarcastic teacher who liked giving his students long-winded problems and then resting, or on some occasions sleeping, in class. On his first day with Gauss, who was 8 years old, the teacher began, as usual, by telling the students to find the sum of all the numbers from 1 to 100. The teacher barely had a chance to sit down before Gauss raised his hand and said "5050." The dumbfounded teacher, who believed Gauss must have heard the problem before and memorized the answer, asked Gauss to explain how he had solved the problem. He replied: "The numbers 1, 2, 3, 4, 5, and so on to 100 can be paired as 1 and 100, 2 and 99, 3 and 98, and so on. Since each pair has a sum of 101, and there are 50 pairs, the total is 5050."

Basic theory is presented in intuitive form, augmented by numerous **worked-out examples** and **annotations** to clearly describe steps and procedures so the student can understand the mechanics of mathematics.

1-3 SUBTRACTING FRACTIONS

To subtract one fraction from another, we follow the same steps as in addition except that we subtract the numerators. Let us examine a few examples, starting with a subtraction that involves two fractions with the same denominator.

■ **EXAMPLE:**

What is $\frac{9}{16} - \frac{4}{16}$?

■ ***Solution:***

$$\frac{9}{16} - \frac{4}{16} = \frac{9-4}{16} = \frac{5}{16}$$

If the two fractions involved in the subtraction are of different types, we will need to determine the lowest common denominator so that we are subtracting one type of fraction from the same type of fraction.

■ **EXAMPLE:**

What is the result of $\frac{6}{10} - \frac{4}{20}$?

■ ***Solution:***

$$\frac{6}{10} - \frac{4}{20} = \frac{12-4}{20} = \frac{8}{20}$$
$$= \frac{8 \div 4}{20 \div 4} = \frac{2}{5}$$

Steps:
Find lowest common denominator (20).
10 into 20 = 2,
2 × 6 = 12.
20 into 20 = 1,
1 × 4 = 4.
12 − 4 = 8.
Reduce result.

1-3-1 *Subtracting Mixed Fractions*

As with addition, the way to subtract one mixed fraction from another is to deal with the whole numbers and fractions separately.

■ **EXAMPLE**

Subtract $2\frac{1}{3}$ from $5\frac{4}{5}$.

■ ***Solution:***

$$5\frac{4}{5} - 2\frac{1}{3} = 3\frac{12-5}{15}$$
$$= 3\frac{7}{15}$$

Steps:
Subtract whole numbers: 5 − 2 = 3.
Determine the lowest common denominator.
3 | 5 3
5 | 5 1
└→ 3 × 5 = 15
5 into 15 = 3, 3 × 4 = 12.
3 into 15 = 5, 5 × 1 = 5.
12 − 5 = 7.

SECTION 1-3 / SUBTRACTING FRACTIONS

Numerous electronic and computer **integrated math applications** instantly demonstrate the value of math as a tool.

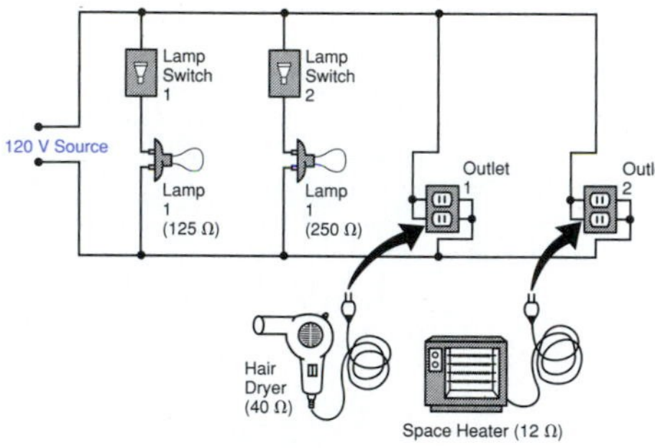

FIGURE 5-3 **Home Electrical System.**

3. Power Formula Example

Another formula frequently used in electricity and electronics is the power formula. We will also study this formula so that we can practice formula transposition again.

Power (Electric)
The amount of energy converted by a component or circuit in a unit of time, normally seconds. It is measured in watts or joules per second.

Electric power is the rate at which electric energy is converted into some other form of energy. Power, which is symbolized by *P*, is measured in *watts* (symbolized by W) in honor of James Watt. We hear watt used frequently in connection with the brightness of a lightbulb: for instance, a 60 watt bulb, a 100 watt bulb, and so on. This designation is a power rating and describes how much electric energy is converted every second. Because a lightbulb generates both heat and light energy, we would describe a 100 watt lightbulb as a device that converts 100 watts of electric energy into heat and light energy every second. Referring to Figure 5-1(b), you can see a lightbulb connected in an electric circuit. The amount of electric power supplied to that lightbulb is dependent on the voltage and the current. It is probably no surprise, therefore, that power is equal to the product of voltage and current. To state the formula:

Power = voltage × current
$P = V \times I$

Quantity	Symbol	Unit	Symbol
Power	P	Watts	W
Voltage	V	Volts	V
Current	I	Amperes	A

This formula states that the amount of power delivered to a device is dependent on the electrical pressure or voltage applied across the device and the electric current flowing through the device.

■ **INTEGRATED MATH APPLICATION:**

What will be the electric current (I) in a circuit if a 60 watt lightbulb is connected across a 120 volt battery?

■ ***Solution:***

Let us begin by inserting the known values in their appropriate places in the power formula.

Power (P) = voltage (V) × current (I)
60 W = 120 V × ? amperes

122 CHAPTER 5 / ALGEBRA, EQUATIONS, AND FORMULAS

Acute Angle
An angle that measures less than 90°.

Obtuse Angle
An angle that measures greater than 90°.

Figure 6-1(d) shows a few more geometric basics. A right angle is one of 90°, and can be symbolized with a small square in the corner. Two lines are perpendicular to one another when the point at which they intersect forms right angles. The angle measure of a straight angle is 180°. The terms **acute angle** and **obtuse angle** describe angles that are relative to a right angle (90°). An acute angle is one that is less than a right angle (<90°), while an obtuse angle is one that is greater than a right angle (>90°).

SELF-TEST EVALUATION POINT FOR SECTION 6-1

Now that you have completed this section, you should be able to:

■ **Objective 1.** *Describe the purpose of geometry, and define many of the basic geometric terms.*

■ **Objective 2.** *Demonstrate how to use a protractor to measure and draw an angle.*

Use the following questions to test your understanding of section 6-1:

1. Using a protractor, measure:
   a. ∠AXB   d. ∠AXE   g. ∠AXH   j. ∠IXG
   b. ∠AXC   e. ∠AXF   h. ∠AXI   k. ∠IXF
   c. ∠AXD   f. ∠AXG   i. ∠IXH   l. ∠IXE

2. Which of the angles in the previous question are right angles, obtuse angles, and acute angles?

3. Draw:
   a. A vertical line   e. An angle of 35°
   b. A horizontal line   f. An angle of 172°
   c. Three parallel lines g. An obtuse angle
   d. A right angle   h. An acute angle

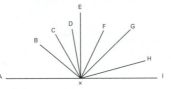

6-2   PLANE FIGURES

A plane figure is a perfectly flat surface, such as a tabletop, floor, wall, or windowpane. In this section we will examine the details of several flat plane surfaces, which, as you will discover, have only two dimensions—length and width.

6-2-1   *Quadrilaterals*

Polygon
A plane geometry figure bounded by 3 or more line segments.

Parallelogram
A quadrilateral where both pairs of opposite sides are parallel.

A **polygon** is a plane geometric figure that is made up of three or more line segments. Quadrilaterals, such as the square, rectangle, rhombus, parallelogram, and trapezoid, are all examples of four-sided polygons.

The information given in Figure 6-2 summarizes the facts and formulas for the square, which has four right angles and four equal length sides, and is a **parallelogram** (which means it has two sets of parallel sides). The box in Figure 6-2 lists the formulas for calculating a square's perimeter, area and side length.

Figures 6-3, 6-4, 6-5 and 6-6 on pp.141–143 summarize the facts and formulas for other quadrilaterals. In addition, these diagrams include, when necessary, a visual breakdown of the formulas so that it is easier to see how the formula was derived.

6-2-2   *Right Angle Triangles*

Right Angle Triangle
A triangle that contains one right or 90° angle.

The **right-angle triangle**, or right triangle shown in Figure 6-7(a) (p. 145), has three sides and three corners. Its distinguishing feature, however, is that two of the sides of this triangle are at right angles (at 90°) to one another. The small square box within the triangle is placed in the corner to show that sides *A* and *B* are *square* or at *right angles* to one another.

Calculator usage and **sequence examples** are given in illustrative format, making the transition to calculator use virtually seamless.

Most calculators have a key specifically for calculating the reciprocal of any number. It is called the reciprocal key and operates as follows.

CALCULATOR KEYS

Name: Reciprocal key

Function: Divides the value on the display into 1.

Example: $\frac{1}{3.2} = ?$

Press keys: [3] [.] [2] [¹⁄ₓ]

Display shows: 0.3125

PI
The sixteenth letter of the Greek alphabet; the symbol π denoting the ratio of the circumference of a circle to its diameter; a transcendental number having a value to eight decimal places of 3.14159265.

The reciprocal of π is used in many different formulas. The Greek lowercase letter pi, which is pronounced "pie" and symbolized by π, is a constant value that describes how much bigger a circle's circumference is than its diameter.

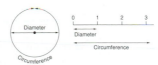

The approximate value of this constant is 3.142, which means that the circumference of any circle is approximately 3 times bigger than its diameter.

Most calculators have a key for quickly obtaining the value of pi. It is called the pi (π) key and operates as follows.

CALCULATOR KEYS

Name: Pi key

Function: Enters the value of pi correct to eight decimal digits.

$\pi = 3.1415927$

Example: $2 \times \pi = ?$

Press keys: [2] [×] [π] [=]

Display shows: 6.2831853

Circuit Analysis Tables train the student to collect circuit facts in an easy-to-read table that enables the student to clearly see the total and individual current, voltage, resistance, and power values and their relationships.

therefore, that the same value of water flow exists throughout a series-connected fluid system. This rule will always remain true, for if the valves were adjusted to double the opposition to flow, then half the flow, or 1 gallon of water per second, would be leaving the pump and flowing throughout the system.

Similarly, with the electronic series circuit shown in Figure 10-4(b), there is a total of 2 A leaving and 2 A arriving at the battery, and so the same value of current exists throughout the series-connected electronic circuit. If the circuit's resistance was changed, a new value of series circuit current would be present throughout the circuit. For example, if the resistance of the circuit was doubled, then half the current, or 1 A, would leave the battery, but that same value of 1 A would flow throughout the entire circuit. This series circuit current characteristic can be stated mathematically as

$$I_T = I_1 = I_2 = I_3 = \cdots$$

Total current = current through R_1 = current through R_2 = current through R_3, and so on.

FIGURE 10-5 **Total Current Example. (a) Schematic. (b) Protoboard Circuit. (c) Circuit Analysis Table.**

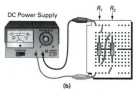

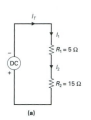

CIRCUIT ANALYSIS TABLE

Resistance $R = V/I$	Current $I = V/R$	Voltage $V = I \times R$	Power $P = V \times I$
$R_1 = 5\ \Omega$	$I_1 = 1\ \text{A}$		
$R_2 = 15\ \Omega$	$I_2 = 1\ \text{A}$		
	$I_T = 1\ \text{A}$		

(c)

Protoboard Pictorials help the beginning student make the transition from the circuit schematic on paper to the constructed lab circuit on a protoboard.

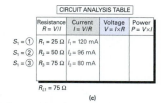

CIRCUIT ANALYSIS TABLE

	Resistance $R = V/I$	Current $I = V/R$	Voltage $V = I \times R$	Power $P = V \times I$
$S_1 = ①$	$R_1 = 25\ \Omega$	$I_1 = 120\ \text{mA}$		
$S_1 = ②$	$R_2 = 50\ \Omega$	$I_2 = 96\ \text{mA}$		
$S_1 = ③$	$R_3 = 75\ \Omega$	$I_3 = 80\ \text{mA}$		

$R_{L1} = 75\ \Omega$

(c)

FIGURE 10-8 **Three-Position Switch Controlling Lamp Brightness. (a) Schematic. (b) Protoboard Circuit. (c) Circuit Analysis Table.**

SELF-TEST EVALUATION POINT FOR SECTION 10-3

Now that you have completed this section, you should be able to:

■ **Objective 5.** *Explain how to calculate total resistance in a series circuit.*

■ **Objective 6.** *Explain how Ohm's law can be applied to calculate current, voltage, and resistance.*

Use the following questions to test your understanding of Section 10-3.

1. State the total resistance formula for a series circuit.
2. Calculate R_T if $R_1 = 2\ \text{k}\Omega$, $R_2 = 3\ \text{k}\Omega$, and $R_3 = 4700\ \Omega$.

10-4   VOLTAGE IN A SERIES CIRCUIT

A potential difference or voltage drop will occur across each resistor in a series circuit when current is flowing. The amount of voltage drop is dependent on the value of the resistor and the amount of current flow. This idea of potential difference or voltage drop is best explained by returning to the water analogy. In Figure 10-9(a), you can see that the high pressure from the pump's outlet is present on the left side of the valve. On the right side of the valve, however, the high pressure is no longer present. The high potential that exists on the left of the valve is not present on the right, so a potential or pressure difference is said to exist across the valve.

28. Calculate the length or magnitude of the resultant vectors in the following vector diagrams.

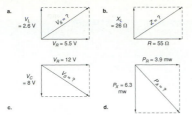

a.   b.

c.   d.

29. Calculate the value of the following trigonometric functions.

a. sin 0°   i. cos 60°
b. sin 30°   j. cos 90°
c. sin 45°   k. tan 0°
d. sin 60°   l. tan 30°
e. sin 90°   m. tan 45°
f. cos 0°   n. tan 60°
g. cos 30°   o. tan 90°
h. cos 45°

30. Calculate angle θ from the function given.

a. $\sin \theta = 0.707, \theta = ?$
b. $\sin \theta = 0.233, \theta = ?$
c. $\cos \theta = 0.707, \theta = ?$
d. $\cos \theta = 0.839, \theta = ?$
e. $\tan \theta = 1.25, \theta = ?$
f. $\tan \theta = 0.866, \theta = ?$

31. Name each of the triangle's sides, and calculate the unknown values.

a.   b.   c.

32. If a cylinder has a radius of 3 cm and a height of 10 cm, calculate its:
 a. Lateral surface area
 b. Total surface area
 c. Volume

33. If a sphere has a diameter of 2 inches, what would be its surface area and volume?

34. What is the lateral surface area of a rectangular prism that has a base perimeter of 40 cm and a height of 12 cm?

35. Calculate the lateral surface area of a pyramid that has a base perimeter of 24 inches and a slant height of 8 inches?

Web Site Questions

Go to the web site http://www.prenhall.com/cook, select the textbook *Mathematics for Electronics and Computers*, select this chapter, and then follow the instruc... the multiple-choice practice problems.

REVIEW QUESTIONS

6-4   SOLID FIGURES

As mentioned previously, a plane figure is a flat surface that has only two dimensions—length and width. A **solid figure**, on the other hand, is a three-dimensional figure having length, width and height.

6-4-1   *Prisms*

A **prism** is a solid figure made up of plane figure faces, and has a least one pair of parallel surfaces. Figure 6-22 details the facts and formulas for the prism. The visual breakdown in Figure 6-22(a) shows how a cube is made up of an upper base, lower base, and a set of sides. The **lateral surface area** (**L**) of the cube is the area of the sides only, as shown in Figure 6-22(b). The **total surface area** (**S**) is the lateral surface area of the cube plus the area of the two bases, as shown in Figure 6-22(c). The holding capacity or **volume** of the cube measures the space inside the prism, and is described visually in Figure 6-22(d). Figure 6-22(e) shows examples of other prism types.

6-4-2   *Cylinders*

A **cylinder** is a solid figure with curved side walls extending between two identical circular bases. Figure 6-23 details the facts and formulas for the cylinder. As with prisms, plane figure formulas are used to calculate the cylinder's lateral surface area and base area, and as expected, the volume of a cylinder is the product of its base area and height.

6-4-3   *Spheres*

A **sphere** is a globular solid figure in which any point on the surface is at an equal distance from the center. Figure 6-24 lists the formulas for a sphere along with a visual breakdown for a sphere's surface area and volume.

6-4-4   *Pyramids and Cones*

Figure 6-25(a) (p. 164) lists the formulas for a **pyramid,** which has only one base and three or more lateral surfaces that taper up to a single point called an **apex.** Figure 6-25(a) lists the formulas for a **cone,** which has a circular base and smooth surface that extends up to an apex. Figures 6-25(b) and 6-26(b) (p. 165), list the formulas for the frustum of a pyramid and the frustum of a cone. A **frustum** is the base section of a solid pyramid or cone, and is formed by cutting off the top of the solid figure so that the upper base and lower base are parallel to one another.

Solid Figure
A 3-dimensional figure having length, width, and height.

Prism
A solid figure made up of plane figure faces; having at least one pair of parallel surfaces.

Lateral Surface Area
The side area of a solid figure.

Total Surface Area
The lateral surface area of a solid figure plus the area of its two bases.

Volume
The space within a solid figure, or its holding capacity.

Cylinder
A solid figure with curved side walls extending between two identical circular bases.

Sphere
A globular solid figure in which any point on the surface is at equal distance from the center.

Pyramid
A solid figure having only one base and 3 or more lateral surfaces that taper up to a single point.

Apex
The uppermost point or tip of a solid figure.

Cone
A solid figure having a circular base and smooth surface that extends up to an apex.

Frustum
The base section of a solid pyramid or cone.

SELF-TEST EVALUATION POINT FOR SECTION 6-4

Now that you have completed this section, you should be able to:

■ **Objective 11.** *Define the term* solid figure, *and identify the different types.*

■ **Objective 12.** *List the formulas for determining the lateral surface area, total surface area, and volume of solid figures.*

Use the following questions to test your understanding of Section 6-4:

1. What is the difference between a plane figure and a solid figure?

2. What is the difference between lateral surface area and total surface area?

3. Sketch the following solid figures and list the formulas for lateral surface area, total surface area and volume.
 a. Prism   d. Pyramid
 b. Cylinder   e. Cone
 c. Sphere

4. What is a frustum?

SECTION 6-4 / SOLID FIGURES   **161**

Contents

PART A
Basic Math

1
Fractions   2
Vignette: There's No Sleeping When He's Around   2
Outline and Objectives   3
Introduction   4
1-1   Representing Fractions   4
1-2   Adding Fractions   7
1-3   Subtracting Fractions   15
1-4   Multiplying Fractions   17
1-5   Dividing Fractions   20
1-6   Canceling Fractions   22
Review Questions   23

2
Decimal Numbers   26
Vignette: An Apple a Day   26
Outline and Objectives   27
Introduction   28
2-1   Counting in Decimal   28
2-2   The Decimal Point   30
2-3   Calculating in Decimal   39
Review Questions   55

3
Positive and Negative Numbers   58
Vignette: Blaise of Genius   58
Outline and Objectives   59
Introduction   60
3-1   Expressing Positive and Negative Numbers   61
3-2   Adding Positive and Negative Numbers   63
3-3   Subtracting Positive and Negative Numbers   65
3-4   Multiplying Positive and Negative Numbers   68
3-5   Dividing Positive and Negative Numbers   70
3-6   Order of Operations   72
Review Questions   74

4
Exponents and the Metric System   76
Vignette: Not a Morning Person   76
Outline and Objectives   77
Introduction   78
4-1   Raising a Base Number to a Higher Power   78
4-2   Powers of 10   85
4-3   The Metric System   90
Review Questions   98

5
Algebra, Equations, and Formulas   100
Vignette: Back to the Future   100
Outline and Objectives   101
Introduction   102
5-1   The Basics of Algebra   102
5-2   Transposition   104
5-3   Substitution   124
5-4   Rules of Algebra—A Summary   131
Review Questions   133

6
Geometry and Trigonometry   136
Vignette: Finding the Question to the Answer!   136
Outline and Objectives   137
Introduction   138
6-1   Basic Geometric Terms   138
6-2   Plane Figures   140
6-3   Trigonometry   151
6-4   Solid Figures   161
Review Questions   165

7
Logarithms and Graphs 168

Vignette: The First Pocket Calculator 168
Outline and Objectives 169
Introduction 170
7-1 Common Logarithms 170
7-2 Graphs 175
Review Questions 183

PART B
Electronics Math

8
Current and Voltage 188

Vignette: Problem Solver 188
Outline and Objectives 189
Introduction 190
8-1 Current 190
8-2 Voltage 196
8-3 Conductors 201
8-4 Insulators 204
Review Questions 206

9
Resistance and Power 208

Vignette: Genius of Chippewa Falls 208
Outline and Objectives 209
Introduction 210
9-1 Resistance 210
9-2 Resistors 212
9-3 Measuring Resistance 222
9-4 Resistor Coding 223
9-5 Protoboards 227
9-6 Energy, Work, and Power 228
Review Questions 237

10
Series DC Circuits 240

Vignette: The First Computer Bug 240
Outline and Objectives 241
Introduction 242
10-1 Components in Series 242
10-2 Currents in a Series Circuit 244
10-3 Resistance in a Series Circuit 246
10-4 Voltage in a Series Circuit 249
10-5 Power in a Series Circuit 265
Review Questions 270

11
Parallel DC Circuits 274

Vignette: Let's Toss for It! 274
Outline and Objectives 275
Introduction 276
11-1 Components in Parallel 276
11-2 Voltage in a Parallel Circuit 278
11-3 Current in a Parallel Circuit 280
11-4 Resistance in a Parallel Circuit 286
11-5 Power in a Parallel Circuit 292
Review Questions 295

12
Series–Parallel DC Circuits 300

Vignette: The Christie Bridge Circuit 300
Outline and Objectives 301
Introduction 302
12-1 Series- and Parallel-Connected Components 302
12-2 Total Resistance in a Series–Parallel Circuit 305
12-3 Voltage Division in a Series–Parallel Circuit 307
12-4 Branch Currents in a Series–Parallel Circuit 309
12-5 Five-Step Method for Series–Parallel Circuit Analysis 310
12-6 Five-Step Method for Series–Parallel Circuit Analysis 311
12-7 Series–Parallel Circuits 316
12-8 Theorems for DC Circuits 325
Review Questions 338

13
Alternating Current (AC) 342

Vignette: The Laser 342
Outline and Objectives 343
Introduction 344
13-1 The Difference Between AC and DC 344
13-2 Why Alternating Current? 347
13-3 AC Wave Shapes 352
13-4 Measuring AC Signals 377
13-5 Decibels 385
Review Questions 394

14
Capacitors 398

Vignette: The Turing Enigma 398
Outline and Objectives 399
14-1 The Unit of Capacitance 400
14-2 Factors Determining Capacitance 402
14-3 Capacitors in Combination 405
14-4 Capacitive Time Constant 410
14-5 Capacitive Reactance 415

14-6 Series *RC* Circuit   417
14-7 Parallel *RC* Circuit   430
Review Questions   433

15

Inductors and Transformers   438

Vignette: The Wizard of Menlo Park   438
Outline and Objectives   439
Introduction   440
15-1 Self-Induction   440
15-2 The Inductor   443
15-3 Factors Determining Inductance   443
15-4 Inductors in Combination   447
15-5 Inductive Time Constant   449
15-6 Inductive Reactance   453
15-7 Series *RL* Circuit   454
15-8 Parallel *RL* Circuit   463
15-9 Mutual Inductance   465
15-10 Basic Transformer   467
15-11 Transformer Loading   468
15-12 Transformer Ratios and Applications   469
15-13 Transformer Ratings   476
Review Questions   477

16

RLC Circuits and Complex Numbers   480

Vignette: The Fairchildren   480
Outline and Objectives   481
Introduction   481
16-1 Series *RLC* Circuit   481
16-2 Parallel *RLC* Circuit   488
16-3 Resonance   491
16-4 Applications of *RLC* Circuits   505
16-5 Complex Numbers   509
Review Questions   520

17

Diodes and Transistors   524

Vignette: Spitting Lightning Bolts   524
Outline and Objectives   525
Introduction   526
17-1 The Junction Diode   526
17-2 The Zener Diode   531
17-3 The Light-Emitting Diode   534
17-4 Introduction to the Transistor   536
17-5 Detailed Transistor Operation   546
Review Questions   582

PART C
Computer Math

18

Analog to Digital   590

Vignette: Moon Walk   590
Outline and Objectives   591
Introduction   592
18-1 Analog and Digital Data and Devices   592
18-2 Analog and Digital Signal Conversion   597
Review Questions   598

19

Number Systems and Codes   600

Vignette: Leibniz's Language of Logic   600
Outline and Objectives   601
Introduction   602
19-1 The Decimal Number System   602
19-2 The Binary Number System   604
19-3 The Hexadecimal Number System   609
19-4 The Octal Number System   616
19-5 Binary Codes   620
Review Questions   626

20

Logic Gates   628

Vignette: The Great Experimenter   628
Outline and Objectives   629
Introduction   630
20-1 Hardware for Binary Systems   630
20-2 Basic Logic Gates   631
20-3 Inverting Logic Gates   636
20-4 Exclusive Logic Gates   639
20-5 IEEE/ANSI Symbols for Logic Gates   642
Review Questions   644

21

Boolean Expressions and Algebra   648

Vignette: From Folly to Foresight   648
Outline and Objectives   649
Introduction   650
21-1 Boolean Expressions for Logic Gates   650
21-2 Boolean Algebra Laws and Rules   660
21-3 From Truth Table to Gate Circuit   667
21-4 Gate Circuit Simplification   669
Review Questions   677

22

Binary Arithmetic   680

Vignette: Woolen Mill Makes Minis   680
Outline and Objectives   681
Introduction   682
22-1   Binary Arithmetic   682
22-2   Representing Positive and Negative Numbers   689
22-3   Two's Complement Arithmetic   695
22-4   Representing Large and Small Numbers   703
Review Questions   704

23

Introduction to Computers and Programming   706

Vignette: Making an Impact   706
Outline and Objectives   707
Introduction   708
23-1   Computer Hardware   708
23-2   Computer Software   710

Review Questions   717

Appendices   721

A   Answers to Self-Test Evaluation Points   721
B   Answers to Odd-Numbered Problems   734

Index   745

Fractions

There's No Sleeping When He's Around!

Carl Friedrich Gauss was born April 30, 1777, to poor, uneducated parents in Brunswick, Germany. He was a child of precocious abilities, particularly in mental computation. In elementary school he soon impressed his teachers, who said that mathematical ability came easier to Gauss than speech.

In secondary school he rapidly distinguished himself in ancient languages and mathematics. At 14, Gauss was presented to the court of the duke of Brunswick, where he displayed his computing skill. Until his death in 1806, the duke generously supported Gauss and his family, encouraging the boy with textbooks and a laboratory.

In the early years of the nineteenth century, Gauss's interest was in astronomy, and his accumulated work on celestial mechanics was published in 1809. In 1828, at a conference in Berlin, Gauss met physicist Wilhelm Weber, who would eventually become famous for his work on electricity. They worked together for many years and became close friends, investigating electromagnetism and the use of a magnetic needle for the measurement of current. In 1833 they constructed an electric telegraph system that could communicate across Göttingen from Gauss's observatory to Weber's physics laboratory. (This telegraph system of communication was later developed independently by U.S. inventor Samuel Morse.)

Gauss conceived almost all of his fundamental mathematical discoveries between the ages of 14 and 17. There are many stories of his genius in his early years, one of which involved a sarcastic teacher who liked giving his students long-winded problems and then resting, or on some occasions sleeping, in class. On his first day with Gauss, who was 8 years old, the teacher began, as usual, by telling the students to find the sum of all the numbers from 1 to 100. The teacher barely had a chance to sit down before Gauss raised his hand and said "5050." The dumbfounded teacher, who believed Gauss must have heard the problem before and memorized the answer, asked Gauss to explain how he had solved the problem. He replied: "The numbers 1, 2, 3, 4, 5, and so on to 100 can be paired as 1 and 100, 2 and 99, 3 and 98, and so on. Since each pair has a sum of 101, and there are 50 pairs, the total is 5050."

Outline and Objectives

VIGNETTE: THERE'S NO SLEEPING WHEN HE'S AROUND!

INTRODUCTION

1-1 REPRESENTING FRACTIONS

Objective 1: Describe what a fraction is, and name its elements.

Objective 2: Demonstrate how fractions are represented.

1-2 ADDING FRACTIONS

Objective 3: Describe how fractions are added.

1-2-1 Method 1 for Finding a Common Denominator

Objective 4: Show two methods used for calculating the lowest common denominator.

1-2-2 Method 2 for Finding a Common Denominator

1-2-3 Improper Fractions and Mixed Fractions

Objective 5: Define a(n):
  a. Proper fraction
  b. Improper fraction
  c. Mixed fraction

1-2-4 Adding Mixed Fractions

Objective 6: Show how mixed fractions are added.

1-2-5 Reducing Fractions

Objective 7: Describe how to reduce fractions.

1-3 SUBTRACTING FRACTIONS

Objective 8: Show how to subtract proper and mixed fractions.

1-3-1 Subtracting Mixed Fractions

1-3-2 Borrowing

1-4 MULTIPLYING FRACTIONS

Objective 9: Show how to multiply proper and mixed fractions.

1-4-1 Multiplying Mixed Fractions

1-5 DIVIDING FRACTIONS

Objective 10: Show how to divide proper and mixed fractions.

1-5-1 Dividing Mixed Fractions

1-6 CANCELING FRACTIONS

Objective 11: Demonstrate how to use cancellation to simplify fraction multiplication or division.

MULTIPLE CHOICE QUESTIONS

COMMUNICATION SKILL QUESTIONS

PRACTICE PROBLEMS

WEB SITE QUESTIONS

Fraction

A numerical representation indicating the quotient of two numbers; a piece, fragment, or portion.

Half

Either of two equal parts into which a thing is divisible.

Quarter

One of four equal parts into which something is divisible; a fourth part.

Fourth

See *quarter*.

Another word for **fraction** is "piece"; therefore, if you have a fraction of something, you have a piece of that something. Having only a fraction or piece of something means that the object has been broken up into pieces, and therefore a fraction is something less than the whole thing. For example, if you wanted half of a pizza, a whole pizza would be cut into two pieces and you would be given one of the halves—a **half** is therefore a fraction. If, in another instance, you wanted only a quarter of an apple pie, a whole apple pie would be cut into four pieces and you would be given one of the quarters—a **quarter** is therefore also a fraction. A quarter (or a **fourth**) is a smaller fraction than a half, and therefore we can say that *the more cuts you make, the smaller the fraction.*

Since this is your first chapter, let me try to put you in the right frame of mind before you begin. Most people have trouble with mathematics because they never mastered the basics. As you proceed through this chapter and the succeeding chapters, it is imperative that you study every section, example, self-test evaluation point, and end-of-chapter question. If you cannot understand a particular section or example, go back and review the material that led up to the problem and make sure that you fully understand all the basics before you continue. Since each chapter builds on previous chapters, you may find that you need to return to an earlier chapter to refresh your understanding before moving on with the current chapter. This process of moving forward and then backtracking to refresh your understanding is very necessary and helps to engrave the basics of mathematics in your mind. Try never to skip a section or chapter because you feel that you already have a good understanding of that material. If it is a basic topic that you have no problem with, read it anyway to refresh your understanding about the steps involved and the terminology because these may be used in a more complex operation in a later chapter.

1-1   REPRESENTING FRACTIONS

Half and *quarter* (or *fourth*) are names that describe a fraction; however, some other system, sign, or symbol is needed to represent or show the fraction. This system will have to show two things:

1. The number of pieces into which the entire object was cut or divided up
2. The number of pieces we have or are concerned with

Numerator

The part of a fraction that is above the line and signifies the number of parts of the denominator taken.

Denominator

The part of a fraction below the line signifying division that functions as the divisor of the numerator.

Vinculum or Fraction Bar

A straight horizontal line placed between two or more numbers of a mathematical expression.

■ **EXAMPLE:**

Write down the proper or written fraction for one-quarter or one-fourth, and label its elements.

■ *Solution:*

Fraction bar
(Its official name is **vinculum**; however, we will refer to it as the **fraction bar.**)

$$\frac{1}{4}$$

Called the **numerator,** this tells us how many of the original pieces we have.

Called the **denominator,** this tells us into how many pieces the original object was cut.

The preceding example fraction represents an amount or quantity. In this example it represents "one-quarter" or "one-fourth" and indicates that the object was cut or divided into four pieces and that we have one of the four pieces.

■ **EXAMPLE:**

First, draw a picture of a pizza cut into six pieces; then show one of the six pieces separately and label it with its fraction.

■ *Solution:*

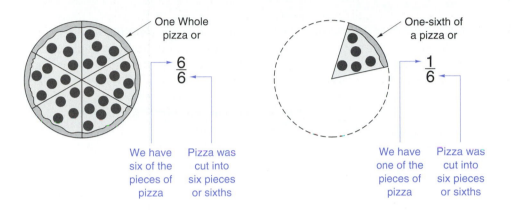

One Whole pizza or

$$\frac{6}{6}$$

We have six of the pieces of pizza

Pizza was cut into six pieces or sixths

One-sixth of a pizza or

$$\frac{1}{6}$$

We have one of the pieces of pizza

Pizza was cut into six pieces or sixths

■ **EXAMPLE:**

A pie is divided into eight pieces. If five of the pieces are eaten and three remain, how would we indicate or represent these fractions?

■ *Solution:*

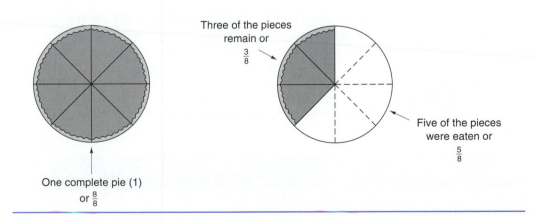

Three of the pieces remain or

$\frac{3}{8}$

Five of the pieces were eaten or

$\frac{5}{8}$

One complete pie (1) or $\frac{8}{8}$

In all technical fields you will have to interpret scales, such as the divisions on the straightedge shown next.

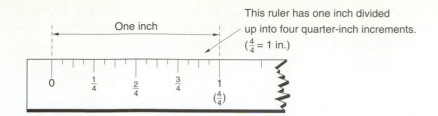

This ruler has one inch divided up into four quarter-inch increments. ($\frac{4}{4}$ = 1 in.)

One inch

Other rulers might have each inch broken up into halves (there would be two halves or $\frac{2}{2}$ in an inch), eighths (there would be eight eighths or $\frac{8}{8}$ in an inch), sixteenths (there would be sixteen sixteenths or $\frac{16}{16}$ in an inch), and so on.

■ INTEGRATED MATH APPLICATION:

Use the ruler to measure the length of the screws shown in the following figure.

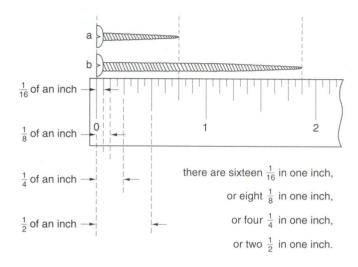

$\frac{1}{16}$ of an inch →

$\frac{1}{8}$ of an inch →

$\frac{1}{4}$ of an inch →

$\frac{1}{2}$ of an inch →

there are sixteen $\frac{1}{16}$ in one inch,

or eight $\frac{1}{8}$ in one inch,

or four $\frac{1}{4}$ in one inch,

or two $\frac{1}{2}$ in one inch.

■ Solution:

a. The smaller screw measures three-fourths of an inch ($\frac{3}{4}$ in.), or six-eighths of an inch ($\frac{6}{8}$ in.), or twelve-sixteenths ($\frac{12}{16}$ in.) of an inch.

b. The second screw is greater than 1 inch long but less than 2 inches long, which means that it is 1 inch plus a fraction of an inch long. Since its tip falls alongside an $\frac{1}{8}$ inch mark on the ruler, we will use the $\frac{1}{8}$ inch scale. The longer screw measures one and seven-eighths of an inch ($1\frac{7}{8}$ in.) long. We could also say that screw b measures $\frac{15}{8}$ of an inch, since the $\frac{8}{8}$ in the first inch, plus the $\frac{7}{8}$ in the second inch, equals a total of $\frac{15}{8}$. Using the sixteenths ruler marks, screw b measures $1\frac{14}{16}$ or $\frac{30}{16}$ of an inch long.

SELF-TEST EVALUATION POINT FOR SECTION 1-1

Now that you have completed this section, you should be able to:

■ **Objective 1.** *Describe what a fraction is, and name its elements.*

■ **Objective 2.** *Demonstrate how fractions are represented.*

Use the following questions to test your understanding of Section 1-1.

1. What fraction is shown shaded?

2. What would be the readings on the following scale?

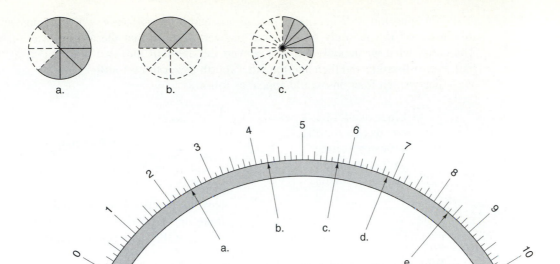

1-2   ADDING FRACTIONS

The *denominator* (number below the fraction bar) determines the fraction type. For example, ³⁄₉ tells us that the object was divided up into nine parts and that we have three of those nine parts. The number nine therefore tells us the type of fraction we are dealing with. When fractions are of the same type (have the same denominators), they are added as simply as adding 5 apples and 3 apples.

■ **EXAMPLE:**

Add the following fractions: $\dfrac{5}{9} + \dfrac{3}{9}$

■ *Solution:*

Since we are dealing with two fractions that are of the same type (ninths), the result is simply the sum or total of the two *numerators* (numbers above the fraction bar).

$$\text{Numerators} \longrightarrow \quad \frac{5}{9} + \frac{3}{9} = \frac{8}{9} \quad \longleftarrow \text{Denominators}$$

Notice that the numerators were added, whereas the denominators remained the same. This is because the type of fraction remained the same; only the number of pieces changed. To use an analogy, we could say:

$$\frac{5}{\text{apples}} + \frac{3}{\text{apples}} = \frac{8}{\text{apples}} \quad \begin{array}{l}\longleftarrow \text{Sum} \\ \longleftarrow \text{Type}\end{array}$$

The next step is to address how we can add different types of fractions. For example, how would we add ½ plus ¼? In this instance we have two different types of fractions, and therefore they are as different as apples and oranges.

Using pictures, we know that ½ plus ¼ equal ¾.

The question is: What mathematical procedure do we follow when the answer is not this obvious? If you study the previous answer, you see that the answer was in fourths. Therefore, what we actually did was convert the one-half into two-fourths ($\frac{1}{2} = \frac{2}{4}$, one-half = two-fourths) and then add the other fourth to get three-fourths.

The written fractions would appear as follows:

Since one-half equals two-fourths, we can subsitute $\frac{2}{4}$ (two-fourths) for $\frac{1}{2}$ (one-half).

$$\frac{1}{2} + \frac{1}{4} = ?$$

$$\frac{2}{4} + \frac{1}{4} = \frac{3}{4}$$

— Sum of numerators

Now the fractions are the same type; therefore the denominator answer indicates the same fraction type.

What we have noticed from this example is that to add fractions that are of different types, we need to find a *denominator that is common to both fractions*. There are two methods for finding a **common denominator** so that we can add different types of fractions.

Common Denominator

A common multiple of the denominators of a number of fractions.

1-2-1   *Method 1 for Finding a Common Denominator*

Probably the easiest way to find a common denominator is to multiply one denominator by another. For example, let us assume that we want to add $\frac{1}{2}$ plus $\frac{1}{3}$.

$$\frac{1}{2} + \frac{1}{3} = ?$$

Our common denominator can be obtained by multiplying one denominator by the other; therefore, $2 \times 3 = 6$.

Now that a common denominator has been found, how many sixths ($\frac{1}{6}$) are in one-half ($\frac{1}{2}$) and one-third ($\frac{1}{3}$)? This is determined in the following way.

$$\frac{1}{2} + \frac{1}{3} = \frac{?}{6}$$

(One-half plus one-third equals how many sixths?)

$$\frac{1}{2} + \frac{1}{3} = \frac{3}{6}$$

Step a: 2 into 6 = 3.
Step b: 3 × 1 = 3.
Step c: Place result in answers numerator.

$$\frac{1}{2} + \frac{1}{3} = \frac{3 + 2}{6}$$

Step a: 3 into 6 = 2.
Step b: 2 × 1 = 2.
Step c: Place result in answers numerator.

$$\frac{1}{2} + \frac{1}{3} = \frac{3 + 2}{6}$$

The last step is to add the two answers for the numerator (3 + 2 = 5).

$$\frac{1}{2} + \frac{1}{3} = \frac{5}{6}$$

■ **EXAMPLE:**

Add $\frac{1}{4}$ and $\frac{2}{3}$.

8   CHAPTER 1 / FRACTIONS

■ **Solution**

Common denominator = denominator × denominator

$$= \quad 4 \quad \times \quad 3$$
$$= \quad 12$$

$$\frac{1}{4} + \frac{2}{3} = \frac{3+8}{12} = \frac{11}{12}$$

Steps:
4 into 12 = 3,
3 × 1 = 3,
3 into 12 = 4,
4 × 2 = 8,
3 + 8 = 11.

1-2-2 *Method 2 for Finding a Common Denominator*

The problem with method 1 is that it often yields a common denominator that is larger than necessary. For example, what is the lowest-value common denominator for the following three-fraction addition?

$$\frac{3}{9} + \frac{1}{4} + \frac{1}{6} = \frac{}{?}$$

Using method 1, would obtain the answer

$$9 \times 4 \times 6 = 216$$

Even though 216 will work as a common denominator, it is a very large number and therefore the fraction addition will be cumbersome.

The steps in the following method will always yield the **lowest common denominator.** Using the previous example, we first place all three denominators in an upside-down division box, as shown.

$$\underline{9 \qquad 4 \qquad 6}$$

Lowest Common Denominator

The lowest-value common denominator.

The next step is to begin with 2 and see if it will divide exactly into any of the three denominators. Once 2 can no longer be used as a divisor, increase to 3, and then 4, and then 5, and so on. Let us follow through these steps in more detail.

2 does not go into 9 exactly.
2 goes into 4, giving 2,
and 2 goes into 6, giving 3.
Using 2 again as a divisor,
we can further reduce because 2
goes into 2, giving 1.
Because 2 will no longer reduce
any of the denominators, we increase
the divisor to 3. 3 goes into
9, giving 3, and into 3, giving 1.
Using 3 again as a divisor, we
can reduce because 3 goes into
3, giving 1.

Divisors to be
used as "factors."

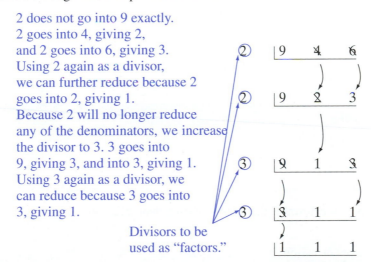

Looking at the previous process, you can see that the point is to keep dividing the denominators until they are all reduced to 1. Because no number is changed when it is multiplied or divided by 1 (for example, 22 × 1 = 22, 115 × 1 = 115, 17 ÷ 1 = 17, 63 ÷ 1 = 63), we can ignore the remaining 1s.

The divisors used in this process are called *factors.* A **factor** *is any number that is multiplied by another number and contributes to the result.* For example, consider 3 × 8. In this multiplication example, 3 is a factor and 8 is a factor that will contribute to the result of 24. Let us

Factor

Any of the numbers or symbols in mathematics that when multiplied together form a product.

now take another example: What factors could contribute to a result of 12? In this case we must determine which small numbers, when multiplied together, will produce 12. The answer is

$$2 \times 6 = 12 \quad \text{or} \quad 3 \times 4 = 12$$

Are these, however, the smallest numbers that when multiplied together will produce 12? The answer is no because the 6 can be broken down into 2×3, and the 4 can be broken down into 2×2. Therefore, $2 \times 3 \times 2$ or $3 \times 2 \times 2$ shows that 12 has the three factors 2, 2, and 3.

Some numbers cannot be reduced or broken down into factors. For example, there are no smaller numbers that when multiplied together will produce 5. This number is called a **prime number,** and there are many, such as 3, 5, 7, 11, 13, and 17.

What we were doing, therefore, in the previous three-fraction addition example was extracting the factors from the denominators.

Prime Number

Any integer other than 0 or 1 that is not divisible without remainder.

These four factors, which were the divisors, can then be used to find the lowest common denominator.

2	9	4	6
2	9	2	3
3	9	1	3
3	3	1	1

(Example: $\dfrac{3}{9} + \dfrac{1}{4} + \dfrac{1}{6} = \dfrac{}{?}$)

$2 \times 2 \times 3 \times 3 = 36$

Denominator factors   Lowest common denominator

Factoring

Resolving into factors; a reducing process that extracts the common factors.

Now that we have completed the *factoring* step and determined the lowest common denominator, we can proceed with the example, which was to add $\tfrac{3}{9} + \tfrac{1}{4} + \tfrac{1}{6}$:

$$\frac{3}{9} + \frac{1}{4} + \frac{1}{6} = \frac{}{36} \quad \longleftarrow \text{ Lowest common denominator}$$

$$\frac{3}{9} + \frac{1}{4} + \frac{1}{6} = \frac{12 +}{36}$$

$$\frac{3}{9} + \frac{1}{4} + \frac{1}{6} = \frac{12 + 9 +}{36}$$

$$\frac{3}{9} + \frac{1}{4} + \frac{1}{6} = \frac{12 + 9 + 6}{36}$$

$$= \frac{12 + 9 + 6}{36} = \frac{27}{36}$$

Steps: 9 into 36 = 4, $4 \times 3 = 12$.

4 into 36 = 9, $9 \times 1 = 9$.

6 into 36 = 6, $6 \times 1 = 6$.

■ **EXAMPLE:**

Add $\dfrac{1}{4} + \dfrac{3}{5} + \dfrac{1}{8}$.

■ *Solution:*

The first step is to determine the factors in each denominator.

Divisors or factors 3 and 4 were ignored because 5 was a prime factor.

2	4	5	8
2	2	5	4
2	1	5	2
5	1	5	1
	1	1	1

$2 \times 2 \times 2 \times 5 = 40$ (Lowest common denominator)

$$\frac{1}{4} + \frac{3}{5} + \frac{1}{8} = \frac{10 + 24 + 5}{40} = \frac{39}{40}$$

1-2-3   *Improper Fractions and Mixed Numbers*

All the fractions discussed so far have a numerator that is smaller than the denominator. These are called **proper fractions.** For example:

$$\text{Proper fractions: } \frac{1}{2} \quad \frac{3}{4} \quad \frac{5}{9} \quad \frac{7}{32} \quad \frac{15}{27} \quad \leftarrow \text{ Smaller numerator}$$
$$\leftarrow \text{ Larger denominator}$$

> **Proper Fraction**
>
> A fraction in which the numerator is less or of lower degree than the denominator.

On some occasions, an addition of two or more proper fractions will produce a fraction with a numerator that is larger than the denominator. To explain this, let us use an example.

■ **EXAMPLE:**

Add $\dfrac{3}{4} + \dfrac{2}{4}$.

■ *Solution:*

$$\frac{3}{4} + \frac{2}{4} = \frac{3 + 2}{4} = \frac{5}{4} \leftarrow \text{Numerator is larger than denominator.}$$

To analyze this answer, $\frac{5}{4}$, we would say that the denominator 4 indicates that the original whole unit was divided into four pieces (quarters) and the numerator indicates that we have five of these pieces. Therefore, with four of the quarters (or fourths) we have one whole unit, leaving one extra fourth.

$$\frac{5}{4} = 1\frac{1}{4}$$

five-fourths = one whole unit and one-fourth

A fraction with a numerator that is larger than the denominator (for example, $\frac{5}{4}$) is called an **improper fraction** and needs to be converted to a **mixed number.** By definition, a mixed fraction or mixed number is a whole number and a fraction (for example, $1\frac{1}{4}$). Since the fraction bar actually indicates the arithmetic operation of division (for example, $\frac{5}{4}$ indicates 5 divided by 4), all we do is perform this division to convert an improper fraction to a mixed number. For example, 4 goes into 5 once with 1 remaining:

> **Improper Fraction**
>
> A fraction whose numerator is equal to or larger than the denominator.
>
> **Mixed Number**
>
> A number composed of an integer and a fraction.

$$\frac{5}{4} = 1\frac{1}{4}$$

four into five = one, with one-fourth remaining

■ EXAMPLE:

Add $\dfrac{2}{3} + \dfrac{4}{6}$.

■ Solution:

$$\frac{2}{3} + \frac{4}{6} = \frac{4 + 4}{6} = \frac{8}{6}$$

Lowest common denominator:

2	3		6
3	3		3

$2 \times 3 = 6$

Because ⁸⁄₆ is an improper fraction, the next step is to convert it to a mixed fraction.

$$\frac{8}{6} = 1\frac{2}{6}$$

six into eight = one, with two-sixths remaining

1-2-4   *Adding Mixed Numbers*

Now that we understand what a mixed number is, let us see how we would add two mixed fractions.

■ EXAMPLE:

Add $5\dfrac{1}{4} + 6\dfrac{3}{8}$.

■ Solution:

The easiest way to add mixed numbers is to deal with the whole numbers and fractions separately.

Step 1: Separate whole number and fractions.

Step 2: Add fractions.

Step 3: Add whole numbers.

Step 4: Combine whole-number result and fraction result.

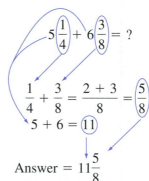

$$5\left(\frac{1}{4}\right) + 6\left(\frac{3}{8}\right) = ?$$

$$\frac{1}{4} + \frac{3}{8} = \frac{2 + 3}{8} = \left(\frac{5}{8}\right)$$

$$5 + 6 = 11$$

$$\text{Answer} = 11\frac{5}{8}$$

■ INTEGRATED MATH APPLICATION:

If two objects measure $6\dfrac{3}{8}$ inches and $2\dfrac{1}{4}$ inches, what is the total length of both objects?

■ Solution:

$$6\frac{3}{8} + 2\frac{1}{4} = 8\frac{3 + 2}{8} = 8\frac{5}{8} \text{ inches}$$

EXAMPLE:

Add $4\dfrac{15}{20} + 5\dfrac{4}{5}$.

■ *Solution:*

$$4\frac{15}{20} + 5\frac{4}{5} = 9\,\frac{15 + 16}{20} = 9\frac{31}{20}$$

Because the fraction $^{31}/_{20}$ is improper, we must convert it by dividing 20 into 31.

$$9\frac{31}{20} = 9 + \left(1\frac{11}{20}\right) \longleftarrow 31 \div 20 = 1\ ^{11}/_{20}$$

$$= 10\frac{11}{20} \longleftarrow \begin{array}{l}\text{9 whole objects plus}\\ \text{1 whole object} = 10\end{array}$$

1-2-5 *Reducing Fractions*

To obtain half of a pizza politely, you could ask for two small quarters because two-quarters equals one-half.

$$\frac{2}{4} = \frac{1}{2}$$

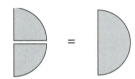

Both fractions therefore indicate the same part of a whole object. In most instances when we have two fractions that describe the same portion, we would use the fraction that has the smallest denominator. For example, the denominator 2 in $^1/_2$ is smaller than the denominator 4 in $^2/_4$, and therefore we would use $^1/_2$ to represent the fraction. The next question is: How do we **reduce** a fraction so that it has the lowest possible denominator? The answer is that we use an algebraic rule that states: *If the same number is divided into both the numerator and denominator of a fraction, the value of the fraction remains the same, only reduced.*

For example, let us divide 2 into both the numerator and denominator of $^2/_4$:

$$\frac{2 \div 2 = 1}{4 \div 2 = 2}$$

To reduce a fraction we must therefore find the largest number that will divide evenly into both its numerator and denominator. This is best achieved by starting with 2, then trying 3, then 4, then 5, and so on. To demonstrate this, let us do the following example.

Reducing

The process of changing to an equivalent but more fundamental expression.

■ **EXAMPLE:**

Reduce the following fractions if possible.

a. $\dfrac{3}{6}$ **c.** $\dfrac{5}{45}$ **e.** $\dfrac{18}{90}$

b. $\dfrac{4}{16}$ **d.** $\dfrac{4}{9}$

■ Solution:

a. $\left(\dfrac{3}{6}\right)\dfrac{\div\ 3}{\div\ 3}=\dfrac{1}{2}$ (three-sixths = one-half)

Will both divide by 2? No. Will both divide by 3? Yes.

b. $\left(\dfrac{4}{16}\right)\dfrac{\div\ 4}{\div\ 4}=\dfrac{1}{4}$ (four-sixteenths = one-fourth)

Will both divide by 2? Yes. Will both divide by 3? No. Will both divide by 4? Yes. We will use 4 because dividing by a larger number will give us a smaller denominator in fewer steps. If we had used 2, we would have ended up with the same answer but would have had to reduce in two steps:

$$\left(\dfrac{4}{16}\right)\dfrac{\div\ 2}{\div\ 2}=\left(\dfrac{2}{8}\right)\dfrac{\div\ 2}{\div\ 2}=\dfrac{1}{4}$$

c. $\left(\dfrac{5}{45}\right)\dfrac{\div\ 5}{\div\ 5}=\dfrac{1}{9}$

Will both divide by 2? No. Will both divide by 3? No. Will both divide by 4? No. Will both divide by 5? Yes.

d. $\left(\dfrac{4}{9}\right)$

Will both divide by 2? No. Will both divide by 3? No. Will both divide by 4? No. This fraction cannot be reduced any further.

e. $\left(\dfrac{18}{90}\right)\dfrac{\div\ 18}{\div\ 18}=\dfrac{1}{5}$

Divide by 2? Yes. Divide by 3? Yes. Divide by 4? No. Divide by 5? No. Divide by 6? Yes. Divide by 7? No. Divide by 8? No. Divide by 9? Yes. . . . Divide by 18? Yes.

SELF-TEST EVALUATION POINT FOR SECTION 1-2

Now that you have completed this section, you should be able to:

■ **Objective 3.** Describe how fractions are added.

■ **Objective 4.** Show two methods used for calculating the lowest common denominator.

■ **Objective 5.** Define a proper fraction, an improper fraction, and a mixed fraction.

■ **Objective 6.** Show how mixed fractions are added.

■ **Objective 7.** Describe how to reduce fractions.

Use the following questions to test your understanding of Section 1-2.

Add the following fractions. Convert to proper fractions and reduce if necessary.

1. $\dfrac{4}{6}+\dfrac{1}{6}=?$

2. $\dfrac{3}{64}+\dfrac{2}{64}+\dfrac{25}{64}=?$

3. $\dfrac{3}{9}+\dfrac{6}{18}=?$

4. $\dfrac{1}{3}+\dfrac{1}{4}+\dfrac{3}{15}=?$

5. $\dfrac{3}{4}+\dfrac{4}{5}=?$

6. $\dfrac{9}{12}+\dfrac{4}{24}+\dfrac{3}{4}=?$

7. $\dfrac{7}{9}+\dfrac{5}{9}+\dfrac{4}{18}=?$

(Convert to proper fractions and/or reduce if necessary.)

8. $\dfrac{15}{6}=?$

9. $\dfrac{4}{24}=?$

10. $\dfrac{25}{100}=?$

1-3　SUBTRACTING FRACTIONS

To subtract one fraction from another, we follow the same steps as in addition except that we subtract the numerators. Let us examine a few examples, starting with a subtraction that involves two fractions with the same denominator.

■ EXAMPLE:

What is $\dfrac{9}{16} - \dfrac{4}{16}$?

■ *Solution:*

$$\frac{9}{16} - \frac{4}{16} = \frac{9-4}{16} = \frac{5}{16}$$

If the two fractions involved in the subtraction are of different types, we will need to determine the lowest common denominator so that we are subtracting one type of fraction from the same type of fraction.

■ EXAMPLE:

What is the result of $\dfrac{6}{10} - \dfrac{4}{20}$?

■ *Solution:*

$$\frac{6}{10} - \frac{4}{20} = \frac{12-4}{20} = \frac{8}{20}$$
$$= \frac{8 \div 4}{20 \div 4} = \frac{2}{5}$$

Steps:
Find lowest common denominator (20).
10 into 20 = 2,
2 × 6 = 12.
20 into 20 = 1,
1 × 4 = 4.
12 − 4 = 8.
Reduce result.

1-3-1　*Subtracting Mixed Fractions*

As with addition, the way to subtract one mixed fraction from another is to deal with the whole numbers and fractions separately.

■ EXAMPLE

Subtract $2\dfrac{1}{3}$ from $5\dfrac{4}{5}$.

■ *Solution:*

$$5\frac{4}{5} - 2\frac{1}{3} = 3\frac{12-5}{15}$$
$$= 3\frac{7}{15}$$

Steps:
Subtract whole numbers: 5 − 2 = 3.
Determine the lowest common denominator.

3	5		3
5	5		1

└→ 3 × 5 = 15
5 into 15 = 3, 3 × 4 = 12.
3 into 15 = 5, 5 × 1 = 5.
12 − 5 = 7.

Borrowing

The process of taking from the next higher order digit so that a subtraction can take place.

1-3-2 *Borrowing*

As is normal in subtraction, we sometimes have problems in which we need to subtract a large number from a small number. In these cases we will need to **borrow**, as shown in the following example.

■ **EXAMPLE:**

What is $6\dfrac{1}{3} - 2\dfrac{2}{3}$?

■ *Solution:*

Borrow 1 from 6 and convert it to ⅔.

$$6\frac{1}{3} - 2\frac{2}{3} = 4\frac{1-2}{3}?$$

$$6\frac{1}{3} = 5 + \left(\frac{3}{3}\right) + \frac{1}{3} = 5\frac{4}{3}$$

$$5\frac{4}{3} - 2\frac{2}{3} = 3\frac{4-2}{3} = 3\frac{2}{3}$$

Steps:
Subtract whole numbers; $6 - 2 = 4$.
Because fractions have the same denominator, subtract numerators.
You cannot subtract 2 from 1.
Borrow whole number from 6, and convert it to a fraction ($1 = $ ⅓).
Add 3 pieces to 1 piece to get 4 pieces, or 4 thirds.
Perform subtraction.

■ **INTEGRATED MATH APPLICATION:**

The interest rate on a car loan has been reduced $2\dfrac{1}{4}$ percent from $6\dfrac{3}{16}$ percent. What is the new interest rate?

■ *Solution:*

$$6\frac{3}{16} - 2\frac{1}{4} = 4\frac{3-4}{16} \longleftarrow \text{You cannot subtract 4 from 3.}$$

$$5 + \frac{16}{16} + \frac{3}{16} - 2\frac{1}{4} =$$

$$5\frac{19}{16} - 2\frac{1}{4} = 3\frac{19-4}{16} = 3\frac{15}{16}$$

SELF-TEST EVALUATION POINT FOR SECTION 1-3

Now that you have completed this section, you should be able to:

■ *Objective 8.* *Show how to subtract proper and mixed fractions.*

Use the following questions to test your understanding of Section 1-3.

1. $\dfrac{4}{32} - \dfrac{2}{32} = ?$

2. $\dfrac{4}{5} - \dfrac{3}{10} = ?$

3. $\dfrac{6}{12} - \dfrac{4}{16} = ?$

4. $15\dfrac{1}{3} - 2\dfrac{4}{12} = ?$

5. $5\dfrac{15}{36} - 2\dfrac{31}{36} = ?$

One of the best ways to understand something new is to use an example that you can imagine easily.

■ **EXAMPLE:**

What is half of a half?

■ *Solution:*

$\frac{1}{2}$ of ◗ = ◗

half of a half = a fourth

$$\frac{1}{2} \times \frac{1}{2} = \frac{1}{4}$$

It is not surprising that the product (multiplication result) of two fractions has a denominator that is smaller than the denominators of the two fractions, because multiplication is calculating the fractional part of a fraction. Let us look at another example.

■ **EXAMPLE:**

What is half of a fourth?

■ *Solution:*

$\frac{1}{2}$ of ◢ = ◿

half of one-fourth = one-eighth

$$\frac{1}{2} \times \frac{1}{4} \qquad = \frac{1}{8}$$

From the last two examples you can probably see that to obtain the product of two fractions, simply multiply the numerators and multiply the denominators, as shown.

$$\frac{1}{2} \times \frac{1}{2} = \frac{1 \times 1 = 1}{2 \times 2 = 4}$$ (one-half of one-half equals one-fourth)

$$\frac{1}{2} \times \frac{1}{4} = \frac{1 \times 1 = 1}{2 \times 4 = 8}$$ (one-half of one-fourth equals one-eighth)

■ **EXAMPLE:**

What is $\frac{3}{4}$ of $\frac{3}{5}$?

■ *Solution:*

$$\frac{3}{4} \times \frac{3}{5} = \frac{3 \times 3 = 9}{4 \times 5 = 20}$$

When adding or subtracting mixed fractions, we deal with the whole numbers and fractions separately. When multiplying mixed fractions, we must first convert the mixed fractions to improper fractions and then perform the multiplication. For example, what is half of one and one-third?

$$\frac{1}{2} \times 1\frac{1}{3} = ?$$

If we convert $1\frac{1}{3}$ to an improper fraction, we will be dealing only with thirds. Because there are three thirds in one whole ($\frac{3}{3} = 1$), these three thirds plus the additional one-third piece equals four-thirds.

$$1\frac{1}{3} = \frac{3}{3} + \frac{1}{3} = \frac{4}{3}$$

Now that the mixed fraction is an improper fraction, we can complete the original problem, which is: What is half of one and one-third, or four-thirds?

$$\frac{1}{2} \times \frac{4}{3} = \frac{4}{6} = \frac{4 \div 2}{6 \div 2} = \frac{2}{3}$$

As expected, half of four-thirds equals two-thirds. As you can see, it was necessary to reduce the first result, $\frac{4}{6}$, to $\frac{2}{3}$. This often occurs due to the larger numbers in the improper fractions.

Before doing another example, let's develop a procedure for converting a mixed fraction to an improper fraction. As an example, we will convert the mixed fraction $2\frac{1}{2}$ to an improper fraction. Because the result will be all halves ($2\frac{1}{2} = \frac{?}{2}$), we simply have to determine how many halves are in 2 whole units and then add this value to the additional half. In summary:

$$2\frac{1}{2} = \frac{4}{2} + \frac{1}{2} = \frac{5}{2}$$

Two and one-half = four-halves + one-half = five-halves

To change a mixed fraction to an improper fraction, therefore, you *multiply the whole number by the fraction's denominator and then add the fraction's numerator.*

$$2\frac{1}{2} = \frac{5}{2}$$

Steps:
Multiply whole number by fraction's denominator: $2 \times 2 = 4$.
Add result to fraction's numerator: $4 + 1 = 5$.

Now that we understand how to convert a mixed fraction to an improper fraction, we can do a multiplication example involving mixed fractions.

EXAMPLE:

What is $\frac{4}{9}$ of $5\frac{3}{4}$?

Solution:

Note that the term *of* implies multiplication. To determine the result, we must first convert the mixed fraction, $5\frac{3}{4}$, to an improper fraction.

$$5\frac{3}{4} = \frac{23}{4}$$

Steps:
$5 \times 4 = 20$.
$20 + 3 = 23$.

The next step is to perform the multiplication using the improper fraction.

$$\frac{4}{9} \times \frac{23}{4} = \frac{\boxed{92} \div 4 = 23}{\boxed{36} \div 4 = 9}$$

↑ Reduce

Once reduced, the answer may be an improper fraction and therefore will have to be converted back into a mixed fraction. With $^{23}\!/_9$, the denominator 9 indicates that the original whole unit was divided into 9 pieces, and the numerator 23 indicates that we have 23 of these pieces. Dividing 9 into 23 will determine how many whole units we can obtain and how many ninths remain.

$$\frac{23}{9} = 2\frac{5}{9} \qquad \begin{array}{l} \textit{Steps:} \\ \text{9 into 23} = 2 \\ \qquad \text{remainder 5} \end{array}$$

■ INTEGRATED MATH APPLICATION:

The formula for converting a temperature in degrees Celcius (°C) to degrees Fahrenheit (°F) is:

$$°F = \left(\frac{9}{5} \times °C\right) + 32$$

If you were traveling in Europe and the temperature was 37 °C, what would that temperature be in degrees Fahrenheit?

$$°F = \left(\frac{9}{5} \times °C\right) + 32$$

$$= \left(\frac{9}{5} \times 37 °C\right) + 32$$

$$= \left(\frac{9}{5} \times \frac{37}{1}\right) + 32$$

$$= \frac{333}{5} + 32$$

$$= 66\frac{3}{5} + 32$$

$$= 98\frac{3}{5} °F$$

SELF-TEST EVALUATION POINT FOR SECTION 1-4

Now that you have completed this section, you should be able to:

■ **Objective 9.** *Show how to multiply proper and mixed fractions.*

Use the following questions to test your understanding of Section 1-4.

1. $\dfrac{1}{3} \times \dfrac{2}{3} = ?$

2. $\dfrac{4}{8} \times \dfrac{2}{9} = ?$

3. $\dfrac{4}{17} \times \dfrac{3}{34} = ?$

4. $\dfrac{1}{16} \times 5\dfrac{3}{7} = ?$

5. $2\dfrac{1}{7} \times 7\dfrac{2}{3} = ?$

1-5   DIVIDING FRACTIONS

Once again let us understand what we are doing by using a simple example: How many quarters are in three-quarters? Stated mathematically, this problem appears as follows:

$$\frac{3}{4} \div \frac{1}{4} = ?$$

The answer is straightforward and requires no mathematical processes or steps because we already know that there are three quarters in three-quarters.

$$\frac{3}{4} \div \frac{1}{4} = 3$$

Since most problems are never this simple, we need to follow a mathematical process to divide fractions. Because addition is the opposite of subtraction, and multiplication is the opposite of division, we can just invert the second fraction in a division problem and then multiply both fractions to obtain the result. Let us try this processs on the previous example:

$$\frac{3}{4} \div \frac{1}{4} = \frac{3 \times 4}{4 \times 1} = \frac{12 \div 4}{4 \div 4} = \frac{3}{1} = 3$$

Invert and multiply   Reduce   1 into 3 = 3

■ **EXAMPLE:**

How many twelfths are in two-thirds $\left(\frac{2}{3} \div \frac{1}{12}\right)$?

■ *Solution:*

$$\frac{2}{3} \div \left(\frac{1}{12}\right) = \frac{2 \times 12}{3 \times 1} = \frac{24 \div 3}{3 \div 3} = \frac{8}{1} = 8$$

Invert and multiply   Reduce

1-5-1   *Dividing Mixed Fractions*

As when we multiply mixed fractions, to divide mixed fractions we must first convert all mixed fractions to improper fractions. Once we have done this, we can divide as usual, which means: Invert the second fraction and multiply. To start, let us use a simple example and test the procedure to see how many halves are in $2\frac{1}{2}$. As you have already determined, the result should indicate that there are five halves in two and one-half.

Steps

Convert mixed fraction to improper fraction:
$2 \times 2 = 4, 4 + 1 = 5.$

$$\left(2\frac{1}{2}\right) \div \frac{1}{2} = ?$$

$$\frac{5}{2} \div \left(\frac{1}{2}\right) = ?$$

Invert second fraction and multiply.

$$\frac{5 \times 2}{2 \times 1} = \frac{10}{2}$$

Reduce.

$$\frac{10 \div 2}{2 \div 2} = \frac{5}{1} = 5$$

■ **EXAMPLE:**

$$6\frac{2}{3} \div 1\frac{1}{3} = ?$$

■ *Solution:*

$$6\frac{2}{3} \div 1\frac{1}{3} = ?$$

Steps:
Convert mixed fractions to improper fractions:
$6 \times 3 = 18; 18 + 2 = 20.$
$1 \times 3 = 3; 3 + 1 = 4.$

$$\frac{20}{3} \div \frac{4}{3} = ?$$

$$\frac{20 \times 3}{3 \times 4} = \frac{60}{12}$$

Invert second fraction and multiply.

$$\frac{60 \div 12}{12 \div 12} = \frac{5}{1} = 5$$

Reduce.

Now let's do an example that does not work out so neatly.

■ **EXAMPLE:**

How many times will $3\frac{5}{9}$ go into $6\frac{1}{2}$?

■ *Solution:*

Note that the term *go into* implies division and is asking how many times the first number will go into the second number.

$$6\frac{1}{2} \div 3\frac{5}{9} = ?$$

Steps:

$$\frac{13}{2} \div \frac{32}{9} = ?$$

Convert mixed fractions to improper fractions.

$$\frac{13 \times 9}{2 \times 32} = \frac{117}{64}$$

Invert second fraction and multiply.

$$\frac{117}{64} = 1\frac{53}{64}$$

Convert improper fraction to mixed fraction.
$117 \div 64 = 1$ remainder 53.

■ **INTEGRATED MATH APPLICATION:**

If you are paying an annual percentage rate of $15\frac{1}{8}$ percent on a credit card, what is the monthly percentage rate?

■ *Solution:*

$$15\frac{1}{8} \div 12 = ?$$

$$\frac{120}{8} \div \frac{12}{1} = ?$$

$$\frac{120 \times 1}{8 \times 12} = \frac{120}{96} = 1\frac{24}{96} = 1\frac{1}{4}\%$$

Now that you have completed this section, you should be able to:

■ **Objective 10.** *Show how to divide proper and mixed fractions.*

Use the following questions to test your understanding of Section 1-5.

1. $\dfrac{2}{3} \div \dfrac{1}{12} = ?$

2. $3\dfrac{1}{4} \div \dfrac{3}{4} = ?$

3. $8\dfrac{16}{20} \div 4\dfrac{4}{20} = ?$

4. $4\dfrac{3}{8} \div 2\dfrac{5}{16} = ?$

1-6   CANCELING FRACTIONS

Canceling

The process of removing a common divisor from numerator and denominator.

Canceling is an operation that can be performed on fractions only when a multiplication symbol exists between the two. Its advantage is a reduction in the size of the numbers in the problem and therefore in time needed to solve, because fewer steps are needed. Let us demonstrate the process with a simple example.

$$\frac{5}{12} \times \frac{4}{20} = ?$$

$$\frac{5 \times 4}{12 \times 20} = \frac{20}{240} \overset{\div\,20}{\underset{\div\,20}{=}} \frac{1}{12}$$

As you can see from this example, multiplying large numbers generates large results that generally need to be reduced. By canceling before we perform the multiplication, we start with smaller fractions and thus simplify the process.

$$\frac{5}{12} \times \frac{4}{20} \qquad = \qquad \frac{5}{3} \times \frac{1}{20} \qquad = \qquad \frac{1}{3} \times \frac{1}{4} = \frac{1}{12}$$

Four divides evenly into 12 and 4.
$12 \div 4 = 3$,
$4 \div 4 = 1$.

Five divides evenly into 5 and 20.
$5 \div 5 = 1$,
$20 \div 5 = 4$.

Multiplication is now easier with no need for reducing.

Up to this point we have "reduced" fractions only by finding a number that will divide evenly into the top and bottom part of the same fraction. In this example you can see that we could not reduce $^5/_{12}$ because there was no number that would evenly divide into 5 and 12; however, we could reduce the numerator of one fraction with the denominator of another fraction. Reducing in this crisscross manner is called cancellation and can be done only when a multiplication sign is present between the two fractions.

■ **EXAMPLE:**

Calculate the following: $\dfrac{4}{9} \div \dfrac{8}{27}$

■ *Solution:*

$$\frac{4}{9} \div \frac{8}{27} = ?$$

Invert second fraction and multiply.

Now that a multiplication symbol exists between the two fractions, we can see if it is possible to crisscross cancel.

$$\frac{\textcircled{4}}{9} \times \frac{27}{\textcircled{8}} = ?$$

Cancellation: $4 \div 4 = 1, 8 \div 4 = 2$.

$$\frac{1}{\textcircled{9}} \times \frac{\textcircled{27}}{2} =$$

Cancellation: $9 \div 9 = 1, 27 \div 9 = 3$.

$$\frac{1}{1} \times \frac{3}{2} = \frac{3}{2} = 1\frac{1}{2}$$

Convert improper fraction to mixed fraction.

■ EXAMPLE:

Calculate the following: $3\frac{5}{7} \times 2\frac{2}{13}$

■ *Solution:*

$$3\frac{5}{7} \times 2\frac{2}{13} = ?$$

$$\frac{26}{7} \times \frac{28}{13} = \frac{2}{7} \times \frac{28}{1} = \frac{2}{1} \times \frac{4}{1}$$

$$\frac{2}{1} \times \frac{4}{1} = \frac{8}{1} = 8$$

Steps:
Convert mixed fractions to improper fractions:
$3 \times 7 = 21, 21 + 5 = 26$.
$2 \times 13 = 26, 26 + 2 = 28$.
Cancellation: $13 \div 13 = 1, 26 \div 13 = 2$.
Cancellation: $28 \div 7 = 4, 7 \div 7 = 1$.

SELF-TEST EVALUATION POINT FOR SECTION 1-6

Now that you have completed this section, you should be able to:

■ *Objective 11.* *Demonstrate how to use cancellation to simplify fraction multiplication or division.*

Use the following questions to test your understanding of Section 1-6.

1. $\dfrac{16}{27} \times \dfrac{27}{240} = ?$

2. $\dfrac{8}{17} \times \dfrac{3}{4} = ?$

3. $\dfrac{15}{32} \div \dfrac{30}{4} = ?$

4. $6\dfrac{2}{3} \times 2\dfrac{7}{10} = ?$

5. $6\dfrac{1}{2} \div \dfrac{9}{32} = ?$

REVIEW QUESTIONS

Multiple Choice Questions

1. A fraction is a piece; therefore if you have a fraction, you have something _____.

   a. Between 1 and 10   c. Less than 1
   b. Greater than 1   d. Between 1 and 100

2. The more cuts you make in the whole unit, the _____ the fraction.

   a. Larger
   b. Smaller

3. The number above the fraction bar is called the _____, while the number below the fraction bar is called the _____.

   a. Denominator, factor
   b. Numerator, denominator
   c. Denominator, numerator
   d. Numerator, factor

4. How many ninths are in a whole unit?

  a. 9
  b. Depends on the value of the numerator
  c. 16
  d. All the above

5. If you had $6\frac{3}{4}$ apples in one pocket and $2\frac{1}{2}$ apples in the other, what would you have?

  a. Big pockets
  b. $8\frac{1}{4}$ apples
  c. $9\frac{3}{4}$ apples
  d. $9\frac{1}{4}$ apples

6. What is the lowest common denominator of 3 and 15?

  a. 45   **c.** 5
  b. 15   **d.** 90

7. A(n) _____ fraction has a numerator that is smaller than the denominator.

  a. Mixed   **c.** Proper
  b. Improper   **d.** All the above

8. A(n) fraction contains both a whole number and a fraction.

  a. Mixed   **c.** Improper
  b. Proper   **d.** All the above

9. A(n) _____ fraction has a numerator that is larger than the denominator.

  a. Mixed   **c.** Improper
  b. Proper   **d.** All the above

10. Reduce the following fraction to its lowest terms: $^{32}/_{144}$.

  a. $\dfrac{4}{18}$ **b.** $\dfrac{2}{9}$ **c.** $\dfrac{1}{6}$ **d.** $\dfrac{8}{32}$

Communication Skill Questions

11. What is a fraction, and what do the two numbers above and below the fraction bar line indicate?   (1-1)

12. Describe the best method for calculating the lowest-value common denominator.   (1-2)

13. What is a prime factor?   (1-2)

14. What is an improper fraction?   (1-2)

15. What is a mixed fraction?   (1-2)

16. Arbitrarily choose values, and describe the steps used to:

  a. Add fractions.   (1-2)
  b. Subtract fractions.   (1-3)
  c. Multiply fractions.   (1-4)
  d. Divide fractions.   (1-5)

17. How can fractions be reduced?   (1-2)

18. Why do you need to sometimes borrow when subtracting fractions?   (1-3)

19. Arbitrarily choose values, and describe the steps used to:

  a. Add mixed fractions.   (1-2)
  b. Subtract mixed fractions.   (1-3)
  c. Multiply mixed fractions.   (1-4)
  d. Divide mixed fractions.   (1-5)

20. How are fractions canceled?   (1-6)

Practice Problems

21. Write the whole number and fraction for each of the following quantities.

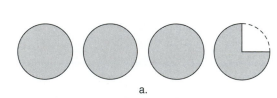

a.

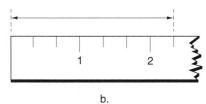

b.

22. Consider the value $5\frac{7}{9}$. State which of the numbers represents the:

  a. Whole number
  b. Denominator
  c. Numerator
  d. Number of original pieces
  e. Number of fractional pieces we have

23. Add the following fractions.

  a. $\dfrac{5}{8} + \dfrac{2}{8} = ?$

  b. $\dfrac{47}{76} + \dfrac{15}{76} + \dfrac{1}{76} = ?$

24. Calculate the lowest common denominator and then add the following fractions.

  a. $\dfrac{4}{9} + \dfrac{1}{3} = ?$

  b. $\dfrac{4}{7} + \dfrac{2}{21} = ?$

  c. $\dfrac{2}{14} + \dfrac{1}{8} + \dfrac{1}{4} = ?$

  d. $\dfrac{4}{10} + \dfrac{1}{5} + \dfrac{2}{15} = ?$

25. Convert the following improper fractions to mixed fractions.

   a. $\dfrac{5}{2} = ?$ **c.** $\dfrac{25}{16} = ?$

   b. $\dfrac{17}{4} = ?$ **d.** $\dfrac{37}{3} = ?$

26. Add the following mixed fractions.

   a. $1\dfrac{3}{4} + 2\dfrac{1}{2} = ?$

   b. $9\dfrac{4}{10} + 6\dfrac{7}{8} = ?$

27. Reduce the following fractions.

   a. $\dfrac{4}{16} = ?$ **c.** $\dfrac{74}{148} = ?$

   b. $\dfrac{16}{18} = ?$ **d.** $\dfrac{28}{45} = ?$

28. Perform the following subtraction of fractions.

   a. $\dfrac{8}{14} - \dfrac{2}{7} = ?$ **c.** $4\dfrac{1}{16} - 2\dfrac{31}{32} = ?$

   b. $3\dfrac{4}{9} - 1\dfrac{1}{3} = ?$ **d.** $5\dfrac{3}{5} - 2\dfrac{4}{15} = ?$

29. Multiply the following. Cancel and reduce if possible.

   a. $\dfrac{1}{9} \times 4 = ?$ **c.** $\dfrac{1}{3} \times 4\dfrac{1}{2} = ?$

   b. $\dfrac{3}{6} \times \dfrac{4}{5} = ?$ **d.** $2\dfrac{3}{4} \times 4\dfrac{4}{11} = ?$

30. Divide the following. Cancel and reduce if possible.

   a. $\dfrac{1}{4} \div 8 = ?$ **c.** $3\dfrac{5}{7} \div 1\dfrac{6}{7} = ?$

   b. $\dfrac{3}{5} \div \dfrac{5}{9} = ?$ **d.** $14\dfrac{8}{9} \div 2\dfrac{4}{5} = ?$

Web Site Questions

Go to the Web site http://www.prenhall.com/cook, select the textbook *Mathematics for Electronics and Computers*, select this chapter, and then follow the instructions when answering the multiple-choice practice problems.

Decimal Numbers

An Apple a Day

Electronics is the branch of science concerned with the use of electricity and electronic devices to manage information signals. As a science it has grown out of electrical engineering, which is concerned with the generation, control, distribution, storage, and applications of electrical power. Today, electronics is a separate science involving numerous scientific disciplines, including physics, chemistry, thermodynamics, quantum mechanics, and mathematics.

One of the early pioneers who laid the foundations of many branches of science was Isaac Newton. He was born in a small farmhouse near Woolsthorpe in Lincolnshire, England, on Christmas Day in 1642. He was an extremely small, premature baby, which worried the midwives who went off to get medicine and didn't expect to find him alive when they came back. Luckily for science, however, he did survive.

Newton's father was an illiterate farmer who died three months before he was born. His mother married the local vicar soon after Newton's birth and left him in the care of his grandmother. This parental absence while he was growing up had a traumatic effect on him and throughout his life affected his relationships with people.

At school in the nearby town of Grantham, Newton showed no interest in classical studies but rather in making working models and studying the world around him. When he was in his early teens, his stepfather died and Newton had to return to the farm to help his mother. Newton proved to be a hopeless farmer; in fact, on one occasion when he was tending sheep he became so engrossed with a stream that he followed it for miles and was missing for hours. Luckily, a schoolteacher recognized Newton's single-minded powers of concentration and convinced his mother to let him return to school, where he performed better and later went off to Cambridge University.

In 1665, in his graduation year, Newton left Cambridge to return home to escape an epidemic of bubonic plague that had spread throughout London. During this time, Newton reflected on his years of seclusion at his mother's cottage and called them the most significant time in his life. It was here on a warm summer's day that Newton saw the apple fall to the ground, leading him to develop his laws of motion and gravitation. It was here that he wondered about the nature of light and later built a prism and proved that white light contains all the colors in a rainbow.

Later, when Newton returned to Cambridge, he demonstrated many of his discoveries but was reluctant to publish the details and did so finally only at the insistence of others. Newton went on to build the first working reflecting astronomical telescope and wrote a paper on optics that was fiercely challenged by the physicist Robert Hooke. Hooke quarreled bitterly with Newton over the years, and there were also heated debates about whether Newton or the German mathematician Gottfried Leibniz invented calculus.

The truth is that many of Newton's discoveries roamed around with him in the English countryside, and even though many of these would, before long, have been put forward by others, it was Newton's genius and skill (and long walks) that tied together all the loose ends.

Outline and Objectives

VIGNETTE: AN APPLE A DAY

INTRODUCTION

2-1   COUNTING IN DECIMAL

Objective 1: Describe the origin of the decimal number system.

2-1-1 Base or Radix

Objective 2: Explain the basic process of counting in decimal and the following terms.
   a. Base or radix
   b. Positional weight
   c. Reset and carry

2-1-2 Positional Weight

2-1-3 Reset and Carry

2-2   THE DECIMAL POINT

Objective 3: Define the function of the decimal point.

2-2-1 Positional Weight

Objective 4: Describe the positional weight of decimal fractions.

2-2-2 No Decimal Point and Extra Zeros

2-2-3 Converting Written Fractions to Decimal Fractions

Objective 5: Demonstrate how to:
   a. Convert written fractions to decimal fractions.
   b. Calculate reciprocals.
   c. Convert decimal fractions to written fractions.

2-2-4 Reciprocals

2-2-5 Converting Decimal Fractions to Written Fractions

2-3   CALCULATING IN DECIMAL

Objective 6: Describe the relationships among the four basic arithmetic operations: addition, subtraction, multiplication, and division.

2-3-1 Decimal Addition

Objective 7: Demonstrate how to perform the following arithmetic operations.
   a. Decimal addition
   b. Decimal subtraction
   c. Decimal multiplication
   d. Decimal division

2-3-2 Decimal Subtraction

2-3-3 Decimal Multiplication

2-3-4 Decimal Division

2-3-5 Ratios

Objective 8: Describe the following mathematical terms and their associated operations.
   a. Ratios
   b. Rounding off
   c. Significant places
   d. Percentages
   e. Averages

2-3-6 Rounding Off

2-3-7 Significant Places

2-3-8 Percentages

2-3-9 Averages

MULTIPLE CHOICE QUESTIONS

COMMUNICATION SKILL QUESTIONS

PRACTICE PROBLEMS

WEB SITE QUESTIONS

Introduction

The *decimal system of counting and calculating* was developed by Hindu mathematicians in India around A.D. 400. The system involved using fingers and thumbs, so it was natural that it would have 10 numerals or *digits,* which is a word meaning "finger." The use of this 10-digit or 10-finger system became widespread and by A.D. 800 was used extensively by Arab cultures. The 10 digits were represented by combinations of the 10 arabic numerals: *0, 1, 2, 3, 4, 5, 6, 7, 8,* and *9.* This is still called the *arabic number system.* The system eventually found its way to England, where it was adopted by nearly all the European countries by A.D. 1200. The Europeans called it the *decimal number system* after the Latin word *decimus,* which means "ten."

In this chapter we will discuss the decimal system of counting and calculating. **Decimal** is a topic with which you probably already feel extremely comfortable because we make use of it every day in countless applications. For example, our money is based on the decimal number system (there are 100 cents in 1 dollar, or 10 dimes in 1 dollar), and the metric system of measurement is also based on the decimal system (there are 10 millimeters in 1 centimeter, and 100 centimeters in 1 meter). Basic as it seems, however, you should still review this chapter because it will discuss all the terminology and procedures regarding how the decimal system is used, and this will serve as our foundation for all subsequent chapters.

Decimal

Numbered or proceeding by tens; based on the number 10.

2-1   COUNTING IN DECIMAL

Counting was probably the first application for a number system such as decimal because people needed some way of keeping track of things. *Decimal is based on the number 10.* Let us first examine what is meant by the *base* of a number system.

2-1-1   *Base or Radix*

Base or Radix

A number equal to the number of units in a given number system. The decimal system uses a base of 10.

Subscript

A distinguishing letter or numeral written immediately below and to the right or left of another character.

The key element that distinguishes one number system from another is the **base** *or* **radix** *of the number system.* The base describes the number of digits that are used. The decimal number system has a base of 10, which means that it makes use of 10 digits (0 through 9) to represent the value of a quantity. A **subscript,** which is a smaller number written to the right or left of and below the main number, is sometimes included to indicate the number's base. For example, all the numbers following have the subscript "$_{10}$" to indicate that they are base 10, or decimal, numbers.

Large numbers indicate value of quantity. 4956_{10} $235{,}163_{10}$ 8_{10} 11_{10} Subscript indicates base of number.

If a subscript is not present, the number is assumed to be a base 10, or decimal, number. For example, the following are all decimal numbers, even though they do not include the subscript 10.

23   101   3,634,987   47

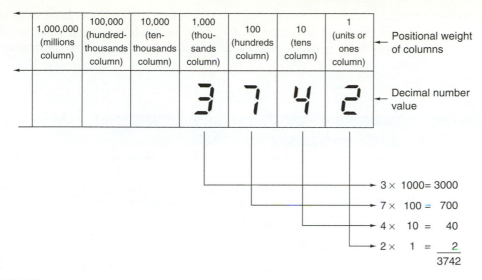

FIGURE 2-1 Positional Weight of Each Decimal Column.

2-1-2 Positional Weight

The position of each digit of a decimal number determines the weight of that digit. For example, a 2 by itself indicates only a value of two, whereas a 2 to the left of three zeros (2000) indicates a value of two thousand. The position of a decimal digit therefore determines its weight, or value, with digits to the left carrying the greater weight. For example, the decimal number 3742 contains three thousands (3 × 1000), seven hundreds (7 × 100), four tens (4 × 10), and two ones (2 × 1). Figure 2-1 shows this decimal number with the positional weight of each decimal column.

You may have noticed that the weight of each decimal (base 10) column increases by a factor of 10 as you move to the left. For example, the units column has a weight of 1. The next column to the left of the units column (tens column) has a weight that is ten times (× 10) larger than the units column—10 × one = 10. The next column to the left of the tens column (hundreds column) has a weight ten times (× 10) larger than the tens column— 10 × ten = 100. Therefore, each column in the decimal number system increases by ten times (× 10) as you move to the left.

2-1-3 Reset and Carry

Figure 2-2 shows the reset and carry action that occurs when we count in decimal. *This* **reset and carry action** *occurs whenever a column reaches the highest decimal digit of 9.* For example, when we begin to count from zero (0), the units column continues to advance by one or **increment** until it reaches 9. An increment beyond this point causes the units column to be reset to zero and a one to be carried into the tens column, producing 10 (ten). As this count continues, the units column will again repeat the cycle going from 0 to 9, which with the one in the tens column will be a count of 10 (ten) to 19 (nineteen). An increase by one beyond 19 will again cause the units column to reset to zero and carry a one into the tens column, resulting in a count of 20 (twenty). The units column will continue to cycle from 0 to 9 and reset and carry into the tens column until the tens column reaches the highest decimal digit of 9. As you can see in Figure 2-2, an advance by one beyond a count of 99 will cause the units column to reset and carry one into the tens column, and because the tens column is at its maximum of 9, it will also reset to zero and carry one into the hundreds column. This "reset to zero and carry one into the next left column action" will continue to occur in

FIGURE 2-2 The Reset and Carry Action That Occurs When Counting in Decimal.

Reset and Carry Action

An operation that occurs when counting, in which a digit resets after advancing beyond its maximum count and carries a one into the next higher order column.

Increment

The action or process of increasing a quantity or value by one.

a right-to-left motion whenever a column reaches a maximum digit of nine. In other words, the units column will reset and carry into the tens column, the tens column will reset and carry into the hundreds column, the hundreds column will reset and carry into the thousands column, and so on.

2-2 THE DECIMAL POINT

Decimal Point

A period or centered dot between the parts of a decimal mixed number, separating the whole number from the fraction.

Decimal Fraction

A decimal value that is less than one.

The **decimal point (dp),** *symbolized by "."*, *is used to separate a whole decimal number from a* **decimal fraction.** Using a monetary example, we ask: What does the amount $12.75 actually mean? The number to the left of the decimal point, which is 12 in this example, indicates how many whole or complete dollars we have. On the other hand, the number to the right of the decimal point indicates the fraction, or the amount that is less than one dollar. Because there are 100 cents in $1, and this example indicates that we have 75 cents, this fraction is actually three-fourths of a dollar. In summary:

The numbers to the left of the decimal point indicate the amount of whole or complete units we have. In this monetary example, we have 12 complete or whole dollars.

Decimal point

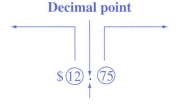

$12 ⁝ 75

The numbers to the right of the decimal point indicate the *decimal fraction*, or amount that is less than one. In this monetary example, we have 75 cents or $^{75}/_{100}$ (which reduces to $^3/_4$).

The decimal point (dp) separates the whole numbers on the left from the fraction on the right. In this monetary example, the decimal point separates the complete dollars on the left from the fraction of a dollar or cents on the right.

Most calculators have a key specifically for entering the decimal point. It operates as follows.

30   CHAPTER 2 / DECIMAL NUMBERS

2-2-1 *Positional Weight*

Most of us know that $0.50 is half a dollar, or 50 cents; $0.25 is a fourth or a quarter of a dollar, or 25 cents; and $0.75 is three-fourths or three-quarters of a dollar, or 75 cents; however, when it comes to explaining why $½ = $0.50, $¼ = $0.25, and $¾ = $0.75, many people have difficulty. Why are these decimal fractions of 0.50, 0.25, and 0.75 equivalent to ½, ¼, and ¾? The answer is in the position that each digit occupies compared to the decimal point. To explain this further, Figure 2-3 repeats our previous example of $12.75, or twelve dollars and seventy-five cents. Once again, each place or position carries its own weight or value, and these positions are all relative to the decimal point. For instance, the digit in the first column to the left of the decimal point indicates how many units or ones are in the value, whereas the digit in the second column to the left of the decimal point indicates how many tens are in the value. Digits to the left of the decimal point all indicate the number of whole numbers in the value. On the other hand, digits to the right of the decimal point indicate the fraction or piece of a whole that is in the value. A value such as 3.5, therefore, indicates that we have three whole units and five-tenths ($^5/_{10}$). Remembering our proper fractions from Chapter 1,

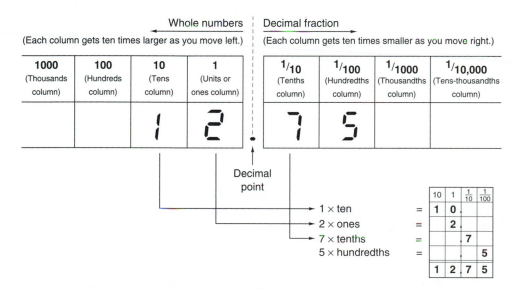

FIGURE 2-3   The Positional Weight of Decimal Fractions.

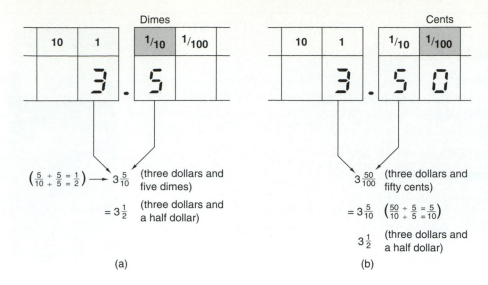

$\left(\frac{5}{10} \div \frac{5}{5} = \frac{1}{2}\right) \longrightarrow 3\frac{5}{10}$ (three dollars and five dimes)

$= 3\frac{1}{2}$ (three dollars and a half dollar)

(a)

$3\frac{50}{100}$ (three dollars and fifty cents)

$= 3\frac{5}{10}$ $\left(\frac{50}{10} \div \frac{5}{5} = \frac{5}{10}\right)$

$3\frac{1}{2}$ (three dollars and a half dollar)

(b)

FIGURE 2-4   **Interpreting the Decimal Fraction.**

we know that $\frac{5}{10}$ means that a whole unit was divided into ten parts and we have five of those parts. Because 5 is half of 10, we have half of a whole unit ($\frac{5}{10}$ can be reduced to $\frac{1}{2}$).

Is there a difference, therefore, between \$3.5 and \$3.50? The answer is no, which can be explained by looking at Figure 2-4. In Figure 2-4(a) we have interpreted the value as 3 dollars and $\frac{5}{10}$ of a dollar. Because $\frac{5}{10} = \frac{1}{2}$, the value in Figure 2-4(a) is 3 dollars and a half dollar. In Figure 2-4(b) we can interpret the decimal fraction as $\frac{5}{10}$ and $\frac{0}{100}$, or $\frac{50}{100}$. Because $\frac{50}{100} = \frac{5}{10} = \frac{1}{2}$, the value in Figure 2-4(b) is also 3 dollars and a half dollar. As far as money is concerned, we tend to refer to the fraction of a dollar in $\frac{1}{100}$ parts because a dollar is divided into 100 parts, and each of these parts is called a cent (there are 100 fractional parts or cents in a dollar).

Looking at this another way, we could label the $\frac{1}{10}$ column as dimes (10¢) and the $\frac{1}{100}$ column as cents (1¢). A value of 3.5 would then be interpreted as 3 dollars and 5 dimes, and a value of 3.50 would be interpreted as 3 dollars and 50 cents, as shown in Figure 2-4(a) and (b).

■ EXAMPLE:

State the individual digit weights of the following number: 7.25.

■ *Solution:*

The first step is to give each column its respective weight or value.

1	$\frac{1}{10}$	$\frac{1}{100}$
7	2	5

It is now easier to see that we have 7 whole units, $\frac{2}{10}$ (two-tenths), and $\frac{5}{100}$ (five-hundredths). Said another way, we have 7 whole units and $\frac{25}{100}$ (twenty-five hundredths). Because $\frac{25}{100}$ can be reduced to $\frac{1}{4}$,

$$\frac{25 \div 25 = 1}{100 \div 25 = 4}$$

we could also interpret this value as seven and one-fourth ($7\frac{1}{4}$).

2-2-2 *No Decimal Point and Extra Zeros*

What happens if a value such as

$$76$$

does not have a decimal point? The answer is: If you do not see a decimal point, always assume that it is to the right of the last digit in the number. In this example, therefore, the decimal point will follow the 6 in the number 76, as follows:

$$76.$$

Another point to remember: Always ignore extra zeros that are added to the left or right of a number. For example, to emphasize that a number such as .33 is a fraction, many people write this number as 0.33. The extra zero in this example is used to frame the decimal point so that it is not missed. You may have noticed that your calculator will add this zero if you press the keys $\boxed{.}\ \boxed{3}\ \boxed{3}$ and will display 0.33. Sometimes, zeros are added to the left of decimal whole numbers as well as to the right of decimal fractions. For example:

$$1367.5$$

Extra zeros → 0293.3
are used to put digits 0006.67
in their correct column. 0049.24

These extra zeros are used to align the digits of each number in the appropriate column. Without them, digits may be placed in the wrong column, as follows:

$$1367.5$$
$$0293.3$$
$$6.67$$
$$49.24$$

In other instances, extra zeros are placed to the right of a decimal fraction to indicate how to interpret the fraction. For example, three dollars and fifty cents would be written as 3.50 instead of 3.5 even though the extra zero after the five has no bearing on the value. This is so the value will be interpreted as three dollars and fifty cents ($3 + \frac{50}{100}$) instead of three dollars and five dimes ($3 + \frac{5}{10}$).

In all instances, extra zeros that are outside the number (to the left of the decimal whole number or to the right of the decimal fraction) do not change the value of the number.

2-2-3 *Converting Written Fractions to Decimal Fractions*

To convert proper fractions (such as $\frac{1}{2}$, $\frac{2}{6}$, $\frac{2}{3}$, $\frac{9}{16}$) to decimal fractions (such as 0.5, 0.3, 0.666, 0.562) we simply perform the operation that is indicated by the fraction bar. For example, when a fraction bar separates a numerator of 1 from a denominator of 2 ($\frac{1}{2}$), the arithmetic operation described by the fraction bar is a division of 2 into 1. Performing this operation, you will convert the proper fraction $\frac{1}{2}$ to its decimal fraction equivalent.

$$\frac{1}{2} = 1 \div 2 = 0.5$$

Calculator sequence:

$\boxed{1}\ \div\ \boxed{2}\ \boxed{=}$
Answer: 0.5

■ **EXAMPLE:**

Convert the following proper fractions to decimal fractions.

   a. $\dfrac{3}{4}$ **b.** $\dfrac{1}{4}$

c. $\dfrac{2}{3}$ **e.** $\dfrac{3}{9}$

d. $\dfrac{8}{16}$ **f.** $\dfrac{1}{8}$

■ *Solution:*

a. $\dfrac{3}{4} = 3 \div 4 = 0.75$

b. $\dfrac{1}{4} = 1 \div 4 = 0.25$

c. $\dfrac{2}{3} = 2 \div 3 = 0.666$

d. $\dfrac{8}{16} = 8 \div 16 = 0.5$ $\quad\left(\dfrac{8 \div 8}{16 \div 8} = \dfrac{1}{2} = 0.5 \right)$

e. $\dfrac{3}{9} = 3 \div 9 = 0.333$ $\quad\left(\dfrac{3 \div 3}{9 \div 3} = \dfrac{1}{3} = 0.3 \right)$

f. $\dfrac{1}{8} = 1 \div 8 = 0.125$

The next question is: How do we convert a mixed fraction such as 3½ to its decimal equivalent? The answer is to do nothing with the whole number except place it in the whole-number decimal columns and then convert the proper fraction as before. Let us do the following example.

■ **EXAMPLE:**

Convert the following mixed fractions to their decimal equivalents.

a. $3\dfrac{1}{2}$ **c.** $27\dfrac{4}{9}$

b. $486\dfrac{1}{4}$ **d.** $31,348\dfrac{8}{9}$

■ *Solution:*

a. $3\dfrac{1}{2} = 3.5$ $\quad(1 \div 2 = 0.5)$

b. $486\dfrac{1}{4} = 486.25$ $\quad(1 \div 4 = 0.25)$

c. $27\dfrac{4}{9} = 27.444$ $\quad(4 \div 9 = 0.444)$

d. $31,348\dfrac{8}{9} = 31,348.888$ $\quad(8 \div 9 = 0.888)$

Finally, let us convert an improper fraction such as ⁴⁄₂ (four halves) to its decimal equivalent. This is the easiest operation to perform, as shown in the following example, because all you do is perform the division on your calculator to obtain a decimal whole number and fraction.

■ **EXAMPLE:**

Convert the following improper fractions to their decimal equivalents.

a. $\dfrac{4}{2}$ **b.** $\dfrac{7}{3}$ **c.** $6\dfrac{7}{2}$ **d.** $\dfrac{18}{16}$

■ *Solution:*

a. $\dfrac{4}{2} = 4 \div 2 = 2.0$

b. $\dfrac{7}{3} = 7 \div 3 = 2.333$

c. $6\dfrac{7}{2} = 6 + (7 \div 2) = 6 + 3.5 = 9.5 \qquad \left(\dfrac{7}{2} = 3\dfrac{1}{2} = 3.5 \right)$

d. $\dfrac{18}{16} = 18 \div 16 = 1.125 \qquad \left(\dfrac{18 \div 2}{16 \div 2} = \dfrac{9}{8} = 1\dfrac{1}{8} \right)$

2-2-4   *Reciprocals*

Whenever you divide a number into 1, you get the **reciprocal** *of that number.* The word *reciprocal* means "inverse" or "opposite relation." This term can best be described by determining the reciprocal of two numbers, 2 and 2000. As just mentioned, to get the reciprocal of these numbers, we simply divide these numbers into 1, as shown:

Original numbers

Reciprocals of original numbers

$$\dfrac{1}{2} = 0.5 \qquad \dfrac{1}{2000} = 0.0005$$

Comparing these two answers, you can see why the reciprocals, or result of dividing the numbers into 1, are inversely related to the original numbers. For instance, the reciprocal of a small number ($\downarrow$), such as 2, results in a large fraction ($\uparrow$), 0.5, whereas the reciprocal of a large number ($\uparrow$), such as 2000, results in a small fraction ($\downarrow$), 0.0005. This inverse relationship is even more evident when you consider that 2000 is one thousand times greater than 2, yet the reciprocal of 2000 is one thousand times smaller than the reciprocal of 2. To summarize:

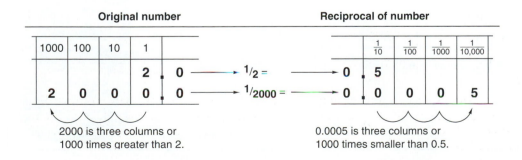

	Original number						Reciprocal of number			
1000	100	10	1				$\frac{1}{10}$	$\frac{1}{100}$	$\frac{1}{1000}$	$\frac{1}{10{,}000}$
			2 .	0	½ =	0 .	5			
2	0	0	0 .	0	½₀₀₀ =	0 .	0	0	0	5

2000 is three columns or 1000 times greater than 2.

0.0005 is three columns or 1000 times smaller than 0.5.

Most calculators have a key specifically for calculating the reciprocal of any number. It is called the reciprocal key and operates as follows.

CALCULATOR KEYS

Name: Reciprocal key

Function: Divides the value on the display into 1.

Example: $\dfrac{1}{3.2} = ?$

Press keys: $\boxed{3}\boxed{.}\boxed{2}\boxed{1/x}$

Display shows: 0.3125

PI

The sixteenth letter of the Greek alphabet; the symbol π denoting the ratio of the circumference of a circle to its diameter; a transcendental number having a value to eight decimal places of 3.14159265.

The reciprocal of π is used in many different formulas. The Greek lowercase letter pi, which is pronounced "pie" and symbolized by π, is a constant value that describes how much bigger a circle's circumference is than its diameter.

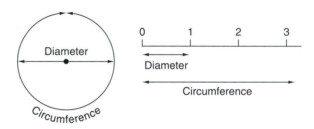

The approximate value of this constant is 3.142, which means that the circumference of any circle is approximately 3 times bigger than its diameter.

Most calculators have a key for quickly obtaining the value of pi. It is called the pi (π) key and operates as follows.

CALCULATOR KEYS

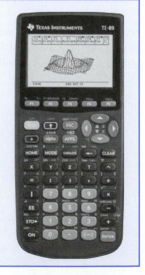

Name: Pi key

Function: Enters the value of pi correct to eight decimal digits.

π = 3.1415927

Example: $2 \times \pi = ?$

Press keys: $\boxed{2}\boxed{\times}\boxed{\pi}\boxed{=}$

Display shows: 6.2831853

The factor π is used frequently in many formulas involved with the analysis of circular motion. The reciprocal $\frac{1}{2}\pi$ (1 over $2 \times \pi$) is frequently used, so we should do a few problems involving the reciprocal and pi keys on the calculator.

■ **EXAMPLE:**

Calculate the reciprocals of the following numbers.

    a. 5   **b.** π **c.** 60   **d.** $2 \times \pi$

■ *Solution:*

    a. The reciprocal of $5 = \dfrac{1}{5} = 1 \div 5 = 0.2$.

        Calculator sequence: $\boxed{5}\boxed{1/x}$; display shows: 0.2

    b. The reciprocal of $\pi = \dfrac{1}{\pi} = \dfrac{1}{3.142} = 0.318$.

        Calculator sequence: $\boxed{\pi}\boxed{1/x}$; display shows: 0.3183098

    c. The reciprocal of $60 = \dfrac{1}{60} = 0.0166$.

        Calculator sequence: $\boxed{6}\boxed{0}\boxed{1/x}$; display shows: 0.0166667

    d. The reciprocal of $2 \times \pi = \dfrac{1}{2 \times \pi} = \dfrac{1}{6.28} = 0.159$.

        Calculator sequence: $\boxed{2}\boxed{\times}\boxed{\pi}\boxed{=}\boxed{1/x}$; display shows: 0.1591549

2-2-5   *Converting Decimal Fractions to Written Fractions*

First, let us discuss why we would want to convert a decimal fraction, such as 0.5, into a written fraction, such as $\frac{1}{2}$. The answer is so that we can convert from a scale of 10 (decimal) to a scale that is something other than 10. For example, imagine that you have a standard ruler with 1 inch divided into 16 parts (there are sixteen $\frac{1}{16}$ in 1 inch). Next, imagine that you have to measure a length of 0.125 of an inch. Since the scale of the ruler is dealing with 16 parts and your decimal value of 0.125 is based on 10 parts, the two are not compatible and a conversion is needed. The first step is to convert $\frac{1}{16}$ into its decimal equivalent:

$$\frac{1}{16} = 1 \div 16 = 0.0625$$

This answer tells us that one-sixteenth of an inch is equivalent to 0.0625 of an inch in decimal. This can be proved by multiplying 0.0625 by 16 to see if we have 16 of these parts in 1 inch.

$$16 \times 0.0625 = 1 \text{ inch}$$

Now that we know that the value 0.0625 inch is equal to $\frac{1}{16}$ inch, the next step is to find out how many 0.0625 parts (or $\frac{1}{16}$) are in our decimal measurement of 0.125.

$$\frac{0.125}{0.0625} = 0.125 \div 0.0625 = 2$$

The answer of 2 tells us that there are two 0.0625 parts (or sixteenths) in 0.125. Therefore, to measure off a length of 0.125 of an inch using a $\frac{1}{16}$ inch scale, simply measure a length of two-sixteenths ($\frac{2}{16}$).

$$0.125 \text{ in.} = \frac{2}{16} \text{ in.}$$

■ **EXAMPLE:**

Convert the following decimal fractions to proper fractions.

a. $0.25 = \dfrac{?}{8}$

b. $0.666 = \dfrac{?}{3}$

c. $0.59375 = \dfrac{?}{32}$

■ *Solution:*

a. 0.25 equals how many eighths? $\frac{1}{8}$ has a decimal equivalent of 0.125 ($1 \div 8 = 0.125$). How many 0.125 parts are in 0.25?

$$\frac{0.25}{0.125} = 0.25 \div 0.125 = 2$$

$$0.25 = \frac{2}{8} = \frac{1}{4}$$

b.

$$\frac{1}{3} = 0.333$$

$$\frac{0.666}{0.333} = 2$$

$$0.666 = \frac{2}{3}$$

c.

$$\frac{1}{32} = 0.03125$$

$$\frac{0.59375}{0.03125} = 19$$

$$0.59375 = \frac{19}{32}$$

CALCULATOR KEYS

Name: Conversion Functions

Display the answer as a fraction (▶ Frac).
Display the answer as a decimal (▶ Dec).

Function:

Value ▶ Frac (display as a fraction) displays an answer as its written fraction equivalent.
Value ▶ Dec (display as a decimal) displays an answer as its decimal equivalent.

Example: $\frac{1}{2} + \frac{1}{3}$ ▶ Frac

$$\frac{5}{6}$$

Ans ▶ Dec
.8333333333

Now that you have completed this section, you should be able to:

■ **Objective 3.** *Define the function of the decimal point.*

■ **Objective 4.** *Describe the positional weight of decimal fractions.*

■ **Objective 5.** *Demonstrate how to convert written fractions to decimal fractions, calculate reciprocals, and convert decimal fractions to written fractions.*

Use the following questions to test your understanding of Section 2-2.

1. What are the individual digit weights of the number 178.649?

2. Convert the following written fractions to decimal fractions.

   a. $1\dfrac{5}{2}$ **b.** $192\dfrac{3}{4}$ **c.** $67\dfrac{6}{9}$

3. Calculate the reciprocal of the following numbers.

   a. 2500   **b.** 0.25   **c.** 0.000025

4. Convert the following decimal fractions to written fractions.

   a. $7.25 = \dfrac{?}{16}$ **b.** $156.90625 = \dfrac{?}{32}$

2-3   CALCULATING IN DECIMAL

In this section we first review the basic principles of decimal addition, subtraction, multiplication, and division. As you will discover, there are really only *two basic operations: addition and subtraction.* If you can perform an addition, you can perform a multiplication because *multiplication is simply repeated addition.* For example, a problem such as 4×2 (4 multiplied by 2) is asking you to calculate the sum of four 2s ($2 + 2 + 2 + 2$). Similarly, *division is simply repeated subtraction.* For example, a problem such as $9 \div 3$ (9 divided by 3) is asking you to calculate how many times 3 can be subtracted from 9 ($9 - 3 = 6, 6 - 3 = 3, 3 - 3 = 0$; therefore, there are three 3s in 9). In summary, therefore, the two basic mathematical operations (addition and subtraction) and their associated mathematical operations (multiplication and division) are as follows:

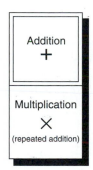

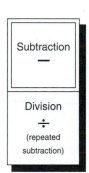

There is another relationship that needs to be discussed, and that is the inverse or opposite nature between addition and subtraction and between multiplication and division. For example, *subtraction is the opposite arithmetic operation of addition.* This can be proved by starting with a number such as 10, then adding 5, then subtracting 5.

$$\text{(10)} + 5 = 15, \quad 15 - 5 = \text{(10)}$$
$$\uparrow \qquad\qquad\qquad\qquad \uparrow$$

    Original number   Back to original number

Similarly, *multiplication is the opposite arithmetic operation of division*. This can be proved by starting with a number such as 4, then multiplying by 3, then dividing by 3.

$$④ \times 3 = 12, \quad 12 \div 3 = ④$$

Original number Back to original number

In summary, therefore, addition is the opposite mathematical operation of subtraction, and multiplication is the opposite mathematical operation of division. This opposite relationship will be made use of in a subsequent chapter.

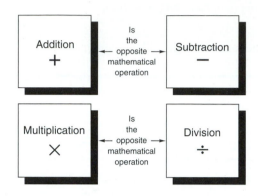

Every calculator has keys for performing these four basic arithmetic operations. The name, function, and procedure for using these keys are as follows.

CALCULATOR KEYS

Name: Decimal point key $\boxed{.}$ (discussed previously)

Name: Equals key $\boxed{=}$

Function: Combines all previously entered numbers and operations. Used to obtain intermediate and final results.

Example: $3 \times 4 + 6 = ?$

Press keys: $\boxed{3}\boxed{\times}\boxed{4}\boxed{=}\boxed{+}\boxed{6}\boxed{=}$

Display shows: 18.0

Name: Add key $\boxed{+}$

Function: Instructs calculator to add the next entered quantity to the displayed number.

Example: Add $31.6 + 2.9 = ?$

Press keys: $\boxed{3}\boxed{1}\boxed{.}\boxed{6}\boxed{+}\boxed{2}\boxed{.}\boxed{9}\boxed{=}$

Displays shows: 34.5

Name: Subtract key $\boxed{-}$

Function: Instructs calculator to subtract the next entered quantity from the displayed number.

2-3-1 *Decimal Addition*

To review decimal **addition,** let us examine the steps involved in adding the following numbers.

$$66.332 + 285 + 0.002 + 182.788 = ?$$

First, list the numbers vertically, making sure that the decimal points are lined up as follows.

Addition

The operation of combining numbers so as to obtain an equivalent quantity.

Addition process:
$(100)\ (10)\ (1)\ .\ (\frac{1}{10})\ (\frac{1}{100})\ (\frac{1}{1000})$

Carry →

	2	1	1	1	1
	0 6	6.	3	3	2
	2 8	5.	0	0	0
	0 0	0.	0	0	2
+1 8	2.	7	8	8

Sum → 5 3 4. 1 2 2

The extra zeros are included to keep the digits in their proper columns.

Steps:
Starting at the right column ($\frac{1}{1000}$), add vertically.

$2 + 0 + 2 + 8 = 12,$ 2 carry 1.
$1 + 3 + 0 + 0 + 8 = 12,$ 2 carry 1.
$1 + 3 + 0 + 0 + 7 = 11,$ 1 carry 1.
$1 + 6 + 5 + 0 + 2 = 14,$ 4 carry 1.
$1 + 6 + 8 + 0 + 8 = 23,$ 3 carry 2.
$2 + 0 + 2 + 0 + 1 = 5.$

Calculator sequence:
$\boxed{6}\boxed{6}\boxed{.}\boxed{3}\boxed{3}\boxed{2}\boxed{+}\boxed{2}\boxed{8}\boxed{5}\boxed{+}\boxed{.}\boxed{0}\boxed{0}\boxed{2}\boxed{+}\boxed{1}\boxed{8}\boxed{2}\boxed{.}\boxed{7}\boxed{8}\boxed{8}\boxed{=}$
Answer: 534.122

There are a few points to note regarding this previous addition. First, the number 285 did not have a decimal point indicated; therefore, it is assumed to be to the right of the last digit (285.0). Second, what is really happening when we **carry** or overflow into the next-left column? The answer is best explained by examining the addition of the units column in the

Carrying

The process of transferring from one column to another during an addition process.

Sum

The result obtained when adding numbers.

previous example. In this column we had a result of $1 + 6 + 5 + 0 + 2 = 14$. As you know from our discussion on positional weight, 14 can be broken down into 1 ten (1×10) and 4 ones (4×1). This is why the 1 ten is carried over to the tens column, and the 4 ones or units are placed in the units sum column. Explained another way, whenever the sum of all the digits in any column exceeds the maximum single decimal digit of 9, a reset and carry action occurs. In the case of the units column with a sum of 14, we exceeded 9 and therefore the units sum was reset to 0 and then advanced to 4, and a 1 was carried into the tens column to make 14.

In summary, therefore, because each column to the left is 10 times larger, a reset and carry action will occur every time the sum of any column is incremented or advanced by one, from 9 to 10.

■ EXAMPLE:

How much is $976.73 + $998.28?

■ *Solution:*

Addend

The number or quantity to be added to an existing quantity called the augend.

Augend

The number to which a new quantity, called the addend, is to be added.

		Steps
Carry:	1111 1	$3 + 8 = 11$, 1 carry 1.
Addend:	0976.73	$1 + 7 + 2 = 10$, 0 carry 1.
Augend:	$+ 0998.28$	$1 + 6 + 8 = 15$, 5 carry 1.
Sum:	$1975.01	$1 + 7 + 9 = 17$, 7 carry 1.
		$1 + 9 + 9 = 19$, 9 carry 1.
		$1 + 0 + 0 = 1$, 1 carry 0.

Calculator sequence: 9 7 6 . 7 3 +
9 9 8 . 2 8 =

Answer: 1975.01

■ INTEGRATED MATH APPLICATION:

Calculate the total cost of the following service station repair invoice:

Parts:	$ 64.95
Labor:	$226.35
Coolant:	$ 4.25
Tax:	$ 22.17

■ *Solution:*

$$
\begin{array}{r}
64.95 \\
226.35 \\
4.25 \\
+ 22.17 \\
\hline
\$317.72
\end{array}
$$

■ INTEGRATED MATH APPLICATION:

How many hours did you work in the previous week if your time card shows the following:

Monday—6 hours
Tuesday—4.5 hours
Wednesday—7.5 hours
Thursday—5 hours
Friday—8 hours

■ *Solution:*

$$6 + 4.5 + 7.5 + 5 + 8 = 31 \text{ hours}$$

2-3-2   *Decimal Subtraction*

To review decimal **subtraction,** let us examine the steps involved in the following example:

$$\$2005.23 - \$192.41 = ?$$

To begin, list the numbers vertically, making sure that the decimal points are lined up as follows:

Subtraction process:

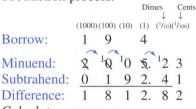

	(1000)	(100)	(10)	(1)	Dimes ($^1/_{10}$)	Cents ($^1/_{100}$)
Borrow:	1	9	4			
Minuend:	2	0	0	5.	2	3
Subtrahend:	0	1	9	2.	4	1
Difference:	1	8	1	2.	8	2

Calculator sequence:

| 2 | 0 | 0 | 5 | . | 2 | 3 | − |

| 1 | 9 | 2 | . | 4 | 1 | = |

Answer: 1812.82
           or $1812.82

Steps:
Starting at the right column ($^1/_{100}$ or cent) subtract vertically:
$3 - 1 = 2$
2 − 4: impossible to subtract 4 from the smaller number 2. Therefore, we will *borrow* $1 from the $5 in the units column. This will leave $4 in the units column; because $1 = 10 dimes, we will be adding 10 dimes to the 2 dimes already present, giving a total of 12 dimes.
$12 - 4 = 8$
$4 - 2 = 2$
0 − 9: impossible to subtract 9 from nothing or zero.
Because we cannot borrow from the tens or hundreds columns, we will have to borrow from the thousands column.
Borrowing 1000 (10×100) gives 10 hundreds; borrowing 1 hundred from 10 hundreds gives 9 hundreds. Borrowing 100 (10×10) gives 10 tens.
$10 - 9 = 1$
$9 - 1 = 8$
$1 - 0 = 1$

As you can see from this example, *borrowing* is the exact opposite of carrying. With addition we sometimes end up with a sum of 10 or more, which results in an overflow or carry into the next-higher or next-left column, whereas with subtraction we sometimes end up with a larger number in the subtrahend than the minuend, and it becomes necessary to borrow a 1 from the next-higher place to the left in the minuend.

■ EXAMPLE:

Perform the following subtraction:

$$27,060.0143 - 963.86$$

■ *Solution:*

	(10,000)	(1000)	(100)	(10)	(1)	($^1/_{10}$)	($^1/_{100}$)	($^1/_{1000}$)	($^1/_{10,000}$)	
		6	9	5	9.	9				*Steps*
	2	7	0	6	0.	0	1 4 3			$3 - 0 = 3$
	−0	0	9	6	3.	8	6 0 0			$4 - 0 = 4$
Difference	2	6	0	9	6.	1	5 4 3			1 − 6: can't do; borrow.

Calculator sequence:

| 2 | 7 | 0 | 6 | 0 | . | 0 | 1 | 4 | 3 | − |

| 9 | 6 | 3 | . | 8 | 6 | = |

Answer: 26,096.1543

Borrowing 1 ten from 6 tens gives 5 tens and 10 ones.
Borrowing 1 one from 10 ones gives 9 ones and 10 tenths.
Borrowing 1 tenth from 10 tenths gives 9 tenths and 10 hundredths.

Subtraction
The operation of deducting one number from another.

Minuend
A number from which the subtrahend is to be subtracted.

Subtrahend
A number that is to be subtracted from a minuend.

Difference
The amount by which things differ in quantity or measure.

10 hundredths + the original 1 hundredth = 11 hundredths:

$11 - 6 = 5$
$9 - 8 = 1$
$9 - 3 = 6$
$5 - 6$: can't do; borrow.

Borrowing 1 thousand from 7 thousands gives 6 thousands and 10 hundreds. Borrowing 1 hundred from 10 hundreds gives 9 hundreds and 10 tens. 10 tens + the original 5 tens = 15 tens:

$15 - 6 = 9$
$9 - 9 = 0$
$6 - 0 = 6$
$2 - 0 = 2$

■ INTEGRATED MATH APPLICATION:

If the regular price of a television is $856.99 and the sale price is $699.99, how much will you save?

■ *Solution:*

$$\begin{array}{r} \$859.99 \\ -\$699.99 \\ \hline \$160.00 \end{array}$$ A savings of $160.00.

■ INTEGRATED MATH APPLICATION:

If your gross weekly salary is $695.00 and your deductions total $212.65, what is your take-home pay?

■ *Solution:*

$$\begin{array}{r} \$695.00 \\ -\$212.65 \\ \hline \$482.35 \end{array}$$ Your weekly take-home pay is $482.35.

2-3-3   *Decimal Multiplication*

Multiplication

The operation of adding an integer to itself a specified number of times.

Multiplicand

The number that is to be multiplied by another.

Multiplier

One that multiplies; a number by which another number is multiplied.

To review decimal **multiplication,** let us examine the steps involved in the following example:

$$27.66 \times 2.4 = ?$$

Unlike in the previous mathematical operations, we can ignore the alignment of the decimal point and simply make sure that the two numbers are flush on the right side.

Multiplicand: 27.66
Multiplier: $\times$ 2.4   Two numbers are listed vertically so that they are flush on the right side.

We can now use the following multiplication steps:

Multiplication process:

Second carry:	11 1
First carry:	132 2
Multiplicand:	27.6 6
Multiplier:	$\times$ 2.4
First partial product:	110 6 4
Second partial product:	5532 0
Final product:	6638 4

Steps:

To begin, multiply each digit in the multiplicand by the first digit of the multiplier (which is 4) to obtain the first partial product. Line up the rightmost digit of the first partial product under the multiplier digit being used (see arrow).

44   CHAPTER 2 / DECIMAL NUMBERS

$4 \times 6 = 24$, 4 carry 2.
$4 \times 6 = 24$, $24 +$ previous carry of $2 = 26$, 6 carry 2.
$4 \times 7 = 28$, $28 +$ carry of $2 = 30$, 0 carry 3.
$4 \times 2 = 8$, $8 +$ carry of $3 = 11$, 1 carry 1; bring 1 down.

The next step is to multiply each digit in the multiplicand by the second digit of the multiplier (which is 2) to obtain the second partial product. Line up the rightmost digit of the second partial product under the multiplier digit being used (see arrow).

$2 \times 6 = 12$, 2 carry 1.
$2 \times 6 = 12$, $12 +$ carry of $1 = 13$, 3 carry 1.
$2 \times 7 = 14$, $14 +$ carry of $1 = 15$, 5 carry 1.
$2 \times 2 = 4$, $4 +$ carry of $1 = 5$.

The next step is to add the first and second partial products to obtain the final product.

$4 + 0 = 4$
$6 + 2 = 8$
$0 + 3 = 3$
$1 + 5 = 6$
$1 + 5 = 6$

Calculator sequence:

[2][7][.][6][6][×][2][.][4][=]

Answer: 66.384

27.6 6 The multiplicand has two digits to the right of the decimal point.
2.4 The multiplier has one digit to the right of the decimal point.
$2 + 1 = 3$ places.

Answer is 66.384

← The final step is to determine the position of the decimal point in the answer by counting the number of digits to the right of the decimal point in the multiplier and multiplicand. There are 3 in this example, so the final product will have three digits to the right of the decimal point.

66.3 8 4

■ **EXAMPLE:**

How much is 0.035 multiplied by 0.005?

■ *Solution:*

In this example we will take a few shortcuts. Since the decimal points do not make a difference at this stage, let us drop them completely. This will produce:

$$\begin{array}{r} 0035 \\ \times\ 0005 \end{array}$$

Second, we learned earlier that extra zeros appearing to the left or right of a number are not needed in the multiplication process, so let us remove these zeros:

$$\begin{array}{r} 35 \\ \times\ 5 \end{array}$$

The multiplication is now a lot easier because it is simply 35×5. Performing the multiplication, we obtain the product:

$$\begin{array}{r} \overset{2}{3}5 \\ \times\ \ 5 \\ \hline 175 \end{array}$$

The final step is to determine the position of the decimal point. The original multiplicand had three digits to the right of the decimal point (0.0 3 5), and the original multiplier also had three digits to the right of the decimal point (0.0 0 5), so the final product should have six digits to the right (3 + 3) of the decimal point.

$$.000175 \text{ or } 0.000175$$

■ INTEGRATED MATH APPLICATION:

If your car averages 26.3 miles per gallon, how far can you travel if your gas tank is filled with 12 gallons of gasoline?

■ *Solution:*

$$26.3 \text{ miles per gallon} \times 12 \text{ gallons} = 315.6 \text{ miles}$$

■ INTEGRATED MATH APPLICATION:

If an inch on a street map is equivalent to 0.5 mile, how many miles will you have to travel if the distance on the map measures 14.5 inches?

■ *Solution:*

$$14.5 \text{ inches} \times 0.5 \text{ mile} = 7.25 \text{ miles}$$

2-3-4   *Decimal Division*

Division

The operation of separating a number into equal parts.

Quotient

The number resulting from the division of one number by another.

Divisor

The number by which a dividend is divided.

Dividend

A number to be divided.

Remainder

The final undivided part after division that is less or of lower degree than the divisor.

To review decimal **division** (or *long division,* as it is sometimes called) let us step through the following division example: 1362 ÷ 12. The procedure is as follows.

Long-division process:

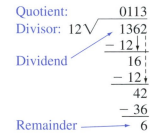

Quotient: 0113
Divisor: 12 √ 1362
            − 12
Dividend   16
            − 12
            42
            − 36
Remainder ————→ 6

Steps:
The first step is to see how many times the divisor (12) can be subtracted from the first digit of the dividend (1). Because the dividend is smaller, the answer or quotient is zero (1 − 12 cannot be done). We place this zero above the dividend 1.

Next, we see how many times the divisor (12) can be subtracted from the first two digits of the dividend (13). 13 − 12 = 1, remainder 1. The answer or quotient of 1 is placed above the next digit of the dividend (3), and the remainder of the subtraction (1) is dropped below the 13 − 12 line.

We carry the next digit of the dividend (6) down to the remainder (1) to address the next part of the dividend, which is 16. Now we repeat the process by seeing how many times the divisor, 12, can be subtracted from these two digits, 16. The answer is a quotient of 1 (which is placed above the 6 dividend) with a remainder of 4 (which is dropped below the 16 − 12 line).

Calculator sequence:

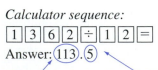

1 3 6 2 ÷ 1 2 =
Answer: 113.5

The whole decimal number is telling you how many times 12 will go

The decimal fraction indicates how much of a fraction of the divisor is remaining.

Bringing down the next dividend digit of 2 and combining it with the previous remainder of 4, we now have the final part of the dividend, which is 42. Because 12 can be subtracted from 42 three times, a quotient of 3 is placed

into 1362.
Therefore,
$1362 \div 12 = $ ⟨113⟩
Answer to $1362 \div 12$ is 113, remainder 6.

As another example, $13 \div 4 = ?$

| 1 | 3 | ÷ | 4 | = |

Answer: ③. ㉕

4 goes into 13 three times.
$13 \div 4 = $ ③

Remainder is $\frac{1}{4}$ or 0.25 of 4; therefore, $0.25 \times 4 = $ ①

Answer to $13 \div 4$ is 3, remainder 1.

Therefore,
$0.5 \times 12 = 6$

above the final digit of the dividend (2), and the remainder of the subtraction, 6, is dropped below the $42 - 36$ line. Therefore, $1362 \div 12 = 113$, remainder 6. Said another way, 12 can be subtracted from 1362 113 times, with 6 remaining.

■ **EXAMPLE:**

Calculate the result of dividing 19,205.628 by 1.56.

■ *Solution:*

$19,205.628 \div 1.56 = 12,311.3$
Calculator sequence: | 1 | 9 | 2 | 0 | 5 | . | 6 | 2 | 8 | ÷ | 1 | . | 5 | 6 | = |
Answer: 12,311.3

■ **INTEGRATED MATH APPLICATION:**

If your round-trip commute to work for five days totals 87.5 miles, how many miles are you driving per day?

■ *Solution:*

$$87.5 \text{ miles} \div 5 \text{ days} = 17.5 \text{ miles per day}$$

■ **INTEGRATED MATH APPLICATION:**

If your gross pay for a week is $536.50 and you worked 37 hours, what is your hourly rate of pay?

■ *Solution:*

$$\$536.50 \div 37 \text{ hours} = \$14.50 \text{ per hour}$$

2-3-5   *Ratios*

A **ratio** is a comparison of one number to another number. For example, a ratio such as 7:4 ("seven to four") is comparing the number 7 to the number 4. This ratio could be written in one of the following three ways:

$\frac{7}{4}$ ← Ratio is written as a fraction.
7:4 ← Ratio is written using the ratio sign (:).
1.75 to 1 ← Ratio is written as a decimal.

Ratio

The relationship in quantity, amount, or size between two or more things.

To express the ratio as a decimal, the number 7 was divided by the number 4, giving a result of 1.75 (7 ÷ 4 = 1.75). This result indicates that the number 7 is 1.75 times (or one and three-quarter times) larger than the number 4.

■ INTEGRATED MATH APPLICATION:

If one store has 360 items and another store has 100 of the same items, express the ratio of 360 to 100:

   a. Using the divide sign
   b. Using the ratio sign
   c. As a decimal

■ *Solution:*

   a. $\dfrac{360}{100} = \dfrac{360 \div 20}{100 \div 20} = \dfrac{18}{5}$ ← 360 to 100 and 18 to 5 are equivalent ratios.

   b. $360 : 100 = 18 : 5$

   c. 3.6 ← The first store has 3.6 times as many items as the other store (18 ÷ 5 = 3.6).

■ EXAMPLE:

Calculate the ratio of the following two like quantities. If necessary, reduce the ratio to its lowest terms.

   a. $0.8 \div 0.2 = ?$
   b. $0.0008 \div 0.0002 = ?$

■ *Solution:*

   a. How many 0.2s are in 0.8? The answer is the same as saying, How many 2s are in 8?

   $$0.8 \div 0.2 = 4 \qquad \dfrac{8 \div 2}{2 \div 2} = \dfrac{4}{1} \qquad \text{(a ratio of 4 to 1)}$$

   b. How many 0.0002s are in 0.0008?

   $$0.0008 \div 0.0002 = 4 : 1$$

The ratio in example (a) is the same as in (b). Another way to describe this is that because the size of each decimal fraction compared to the other decimal fraction (or the *ratio*) remained the same, the answer or quotient remained the same.

   8 is four times larger than 2, or 2 is four times smaller than 8.

   0.8 is four times larger than 0.2, or 0.2 is four times smaller than 0.8.

   0.0008 is four times larger than 0.0002, or 0.0002 is four times smaller than 0.0008. Therefore, all have the same 4 : 1 ratio.

■ INTEGRATED MATH APPLICATION:

In some instances, a ratio will be used to describe how a quantity is to be divided. For example, if 600 textbooks are to be divided between two schools in the ratio of 1 to 2 (1:2), how many textbooks will each school get? To calculate this, you should first add the terms of the ratio:

■ *Solution:*

   Step 1: The ratio is 1:2, so 1 + 2 = 3 (this means that the total is to be divided into 3 parts).

Next, you should multiply the total by each fractional part:

Step 2: The first school will receive ⅓ of the total 600 textbooks, which is 200 texts (⅓ × 600 = 200). The second school will receive ⅔ of the total 600 textbooks, which is 400 texts (⅔ × 600 = 400).

■ **INTEGRATED MATH APPLICATION:**

If three restaurants receive 216 hamburgers in the ratio of 6:14:7, how many hamburgers will each restaurant get?

The first step is to add the terms of the ratio:

■ *Solution:*

Step 1: The ratio is 6:14:7, so 6 + 14 + 7 = 27 (this means that the total is to be divided into 27 parts).

Next, you should multiply the total by each fractional part:

Step 2: The first restaurant will receive ⁶⁄₂₇ of the total 216 hamburgers, which is 48 (⁶⁄₂₇ × 216 = 48). The second restaurant will receive ¹⁴⁄₂₇ of the total 216 hamburgers, which is 112 (¹⁴⁄₂₇ × 216 = 112). The third restaurant will receive ⁷⁄₂₇ of the total 216 hamburgers, which is 56 (⁷⁄₂₇ × 216 = 56).

2-3-6   *Rounding Off*

In many cases we will **round off** *decimal numbers because we do not need the accuracy indicated by a large number of decimal digits.* For example, it is usually unnecessary to have so many decimal digits in a value such as 74.139896428. If we were to round off this value to the nearest hundredths place, we would include two digits after the decimal point, which is 74.13. This is not accurate, however, since the digit following 74.13 was a 9 and therefore one count away from causing a reset and carry action into the hundredths column. To take into account the digit that is to be dropped when rounding off, therefore, we follow this basic rule: *When the first digit to be dropped is 6 or more, increase the previous digit by 1. On the other hand, when the first digit to be dropped is 4 or less, do not change the previous digit.* Therefore, the value 74.139896428, rounded off to the nearest hundredths place, would equal:

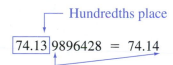

┌── Hundredths place

74.13│9896428 = 74.14

First digit to be dropped is greater than 6, so previous digit should be increased by 1.

■ **EXAMPLE:**

Round off the value 74.139896428 to the nearest ten, whole number, tenth, hundredth, thousandth, and ten-thousandth.

■ *Solution:*

[7]4.139896428 rounded off to the nearest ten = 70
    └─ First digit to be dropped is 4 or less; therefore, do not change previous digit.

[74.]139896428 rounded off to the nearest whole number = 74
    └─ First digit to be dropped is 4 or less; therefore, do not change previous digit.

$\boxed{74.1}$ 39896428 rounded off to the nearest tenth = 74.1

    └─ First digit to be dropped is 4 or less; therefore, do not change previous digit.

$\boxed{74.13}$ 9896428 rounded off to the nearest hundredth = 74.14

    └─ First digit to be dropped is 6 or greater; therefore, increase previous digit by 1 (3 to 4).

$\boxed{74.139}$ 896428 rounded off to the nearest thousandth = 74.140

    └─ First digit to be dropped is 6 or greater; therefore, increase previous digit by 1. Since previous digit is 9, allow reset and carry action to occur. The steps are:

    74.1398 ← 8 carries 1 into thousandths column.

    74.130 ← 9 resets to 0, and carries 1 into hundredths column.

    74.140 ← Hundredths digit is increased by 1.

$\boxed{74.1398}$ 96428 rounded off to the nearest ten-thousandth = 74.1399

    └─ First number to be dropped is 6 or greater; therefore, increase previous number by 1.

In the preceding example we discovered how to round off if a digit was over halfway up the decimal scale (greater than 6), or if a digit was below the halfway point on the decimal scale (less than 4). The next question, therefore, is: How do we round off when the digit to be dropped is a 5 and therefore midway in the decimal scale of 10? In this instance, we follow this rule: *When the 5 is followed by a digit that is more than zero, increase the previous digit by 1. On the other hand, when the 5 is followed by zero, do not change the previous digit.* Let us apply this to an example:

$\boxed{79.5}$ 3 rounded off to the nearest whole number = 80

Let us examine why this is correct. Because the digit following the 5 is more than zero, the decimal fraction (0.53) is in this case more than one-half or 0.5 (it is actually 5 tenths or ½ plus 3 hundredths), and therefore the 9 should be changed to 0 and the 7 increased to 8.

■ **EXAMPLE:**

Round off the value 31,520.565 to the nearest ten thousand, thousand, hundred, whole number, tenth, and hundredth.

■ *Solution:*

$\boxed{3}$ 1,520.565 rounded off to the nearest ten thousand = 30,000

    └ First digit to be dropped is 4 or less; therefore, do not change previous digit.

$\boxed{31,}$ 520.565 rounded off to the nearest thousand = 32,000

    ↖ The 5 is followed by a digit that is more than zero; therefore, increase previous digit by 1.

$\boxed{31,5}$ 20.565 rounded off to the nearest hundred = 31,500

    └ First digit to be dropped is 4 or less; therefore, do not change previous digit.

$\boxed{31,520}$ .565 rounded off to the nearest whole number = 31,521

    ↖ The 5 is followed by a digit that is more than zero; therefore, increase previous digit by 1.

$\boxed{31,520.5}$ 65 rounded off to the nearest tenth = 31,520.6

    └ Digit to be dropped is 6 or more; therefore, increase previous digit by 1.

31,520.56 5 rounded off to the nearest hundredth = 31,520.56

                  └ Because no digit follows 5, it is assumed to be a zero; therefore, do not change previous digit.

CALCULATOR KEYS

Name: Round function
round (*value*[,*#decimals*])

Function: Value is rounded to the number of decimals specified.

Example:
round (π, 4)
        3.1416

2-3-7 *Significant Places*

The number of significant places describes how many digits are in the value and how many digits in the value are accurate after rounding off. For example, a number such as 347.63 is a five-significant-place (or significant-figure) value because it has five digits in five columns. If we were to round off this value to a whole number, we would get 348.00. This value would still be a five-significant-place number; however, it would now be accurate to only three significant places. Let us examine a few problems to practice using the terms *significant places* and *accurate to significant places*.

■ **EXAMPLE:**

On a calculator, the value of π will come up as 3.141592654.

   a. Write the number π to six significant places.

   b. Give π to five significant places, and also round off π to ten-thousandths.

   c. 3.14159000 is the value of π to _____ significant places; however, it is accurate to only _____ significant places.

■ *Solution:*

   a. The value π to six significant places or figures is 3.14159 27

         3.14159

              └ Uses six digits or columns.

   b. The value π to five significant places.
     3.1415 92 rounded off to ten-thousandths = 3.1416

   c. 3.14159000 is the value of π to nine significant places (has nine digits); however, it is accurate to only six significant places (because the zeros to the right are just extra, and the value 3.14159 has only six digits).

2-3-8 *Percentages*

Percent

In the hundred; of each hundred.

The percent sign (%) means hundredths, which as a proper fraction is $\frac{1}{100}$ or as a decimal fraction is 0.01. For example, 50% means 50 hundredths ($\frac{50}{100} \div \frac{50}{} \div 50 = \frac{1}{2}$), which is one-half. In decimal, 50% means 50 hundredths (50 × 0.01 = 0.5), which is also one-half. The following shows how some of the more frequently used percentages can be expressed in decimal.

$$1\% = 1 \times 0.01 = 0.01 \quad \text{(For example, 1\% is expressed mathematically}$$
$$\text{in decimal as 0.01.)}$$
$$5\% = 5 \times 0.01 = 0.05$$
$$10\% = 10 \times 0.01 = 0.10$$
$$25\% = 25 \times 0.01 = 0.25$$
$$50\% = 50 \times 0.01 = 0.50$$
$$75\% = 75 \times 0.01 = 0.75$$

Most calculators have a percent key that automatically makes this conversion to decimal hundredths. It operates as follows.

CALCULATOR KEYS

Name: Percent key

Function: Converts the displayed number from a percentage to a decimal fraction.

Example: 18.6% = ?

Press keys: $\boxed{1}\boxed{8}\boxed{.}\boxed{6}\boxed{\%}$

Display shows: 0.186

Now that we understand that the percent sign stands for hundredths, what does the following question actually mean: What is 50% of 12? We now know that 50% is 50 hundredths ($\frac{50}{100}$), which is one-half. This question is actually asking: What is half of 12? Expressed mathematically as both proper fractions and decimal fractions, it would appear as follows:

Proper Fractions	Decimal Fractions
50% of 12 = ?	50% of 12 = ?
$= \dfrac{50}{100} \times 12$	$= (50 \times 0.01) \times 12$
$= \dfrac{1}{2} \times 12$	$= 0.5 \times 12$
$= 6$	$= 6$

Explaining this another way, we could say that percentages are fractions in hundredths. In the preceding example, therefore, the question is actually asking us to divide 12 into 100 parts and then to determine what value we would have if we had 50 of those 100 parts. Therefore, if we were to split 12 into 100 parts, each part would have a value of 0.12 ($12 \div 100 = 0.12$). Having 50 of these 100 parts would give us a total of $50 \times 0.12 = 6$.

■ **INTEGRATED MATH APPLICATION:**

If sales tax is 8% and the merchandise price is $256.00, what is the tax on the value, and what will be the total price?

■ *Solution:*

First, determine the amount of tax to be paid.

$$8\% \text{ of } \$256 =$$
$$0.08 \times 256 = \$20.48$$

Calculator sequence: 8 % × 2 5 6 =
Answer: 20.48
The total price paid will therefore equal

$$\$256.00 + \$20.48 = \$276.48$$

■ **INTEGRATED MATH APPLICATION:**

Which is the better buy—a $599.95 refrigerator at 25% off or a $499.95 refrigerator with a $50 manufacturer's rebate?

■ *Solution:*

$$25\% \text{ of } \$599.95 = \$150$$
$$\text{Refrigerator A: } \$599.95 - \$150.00 = \$449.95$$
$$\text{Refrigerator B: } \$499.95 - \$50.00 = \$449.95$$

The refrigerators are the same price.

2-3-9   *Averages*

*A mean **average** is a value that summarizes a set of unequal values.* This value is equal to the sum of all the values divided by the number of values. For example, if a team has scores of 5, 10, and 15, what is their average score? The answer is obtained by adding all the scores ($5 + 10 + 15 = 30$) and then dividing the result by the number of scores ($30 \div 3 = 10$).

Average

A single value that summarizes or represents the general significance of a set of unequal values.

■ **INTEGRATED MATH APPLICATION:**

The voltage at the wall outlet in your home was measured at different times in the day and equaled 108.6, 110.4, 115.5, and 123.6 volts (V). Find the average voltage.

■ *Solution:*

Add all the voltages.

$$108.6 + 110.4 + 115.5 + 123.6 = 458.1$$

Divide the result by the number of readings.

$$\frac{458.1}{4} = 114.525 \text{ V}$$

The average voltage present at your home was therefore 114.525 volts.

Statistics
A branch of mathematics dealing with the collection, analysis, interpretation, and presentation of masses of numerical data.

Statistics are generally averages and can misrepresent the actual conditions. For example, if five people earn an average salary of $204,600 a year, one would believe that all five are very well paid. Studying the individual figures, however, we find out that their yearly earnings are $8000, $6000, $4000, $20,000, and $985,000 per year. Therefore,

$$8000 + 6000 + 4000 + 20,000 + 985,000 = 1,023,000$$
$$\frac{1,023,000}{5} = 204,600$$

If one value is very different from the others in a group, this value will heavily influence the result, creating an average that misrepresents the actual situation.

■ **INTEGRATED MATH APPLICATION:**

A football player's "average number of yards gained per carry" ranking is calculated by dividing the total number of yards gained by the total number of times the player carried the football. Which of the following players had the higher average number of yards gained per carry?

Player	Total Yards	Total Carries
Payton	16,726	3,838
Brown	12,312	2,359

■ *Solution:*

$$\text{Payton: } \frac{16,726}{3,838} = 4.4 \text{ yards per carry average}$$

$$\text{Brown: } \frac{12,312}{2,359} = 5.2 \text{ yards per carry average}$$

On average, Brown gained a higher number of yards per carry.

SELF-TEST EVALUATION POINT FOR SECTION 2-3

Now that you have completed this section, you should be able to:

■ *Objective 6.* *Describe the relationships among the four basic arithmetic operations: addition, subtraction, multiplication, and division.*

■ *Objective 7.* *Demonstrate how to perform the following arithmetic operations: decimal addition,*

decimal subtraction, decimal multiplication, and decimal division.

■ *Objective 8.* *Describe the following mathematical terms and their associated operations: ratios, rounding off, significant places, percentages, and averages.*

Use the following questions to test your understanding of Section 2-3.

1. The two basic arithmetic operations are _____ and _____ .

2. Give the opposite of the following arithmetic operations.
   a. Addition
   b. Subtraction
   c. Multiplication
   d. Division

3. Perform the following arithmetic operations by hand, and then confirm your answers with a calculator.
   a. $26.443 + 197.1 + 2.1103 + 0.004 = ?$
   b. $19,637.224 - 866.43 = ?$
   c. $894.357 \times 8.6 = ?$
   d. $1.3397 \div 0.015 = ?$

4. Express a 176 meter to 8 meter ratio as a fraction, using the ratio sign and as a decimal.

5. Round off the following values to the nearest hundredth.
   a. 86.43760
   b. 12,263,415.00510
   c. 0.176600

6. Referring to the values in Question 5, describe:
   a. Their number of significant places
   b. To how many significant places they are accurate

7. Calculate the following.
   a. 15% of 0.5 = ?
   b. 22% of 1000 = ?
   c. 2.35% of 10 = ?
   d. 96% of 20 = ?

8. Calculate the average of the following values.
   a. 20, 30, 40, and 50
   b. 4000, 4010, 4008, and 3998

REVIEW QUESTIONS

Multiple Choice Questions

1. What is the base or radix of the decimal number system?
   a. 100
   b. 10
   c. 0 through 9
   d. 10, 100, 1000, 10,000

2. Because each column in the decimal number system increases by _____ times as you move left, two columns to the left will be _____ times greater.
   a. 100, 10
   b. 10, 1000
   c. 1000, 10
   d. 10, 100

3. The decimal point is used to:
   a. Separate the whole numbers from the fraction
   b. Separate the tens column from the units column
   c. Identify the decimal number system
   d. All the above

4. What is the positional weight of the second column to the right of the decimal point?
   a. Tenths
   b. Tens
   c. Units
   d. Hundredths

5. The Greek letter pi represents a constant value that describes how much bigger a circle's _____ is compared with its _____ , and it is frequently used in many formulas involved with the analysis of circular motion.
   a. Circumference, radius
   b. Circumference, diameter
   c. Diameter, radius
   d. Diameter, circumference

6. _____ is the opposite arithmetic operation of _____ , and _____ is the opposite arithmetic operation of _____ .
   a. Multiplication, division, subtraction, addition
   b. Division, addition, multiplication, subtraction
   c. $+, \div, -, \times$
   d. $\times, -, \div, +$

7. Is π a ratio?
   a. Yes
   b. No

8. Round off the following number to the nearest hundredth: 74.8552.
   a. 74.85
   b. 74.86
   c. 74.84
   d. 74.855

9. The number 27.0003 is a _____ significant-place number and is accurate to _____ significant places.
   a. 6, 2
   b. 2, 6
   c. 6, 6
   d. 4, 6

10. Express 29.63% mathematically as a decimal.
    a. 29.63
    b. 2.963
    c. 0.02963
    d. 0.2963

Communication Skill Questions

11. Describe the base, positional weight, and reset and carry action of the decimal number system. (2–1)

12. How are whole numbers and fractions represented in decimal? (2–2)

13. Describe how written fractions are converted to decimal fractions and, conversely, how decimal fractions are converted to written fractions. (2–2)

14. How do you obtain the reciprocal of a number, and what is its relationship to the original number? (2–2)

15. Arbitrarily choose values, and describe the steps involved in: (2–2)
    a. Decimal addition
    b. Decimal subtraction
    c. Decimal multiplication
    d. Decimal division

16. What is a ratio, and how is it expressed? (2–3)

17. Briefly describe the rules for rounding off. (2–3)

18. What is a percentage? (2–3)

19. How is an average calculated? (2–3)

20. What is the approximate value of the constant pi (π), and what does it describe? (2–2)

Practice Problems

21. Indicate the positional weight of each of the following digits: 96,237.

22. Write each of the following values as a number.
    a. Two hundred and seventy-six
    b. Eight thousand and seven tens
    c. Twenty thousand, four tens, and nine units

23. Which columns will reset and carry when each of the following values is advanced or incremented by 1?
    a. 199
    b. 409
    c. 29,909
    d. 49,999

24. What are the individual digit weights of 2.7463?

25. Write each of the following values as a number.
    a. Two units and three-tenths
    b. Five-tenths and seven-thousandths
    c. Nine thousand, three tens, and four-hundredths

26. Indicate which zeros can be dropped in the following values (excluding the zero in the units place).
    a. 02.017
    b. 375.011020
    c. 0.0102000

27. Convert the following written fractions to their decimal equivalents.
    a. $\dfrac{16}{32}$
    b. $3\dfrac{8}{9}$
    c. $4\dfrac{9}{8}$
    d. $195\dfrac{7}{3}$

28. Calculate the reciprocal of the following values.
    a. π
    b. 0.707
    c. 0.3183

29. Convert the following decimal fractions to written fractions.
    a. $0.777 = \dfrac{?}{9}$
    b. $0.6149069 = \dfrac{?}{161}$
    c. $43.125 = \dfrac{?}{8}$

30. Perform the following arithmetic operations by hand, and then confirm your answers with a calculator.
    a. $764 + 37 + 2199 = ?$
    b. $22,763 + 4 + 56,003 = ?$
    c. $451 + 19 + 17 = ?$
    d. $5555 + 3333 + 9999 = ?$
    e. $83.57 + 4 + 0.663 = ?$
    f. $898 - 457 = ?$
    g. $7,660,232 - 23,000 = ?$
    h. $4.9 - 3.7 = ?$
    i. $22 - 14.7 = ?$
    j. $2.01 - 1.1 = ?$
    k. $4 \times 16 = ?$
    l. $488 \times 14.0 = ?$
    m. $0.3 \times 99 = ?$
    n. $22,000 \times 14,000 = ?$
    o. $0.5 \times 10,000 = ?$
    p. $488 \div 4 = ?$
    q. $64.8 \div 8.1 = ?$
    r. $99,000 \div 10 = ?$
    s. $123 \div 17.5 = ?$
    t. $0.662 \div 0.002 = ?$

31. Express the ratio of the following in their lowest terms. (Remember to compare like quantities.)
    a. The ratio of 20 feet to 5 feet
    b. The ratio of $2\frac{1}{2}$ minutes to 30 seconds

32. Round off as indicated.
    a. 48.36 to the nearest whole number
    b. 156.3625 to the nearest tenth
    c. 0.9254 to the nearest hundredth

33. Light travels in a vacuum at a speed of 186,282.3970 miles per second. This _____-significant-place value is accurate to _____ significant places.

34. Calculate the following percentages.

   a. 23% of 50 = ?

   b. 78% of 10 = ?

   c. 3% of 1.5 = ?

   d. 20% of 3300 = ?

   e. 5% of 10,000 = ?

35. Calculate the average value of the following groups of values (include units).

   a. 4, 5, 6, 7, and 8

   b. 15 seconds, 20 seconds, 18 seconds, and 16 seconds

   c. 110 volts, 115 volts, 121 volts, 117 volts, 128 volts, and 105 volts

   d. 33 ohms, 17 ohms, 1000 ohms, and 973 ohms

   e. 150 meters, 160 meters, and 155 meters

Web Site Questions

Go to the web site http://www.prenhall.com/cook, select the textbook *Mathematics for Electronics and Computers,* select this chapter, and then follow the instructions when answering the multiple-choice practice problems.

Positive and Negative Numbers

Blaise of Genius

Blaise Pascal, the son of a regional tax official, was born in France in 1623. His father realized early that his son was a genius when his son's understanding of mathematics at the age of 8 exceeded his own. By the age of 16, Pascal had published several mathematical papers on conic sections and hydraulics. While still a youth, he invented, among other things, the syringe, the hydraulic press, and the first public bus system in Paris.

At the age of 19 he began working on an adding machine in an attempt to reduce the computational drudgery of his father's job. Pascal's machine, which became known as the *Pascaline,* became the rage of European nobility and is recognized today as one of the first computers. The operator loaded a number into a machine by setting a series of wheels. Each wheel, marked with the digits zero through nine, stood for each of the decimal columns: 1s, 10s, 100s, and so on. A carry was achieved when a total greater than nine caused a complete revolution of a wheel, which advanced the wheel to the left by one digit.

Although widely praised, the Pascaline, whose adding principle of operation would remain the standard for the next 300 years, did not make Pascal rich. Like many geniuses, he was a troubled person whose ideas and problems would plague him until he could find solutions. At the age of 37, he dropped out of society to join a religious monastery that refrained from all scientific pursuits. He died just two years later at the age of 39.

In his lifetime he became famous as a mathematician, physicist, writer, and philosopher, and in honor of his achievements one of today's computer programming languages has been named after him—Pascal.

Outline and Objectives

VIGNETTE: BLAISE OF GENIUS

INTRODUCTION

Objective 1: Define the difference between a positive number and a negative number.

Objective 2: Describe some applications in which positive and negative numbers are used.

3-1 EXPRESSING POSITIVE AND NEGATIVE NUMBERS

Objective 3: Explain how to express positive and negative numbers.

3-2 ADDING POSITIVE AND NEGATIVE NUMBERS

Objective 4: Describe the three basic rules of positive and negative numbers as they relate to:
    a. Addition
    b. Subtraction
    c. Multiplication
    d. Division

3-3 SUBTRACTING POSITIVE AND NEGATIVE NUMBERS

3-4 MULTIPLYING POSITIVE AND NEGATIVE NUMBERS

3-5 DIVIDING POSITIVE AND NEGATIVE NUMBERS

3-6 DEALING WITH STRINGS OF POSITIVE AND NEGATIVE NUMBERS

Objective 5: Explain how to combine strings of positive and negative numbers.

MULTIPLE CHOICE QUESTIONS

COMMUNICATION SKILL QUESTIONS

PRACTICE PROBLEMS

WEB SITE QUESTIONS

Introduction

Until now we have been dealing only with **positive numbers.** In fact, all the values we have discussed so far have been positive. For example, even though 12 does not have a positive (+) sign in front of it, it is still a positive number and therefore could be written as +12. We do not normally include the positive sign in front of a positive value unless we have other values that are negative. In instances when a **negative number** is present, we include the positive sign (+) in front of positive numbers to distinguish them from negative numbers that have the negative sign (−) in front of them. For example, if a list contained both positive and negative values, we would have to include their signs, as follows:

$$-4 \qquad +12 \qquad -6 \qquad +156 \qquad -198{,}765 \qquad +1{,}000{,}000$$

On the other hand, if we were to list only the positive values, there would be no need to include the positive sign because all values are positive.

$$12 \qquad 156 \qquad 1{,}000{,}000 \qquad \text{or} \qquad +12 \qquad +156 \qquad +1{,}000{,}000$$

By definition, therefore, *a positive number is any value that is greater than zero, whereas a negative number is any value that is less than zero,* as shown in Figure 3-1.

The fractions discussed so far, which are pieces of a whole, are less than 1 and can be positive values or negative values. For example, some of the following fractions are positive (have a value between zero and positive one), and some are negative (have a value between zero and negative one).

$$-\frac{1}{2} \qquad +0.5 \qquad +\frac{1}{8} \qquad -0.0016$$

The next important question is: *Why do we need both positive and negative numbers? The answer is: to indicate values that are above and below a reference point.* For example, consider the thermometer shown in Figure 3-2, which measures temperature in degrees Celsius (symbolized °C). This scale is used to indicate the amount of heat. On this scale, a reference point of 0 °C (zero degrees Celsius) was assigned to the temperature at which water freezes or ice melts, and this is an easy point to remember. For example, let us assume that the ice cubes inside your freezer are at a temperature of −15 °C (negative 15 degrees Celsius), and you remove several ice cubes and place them in a glass on the kitchen counter. At the warmer room temperature the ice cubes will absorb heat, and their temperature value will increase, or become less negative, as follows: −15 °C, −12 °C, −10 °C, −6 °C, and so on.

As we cross this temperature scale's reference point of 0 °C, the ice cubes change from a solid (ice cubes) to a liquid (water). If we then pour the water into a kettle and heat it even further, its temperature value will increase or become more positive, as follows: +5 °C, +12 °C, +22 °C, and so on.

Negative and positive numbers therefore are used to designate some quantity or value relative to, or compared with, a zero reference point. Temperature is not the only application for these numbers. For example, positive values can be used to express money that you have, whereas negative values can be used to express money that you owe, or have paid out. Similarly, a zero reference point can be used to indicate the average rainfall or snowfall, with positive values indicating the amount of rain or snow above the average, and negative values indicating the amount of rain or snow below the average.

In Chapter 3 we will discuss how to add, subtract, multiply, and divide positive and negative numbers, or signed numbers (numbers with signs).

CHAPTER 3 / POSITIVE AND NEGATIVE NUMBERS

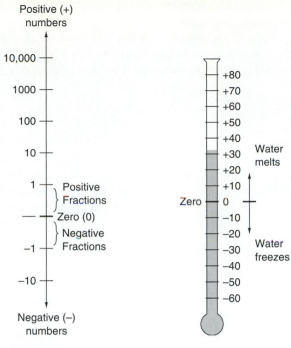

FIGURE 3-1 The Positive and Negative Number Line.

FIGURE 3-2 The Celsius Temperature Scale.

3-1   EXPRESSING POSITIVE AND NEGATIVE NUMBERS

Let us first discuss how we will mathematically express an addition, subtraction, multiplication, or division involving positive and negative numbers. As an example, let us assume that the rainfall over the last 2 years has been 3 and 9 graduations (marks on the scale) above the average reference point of zero. Because these two values of 3 and 9 are above the zero reference, they are positive numbers and can therefore be written as

$$3 \text{ and } 9 \quad \text{or} \quad +3 \text{ and } +9$$

To calculate the total amount of rainfall over the last 2 years, therefore, we must add these two values, as follows:

$$3 + 9 = 12 \quad \text{or} \quad +3 + (+9) = +12$$

If each graduation or mark on the scale was equivalent to 1 inch, over the past 2 years we would have had 12 inches of rain (+12) above the average of zero. Both preceding statements or expressions are identical, although the positive signs precede the values or numbers in the statement on the right. You can also see how the parentheses were used to separate the *plus sign* (indicating mathematical addition) from the *positive sign* (indicating that 9 is a positive number). If the parentheses had not been included, you might have found the statement confusing, because it would have appeared as follows:

$$+3 + +9 = +12$$

Plus sign ⌐   ⌐ Positive sign

The parentheses are included to isolate the sign of a number from a mathematical operation sign and therefore to prevent confusion.

As another example, let us imagine that last year the rainfall was 7 inches or graduations above the average, and the year before that there was a bit of a drought and the rainfall was 5 graduations or inches below the average. The total amount of rainfall for the last two years was therefore

$+7 + (-5) = +2$ ← Expression or meaningful combination of symbols

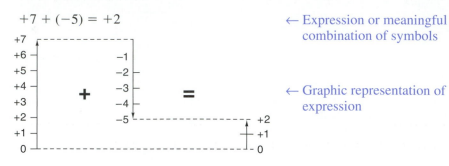

← Graphic representation of expression

Once again, notice how the parentheses were used to separate the addition symbol (plus sign) from the sign of the number (negative 5). This example is probably best understood by referring to the graphic representation below the expression, which shows how the −5 arrow cancels 5 graduations of the +7 arrow, resulting in a sum of 2 positive graduations (+2).

Before addressing the details of adding positive and negative numbers, let us discuss the *positive/negative* or *change-sign key* available on most calculators and used to change the sign of an inputted number.

CALCULATOR KEYS

Name: Change-sign, positive/negative, or negation key

Function: Generally used after a number has been entered to change the sign of the number. Numbers with no sign indicated are always positive. Normally you press the negation key $\boxed{(-)}$ before you enter the value.

Example: $\boxed{+/-}$ $+7 + (-5) = ?$

Press keys: $\boxed{7}\boxed{+}\boxed{5}\boxed{+/-}\boxed{=}$

Answer: 2 or +2

Example: $\boxed{(-)}$ $-3 + (-15) = ?$

Press keys: $\boxed{(-)}\boxed{3}\boxed{+}\boxed{(-)}\boxed{1}\boxed{5}\boxed{=}$

Answer: −18

Note: If the subtraction key $\boxed{-}$ is used in place of the negation key $\boxed{(-)}$, or vice versa, an error occurs.

▮ SELF-TEST EVALUATION POINT FOR SECTION 3-1

Now that you have completed this section, you should be able to:

■ *Objective 1.* *Define the difference between a positive number and a negative number.*

■ *Objective 2.* *Describe some applications in which positive and negative numbers are used.*

■ *Objective 3.* *Explain how to express positive and negative numbers.*

Use the following questions to test your understanding of Section 3-1.

1. Use parentheses to separate the arithmetic operation sign from the sign of the number, and then determine the result.

   a. $+3 - -4 =$
   b. $12 \times +4 =$
   c. $-5 \div -7 =$
   d. $-0.63 \times +6.4 =$

2. Express the following statements mathematically.

   a. Positive six plus negative seven
   b. Negative zero point seven five multiplied by eighteen

3. Write the calculator key sequences to input the following problems.

   a. $-7.5 + (-4.6) =$
   b. $+5 \times (-2) =$
   c. $+316.629 \div (-1.44) =$

3-2   ADDING POSITIVE AND NEGATIVE NUMBERS

To begin, we will review how to add one positive number $(+)$ to another positive number $(+)$.

RULE 1: $(+) + (+) = (+)$

Description: A positive number $(+)$ plus a positive number $(+)$ equals a positive number $(+)$.

Example:

$$5 + 15 = ?$$
$$+5 + (+15) = +20$$

Explanation:

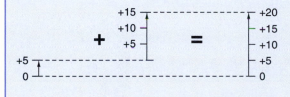

A positive 5-unit arrow plus a positive 15-unit arrow equals a positive 20-unit arrow.

Calculator sequence: [5][+][1][5][=]
Answer: $20 (+20)$

Other Examples:

$$+9 + 6 = ?$$
$$+9 + (+6) = +15$$
$$81 + +16 = ?$$
$$+81 + (+16) = +97$$
$$+(15) + +15 = ?$$
$$+15 + (+15) = +30$$

The preceding box helps to collect all the facts about this rule of addition. As you know, this basic rule of addition was covered in Chapter 2.

RULE 2: $(-) + (+)$ or $(+) + (-)$ = (sign of larger number) and difference

Description: A negative number $(-)$ plus a positive number $(+)$ or a positive number $(+)$ plus a negative number $(-)$ equals the sign of the larger number and the difference.

Two Examples:

$$-8 + (+15) = ? \qquad \text{or} \qquad +15 + (-8) = ?$$
$$-8 + (+15) = +7 \qquad\qquad\qquad +15 + (-8) = +7$$

Because 15 is the larger number, sign is positive.　Difference between 8 and 15 is 7.

Explanation:

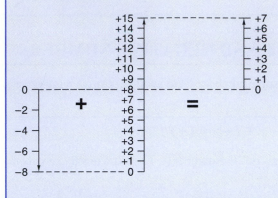

A negative 8-unit arrow plus a positive 15-unit arrow equals a positive 7-unit arrow.

Calculator sequence: [8] [⁺/₋] [+] [1] [5] [=]
Answer: 　　　　7 (+7)

Other Examples:

$$6 + (-3) = ?$$
$$+6 + (-3) = +3 \qquad \text{(larger number is +, difference = 3)}$$
$$-112 + +59 = ?$$
$$-112 + (+59) = -53 \qquad \text{(larger number is −, difference = 53)}$$
$$-18 + +12 = ?$$
$$-18 + (+12) = -6 \qquad \text{(larger number is −, difference = 6)}$$

RULE 3: $(-) + (-) = (-)$

Description: A negative number $(-)$ plus a negative number $(-)$ equals a negative number $(-)$.

Example:

$$-16 + (-4) = ?$$
$$-16 + (-4) = -20$$

Explanation:

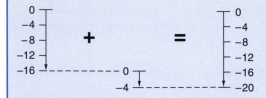

A negative 16-unit arrow plus a negative 4-unit arrow equals a negative 20-unit arrow.

Other Examples:

$$-15 + -25 = ?$$
$$-15 + (-25) = -40$$
$$(-1063) + (-1000) = ?$$
$$-1063 + (-1000) = -2063$$

SELF-TEST EVALUATION POINT FOR SECTION 3-2

Now that you have completed this section, you should be able to:

■ **Objective 4.** *Describe the three basic rules of positive and negative numbers as they relate to addition.*

Use the following questions to test your understanding of Section 3-2.

Add the following values.

1. $+15 + (+4) = ?$
2. $-3 + (+45) = ?$
3. $+114 + (-111) = ?$
4. $-357 + (-74) = ?$
5. $17 + 15 = ?$
6. $-8 + (-177) = ?$
7. $+4600 + (-3400) = ?$
8. $-6.25 + (+0.34) = ?$

3-3 SUBTRACTING POSITIVE AND NEGATIVE NUMBERS

In Chapter 2 we performed a basic subtraction in the following way:

$$19 - 12 = ?$$

$$\begin{array}{r} 19 \\ -12 \\ \hline 07 \end{array}$$

Steps:
$9 - 2 = 7.$
$1 - 1 = 0.$

Therefore, $19 - 12 = 7$.

If we wanted to, however, we could perform the same subtraction simply by changing the sign of the second number and then adding the two numbers. Let us apply this to the previous example of $19 - 12$ and see if it works.

$$19 - 12 = ? \quad \leftarrow \text{Original example}$$
$$+19 - (+12) = ? \quad \leftarrow \text{Values written as signed numbers}$$

——— *Step 1:* Change sign of second number.

——— *Step 2:* Add the two numbers.

$$+19 + (-12) = +7 \quad \leftarrow \text{Same result is achieved as performing the operation } 19 - 12.$$

Previous rule of addition
$(+) + (-) = $ (sign of larger number) and difference

In summary, therefore, *to subtract one number from another, we can change the sign of the second number and then add the two numbers together.* If you apply this "general rule of subtraction" to all the following positive and negative number subtraction examples, you will find that it simplifies the operation because all you have to do is change the sign of the second number and then apply the rules of addition previously discussed.

RULE 1: $(+) - (+)$ or $(-) - (-) =$ (sign of larger number) and difference

Description: A positive number $(+)$ minus a positive number $(+)$, or a negative number $(-)$ minus a negative number $(-)$, equals the sign of the larger number and the difference.

Two Examples:

$$+15 - (+8) = ? \qquad\qquad -10 - (-2) = ?$$
$$+15 + (-8) = ? \qquad\qquad -10 + (+2) = ?$$

Apply rule of subtraction:
Step 1: Change sign of second number.
Step 2: Add the two numbers.

$$+15 + (-8) = +7 \qquad\qquad -10 + (+2) = -8$$

Apply rule of addition:
$(+) + (-)$ or $(-) + (+) =$
(sign of larger number)
and difference

Explanation:

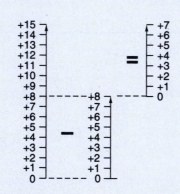
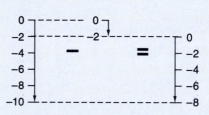

Calculator sequence: 1 5 − 8 =   1 0 ⁺⁄₋ − 2 ⁺⁄₋ =
Answer: $7(+7)$ -8

Other Examples:

$$-15 - (+8) = ?$$
$$-15 + (-8) = -23 \quad\longleftarrow\quad \text{(Change sign of second}$$
number and then add.)

$$-2 - (-10) = ?$$
$$-2 + (+10) = +8 \quad\longleftarrow\quad \text{(Change sign of second}$$
number and then add.)

$$-654 - (-601) = ?$$
$$-654 + (+601) = -53 \quad\longleftarrow\quad \text{(Change sign of second}$$
number and then add.)

RULE 2: $(+) - (-) = (+)$

Description: A positive number $(+)$ minus a negative number $(-)$ will always equal a positive number $(+)$.

Example:

$$+6 - (-5) = ?$$ ⟵ Apply rule of subtraction:

Step 1: Change sign of second number.
Step 2: Add two numbers.

$$+6 + (+5) = +11$$ Apply rule of addition:

$$(+) + (+) = +.$$

Explanation: As you can see from the preceding example, subtracting a negative number is the same as adding a positive number, so the result will always be positive. To explain this further, the minus sign is immediately followed by a negative sign, causing two reversals in direction, and therefore the two numbers are combined.

   This *double negative* (minus sign followed by a negative sign) can also be explained by using an English language analogy. If you were to use the expression "I don't have no money," you would be using two negatives (don't and no) or reversals and therefore be saying the opposite. If you don't have "no money," you must have some money.

Calculator sequence: $\boxed{6}\ \boxed{-}\ \boxed{5}\ \boxed{^+/_-}\ \boxed{=}$
Answer: 11 $(+11)$

Other Examples:

$$+63 - (-14) = ?$$ ⟵ (Change sign of second number and then add.)
$$+63 + (+14) = +77$$
$$1 - -2 = ?$$
$$+1 - (-2) = ?$$ ⟵ (Change sign of second number and then add.)
$$+1 + (+2) = +3$$
$$(18) - -14 = ?$$
$$+(18) - (-14) = ?$$ ⟵ (Change sign of second number and then add.)
$$+18 + (+14) = +32$$

RULE 3: $(-) - (+) = (-)$

Description: A negative number $(-)$ minus a positive number $(+)$ equals a negative number $(-)$.

Example:

$$-7 - (+4) = ?$$ ⟵ Apply rule of subtraction:

Step 1: Change sign of second number.
Step 2: Add two numbers.

$$-7 + (-4) = -11$$ ⟵ Apply rule of addition:
$$(-) + (-) = -.$$

Explanation: A negative number minus a positive number $[(-) - (+) = (-)]$ is the same as a negative number plus a negative number $[(-) + (-) = (-)]$. The result will always be negative because the arithmetic operation is designed to calculate the sum of two negatives.

Calculator sequence: [7] [⁺/₋] [−] [4] [=]

Answer: -11

Other Examples:

$$-63 - (+15) = ?$$
$$-63 + (-15) = -78 \quad \longleftarrow \text{(Change sign of second number and then add.)}$$
$$-(3) - (+15) = ?$$
$$-3 - (+15) = ?$$
$$-3 + (-15) = -18 \quad \longleftarrow \text{(Change sign of second number and then add.)}$$

SELF-TEST EVALUATION POINT FOR SECTION 3-3

Now that you have completed this section, you should be able to:

■ *Objective 4. Describe the three basic rules of positive and negative numbers as they relate to subtraction.*

Use the following questions to test your understanding of Section 3-3.

Subtract the following values.

1. $+18 - (+7) = ?$
2. $-3.4 - (-5.7) = ?$
3. $19,665 - (-5031) = ?$
4. $-8 - (+5) = ?$
5. $467 - 223 = ?$
6. $-331 - (-2.6) = ?$
7. $8 - (+25) = ?$
8. $-0.64 - (-0.04) = ?$

3-4   MULTIPLYING POSITIVE AND NEGATIVE NUMBERS

Let us begin by reviewing how to multiply one positive number by another positive number. This basic rule of multiplication was covered in Chapter 2.

RULE 1: $(+) \times (+) = (+)$

Description: A positive number $(+)$ multiplied by a positive number $(+)$ will always yield a positive product or result $(+)$.

Example:

$$+6 \times (+3) = ? \quad \leftarrow \text{(same as } 6 \times 3 = ?)$$
$$+6 \times (+3) = +18$$

Explanation: Multiplication is simply repeated addition, so adding a series of positive numbers will always produce a positive result.

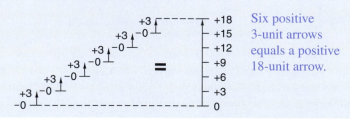

Six positive 3s

Six positive 3-unit arrows equals a positive 18-unit arrow.

Calculator sequence: $\boxed{6}\,\boxed{\times}\,\boxed{3}\,\boxed{=}$

Answer: 18 (+18)

Other Examples:

$$24 \times 2 = ?$$
$$+24 \times (+2) = +48$$
$$+3 \times +7 = ?$$
$$+3 \times (+7) = +21$$
$$+26{,}397 \times +31 = ?$$
$$+26{,}397 \times (+31) = +818{,}307$$

RULE 2: $(+) \times (-)$ or $(-) \times (+) = (-)$

Description: A positive number $(+)$ multiplied by a negative number $(-)$, or a negative number $(-)$ multiplied by a positive number $(+)$, will always yield a negative result $(-)$.

Example:

$$+6 \times (-4) = ? \qquad \text{or} \qquad -4 \times (+6) = ?$$
$$+6 \times (-4) = -24 \qquad \qquad\quad -4 \times (+6) = -24$$

Explanation: The result will always be negative because multiplication is simply repeated addition, and therefore adding a series of negative numbers will always result in a negative answer.

Calculator sequence: $\boxed{6}\,\boxed{\times}\,\boxed{4}\,\boxed{+/-}\,\boxed{=}$ or $\boxed{4}\,\boxed{+/-}\,\boxed{\times}\,\boxed{6}\,\boxed{=}$

Answer: -24 -24

Other Examples:

$$18 \times (-2) = ?$$
$$+18 \times (-2) = -36$$
$$-4 \times +3 = ?$$
$$-4 \times (+3) = -12$$
$$+21.3 \times -0.2 = ?$$
$$+21.3 \times (-0.2) = -4.26$$

RULE 3: $(-) \times (-) = (+)$

Description: A negative number $(-)$ multiplied by a negative number $(-)$ equals a positive number $(+)$.

Example:

$$-3 \times (-5) = ?$$
$$-3 \times (-5) = +15$$

Explanation: The explanation for this is once again the double negative. After a series of negative numbers are added repeatedly, the negative answer is reversed to a positive. In summary, the reversals of direction in the operation coupled with the two negative numbers reverse the result to a positive value.

Calculator sequence: $\boxed{3}\,\boxed{^+/_-}\,\boxed{\times}\,\boxed{4}\,\boxed{^+/_-}\,\boxed{=}$

Answer: 15 (+15)

Other Examples:

$$-30 \times (-2) = ?$$
$$-30 \times (-2) = +60$$
$$-(2) \times -3 = ?$$
$$-2 \times (-3) = +6$$
$$-76.44 \times -1.3 = ?$$
$$-76.44 \times (-1.3) = +99.37$$

SELF-TEST EVALUATION POINT FOR SECTION 3-4

Now that you have completed this section, you should be able to:

■ *Objective 4.* *Describe the three basic rules of positive and negative numbers as they relate to multiplication.*

Use the following questions to test your understanding of Section 3-4.

Multiply the following values.

1. $4 \times (+3) = ?$

2. $+17 \times (-2) = ?$

3. $-8 \times 16 = ?$

4. $-8 \times (-5) = ?$

5. $+12.6 \times 15 = ?$

6. $-3.3 \times +1.4 = ?$

7. $0.3 \times (-4) = ?$

8. $-4.6 \times -3.3 = ?$

3-5 DIVIDING POSITIVE AND NEGATIVE NUMBERS

Let us begin by reviewing how to divide one positive number by another positive number. This basic rule of division was covered in Chapter 2.

RULE 1: $(+) \div (+) = (+)$

Description: A positive number $(+)$ divided by a positive number $(+)$ will always result in a positive number $(+)$.

Example:

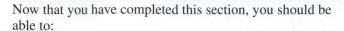

or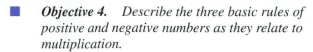

$$+12 \div (+3) = +4$$

Explanation: Because division is simply repeated subtraction, subtracting a series of positive numbers from a positive number will yield a positive answer.

Calculator sequence: $\boxed{1}\,\boxed{2}\,\boxed{\div}\,\boxed{3}\,\boxed{=}$

Answer: 4 (+4)

Other Examples:

$$+36 \div (+2) = ?$$
$$+36 \div (+2) = +18$$
$$\frac{+504}{(+8)} = ?$$
$$+504 \div (+8) = +63$$
$$+22.7 \div +3.6 = ?$$
$$+22.7 \div (+3.6) = +6.31$$

---

70 CHAPTER 3 / POSITIVE AND NEGATIVE NUMBERS

RULE 2: $(+) \div (-)$ or $(-) \div (+) = (-)$

Description: A positive number $(+)$ divided by a negative number $(-)$ or a negative number $(-)$ divided by a positive number $(+)$ will always yield a negative result $(-)$.

Example:

Dividend is larger than divisor:

$$+16 \div (-4) = ? \qquad \text{or} \qquad -16 \div (+4) = ?$$
$$+16 \div (-4) = -4 \qquad\qquad -16 \div (+4) = -4$$

Dividends (16) are larger than divisors (4).

Dividend is smaller than divisor:

$$+4 \div (-16) = -0.25 \qquad \text{or} \qquad -4 \div (+16) = -0.25$$

Even if dividends (4) are smaller than divisors (16), the answer is still negative.

Explanation: The result will always be negative because division is simply repeated subtraction, and therefore subtracting a series of negative numbers will always result in a negative answer.

Calculator sequence:

(Dividend is larger) → $\boxed{1}\boxed{6}\boxed{\div}\boxed{4}\boxed{+/_}\boxed{=}$ or $\boxed{1}\boxed{6}\boxed{+/_}\boxed{\div}\boxed{4}\boxed{=}$

Answer: -4 $\qquad\qquad\qquad$ -4

(Divisor is larger) → $\boxed{4}\boxed{\div}\boxed{1}\boxed{6}\boxed{+/_}\boxed{=}$ or $\boxed{4}\boxed{+/_}\boxed{\div}\boxed{1}\boxed{6}\boxed{=}$

Answer: -0.25 $\qquad\qquad$ -0.25

Other Examples:

$$+48 \div (-4) = ?$$
$$+48 \div (-4) = -12$$
$$-4 \div +12 = ?$$
$$-4 \div (+12) = -0.333$$
$$\frac{+677.25}{-3.66} = ?$$
$$+677.25 \div (-3.66) = -185.04$$

RULE 3: $(-) \div (-) = (+)$

Description: A negative number $(-)$ divided by a negative number $(-)$ equals a positive number $(+)$.

Example:

$$-15 \div (-5) = ?$$
$$-15 \div (-5) = +3$$

Explanation: The explanation is once again the double negative. After a series of negative numbers are repeatedly subtracted, the negative answer is reversed to a positive. In summary, the reversals of direction in the operation coupled with the two negative numbers reverse the result to a positive value.

Calculator sequence: [1] [5] [+/−] [÷] [5] [+/−] [=]
Answer: 3 (+3)

Other Examples:

$$-4 \div (-2) = ?$$
$$-4 \div (-2) = +2$$
$$\frac{-96}{-3} = ?$$
$$-96 \div (-3) = +32$$
$$-2.2 \div -11.0 = ?$$
$$-2.2 \div (-11.0) = +0.2$$

SELF-TEST EVALUATION POINT FOR SECTION 3-5

Now that you have completed this section, you should be able to:

■ ***Objective 4.*** *Describe the three basic rules of positive and negative numbers as they relate to division.*

Use the following questions to test your understanding of Section 3-5.

Divide the following values.

1. $\frac{+16.7}{+2.3} = ?$

2. $18 \div (+6) = ?$

3. $-6 \div (+2) = ?$

4. $+18 \div (-4) = ?$

5. $2 \div (-8) = ?$

6. $-8 \div +5 = ?$

7. $\frac{-15}{-5} = ?$

8. $0.664 \div (-0.2) = ?$

3-6 ORDER OF OPERATIONS

For easy access, Figure 3-3 summarizes all the rules for signed number addition, subtraction, multiplication, and division. Remember that it is more important to understand these rules rather than just to memorize them. Returning to the detailed in-chapter explanation for each of the end-of-chapter problems is not a waste of time but an investment, because you will be developing logic and reasoning skills that can be applied to all aspects of science and technology.

Addition	Subtraction
(+) + (+) = (+) (−) + (+) or (+) + (−) = (sign of larger number) and difference (−) + (−) = (−)	(+) − (+) or (−) − (−) = (sign of larger number) and difference (+) − (−) = (+) (−) − (+) = (−)
Multiplication	**Division**
(+) × (+) = (+) (+) × (−) or (−) × (+) = (−) (−) × (−) = (+)	(+) ÷ (+) = (+) (+) ÷ (−) or (−) ÷ (+) = (−) (−) ÷ (−) = (+)

FIGURE 3-3 Positive and Negative Number Rules of Addition, Subtraction, Multiplication, and Division.

When a string of more than two positive and negative numbers has to be combined by some arithmetic operation, the order in which the mathematical operations should be combined is as follows:

ORDER OF OPERATIONS	MEMORY AID
First: **Parentheses**	**Please**
Second: **Exponents**	**Excuse**
Third: **Multiplication**	**My**
Fourth: **Division**	**Dear**
Fifth: **Addition**	**Aunt**
Sixth: **Subtraction**	**Sally**

Order of Operations
The sequence in which a string of two or more signed numbers should be combined.

This **order of operations** can more easily be remembered with the phrase "Please Excuse My Dear Aunt Sally."

■ **EXAMPLE:**

Calculate the following:
  a. $3 + 6 \cdot 4 = ?$
  b. $4(4 + 3) = ?$
  c. $8^2 - 5 = ?$

■ *Solution:*

a. Problem has multiplication and addition operations. According to the order of operations (PEMDAS), multiplication should be performed first and then addition second.

$$3 + 6 \cdot 4 = ?$$
$$3 + 24 = \qquad (6 \cdot 4 = 24)$$
$$27 \qquad (3 + 24 = 27)$$

b. $4(4 + 3) = ?$ Parentheses operation should be performed first, multiplication second.

$$4(4 + 3) = ?$$
$$4(7) =$$
$$4 \cdot 7 =$$
$$28$$

c. $8^2 - 5 = ?$ Exponents first, subtraction second.

$$8^2 - 5 = ?$$
$$64 - 5 =$$
$$59$$

CALCULATOR KEYS

Name: Order of evaluation key

Function: Most calculators define the order in which functions in expressions are entered and evaluated. This order is as follows:

CALCULATOR KEYS *(continued)*

ORDER NUMBER	FUNCTION
1	Functions that precede the argument, such as $\sqrt{\ }$ (, **sin**(, or **log**(
2	Functions that are entered after the argument, such as ^{2}, $^{-1}$, $^\circ$, ^{r}, and conversions
3	Powers and roots, such as **2^5** or **5$^x\sqrt{32}$**
4	Permutations (**nPr**) and combinations (**nCr**)
5	Multiplication, implied multiplication, and division
6	Addition and subtraction
7	Relational functions, such as $>$ or $\leq$
8	Logic operator **and**
9	Logic operators **or** and **xor**

Example: Calculate $3.76 \div (-7.9 + \sqrt{5}) + 2 \log 45$.

Press keys: [3] [.] [76] [÷] [(] [(-)] [7] [.] [9] [+] [2nd] [$\sqrt{\ }$]
[5] [)] [+] [2] [×] [LOG] [(] [45] [)] | 3.76/(-7.9+√ 5)
[ENTER] | +2log(45)
                                           | 2.642575252

SELF-TEST EVALUATION POINT FOR SECTION 3-6

Now that you have completed this section, you should be able to:

■ **Objective 5.** *Explain how to combine strings of positive and negative numbers.*

Use the following questions to test your understanding of Section 3-6.

1. $+6 + (+3) + (-7) + (-5) = ?$

2. $+9 - (+2) - (-13) - (-4) = ?$

3. $-6 \times (-4) \times (-5) = ?$

4. $+8 \div (+2) \div (-5) = ?$

or $\dfrac{\left(\dfrac{+8}{+2}\right)}{(-5)} = ?$

5. $-4 \div (-2) \times (+8) = ?$ or $\dfrac{-4}{-2} \times (+8) = ?$

6. $-9 + (+5) - (-7) = ?$

REVIEW QUESTIONS

Multiple Choice Questions

1. A _____ number is any value greater than zero, whereas a _____ number is any value less than zero.

   a. Positive, +   **c.** +, negative
   b. −, +   **d.** −, −

2. The fraction 0.3417 is a positive number.

   a. True   **b.** False

3. Negative and positive numbers are used to designate some value relative to a _____ reference point.

   a. 0   **c.** 10 (decimal)
   b. 0.5   **d.** π

4. Which of the following calculator sequences would be used to input the following expression? $+5 - (-15) =$

   a. [5] [+/−] [−] [1] [5] [=]
   b. [+/−] [5] [−] [−] [1] [5] [=]
   c. [5] [−] [−] [1] [5] [=]
   d. [5] [−] [1] [5] [+/−] [=]

5. $(+) + (+) = ?$

   a. $(+)$
   b. $(-)$
   c. (Sign of larger number) and difference
   d. None of the above

6. $(+) + (-) = ?$

   a. $(+)$

   b. $(-)$

   c. (Sign of larger number) and difference

   d. None of the above

7. $(-) + (-) = ?$

   a. $(+)$

   b. $(-)$

   c. (Sign of larger number) and difference

   d. None of the above

8. $(+) - (-)$ is equivalent to?

   a. $(-) + (-)$ **c.** $(+) + (+)$

   b. $(-) + (+)$ **d.** $(+) + (-)$

9. $(-) - (-)$ is equivalent to?

   a. $(-) + (+)$ **c.** $(-) + (-)$

   b. $(+) + (+)$ **d.** $(+) + (-)$

10. $(-) - (+)$ is equivalent to?

   a. $(-) + (+)$ **c.** $(+) + (+)$

   b. $(-) + (-)$ **d.** $(+) + (-)$

Communication Skill Questions

11. Define the following: (Introduction)

   a. Positive number   **c.** Positive fraction

   b. Negative number   **d.** Negative fraction

12. What method is used to show whether the number is positive or negative? (3-1)

13. Describe the three rules of adding positive and negative numbers. (3-2)

14. How can you achieve subtraction through addition? (3-3)

15. Describe the three rules of subtracting positive and negative numbers. (3-3)

16. What are the three rules of multiplying positive and negative numbers? (3-4)

17. List and describe the three rules of positive and negative number division. (3-5)

18. Arbitrarily choose values, and then describe the following: (3-2)

   a. $(+) + (-) = ?$

   b. $(-) + (-) = ?$

   c. $(+) + (+) = ?$

19. Arbitrarily choose values, and then describe the following: (3-3)

   a. $(+) - (+) = ?$

   b. $(-) - (+) = ?$

   c. $(+) - (-) = ?$

20. Arbitrarily choose values, and then describe the following: (3-4 and 3-5)

   a. $(+) \times (-) = ?$ **c.** $(+) \div (-) = ?$

   b. $(-) \times (-) = ?$ **d.** $(-) \div (-) = ?$

Practice Problems

21. Express the following statements mathematically.

   a. Positive eight minus negative five

   b. Negative zero point six times thirteen

   c. Negative twenty-two point three divided by negative seventeen

   d. Positive four divided by negative nine

22. Add the following values.

   a. $+8 + (+9) = ?$ **c.** $-6 + (+9) = ?$

   b. $+9 + (-6) = ?$ **d.** $-8 + (-9) = ?$

23. Subtract the following values.

   a. $+6 - (+8) = ?$ **c.** $-75 - (+62) = ?$

   b. $+9 - (-6) = ?$ **d.** $-39 - (-112) = ?$

24. Multiply the following values.

   a. $+18 \times (+4) = ?$ **c.** $-20 \times (+3) = ?$

   b. $+8 \times (-2) = ?$ **d.** $-16 \times (-10) = ?$

25. Divide the following values.

   a. $+19 \div (+3) = ?$ **c.** $-80 \div (+5) = ?$

   b. $+36 \div (-3) = ?$ **d.** $-44 \div (-2) = ?$

Calculate the answers to the following problems.

26. $+0.75 + (-0.25) - (+0.25) = ?$

27. $\left(\dfrac{+15}{+5} \right) \times (-3.5) = +15 \div (+5)$

   $\times (-3.5) = ?$

28. $\dfrac{+2 \times (-4) \times (-7)}{+3} = +2 \times (-4) \times (-7)$

   $\div (+3) = ?$

29. $-6 \div (-4) \div (-3) \times (-15)$

   $= \dfrac{\left(\dfrac{-6}{-4} \right)}{-3} \times (-15) = ?$

30. $[+15 - (-4) - (+5) - (-10)] \div 4 =$

   $\dfrac{+15 - (-4) - (+5) - (-10)}{4} = ?$

Web Site Questions

Go to the web site http://www.prenhall.com/cook, select the textbook *Mathematics for Electronics and Computers,* select this chapter, and then follow the instructions when answering the multiple-choice practice problems.

Exponents and the Metric System

Not a Morning Person

René Descartes was born in 1596 in Brittany, France, and at a very early age began to display an astonishing analytical genius. At the age of 8 he was sent to Jesuit College in La Flèche, then one of the most celebrated institutions in Europe, where he studied several subjects, none of them interesting him as much as mathematics. His genius impressed all his professors, who gave him permission, because of his delicate health, to study in bed until midday—a practice that he retained throughout his life. In 1612 he left La Flèche to go to the University of Poitiers, where he graduated in law in 1616, a profession he never practiced.

Wanting to see the world, he joined the army, which made use of his mathematical ability in military engineering. In 1619 he met Dutch philosopher Isaac Beeckman, who convinced Descartes to "turn his mind back to science and worthier occupations." After leaving the army, Descartes traveled to Neuberg, Germany, where he shut himself in a well-heated room for the winter. On the eve of St. Martin's Day (November 10, 1619), Descartes described a vivid dream that determined all his future endeavors. The dream clarified his purpose and showed him that physics and all sciences could be reduced to geometry, and therefore all were interconnected "as by a chain." From this point on, his genius was displayed in his invention of coordinate geometry and in contributions to theoretical physics, methodology, and metaphysics.

In his time he was heralded as an analytical genius, a reputation that has lasted to this day. His fame was so renowned that he was asked frequently to tutor royalty. When in Paris in 1649, Descartes was asked to tutor Queen Christina of Sweden. He did not want the opportunity to, in his words, "live in the land of bears among rock and ice, and lose my independence," but he was persuaded to do so by the French ambassador a month later. The queen chose five o'clock in the morning for her lessons, and on his travels one bitter morning Descartes caught a severe chill and died within two weeks.

Descartes's problem-solving ability was incredible. You may find his four-step method to solving a problem helpful:

1. Never accept anything as true unless it is clear and distinct enough to exclude all doubt from your mind.
2. Divide the problem into as many parts as necessary to reach a solution.
3. Start with the simplest things and proceed step by step toward the complex.
4. Review the solution so completely and generally that you are sure nothing was omitted.

Outline and Objectives

VIGNETTE: NOT A MORNING PERSON

INTRODUCTION

Objective 1: Define the term *exponent*.

4-1 RAISING A BASE NUMBER TO A HIGHER POWER

Objective 2: Describe what is meant by raising a number to a higher power.

4-1-1 Square of a Number

Objective 3: Explain how to find the square and root of a number.

4-1-2 Root of a Number

4-1-3 Powers and Roots in Combination

Objective 4: Calculate the result of problems with powers and roots in combination.

4-2 POWERS OF 10

Objective 5: Explain the powers-of-ten method and how to convert to powers of ten.

4-2-1 Converting to Powers of 10

4-2-2 Powers of 10 in Combination

Objective 6: Calculate the result of problems with powers of ten in combination.

4-2-3 Scientific and Engineering Notation

Objective 7: Describe the two following floating-point number systems:
    a. Scientific notation
    b. Engineering notation

4-3 THE METRIC SYSTEM

Objective 8: Define and explain the purpose of the metric system.

4-3-1 Metric Units and Prefixes

Objective 9: List the metric prefixes and describe the purpose of each.

4-3-2 Metric Units of Length

Objective 10: Describe the following metric units for:
    a. Length
    b. Area
    c. Weight
    d. Volume
    e. Temperature

Objective 11: Convert U.S. customary units to metric units, and metric units to U.S. customary units.

4-3-3 Metric Units of Area

4-3-4 Metric Units of Weight

4-3-5 Metric Units of Volume

4-3-6 Metric Unit of Temperature

4-3-7 Electrical Units, Prefixes, and Conversions

Objective 12: List and describe many of the more frequently used electrical units and prefixes, and describe how to interconvert prefixes.

MULTIPLE CHOICE QUESTIONS

COMMUNICATION SKILL QUESTIONS

PRACTICE PROBLEMS

WEB SITE QUESTIONS

Introduction

Like many terms used in mathematics, the word **exponent** sounds as if it will be complicated; however, once you find out that exponents are simply a sort of "math shorthand," the topic loses its intimidation. Many of the values used in science and technology contain numbers that have exponents. An exponent is a number in a smaller type size that appears to the right of and slightly higher than another number, for example:

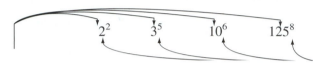

$$2^2 \qquad 3^5 \qquad 10^6 \qquad 125^8$$

The number in the larger type size in these examples is called the *base number.*

All these numbers in the smaller type size are examples of an *exponent.*

However, what does a term like 2^3 mean? It means that the base 2 is to be used as a factor 3 times; therefore,

$$2^3 = 2 \times 2 \times 2 = 8$$

Similarly, 3^5 means that the base 3 is to be used as a factor 5 times; therefore,

$$3^5 = 3 \times 3 \times 3 \times 3 \times 3 = 243$$

As you can see, it is much easier to write 3^5 than to write $3 \times 3 \times 3 \times 3 \times 3$, and since both mean the same thing and equal the same amount (which is 243), exponents are a quick and easy math shorthand.

In this chapter we will examine the details relating to squares, roots, exponents, scientific and engineering notation, and prefixes.

4-1 RAISING A BASE NUMBER TO A HIGHER POWER

A base number's exponent indicates how many times the base number must be multiplied by itself. This is called *raising a number to a higher power.* For example, $5 \times 5 \times 5 \times 5$ can be written as 5^4, which indicates that the base number 5 is raised to the fourth power by the exponent 4. As another example, 3×3 can be written as 3^2, which indicates that the base number 3 is raised to the second power by the exponent 2. The second power is also called the **square** of the base number, and therefore 3^2 can be called "three squared" or "three to the second power."

What does 10^3 mean?

■ *Solution:*

10^3 indicates that the base number 10 is raised to the third power by the exponent 3. Described another way, it means that the base 10 is to be used as a factor 3 times; therefore,

$$10^3 = 10 \times 10 \times 10 = 1000$$

Thus, instead of writing $10 \times 10 \times 10$, or 1000, you could simply write 10^3 (pronounced "ten to the three," "ten to the third power," or "ten cubed").

Because the square of a base number is used very frequently, let us begin by discussing raising a base number to the second power.

4-1-1 *Square of a Number*

The square of a base number means that the base number is to be multiplied by itself. For example, 4^2, which is pronounced "four squared" or "four to the second power," means 4×4. The squares of the first ten base numbers are used very frequently in numerical problems and are as follows:

$$0^2 = 0 \times 0 = 0$$
$$1^2 = 1 \times 1 = 1$$
$$2^2 = 2 \times 2 = 4$$
$$3^2 = 3 \times 3 = 9$$
$$4^2 = 4 \times 4 = 16$$
$$5^2 = 5 \times 5 = 25$$
$$6^2 = 6 \times 6 = 36$$
$$7^2 = 7 \times 7 = 49$$
$$8^2 = 8 \times 8 = 64$$
$$9^2 = 9 \times 9 = 81$$
$$10^2 = 10 \times 10 = 100$$

Many people get confused with the first three of the squares. Be careful to remember that $0^2 = 0$ because nothing × nothing equals nothing; $1^2 = 1$ because one times one is still one; and 2^2 means 2×2, not $2 + 2$, even though the answer works out both ways to be 4.

■ **EXAMPLE:**

Give the square of the following.
   a. 12^2 b. 7^2

■ *Solution:*

   a. $12^2 = 12 \times 12 = 144$
   b. $7^2 = 7 \times 7 = 49$

Finding the square of a base number is done so frequently in science and technology that most calculators have a special key just for that purpose. It is called the *square key* and operates as follows.

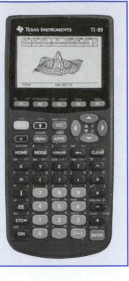
Most calculators also have a key for raising a base number of any value to any power. This is called the *y to the x power key,* and it operates as follows.

4-1-2   *Root of a Number*

What would we do if we had the result 64 and we didn't know the value of the number that was multiplied by itself to get 64? In other words, we wanted to *find the source or root number that was squared to give us the result.* This process, called **square root,** uses a special symbol called a *radical sign* and a smaller number called the *index.*

$$\sqrt[2]{64} = ?$$

Smaller number is called the *index.* It indicates how many times a number was multiplied by itself to get the value shown inside the radical sign. A 2 index is called the *square root.*

Radical sign ($\sqrt{\ }$) indicates that the value inside is the result of a multiplication of a number two or more times.

Square Root

A factor of a number that when squared gives the number.

In this example the index is 2, which indicates that the number we are trying to find was multiplied by itself two times. Of course, in this example we already know that the answer is 8 because $8 \times 8 = 64$.

If squaring a number takes us forward ($8^2 = 8 \times 8 = 64$), taking the square root of a number must take us backward ($\sqrt[2]{64} = 8$). Almost nobody extracts the squares from square root problems by hand because calculators make this process more efficient. Most calculators have a special key just for determining the square root of a number. Called the *square root key*, it operates as follows.

CALCULATOR KEYS

Name: Square root key

Function: Calculates the square root of the number in the display.

Example: $\sqrt{81}$ = ?

Press keys: $\boxed{8}\ \boxed{1}\ \boxed{\sqrt{x}}$

Display shows: 9

Some of the more frequently used *square root* values are as follows:

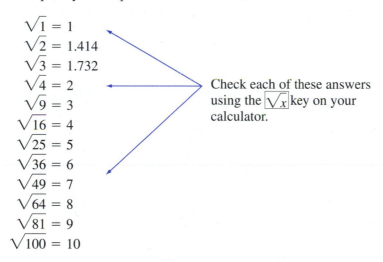

$$\sqrt{1} = 1$$
$$\sqrt{2} = 1.414$$
$$\sqrt{3} = 1.732$$
$$\sqrt{4} = 2$$
$$\sqrt{9} = 3$$
$$\sqrt{16} = 4$$
$$\sqrt{25} = 5$$
$$\sqrt{36} = 6$$
$$\sqrt{49} = 7$$
$$\sqrt{64} = 8$$
$$\sqrt{81} = 9$$
$$\sqrt{100} = 10$$

Check each of these answers using the $\boxed{\sqrt{x}}$ key on your calculator.

In some cases the index may be 3, meaning that some number was multiplied by itself three times to get the value within the radical sign; for example,

$$\sqrt[3]{125} = ?$$

Using the index 3 instead of the index 2 is called taking the **cube root.** The answer to this problem is 5 because 5 multiplied by itself three times equals 125.

$$\sqrt[3]{125} = 5$$

Cube Root

A factor of a number that when multiplied by itself three times gives the number.

Most calculators also have a key for calculating the root of any number with any index. It operates as follows.

CALCULATOR KEYS

Name: *x*th root of *y* key

Function: Calculate the *x*th root of the displayed value *y*.

Example: $\sqrt[3]{512}$ = ?

Press keys: $\boxed{5}\ \boxed{1}\ \boxed{2}\ \boxed{\sqrt[x]{y}}\ \boxed{3}\ \boxed{=}$

Display shows: 8

In most cases a radical sign will not have an index, in which case you can assume that the index is 2, or square root; for example,

$$\sqrt{16} = 4$$
Square root of sixteen = four

4-1-3   *Powers and Roots in Combination*

In some instances you may have several operations to perform before the result can be determined.

■ EXAMPLE:

What is the sum of $4^2 + 3^3$ (four squared plus three cubed)?

$$4^2 + 3^3 = ?$$

■ *Solution:*

The first step in a problem like this is to raise the base numbers to the power indicated by the exponent and then add the two values.

$4^2 + 3^3 =$
$16 + 27 = 43$

Steps:
$4^2 = 4 \times 4 = 16$
$3^3 = 3 \times 3 \times 3 = 27$
$16 + 27 = 43$
Calculator sequence: $\boxed{4}\ \boxed{x^2}\ \boxed{+}\ \boxed{3}\ \boxed{y^x}\ \boxed{3}\ \boxed{=}$

Here are some other examples in which a mathematical operation needs to be performed on numbers with exponents to determine an answer or result.

■ EXAMPLE A:

$$(2 + 9)^2 = 11^2 = 121$$

Exponent 2 indicates that the sum within the parentheses is to be squared.

Parentheses are used to group term $2 + 9$.

Steps:
$2 + 9 = 11$
$11^2 = 11 \times 11 = 121$
Calculator sequence: $\boxed{2}\boxed{+}\boxed{9}\boxed{=}\boxed{x^2}$

■ EXAMPLE B:

$$(8^2)^2 = 64^2 = 4096$$

Steps:
$8^2 = 8 \times 8 = 64$
$64^2 = 64 \times 64 = 4096$
Calculator sequence: $\boxed{8}\boxed{x^2}\boxed{x^2}$

■ EXAMPLE C:

$$(5^2 \times 3^3)^2 = (25 \times 27)^2$$
$$= 675^2 = 455{,}625$$

Steps:
$5^2 = 5 \times 5 = 25$
$3^3 = 3 \times 3 \times 3 = 27$
$25 \times 27 = 675, 675 \times 675 = 455{,}625$
Calculator sequence:
$\boxed{5}\boxed{x^2}\boxed{\times}\boxed{3}\boxed{y^x}\boxed{3}\boxed{=}\boxed{x^2}$

■ EXAMPLE D:

$$(5^2 - 15)^3 = (25 - 15)^3$$
$$= 10^3 = 1000$$

Steps:
$5^2 = 5 \times 5 = 25$
$25 - 15 = 10$
$10^3 = 10 \times 10 \times 10 = 1000$
Calculator sequence:
$\boxed{5}\boxed{x^2}\boxed{-}\boxed{1}\boxed{5}\boxed{=}\boxed{y^x}\boxed{3}$

■ EXAMPLE E:

$$\left(\frac{4^2}{2}\right)^2 + 7^2 \text{ or } (4^2 \div 2)^2 + 7^2$$
$$= (16 \div 2)^2 + 7^2$$
$$= 8^2 + 7^2$$
$$= 64 + 49 = 113$$

Steps:
$4^2 = 4 \times 4 = 16$
$16 \div 2 = 8$
$8^2 = 8 \times 8 = 64$
$7^2 = 7 \times 7 = 49$
$64 + 49 = 113$
Calculator sequence: $\boxed{4}\boxed{x^2}\boxed{\div}\boxed{2}\boxed{=}$
$\boxed{x^2}\boxed{+}\boxed{7}\boxed{x^2}\boxed{=}$

Now let us examine a few examples involving powers and roots.

EXAMPLE A:

No index means square root.

$$\sqrt{3 + 7 + 6} = \sqrt{16} = 4$$

Radical sign extends
over terms $3 + 7 + 6$.

Steps:
$3 + 7 + 6 = 16$
$\sqrt{16} = 4$
Calculator sequence: $\boxed{3}\boxed{+}\boxed{7}\boxed{+}\boxed{6}\boxed{=}\boxed{\sqrt{x}}$

EXAMPLE B:

$$\sqrt{14^2} = \sqrt{196} = 14$$

(Because square is the opposite
of square root, the square root
of fourteen squared $= 14$.)

Steps:
$14^2 = 14 \times 14 = 196$
$\sqrt{196} = 14$
Calculator sequence: $\boxed{1}\boxed{4}\boxed{x^2}\boxed{\sqrt{x}}$

EXAMPLE C:

$$\sqrt[2]{\frac{8}{9}} \text{ or } \sqrt[2]{8 \div 9}$$
$$= \sqrt{0.888} = 0.943$$

Steps:
$8 \div 9 = 0.889$
Calculator sequence: $\boxed{8}\boxed{\div}\boxed{9}\boxed{=}\boxed{\sqrt{x}}$

EXAMPLE D:

$$\sqrt[3]{5^2 \times 2^2} = \sqrt[3]{25 \times 4}$$
$$= \sqrt[3]{100} = 4.6416$$

Steps:
$5^2 = 25$
$2^2 = 4$
$25 \times 4 = 100$
$\sqrt[3]{100} = 4.6416$
Calculator sequence: $\boxed{5}\boxed{x^2}\boxed{\times}\boxed{2}\boxed{x^2}\boxed{=}\boxed{\sqrt[x]{y}}\boxed{3}\boxed{=}$

EXAMPLE E:

$$\sqrt{6^2 - 5} + \sqrt{121}$$
$$= \sqrt{36 - 5} + \sqrt{121}$$
$$= \sqrt{31} + \sqrt{121}$$
$$= 5.568 + 11 = 16.568$$

Steps:
$6^2 = 6 \times 6 = 36$
$36 - 5 = 31$
$\sqrt{31} = 5.568$
$\sqrt{121} = 11$
$5.568 + 11 = 16.568$
Calculator sequence: $\boxed{6}\boxed{x^2}\boxed{-}\boxed{5}\boxed{=}$
$\boxed{\sqrt{x}}\boxed{+}\boxed{1}\boxed{2}\boxed{1}\boxed{\sqrt{x}}\boxed{=}$

Now that you have completed this section, you should be able to:

- **Objective 1.** *Define the term* exponent.
- **Objective 2.** *Describe what is meant by raising a number to a higher power.*
- **Objective 3.** *Explain how to find the square and root of a number.*
- **Objective 4.** *Calculate the result of problems with powers and roots in combination.*

Use the following questions to test your understanding of Section 4-1.

1. Raise the following base numbers to the power indicated by the exponent, and give the answer.
   a. $16^4 = ?$
   b. $32^3 = ?$
   c. $112^2 = ?$
   d. $15^6 = ?$
   e. $2^3 = ?$
   f. $3^{12} = ?$

2. Give the following roots.
   a. $\sqrt[2]{144} = ?$
   b. $\sqrt[3]{3375} = ?$
   c. $\sqrt{20} = ?$
   d. $\sqrt[3]{9} = ?$

3. Calculate the following.
   a. $(9^2 + 14^2)^2 - \sqrt[3]{3 \times 7} = ?$
   b. $\sqrt{3^2 \div 2^2} + \dfrac{(151 - 9^2)}{3.5^2} = ?$

4-2   POWERS OF 10

Many of the sciences deal with numbers that contain a large number of zeros, for example:

$$14,000$$
$$0.000032$$

By using exponents, we can eliminate the large number of zeros to obtain a shorthand version of the same number. This method is called *powers of 10*.

As an example, let us remove all the zeros from the number 14,000 until we are left with simply 14. However, this number (14) is not equal to the original number (14,000), and therefore simply removing the zeros is not an accurate shorthand. Another number needs to be written with the 14 to indicate what has been taken away—this is called a *multiplier*. The multiplier must indicate what you have to multiply 14 by to get back to 14,000; therefore,

$$14,000 = 14 \times 1000$$

As you know from our discussion on exponents, we can replace the 1000 with 10^3 because $1000 = 10 \times 10 \times 10$. Therefore, the powers-of-10 notation for 14,000 is 14×10^3. To convert 14×10^3 back to its original form (14,000), simply remember that each time a number is multiplied by 10, the decimal place is moved one position to the right. In this example 14 is multiplied by 10 three times (10^3, or $10 \times 10 \times 10$), and therefore the decimal point will have to be moved three positions to the right.

$$14 \times 10^3 = 14 \times 10 \times 10 \times 10 = 14\overset{\frown}{_0}\overset{\frown}{_0}\overset{\frown}{_0}0. = 14,000$$

As another example, what is the powers-of-10 notation for the number 0.000032? If we once again remove all the zeros to obtain the number 32, we will again have to include a multiplier with 32 to indicate what 32 has to be multiplied by to return it to its original form. In this case 32 will have to be multiplied by 1/1,000,000 (one millionth) to return it to 0.000032.

$$32 \times \frac{1}{1,000,000} = 0.000032$$

This can be verified because when you divide any number by 10, you move the decimal point one position to the left. Therefore, to divide any number by 1,000,000, you simply move the decimal point six positions to the left.

$$32 \times \frac{1}{1,000,000} = \frac{32}{1,000,000} = \frac{32}{10 \times 10 \times 10 \times 10 \times 10 \times 10} = 0.0000032$$

Once again an exponential expression can be used in place of the 1/1,000,000 multiplier, namely,

$$\frac{1}{1,000,000} = \frac{1}{10^6} = 0.000001 = 10^{-6}$$

Whenever you divide a number into 1, you get the *reciprocal* of that number. In this example, when you divide 1,000,000 into 1, you get 0.000001, which is equal to power-of-10 notation with a negative exponent of 10^{-6}. The multiplier 10^{-6} indicates that the decimal point must be moved back (to the left) by six places, and therefore

$$32 \times \frac{1}{1,000,000} = 32 \times 0.000001 = 32 \times 10^{-6} = 0.0000032$$

Now that you know exactly what a multiplier is, you have only to remember these simple rules:

1. A *negative exponent* tells you how many places *to the left* to move the decimal point.
2. A *positive exponent* tells you how many places *to the right* to move the decimal point.

Remember one important point: a negative exponent does not indicate a negative number; it simply indicates a fraction. For example, 4×10^{-3} meter means that 1 meter has been broken up into 1000 parts and we have 4 of those pieces, or 4/1000.

4-2-1   *Converting to Powers of 10*

To reinforce our understanding, let us try converting a few numbers to powers-of-10 notation.

■ **EXAMPLE A:**

$$230,000,000 = 23\,0,000,000 = 23 \times 10^7$$

■ **EXAMPLE B:**

$$760,000 = 760,000 = 76 \times 10^4$$

■ **EXAMPLE C:**

$$0.0019 = 0.0019 = 19 \times 10^{-4}$$

CHAPTER 4 / EXPONENTS AND THE METRIC SYSTEM

$$1\ 2\ 3\ 4\ 5$$
$$0.00085 = 0.00085. = 85 \times 10^{-5}$$

Most calculators have a key specifically for entering powers of 10. It is called the *exponent key* and operates as follows.

CALCULATOR KEYS

Name: Exponent entry key [EXP] or [EE]

Function: Prepares calculator to accept next digits entered as a power-of-10 exponent. The sign of the exponent can be changed by using the change-sign key [+/−].

Example: $76 \text{ [EXP]} ^4$

Press keys: [7] [6] [EXP] [4]

Display shows: [76. 04]

Example: 85×10^{-5}

Press keys: [8] [5] [EXP] [5] [+/−]

Display shows: [85. −05]

4-2-2   *Powers of 10 in Combination*

A power-of-10 multiplier raises the base number to the power of 10 indicated by the exponent. To get you used to working with powers of 10, let us do a few examples and include the calculator sequences.

■ EXAMPLE A:

$(2 \times 10^2) + (3 \times 10^3)$
$= 200 + 3000 = 3200$

Steps:
$2 \times 10^2 = 2 \times 10 \times 10 = 200$
$3 \times 10^3 = 3 \times 10 \times 10 \times 10 = 3000$
$200 + 3000 = 3200$
Calculator sequence: [2] [EXP] [2] [+] [3] [EXP] [3] [=]
Answer: 3200 or
[32. 02] (32×10^2 or 3.2×10^3)

■ EXAMPLE B:

$(3 \times 10^6) - (2 \times 10^4)$
$= 3,000,000 - 20,000$
$= 2,980,000$

Steps:
$3 \times 10^6 = 3 \times 10 \times 10 \times 10 \times 10 \times 10 \times 10$
$\quad = 3,000,000$
$2 \times 10^4 = 2 \times 10 \times 10 \times 10 \times 10$
$\quad = 20,000$
$3,000,000 - 20,000 = 2,980,000$
Calculator sequence: [3] [EXP] [6] [−] [2] [EXP] [4] [=]
Answer: 2,980,000 or
[298. 04] (2.98×10^6)

EXAMPLE C:

$$\frac{1.6 \times 10^4}{4 \times 10^2} \text{ or}$$
$$(1.6 \times 10^4) \div (4 \times 10^2)$$
$$= 16{,}000 \div 400$$
$$= 40$$

Steps:
$1.6 \times 10^4 = 1.6 \times 10 \times 10$
$\qquad\qquad \times 10 \times 10 = 16{,}000$
$4 \times 10^2 = 4 \times 10 \times 10$
$\qquad\qquad = 400$
$16{,}000 \div 400 = 40$
Calculator sequence: $\boxed{1}\,\boxed{.}\,\boxed{6}\,\boxed{\text{EXP}}\,\boxed{4}\,\boxed{\div}$
$\boxed{4}\,\boxed{\text{EXP}}\,\boxed{2}\,\boxed{=}$

Answer: 40

EXAMPLE D:

$$(7.5 \times 10^{-6}) \times (1.86 \times 10^{-3})$$
$$= 0.0000075 \times 0.00186$$
$$= 0.00000001395 \text{ or}$$
$$1.395 \times 10^{-8}$$

Steps:
$7.5 \times 10^{-6} = 0.00000075$
$1.86 \times 10^{-3} = 0.001.86$
$0.0000075 \times 0.00186 =$
0.00000001395
Calculator sequence: $\boxed{7}\,\boxed{.}\,\boxed{5}\,\boxed{\text{EXP}}\,\boxed{6}$
$\boxed{+/-}\,\boxed{\times}\,\boxed{1}\,\boxed{.}\,\boxed{8}$
$\boxed{6}\,\boxed{\text{EXP}}\,\boxed{3}\,\boxed{+/-}\,\boxed{=}$

Answer: 1.395×10^{-8}

4-2-3   *Scientific and Engineering Notation*

As mentioned previously, powers of 10 are used in science and technology as a shorthand due to the large number of zeros in many values. There are basically two systems or notations used, involving values that have exponents that are a power of ten. They are called **scientific notation** and **engineering notation.**

A number in *scientific notation* is expressed as a base number between 1 and 10 multiplied by a power of 10. In the following examples, the values on the left have been converted to scientific notation.

Scientific Notation

A widely used floating-point system in which numbers are expressed as products consisting of a number between 1 and 10 multiplied by an appropriate power of 10.

Engineering Notation

A widely used floating-point system in which numbers are expressed as products consisting of a number that is greater than 1 multiplied by a power of 10 that is some multiple of 3.

EXAMPLE A:

$$32{,}000 = 3.2000_{\circ} = 3.2 \times 10^4 \qquad \text{— Scientific notation}$$

Decimal point is moved to a position that results in a base number between 1 and 10. If decimal point is moved left, exponent is positive. If decimal point is moved right, exponent is negative.

EXAMPLE B:

$$0.0019 = 0._{\circ}001.9 = 1.9 \times 10^{-3} \qquad \text{— Scientific notation}$$

EXAMPLE C:

$$114{,}300{,}000 = 1.143\,0\,0\,0\,0\,0_{\circ} = 1.143 \times 10^8 \qquad \text{— Scientific notation}$$

■ **EXAMPLE D:**

$$0.26 = 0_{\curvearrowleft}2.6 = 2.6 \times 10^{-1}$$

$\llcorner$ Scientific notation

As you can see from the preceding examples, the decimal point floats backward and forward, which explains why scientific notation is called a **floating-point number system.** Although scientific notation is used in science and technology, the engineering notation system, discussed next, is used more frequently.

In *engineering notation* a number is represented as a base number that is greater than 1 multiplied by a power of 10 that is some multiple of 3. In the following examples, the values on the left have been converted to engineering notation.

Floating-Point Number System

A system in which numbers are expressed as products consisting of a number and a power-of-10 multiplier.

■ **EXAMPLE A:**

$$32{,}000 = 32.\overset{\frown}{0\,0\,0}_{\circ} = 32 \times 10^{3}$$

Decimal point is moved to a position that results in a base number that is greater than 1, and a power-of-10 exponent that is some multiple of 3.

■ **EXAMPLE B:**

$$0.0019 = 0_{\circ}\overset{\frown}{0\,0\,1}.9 = 1.9 \times 10^{-3}$$

■ **EXAMPLE C:**

$$114{,}300{,}000 = 114.\overset{\frown}{3\,0\,0\,0\,0\,0}_{\circ} = 114.3 \times 10^{6}$$

■ **EXAMPLE D:**

$$0.26 = 0_{\circ}\overset{\frown}{2\,6\,0}. = 260 \times 10^{-3}$$

CALCULATOR KEYS

Name: Normal, scientific, engineering modes

Function: Most calculators have different notation modes that affect the way an answer is displayed on the calculator's screen. Numeric answers can be displayed with up to 10 digits and a two-digit exponent. You can enter a number in any format.

Example: Normal notation mode is the usual way we express numbers, with digits to the left and right of the decimal, as in 12345.67.

Sci (scientific) notation mode expresses numbers in two parts. The significant digits display with one digit to the left of the decimal. The appropriate power of 10 displays to the right of E, as in 1.234567E4.

Eng (engineering) notation mode is similar to scientific notation; however, the number can have one, two, or three digits before the decimal, and the power-of-10 exponent is a multiple of three, as in 12.34567E3.

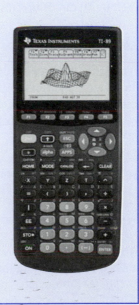

Now that you have completed this section, you should be able to:

■ **Objective 5.** *Explain the powers-of-10 method and how to convert to powers of 10.*

■ **Objective 6.** *Calculate the result of problems with powers of 10 in combination.*

■ **Objective 7.** *Describe the two floating-point number systems: scientific notation and engineering notation.*

Use the following questions to test your understanding of Section 4-2.

1. Convert the following to powers of 10.

    a. $100 = ?$
    b. $1 = ?$
    c. $10 = ?$
    d. $1,000,000 = ?$
    e. $\dfrac{1}{1,000} = ?$

    f. $\dfrac{1}{1,000,000} = ?$

2. Convert the following to common numbers without exponents.

    a. $6.3 \times 10^3 = ?$
    b. $114,000 \times 10^{-3} = ?$
    c. $7,114,632 \times 10^{-6} = ?$
    d. $6624 \times 10^6 = ?$

3. Perform the indicated operation on the following.

    a. $\sqrt{3} \times 10^6 = ?$
    b. $(2.6 \times 10^{-6}) - (9.7 \times 10^{-9}) = ?$
    c. $\dfrac{(4.7 \times 10^3)^2}{3.6 \times 10^6} = ?$

4. Convert the following common numbers to engineering notation.

    a. $47,000 = ?$
    b. $0.00000025 = ?$
    c. $250,000,000 = ?$
    d. $0.0042 = ?$

4-3   THE METRIC SYSTEM

Metric System

A decimal system of weights and measures based on the meter and the kilogram.

The **metric system** *is a decimal system of weights and measures.* This system was developed to make working with values easier. Comparing two examples, you will see that working with the metric system, which uses multiples of 10 (decimal), is much easier than working with the *U.S. customary system of units.*

■ **EXAMPLE: THE METRIC SYSTEM**

The U.S. monetary system is like the metric system. It is a decimal system, which means that it is based on multiples of 10. For instance, there are 100 cents in 1 dollar ($100 \times 1¢ = \$1$), or 10 dimes in 1 dollar ($10 \times 10¢ = \$1$).

■ **EXAMPLE: THE U.S. CUSTOMARY SYSTEM**

The United States still has not adopted the metric system of measurements. For instance, there are 12 inches in 1 foot (12 in. = 1 ft), and 3 feet in 1 yard (3 ft = 1 yd). None of these values are multiples of 10. To add even more to the confusion of this system, inches are divided up into strange fractions such as fourths, sixteenths, thirty-seconds, and sixty-fourths.

From these two examples you can see that it is easier to count, represent, and perform mathematical operations on a system based on multiples of 10.

4-3-1   *Metric Units and Prefixes*

Unit

A determinate quantity adopted as a standard of measurement.

To help explain what we mean by units and prefixes, let us begin by examining how we measure length using the metric system. The standard **unit** of length in the metric system is the *meter* (abbreviated m). You have probably not seen the unit "meter" used a lot on its own. More frequently you have heard and seen the terms *centimeter, millimeter,* and *kilometer.* All these words have two parts: a *prefix name* and *unit*. For example, with the name *centimeter, centi* is the prefix, and *meter* is the unit. Similarly, with the names *millimeter* and *kilometer, milli* and *kilo* are the prefixes, and *meter* is the unit. The next question, therefore, is: What are these prefixes? A **prefix** is *simply a power of 10 or multiplier that precedes the unit.* Figure 4-1 shows the names, symbols, and values of the most frequently used metric prefixes.

What does *centimeter* mean? If you look up the prefix *centi* in Figure 4-1, you can see that it is a prefix indicating a fraction (less than one). Its value or power of 10 is 10^{-2}, or one hundredth ($\frac{1}{100}$). This means that one meter has been divided up into 100 pieces, and each of these pieces is a centimeter or a hundredth of a meter. Millimeters have been used in photography for many years because the width of film negatives is measured in millimeters (for example, 35 mm film). Looking up the prefix *milli* in Figure 4-1, you see that its power of 10 is 10^{-3}, or one thousandth ($\frac{1}{1000}$). This prefix is used to indicate smaller fractions because *milli* measures length in thousandths of a meter, whereas *centi* measures length in hundredths of a meter. When the length of some object is less than 1 meter, therefore, the unit *meter* will have a prefix indicating the fractional multiplier, and this power of 10 will have a negative exponent value. Remember that these negative exponents do not indicate negative numbers; they simply indicate a fractional multiplier.

Looking at the number scale in Figure 4-1, you can see that when values are between 1 and 99, we do not need to use a prefix; for example, 1 meter, 62 meters, 84 meters, and so on. A value like 4500 meters, however, would be shortened to include the 1000 (10^3) prefix *kilo,* as follows:

$$4500 \text{ m} = 4.5 \times 1000 \text{ m} \quad \leftarrow (4.\overset{\frown}{5\,0\,0}.)$$
$$= 4.5 \times 10^3 \text{ m} \quad \leftarrow (\text{Because the 1000}$$
The value 4500 $\longrightarrow = 4.5 \text{ km}$ or 10^3 multiplier is *kilo,* we can
is shortened to 4.5 substitute *kilo,* or k, for 10^3.)
with the prefix kilo.

Similarly, a length of 4,600,000 meters is shortened to

$$4,600,000 \text{ m} = 4.6 \times 1,000,000 \text{ m} \quad \leftarrow (4.\overset{\frown}{6\,0\,0\,0\,0\,0}.)$$
$$= 4.6 \times 10^6 \text{ m} \quad \leftarrow (10^6 = \text{mega, or M})$$
$$= 4.6 \text{ Mm}$$

Now that you have an understanding of metric prefixes, let us discuss some of the metric units and compare them with our known U.S. customary units.

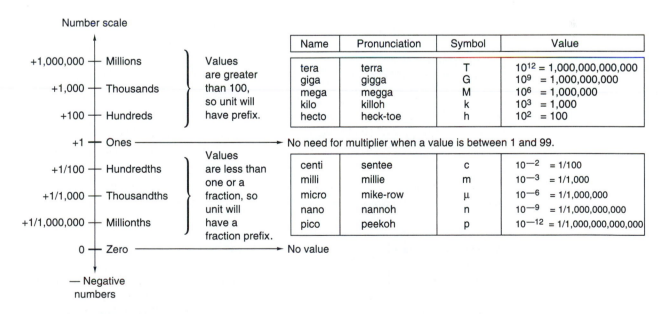

Number scale

Name	Pronunciation	Symbol	Value
tera	terra	T	$10^{12} = 1,000,000,000,000$
giga	gigga	G	$10^9\ = 1,000,000,000$
mega	megga	M	$10^6\ = 1,000,000$
kilo	killoh	k	$10^3\ = 1,000$
hecto	heck-toe	h	$10^2\ = 100$

No need for multiplier when a value is between 1 and 99.

Name	Pronunciation	Symbol	Value
centi	sentee	c	$10^{-2}\ = 1/100$
milli	millie	m	$10^{-3}\ = 1/1,000$
micro	mike-row	μ	$10^{-6}\ = 1/1,000,000$
nano	nannoh	n	$10^{-9}\ = 1/1,000,000,000$
pico	peekoh	p	$10^{-12} = 1/1,000,000,000,000$

No value

FIGURE 4-1 **Metric Prefixes.**

Prefix
An affix attached to the beginning of a word, base, or phrase.

4-3-2 *Metric Units of Length*

As already mentioned, the standard unit of length in the metric system is the meter. Table 4-1 shows how to convert a metric unit of length (millimeters, centimeters, meters, kilometers, and so on) to a U.S. customary unit of length (inches, feet, yards, and miles).

■ **INTEGRATED MATH APPLICATION:**

Use Table 4-1 to find how many feet are in 1 meter.

$$1 \text{ m} = ? \text{ feet}$$

■ *Solution:*

Referring to Table 4-1, you can see that "When You Know" meters, "Multiply by" 3.3 "To Find" feet (ft). Therefore,

$$1 \text{ m} \times 3.3 = 3.3 \text{ ft}$$

Other frequently used conversions are

$$1 \text{ cm} = 0.4 \text{ in.}$$
$$1 \text{ km} = 0.6 \text{ mi}$$

Table 4-2 reverses the process by converting U.S. customary units to metric units. For example,

$$1 \text{ in.} = 2.54 \text{ cm}$$
$$1 \text{ ft} = 30 \text{ cm}$$
$$1 \text{ mi} = 1.6 \text{ km}$$

TABLE 4-1 Converting Metric Units to U.S. Customary Units (Metric → U.S.)

WHEN YOU KNOW	MULTIPLY BY	TO FIND
Length		
millimeters (mm)	0.04	inches (in.)
centimeters (cm)	0.4	inches (in.)
meters (m)	3.3	feet (ft)
meters (m)	1.1	yards (yd)
kilometers (km)	0.6	miles (mi)
Area		
sq. centimeters (cm^2)	0.16	sq. inches (in.^{2})
sq. meters (m^2)	1.2	sq. yards (yd^2)
sq. kilometers (km^2)	0.4	sq. miles (mi^2)
hectares (ha) (10,000 m^2)	2.5	acres
Weight		
grams (g)	0.035	ounces (oz)
kilograms (kg)	2.2	pounds (lb)
tonnes (1000 kg) (t)	1.1	short tons
Volume		
milliliters (mL) ⎫	0.03	fluid ounces (fl oz)
liters (L) ⎪ Liquid volume	2.1	pints (pt)
liters (L) ⎬	1.06	quarts (qt)
liters (L) ⎭	0.26	gallons (gal)
cubic meters (m^3) ⎫ Dry volume	35	cubic feet (ft^3)
cubic meters (m^3) ⎭	1.3	cubic yards (yd^3)
Temperature		
Celsius (°C)	$\left(\dfrac{9}{5} \times °C\right) + 32$	Fahrenheit (°F)

TABLE 4-2 **Converting U.S. Customary Units to Metric Units (U.S. → Metric)**

WHEN YOU KNOW	MULTIPLY BY	TO FIND
Length		
inches (in.)	2.54	centimeters (cm)
feet (ft)	30	centimeters (cm)
yards (yd)	0.9	meters (m)
miles (mi)	1.6	kilometers (km)
Area		
sq. inches (in.^{2})	6.5	sq. centimeters (cm^2)
sq. feet (ft^2)	0.09	sq. meters (m^2)
sq. yards (yd^2)	0.8	sq. meters (m^2)
sq. miles (mi^2)	2.6	sq. kilometers (km^2)
acres	0.4	hectares (ha)
Weight		
ounces (oz)	28	grams (g)
pounds (lb)	0.45	kilograms (kg)
short tons (2000 lb)	0.9	tonnes (t)
Volume		
teaspoons (tsp)	5	milliliters (mL)
tablespoons (tbsp)	15	milliliters (mL)
fluid ounces (fl oz)	30	milliliters (mL)
cups (c)	0.24	liters (L)
pints (pt)	0.47	liters (L)
quarts (qt)	0.95	liters (L)
gallons (gal)	3.8	liters (L)
cubic feet (ft^3)	0.03	cubic meters (m^3)
cubic yards (yd^3)	0.76	cubic meters (m^3)
Temperature		
Fahrenheit (°F)	$\frac{5}{9} \times (°F - 32)$	Celsius (°C)

(Liquid volume: teaspoons, tablespoons, fluid ounces, cups, pints, quarts, gallons)
(Dry volume: cubic feet, cubic yards)

4-3-3   *Metric Units of Area*

The metric unit of area is the *square meter* (abbreviated m^2). For example, an area measuring 4 meters by 3 meters contains 12 square meters (12 m^2), as shown.

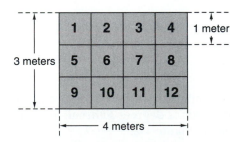

4 m by 3 m = ?
4 m × 3 m = 12 square meters or 12 m^2

■   **INTEGRATED MATH APPLICATION:**

How many square yards are in 12 square meters?

■ *Solution:*

If we wanted to find out how many square yards (yd^2) are in 12 square meters, we would use Table 4-1 (metric → U.S.).

$$12 \text{ m}^2 = ? \text{ yd}^2$$
$$12 \text{ m}^2 \times 1.2 = 14.4 \text{ yd}^2$$

4-3-4   *Metric Units of Weight*

In the metric system, objects are weighed in *grams*. Referring to Table 4-1, you can see that a gram is not very heavy, only 0.035 of an ounce (oz). This is why we usually measure weight in kilograms (kilo = 10^3; therefore, 1 kilogram = 1000 grams). Using Table 4-1, you can see that

$$1 \text{ kilogram (kg)} = 2.2 \text{ pounds (lb)}$$

4-3-5   *Metric Units of Volume*

Volume
The amount of space occupied by a three-dimensional object as measured in cubic units.

Area is a space measured in two dimensions, whereas **volume** is a space measured in three dimensions, as follows:

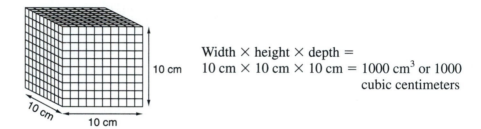

Width $\times$ height $\times$ depth =
10 cm $\times$ 10 cm $\times$ 10 cm = 1000 cm^3 or 1000 cubic centimeters

The cube shown contains 1000 smaller 1 centimeter $\times$ 1 centimeter $\times$ 1 centimeter cubes. In the metric system a volume of 1000 cubic centimeters (1000 cm^3) is equal to 1 *liter* (L), which is the metric unit of *liquid volume*. Referring to Table 4-1, you can see that 1 liter is approximately equal to 1 quart (qt). This means that the next time you are at the gas station and get 2 gallons of gasoline or 8 quarts (4 quarts in 1 gallon), you have actually filled your tank with approximately 8 liters. *Dry volume* is also measured in the metric unit *meters cubed* (m^3).

Degrees Celsius
The international thermometer scale on which the interval between the melting point of ice (0 °C) and the boiling point of water (+100 °C) is divided into 99.9 degrees.

Degrees Fahrenheit
A thermometer scale on which the boiling point of water is at 212 degrees, and the freezing point of ice is at 32 degrees.

4-3-6   *Metric Unit of Temperature*

The metric unit of temperature is **degrees Celsius** (symbolized °C). Most people are more familiar with the U.S. customary temperature unit, which is **degrees Fahrenheit** (symbolized °F). These two temperature scales are compared in Figure 4-2.

As you can see by comparing the scales, the metric Celsius scale has two easy points to remember: 0 °C is the freezing or melting point of water, and 100 °C is the boiling point of water. These same points on the Fahrenheit scale are not as easy to remember (the melting or freezing point of water is 32 °F, and the boiling point of water is 212 °F).

Table 4-1 shows how to convert a °C temperature to °F, and Table 4-2 shows how to convert a °F reading to °C.

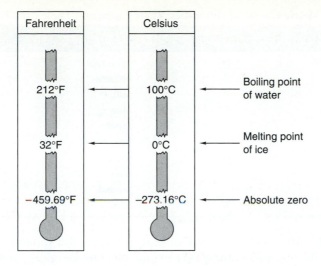

FIGURE 4-2 Comparison of Fahrenheit and Celsius Temperature Scales

---

■ **INTEGRATED MATH APPLICATION:**

If a thermometer reads 29 °C, what would this temperature be in degrees Fahrenheit?

■ *Solution:*

From Table 4-1, you can see that to convert a metric Celsius temperature to a U.S. customary temperature, you multiply the Celsius reading by $\frac{9}{5}$ and then add 32.

$$°F = \left(\frac{9}{5} \times °C\right) + 32$$
$$= (1.8 \times 29) + 32$$
$$= 52.2 + 32 = 84.2 \; °F$$

Steps:
$9 \div 5 = 1.82$
$1.8 \times 29 = 52.2$
$52.2 + 32 = 84.2 \; °F$

■ **INTEGRATED MATH APPLICATION:**

Convert 63 °F to Celsius.

■ *Solution:*

Table 4-2 indicates that to convert °F to °C, you must subtract 32 from the Fahrenheit reading and then multiply the remainder by $\frac{5}{9}$.

$$°C = \frac{5}{9} \times (°F - 32)$$
$$= 0.55 \times (63 - 32)$$
$$= 0.55 \times 31 = 17.05 \; °C$$

Steps:
$5 \div 9 = 0.55$
$63 - 32 = 31 \; °C$

4-3-7   *Electrical Units, Prefixes, and Conversions*

Table 4-3 lists many of the more commonly used metric electrical quantities with their units and symbols. For example, electrical current (symbolized by I) is measured in amperes or amps (symbolized by A). As another example, electrical resistance (R) is measured in ohms (symbolized by the Greek capital letter omega, Ω).

If electrical values are much larger than one or are a fraction of one, the unit of electrical quantity is preceded by a prefix. Most electrical and electronic applications use engineering notation, so these prefixes are power-of-10 exponents that are some multiple of

TABLE 4-3 Metric Electrical Quantities, Their Units, and Their Unit Symbols

QUANTITY	SYMBOL	UNIT	SYMBOL
Charge	Q	coulomb	C
Current	I	ampere	A
Voltage	V	volt	V
Resistance	R	ohm	Ω
Capacitance	C	farad	F
Inductance	L	henry	H
Energy, work	W	joule	J
Power	P	watt	W
Time	t	second	s
Frequency	f	hertz	Hz

three (10^3, 10^6, 10^9, 10^{12}, and so on). As an example, Table 4-4 shows how these prefixes precede ampere, which is the electrical current unit. To help our understanding of these prefixes, let us do a few conversion problems.

■   INTEGRATED MATH APPLICATION:

Convert the following.

a. 0.003 A = _____ mA (milliamperes)

b. 0.07 mA = _____ μA (microamperes)

c. 7333 mA = _____ A (amperes)

d. 1275 μA = _____ mA (milliamperes)

■   *Solution:*

a. 0.003 A = _____ mA. In this example, 0.003 A has to be converted so that it is represented in milliamperes (10^{-3}, or $\frac{1}{1000}$ of an ampere). The basic algebraic rule to remember is that both expressions on either side of the equals sign must be the same.

Left		Right	
Base	Exponent	Base	Exponent
0.003×10^0		$=$	_____ $\times\ 10^{-3}$

TABLE 4-4 Electrical Current Prefixes

NAME	ABBREVIATION	VALUE
Teraampere	TA	$10^{12} = 1{,}000{,}000{,}000{,}000$
Gigaampere	GA	$10^9 = 1{,}000{,}000{,}000$
Megaampere	MA	$10^6 = 1{,}000{,}000$
Kiloampere	kA	$10^3 = 1000$
Ampere	A	$10^0 = 1$
Milliampere	mA	$10^{-3} = \dfrac{1}{1000}$
Microampere	μA	$10^{-6} = \dfrac{1}{1{,}000{,}000}$
Nanoampere	nA	$10^{-9} = \dfrac{1}{1{,}000{,}000{,}000}$
Picoampere	pA	$10^{-12} = \dfrac{1}{1{,}000{,}000{,}000{,}000}$

The exponent on the right in this problem will be decreased 1000 times (10^0 to 10^{-3}), so for the statement to balance, the number on the right will have to be increased 1000 times; that is, the decimal point will have to be moved to the right three places (0.003 to 3). Therefore,

$$0.003 \times 10^0 = 3 \times 10^{-3}$$

or $\quad$ $0.003 \text{ A} = 3 \times 10^{-3} \text{ A}$ $\quad$ or $\quad$ 3 mA

b. 0.07 mA = _____ μA. In this example the exponent is going from milliamperes to microamperes (10^{-3} to 10^{-6}), or 1000 times smaller, so the number must be made 1000 times greater.

$$0.070. \quad \text{or} \quad 70.0$$

Therefore, 0.07 mA = 70 μA.

c. 7333 mA = _____ A. The exponent is going from milliamperes to amperes, increasing 1000 times, so the number must decrease 1000 times.

$$7333. \quad \text{or} \quad 7.333$$

Therefore, 7333 mA = 7.333 A.

d. 1275 μA = _____ mA. The exponent is changing from microamperes to milliamperes, an increase of 1000 times, so the number must decrease by the same factor.

$$1.2750 \quad \text{or} \quad 1.275$$

Therefore, 1275 μA = 1.275 mA.

CALCULATOR KEYS

Name: Convert key $\blacktriangleright$

Function: Some calculators have a function that converts an expression from one unit to another.

$$expression\_unit \; \blacktriangleright \;\_unit2 \quad \Rightarrow \quad expression\_unit2$$

Examples:

How many feet are in 3 meters?

Press keys: 3_m $\blacktriangleright$ _ft ENTER

Answer: 9.842_ft

How many kilometers are in 4 light-years?

Press keys: 4_ltyr $\blacktriangleright$ _km

Answer: 3.78421E13_km

How many ohms are in 4 kiloohms?

Press keys: 4_kΩ $\blacktriangleright$ _Ω

Answer: 4000_Ω

To convert a temperature value, you must use tmpCnv( ) instead of the $\blacktriangleright$ operator.

Example:

Press keys: tmpCnv(100_°C to _°F)

Answer: 212_°F

Now that you have completed this section, you should be able to:

■ **Objective 8.** *Define and explain the purpose of the metric system.*

■ **Objective 9.** *List the metric prefixes and describe the purpose of each.*

■ **Objective 10.** *Describe the metric units for length, area, weight, volume, and temperature.*

■ **Objective 11.** *Convert U.S. customary units to metric units, and metric units to U.S. customary units.*

■ **Objective 12.** *List and describe many of the more frequently used electrical units and prefixes, and describe how to interconvert prefixes.*

1. Give the power-of-10 value for the following prefixes.

   a. kilo   **d.** mega
   b. centi   **e.** micro
   c. milli

2. Convert the following metric units to U.S. customary units.

   a. 15 cm to inches
   b. 23 kg to pounds
   c. 37 L to gallons
   d. 23 °C to degrees Fahrenheit

3. Convert the following U.S. customary units to metric.

   a. 55 miles per hour (mph) to kilometers per hour (km/h)
   b. 16 gal to liters
   c. 3 yd^2 to square meters
   d. 92 °F to degrees Celsius

4. List the names of the metric units for the following quantities.

a. Length	**f.** Current
b. Weight	**g.** Voltage
c. Temperature	**h.** Volume (liquid)
d. Time	**i.** Energy
e. Power	**j.** Resistance

5. Convert the following.

   a. 25,000 V = _____ kilovolts
   b. 0.014 W = _____ milliwatts
   c. 0.000016 μF = _____ nanofarads
   d. 3545 kHz = _____ megahertz

REVIEW QUESTIONS

Multiple Choice Questions

1. A base number's exponent indicates how many times the base number must be multiplied by:

   a. 2
   b. 3
   c. Itself
   d. Both (a) and (c)

2. What does 16^3 mean?

   a. 16×3
   b. $16 \times 16 \times 16$
   c. $16 \times (3 \times 3 \times 3)$
   d. $3 \times 3 \times 3 \times 3 \times 3 \times 3 \times 3 \times 3 \times 3 \times 3 \times 3 \times 3 \times 3 \times 3 \times 3 \times 3$

3. If $x^2 = 16$, then $\sqrt[2]{16} = x$. What is the value of x?

   a. 4
   b. 2
   c. 16
   d. 8

4. Calculate $\sqrt[5]{32}$.

   a. 2
   b. 1.3
   c. 4
   d. 6.4

5. Raise the following number to the power of 10 indicated by the exponent, and give the answer: 3.6×10^2.

   a. 3600   **c.** 0.036
   b. 36   **d.** 360

6. Convert the following number to powers of 10: 0.00029.

   a. 0.29×10^{-3}
   b. 2.9×10^{-4}
   c. 29×10^{-5}
   d. All the above

7. What is the power-of-10 value for the metric prefix *milli*?

   a. 10^{-3} **c.** 10^3
   b. 10^{-2} **d.** All the above

8. What is the metric unit for temperature?

   a. Fahrenheit
   b. Borg
   c. Celsius
   d. Rankine

9. Convert the following value of inductance to engineering notation: 0.016 henry (H).

   a. 1.6×10^2 H   **c.** 1.6 cH
   b. 16 mH   **d.** 16 kH

10. How many centimeters are in 8 inches?

    a. 2.03 m   **c.** 8 cm
    b. 3.2 cm   **d.** 20.32 cm

Communication Skill Questions

11. What is an exponent? (Introduction)

12. Define the following terms: (4-1)
    a. Square of a number
    b. Square root of a number

13. What is a power-of-10 exponent? (4-2)

14. Describe the scientific and engineering notation systems. (4-2)

15. What is the metric system? (4-3)

16. List the metric prefixes and their value. (4-3)

17. Explain how to convert the following: (4-3)
    a. Centimeters to inches
    b. Ounces to grams
    c. Yards to meters
    d. Miles to kilometers

18. What is the metric unit of: (4-3)
    a. Area
    b. Length
    c. Weight
    d. Volume
    e. Temperature

19. Give the units for the following electrical quantities: (4-3)
    a. Current   c. Resistance
    b. Length   d. Power

20. Arbitrarily choose values, and then describe the following conversions: (4-3)
    a. Amps to milliamps
    b. Volts to kilovolts

Practice Problems

21. Determine the square of the following values.
    a. 9^2 d. 0^2
    b. 6^2 e. 1^2
    c. 2^2 f. 12^2

22. Determine the square root of the following values.
    a. $\sqrt{81}$ d. $\sqrt{36}$
    b. $\sqrt{4}$ e. $\sqrt{144}$
    c. $\sqrt{0}$ f. $\sqrt{1}$

23. Raise the following base numbers to the power indicated by the exponent, and give the answer.
    a. 9^3 c. 4^6
    b. 10^4 d. 2.5^3

24. Determine the roots of the following values.
    a. $\sqrt[3]{343} = ?$ c. $\sqrt[4]{760} = ?$
    b. $\sqrt{1296} = ?$ d. $\sqrt[6]{46,656} = 5 ?$

25. Convert the following values to powers of 10.
    a. $\dfrac{1}{100}$ c. $\dfrac{1}{1000}$
    b. 1,000,000,000   d. 1000

26. Convert the following to proper fractions and decimal fractions.
    a. 10^{-4} c. 10^{-6}
    b. 10^{-2} d. 10^{-3}

27. Convert the following values to powers of 10 in both scientific and engineering notation.
    a. 475   c. 0.07
    b. 8200   d. 0.00045

28. Convert the following to whole-number values with a metric prefix.
    a. 0.005 A
    b. 8000 m
    c. 15,000,000 Ω
    d. 0.000016 s

29. Convert the following metric units to U.S. customary units.
    a. 100 km   c. 67 kg
    b. 29 m^2 d. 2 L

30. Convert the following U.S. customary units to metric.
    a. 4 in.   c. 4 oz
    b. 2 mi^2 d. 10 gal

31. Convert the following temperatures from °F to °C.
    a. 32 °F   b. 72 °F

32. Convert the following temperatures from °C to °F.
    a. 6 °C   b. 32 °C

33. State the metric unit of:
    a. Length   d. Dry volume
    b. Area   e. Liquid volume
    c. Weight   f. Temperature

34. What is the metric unit of time?

35. Convert the following.
    a. 8000 ms = _____ microseconds
    b. 0.02 MV = _____ kilovolts
    c. 10 km = _____ meters
    d. 250 mm = _____ centimeters

Web Site Questions

Go to the web site http://www.prenhall.com/cook, select the textbook *Mathematics for Electronics and Computers,* select this chapter, and then follow the instructions when answering the multiple-choice practice problems.

Algebra, Equations, and Formulas

Back to the Future

Each part of a computer system is designed to perform a specific task. You would think that all these units were first thought of by some recent pioneer in the twentieth century, but in fact, two of these elements were first described in 1833.

Born in England in 1791, Charles Babbage became very well known for both his mathematical genius and eccentric personality. For many years Babbage occupied the Cambridge chair of mathematics, once held by Isaac Newton, and although he never delivered a single lecture, he wrote several papers on numerous subjects, ranging from politics to manufacturing techniques. He also helped develop several practical devices, including the tachometer and the railroad cowcatcher.

Babbage's ultimate pursuit, however, was that of mathematical accuracy. He delighted in spotting errors in everything from log tables (used by astronomers, mathematicians, and navigators) to poetry. In fact, he once wrote to poet Alfred Lord Tennyson, pointing out an inaccuracy in his line "Every moment dies a man—every moment one is born." Babbage explained to Tennyson that since the world population was actually increasing and not, as he indicated, remaining constant, the line should be rewritten to read "Every moment dies a man—every moment one and one-sixteenth is born."

In 1822, Babbage described in a paper and built a model of what he called "a difference engine," which could be used to calculate mathematical tables. The Royal Society of Scientists described his machine as "highly deserving of public encouragement," and a year later the government awarded Babbage £1500 for his project. Babbage originally estimated that the project should take 3 years; however, the design had its complications, and after 10 years of frustrating labor, in which the government grants increased to £17,000, Babbage was still no closer to completion. Finally, the money stopped and Babbage reluctantly decided to let his brainchild go.

In 1833, Babbage developed an idea for a much more practical machine, which he named "the analytical engine." It was to be a more general machine that could be used to solve a variety of problems, depending on instructions supplied by the operator. It would include two units, called a "mill" and a "store," both of which would be made of cogs and wheels. The store, which was equivalent to a modern-day computer memory, could hold up to 100 forty-digit numbers. The mill, which was equivalent to a modern computer's arithmetic and logic unit (ALU), could perform both arithmetic and logic operations on variables or numbers retrieved from the store, and the result could be stored in the store and then acted upon again or printed out. The program of instructions directing these operations would be fed into the analytical engine in the form of punched cards.

The analytical engine was never built. All that remains are the volumes of descriptions and drawings, and a section of the mill and printer built by Babbage's son, who also had to concede defeat. It was, unfortunately for Charles Babbage, a lifetime of frustration to have conceived the basic building blocks of the modern computer a century before the technology existed to build it.

Outline and Objectives

VIGNETTE: BACK TO THE FUTURE

INTRODUCTION

5-1 THE BASICS OF ALGEBRA

Objective 1: Define the following terms.
 a. Algebra   c. Formula
 b. Equation   d. Literal numbers

Objective 2: Describe the rules regarding the equality on both sides of the equals sign.

5-1-1 The Equality on Both Sides of the Equals Sign

Objective 3: Demonstrate how to remove a quantity by performing the same arithmetic operation on both sides of an equation or formula.

5-1-2 Treating Both Sides Equally

5-2 TRANSPOSITION

5-2-1 Transposing Equations

Objective 4: Explain the steps involved in transposing an equation to determine the unknown.

1. Example with an Unknown Above the Fraction Bar

2. Example with an Unknown Below the Fraction Bar

3. Example with an Unknown on Both Sides of the Equals Sign

4. Story Problems

Objective 5: Demonstrate how to develop an equation from a story problem.

5. Factoring and Removing Parentheses

Objective 6: Describe the process of factoring and how to remove parentheses within a formula or expression.

6. Square and Square Root

5-2-2 Transposing Formulas

Objective 7: Demonstrate the process of transposing a formula.

1. Directly Proportional and Inversely Proportional

Objective 8: Describe how the terms *directly proportional* and *inversely proportional* are used to show the relationship between quantities in a formula.

2. Ohm's Law Formula Example

Objective 9: Explain the proportionality among the three quantities in the Ohm's law formula and describe how to transpose the formula.

3. Power Formula Example

Objective 10: Describe the electric power formula and transpose the formula to develop alternative procedures.

5-3 SUBSTITUTION

Objective 11: Demonstrate how to use substitution to develop alternative formulas.

5-3-1 Circle Formula Example

5-3-2 Power Formula Example

5-4 RULES OF ALGEBRA—A SUMMARY

Objective 12: Summarize the terms, rules, properties and mathematical operations of algebra.

MULTIPLE CHOICE QUESTIONS

COMMUNICATION SKILL QUESTIONS

PRACTICE PROBLEMS

WEB SITE QUESTIONS

Introduction

I contemplated calling this chapter "using letters in mathematics" because the word *algebra* seems to make many people back away. If you look up *algebra* in the dictionary, it states that it is "a branch of mathematics in which letters representing numbers are combined according to the rules of mathematics." In this chapter you will discover how easy algebra is to understand and then see how we can put it to some practical use, such as rearranging formulas. As you study this chapter you will find that algebra is quite useful, and because it is used in conjunction with formulas, an understanding is essential for anyone entering a technical field.

5-1 THE BASICS OF ALGEBRA

Algebra
A generalization of arithmetic in which letters representing numbers are combined according to the rules of arithmetic.

Formula
A general, fact, rule, or principle expressed usually in mathematical symbols.

As mentioned briefly in the introduction to this chapter, **algebra** by definition is a branch of mathematics in which letters representing numbers are combined according to the rules of mathematics. The purpose of using letters instead of values is to develop a general statement or **formula** that can be used for any values. For example, distance (d) equals velocity (v) multiplied by time (t), or

$$d = v \times t$$

where d = distance in miles
v = velocity or speed in miles per hour
t = time in hours

Using this formula, we can calculate how much distance was traveled if we know the speed or velocity at which we were traveling and the time for which we traveled at that speed. Therefore, if I were to travel at a speed of 20 miles per hour (20 mph) for 2 hours, how far, or how much distance, would I travel? Replacing the letters in the formula with values converts the problem from a formula to an **equation,** as shown:

Equation
A formal statement of the equality or equivalence of mathematical or logical expressions.

$$d = v \times t \leftarrow \text{(Formula)}$$
$$d = 20 \text{ mph} \times 2 \text{ h} \leftarrow \text{(Equation)}$$
$$d = 40 \text{ mi} \leftarrow \text{(Answer)}$$

Literal Number
A number expressed as a letter.

Letters such as d, v, t and a, b, c are called **literal numbers** (letter numbers), and are used, as we have just seen, in general statements showing the relationship between quantities. They are also used in equations to signify an *unknown quantity,* as will be discussed in the following section.

5-1-1 *The Equality on Both Sides of the Equals Sign*

All equations or formulas can basically be divided into two sections that exist on either side of an equals sign, as shown:

Left half   Right half

Fraction bar ⟶ $\dfrac{8 \times x}{2}$ = $\dfrac{16}{1}$ ⟵ Fraction bar

Equals sign

Everything in the left half of the equation is equal to everything in the right half of the equation. This means that 8 times *x* (which is an unknown value) divided by 2 equals 16 divided by 1. Generally, it is not necessary to put 1 under the 16 in the right section because any number divided by 1 equals the same number ($^{16}/_1 = 16 \div 1 = 16$); however, it was included in this introduction to show that each section has both a top and a bottom. If the equation is written without the fraction bar or 1 in the denominator position, it appears as follows:

$$\frac{8 \times x}{2} = 16$$

Although the fraction bar is not visible, you should always assume that a number on its own on either side of the equals sign is above the fraction bar, as shown:

$$\frac{8 \times x}{2} = \frac{16}{}$$

Now that we understand the basics of an equation, let us see how we can manipulate it yet keep both sides equal to one another.

5-1-2   *Treating Both Sides Equally*

If you do exactly the same thing to both sides of an equation or formula, the two halves remain exactly equal, or in balance. This means that as long as you add, subtract, multiply, or divide both sides of the equation by the same number, the equality of the equation is preserved. For example, let us try adding, subtracting, multiplying, and dividing both sides of the following equation by 4 and see if both sides of the equation are still equal.

$$\boxed{2 \times 4 = 8} \quad \leftarrow \text{(Original equation)}$$

1. Add 4 to both sides of the equation:

$$2 \times 4 = 8 \quad \leftarrow \text{(Original equation)}$$
$$(2 \times 4) + 4 = 8 + 4 \quad \leftarrow \text{(Add 4 to both sides.)}$$
$$8 + 4 = 8 + 4$$
$$12 = 12$$

Both sides of the equation remain equal.

2. Subtract 4 from both sides of the equation:

$$2 \times 4 = 8 \quad \leftarrow \text{(Original equation)}$$
$$(2 \times 4) - 4 = 8 - 4 \quad \leftarrow \text{(Subtract 4 from both sides.)}$$
$$8 - 4 = 8 - 4$$
$$4 = 4$$

Both sides of the equation remain equal.

3. Multiply both sides of the equation by 4:

$$2 \times 4 = 8 \quad \leftarrow \text{(Original equation)}$$
$$(2 \times 4) \times 4 = 8 \times 4 \quad \leftarrow \text{(Multiply both sides by 4.)}$$
$$8 \times 4 = 8 \times 4$$
$$32 = 32$$

Both sides of the equation remain equal.

4. Divide both sides of the equation by 4:

$$2 \times 4 = 8 \quad \leftarrow \text{(Original equation)}$$
$$(2 \times 4) \div 4 = 8 \div 4 \quad \leftarrow \text{(Divide both sides by 4.)}$$
$$\frac{2 \times 4}{4} = \frac{8}{4}$$
$$\frac{8}{4} = \frac{8}{4}$$
$$2 = 2$$

Both sides of the equation remain equal.

As you can see from the four preceding procedures, *if you add, subtract, multiply, or divide both halves of an equation by the same number, the equality of the equation is preserved.* In the next section we will see how these operations can serve some practical purpose.

SELF-TEST EVALUATION POINT FOR SECTION 5-1

Now that you have completed this section, you should be able to:

- **Objective 1.** *Define the terms* algebra, equation, formula, *and* literal numbers.

- **Objective 2.** *Describe the rules regarding the equality on both sides of the equals sign.*

- **Objective 3.** *Demonstrate how to remove a quantity by performing the same arithmetic operation on both sides of an equation or formula.*

1. Is the following equation true?

$$\frac{\frac{56}{14}}{2} = 2 \text{ or } \frac{56 \div 14}{2} = 2$$

2. If we were to multiply the left side of the equation in Question 1 by 3 and the right side of the equation by 2, would the two sides be equal to each other?

3. Fill in the missing values.

$$\frac{144}{12} \times \boxed{} = \frac{36}{6} \times 2 \times \boxed{}$$
$$60 = 60$$

4. Determine the result and state whether or not the equation is equal.

$$\frac{(8 - 4) + 26}{5} = \frac{81 - 75}{2}$$

5. Is the following equation balanced?

$$3.2 \text{ k}\Omega = 3200 \ \Omega$$

5-2   TRANSPOSITION

Transposition

The transfer of any term of an equation from one side over to the other side with a corresponding change of the sign.

It is important to know how to *transpose,* or rearrange, equations and formulas so that you can determine the unknown quantity. This process of rearranging, called **transposition,** is discussed in this section.

5-2-1   *Transposing Equations*

As an example, let us return to the original problem introduced at the beginning of this chapter and try to determine the value of the unknown quantity *x*.

$$\frac{8 \times x}{2} = 16$$

To transpose the equation we must follow two steps:

Step 1. Move the unknown quantity so that it is above the fraction bar on one side of the equals sign.

Step 2. Isolate the unknown quantity so that it stands by itself on one side of the equals sign.

Looking at the first step, let us see if our unknown quantity is above the fraction bar on either side of the equals sign.

The unknown quantity x is above the fraction bar.

$$\frac{8 \times x}{2} = 16$$

Since step 1 is done, we can move on to step 2. Looking at the equation, you can see that x does not stand by itself on one side of the equals sign. To satisfy this step, we must somehow move the 8 above the fraction bar, and the 2 below the fraction bar away from the left side of the equation, so that x is on its own.

Let us begin by removing the 2. To remove a letter or number from one side of a formula or equation, simply remember this rule: *To move a quantity, simply apply to both sides the arithmetic opposite of that quantity.* Multiplication is the opposite of division, so to remove a "divide by 2" ($\div 2$), simply multiply both sides by 2 ($\times 2$), as follows:

$$\frac{8 \times x}{2} = 16 \qquad \leftarrow \text{(Original equation)}$$

$$\frac{8 \times x}{2} \times 2 = 16 \times 2 \qquad \leftarrow \text{(Multiply both sides by 2.)}$$

$$\frac{8 \times x}{2} \times \frac{2}{1} = 16 \times 2 \qquad \leftarrow \text{(Because } \frac{2}{2} = 1 \text{, the two 2s on the left side of the equation cancel.)}$$

$$(8 \times x) \times 1 = 16 \times 2 \qquad \text{(Because anything multiplied by 1 equals the same number, the 1 on}$$

$$8 \times x = 16 \times 2 \qquad \text{the left side of the equation can be removed: } 8x \times 1 = 8x.)$$

Looking at the result so far, you can see that by multiplying both sides by 2, we effectively moved the 2 that was under the fraction bar on the left side to the right side of the equation above the fraction bar.

$$\frac{8 \times x}{2} = 16 \qquad \leftarrow \text{(Original equation)}$$

$$8 \times x = 16 \times 2 \qquad \leftarrow \text{(Result after both sides of the equation were multiplied by 2)}$$

However, we have still not completed step 2, which was to isolate x on one side of the equals sign. To achieve this we need to remove the 8 from the left side of the equation. Once again we will do the opposite: Because the opposite of multiply is divide, to remove a "multiply by 8" we must divide both sides by 8 ($\div 8$), as follows:

$$\frac{8 \times x}{2} = 16 \qquad \leftarrow \text{(Original equation)}$$

$$8 \times x = 16 \times 2 \qquad \leftarrow \text{(Equation after both sides were multiplied by 2)}$$

$$\frac{8 \times x}{8} = \frac{16 \times 2}{8} \qquad \leftarrow \text{(Divide both sides by 8.)}$$

$$\frac{8 \times x}{8} = \frac{16 \times 2}{8} \qquad \text{(Because } \frac{8}{8} = 1 \text{, the two 8s on the left side of the equation cancel.)}$$

$$1 \times x = \frac{16 \times 2}{8} \qquad \text{(Anything multiplied by 1 equals the same number, so the 1 on the left side of the equation can be removed: } 1 \times x = x.)$$

$$x = \frac{16 \times 2}{8}$$

Now that we have completed step 2, which was to isolate the unknown quantity so that it stands by itself on one side of the equation, we can calculate the value of the unknown x by performing the arithmetic operations indicated on the right side of the equation.

$$x = \frac{16 \times 2}{8} \qquad (16 \times 2 = 32)$$

$$x = \frac{32}{8} \qquad (32 \div 8 = 4)$$

$$x = 4$$

To double-check this answer, let us insert this value into the original equation to see if it works.

$$\frac{8 \times x}{2} = 16 \qquad (\text{Replace } x \text{ with 4, or substitute 4 for } x.)$$

$$\frac{8 \times 4}{2} = 16 \qquad (8 \times 4 = 32)$$

$$\frac{32}{2} = 16 \qquad (32 \div 2 = 16)$$

$$16 = 16 \qquad (\text{Answer checks out because } 16 = 16.)$$

1. Example with an Unknown Above the Fraction Bar

■ EXAMPLE:

Determine the value of the unknown a in the following equation.

$$(2 \times a) + 5 = 23$$

■ Solution:

Step 1. Is the unknown quantity above the fraction bar? Yes.

Step 2. Is the unknown quantity isolated on one side of the equals sign? No.

Use the following steps to isolate the unknown, a.

	Steps:
$(2 \times a) + 5 = 23$	To remove the "+ 5," do the opposite: subtract 5 from both sides.
$(2 \times a) + \cancel{5} - \cancel{5} = 23 - 5$	$5 - 5 = 0$; any number plus 0 equals the same number: $(2 \times a) + 0 = 2 \times a$
$2 \times a = 23 - 5$	To remove the "2 ×," do the opposite: divide both sides by 2.
$\dfrac{\cancel{2} \times a}{\cancel{2}} = \dfrac{23 - 5}{2}$	$2 \div 2 = 1$; any number multiplied by 1 equals the same number: $1 \times a = a$
$1 \times a = \dfrac{23 - 5}{2}$	
$a = \dfrac{23 - 5}{2}$	Perform the arithmetic operations indicated to determine the value a.
	$23 - 5 = 18 \qquad 18 \div 2 = 9$
$a = \dfrac{18}{2}$	
$a = 9$	

Double-check your answer by inserting it into the original equation.

$$(2 \times a) + 5 = 23 \quad \text{(Replace } a \text{ with 9.)}$$
$$(2 \times 9) + 5 = 23$$
$$18 + 5 = 23$$
$$23 = 23 \quad \text{(Answer checks out because } 23 = 23.\text{)}$$

2. Example with an Unknown Below the Fraction Bar

■ **EXAMPLE:**

Determine the value of the unknown x in the following equation:

$$\frac{72}{x} = 12$$

■ *Solution:*

Step 1. Is the unknown quantity above the fraction bar? No.

Use the following steps to move x above the fraction bar.

Steps:

$$\frac{72}{x} = 12$$

To move the x do the opposite: the opposite of "divide by x" is multiply by x.

$$\frac{72}{x} \left(\times x \right) = 12 \left(\times x \right)$$

Any number divided by itself = 1, so $x \div x = 1$ and therefore the two x's on the left side of the equation cancel.

$$72 \times 1 = 12 \times x$$

$$72 = 12 \times x \qquad 72 \times 1 = 72$$

Now that the unknown quantity x is above the fraction bar, we can proceed to step 2.

Step 2. Isolate the unknown quantity on one side of the equals sign.

$$72 = 12 \times x$$

To remove the "12 ×," simply do the opposite: divide both sides by 12.
$12 \div 12 = 1$, $1 \times x = x$; therefore, the 12s cancel.

$$\frac{72}{12} = \frac{12 \times x}{12}$$

$$\frac{72}{12} = 1 \times x$$

$$\frac{72}{12} = x \qquad 72 \div 12 = 6, x = 6$$

$$6 = x \qquad \text{or} \qquad x = 6$$

To double-check your answer, replace x with 6 in the original equation.

$$\frac{72}{x} = 12 \qquad \text{(Original equation)}$$

$$\frac{72}{6} = 12$$

$$12 = 12$$

3. Example with an Unknown on Both Sides of the Equals Sign

In this section we will see how to deal with equations containing an unknown on both sides of the equals sign. In all equations like this, the same letter or literal number will have the same value no matter where it appears.

■ **EXAMPLE:**

Determine the value of the unknown y in the following equation:

$$(6 \times y) + 7 = y + 27$$

■ *Solution:*

Step 1. Is the unknown above the fraction bar? Yes.

Step 2. Is the unknown isolated on one side of the equals sign? No.

To fulfill step 2 we need to isolate the unknown on one side of the equals sign. Since y exists on both sides, we must remove one. Studying the left side of the equation you can see that it has six y's ($6 \times y$), whereas the right side of the equation has only one y (y). The first step could therefore be to subtract y ($-y$) from both sides.

$$(6 \times y) + 7 = y + 27 \qquad \leftarrow \text{(Original equation)}$$

Steps:

$$(6 \times y) + 7 \,\boxed{- y} = y + 27 \,\boxed{- y} \qquad \text{Subtract } y \text{ from both sides:}$$
$$(5 \times y) + 7 = 0 + 27 \qquad\qquad (6 \times y) - y = 5 \times y,$$
$$(5 \times y) + 7 = 27 \qquad\qquad y - y = 0, 0 + 27 = 27$$
$$(5 \times y) + 7 \,\boxed{- 7} = 27 \,\boxed{- 7} \qquad \text{To remove } + 7, \text{ subtract 7 from both sides.}$$
$$\qquad\qquad\qquad\qquad 7 - 7 = 0, \quad 27 - 7 = 20$$
$$5 \times y = 20 \qquad\qquad \text{The opposite of multiply is}$$
$$\qquad\qquad\qquad\qquad \text{divide. To remove ``5 } \times \text{,''}$$
$$\qquad\qquad\qquad\qquad \text{divide both sides by 5.}$$

$$\frac{\cancel{5} \times y}{\cancel{5}} = \frac{20}{5} \qquad\qquad 5 \div 5 = 1$$

$$1 \times y = \frac{20}{5} \qquad\qquad 1 \times y = y$$

$$y = \frac{20}{5} \qquad\qquad 20 \div 5 = 4; \text{ therefore, } y = 4.$$

$$y = 4$$

To double-check your answer, replace y with 4 in the original equation.

$$(6 \times y) + 7 = y + 27 \quad \leftarrow \text{(Original equation)}$$
$$(6 \times 4) + 7 = 4 + 27 \quad \leftarrow \text{(Replace } y \text{ with 4.)}$$
$$24 + 7 = 4 + 27$$
$$31 = 31$$

4. Story Problems

Story problems are probably the best practice because they are connected to our everyday life and help us to see more clearly how we can apply our understanding of mathematics. I have found the following steps helpful with story problems:

a. Determine what is unknown and assign it a literal number such as x.

b. Determine how many other elements are involved and how they relate to the unknown.

c. Construct an equation using the unknown and the other elements with their associated arithmetic operations.

d. Calculate the unknown using transposition if necessary.

■ **INTEGRATED MATH APPLICATION:**

If you multiply the number of dollars in your wallet by 6 and then subtract $23, you would still end up with one fourth of $100. How many dollars do you have in your wallet?

■ *Solution:*

To start with step a, we must determine the unknown. The last sentence, "How many dollars do you have in your wallet?" states quite clearly that this is our unknown quantity. We will let x represent the number of dollars in our wallet.

Step b says to determine how many other elements are involved and how they relate. If we read again from the beginning, the example says:

"If you multiply the number of dollars in your wallet by 6 . . .	This part of the story indicates that our unknown x should be multiplied by 6; $x \times 6$.
. . . and then subtract \$23, . . .	This part states that after our unknown has been multiplied by 6, it should have \$23 subtracted; $(x \times 6) - 23$.
. . . you would still end up with . . .	This statement precedes a result and therefore an equals sign should follow our equation at this point; $(x \times 6) - 23 =$.
. . . one-fourth of \$100."	This is the result we end up with: ¼ of 100 or ¼ $\times$ 100.

Now, to complete step c, we must construct an equation using the unknown and the other elements with their associated arithmetic operations.

$$(x \times 6) - 23 = \frac{1}{4} \times 100$$

Now that we have an equation, we can move on to step d. Since x is not isolated on one side of the equals sign, we will have to transpose or rearrange the equation.

Step 1: Is the unknown above the fraction bar? Yes.

Step 2: Is the unknown isolated on one side of the equals sign? No.

$(x \times 6) - 23 = \frac{1}{4} \times 100$

Steps:
Rather than multiply both sides by 4, let us reduce the fraction ¼ of 100. 100 is above the fraction bar, therefore it can be written

$(x \times 6) - 23 = \frac{1 \times 100}{4}$

$1 \times 100 = 100$

$(x \times 6) - 23 = \frac{100}{4}$

$100 \div 4 = 25$; the fraction is now eliminated.

$(x \times 6) - 23 = 25$
$(x \times 6) - 23 \,\boxed{+ 23} = 25 \,\boxed{+ 23}$

Add 23 to both sides to remove -23.
$-23 + 23 = 0$,
$(x \times 6) - 0 = x \times 6$,
$25 + 23 = 48$

$x \times 6 = 48$

$\dfrac{x \times 6}{\boxed{6}} = \dfrac{48}{\boxed{6}}$

Divide both sides by 6 to remove "multiply by 6".
$6 \div 6 = 1, 1 \times x = x$

$x = \dfrac{48}{6}$

$x = 8$

You therefore started off with \$8 in your pocket. To double-check this answer, replace x with 8 in the original equation.

$$(x \times 6) - 23 = \frac{1}{4} \times 100 \qquad \left(\frac{1}{4} \text{ of } 100 = 25\right)$$
$$(8 \times 6) - 23 = 25 \qquad (8 \times 6 = 48)$$
$$48 - 23 = 25 \qquad (48 - 23 = 25)$$
$$25 = 25$$

■ **INTEGRATED MATH APPLICATION:**

Store owner A has four times as many nails as store owner B, who has 12,000. Combined, A and B have 60,000 nails. How many nails does store owner A have?

■ *Solution:*

A is the unknown, and as stated in the question, A is four times larger than B.

$$A = 4 \times B$$

If B has 12,000 nails, we can substitute 12,000 for B.

$$A = 4 \times 12,000$$

To calculate how many nails store owner A has, all we have to do is a simple multiplication.

$$A = 4 \times 12,000$$
$$A = 48,000$$

To prove this answer is true, we can return to the part of the question that states "combined, both A and B have 60,000 nails."

$$A + B = 60,000$$

We know A has 48,000 nails and B has 12,000 nails, so we can replace A and B with their equivalent values as follows:

$$A + B = 60,000$$
$$48,000 + 12,000 = 60,000$$
$$60,000 = 60,000$$

5. Factoring and Removing Parentheses

A factor is any number multiplied by another number and contributing to the result. For example, consider 3×8. In this multiplication example, 3 is a factor and 8 is a factor, and both will contribute to a result of 24.

Now let us reverse the process. What factors contributed to a result of 12? In this case we must determine which small numbers, when multiplied together, will produce 12. The answer could be

$$2 \times 6 = 12 \qquad \text{or} \qquad 3 \times 4 = 12$$

Are these, however, the smallest numbers that, when multiplied together, will produce 12? The answer is no because the 6 (in 2×6) can be broken into 2×3, and the 4 (in 3×4) can be broken into 2×2. Therefore, 12 has the factors $2 \times 2 \times 3$.

Now consider the equation $12 + 9$. What smaller number is a factor of both these numbers? The answer is 3 because you can get results of 12 and 9 by multiplying some other number by 3. Therefore, both 12 and 9 contain the factor 3. If we now remove the common factor 3 from both the numbers 12 and 9, we end up with the following equation:

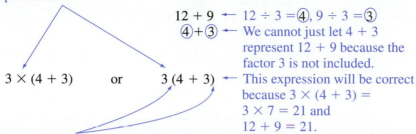

If no symbol appears, it is assumed that it is a multiplication.

$12 + 9$ ← $12 \div 3 = ④, 9 \div 3 = ③$
$④ + ③$ ← We cannot just let $4 + 3$ represent $12 + 9$ because the factor 3 is not included.

$3 \times (4 + 3)$ or $3 (4 + 3)$ ← This expression will be correct because $3 \times (4 + 3) = 3 \times 7 = 21$ and $12 + 9 = 21$.

Parentheses are used to group the addition $4 + 3$. The factor 3 outside the parentheses applies to everything inside the parentheses.

Factoring is therefore a reducing process that extracts the common factor from two or more larger numbers. As another example, remove the factor from the following equation:

Steps:

$27x - 9x$ 9 is a factor common to both 27 and 9.
                      $27x \div = 9 = 3x, 9x \div 9 = 1x$
$= 9(3x - 1x)$ Place 9 outside the parentheses and the results of the division inside the parentheses.

We can check if $9(3x - 1x) = 27x - 9x$ simply by performing the arithmetic operations indicated:

$27x - 9x = ⟨18x⟩$

$1x = x$

$9(3x - 1x)$
$= 9(3x - x)$ ← $3x - x = 2x$
$= 9 \times 2x$
$⟨= 18x⟩$

Results are the same.

CALCULATOR KEYS

Name: Factor function

Function: Some calculators have a factor function that returns an expression factored with respect to all of its variables.

factor (expression)

Examples: Extract the common factor from $(64x - 8y)$.

Press keys: Factor $(64x - 8y)$
Answer: $8(8x - y)$

Press keys: Factor $(16a - 4)$
Answer: $4(4a - 1)$

To reverse the factoring process and convert $9(3x - 1x)$ back to the original equation $27x - 9x$, simply reverse the arithmetic steps. Because we began by dividing both numbers by the common factor, simply do the opposite and multiply both values within the parentheses by the factor 9, as follows:

$$9(3x - 1x)$$
$$= 27x - 9x$$

Steps:
$9 \times 3x = 27x,$
$9 \times 1x = 9x$

To make any equation with parentheses easier to solve, always begin by removing the parentheses before doing any other operation.

CALCULATOR KEYS

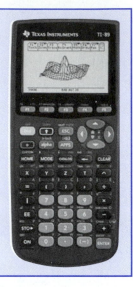

Name: Expand function

Function: Some calculators have a reverse factoring function that removes the parentheses.

Expand (expression)

Examples:

Press keys: Expand $(8(8x - y))$
Answer: $64x - 8y$

Press keys: Expand $(4(7a + 2b - 4c))$
Answer: $28a + 8b - 16c$

■ EXAMPLE A:

$$5(7x + 3x) \leftarrow 5 \times 7x = 35x, 5 \times 3x = 15x$$
$$= 35x + 15x$$

■ EXAMPLE B:

$$6(4a - 3) \quad 6 \times 4a = 24a, \quad 6 \times 3 = 18$$
$$= 24a - 18$$

To summarize, we now know about two new processes called *factoring* and *removing parentheses*. To show how these two are the reverse of one another, let us do a few examples.

■ EXAMPLE A:

Factoring:

$$12 + 18$$
$$= 6(2 + 3)$$

Common factor is 6.
$12 \div 6 = 2, 18 \div 6 = 3$

Removing parentheses:

$$6(2 + 3)$$
$$= 12 + 18$$

$6 \times 2 = 12, 6 \times 3 = 18$

$$22y - 11$$
$$= 11(2y - 1)$$

Factoring:
Common factor is 11.
$22y \div 11 = 2y, 11 \div 11 = 1$

$$11(2y - 1)$$
$$= \quad 22y - 11$$

Removing parentheses:
$11 \times 2y = 22y, 11 \times 1 = 11$

Now that we know how to perform both of these operations, let us look at a practical application of them.

■ INTEGRATED MATH APPLICATION:

A motorbike uses a certain number of gallons of gasoline a week, a car uses twice that amount, and a truck uses four times as much as the motorbike and car combined. If all three use 100 gallons of gasoline a week, what does each vehicle consume per week?

■ *Solution:*

Reading the problem carefully, you can see that each vehicle's consumption is related to the amount consumed by the motorbike. Therefore, if we can find the amount of gasoline consumed by the motorbike, we can calculate the amount consumed by the car and truck. We will represent the unknown quantity of gasoline that the bike consumes by x.

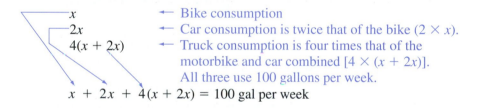

x ← Bike consumption
$2x$ ← Car consumption is twice that of the bike ($2 \times x$).
$4(x + 2x)$ ← Truck consumption is four times that of the motorbike and car combined [$4 \times (x + 2x)$]. All three use 100 gallons per week.

$$x + 2x + 4(x + 2x) = 100 \text{ gal per week}$$

Now that we have an equation, the next step is to use transportation to determine the unknown value x.

$$x + 2x + 4(x + 2x) = 100$$

Steps:
Remove parentheses:
$4 \times x = 4x, 4 \times 2x = 8x$

$$x + 2x + 4x + 8x = 100$$

Combine all the unknowns:
$x + 2x + 4x + 8x = 15x$

$$15x = 100$$

Divide both sides by 15 to remove the "multiply by 15."

$$\frac{\cancel{15}x}{\cancel{15}} = \frac{100}{15}$$

$15 \div 15 = 1, 1 \times x = x$

$$x = \frac{100}{15}$$

$100 \div 15 = 6.67$

$$x = 6.67$$

Therefore,

$$\text{Motorbike consumption} = x = 6.67 \text{ gal per week}$$

The car consumes twice that amount:

$$\text{Car consumption} = 2 \times x = 2 \times 6.67 = 13.34 \text{ gal per week}$$

The truck consumes four times as much as both combined:

$$
\begin{aligned}
\text{Truck consumption} &= 4(x + 2x) = 4(6.67 + 2 \times 6.67) \\
&= 4(6.67 + 13.34) \qquad 4 \times 6.67 = 26.68 \\
&\qquad\qquad\qquad\qquad\qquad\ 4 \times 13.34 = 53.36 \\
&= 26.68 + 53.36 \\
&= 80.0 \text{ gal per week}
\end{aligned}
$$

To double-check our answer, let us add all the individual amounts to see if they equal the total:

$$
\begin{aligned}
\text{Motorbike} + \text{car} + \text{truck} &= 100 \text{ gal per week} \\
6.67 + 13.34 + 80.0 &= 100 \text{ gal per week}
\end{aligned}
$$

Referring to the preceding example, you may have noticed that *you used factoring to develop the equation:*

$$\text{Truck consumption is 4 times that of bike and car} = 4(x + 2x)$$

and *you removed parentheses to transpose the equation:*

$$4(x + 2x) = 4x + 8x$$

6. Square and Square Root

How do we remove the square or square root on one side of an equation? The answer once again is: Perform the opposite function. For example, if we square a number and then find its square root, we end up with the number with which we started:

$$4 \text{ squared} = 4^2 = 4 \times 4 = 16$$
$$\text{Square root of } 16 = \sqrt{16} = 4$$

■   **EXAMPLE:**

Determine the value of the unknown I in the following equation:

$$I^2 = 81$$

■   *Solution:*

To remove the square from the unknown quantity I, we simply take the square root of both sides.

$$
\begin{aligned}
I^2 &= 81 \\
\sqrt{I^2} &= \sqrt{81} \quad \leftarrow \text{(Take square root of both sides.)} \\
I &= \sqrt{81} \quad \leftarrow \text{(Square and square root cancel.)} \\
I &= 9 \qquad\ \leftarrow (\sqrt{81} = 9, \text{ since } 9 \times 9 = 81)
\end{aligned}
$$

CHAPTER 5 / ALGEBRA, EQUATIONS, AND FORMULAS

■ EXAMPLE:

Determine the value of the unknown x in the following equation.

$$\sqrt{x} = 2 + 6$$

■ Solution:

$\sqrt{x} = 2 + 6$ ← (To remove the square root, square both sides.)
$\sqrt{x}^2 = (2 + 6)^2$ ← (Square and square root cancel.)
$x = (2 + 6)^2$
$x = 8^2$ $\quad\quad (8 \times 8 = 64)$
$x = 64$

5-2-2 *Transposing Formulas*

As mentioned previously, *a formula is a general statement using letters and sometimes numbers that enables us to calculate the value of an unknown quantity.* Some of the simplest relationships are formulas. For example, to calculate the distance traveled, you can use the following formula, which was discussed earlier.

$$d = v \times t$$

where d = distance in miles
$\quad\quad\quad v$ = velocity in miles per hour
$\quad\quad\quad t$ = time in hours

Using this formula we can calculate distance if we know the speed and time. If we need only to calculate distance, this formula is fine. But what if we want to know what speed to go ($v = ?$) to travel a certain distance in a certain amount of time, or what if we want to calculate how long it will take for a trip ($t = ?$) of a certain distance at a certain speed? The answer is: Transpose or rearrange the formula so that we can solve for any of the quantities in the formula.

■ INTEGRATED MATH APPLICATION:

As an example, calculate how long it will take to travel 250 miles when traveling at a speed of 50 miles per hour (mph). To solve this problem, let us place the values in their appropriate positions in the formula.

$$d = v \times t$$
$$250 \text{ mi} = 50 \text{ mph} \times t$$

The next step is to transpose just as we did before with equations to determine the value of the unknown quantity t.

	Steps:
$250 = 50 \times t$	To isolate t, divide both sides by 50.
$\dfrac{250}{50} = \dfrac{\cancel{50} \times t}{\cancel{50}}$	$50 \div 50 = 1, 1 \times t = t$
$\dfrac{250}{50} = t$	$250 \div 50 = 5$
$t = 5$	

It will therefore take 5 hours to travel 250 miles at 50 miles per hour ($250 \text{ mi} = 50 \text{ mph} \times 5 \text{ h}$).

All we have to do, therefore, is transpose the original formula so that we can obtain formulas for calculating any of the three quantities (distance, speed, or time) if two values are known. This time, however, we will transpose the formula instead of the inserted values.

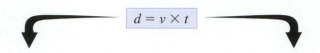

$$d = v \times t$$

Solve for v:

Steps:

$d = v \times t$ To isolate v, divide

$\dfrac{d}{t} = \dfrac{v \times t}{t}$ both sides by t.

$\dfrac{d}{t} = v$ $t \div t = 1$,

$1 \times v = v$

$$\text{Velocity} = \frac{\text{distance}}{\text{time}} \qquad v = \frac{d}{t}$$

Solve for t:

Steps:

$d = v \times t$ To isolate t, divide both

$\dfrac{d}{v} = \dfrac{v \times t}{v}$ sides by v.

$\dfrac{d}{v} = t$ $v \div v = 1$,

$1 \times t = t$

$$t = \frac{d}{v} \qquad \text{Time} = \frac{\text{distance}}{\text{velocity}}$$

$$d = v \times t \qquad\qquad v = \frac{d}{t} \qquad\qquad t = \frac{d}{v}$$

$$250 \text{ mi} = 50 \text{ mph} \times 5 \text{ h} \qquad 50 \text{ mph} = \frac{250 \text{ mi}}{5 \text{ h}} \qquad 5 \text{ h} = \frac{250 \text{ mi}}{50 \text{ mph}}$$
$$(250 = 250) \qquad\qquad\qquad (50 = 50) \qquad\qquad\qquad (5 = 5)$$

1. Directly Proportional and Inversely Proportional

The word **proportional** is used to describe a relationship between two quantities. For example, reconsider the formula

$$\text{Time } (t) = \frac{\text{distance } (d)}{\text{velocity } (v)}$$

With this formula we can say that the time (t) it takes for a trip is directly proportional (symbolized by $\propto$) to the amount of distance (d) that needs to be traveled. In symbols, the relationship appears as follows:

$t \propto d$ (Time is directly proportional to distance.)

The term **directly proportional** can therefore be used to describe the relationship between time and distance because an increase in distance will cause a corresponding increase in time ($d \uparrow$ causes $t \uparrow$), and similarly, a decrease in distance will cause a corresponding decrease in time ($d \downarrow$ causes $t \downarrow$).

Proportional

The relation of one part to another or to the whole with respect to magnitude, quantity, or degree.

Directly Proportional

The relation of one part to one or more other parts in which a change in one causes a similar change in the other.

116

CHAPTER 5 / ALGEBRA, EQUATIONS, AND FORMULAS

Using the previous example,

$$5 \text{ h } (t) = \frac{250 \text{ mi } (d)}{50 \text{ mph } (v)}$$

let us try doubling the distance and then halving the distance to see what effect it has on time. If time and distance are directly proportional to each other, they should change in proportion.

Original example → $\text{Time} = \dfrac{\text{distance}}{\text{velocity}} = \dfrac{250 \text{ mi}}{50 \text{ mph}} = 5 \text{ h}$

Doubling the distance from 250 mi to 500 mi should double the time it takes for the trip, from 5 h to 10 h.

$$t = \frac{d}{v} = \frac{500 \text{ mi}}{50 \text{ mph}} = 10 \text{ h}$$

Doubling the distance did double the time ($d \uparrow, t \uparrow$).

Halving the distance from 250 mi to 125 mi should halve the time it takes for the trip, from 5 h to 2½ h.

$$t = \frac{d}{v} = \frac{125 \text{ mi}}{50 \text{ mph}} = 2.5 \text{ h}$$

Halving the distance did halve the time ($d \downarrow, t \downarrow$).

As you can see from this exercise, time is directly proportional to distance ($t \propto d$). Therefore, *formula quantities are always directly proportional to one another when both quantities are above the fraction bars on opposite sides of the equals sign.*

On the other hand, **inversely proportional** means that the two quantities compared are opposite in effect. For example, consider again our time formula:

$$t = \frac{d}{v}$$

Inversely Proportional
The relation of one part to one or more other parts in which a change in one causes an opposite change in the other.

Observing the relationship between these quantities again, we can say that the time (t) it takes for a trip is *inversely proportional* (symbolized by $1/\propto$) to the speed or velocity (v) we travel. In symbols, the relationship appears as follows:

$$t \frac{1}{\propto} v \quad \text{(Time is inversely proportional to velocity.)}$$

The term *inversely proportional* can therefore be used to describe the relationship between time and velocity because an increase in velocity will cause a corresponding decrease in time ($v \uparrow$ causes $t \downarrow$), and similarly, a decrease in velocity will cause a corresponding increase in time ($v \downarrow$ causes $t \uparrow$).

■ INTEGRATED MATH APPLICATION:

Using the previous example,

$$5 \text{ h}(t) = \frac{250 \text{ mi } (d)}{50 \text{ mph } (v)}$$

let us try doubling the speed we travel, or velocity, and then halving the velocity to see what effect it has on time. If time and velocity are inversely proportional to each other, a velocity change should have the opposite effect on time.

$$\text{Original example } \text{j} \quad \text{Time} = \frac{\text{distance}}{\text{velocity}} = \frac{250 \text{ mi}}{50 \text{ mph}} = 5 \text{ h}$$

Doubling the speed from 50 mph to 100 mph should halve the time it takes to complete the trip, from 5 h to 2½ h.

$$\longrightarrow \quad t = \frac{d}{v} = \frac{250 \text{ mi}}{\boxed{100 \text{ mph}}} = \boxed{2.5 \text{ h}}$$

Doubling the speed did halve the time ($v \downarrow, t \uparrow$).

Halving the speed from 50 mph to 25 mph should double the time it takes to complete the trip, from 5 h to 10 h.

$$\longrightarrow \quad t = \frac{d}{v} = \frac{250 \text{ mi}}{\boxed{25 \text{ mph}}} = \boxed{10 \text{ h}}$$

Halving the speed did double the time ($v \downarrow, t \uparrow$).

As you can see from this exercise, time is inversely proportional to velocity ($t \frac{1}{\propto} v$). Therefore, *formula quantities are always inversely proportional to one another when one of the quantities is above the fraction bar and the other is below the fraction bar on opposite sides of the equals sign.*

2. Ohm's Law Formula Example

In this section we will examine a law that was first discovered by Georg Ohm in 1826. It states that the **electric current** in a circuit is directly proportional to the **voltage** source *applied and inversely proportional to the* **resistance** *in the circuit.* Stated as a formula, it appears as follows:

Ohm's law formula: $\boxed{\text{Current} = \dfrac{\text{voltage}}{\text{resistance}}}$

Quantity	Symbol	Unit	Symbol
Current	I	Amperes	A
Voltage	V	Volts	V
Resistance	R	Ohms	Ω

To explain this formula, let us use the fluid analogy shown in Figure 5-1(a). An object will move only when a force is applied, and as you can see in Figure 5-1(a), the greater the pressure applied to the piston, the greater the amount of water flow through the pipe. Water flow is therefore directly proportional to pressure applied:

Water flow $\propto$ pressure applied

In a similar way, the electric current (I) shown in Figure 5-1(b) is directly dependent on the voltage (V) force applied by the battery. For instance, a smaller voltage ($V \downarrow$) will cause a corresponding smaller circuit current ($I \downarrow$), and similarly, a larger voltage ($V \uparrow$) will result in a larger circuit current ($I \uparrow$). It can therefore be said that electric current, which is the flow of electrons, is directly proportional to electric voltage, which is the force that moves electrons:

Current (I) $\propto$ voltage (V)

Now, referring to the Ohm's law formula, you can see that current (I) and voltage (V) are directly proportional because these quantities are above the fraction bars on opposite sides of the equals sign.

$$\boxed{\text{Current } (I)} = \frac{\boxed{\text{voltage } (V)}}{\text{resistance } (R)}$$

The current (I) in an electric circuit is also inversely proportional to the opposition or resistance in the circuit. This can best be explained by the fluid analogy in Figure 5-2(a). If the valve is adjusted to offer a small opposition, a large amount of water will flow, whereas if the valve is adjusted to introduce a large amount of opposition, only a small amount of

Current (electric)

The flow of electronics through a conductor. It is measured in amperes or amps.

Voltage

Term used to designate electrical power or the force that causes current flow.

Resistance

The opposition to current flow with the dissipation of energy in the form of heat. It is measured in ohms.

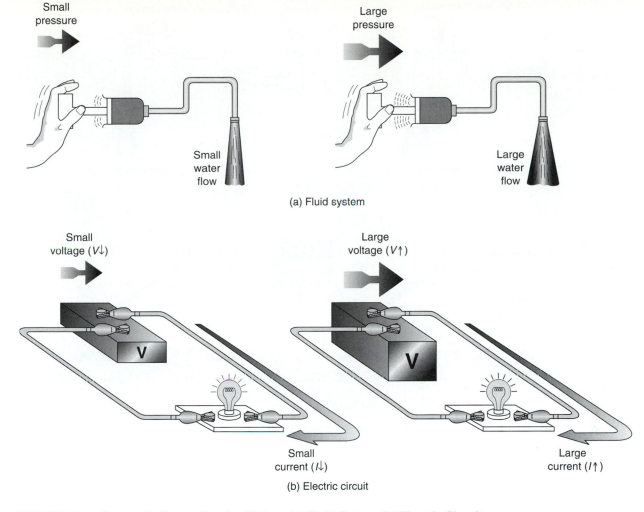

Small
pressure

Large
pressure

Small
water
flow

Large
water
flow

(a) Fluid system

Small
voltage ($V\downarrow$)

Large
voltage ($V\uparrow$)

V

V

Small
current ($I\downarrow$)

Large
current ($I\uparrow$)

(b) Electric circuit

FIGURE 5-1 **Current Is Proportional to Voltage (a) Fluid System (b) Electric Circuit.**

water will be permitted to flow. Water flow is therefore inversely proportional to the valve's opposition:

$$\text{Water flow} \frac{1}{\propto} \text{valve opposition}$$

In a similar way, the electric current (I) shown in Figure 5-2(b) is inversely dependent on the opposition or resistance (R) offered by the resistor in the circuit. For instance, a small-value resistor will offer only a small opposition or resistance ($R \downarrow$), which will permit a large circuit current ($I \uparrow$), whereas a large-value resistor will offer a large opposition or resistance ($R \uparrow$), which will allow only a small circuit current ($I \downarrow$). It can be said, therefore, that electric current, which is the flow of electrons, is inversely proportional to electric resistance, which is the restriction to current flow:

$$\text{Current} (I) \frac{1}{\propto} \text{resistance} (R)$$

Once again, referring to the Ohm's law formula, you can see that current (I) and resistance (R) are inversely proportional to each other because one quantity is above the fraction bar and the other is below the fraction bar on opposite sides of the equals sign:

$$\text{Current} (I) = \frac{\text{voltage} (V)}{\text{resistance} (R)}$$

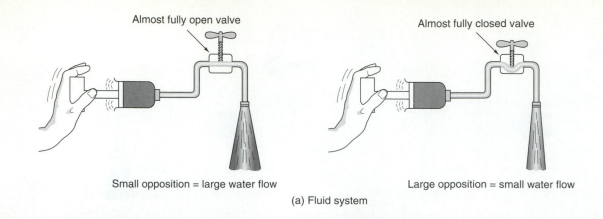

Small opposition = large water flow   Large opposition = small water flow

(a) Fluid system

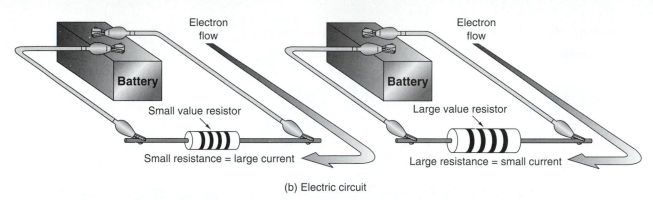

(b) Electric circuit

FIGURE 5-2 **Current Is Inversely Proportional to Resistance (a) Fluid System (b) Electric Circuit.**

■ INTEGRATED MATH APPLICATION:

Calculate what voltage would cause 2 amperes (A) of electric current in a circuit having an opposition or resistance of 3 ohms (Ω).

■ *Solution:*

To solve this problem, begin by placing the values in their appropriate position within the formula. In this example,

$$\text{Current } (I) = \frac{\text{voltage } (V)}{\text{resistance } (R)}$$

Therefore,

$$2\,A = \frac{V?}{3\,\Omega}$$

or

$$2 = \frac{V}{3}$$

As you can see, we will first need to transpose the current formula ($I = V/R$) so that we can find a formula for voltage (V), because V is the unknown in this example.

$$\text{Current } (I) = \frac{\text{voltage } (V)}{\text{resistance } (R)}$$

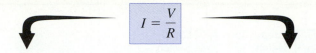

$$I = \frac{V}{R}$$

Solve for V	*Solve for R*
$\dfrac{I}{} = \dfrac{V}{R}$	$\dfrac{I}{} = \dfrac{V}{R}$
$\dfrac{R \times I}{} = \dfrac{V \times R}{R}$ (Multiply both sides by R.)	$\dfrac{R \times I}{} = \dfrac{V \times R}{R}$ (Multiply both sides by R to bring R above the line.)
$\dfrac{R \times I}{} = \dfrac{V \times \cancel{R}}{\cancel{R}}$ ($R \div R$ cancels.)	$\dfrac{R \times I}{} = \dfrac{V \times \cancel{R}}{\cancel{R}}$ ($R \div R$ cancels.)
$R \times I = V$	$\dfrac{R \times I}{I} = \dfrac{V}{I}$ (Divide both sides by I to isolate R.)
$\boxed{V = I \times R}$	$\dfrac{R \times \cancel{I}}{\cancel{I}} = \dfrac{V}{I}$ ($I \div I$ cancels.)
	$\boxed{R = \dfrac{V}{I}}$

Now that we have a formula for voltage ($V = I \times R$), we can complete this problem:

$$V = I \times R \quad \text{(Voltage = current} \times \text{resistance)}$$
$$= 2\,\text{A} \times 3\,\Omega$$
$$= 6\,\text{V}$$

This equation means that a voltage or electrical pressure of 6 volts would cause 2 amperes of electric current in a circuit having 3 ohms of resistance.

To test if these three formulas are correct, we can now plug into each formula the values of the preceding example, as follows:

$$I = \frac{V}{R} \qquad\qquad V = I \times R \qquad\qquad R = \frac{V}{I}$$

$$2\,A = \frac{6\,V}{3\,\Omega} \qquad 6\,V = 2\,A \times 3\,\Omega \qquad 3\,\Omega = \frac{6\,V}{2\,A}$$

As you can see from these formulas, the transposition has been successful.

■ INTEGRATED MATH APPLICATION:

Figure 5-3 shows a basic circuit for a home electrical system. If each household light and appliance has 120 volts connected across it, how much current will pass through lamp 1, lamp 2, the hair dryer, and the space heater?

■ Solution:

$$\text{Current through lamp 1 } (I_{L1}) = \frac{V_{L1}}{R_{L1}} = \frac{120\,V}{125\,\Omega} = 960\,\text{mA}$$

$$\text{Current through lamp 2 } (I_{L2}) = \frac{V_{L2}}{R_{L2}} = \frac{120\,V}{250\,\Omega} = 480\,\text{mA}$$

$$\text{Current through hair dryer } (I_{HD}) = \frac{V_{HD}}{R_{HD}} = \frac{120\,V}{40\,\Omega} = 3\,\text{A}$$

$$\text{Current through space heater } (I_{SH}) = \frac{V_{SH}}{R_{SH}} = \frac{120\,V}{12\,\Omega} = 10\,\text{A}$$

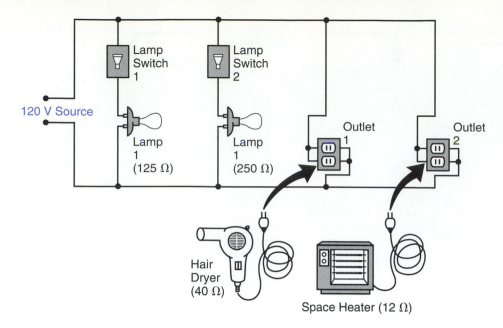

FIGURE 5-3   Home Electrical System.

3. Power Formula Example

Another formula frequently used in electricity and electronics is the power formula. We will also study this formula so that we can practice formula transposition again.

Electric power is the rate at which electric energy is converted into some other form of energy. Power, which is symbolized by *P,* is measured in *watts* (symbolized by W) in honor of James Watt. We hear watt used frequently in connection with the brightness of a lightbulb: for instance, a 60 watt bulb, a 100 watt bulb, and so on. This designation is a power rating and describes how much electric energy is converted every second. Because a lightbulb generates both heat and light energy, we would describe a 100 watt lightbulb as a device that converts 100 watts of electric energy into heat and light energy every second. Referring to Figure 5-1(b), you can see a lightbulb connected in an electric circuit. The amount of electric power supplied to that lightbulb is dependent on the voltage and the current. It is probably no surprise, therefore, that power is equal to the product of voltage and current. To state the formula:

Power = voltage × current

$P = V \times I$

Quantity	Symbol	Unit	Symbol
Power	*P*	Watts	W
Voltage	*V*	Volts	V
Current	*I*	Amperes	A

This formula states that the amount of power delivered to a device is dependent on the electrical pressure or voltage applied across the device and the electric current flowing through the device.

■ INTEGRATED MATH APPLICATION:

What will be the electric current (*I*) in a circuit if a 60 watt lightbulb is connected across a 120 volt battery?

■ *Solution:*

Let us begin by inserting the known values in their appropriate places in the power formula.

Power (*P*) = voltage (*V*) × current (*I*)
60 W = 120 V × ? amperes

120 V Source

Lamp Switch 1

Lamp Switch 2

Lamp 1 (125 Ω)

Lamp 1 (250 Ω)

Outlet 1

Outlet 2

Hair Dryer (40 Ω)

Space Heater (12 Ω)

As you can see, we first need to transpose the power formula ($P = V \times I$) so that we can find a formula for current (I), because I is the unknown in this example.

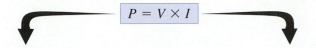

$$P = V \times I$$

Solve for V:

$P = V \times I$

$\dfrac{P}{I} = \dfrac{V \times I}{I}$ (Divide both sides by I.)

$\dfrac{P}{I} = \dfrac{V \times \cancel{I}}{\cancel{I}}$ ($I \div I = 1$, and $V \times 1 = V$.)

$\dfrac{P}{I} = V$

$$\boxed{V = \dfrac{P}{I}}$$

This formula can be used to calculate V when P and I are known.

Solve for I:

$P = V \times I$

$\dfrac{P}{V} = \dfrac{V \times I}{V}$ (Divide both sides by V.)

$\dfrac{P}{V} = \dfrac{\cancel{V} \times I}{\cancel{V}}$ ($V \div V = 1$, and $I \times 1 = I$.)

$\dfrac{P}{V} = I$

$$\boxed{I = \dfrac{P}{V}}$$

This formula can be used to calculate I when P and V are known.

Now that we have a formula for current ($I = P/V$), we can complete this problem:

$$I = \dfrac{P}{V} \quad \left(\text{Current} = \dfrac{\text{power}}{\text{voltage}} \right)$$
$$= \dfrac{60\ \text{W}}{120\text{V}}$$
$$= 0.5\ \text{A} \qquad \text{or} \qquad 500 \times 10^{-3}\ \text{A} \;\leftarrow\; (0.500.0)$$
$$= 500\ \text{mA}$$

This result means that a current of 500 milliamperes will exist in an electric circuit if a 120 volt battery is connected across a 60 watt light bulb.

CALCULATOR KEYS

Name: Solve function key

Function: Some calculators have a function that solves an expression for a specified variable.

Solve (equation, variable)

Example: Solve for R, when $I = V/R$
Press keys: Solve($I = V/R,R$)
Answer: $R = V/I$

CALCULATOR KEYS *(continued)*

Solve for *t*, when $250 = 50 \times t$.
Press keys: Solve ($250 = 50 \times t,t$)
Answer: $t = 5$

Solve for *y*, when $(6 \times y) + 27 = y + 27$.
Press keys: Solve$((6 \times y) + 7 = y + 27, y)$
Answer: $y = 4$

SELF-TEST EVALUATION POINT FOR SECTION 5-2

Now that you have completed this section, you should be able to:

■ **Objective 4.**   *Explain the steps involved in transposing an equation to determine the unknown.*

■ **Objective 5.**   *Demonstrate how to develop an equation from a story problem.*

■ **Objective 6.**   *Describe the process of factoring and how to remove parentheses within a formula or expression.*

■ **Objective 7.**   *Demonstrate the process of transposing a formula.*

■ **Objective 8.**   *Describe how the terms* directly proportional *and* inversely proportional *are used to show the relationship between quantities in a formula.*

■ **Objective 9.**   *Explain the proportionality among the three quantities in the Ohm's law formula, and describe how to transpose the formula.*

■ **Objective 10.**   *Describe the electric power formula, and transpose the formula to develop alternative procedures.*

Use the following questions to test your understanding of Section 5-2.

1. Calculate the result of the following arithmetic operations involving literal numbers.

   a. $x + x =$ **d.** $2x - x =$

   b. $x \times x =$ **e.** $\dfrac{x}{x} =$

   c. $7a + 4a =$ **f.** $x - x =$

2. Transpose the following equations to determine the unknown quantity.

   a. $x + 14 = 30$

   b. $8 \times x = \dfrac{80 - 40}{10} \times 12$

   c. $y - 4 = 8$

   d. $(x \times 3) - 2 = \dfrac{26}{2}$

   e. $x^2 + 5 = 14$

   f. $2(3 + 4x) = (2x + 13)$

3. Transpose the following formulas.

   a. $x + y = z, y = ?$
   b. $Q = C \times V, C = ?$
   c. $x_L = 2 \times \pi \times f \times L, L = ?$
   d. $V = I \times R, R = ?$

4. Determine the unknown in the following equations by using transposition.

   a. $I^2 = 9$ **b.** $\sqrt{z} = 8$

5-3   SUBSTITUTION

Substitution

The replacement of one mathematical entity by another of equal value.

Substitution is a mathematical process used to develop alternative formulas by replacing or substituting one mathematical term with an equivalent mathematical term. To explain this process, let us try a couple of examples.

Circle

A closed plane curve, every point of which is equidistant from a fixed point within the curve.

5-3-1   *Circle Formula Example*

A **circle** is a perfectly round figure in which every point is at an equal distance from the center.

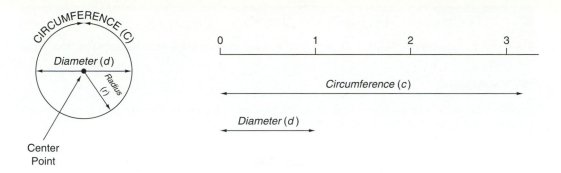

Center
Point

Some of the elements of a circle are as follows.

a. The **circumference** (*C*) of a circle is the distance around the perimeter of the circle.

b. The **radius** (*r*) of a circle is a straight line drawn from the center of the circle to any point of the circumference.

c. The **diameter** (*d*) of a circle is a straight line drawn through the center of the circle to opposite sides of the circle. The diameter of a circle is therefore equal to twice the circle's radius ($d = 2 \times r$).

To review, *pi* (symbolized by π) is the name given to the ratio, or comparison, of a circle's circumference to its diameter. This constant will always be equal to approximately 3.14 because the circumference of any circle will always be about 3 times larger than the same circle's diameter. Stated mathematically:

$$\pi = \frac{\text{circumference } (C)}{\text{diameter } (d)} = 3.14$$

Circumference
The perimeter of a circle.

Radius
A line segment extending from the center of a circle or sphere to the circumference.

Diameter
The length of a straight line passing through the center of an object.

■ **INTEGRATED MATH APPLICATION:**

Determine the circumference of a bicycle wheel that is 26 inches in diameter.

■ *Solution:*

To calculate the wheel's circumference, we will have to transpose the following formula to isolate *C*.

$$\pi = \frac{C}{d} \qquad \text{Multiply both sides by } d.$$

$$\pi \times d = \frac{C}{d} \times d \qquad d \div d = 1, \text{ and } C \times 1 = C.$$

$$\pi \times d = C$$

The circumference of the bicycle wheel is therefore:

$$C = \pi \times d$$
$$C = 3.14 \times 26 \text{ in.} = 81.64 \text{ in.}$$

In some examples, only the circle's radius will be known. In situations like this, we can substitute the expression $\pi = C/d$ to arrive at an equivalent expression comparing a circle's circumference to its radius.

$$\pi = \frac{C}{d} \qquad \text{Because } d = 2 \times r, \text{ we can substitute } 2 \times r \text{ for } d.$$

$$\pi = \frac{C}{d} \qquad \text{Therefore, this is equivalent to } \pi = \frac{C}{2 \times r}$$

How much decorative edging will you need to encircle a garden that measures 6 feet from the center to the edge?

■ *Solution:*

In this example r is known, so to calculate the garden's circumference, we will have to transpose the following formula to isolate C:

$$\pi = \frac{C}{2 \times r}$$ Multiply both sides by $2 \times r$.

$$\pi \times (2 \times r) = \frac{C}{2 \times r} \times (2 \times r)$$ $(2 \times r) \div (2 \times r) = 1$, and $C \times 1 = C$.

$$\pi \times (2 \times r) = C$$

The edging needed for the circumference of the garden will therefore be:

$$C = \pi \times (2 \times r)$$
$$C = 3.14 \times (2 \times 6 \text{ ft}) = 37.68 \text{ ft}$$

To continue our example of the circle, how could we obtain a formula for calculating the area within a circle? The answer is best explained with the diagram shown in Figure 5-4. In Figure 5-4(a) a circle has been divided into eight triangles, and then these triangles have been separated as shown in Figure 5-4(b). Figure 5-4(c) shows how each of these triangles can be thought of as having an A and a B section that, if separated and rearranged, will form a square. The area of a square is easy to calculate because it is simply the product of the square's width and height (area = $w \times h$). The area of the triangles shown in Figure 5-4(c), therefore, will be

$$\text{Area} = \frac{1}{2} \times (w \times h)$$

Referring to Figure 5-4(d), you can see that the width of all eight triangles within a circle is equal to the circle's circumference ($w = C$), and the height of the triangles is equal to the circle's radius ($h = r$). The area of the circle, therefore, can be calculated with the following formula:

$$\text{Area of a circle} = \frac{1}{2} \times (C \times r)$$

Using both substitution and transposition, let us now try to reduce this formula with $C = \pi \times (2 \times r)$.

$$\text{Area of a circle} = \frac{1}{2} \times (C \times r)$$ ← Because $C = \pi \times (2 \times r)$, we can replace C with $\pi \times (2 \times r)$.

$$= \frac{1}{2} \times [\pi \times (2 \times r) \times r]$$ ← We can now use transposition to reduce.

$$= \frac{1 \times \pi \times 2 \times r \times r}{2}$$ ← $2 \div 2 = 1$

$$= 1 \times \pi \times r \times r$$ ← $r \times r = r^2$
$$= 1 \times \pi \times r^2$$ ← $1 \times (\pi \times r^2) = \pi \times r^2$

$$\text{Area of a circle} = \pi \times r^2$$

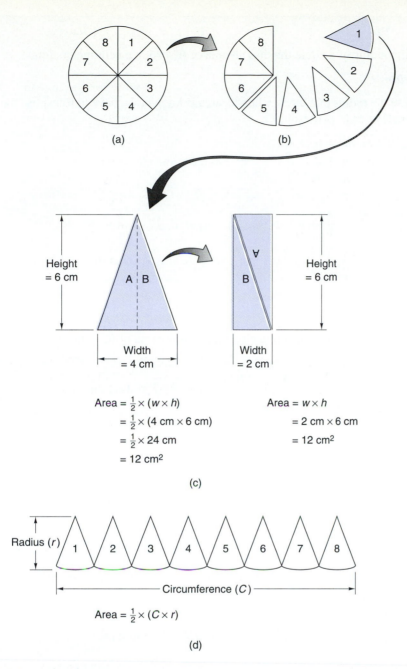

FIGURE 5-4   **Calculating the Area of a Circle.**

■   **EXAMPLE:**

Calculate the area of a circle that has a radius of 14 inches.

■   *Solution:*

$$\text{Area of a circle} = \pi \times r^2$$
$$= 3.14 \times (14 \text{ in.})^2$$
$$= 3.14 \times 196 \text{ in.}^2$$
$$= 615.4 \text{ in.}^2$$

■ **EXAMPLE:**

Calculate the radius of a circle that has an area of 624 square centimeters (cm^2).

■ *Solution:*

To determine the radius of a circle when its area is known, we have to transpose the formula area of a circle $= \pi \times r^2$ to isolate r.

Area of a circle $= \pi \times r^2$

$A = \pi \times r^2$ ← Divide both sides by π.

$\dfrac{A}{\pi} = \dfrac{\pi \times r^2}{\pi}$ ← $\pi \div \pi = 1$, $1 \times r^2 = r^2$

$\dfrac{A}{\pi} = r^2$ ← Take the square root of both sides.

$\sqrt{\dfrac{A}{\pi}} = \sqrt{r^2}$ ← $\sqrt{r^2} = r$

$\sqrt{\dfrac{A}{\pi}} = r$

Now we can calculate the radius of a circle that has an area of 624 cm^2.

$$r = \sqrt{\dfrac{A}{\pi}} = \sqrt{\dfrac{624}{3.14}} = \sqrt{198.7} = 14.1 \text{ cm}$$

5-3-2 *Power Formula Example*

In Section 5-2 you were introduced to two electrical formulas: the Ohm's law formula, which states

$$\text{Current } (I) = \frac{\text{voltage } (V)}{\text{resistance } (R)}$$

and the power formula, which states

$$\text{Power } (P) = \text{voltage } (V) \times \text{current } (I)$$

Using transposition we were able to develop the following alternative formulas that allowed us to calculate any quantity if two quantities were known.

OHM'S LAW FORMULA			POWER FORMULA		
FORMULA	CALCULATES	WHEN THESE ARE KNOWN:	FORMULA	CALCULATES	WHEN THESE ARE KNOWN:
$I = \dfrac{V}{R}$	Current (I) in amperes (A)	Voltage (V) and resistance (R)	$P = V \times I$	Power (P) in watts (W)	Voltage (V) and current (I)
$V = I \times R$	Voltage (V) in volts (V)	Current (I) and resistance (R)	$V = \dfrac{P}{I}$	Voltage (V) in volts (V)	Power (P) and current (I)
$R = \dfrac{V}{I}$	Resistance (R) in ohms (Ω)	Voltage (V) and current (I)	$I = \dfrac{P}{V}$	Current (I) in amperes (A)	Power (P) and voltage (V)

Although it seems from this table that we have covered all possible alternatives, there are a few missing. For example, consider the power formula $P = V \times I$. When trying to calculate power, we may not always have voltage (V) and current (I). We may, for example, have values only for voltage (V) and resistance (R), or current (I) and resistance (R). In situations such as this the mathematical process called *substitution* can be used to develop alternative power formulas by replacing or substituting one mathematical term for an equivalent mathematical term. The following process shows how we can substitute terms in the $P = V \times I$ formula to arrive at alternative power formulas for wattage calculations.

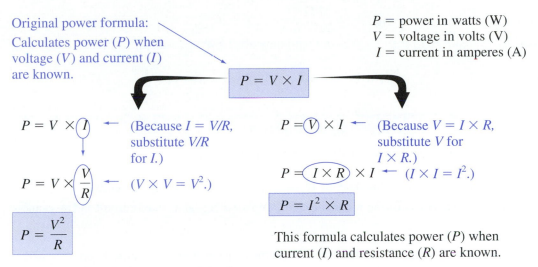

Original power formula:

Calculates power (P) when voltage (V) and current (I) are known.

$P = $ power in watts (W)
$V = $ voltage in volts (V)
$I = $ current in amperes (A)

$$P = V \times I$$

$P = V \times \boxed{I}$ ← (Because $I = V/R$, substitute V/R for I.)

$P = V \times \dfrac{V}{R}$ ← ($V \times V = V^2$.)

$$P = \dfrac{V^2}{R}$$

This formula calculates power (P) when voltage (V) and resistance (R) are known.

$P = \boxed{V} \times I$ ← (Because $V = I \times R$, substitute V for $I \times R$.)

$P = \boxed{I \times R} \times I$ ← ($I \times I = I^2$.)

$$P = I^2 \times R$$

This formula calculates power (P) when current (I) and resistance (R) are known.

Let us now look at two applications to see how we can use these formulas to calculate unknown quantities.

■ INTEGRATED MATH APPLICATION:

The household heater shown here has a resistance of 7 ohms and is connected to the 120 volt source at the wall outlet. Calculate the amount of power dissipated by the heater.

Electric heater

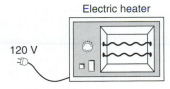

120 V

■ *Solution:*

In this example, both voltage (120 V) and resistance (7 Ω) are known, and the unknown to be calculated is power (P). We must therefore look at all the power formulas to select one that includes voltage (V) and resistance (R).

$$P = V \times I$$

$P = \dfrac{V^2}{R}$ ← This power formula is the one to use because it includes V and R.

$$P = I^2 \times R$$

$$\text{Power } (P) = \frac{\text{voltage}^2 \ (V^2)}{\text{resistance } (R)}$$

Replace the literal numbers *V* and *R* with values.

$$= \frac{(120 \ V)^2}{7 \ \Omega}$$

$120^2 = 14{,}400$

$$= \frac{14{,}400}{7 \ \Omega}$$

$$= 2057 \ W$$

$2.\overset{\frown}{057}$

or approximately 2 kW

■ **INTEGRATED MATH APPLICATION:**

The lightbulb shown here is rated at 6 volts, 60 milliamps.

a. Using Ohm's law, calculate the resistance of the electric lamp or lightbulb.

b. Using the power formula, calculate the lamp's wattage at its rated current and resistance.

■ *Solution:*

Part (a) asks us to use Ohm's law to calculate the resistance of the electric lamp.

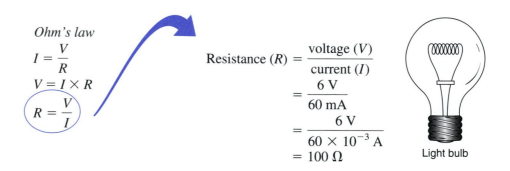

Ohm's law

$$I = \frac{V}{R}$$

$$V = I \times R$$

$$\boxed{R = \frac{V}{I}}$$

$$\text{Resistance } (R) = \frac{\text{voltage } (V)}{\text{current } (I)}$$

$$= \frac{6 \ V}{60 \ mA}$$

$$= \frac{6 \ V}{60 \times 10^{-3} \ A}$$

$$= 100 \ \Omega$$

Light bulb

Part (b) directs us to use the power formula (*P*) to calculate the lamp's wattage at its rated current (*I*) and resistance (*R*). *I* and *R* are known, so we will use the following formula:

Power formula

$$P = V \times I$$

$$P = \frac{V^2}{R}$$

$$\boxed{P = I^2 \times R}$$

$$\text{Power } (P) = \text{current}^2 (I^2) \times \text{resistance}$$

$$= (60 \ mA)^2 \times 100 \ \Omega$$

$$= (60 \times 10^{-3})^2 \times 100 \ \Omega$$

$$= 0.0036 \times 100$$

$$= 0.36 \ W \ (0.\overset{\frown}{360.})$$

or $360 \times 10^{-3} \ W$

or 360 mW

As a review of the Ohm's law and power formula combinations, Figure 5-5 shows all the possible formulas for the four electrical properties.

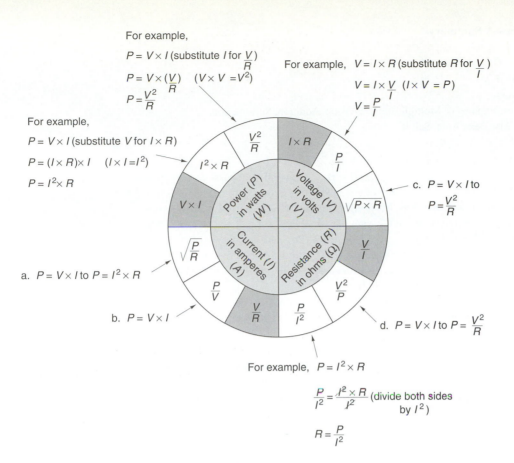

For example,

$P = V \times I$ (substitute I for $\frac{V}{R}$)

$P = V \times (\frac{V}{R})$ $(V \times V = V^2)$

$P = \frac{V^2}{R}$

For example, $V = I \times R$ (substitute R for $\frac{V}{I}$)

$V = I \times \frac{V}{I}$ $(I \times V = P)$

$V = \frac{P}{I}$

For example,

$P = V \times I$ (substitute V for $I \times R$)

$P = (I \times R) \times I$ $(I \times I = I^2)$

$P = I^2 \times R$

c. $P = V \times I$ to

$P = \frac{V^2}{R}$

a. $P = V \times I$ to $P = I^2 \times R$

b. $P = V \times I$

d. $P = V \times I$ to $P = \frac{V^2}{R}$

For example, $P = I^2 \times R$

$\frac{P}{I^2} = \frac{I^2 \times R}{I^2}$ (divide both sides by I^2)

$R = \frac{P}{I^2}$

FIGURE 5-5 The Ohm's Law and Power Formula Circle.

SELF-TEST EVALUATION POINT FOR SECTION 5-3

Now that you have completed this section, you should be able to:

■ *Objective 11. Demonstrate how to use substitution to develop alternative formulas.*

Use the following questions to test your understanding of Section 5-3.

1. Use substitution to calculate the value of the unknown.
   $x = y \times z$ and $a = x \times y$
   Calculate a if $y = 14$ and $z = 5$.

2. If $I = V \times R$ and $P = V \times I$, show the transposition and substitution steps that were followed to develop the formula circle shown in Figure 5-5. (The development follows parts (a) to (d) in the figure.)

3. What percentage of 12 gives 2.5?
   (x% of 12 = 2.5)

4. 60 is 22% of what number?
   (22% of x = 60)

5-4   RULES OF ALGEBRA—A SUMMARY

As a review of the lessons learned in this chapter, Table 5-1 summarizes the rules of algebra with many examples. Study every problem, since these will provide you with excellent practice for the electronics and computer applications that follow.

TABLE 5-1   Rules of Algebra—A Summary

Monomial (One Term): $2a$

Binomial (Two Terms): $2a + 3b$

Trinomial (Three Terms): $2a + 3b + c$

Polynomial (Any Number of Terms).

Order of Operations: Parentheses, Exponents, Multiplication, Division, Addition, Substraction.
   (Memory Aid: Please, Excuse, My, Dear, Aunt, Sally)

Basic Rules:

$a + a = 2a$	$a - a = 0$	$a \cdot a = a^2$	$a \div a = 1$	$\sqrt{a^2} = a$
$a + 1 = a + 1$	$a - 1 = a - 1$	$a \cdot 1 = a$	$a \div 1 = a$	
$a + 0 = a$	$a - 0 = a$	$a \cdot 0 = 0$	$a \div 0 = $ Impossible	$0 \div a = 0$

Parentheses First: $(3 + 6) \cdot 2 = 9 \cdot 2 = 18$ $\qquad 9 - (2 \cdot 3) = 9 - 6 = 3 \qquad 6 + (9 - 7) = 6 + 2 = 8$

No Parentheses—Step 1:   Multiplications and Divisions, left to right. $\qquad 3 + 6 \cdot 4 = 3 + 24 = 27$

$\qquad\qquad\qquad$ Step 2:   Additions and Subtractions, left to right. $\qquad 4 \cdot 2 + 3 - 4 \div 2 = 8 + 3 - 2 = 9$

Combine Only Like Terms: $\qquad 2x + x + y = 3x + y \qquad 3a + 4b - a = 2a + 4b$

Commutative Property of Addition—order of terms makes no difference: $\qquad a + b = b + a$

Associative Property of Addition—grouping of terms makes no difference: $\qquad (a + b) + c = a + (b + c)$

Coefficients: $\qquad x + x + x = 3x \qquad x + y + x + y = 2x + 2y$

Exponents: $\qquad x \cdot x \cdot x = x^3 \qquad x \cdot y \cdot x \cdot y = x^2 \cdot y^2 = x^2 y^2$

Commutative Property of Multiplication—order of factors makes no difference: $\qquad a \cdot b = b \cdot a$

Associative Property of Multiplication—grouping of factors makes no difference: $\qquad (ab)c = a(bc)$

Distributive Property of Multiplication—$a(b + c) = ab + ac \qquad 2(a - 3b) = 2 \cdot a - 2 \cdot 3b = 2a - 6b$

Transposition—Step 1: Ensure unknown is above fraction bar. $\qquad x \cdot 8 = 16$ (divide both sides by 8, $\div 8$)

$\qquad\qquad\qquad$ Step 2: Isolate unknown. $\qquad (x \cdot 8) \div 8 = (16) \div 8, \qquad x = 2$

Unknown Above Fraction Bar	Unknown Below Fraction Bar	Unknown on Both Sides	Combining Unknowns
$(2 \cdot a) + 5 = 23 \quad (-5)$	$\dfrac{72}{x} = 12 \qquad (\cdot x)$	$(6 \cdot y) + 7 = y + 27 \quad (-7)$	$x + 2x + 4(x + 2x) = 100$
$2 \cdot a = 18 \quad (\div 2)$	$72 = 12 \cdot x \quad (\div 12)$	$6y = y + 20 \quad (-y)$	$x + 2x + 4x + 8x = 100$
$a = 9$	$6 = x$	$5y = 20 \qquad (\div 5)$	$15x = 100$
		$y = 4$	$x = 6.67$

Positive and Negative Number Rules

Addition $\oplus$ $\qquad (+) + (+) = (+), (-) + (-) = (-)$
$\qquad\qquad\qquad (+) + (-)$ or $(-) + (+) = $ Difference

Subtraction $\ominus$ $\qquad$ Change sign of second number, and then add.

Multiplication $\otimes$ $\qquad (+) \cdot (+) = (+), (-) \cdot (-) = (+)$
$\qquad\qquad\qquad (+) \cdot (-)$ or $(-) \cdot (+) = (-)$

Division $\oslash$ $\qquad (+) \div (+) = (+), (-) \div (-) = (+)$
$\qquad\qquad\qquad (+) \div (-)$ or $(-) \div (+) = (-)$

Positive and Negative Numbers

Addition $\oplus$ $\qquad 2x + 5x = 7x \qquad -6a + (+3a) = -3a$

Subtraction $\ominus$ $\qquad -3y^2 - (-2y^2) = -3y^2 + (+2y^2) = -y^2$

Multiplication $\otimes$ $\qquad -3(6a) = -3 \cdot (+6a) = -18a$

Division $\oslash$ $\qquad 14a \div (-7) = -2a$

Adding Polynomials: $\qquad (4a^2 + 6b^2 + x - 2) + (7a^2 - b^2 + 3)$

$$
\begin{array}{r}
4a^2 + 6b^2 + x - 2 \\
+ \quad 7a^2 - b^2 \quad\;\; + 3 \\
\hline
11a^2 + 5b^2 + x + 1
\end{array}
$$

Subtracting Polynomials: $\qquad (6x^2 + y^2 - 4) - (-4x^2 + 2y^2 + 6)$

$$
\begin{array}{r}
6x^2 + y^2 - 4 \\
- \quad -4x^2 + 2y^2 + 6 \\
\end{array}
$$

Rule: Add the Opposite $\longrightarrow$

$$
\begin{array}{r}
6x^2 + y^2 - 4 \\
+ \quad 4x^2 - 2y^2 - 6 \\
\hline
10x^2 - y^2 - 10
\end{array}
$$

Multiplying Monomials, Binomials

$x^2 \cdot x^3 \cdot x = x \cdot x \cdot x \cdot x \cdot x \cdot x = x^{2+3+1} = x^6$

$(x^3)^2 = (x \cdot x \cdot x) \cdot (x \cdot x \cdot x) = x^{3 \cdot 2} = x^6$

$2(a + 2b) = 2a + 4b$
$\quad + 6b$

$4x^2 \cdot x^3 = 4 \cdot x \cdot x \cdot x \cdot x \cdot x = 4x^{2+3} = 4x^5$

$(-6y)^2 \cdot y^3 = (-6)^2 \cdot y^2 \cdot y^3 = 36y^5$

$-2(2a - 3b) = (-2 \cdot 2a) - (-2 \cdot 3b) = -4a - (-6b) = \quad$ p$-4a$

Multiplying Polynomials

$(x^2 + 2x - 1) \cdot (x - 3) = ?$

$$
\begin{array}{r}
x^2 + 2x - 1 \\
\times \qquad\quad x - 3 \\
\hline
-3x^2 - 6x + 3 \;\leftarrow\; -3(x^2 + 2x - 1) \\
x^3 + 2x^2 - x \qquad\quad\; \leftarrow\; x(x^2 + 2x - 1) \\
\hline
x^3 - x^2 - 7x + 3
\end{array}
$$

$(a - 3)^2 = ?$

$$
\begin{array}{r}
a - 3 \\
\times \quad a - 3 \\
\hline
-3a + 9 \;\leftarrow\; -3(a - 3) \\
a^2 - 3a \qquad\;\; \leftarrow\; a(a - 3) \\
\hline
a^2 - 6a + 9
\end{array}
$$

Dividing Polynomials

$\dfrac{a^4}{a} = \dfrac{\overset{1}{\cancel{a}} \cdot a \cdot a \cdot a}{\underset{1}{\cancel{a}}} = a^3$

$\dfrac{-6ab^3}{2ab} = -\dfrac{\overset{-3}{\cancel{6}} \cdot \cancel{a} \cdot \cancel{b} \cdot b \cdot b}{\underset{1}{\cancel{2} \cdot \cancel{a} \cdot \cancel{b}}} = -3b^2$

$\dfrac{2ab - 4a^2b^2}{2a} = \dfrac{2ab}{2a} - \dfrac{4a^2b^2}{2a} = b - 2ab^2$

$\dfrac{2a^2 + 6a + 8ab}{2a} = \dfrac{2a^2}{2a} + \dfrac{6a}{2a} + \dfrac{8ab}{2a} = a + 3 + 4b$

Now that you have completed this section, you should be able to:

■ **Objective 12.** *Summarize the terms, rules, properties, and mathematical operations of algebra.*

Use the following questions to test your understanding of Section 5-4.

1. What is the order of operations?

2. Give an example of a:

   a. Monomial   **c.** Trinomial
   b. Binomial   **d.** Polynomial

3. Determine the following:

   a. $a \div 1 = ?$
   b. $a + a = ?$
   c. $a - a = ?$
   d. $a \times a = ?$
   e. $3x + b - 2x = ?$

4. Determine the unknown:

   a. $16 + a = 5 \times a, a = ?$
   b. $\dfrac{42}{b} = 3 \times 7$

REVIEW QUESTIONS

Multiple Choice Questions

1. If you perform exactly the same mathematical operation on both sides of an equation, the equality is preserved.

   a. True
   b. False

2. Are the two sides of the following equation, $(3 - 2) \times 16 = 32 \div 2$, equal?

   a. Yes
   b. No
   c. Only if the right half is doubled
   d. Only if the left half is doubled

3. $x + x =$ _____, $x \times x =$ _____, $x \div x =$ _____, and $x - x$ _____.

   a. $x^2, 2x, 1, 0$
   b. $2x, \sqrt{x}, 0, 1$
   c. $1x, 2x, 1, 0$
   d. $2x, x^2, 1, 0$

4. What is the value of the unknown in the following equation: $3a + 9 = 2a + 10$?

   a. 2   **c.** 1
   b. 10   **d.** 3

5. Calculate the value x for the following equation: $x + y^2 = y^2 + 9$.

   a. 3   **c.** 9
   b. 6   **d.** 12

6. What is the value of the unknown, s, in the following equation: $\sqrt{s^2} + 14 = 26$?

   a. 12   **c.** 4
   b. 26   **d.** 5

7. What is the correct transposition for the following formula:

   $a = \dfrac{b - c}{x}, x = ?$

   a. $x = \dfrac{b - a}{c}$ **c.** $x = \dfrac{b - c}{a}$
   b. $x = (b - a) \times c$ **d.** $x = x(b - c)$

8. If $V = I \times R$ and $P = V \times I$, develop a formula for P when I and R are known.

   a. $P = \dfrac{I}{R}$
   b. $P = I^2 \times R^2$
   c. $P = I^2 \times R$
   d. $P = \sqrt{\dfrac{I}{R}}$

9. If $E = M \times C^2, M = ?$ and $C = ?$

   a. $\sqrt{\dfrac{E}{C}}, \dfrac{E^2}{M}$
   b. $\dfrac{E}{C^2}, \sqrt{\dfrac{E}{M}}$
   c. $E \times C^2, \sqrt{C^2 \times E}$
   d. $\dfrac{C^2}{E}, \sqrt{\dfrac{M}{E}}$

10. If $P = \dfrac{V^2}{R}$, transpose to obtain a formula to solve for V.

   a. $V = \sqrt{P \times R}$
   b. $V = \dfrac{P^2}{R}$
   c. $V = P \times R^2$
   d. $V = \sqrt{P} \times R$

Communication Skill Questions

11. Define the following terms: (5-1)

    a. Algebra   c. Equation
    b. Formula   d. Literal number

12. Why is it important to treat both sides of an equation equally?   (5-1)

13. Describe how the following three factors can be combined in a formula: distance (d), velocity (v), and time (t).   (5-2)

14. What two steps must be followed to transpose an equation? (5-2)

15. Describe the factoring process.   (5-2)

16. Define the terms *proportional and inversely proportional*. (5-2)

17. Explain how current (I), voltage (V), and resistance (R) can be combined in a formula.   (5-2)

18. Transpose the formula from question 22 to solve for I, V, and R.   (5-2)

19. If $P = V \times I$, describe how you would solve for V and I.   (5-2)

20. Define substitution.   (5-3)

21. If $\pi = C/d$, and area $= \frac{1}{2} \times (C \times r)$, describe how substitution and transposition can be employed to obtain a formula for area when only the radius of a circle is known.   (5-3)

22. Explain how the parentheses can be removed from the following examples: (5-2)
    a. $4(3b - 1b)$ b. $2(3x \times 2x)$

23. If $a = b \times c$, explain how you would solve for b and c.   (5-2)

24. Describe the relationship between square and square root. (5-2)

25. Why is formula manipulation important?   (5-1)

Practice Problems

26. Calculate the result of the following arithmetic operations involving literal numbers.

    a. $x + x + x = ?$ g. $4y \times 3y^2 = ?$
    b. $5x + 2x = ?$ h. $a \times a = ?$
    c. $y - y = ?$ i. $2y \div y = ?$
    d. $2x - x = ?$ j. $6b \div 3b = ?$
    e. $y \times y = ?$ k. $\sqrt{a^2} = ?$
    f. $2x \times 4x = ?$ l. $(x^3)^2 = ?$

27. Transpose the following equations to determine the unknown value.

    a. $4x = 11$
    b. $6a + 4a = 70$
    c. $5b - 4b = \dfrac{7.5}{1.25}$
    d. $\dfrac{2z \times 3z}{4.5} = 2z$

28. Apply transposition and substitution to Newton's second law of motion.
    a. $F = ma$ [Force (F) = mass (m) $\times$ acceleration (a)] If $F = m \times a$, $m = ?$ and $a = ?$
    b. If acceleration (a) equals change in speed (Δv) divided by time (t), how would the $F = ma$ formula appear if acceleration (a) was substituted for $\dfrac{\Delta v}{t}$?

29. a. Using the formula power (P) = voltage (V) $\times$ current (I), calculate the value of electric current through a hairdryer that is rated as follows: power = 1500 watts and voltage = 120 volts.

$$\boxed{P = V \times I}$$

    where P = power in watts
          V = voltage in volts
          I = current in amperes

b. Using the Ohm's formula voltage (V) = current (I) $\times$ resistance (R), calculate the resistance or opposition to current flow offered by the hairdryer in Question 14a.

30. The following formula is used to calculate the total resistance (R_T) in a parallel circuit containing two parallel resistances (R_1 and R_2).

$$R_T = \dfrac{1}{\dfrac{1}{R_1} + \dfrac{1}{R_2}}$$

    a. Calculate R_T (total resistance) if the resistance of $R_1 = 6$ ohms and the resistance of $R_2 = 12$ ohms.
    b. Indicate the calculator sequence used for part (a).
    c. If R_2 and R_T were known and we needed to calculate the value of R_1, we would have to transpose the total parallel resistance formula to solve for R_1. The following steps show how to transpose the total resistance formula to calculate R_1 when R_T and R_2 are known. Describe what has occured in each of the steps.

$$R_T = \dfrac{1}{\dfrac{1}{R_1} + \dfrac{1}{R_2}} \qquad \text{(Original formula)}$$

1. $R_T \times \left(\dfrac{1}{R_1} + \dfrac{1}{R_2}\right) = \dfrac{1}{\dfrac{1}{R_1} + \dfrac{1}{R_2}} \times \left(\dfrac{1}{R_1} + \dfrac{1}{R_2}\right)$

$$R_T \times \left(\dfrac{1}{R_1} + \dfrac{1}{R_2}\right) = 1$$

2. $\dfrac{R_T \times \left(\dfrac{1}{R_1} + \dfrac{1}{R_2}\right)}{R_T} = \dfrac{1}{R_T}$

$$\dfrac{1}{R_1} + \dfrac{1}{R_2} = \dfrac{1}{R_T}$$

or $\quad \dfrac{1}{R_T} = \dfrac{1}{R_1} + \dfrac{1}{R_2}$

3. $\dfrac{1}{R_T} - \dfrac{1}{R_2} = \dfrac{1}{R_1} + \dfrac{\cancel{1}}{\cancel{R_2}} - \dfrac{\cancel{1}}{\cancel{R_2}}$

$\qquad \dfrac{1}{R_T} - \dfrac{1}{R_2} = \dfrac{1}{R_1}$

$\quad$ or $\quad \dfrac{1}{R_1} = \dfrac{1}{R_T} - \dfrac{1}{R_2}$

4. $\dfrac{1}{R_1} = \boxed{\dfrac{1}{R_T} - \dfrac{1}{R_2}}$

$\qquad \dfrac{1}{R_T} - \dfrac{1}{R_2} = \dfrac{}{} \quad$ *lowest common denominator (LCD)*

$\qquad \dfrac{1}{R_T} - \dfrac{1}{R_2} = \dfrac{}{R_T \times R_2} \qquad$ (LCD) method 1: fractions

$\qquad \qquad \qquad \qquad \qquad \qquad \qquad R_1 \times R_2 = R_1 \times R_2$

$\qquad \dfrac{1}{R_T} - \dfrac{1}{R_2} = \dfrac{R_2 - R_T}{R_T \times R_2} \qquad$ *Steps:*

$\qquad \qquad \qquad \qquad \qquad \qquad (R_T \times R_2) \div R_T = R_2,$

$\qquad \dfrac{1}{R_T} - \dfrac{1}{R_2} = \dfrac{\boxed{R_2 - R_T}}{R_T \times R_2} \qquad R_2 \times 1 = R_2$

$\qquad \dfrac{1}{R_1} = \dfrac{R_2 - R_T}{R_T \times R_2} \qquad \qquad (R_T \times R_2) \div R_2 = R_T,$

$\qquad \qquad \qquad \qquad \qquad \qquad R_T \times 1 = R_T$

5. $\dfrac{1}{\cancel{R_1}} \times \cancel{R_1} = \dfrac{R_2 - R_T}{R_T \times R_2} \times R_1$

$\qquad \qquad 1 = \dfrac{R_2 - R_T}{R_T \times R_2} \times R_1$

6. $1 \times R_T \times R_2 = \dfrac{R_2 - R_T}{\cancel{R_T} \times \cancel{R_2}} \times R_1 \times \cancel{R_T} \times \cancel{R_2}$

$\qquad \quad R_T \times R_2 = (R_2 - R_T) \times R_1$

7. $\dfrac{R_T \times R_2}{R_2 - R_T} = \dfrac{(\cancel{R_2} - \cancel{R_T}) \times R_1}{\cancel{R_2} - \cancel{R_T}}$

$\qquad \dfrac{R_T \times R_2}{R_2 - R_T} = R_1 \quad$ or $\quad R_1 = \dfrac{R_T \times R_2}{R_2 - R_T}$

d. Using the values from part (a), double-check the new formula.

e. If we wanted to obtain a formula for calculating R_2 when both R_T and R_1 are known, which of the steps would we change, and in what way?

f. Give the formula for calculating R_2 when both R_T and R_1 are known.

g. Using the values from part (a), double-check this new formula.

Web Site Questions

Go to the web site http://www.prenhall.com/cook, select the textbook *Mathematics for Electronics and Computers,* select this chapter, and then follow the instructions when answering the multiple-choice practice problems.

Geometry and Trigonometry

Finding the Question to the Answer!

More than 350 years ago in 1637, Pierre de Fermat, a French mathematician and physicist, wrote an apparently simple theorem in the margins of a book. He also added that he had discovered marvelous proof of it but lacked enough space to include it in the margin. Later that night he died and took with him a mystery that mathematicians have been trying to solve ever since. Many of the brightest minds in mathematics have struggled to find the proof, and many have concluded that Fermat must have been mistaken despite his considerable mathematical ability.

Fermat's last theorem, as it has been appropriately named, has to do with equations of the form x to the nth power, plus y to the nth power = z to the nth power.

$$x^n + y^n = z^n$$

A power of 2 (square) is familiar as the Pythagorean theorem, which states that the sum of the squares of the lengths of two sides of a right-angle triangle is equal to the square of the length of the other side.

$$c^2 = a^2 + b^2$$

For example,

$$5^2 = 3^2 + 4^2$$
$$25 = 9 + 16$$

Fermat's last theorem seemed not to work because, for example,

$$? = 3^3 + 4^3$$
$$91 = 27 + 64$$

and there is not a whole number that can be cubed to equal 91.

Now, after thousands of claims of success that proved untrue, mathematicians say the daunting challenge, perhaps the most famous of unsolved mathematical problems, has been surmounted. The conqueror is Andrew Wiles, 40, a British mathematician who works at Princeton University. Wiles announced the result at the last of three lectures at Cambridge University in England. Within a few minutes of the conclusion of his final lecture, computer mail messages were winging around the world, sending mathematicians into a frenzy. It had apparently taken 7 years for Wiles to solve the problem. He continued a chain of ideas that was begun in 1954 by several other mathematicians.

After being criticized by many in the eighteenth and nineteenth centuries and ridiculed in the twentieth century by mathematicians with supercomputers, Fermat was elevated finally in 1993 to the status of "a genius before his time."

6

Outline and Objectives

VIGNETTE: FINDING THE QUESTION TO THE ANSWER!

INTRODUCTION

6-1   BASIC GEOMETRY TERMS

Objective 1: Describe the purpose of geometry, and define many of the basic geometric terms.

Objective 2: Demonstrate how to use a protractor to measure and draw an angle.

6-2   PLANE FIGURES

Objective 3: Define the term *plane figure,* and identify the different types of polygons.

Objective 4: List the formulas for determining the perimeter and area of plane figures.

6-2-1 Quadrilaterals

Objective 5: Explain how the Pythagorean theorem relates to right-angle triangles.

Objective 6: Determine the length of one side of a right-angle triangle when the lengths of the other two sides are known.

Objective 7: Describe how vectors are used to represent the magnitude and direction of physical quantities and how they are arranged in a vector diagram.

Objective 8: Explain how the Pythagorean theorem can be applied to a vector diagram to calculate the magnitude of a resultant vector.

6-2-2 Right Angle Triangles

6-2-3 Vectors and Vector Diagrams

6-2-4 Other Triangles

6-2-5 Circles

6-2-6 Other Polygons

6-3   TRIGONOMETRY

Objective 9: Define the trigonometric terms:
  a. Opposite
  b. Adjacent
  c. Hypotenuse
  d. Theta
  e. Sine
  f. Cosine
  g. Tangent

Objective 10: Demonstrate how the sine, cosine, and tangent trigonometric functions can be used to calculate:
  a. The length of an unknown side of a right-angle triangle if the length of another side and the angle theta are known
  b. The angle theta of a right-angle triangle if the lengths of two sides of the triangle are known

6-3-1 Sine of Theta

6-3-2 Cosine of Theta

6-3-3 Tangent of Theta

6-3-4 Summary

6-4   SOLID FIGURES

Objective 11: Define the term *solid figure,* and identify the different types.

Objective 12: List the formulas for determining the lateral surface area, total surface area, and volume of solid figures.

6-4-1 Prisms

6-4-2 Cylinders

6-4-3 Spheres

6-4-4 Pyramids and Cones

Geometry is one of the oldest branches of mathematics, and deals with the measurement, properties, and relationships of points, lines, angles, plane figures or surfaces, and solid figures. The practical attributes of geometry were used by the ancient Egyptians to build the pyramids, and today, it is widely used in almost every branch of engineering technology. Unlike algebra, which deals with unknowns that are normally represented by a letter, geometry deals with a physical point, line, surface, or figure.

The first section of this chapter deals with the basic elements of geometry, the second section covers plane figures, the third trigonometry, and the final section examines solid figures.

6-1   BASIC GEOMETRIC TERMS

Figure 6-1(a) shows some examples of line segments. In the first example you can see that a line segment exists between ending points A and B. An abbreviation is generally used to represent a line segment, in which a bar is placed over the two end-point letters, and so "line segment AB" could be written as $\overline{AB}$.

Figure 6-1(a) also shows an example of a vertical line that extends straight up and down, a horizontal line that extends from left to right, and two parallel lines that extend in the same direction while remaining at the same distance from one another.

Sides

The lines that form an angle.

Vertex

A point where two lines meet.

In Figure 6-1(b) an angle is formed between two intersecting line segments or lines, which are also called **sides.** The common point at which they connect is called the **vertex.** The angle symbol $\angle$ is generally used as an abbreviation, and so "angle ABC" could be written as $\angle ABC$. The unit of measurement for the opening between the sides, or angle, is the degree (symbolized " ° ").

Figure 6-1(c) shows how a protractor can be used to measure angles. An image of a protractor has been printed in the back of this text, and can, if copied, be used for measuring and drawing angles. In this example, we are using the protractor to measure $\angle ABC$, from Figure 6-1(b). The 3-step procedure for measuring an angle using a protractor is as follows:

MEASURING AN ANGLE WITH A PROTRACTOR

Step 1:   Place protractor's center mark on vertex.

Step 2:   Rotate protractor until zero-degree mark is on one line.

Step 3:   Read the measure in degrees where the line intersects the protractor's scale.

You may have noticed that the protractor has a reversing upper and lower scale, so that an angle can be read either clockwise or counter-clockwise. And so, if you were to measure the angle in Figure 6-1(b), you would use the lower scale of the protractor as shown in Figure 6-1(c), moving counterclockwise from 0°. On the other hand, if the angle was in a reverse direction as shown with the 90° angle in Figure 6-1(d), you would use the upper scale of the protractor moving clockwise from 0°.

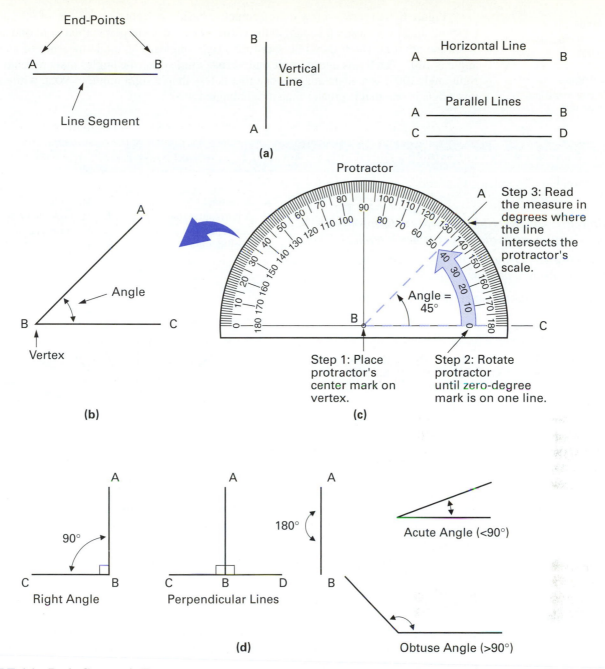

FIGURE 6-1   Basic Geometric Terms.

The protractor can also be used to draw an angle. The 3-step procedure for drawing an angle using a protractor is as follows:

DRAWING AN ANGLE WITH A PROTRACTOR

Step 1: Draw a straight line on the paper representing one side of the angle.

Step 2: Place the protractor's center mark on the end of the line that is to serve as the angle's vertex.

Step 3: Draw a dot on the paper next to the protractor's scale equal to the measure of the desired angle, and then draw a line connecting this dot to the vertex.

Acute Angle

An angle that measures less than 90°.

Obtuse Angle

An angle that measures greater than 90°.

Figure 6-1(d) shows a few more geometric basics. A right angle is one of 90°, and can be symbolized with a small square in the corner. Two lines are perpendicular to one another when the point at which they intersect forms right angles. The angle measure of a straight angle is 180°. The terms **acute angle** and **obtuse angle** describe angles that are relative to a right angle (90°). An acute angle is one that is less than a right angle (<90°), while an obtuse angle is one that is greater than a right angle (>90°).

SELF-TEST EVALUATION POINT FOR SECTION 6-1

Now that you have completed this section, you should be able to:

■ *Objective 1.* *Describe the purpose of geometry, and define many of the basic geometric terms.*

■ *Objective 2.* *Demonstrate how to use a protractor to measure and draw an angle.*

Use the following questions to test your understanding of section 6-1:

1. Using a protractor, measure:

a. ∠AXB	**d.** ∠AXE	**g.** ∠AXH	**j.** ∠IXG
b. ∠AXC	**e.** ∠AXF	**h.** ∠AXI	**k.** ∠IXF
c. ∠AXD	**f.** ∠AXG	**i.** ∠IXH	**l.** ∠IXE

2. Which of the angles in the previous question are right angles, obtuse angles, and acute angles?

3. Draw:

a. A vertical line	**e.** An angle of 35°
b. A horizontal line	**f.** An angle of 172°
c. Three parallel lines	**g.** An obtuse angle
d. A right angle	**h.** An acute angle

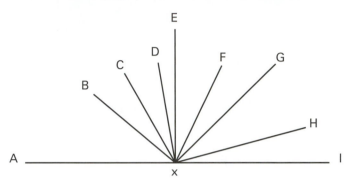

6-2 PLANE FIGURES

A plane figure is a perfectly flat surface, such as a tabletop, floor, wall, or windowpane. In this section we will examine the details of several flat plane surfaces, which, as you will discover, have only two dimensions—length and width.

6-2-1 *Quadrilaterals*

Polygon

A plane geometry figure bounded by 3 or more line segments.

Parallelogram

A quadrilateral where both pairs of opposite sides are parallel.

A **polygon** is a plane geometric figure that is made up of three or more line segments. Quadrilaterals, such as the square, rectangle, rhombus, parallelogram, and trapezoid, are all examples of four-sided polygons.

The information given in Figure 6-2 summarizes the facts and formulas for the square, which has four right angles and four equal length sides, and is a **parallelogram** (which means it has two sets of parallel sides). The box in Figure 6-2 lists the formulas for calculating a square's perimeter, area and side length.

Figures 6-3, 6-4, 6-5 and 6-6 on pp.141–143 summarize the facts and formulas for other quadrilaterals. In addition, these diagrams include, when necessary, a visual breakdown of the formulas so that it is easier to see how the formula was derived.

6-2-2 *Right Angle Triangles*

Right Angle Triangle

A triangle that contains one right or 90° angle.

The **right-angle triangle,** or right triangle shown in Figure 6-7(a) (p. 145), has three sides and three corners. Its distinguishing feature, however, is that two of the sides of this triangle are at right angles (at 90°) to one another. The small square box within the triangle is placed in the corner to show that sides *A* and *B* are *square* or at *right angles* to one another.

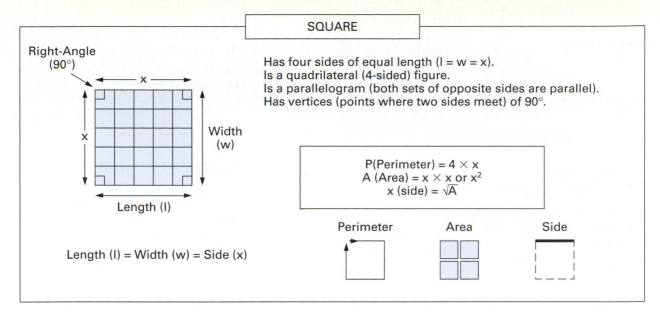

FIGURE 6-2 **The Square.**

If you study this right triangle you may notice two interesting facts about the relative lengths of sides *A*, *B*, and *C*. These observations are as follows.

1. Side *C* is always longer than side *A* or side *B*.
2. The total length of side *A* and side *B* is always longer than side *C*.

The right triangle in Figure 6-7(b) has been drawn to scale to demonstrate another interesting fact about this triangle. If side *A* were to equal 3 cm and side *B* were to equal 4 cm,

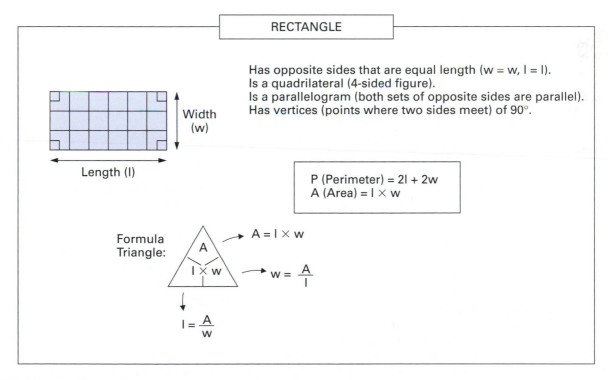

FIGURE 6-3 **The Rectangle.**

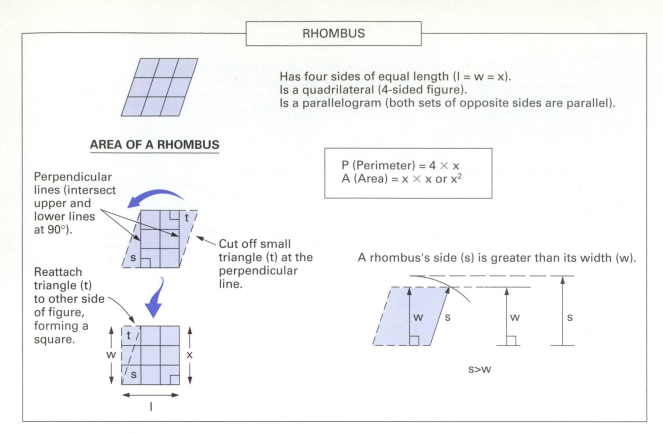

Has four sides of equal length (l = w = x).
Is a quadrilateral (4-sided figure).
Is a parallelogram (both sets of opposite sides are parallel).

AREA OF A RHOMBUS

P (Perimeter) = 4 $\times$ x
A (Area) = x $\times$ x or x^2

Perpendicular lines (intersect upper and lower lines at 90°).

Cut off small triangle (t) at the perpendicular line.

Reattach triangle (t) to other side of figure, forming a square.

A rhombus's side (s) is greater than its width (w).

s>w

FIGURE 6-4   The Rhombus.

PARALLELOGRAM

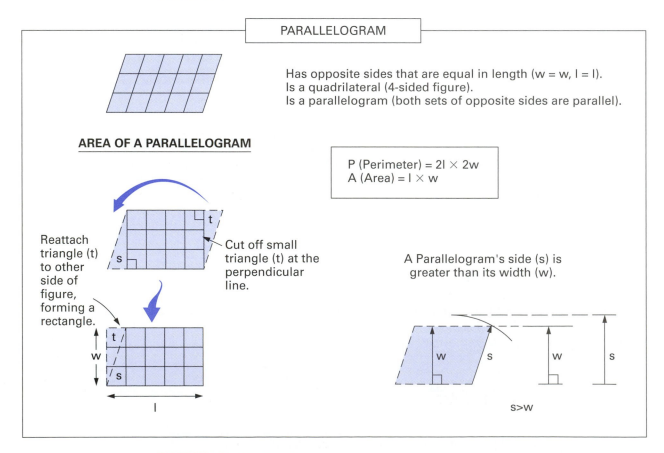

Has opposite sides that are equal in length (w = w, l = l).
Is a quadrilateral (4-sided figure).
Is a parallelogram (both sets of opposite sides are parallel).

AREA OF A PARALLELOGRAM

P (Perimeter) = 2l $\times$ 2w
A (Area) = l $\times$ w

Reattach triangle (t) to other side of figure, forming a rectangle.

Cut off small triangle (t) at the perpendicular line.

A Parallelogram's side (s) is greater than its width (w).

s>w

FIGURE 6-5   The Parallelogram.

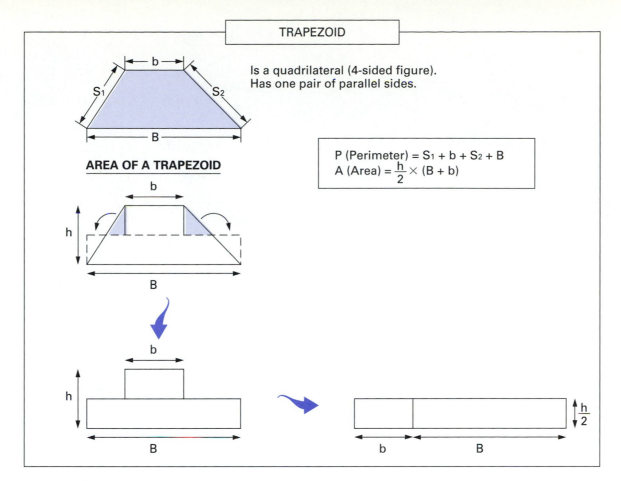

FIGURE 6-6 The Trapezoid.

side *C* would equal 5 cm. This demonstrates a basic relationship among the three sides and accounts for why this right triangle is sometimes referred to as a *3, 4, 5 triangle.* As long as the relative lengths remain the same, it makes no difference whether the sides are 3 cm, 4 cm, and 5 cm or 30 km, 40 km, and 50 km. Unfortunately, in most applications our lengths of *A, B,* and *C* will not work out as easily as 3, 4, 5.

Pythagoras's Theorem

When a relationship exists among three quantities, we can develop a formula to calculate an unknown when two of the quantities are known. It was Pythagoras who first developed this basic formula or equation (known as the **Pythagorean theorem**), which states that *the square of the length of the hypotenuse (side C) of a right triangle equals the sum of the squares of the lengths of the other two sides:*

$$C^2 = A^2 + B^2 \qquad \begin{pmatrix} 5^2 = 3^2 + 4^2 \\ 25 = 9 + 16 \end{pmatrix}$$

By using the rules of algebra, we can transpose this formula to derive formulas for sides *C, B,* and *A.*

Pythagorean Theorem

A theorem in geometry: The square of the length of the hypotenuse of a right triangle equals the sum of the squares of the lengths of the other two sides.

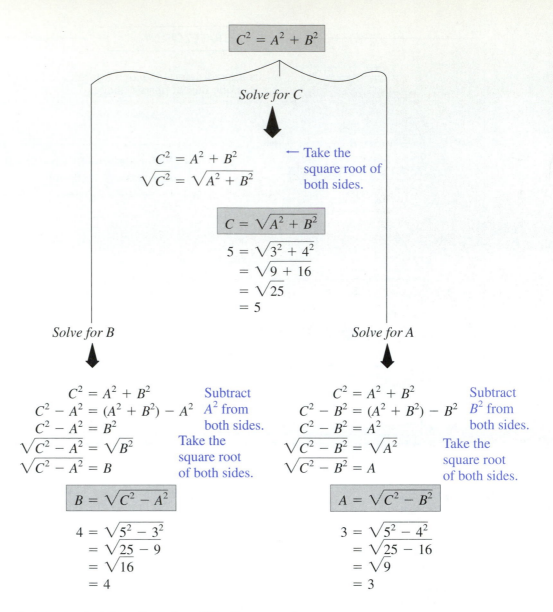

$$C^2 = A^2 + B^2$$

Solve for C

$$C^2 = A^2 + B^2$$
$$\sqrt{C^2} = \sqrt{A^2 + B^2}$$

← Take the square root of both sides.

$$C = \sqrt{A^2 + B^2}$$

$$5 = \sqrt{3^2 + 4^2}$$
$$= \sqrt{9 + 16}$$
$$= \sqrt{25}$$
$$= 5$$

Solve for B

$$C^2 = A^2 + B^2$$
$$C^2 - A^2 = (A^2 + B^2) - A^2$$
$$C^2 - A^2 = B^2$$
$$\sqrt{C^2 - A^2} = \sqrt{B^2}$$
$$\sqrt{C^2 - A^2} = B$$

Subtract A^2 from both sides.
Take the square root of both sides.

$$B = \sqrt{C^2 - A^2}$$

$$4 = \sqrt{5^2 - 3^2}$$
$$= \sqrt{25 - 9}$$
$$= \sqrt{16}$$
$$= 4$$

Solve for A

$$C^2 = A^2 + B^2$$
$$C^2 - B^2 = (A^2 + B^2) - B^2$$
$$C^2 - B^2 = A^2$$
$$\sqrt{C^2 - B^2} = \sqrt{A^2}$$
$$\sqrt{C^2 - B^2} = A$$

Subtract B^2 from both sides.
Take the square root of both sides.

$$A = \sqrt{C^2 - B^2}$$

$$3 = \sqrt{5^2 - 4^2}$$
$$= \sqrt{25 - 16}$$
$$= \sqrt{9}$$
$$= 3$$

Let us now test these formulas with a few examples.

■ INTEGRATED MATH APPLICATION:

A 4-foot ladder has been placed in a position 2 feet from a wall. How far up the wall will the ladder reach?

■ *Solution:*

In this example, A is unknown, and therefore

$$A = \sqrt{C^2 - B^2}$$
$$= \sqrt{4^2 - 2^2}$$
$$= \sqrt{16 - 4}$$
$$= \sqrt{12}$$
$$= 3.46 \quad \text{or} \quad \approx 3.5 \text{ ft}$$

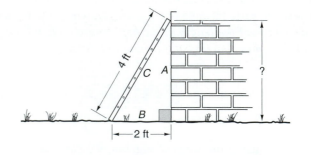

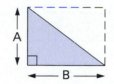

Is a triangle (3-sided figure).
Has one right angle (indicated by square).

(a) AREA OF A RIGHT TRIANGLE

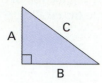

P (Perimeter) = A + B + C

A (Area) = $\dfrac{A \times B}{2}$

Sides (A, B, C)
Pythagorean theorem: $C^2 = A^2 + B^2$

$C = \sqrt{A^2 + B^2}$ $B = \sqrt{C^2 - A^2}$ $A = \sqrt{C^2 - B^2}$

(b) SIDES OF A RIGHT TRIANGLE

Pythagorean theorem: $A^2 + B^2 = C^2$
Square of side A + Square of side B = Square of side C.
Example: 9 + 16 = 25

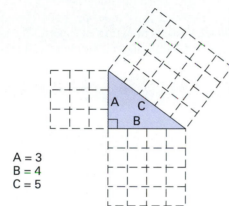

A = 3
B = 4
C = 5

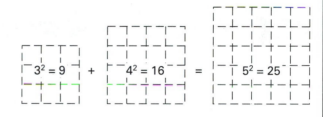

$3^2 = 9$ + $4^2 = 16$ = $5^2 = 25$

FIGURE 6-7   The Right Triangle.

■ INTEGRATED MATH APPLICATION:

A 16-foot flagpole is casting a 12-foot shadow. What is the distance from the end of the shadow to the top of the flagpole?

■ *Solution:*

In this example, *C* is unknown, and therefore

$$C = \sqrt{A^2 + B^2}$$
$$= \sqrt{16^2 + 12^2}$$
$$= \sqrt{256 + 144}$$
$$= \sqrt{400}$$
$$= 20 \text{ ft}$$

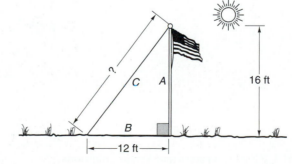

■ INTEGRATED MATH APPLICATION:

In Figure 6-2(c), a 200-foot piece of string is stretched from the top of a 125-foot cliff to a point on the beach. What is the distance from this point to the cliff?

■ *Solution:*

In this example, *B* is unknown, and therefore

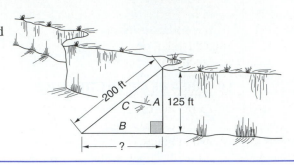

$$B = \sqrt{C^2 - A^2}$$
$$= \sqrt{200^2 - 125^2}$$
$$= \sqrt{40{,}000 - 15{,}625}$$
$$= \sqrt{24{,}375}$$
$$= 156 \text{ ft}$$

6-2-3   *Vectors and Vector Diagrams*

Vector or Phasor

A quantity that has magnitude and direction and that is commonly represented by a directed line segment whose length represents the magnitude and whose orientation in space represents the direction.

A **vector** or **phasor** is an arrow used to represent the magnitude and direction of a quantity. Vectors are generally used to represent a physical quantity that has two properties. For example, Figure 6-8(a) shows a motorboat heading north at 12 miles per hour. Figure 6-8(b) shows how vector **A** could be used to represent the vessel's direction and speed. The size of the vector represents the speed of 12 mph by being 12 cm long, and because we have made the top

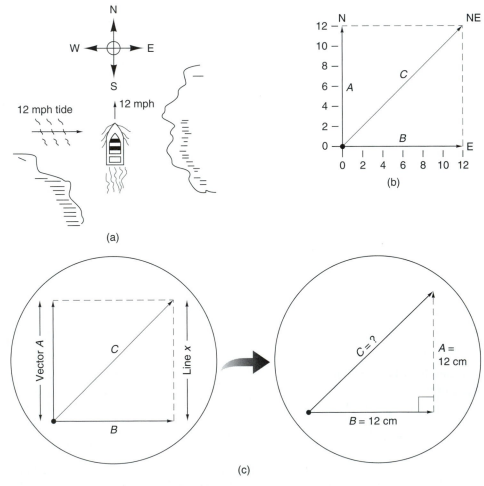

(a)

(b)

(c)

FIGURE 6-8   **Vectors and Vector Diagrams.**

146   CHAPTER 6 / GEOMETRY AND TRIGONOMETRY

of the page north, vector **A** should point straight up so that it represents the vessel's direction. Referring to Figure 6-8(a) you can see that there is another factor that also needs to be considered, a 12 mile per hour easterly tide. This tide is represented in our vector diagram in Figure 6-8(b) by vector **B,** which is 12 centimeters long and pointing east. Since the motorboat is pushing north at 12 miles per hour and the tide is pushing east at 12 miles per hour, the resultant course will be northeast, as indicated by vector **C** in Figure 6-8(b). Vector **C** was determined by **vector addition** (as seen by the dashed lines) of vectors **A** and **B**. The result, vector **C,** is called the **resultant vector.** This resultant vector indicates the motorboat's course and speed. The direction or course, we can see, is northeast because vector **C** points in a direction midway between north and east. The speed of the motorboat, however, is indicated by the length or magnitude of vector **C.** This length, and therefore the motorboat's speed, can be calculated by using the Pythagorean theorem. Probably your next question is: How is the **vector diagram** in Figure 6-8(b) similar to a right-angle triangle? The answer is to redraw Figure 6-8(b), as shown in Figure 6-8(c). Because the dashed line *x* is equal in length to vector **A,** vector **A** can be put in the position of line *x* to form a right-angle triangle with vector **B** and vector **C.** Now that we have a right-angle triangle with two known lengths and one unknown length, we can calculate the unknown vector's length and therefore the motorboat's speed.

$$
\begin{aligned}
C &= \sqrt{A^2 + B^2} \\
&= \sqrt{12^2 + 12^2} \\
&= \sqrt{144 + 144} \\
&= \sqrt{288} = 16.97
\end{aligned}
$$

The length of vector **C** is 16.97 cm, and because each 1 mile per hour of the motorboat was represented by 1 centimeter, the motorboat will travel at a speed of 16.97 miles per hour in a northeasterly direction.

■ INTEGRATED MATH APPLICATION:

An electrical circuit has two elements that oppose current flow; one form of opposition is called *resistance* (symbolized by *R*), and the other form of opposition is called *reactance* (symbolized by *X*). These two oppositions are out of sync, or not in phase, with each other, as seen by the vector diagram in Figure 6-9(a), and therefore cannot simply be added. Using the Pythagorean theorem, vectorially add the two forms of opposition (*R* and *X*) to determine the total opposition to current flow, which is called *impedance* (symbolized by *Z*).

■ *Solution:*

First, redraw the vector diagram to produce a right-angle triangle, as shown in Figure 6-9(b). Then calculate Z:

$$
\begin{aligned}
Z &= \sqrt{R^2 + X^2} \\
&= \sqrt{30^2 + 16^2} \\
&= \sqrt{900 + 256} \\
&= \sqrt{1156} \\
&= 34 \ \Omega
\end{aligned}
$$

FIGURE 6-9   An Electrical Application of the Pythagorean Theorem.

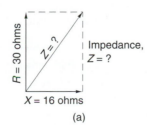 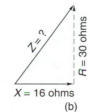

(a)   (b)

The total opposition or impedance (Z) to current flow is therefore 34 ohms.

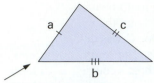

SCALENE TRIANGLE

Is a triangle (3-sided figure).
Has no sides that are equal.

A different number of
check marks on the
sides of the triangle,
are used to indicate
sides of different sizes.

$$P \text{ (Perimeter)} = a + b + c$$
$$A \text{ (Area)} = \frac{b \times h}{2}$$

If the length of the sides are known:

$$A = \sqrt{s \times (s{-}a) \times (s{-}b) \times (s{-}c)}$$

In which s (half perimeter) $= \dfrac{a + b + c}{2}$

AREA OF A SCALENE TRIANGLE

Scalene triangle occupies half the
area of the rectangle below.

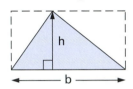

FIGURE 6-10   **The Scalene Triangle.**

6-2-4   *Other Triangles*

Figures 6-10, 6-11 and 6-12 summarize the facts and formulas for other triangle types.

6-2-5   *Circles*

Details relating to the circle have been discussed in previous chapters. These facts and
formulas are summarized in Figure 6-13.

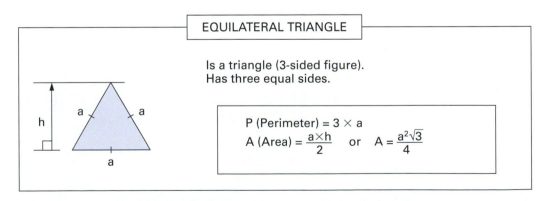

EQUILATERAL TRIANGLE

Is a triangle (3-sided figure).
Has three equal sides.

$$P \text{ (Perimeter)} = 3 \times a$$
$$A \text{ (Area)} = \frac{a \times h}{2} \quad \text{or} \quad A = \frac{a^2 \sqrt{3}}{4}$$

FIGURE 6-11   **The Equilateral Triangle.**

ISOSCELES TRIANGLE

Is a triangle (3-sided figure).
Has two equal sides.

$$P \text{ (Perimeter)} = 2a + b$$
$$A \text{ (Area)} = \frac{b \times h}{2} \text{ or } A = \frac{b}{2}\sqrt{a^2 - \left(\frac{b}{2}\right)^2}$$

FIGURE 6-12 The Isosceles Triangle.

CIRCLE

Radius
(r)

Is a perfectly round figure.
Every point is at an equal distance from the center.

Circumference (C)

Diameter
(d)

$$\text{Diameter (d)} = 2 \times r$$
$$\text{Circumference (c)} = \pi \times d \text{ or } c = \pi \times (2r)$$
$$\text{Area (A)} = \pi r^2$$

Every circle circumference is 3.14 times larger than its diameter. This constant is called pi (symbolized π).

AREA OF A CIRCLE

$$\pi = \frac{\text{Circumference (c)}}{\text{Diameter (d)}} = 3.14$$

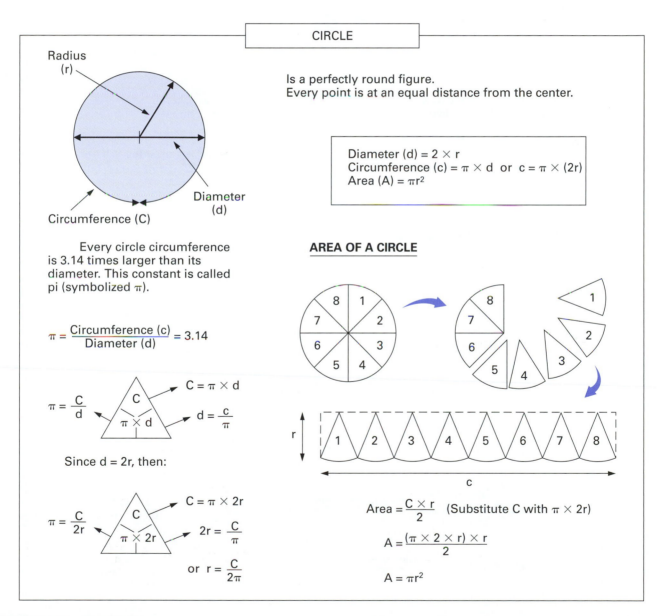

$$\pi = \frac{C}{d} \qquad C = \pi \times d \qquad d = \frac{c}{\pi}$$

Since d = 2r, then:

$$\pi = \frac{C}{2r} \qquad C = \pi \times 2r \qquad 2r = \frac{C}{\pi}$$

$$\text{or } r = \frac{C}{2\pi}$$

$$\text{Area} = \frac{C \times r}{2} \quad \text{(Substitute C with } \pi \times 2r\text{)}$$

$$A = \frac{(\pi \times 2 \times r) \times r}{2}$$

$$A = \pi r^2$$

FIGURE 6-13   The Circle.

6-2-6   *Other Polygons*

Figure 6-14(a) shows that if lines are drawn from the center of any polygon to each vertex, equivalent triangles are formed, which enables us to determine the polygon's area. In Figure 6-14(b), the formula given allows us to determine the number of degrees in the angles for each polygon type. From the examples shown we can see that each angle in an equilateral triangle is 60°, for a square each angle is 90° as expected, for a pentagon each angle is 108°, a hexagon has six 120° angles, and an octagon has eight 135° vertices.

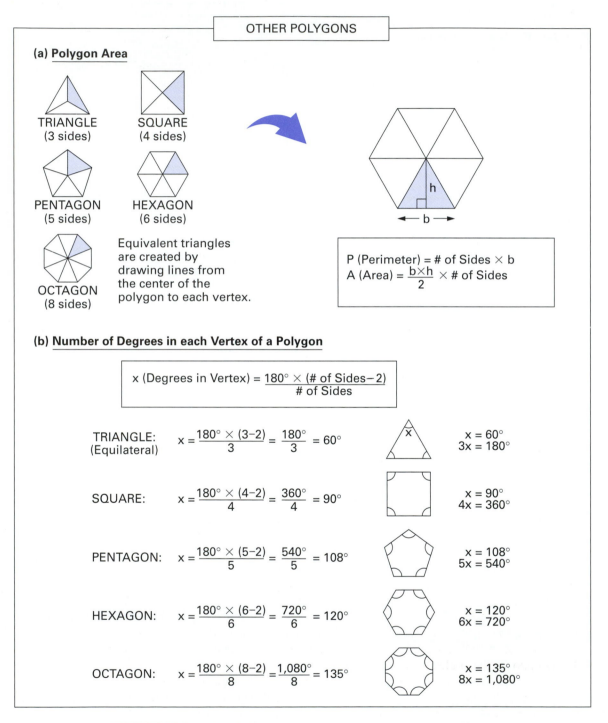

FIGURE 6-14   Other Polygons.

Now that you have completed this section, you should be able to:

■ **Objective 3.** *Define the term* plane figure, *and identify the different types of polygons.*

■ **Objective 4.** *List the formulas for determining the perimeter and area of plane figures.*

■ **Objective 5.** *Explain how the Pythagorean theorem relates to right-angle triangles.*

■ **Objective 6.** *Determine the length of one side of a right-angle triangle when the lengths of the other two sides are known.*

■ **Objective 7.** *Describe how vectors are used to represent the magnitude and direction of physical quantities and how they are arranged in a vector diagram.*

■ **Objective 8.** *Explain how the Pythagorean theorem can be applied to a vector diagram to calculate the magnitude of a resultant vector.*

Use the following questions to test your understanding of Section 6-2:

1. Define the terms:

  a. Plane figure
  b. Polygon

2. Sketch the following and list the formulas for perimeter and area:

  a. Square
  b. Rectangle
  c. Rhombus
  d. Parallelogram
  e. Trapezoid
  f. Right triangle
  g. Scalene triangle

  h. Equilateral triangle
  i. Isosceles triangle
  j. Circle

3. Calculate the lengths of the unknown sides in the following triangles.

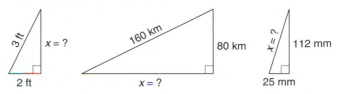

4. Calculate the magnitude of the resultant vectors in the following vector diagrams.

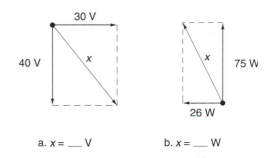

  a. x = ___ V   b. x = ___ W

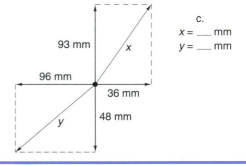

c.
x = ___ mm
y = ___ mm

6-3   TRIGONOMETRY

In the previous section you saw how Pythagoras's theorem could be used to find the length of one side of a right-triangle, when the length of the other two sides are known. **Trigonometry** is the study of the properties of triangles, and in this section we will examine the right-angle triangle in a little more detail, and see how angles and sides are related.

    Until this point we have called the three sides of our right triangle *A, B,* and *C.* Figure 6-15 shows the more common names given to the three sides of a right-angle triangle. Side *C,* called the **hypotenuse,** is always the longest of the three sides. Side *B,* called the **adjacent side,** always extends between the hypotenuse and the vertical side. An angle called **theta** (θ, a Greek letter) is formed between the hypotenuse and the adjacent side. The side that is always opposite the angle θ is called the **opposite side.**

    Angles are always measured in degrees because all angles are part of a circle, like the one shown in Figure 6-16(a). A circle is divided into 360 small sections called **degrees.** In Figure 6-16(b) our right triangle has been placed within the circle. In this example the triangle occupies a 45° section of the circle and therefore theta equals 45 degrees ($\theta = 45°$). Mov-

Trigonometry
The study of the properties of triangles.

Hypotenuse
The side of a right-angle triangle that is opposite the right angle.

Adjacent
The side of a right-angle triangle that has a common endpoint, in that it extends between the hypotenuse and the vertical side.

FIGURE 6-15 **Names Given to the Sides of a Right Triangle.**

Theta

The eighth letter of the Greek alphabet, used to represent an angle.

Opposite

The side of a right-angle triangle that is opposite the angle theta.

Degree

A unit of measure for angles equal to an angle with its vertex at the center of a circle and its sides cutting off $1/360$ of the circumference.

Sine

The trigonometric function that for an acute angle is the ratio between the leg opposite the angle when it is considered part of a right triangle, and the hypotenuse.

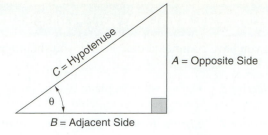

ing from left to right in Figure 6-16(c) you will notice that angle θ increases from 5° to 85°. The length of the hypotenuse (H) in all these examples remains the same; however, the adjacent side's length decreases ($A \downarrow$) and the opposite side's length increases ($O \uparrow$) as angle θ is increased ($\theta \uparrow$). This relationship between the relative length of a triangle's sides and theta means that we do not have to know the length of two sides to calculate a third. If we have the value of just one side and the angle theta, we can calculate the length of the other two sides.

6-3-1   *Sine of Theta (Sin θ)*

In the preceding section we discovered that a relationship exists between the relative length of a triangle's sides and the angle theta. **Sine** is a comparison between the length of the opposite side and the length of the hypotenuse. Expressed mathematically,

$$\text{Sine of theta (sin } \theta) = \frac{\text{opposite side } (O)}{\text{hypotenuse } (H)}$$

Because the hypotenuse is always larger than the opposite side, the result will always be less than 1 (a decimal fraction). Let us use this formula in a few examples to see how it works.

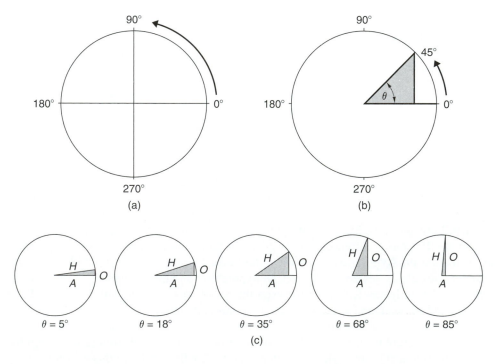

FIGURE 6-16 **Angle Theta (θ).**

■ INTEGRATED MATH APPLICATION:

In Figure 6-17(a), angle theta is equal to 41°, and the opposite side is equal to 20 centimeters in length. Calculate the length of the hypotenuse.

■ *Solution:*

Inserting these values in our formula, we obtain the following:

$$\sin \theta = \frac{O}{H}$$

$$\sin 41° = \frac{20 \text{ cm}}{H}$$

By looking up 41° in a sine trigonometry table or by using a scientific calculator that has all the trigonometry tables stored permanently in its memory, you will find that the sine of 41° is 0.656. This value describes the fact that when $\theta = 41°$, the opposite side will be 0.656, or 65.6%, as long as the hypotenuse side. By inserting this value into our formula and

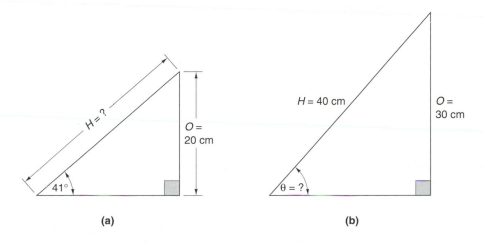

(a)   (b)

FIGURE 6-17 Sine of Theta.

transposing the formula according to the rules of algebra, we can determine the length of the hypotenuse.

$$\sin 41° = \frac{20 \text{ cm}}{H}$$

Calculator sequence: 4 1 SIN

$$0.656 = \frac{20 \text{ cm}}{H}$$

$$0.656 \times H = \frac{20 \text{ cm} \times \cancel{H}}{\cancel{H}}$$

Multiply both sides by *H.*

$$\frac{\cancel{0.656} \times H}{\cancel{0.656}} = \frac{20 \text{ cm}}{0.656}$$

Divide both sides by 0.656.

$$H = \frac{20 \text{ cm}}{0.656} = 30.5 \text{ cm}$$

■ **INTEGRATED MATH APPLICATION:**

Figure 6-17(b) illustrates another example; however, in this case the lengths of sides *H* and *O* are known but θ is not.

■ *Solution:*

$$\sin \theta = \frac{O}{H}$$

$$\sin \theta = \frac{30 \text{ cm}}{40 \text{ cm}}$$

$$= 0.75$$

The ratio of side *O* to side *H* is 0.75, or 75%, which means that the opposite side is 75%, or 0.75, as long as the hypotenuse. To calculate angle θ we must isolate it on one side of the equation. To achieve this we must multiply both sides of the equation by **arcsine,** or **inverse sine** (invsin), which does the opposite of sine.

Arcsine or Inverse Sine
The inverse function to the sine. (If *y* is the sine of θ, then θ is the arcsine of *y*.)

$$\sin \theta = 0.75$$

$$\cancel{\text{invsin}} \ (\cancel{\sin} \ \theta) = \text{invsin } 0.75$$

$$\theta = \text{invsin } 0.75$$

$$= 48.6°$$

Take the inverse sine of 0.75.
Calculator sequence: . 7 5 INV SIN

In summary, therefore, the sine trig functions take an angle θ and give you a number *x.* The inverse sine (arcsin) trig functions take a number *x* and give you an angle θ. In both cases, the number *x* is the ratio of the opposite side to the hypotenuse.

Sine: angle $\theta \rightarrow$ number *x*
Inverse sine: number *x* $\rightarrow$ angle θ

6-3-2   *Cosine of Theta (Cos θ)*

Cosine
The trigonometric function that for an acute angle is the ratio between the leg adjacent to the angle when it is considered part of a right triangle, and the hypotenuse.

Sine is a comparison between the opposite side and the hypotenuse, and **cosine** is a comparison between the adjacent side and the hypotenuse.

$$\text{Cosine of theta (cos } \theta) = \frac{\text{adjacent } (A)}{\text{hypotenuse } (H)}$$

■ **INTEGRATED MATH APPLICATION:**

Figure 6-18(a) illustrates a right triangle in which the angle θ and the length of the hypotenuse are known. From this information, calculate the length of the adjacent side.

■ *Solution:*

$$\cos \theta = \frac{A}{H}$$

$$\cos 30° = \frac{A}{40 \text{ cm}} \qquad \textit{Calculator sequence: } \boxed{3}\boxed{0}\boxed{\text{COS}}$$

$$0.866 = \frac{A}{40 \text{ cm}}$$

Looking up the cosine of 30° you will obtain the fraction 0.866. This value states that when $\theta = 30°$, the adjacent side will always be 0.866, or 86.6%, as long as the

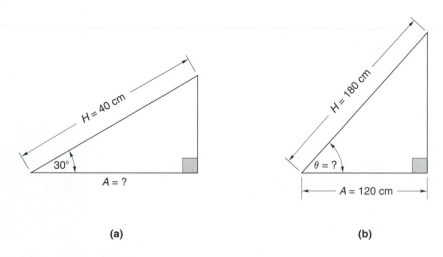

(a)   (b)

FIGURE 6-18 **Cosine of Theta.**

hypotenuse. By transposing this equation we can calculate the length of the adjacent side:

$$0.866 = \frac{A}{40 \text{ cm}}$$

$$40 \times 0.866 = \frac{A}{40 \text{ cm}} \times 40 \qquad \text{\textcolor{blue}{Multiply both sides by 40.}}$$

$$A = 0.866 \times 40$$

$$= 34.64 \text{ cm}$$

■ INTEGRATED MATH APPLICATION:

Calculate the angle θ in Figure 6-18(b) if $A = 120$ centimeters and $H = 180$ centimeters.

■ *Solution:*

$$\cos \theta = \frac{A}{H}$$

$$= \frac{120 \text{ cm}}{180 \text{ cm}}$$

$$= 0.667 \qquad \text{\textcolor{blue}{\textit{A} is 66.7\% as long as \textit{H}.}}$$

$$\text{invcos}(\cos \theta) = \text{invcos } 0.667 \qquad \text{\textcolor{blue}{Multiply both sides by invcos.}}$$

$$\theta = \text{invcos } 0.667 \qquad \text{\textcolor{blue}{\textit{Calculator sequence:}}} \boxed{0}\boxed{.}\boxed{6}\boxed{6}\boxed{7}\boxed{\text{INV}}\boxed{\text{COS}}$$

$$= 48.2°$$

The inverse cosine trig function performs the reverse operation of the cosine function:

Cosine: angle $\theta \rightarrow$ number x

Inverse cosine: number $x \rightarrow$ angle θ

6-3-3   *Tangent of Theta (Tan θ)*

Tangent

The trigonometric function that for an acute angle is the ratio between the leg opposite the angle when it is considered part of a right triangle, and the leg adjacent.

Tangent is a comparison between the opposite side of a right triangle and the adjacent side.

$$\text{Tangent of theta (tan } \theta) = \frac{\text{opposite } (O)}{\text{adjacent } (A)}$$

CALCULATOR KEYS

Name: Tangent key

Function: Instructs the calculator to find the tangent of the displayed value (angle → value).

Example: sin 73° = ?

Press keys: $\boxed{7}\boxed{3}\boxed{\text{TAN}}$

Answer: 3.2709

Name: Arctangent (tan^{-1}) or inverse tangent sequence.

Function: Calculates the smallest angle whose tangent is in the display (value → angle).

Example: invtan 0.95 = ?

Press keys: $\boxed{.}\boxed{9}\boxed{5}\boxed{\text{INV}}\boxed{\text{TAN}}$

Answer: 43.5312

INTEGRATED MATH APPLICATION:

Figure 6-19(a) illustrates a right triangle in which $\theta = 65°$ and the opposite side is 43 centimeters. Calculate the length of the adjacent side.

Solution:

$$\tan \theta = \frac{O}{A}$$

$$\tan 65° = \frac{43 \text{ cm}}{A}$$

Calculator sequence: 6 5 TAN

$$2.14 = \frac{43 \text{ cm}}{A}$$

Whenever $\theta = 65°$, the opposite side will be 2.14 times longer than the adjacent side.

$$2.14 \times A = \frac{43 \text{ cm}}{A} \times A$$

Multiply both sides by A.

$$\frac{\cancel{2.14} \times A}{\cancel{2.14}} = \frac{43 \text{ cm}}{2.14}$$

Divide both sides by 2.14.

$$A = \frac{43 \text{ cm}}{2.14}$$

$$= 20.1 \text{ cm}$$

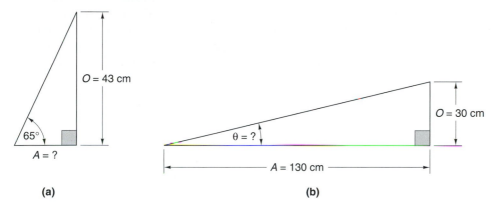

(a)   **(b)**

FIGURE 6-19   **Tangent of Theta.**

INTEGRATED MATH APPLICATION:

Calculate the angle θ in Figure 6-19(b) if $O = 30$ centimeters and $A = 130$ centimeters.

Solution:

$$\tan \theta = \frac{O}{A}$$

$$= \frac{30 \text{ cm}}{130 \text{ cm}}$$

$$= 0.231$$

O is 0.231 or 23.1% as long as A.

$$\text{invtan } (\tan \theta) = \text{invtan } 0.231$$

Multiply both sides by invtan.

$$\theta = \text{invtan } 0.231$$

Calculator sequence: 0 . 2 3 1 INV TAN

$$= 12.99 \text{ or } 13°$$

6-3-4   *Summary*

As you have seen, trigonometry involves the study of the relationships among the three sides (O, H, A) of a right triangle and also the relationships among the sides of the right triangle and the number of degrees contained in the angle theta (θ).

If the lengths of two sides of a right triangle are known, and the length of the third side is needed, remember that

$$H^2 = O^2 + A^2$$

If the angle θ is known along with the length of one side, or if angle θ is needed and the lengths of the two sides are known, one of the three formulas can be chosen based on which variables are known and what is needed.

$$\sin \theta = \frac{O}{H} \quad \cos \theta = \frac{A}{H} \quad \tan \theta = \frac{O}{A}$$

When I was introduced to trigonometry, my mathematics professor spent 15 minutes having the whole class practice what he described as an old Asian war cry that went like this: "SOH CAH TOA." After he explained that it wasn't a war cry but in fact a memory aid to help us remember that SOH was in fact sin θ = O/H, CAH was cos θ = A/H, and TOA was tan θ = O/A, we understood the method in his madness.

■ INTEGRATED MATH APPLICATION:

Sine Wave

A waveform that represents periodic oscillations in which the amplitude of displacement at each point is proportional to the sine of the phase angle of the displacement and that is visualized as a sine curve.

The **sine wave** is the most common type of electrical waveform. It is the natural output of a generator that converts a mechanical input, in the form of a rotating shaft, into an electrical output in the form of a sine wave. In fact, for one cycle of the input shaft, the generator will produce one sinusoidal ac voltage waveform, as shown in Figure 6-20. When the input shaft of the generator is at 0°, the ac output is 0 volts. As the shaft is rotated through 360° the ac output voltage will rise to a maximum or peak positive voltage at 90°, fall back to 0 volts at 180°, then reach a maximum or peak negative voltage at 270°, and finally return to 0 volts at 360°. If this ac voltage is applied across a closed circuit, it produces a current that continually reverses or alternates in direction. (It switches from being a positive voltage to a negative voltage, back to a positive voltage, and so on.) The sine wave is the most common type of waveform shape, and why the name *sine* was given to this wave needs to be explained further.

Figure 6-21(a) shows the correlation between the 360° of a circle and the 360° of a sine wave. Within the circle a triangle has been drawn to represent the right-angle triangle

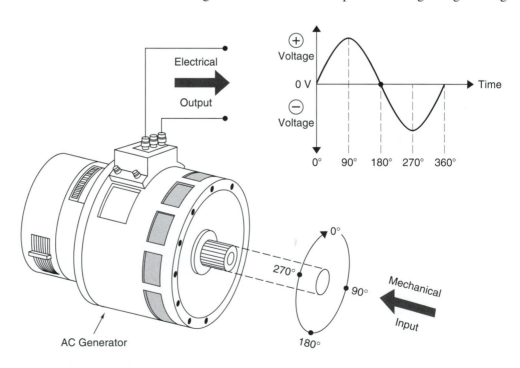

FIGURE 6-20 **Degrees of a Sine Wave.**

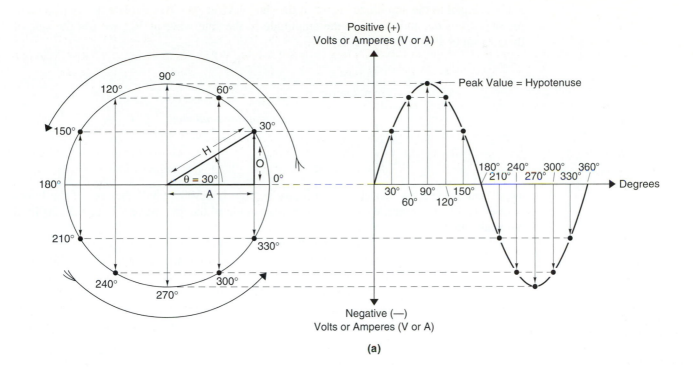

(a)

Angle	Sine
0…°	0.000
15	0.259
30	0.500
45	0.707
60	0.866
75	0.966
90	1.000

(b)

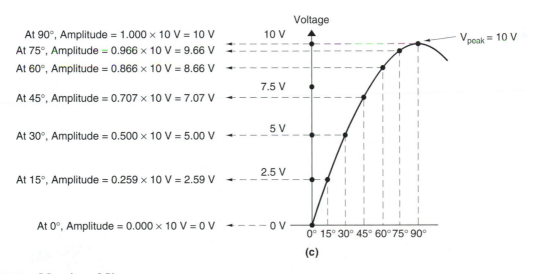

At 90°, Amplitude = 1.000 × 10 V = 10 V
At 75°, Amplitude = 0.966 × 10 V = 9.66 V
At 60°, Amplitude = 0.866 × 10 V = 8.66 V
At 45°, Amplitude = 0.707 × 10 V = 7.07 V
At 30°, Amplitude = 0.500 × 10 V = 5.00 V
At 15°, Amplitude = 0.259 × 10 V = 2.59 V
At 0°, Amplitude = 0.000 × 10 V = 0 V

(c)

FIGURE 6-21   **Meaning of Sine.**

formed at 30°. The hypotenuse side will always remain the same length throughout 360°, and will equal the maximum voltage or peak of the sine wave. The opposite side of the triangle is equal to the amplitude vector of the sine wave at 30°. To calculate the amplitude of the opposite side, and therefore the amplitude of the sine wave at 30°, we use the sine of theta formula discussed previously:

$$\sin \theta = \frac{\text{opposite}}{\text{hypotenuse}}$$

$$\sin 30° = \frac{O}{H}$$ Calculator sequence: [3] [0] [SIN]

$$0.500 = \frac{O}{H}$$

This equation tells us that at 30°, the opposite side is 0.500 or half the size of the hypotenuse. At 30°, therefore, the amplitude or magnitude of the sine wave will be 0.5 (50%) of the peak value.

Figure 6-21(b) lists the sine values at 15° increments. Figure 6-21(c) shows an example of a 10 volt peak sine wave. At 15°, a sine wave will always be at 0.259 (sine of 15°) of the peak value, which for a 10 volt sine wave will be 2.59 volts (0.259 × 10 volts = 2.59 volts). At 30°, a sine wave will have increased to 0.500 (sine of 30°) of the peak value. At 45°, a sine wave will be at 0.707 of the peak, and so on. The sine wave is called a sine wave because it changes in amplitude at the same rate as the sine trigonometric function.

SELF-TEST EVALUATION POINT FOR SECTION 6-3

Now that you have completed this section, you should be able to:

■ **Objective 9.** *Define the trigonometric terms opposite, adjacent, hypotenuse, theta, sine, cosine, and tangent.*

■ **Objective 10.** *Demonstrate how the sine, cosine, and tangent trigonometric functions can be used to*

calculate (1) the length of an unknown side of a right-angle triangle if the length of another side and the angle theta are known or (2) the angle theta of a right-angle triangle if the lengths of two sides of the triangle are known.

Use the following questions to test your understanding of Section 6-3.

1. Calculate the length of the unknown side in the following triangles.

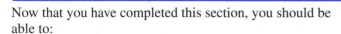

2. Calculate the angle θ for the following right-angle triangles.

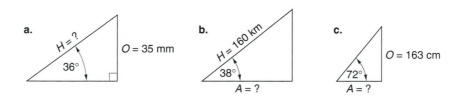

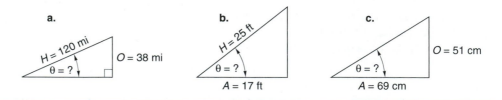

6-4   SOLID FIGURES

As mentioned previously, a plane figure is a flat surface that has only two dimensions—length and width. A **solid figure**, on the other hand, is a three-dimensional figure having length, width and height.

6-4-1   *Prisms*

A **prism** is a solid figure made up of plane figure faces, and has a least one pair of parallel surfaces. Figure 6-22 details the facts and formulas for the prism. The visual breakdown in Figure 6-22(a) shows how a cube is made up of an upper base, lower base, and a set of sides. The **lateral surface area (L)** of the cube is the area of the sides only, as shown in Figure 6-22(b). The **total surface area (S)** is the lateral surface area of the cube plus the area of the two bases, as shown in Figure 6-22(c). The holding capacity or **volume** of the cube measures the space inside the prism, and is described visually in Figure 6-22(d). Figure 6-22(e) shows examples of other prism types.

6-4-2   *Cylinders*

A **cylinder** is a solid figure with curved side walls extending between two identical circular bases. Figure 6-23 details the facts and formulas for the cylinder. As with prisms, plane figure formulas are used to calculate the cylinder's lateral surface area and base area, and as expected, the volume of a cylinder is the product of its base area and height.

6-4-3   *Spheres*

A **sphere** is a globular solid figure in which any point on the surface is at an equal distance from the center. Figure 6-24 lists the formulas for a sphere along with a visual breakdown for a sphere's surface area and volume.

6-4-4   *Pyramids and Cones*

Figure 6-25(a) (p. 164) lists the formulas for a **pyramid,** which has only one base and three or more lateral surfaces that taper up to a single point called an **apex.** Figure 6-25(a) lists the formulas for a **cone,** which has a circular base and smooth surface that extends up to an apex. Figures 6-25(b) and 6-26(b) (p. 165), list the formulas for the frustum of a pyramid and the frustum of a cone. A **frustum** is the base section of a solid pyramid or cone, and is formed by cutting off the top of the solid figure so that the upper base and lower base are parallel to one another.

Solid Figure
A 3-dimensional figure having length, width, and height.

Prism
A solid figure made up of plane figure faces; having at least one pair of parallel surfaces.

Lateral Surface Area
The side area of a solid figure.

Total Surface Area
The lateral surface area of a solid figure plus the area of its two bases.

Volume
The space within a solid figure, or its holding capacity.

Cylinder
A solid figure with curved side walls extending between two identical circular bases.

Sphere
A globular solid figure in which any point on the surface is at equal distance from the center.

Pyramid
A solid figure having only one base and 3 or more lateral surfaces that taper up to a single point.

Apex
The uppermost point or tip of a solid figure.

Cone
A solid figure having a circular base and smooth surface that extends up to an apex.

Frustum
The base section of a solid pyramid or cone.

SELF-TEST EVALUATION POINT FOR SECTION 6-4

Now that you have completed this section, you should be able to:

■ **Objective 11.** *Define the term* solid figure, *and identify the different types.*

■ **Objective 12.** *List the formulas for determining the lateral surface area, total surface area, and volume of solid figures.*

Use the following questions to test your understanding of Section 6-4:

1. What is the difference between a plane figure and a solid figure?

2. What is the difference between lateral surface area and total surface area?

3. Sketch the following solid figures and list the formulas for lateral surface area, total surface area and volume.

   a. Prism   **d.** Pyramid
   b. Cylinder **e.** Cone
   c. Sphere

4. What is a frustum?

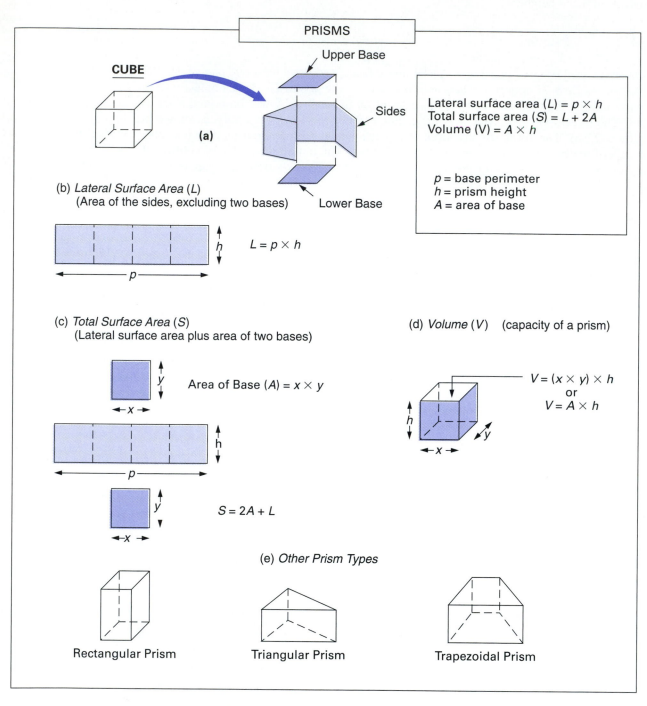

FIGURE 6-22   Prisms.

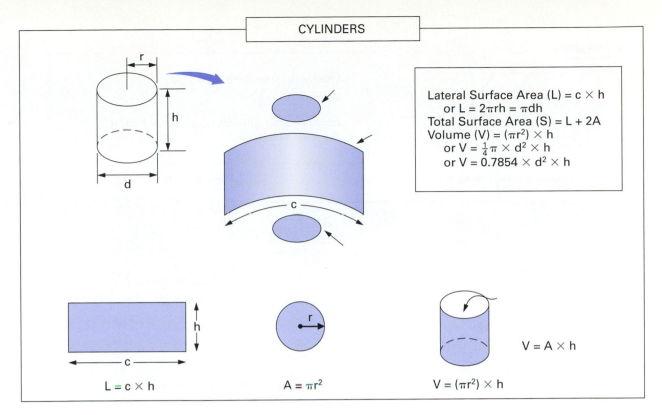

FIGURE 6-23 Cylinders.

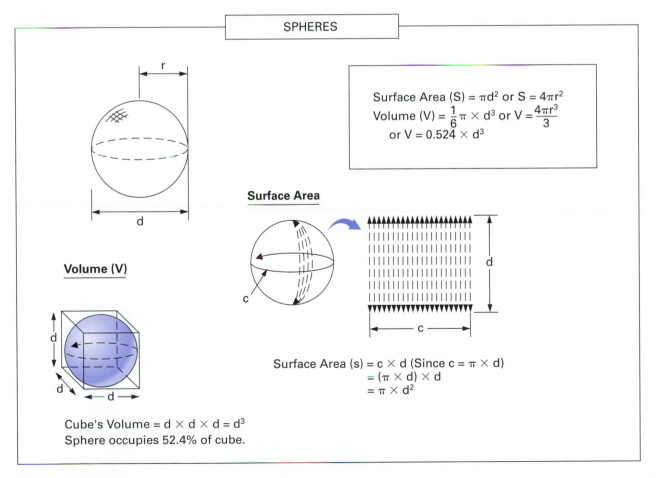

FIGURE 6-24 Spheres.

(a) PYRAMID

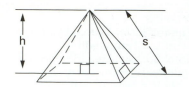

Lateral Surface Area (L) = $\frac{1}{2}$ ps

Volume (V) = $\frac{1}{3}$ Ah

p = perimeter of base   s = slant height
A = area of base   h = height

Lateral Surface Area (L)

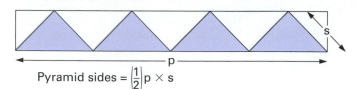

Pyramid sides = $\left(\frac{1}{2}\right)$p × s

Volume (V)

Cube Volume = A × h.
Pyramid occupies $\frac{1}{3}$ of cube.

(b) FRUSTUM OF PYRAMID

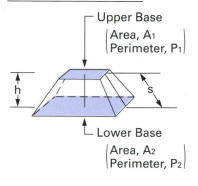

Upper Base
$\left(\begin{array}{l}\text{Area, A}_1 \\ \text{Perimeter, P}_1\end{array}\right)$

Lower Base
$\left(\begin{array}{l}\text{Area, A}_2 \\ \text{Perimeter, P}_2\end{array}\right)$

Lateral Surface Area (L) = $\frac{1}{2}$ (P$_1$ + P$_2$) s

Volume (V) = $\frac{1}{3}$ h (A$_1$ + A$_2$ + $\sqrt{A_1 \times A_2}$)

FIGURE 6-25   **Pyramids and Frustums of Pyramids.**

(a) CONE

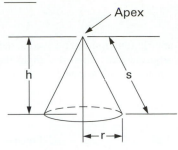

Apex

h

s

r

Lateral Surface Area (L) = πrs or $L = \frac{1}{2}\pi ds$

Volume (V) = $\frac{1}{3}\pi r^2 h$ or $V = \frac{1}{12}\pi d^2 h$

r = base radius   h = height
s = slant height   d = base diameter

(b) FRUSTUM OF CONE

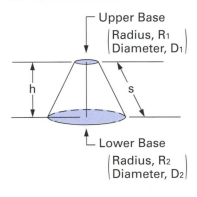

Upper Base
$\left(\begin{array}{l}\text{Radius, } R_1 \\ \text{Diameter, } D_1\end{array}\right)$

h

s

Lower Base
$\left(\begin{array}{l}\text{Radius, } R_2 \\ \text{Diameter, } D_2\end{array}\right)$

Lateral Surface Area (L) = $\pi s(r_1 + r_2)$

or $L = \frac{1}{2}\pi s(d_1 + d_2)$

Volume (V) = $\frac{1}{3}\pi h(r_1^2 + r_2^2 + (r_1 r_2))$

or $V = \frac{1}{12}\pi h(d_1^2 + d_2^2 + (d_1 d_2))$

FIGURE 6-26   **Cones and Frustums of Cones.**

REVIEW QUESTIONS

Multiple Choice Questions

1. What instrument can be used to measure an angle?

   a. Compass   c. Ruler
   b. Protractor   d. None of the above

2. A right angle is equal to:

   a. 45°   c. 180°
   b. 100°   d. 90°

3. A protractor can be used to:

   a. Measure an angle   c. Both (a) and (b)
   b. Draw an angle   d. None of the above

4. A _____ is a quadrilateral parallelogram, that has four sides that are all of equal length.

   a. Square   c. Trapezoid
   b. Rectangle   d. Prism

5. Trigonometry is the study of:

   a. Triangles   c. Trigonometric functions
   b. Angles   d. All the above

6. The Pythagorean theorem states that the _____ of the length of the hypotenuse of a right-angle triangle equals the sum of the _____ of the lengths of the other two sides.

   a. Square, square roots   c. Square, squares
   b. Square root, squares   d. Square root, square roots

7. A vector is an arrow whose length is used to represent the _____ of a quantity and whose point is used to indicate the same quantity's _____.

   a. Magnitude, size   c. Direction, size
   b. Magnitude, direction   d. Direction, magnitude

8. If two vectors in a vector diagram are not working together, or are out of phase with one another, they must be _____ to obtain a _____ vector.

   a. Added, resultant
   b. Multiplied by one another, parallel
   c. Vectorially added, resultant
   d. Both (a) and (b)

9. The _____ of a right-angle triangle is always the longest side.

   a. Opposite side
   b. Adjacent side
   c. Hypotenuse
   d. Angle theta

10. The _____ of a right-angle triangle is always across from the angle theta.

   a. Opposite side
   b. Adjacent side
   c. Hypotenuse
   d. Angle theta

11. The _____ of a right-angle triangle is always formed between the hypotenuse and the adjacent side.

   a. Opposite side   **c.** Hypotenuse
   b. Adjacent side   **d.** Angle theta

12. Which trigonometric function can be used to determine the length of the hypotenuse if the angle theta and the length of the opposite side are known?

   a. Tangent
   b. Sine
   c. Cosine

13. Which trigonometric function can be used to determine the angle theta when the lengths of the opposite and adjacent sides are known?

   a. Tangent
   b. Sine
   c. Cosine

14. Which trigonometric function can be used to determine the length of the adjacent side when the angle theta and the length of the hypotenuse are known?

   a. Tangent
   b. Sine
   c. Cosine

15. A plane figure has _____ dimensions, while a solid figure has _____ dimensions.

   a. 3, 4   **c.** 2, 3
   b. 2, 4   **d.** 3, 2

Communication Skill Questions

16. Define the following terms: (6-1 to 6-4)

   a. Geometry   **c.** Plane figure
   b. Trigonometry   **d.** Solid figure

17. Describe how to use a protractor to draw and measure an angle. (6-1)

18. Sketch the following plane figures and list their associated formulas for calculating perimeter and area: (6-2)

   a. Square   **f.** Right triangle
   b. Rectangle   **g.** Scalene triangle
   c. Rhombus   **h.** Equilateral triangle
   d. Parallelogram   **i.** Isosceles triangle
   e. Trapezoid   **j.** Circle

19. What are the number of degrees in each vertex of a: (6-2-5)

   a. Square   **d.** Hexagon
   b. Equilateral triangle   **e.** Octagon
   c. Pentagon

20. Describe the purpose of Pythagoras's theorem. (6-2-2)

21. Demonstrate how the sine, cosine and tangent trigonometric functions can be used to calculate: (6-3)

   a. The length of an unknown side of a right triangle when the length of another side is known and the angle theta is known.

   b. The angle theta of a right triangle if the lengths of two sides of the triangle are known.

22. Sketch the following solid figures and list their associated formulas: (6-4)

   a. Prism   **d.** Pyramid
   b. Cylinder   **e.** Cone
   c. Sphere

23. Why is it necessary to have inverse trigonometric functions? (6-3)

24. What is a sine wave? (6-3)

25. What is a frustum? (6-4)

Practice Problems

26. Sketch the following:

   a. A line segment.   **e.** Two perpendicular lines.
   b. Two parallel lines.   **f.** An angle of 45°.
   c. A right angle.   **g.** A vertical line.
   d. An acute angle.   **h.** An obtuse angle.

27. Calculate the length of the hypotenuse (C side) of the following right triangles.

   a. $A = 20$ mi, $B = 53$ mi
   b. $A = 2$ km, $B = 3$ km
   c. $A = 4$ in., $B = 3$ in.
   d. $A = 12$ mm, $B = 12$ mm

28. Calculate the length or magnitude of the resultant vectors in the following vector diagrams.

a.

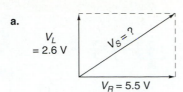

$V_L = 2.6\ V$ $V_S = ?$ $V_R = 5.5\ V$

b.

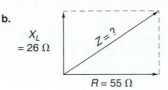

$X_L = 26\ \Omega$ $Z = ?$ $R = 55\ \Omega$

c.

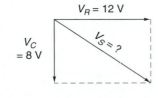

$V_R = 12\ V$ $V_C = 8\ V$ $V_S = ?$

d.

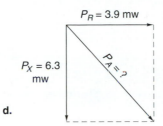

$P_R = 3.9\ mw$ $P_X = 6.3\ mw$ $P_A = ?$

29. Calculate the value of the following trigonometric functions.

a. sin 0°	**i.** cos 60°
b. sin 30°	**j.** cos 90°
c. sin 45°	**k.** tan 0°
d. sin 60°	**l.** tan 30°
e. sin 90°	**m.** tan 45°
f. cos 0°	**n.** tan 60°
g. cos 30°	**o.** tan 90°
h. cos 45°	

30. Calculate angle θ from the function given.

a. $\sin \theta = 0.707, \theta = ?$
b. $\sin \theta = 0.233, \theta = ?$
c. $\cos \theta = 0.707, \theta = ?$
d. $\cos \theta = 0.839, \theta = ?$
e. $\tan \theta = 1.25, \theta = ?$
f. $\tan \theta = 0.866, \theta = ?$

31. Name each of the triangle's sides, and calculate the unknown values.

a.

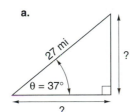

27 mi ? $\theta = 37°$?

b.

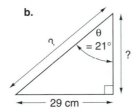

? $\theta = 21°$? 29 cm

c.

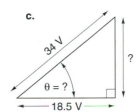

34 V $\theta = ?$? 18.5 V

32. If a cylinder has a radius of 3 cm and a height of 10 cm, calculate its:

a. Lateral surface area
b. Total surface area
c. Volume

33. If a sphere has a diameter of 2 inches, what would be its surface area and volume?

34. What is the lateral surface area of a rectangular prism that has a base perimeter of 40 cm and a height of 12 cm?

35. Calculate the lateral surface area of a pyramid that has a base perimeter of 24 inches and a slant height of 8 inches?

Web Site Questions

Go to the web site http://www.prenhall.com/cook, select the textbook *Mathematics for Electronics and Computers,* select this chapter, and then follow the instructions when answering the multiple-choice practice problems.

Logarithms and Graphs

The First Pocket Calculator

During the seventeenth century, European thinkers were obsessed with any device that could help them with mathematical calculation. Scottish mathematician John Napier decided to meet this need, and in 1614 he published his new discovery of logarithms. In this book, consisting mostly of tediously computed tables, Napier stated that a logarithm is the exponent of a base number. For example:

The common logarithm (base 10) of 100 is 2 ($100 = 10^2$).

The common logarithm of 10 is 1 ($10 = 10^1$).

The common logarithm of 27 is 1.43136 ($27 = 10^{1.43136}$).

The common logarithm of 6 is 0.77815 ($6 = 10^{0.77815}$).

Any number, no matter how large or small, can be represented by or converted to a logarithm. Napier also outlined how the multiplication of two numbers could be achieved by simply adding the numbers' logarithms. For example, if the logarithm of 2 (which is 0.30103) is added to the logarithm of 4 (which is 0.60206), the result will be 0.90309, which is the logarithm of the number 8 ($0.30103 + 0.60206 = 0.90309$, $2 \times 4 = 8$). Therefore, the multiplication of two large numbers can be achieved by looking up the logarithms of the two numbers in a log table, adding them together, and then finding the number that corresponds to the sum in an antilog (reverse log) table. In this example, the antilog of 0.90309 is 8.

Napier's table of logarithms was used by William Oughtred, who developed, just 10 years after Napier's death in 1617, a handy mechanical device that could be used for rapid calculation. This device, considered the first pocket calculator, was the slide rule.

From *Portraits of Eminent Mathematicians*, by David Eugene Smith, published by Pictorial Mathematics, New York, 1936.

7

••••••••••••••••••••••

Outline and Objectives

VIGNETTE: THE FIRST POCKET CALCULATOR

INTRODUCTION

7-1   COMMON LOGARITHMS

Objective 1: Define the following terms:
a. Common logarithm
b. Exponential expression
c. Logarithmic expression
d. Natural logarithm

Objective 2: Describe how to use the calculator's stored table of common logarithms to find the log of any number.

Objective 3: Explain how logarithms can be used to simplify multiplication and division.

Objective 4: Describe how to use the calculator's stored table of common antilogarithms to convert the log answer back to a number.

7-1-1 The Calculator's Table of Common Logarithms

7-1-2 Numbers That Are Not Multiples of Ten

7-1-3 Using Logarithms to Simplify Multiplication

7-1-4 Using Logarithms to Simplify Division

7-2   GRAPHS

Objective 5: Define the purpose of a graph.

Objective 6: Describe how to interpret the data displayed in a graph.

Objective 7: Explain the differences among the:
a. Line graph
b. Bar graph
c. Circle graph or pie chart

Objective 8: Show how data collected from an experiment are plotted on a graph.

Objective 9: Describe the difference between a logarithmic response and a linear response.

Objective 10: Explain how the graphing calculator can be used to find the solution to a problem.

7-2-1 Reading Graphs

7-2-2 Types of Graphs

7-2-3 Plotting a Graph

7-2-4 The Graphing Calculator

MULTIPLE CHOICE QUESTIONS

COMMUNICATION SKILL QUESTIONS

PRACTICE PROBLEMS

WEB SITE QUESTIONS

Introduction

Logarithm

The exponent that indicates the power to which a number is raised to produce a given number.

In the first half of this chapter, we will discuss **logarithms.** Although this seems like a new topic, it is simply a continuation of our discussion of exponents from Chapter 4. For example, you know that a distance of 1 kilometer is equal to 1×10^3 meters (or 1000 meters). The power of 10, exponent 3 in this example, is the logarithm of 1000 (log 1000 = 3). Logarithms can be used to simplify multiplication to addition, division to subtraction, raising to a power to multiplication, and extracting square roots to division. Today, multiplication, division, raising to a power, and extracting square roots can be accomplished more easily with a calculator, and so the use of logarithms to simplify these operations has decreased. Logarithms have found new applications, however, in many areas of science and technology, and so it is important to have a clear understanding of them.

In the second half of this chapter, we will study graph interpretation, types, and plotting. A graph is a diagram used frequently in all areas of science and technology to show the relationship between two or more factors. Graphs actually make it easier for us to understand information because they convert long lists of recorded data into an easy-to-understand visual representation.

7-1   COMMON LOGARITHMS

Common Logarithm

A logarithm whose base is 10.

Stated simply, a **common logarithm** is a base 10 exponent. So far, you have already had considerable experience with base 10 exponents, or as they are officially called, *power of 10.* For example, you know that a computer having 1 megabyte of memory means that the memory can store 1×10^6 bytes (or 1,000,000 bytes), and that a distance of 1 kilometer is equal to 1×10^3 meters (or 1000 meters). The power-of-10 exponent used in these examples is the logarithm; for the first example, *the common logarithm (base 10) of 1,000,000 is 6,* and for the second example, *the common logarithm (base 10) of 1000 is 3.* If we were to state the first of these examples in mathematical form, it would appear as follows:

$$\overset{\text{Exponent}}{10^{6}} = \underset{\text{Number}}{1,000,000} \qquad \log_{10} \underset{\text{Base}}{} \underset{\text{Number}}{1,000,000} = \overset{\text{Logarithm}}{6}$$

The first expression ($10^6 = 1,000,000$) is an *exponential expression* because the quantity is expressed in *exponential form.* The second expression ($\log_{10} 1,000,000 = 6$) is a *logarithmic expression* because the quantity is expressed in *logarithmic form.*

Natural Logarithm

A logarithm with *e* as a base, in which *e* is a transcendental number having a value to eight decimal places of 2.71828183.

There are two basic types of logarithms: **natural logarithms,** which use a base of 2.71828, and *common logarithms,* which use a base of 10. The base of a logarithm should be indicated when there is any chance of confusion; however, when using only common logs, it is not necessary to indicate or say the base because it is assumed to be 10. Therefore, $\log_{10}$ 1,000,000 = 6 can be simplified to log 1,000,000 = 6.

As another example: What is the log of 1000? To find the answer, we simply convert 1000 to its power-of-10 equivalent, which in exponential form is 10^3. Because the common logarithm of any number is its base 10 exponent, in logarithmic form, log 1000 = 3. Here is a list of logs for numbers that are greater than 1:

Exponential Form	Logarithmic Form
$10^0 = 1$	log 1 = 0
$10^1 = 10$	log 10 = 1
$10^2 = 100$	log 100 = 2
$10^3 = 1000$	log 1000 = 3
$10^4 = 10,000$	log 10,000 = 4
$10^5 = 100,000$	log 100,000 = 5
$10^6 = 1,000,000$	log 1,000,000 = 6

What do we do when we are dealing with numbers that are a fraction of 1, for example, 0.001, 0.000001, and so on? The answer is to follow exactly the same procedure. First, express the number or value in exponential form ($10^{-3} = 0.001$). Then, because the logarithm is the base 10 exponent, express the number in logarithmic form (log 0.001 = −3). Here is a list of logs for numbers that are a fraction of 1:

Exponential Form	Logarithmic Form
$10^0 = 1$	log 1 = 0
$10^{-1} = 0.1$	log 0.1 = −1
$10^{-2} = 0.01$	log 0.01 = −2
$10^{-3} = 0.001$	log 0.001 = −3
$10^{-4} = 0.0001$	log 0.0001 = −4
$10^{-5} = 0.00001$	log 0.00001 = −5
$10^{-6} = 0.000001$	log 0.000001 = −6

7-1-1   *The Calculator's Table of Common Logarithms*

Your calculator contains a table of common logarithms. It operates as follows.

CALCULATOR KEYS

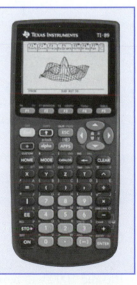

Name: Common logarithm key

Function: Calculates the common logarithm (base 10) of the displayed number.

Example: log 35.5 = ?

Press keys: 3 5 . 5 LOG

Answer: 1.5502283

Therefore, $35.5 = 10^{1.5502283}$.

■ EXAMPLE:

Using a calculator, determine the log of 1000.

■ *Solution:*

To determine the log of 1000 using a calculator, you input the value 1000 and then press the log calculator key, as follows:

Log 1000 = ?
Press keys: $\boxed{1}\,\boxed{0}\,\boxed{0}\,\boxed{0}\,\boxed{\text{LOG}}$
Answer: 3

■ EXAMPLE:

Determine the log of 0.00001.

■ *Solution:*

To determine the log of 0.00001 using the calculator, you simply input the value 0.00001 and then press the log calculator key, as follows:

Log 0.00001 = ?
Press keys: $\boxed{0}\,\boxed{.}\,\boxed{0}\,\boxed{0}\,\boxed{0}\,\boxed{0}\,\boxed{1}\,\boxed{\text{LOG}}$
Answer: −5

7-1-2   *Numbers That Are Not Multiples of 10*

Only numbers that are powers of 10, for example, 0.001, 0.1, 100, 1,000,000,000, and so on, have an exponent or logarithm that is a whole number. In the following examples, you will see that any number between two multiples of 10 will have a decimal fraction. Using your calculator, check the logs (which have been rounded off to 4 decimal places) for the following numbers:

Logarithmic Form	Exponential Form
log 0.000089 = −4.0506	$10^{-4.0506} = 0.000089$
log 0.28 = −0.5528	$10^{-0.5528} = 0.28$

Because the example numbers are a fraction of 1, the log or exponent will be negative.

log 2.5 = 0.3979	$10^{0.3979} = 2.5$
log 6 = 0.7782	$10^{0.7782} = 6$
log 9.75 = 0.9890	$10^{0.9890} = 9.75$

Because the log of 1 is 0, and the log of 10 is 1, all the example numbers between 1 and 10 will have logs or exponents between 0 and 1.

log 12 = 1.0792	$10^{1.0792} = 12$
log 76.4 = 1.8831	$10^{1.8831} = 76.4$

Because the log of 10 is 1 and the log of 100 is 2, all the example numbers between 10 and 100 will have logs or exponents between 1 and 2.

log 135.54 = 2.132	$10^{2.132} = 135.54$
log 873 = 2.941	$10^{2.941} = 873$

Because the log of 100 is 2 and the log of 1000 is 3, all the example numbers between 100 and 1000 will have logs or exponents between 2 and 3.

7-1-3 *Using Logarithms to Simplify Multiplication*

Now that we know what logarithms are and how to find them, let us see what we can do with them. As mentioned previously, one reason for converting numbers to logarithms is to simplify calculations. For example, using logarithms, *we can reduce the multiplication of two numbers to simple addition.* This advantage was first shown in this chapter's opening vignette; however, let us study the steps involved in a little more detail. To find the product of two numbers using logs, you have to follow these four steps:

STEP 1. Find the log of each number.
STEP 2. Change each number to base 10.
STEP 3. Add the base 10 exponents.
STEP 4. Convert the log answer back to a number using the antilog.

■ **INTEGRATED MATH APPLICATION:**

Let us use logarithms to calculate the product of 26×789.

■ *Solution:*

STEP 1: Find the log of each number:

$$\log 26 = 1.41497$$
$$\log 789 = 2.89708$$

STEP 2: Change each number to base 10. Because the log of a number is the base 10 exponent, we have the following:

$$26 = 10^{1.41497}$$
$$789 = 10^{2.89708}$$

STEP 3: The answer to 26×789 can be obtained by simply adding the exponents from step 2.

$$10^{1.41497} \times 10^{2.89708} = 10^{1.41497 + 2.89708} = 10^{4.31205}$$

Note: This law has always applied to exponents; for example, $100 \times 1000 = 10^2 \times 10^3 = 100{,}000 = 10^5$. When multiplying like bases, the exponents or number of zeros are added.

STEP 4: Once you have converted all your numbers to logarithms and finished all your calculations, you then have to convert the log answer back to a number using a **common antilogarithm** table. This reverse log table is also stored in your calculator and operates as follows.

Antilogarithm
The number corresponding to a given logarithm.

CALCULATOR KEYS

Name: Common antilogarithm (10^x) key

Function: Calculates the common antilogarithm of the displayed value.

Example: $10^{1.5502283}$

Press keys: [1][.][5][5][0][2][2][8][3][10^x]

Answer: 35.5

Therefore, $10^{1.5502283} = 35.5$.

CALCULATOR KEYS *(continued)*

To continue with our problem, let us now find the antilog of our answer, $10^{4.31205}$. To find the reverse log of 4.31205, we press the following calculator keys:

Press keys: [4] [.] [3] [1] [2] [0] [5] [10^x]
Answer: 20513.98

Using the calculator to double-check this answer, we find that $26 \times 789 = 20{,}514$.

7-1-4 *Using Logarithms to Simplify Division*

Using logarithms, *we can reduce the division of one number by another number to simple subtraction.* The steps for calculating the result of a division using logs are:

STEP 1. Find the log of the dividend and divisor.
STEP 2. Change each number to base 10.
STEP 3. Subtract the log of the divisor from the log of the dividend.
STEP 4. Convert the log answer back to a number using the antilog.

■ **INTEGRATED MATH APPLICATION:**

Let us use logarithms to calculate the quotient of $8992.5 \div 165$.

■ *Solution:*

STEP 1: Find the log of both the dividend (8992.5) and the divisor (165):

$$\log 8992.5 = 3.95388$$
$$\log 165 = 2.21748$$

STEP 2: Change each number to base 10. Because the log of a number is the base 10 exponent, we have the following:

$$8992.5 = 10^{3.95388}$$
$$165 = 10^{2.21748}$$

STEP 3: The answer to $8992.5 \div 165$ can be obtained by simply subtracting the log or exponent of the divisor from the log of the dividend.

$$10^{3.95388} \div 10^{2.21748} = 10^{3.95388 - 2.21748} = 10^{1.7364}$$

Note: This law has always applied to exponents; for example, $10{,}000 \div 10 = 10^4 \div 10^1 = 10^{4-1} = 10^3 = 1000$. When like bases are divided, the exponents are subtracted.

STEP 4: Convert the log answer back to a number using the antilog table in your calculator.

$$\text{antilog } 1.7364 = 54.5$$

Using the calculator to double-check this answer, we find that $8992.5 \div 165 = 54.5$.

SELF-TEST EVALUATION POINT FOR SECTION 7-1

Now that you have completed this section, you should be able to:

■ *Objective 1.* *Define the terms* common logarithm, exponential expression, logarithmic expression, *and* natural logarithm.

■ *Objective 2.* *Describe how to use the calculator's stored table of common logarithms to find the log of any number.*

■ *Objective 3.* *Explain how logarithms can be used to simplify multiplication and division.*

Objective 4. *Describe how to use the calculator's stored table of common antilogarithms to convert the log answer to a number.*

Use the following questions to test your understanding of Section 7-1.

1. A common logarithm is a base 10 _____.
2. The logarithm of 10,000 is _____.

   a. 2   **c.** 4
   b. 40   **d.** 100

3. Determine the log of each of the following.

   a. 10   **e.** 10,500
   b. 100   **f.** 3746
   c. 25   **g.** 0.35
   d. 150   **h.** 1.75

4. Convert the following logs to their numerical value.

   a. 1.75   **e.** 5.5
   b. 2.35   **f.** 0.25
   c. 5   **g.** 0.1
   d. 1   **h.** 10

7-2   GRAPHS

A **graph** is a diagram showing the relationship between two or more factors. As an example, the graph in Figure 7-1 shows how new technologies, such as electric light, radio, the telephone and television, have been adopted in homes in the United States. This type of graph is called a **line graph** because the data are represented by points joined by line segments. The vertical scale in this graph indicates U.S. households in percent (0% to 100%), while the horizontal scale indicates the year (1900 to 1980). The four plotted lines in this graph show how quickly these new technologies found their way into the home. The data used to create the four plotted lines was obtained by recording what percentage of homes in the United States had electric light, radio, telephone, and television every year from 1900 to 1980.

Graph
A diagram (as a series of one or more points, lines, line segments, curves, or areas) that represents the variation of a variable in comparison with that of one or more other variables.

Line Graph
A graph in which points representing values of a variable are connected by a broken line.

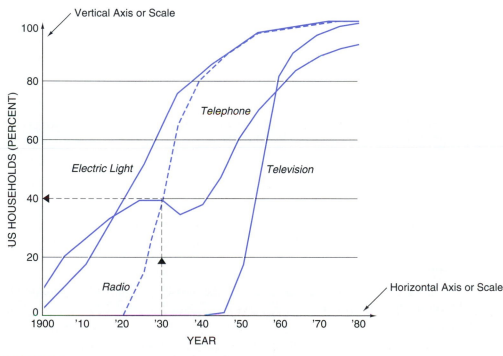

FIGURE 7-1   A Graph Showing How New Technologies Have Been Adopted in the United States.

7-2-1 *Reading Graphs*

If the yearly facts for the graph shown in Figure 7-1 were listed in columns instead of illustrated graphically, it would be difficult to see the characteristics of each individual line graph, or to compare the characteristics of one technology with those of another. To explain this point, let us try an example.

■ INTEGRATED MATH APPLICATION:

From the graph shown in Figure 7-1, determine the following:

a. Approximately what percentage of U.S. households had radio in 1930?
b. Approximately what percentage of U.S. households had telephones in 1960?
c. Which technology was the slowest to be adopted into the home, and which was the fastest?

■ *Solution:*

a. To find this point on the graph, move along the horizontal scale until you reach 1930, and then proceed directly up until you connect with the plotted radio line graph (shown as a dashed line in Figure 7-1). Comparing this point with the vertical scale, you can see that it corresponds to about 40%. In 1930, therefore, approximately 40% of U.S. households had radio.

b. To find this point on the graph, move along the horizontal scale until you reach 1960, and then proceed directly up until you connect with the plotted telephone line graph. Comparing this point with the vertical scale, you can see that it corresponds to about 80%. In 1960, therefore, approximately 80% of U.S. households had a telephone.

c. Comparing the four line graphs shown in Figure 7-1, you can see that the telephone seemed to be the slowest technology to be adopted, while the television was the fastest.

Questions (a) and (b) in the previous example show how a graph can be used to analyze the characteristics of each individual line graph. Question (c) showed how a graph can be used to compare the characteristics of several line graphs.

■ INTEGRATED MATH APPLICATION:

From the graph shown in Figure 7-2, determine the following:

a. Which camera type sells the most, and approximately how many were sold in 1992?
b. Which camera type came on and then went off the market?
c. Compare the sales of 35mm cameras in 1984 with sales in 1994.

■ *Solution:*

a. Disposable or one-time-use cameras sell the most, and in 1992 approximately 22 million units were sold.
b. Disc cameras.
c. In 1984 approximately 5 million 35mm cameras were sold, whereas in 1994 approximately 11 million 35mm cameras were sold.

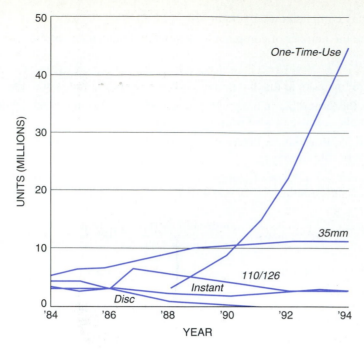

FIGURE 7-2   A Graph Showing Camera Sales.

7-2-2   *Types of Graphs*

There are three types of graphs: line graphs, bar graphs, and circle graphs. *Line graphs,* like the examples shown in Figures 7-1 and 7-2, are ideal for displaying historical data in which a factor is constantly changing.

   Bar graphs use parallel bars of different sizes to compare one variable with others. For example, Figure 7-3(a) shows how a bar graph can be used to compare the sales records of four computer salespersons for the month of May.

■   INTEGRATED MATH APPLICATION:

From the graph shown in Figure 7-3(a), determine the following:

a. How many sales representatives are there?

b. Which sales rep sold the most computers in the month of May, and which sales rep sold the least?

Bar Graph or Bar Chart

A graphic means of quantitative comparison by rectangles with lengths proportional to the measure of the data being compared.

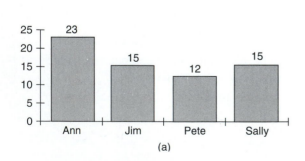

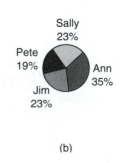

FIGURE 7-3   Bar Graphs and Circle Graphs.

c. How many computers were sold in May?

d. Calculate the number of computers Ann sold as a percentage of the total sold.

■ *Solution:*

a. There are four bars in this bar graph, one for each of the four sales reps.

b. Ann has the largest bar in the graph because she sold 23 computers, while Pete has the smallest bar because he sold only 12 computers.

c. Ann sold 23, Jim sold 15, Pete sold 12, and Sally sold 15. Therefore, a total of 65 computers were sold in May.

d. To calculate the number of computers Ann sold as a percentage of the total sold, use the following formula;

$$\text{Percentage} = \frac{\text{units sold by rep}}{\text{total units sold}} \times 100$$

$$\text{Percentage} = \frac{23}{65} \times 100 \qquad (23 \div 65 = 0.35)$$

$$= 0.35 \times 100 = 35\%$$

Pie Chart or Circle Graph

A circular chart cut by radii into segments illustrating relative magnitudes.

Figure 7-3(b) shows an example of a **circle graph,** which uses different size segments within a circle to compare one variable with others. These circle graphs are often called **pie charts** because the circle resembles a pie and the segments look like different slices in the pie. The size of the segments within a circle graph is usually given as a percentage, as shown in the example in Figure 7-3(b), with the whole circle equal to 100%. Like the bar graph in Figure 7-3(a), the circle graph in Figure 7-3(b) compares the sales records of four computer salespersons for the month of May. Because the data represented in these two graphs are the same, the only difference between these two graphs are the presentation format.

■ **INTEGRATED MATH APPLICATION:**

From the graph shown in Figure 7-3(b), determine the following:

a. How many sales representatives are there?

b. Which sales rep sold the most computers in the month of May, and which sales rep sold the least?

c. If Ann sold 23 units, Jim sold 15 units, Pete sold 12 units, and Sally sold 15 units, calculate each sales rep's percentage of total sales.

d. If a total of 87 computers were sold in the month of June and Jim again sold 23%, how many did he sell?

■ *Solution:*

a. There are four segments in this circle graph, one for each of the four sales reps.

b. The largest segment is assigned to Ann because she sold 23 computers, and the smallest segment is assigned to Pete, who sold only 12 computers.

c. Each sales rep's percentage was calculated as follows:

$$\text{Ann's percentage} = \frac{\text{units sold by Ann}}{\text{total units sold}} \times 100 = \frac{23}{65} \times 100 = 0.35 \times 100 = 35\%$$

$$\text{Jim's percentage} = \frac{\text{units sold by Jim}}{\text{total units sold}} \times 100 = \frac{15}{65} \times 100 = 0.23 \times 100 = 23\%$$

$$\text{Pete's percentage} = \frac{\text{units sold by Pete}}{\text{total units sold}} \times 100 = \frac{12}{65} \times 100 = 0.19 \times 100 = 19\%$$

$$\text{Sally's percentage} = \frac{\text{units sold by Sally}}{\text{total units sold}} \times 100 = \frac{15}{65} \times 100 = 0.23 \times 100 = 23\%$$

d. $23\% \times 87 = 0.23 \times 87 = 20$ Jim sold 20 computers in June.

7-2-3   *Plotting a Graph*

As the expression goes, "a picture is worth a thousand words," and graphs are pictures that can convert a thousand lines of recorded data into an easy-to-understand visual representation. To understand the graph better, let us collect data from an experiment and then plot these data on a graph.

It has been proven through experimentation that the human ear responds logarithmically to changes in sound level. To explain this in more detail, refer to the illustration in Figure 7-4(a), which shows a tone generator connected to a speaker. As the volume control of the generator is increased the power out of the speaker is gradually increased from 1 watt to 100 watts. The listener is asked to signal the points at which a change in volume is detected. These points are listed in the table in Figure 7-4(b). Studying this table, you can see that the power has to be increased from its start point of 1 watt (1.0 W) to approximately 1.26 W before the ear can detect an increase in volume. As the power was increased further, a change in volume was detected at 1.58 W, 2 W, 2.5 W, 3.17 W, and so on, as shown in the table in Figure 7-4(b).

Studying the table in Figure 7-4(b), you can see that the human ear is rather slow to respond to increased stimulation. For example, even when the sound intensity is doubled to twice its original value (1 W to 2 W), it is not interpreted by the listener as being twice as loud. In fact the listener seems to detect only a small change in loudness. To make the tone sound twice as loud to the listener, the power must be increased ten times (1 W to 10 W), and to make the tone sound three times as loud, the power must be increased one hundred times (1 W to 100 W). Let us see how these data can be put to some practical purpose.

■   **INTEGRATED MATH APPLICATION:**

You are considering the purchase of a new amplifier for your home music system and cannot decide whether to buy a 50 W or 100 W unit. If the 100 W unit is twice the price, will you get twice the benefit?

■   *Solution:*

Although the 100 W amplifier delivers twice the output power, your ears will perceive this as only a small difference in volume. For your ears to detect twice the volume, therefore, you would need an amplifier that can deliver 10 times the power. Compared with a 50 W amplifier, a 500 W amplifier will sound as though you are getting twice your money's worth.

The graph in Figure 7-4(c) shows the **logarithmic response** of the human ear by plotting the information from Figure 7-4(b). At point 10, the sound is sensed by the human ear as being twice (2 times) as loud, although power has actually increased from 1 W to 10 W (10 times). At point 20, the sound is sensed as being three times (3 times) as loud, although power has actually increased from 1 W to 100 W (100 times). This response is called logarithmic because a multiple-of-10 change was needed in the input sound level (power had to be increased from 1 W to 10 W to 100 W) to be detected as a change in the sensed output sound (twice as much, or three times as much, respectively).

Logarithmic Response

Having or being a response or output that is logarithmic.

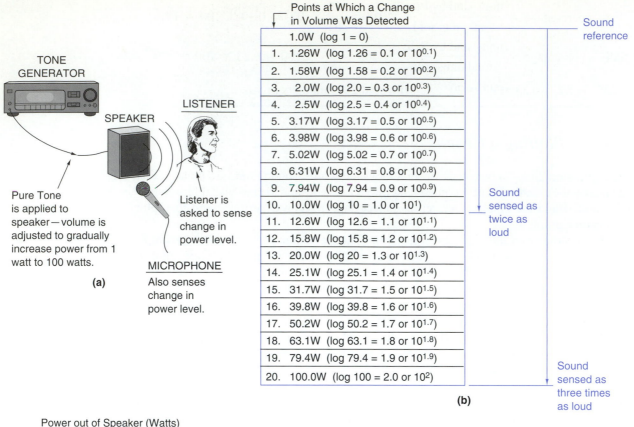

Points at Which a Change in Volume Was Detected

	1.0W (log 1 = 0)
1.	1.26W (log 1.26 = 0.1 or $10^{0.1}$)
2.	1.58W (log 1.58 = 0.2 or $10^{0.2}$)
3.	2.0W (log 2.0 = 0.3 or $10^{0.3}$)
4.	2.5W (log 2.5 = 0.4 or $10^{0.4}$)
5.	3.17W (log 3.17 = 0.5 or $10^{0.5}$)
6.	3.98W (log 3.98 = 0.6 or $10^{0.6}$)
7.	5.02W (log 5.02 = 0.7 or $10^{0.7}$)
8.	6.31W (log 6.31 = 0.8 or $10^{0.8}$)
9.	7.94W (log 7.94 = 0.9 or $10^{0.9}$)
10.	10.0W (log 10 = 1.0 or 10^{1})
11.	12.6W (log 12.6 = 1.1 or $10^{1.1}$)
12.	15.8W (log 15.8 = 1.2 or $10^{1.2}$)
13.	20.0W (log 20 = 1.3 or $10^{1.3}$)
14.	25.1W (log 25.1 = 1.4 or $10^{1.4}$)
15.	31.7W (log 31.7 = 1.5 or $10^{1.5}$)
16.	39.8W (log 39.8 = 1.6 or $10^{1.6}$)
17.	50.2W (log 50.2 = 1.7 or $10^{1.7}$)
18.	63.1W (log 63.1 = 1.8 or $10^{1.8}$)
19.	79.4W (log 79.4 = 1.9 or $10^{1.9}$)
20.	100.0W (log 100 = 2.0 or 10^{2})

TONE GENERATOR

SPEAKER

LISTENER

Pure Tone is applied to speaker—volume is adjusted to gradually increase power from 1 watt to 100 watts.

(a)

Listener is asked to sense change in power level.

MICROPHONE
Also senses change in power level.

Sound reference

Sound sensed as twice as loud

Sound sensed as three times as loud

(b)

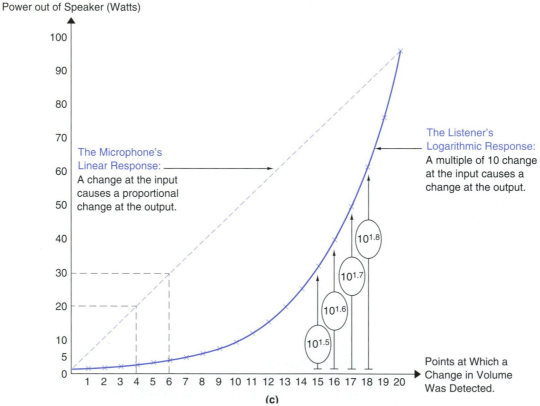

Power out of Speaker (Watts)

The Microphone's Linear Response: A change at the input causes a proportional change at the output.

The Listener's Logarithmic Response: A multiple of 10 change at the input causes a change at the output.

$10^{1.8}$

$10^{1.7}$

$10^{1.6}$

$10^{1.5}$

Points at Which a Change in Volume Was Detected.

(c)

FIGURE 7-4   The Human Ear's Logarithmic Response to Changes in Sound Level.

FIGURE 7-5   **A Graphing Calculator.**

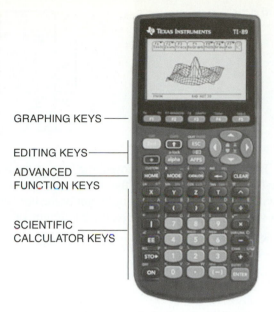

GRAPHING KEYS

EDITING KEYS

ADVANCED
FUNCTION KEYS

SCIENTIFIC
CALCULATOR KEYS

The logarithmic response of the human ear is very different from the **linear response** of the microphone, which is also shown in the graph in Figure 7-4(c). By definition, a linear change at the input causes a proportional change at the output. With a linear response, each change in the sound level applied to the microphone input will result in a proportional change in the microphone's output signal. For example, when the sound level is increased from 20 W to 30 W, the increase on the horizontal scale will be from point 4 to point 6 (a 10 W change at the input results in a 2-point change at the output). Similarly, when the sound level is increased from 70 W to 80 W, the increase on the horizontal scale will be from point 14 to point 16 (a 10 W change at the input will again result in a 2-point change at the output).

Linear Response

Having or being a response or output that is directly proportional to the input.

7-2-4 *The Graphing Calculator*

Up to this point we have been using a standard scientific calculator to perform mathematical operations. Graphing calculators possess all the functions of the scientific calculator, with the added ability that they can be used to graph functions. An example of a typical graphing calculator is shown in Figure 7-5.

The keyboard is divided into four zones: *graphing keys* access the calculator's interactive graphing features, *editing keys* allow you to edit expressions and values, *advanced function keys* display menus that access the advanced functions, and *scientific calculator keys* access the standard capabilities discussed previously.

To explain why we would need graphing capabilities, let us go straight to an application example.

■ INTEGRATED MATH APPLICATION:

Figure 7-6 shows a 20-cm $\times$ 25-cm sheet of cardboard with two $X \times X$ corners cut out of the left side, and two $X \times$ 12.5-cm rectangles cut out of the right side. The cardboard can then be folded into a box with a lid that will have a volume that is equal to:

$$\boxed{V = A \times B \times X}$$

Volume = length $\times$ width $\times$ height

where $A = 20 - 2X$
$B = 25/2 - X$

FIGURE 7-6 A Cardboard Box.

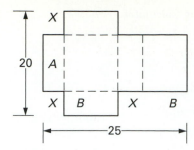

If a small value of X is chosen, the base of the box will be large, but the sides or height of the box would be small. On the other hand, if a large value of X is chosen, the base of the box will be small, but the sides or height of the box will be large. The question, therefore, is, What value of X will give the box the maximum value of volume?

■ *Solution:*

If $V = A \times B \times X$, and $A = 20 - 2X$, and $B = 25/2 - X$, then

$$V = (20 - 2X) \times (25/2 - X) \times (X)$$

To enter this equation into the graphing calculator, you would use the "$Y = $ Editor" function and input the formula as shown in Figure 7-7(a). Now, selecting the calculator's "Table" function will determine the volume of the box ($Y1$) for every possible value of X, as shown in Figure 7-7(b). Scanning down the table shown in Figure 7-7(c), you can see that the maximum value of $Y1$ or volume is 410.26, which occurs when $X = 3.7$. Pressing the calculator's "Graph" function will display the plotted data in graphical rather than tabular form, as seen in Figure 7-7(d).

This ability of the graphing calculator to display every possible output result for every possible input value can be extremely helpful in the analysis of a problem.

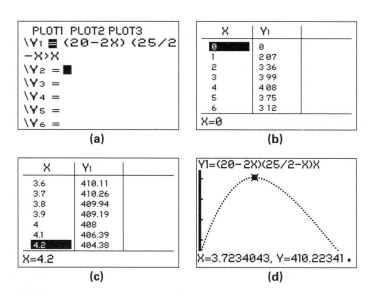

FIGURE 7-7 **Display Results for Box Volume.**

Now that you have completed this section, you should be able to:

- **Objective 5.** *Define the purpose of a graph.*
- **Objective 6.** *Describe how to interpret the data displayed in a graph.*
- **Objective 7.** *Explain the differences among a line graph, a bar graph, and a circle graph or pie chart.*
- **Objective 8.** *Show how the data collected from an experiment are plotted on a graph.*
- **Objective 9.** *Describe the difference between a logarithmic response and a linear response.*
- **Objective 10.** *Explain how the graphing calculator can be used to find the solution to a problem.*

Use the following questions to test your understanding of Section 7-2.

1. Plot a line graph to display the following data:

The Number of Indoor and Outdoor Movie Screens

YEAR	INDOOR	OUTDOOR
1975	12,000	4000
1980	14,000	4000
1985	18,000	3000
1990	23,000	2000

2. Draw a circle graph and bar graph to indicate the following data: Each 24 h day Tom works for 8 h, studies for 3 h, showers and dresses for 2 h, eats for 1.5 h, sleeps for 8 h, and exercises for 1.5 h.

3. What is the difference between a standard scientific calculator and a graphing calculator?

4. Describe the purpose for each of the following graphing calculator steps:
   a. $Y =$ Editor
   b. Table
   c. Graph

REVIEW QUESTIONS

Multiple Choice Questions

1. Logarithms were first discovered by:
   a. Carl Gauss
   c. Isaac Newton
   b. John Napier
   d. Blaise Pascal

2. Natural logarithms use a base of _____, while common logarithms use a base of _____.
   a. 10, 100
   b. 10, 2.71828
   c. 2.71828, 100
   d. 2.71828, 10

3. What is the log of 10,000?
   a. 9.21
   c. 4
   b. 6
   d. 3

4. What is the log of 0.32?
   a. −0.5
   c. −1.5
   b. −1.1
   d. −0.15

5. What is the common antilogarithm of 10^6?
   a. 1,000,000
   c. 1000
   b. 1000.000
   d. 100,000

6. A_____ is a diagram showing the relationship between two or more factors.
   a. Scale
   c. Graph
   b. Logarithm
   d. Exponent

7. In a _____ _____, parallel columns of different sizes are used to compare one variable with others.
   a. Line graph
   c. Circle graph
   b. Bar graph
   d. All the above

8. In a _____ _____, the data are represented by points joined by line segments.
   a. Line graph
   c. Circle graph
   b. Bar graph
   d. All the above

9. In a _____ _____, different size segments within a circle are used to compare one variable with others.
   a. Line graph
   c. Pie chart
   b. Bar graph
   d. All the above

10. With a _____ response, a multiple-of-10 change at the input causes a change at the output, whereas with a _____ response, a change at the input causes a proportional change at the output.
   a. Linear, logarithmic
   b. Logarithmic, linear
   c. Logarithmic, exponential
   d. Linear, exponential

Communication Skill Questions

11. What is a common logarithm? (7–1)

12. What is the difference between a natural logarithm and a common logarithm? (7–1)

13. What is a common antilogarithm? (7–1)

14. State the four-step procedure you would use to simplify multiplication using logarithms. (7–1)

15. Arbitrarily choose values, and then describe the following:
  a. Multiplication using logarithms
  b. Division using logarithms

16. What is a graph? (7–2)

17. List some of the different types of graphs. (7–2)

18. Sketch a diagram and describe the different types of graphs. (7–2)

19. What is a linear response? (7–2)

20. What is a logarithmic response? (7–2)

Practice Problems

21. Determine the log of the following:

a. 10	**g.** 0.5
b. 10,000	**h.** 0.75
c. 1,234,567	**i.** 0.125
d. 50.5	**j.** 0.95
e. 1.5	**k.** 12.25
f. 0.25	**l.** 75.75

22. Determine the antilog of the following:

a. 1	**g.** 2.5
b. 2	**h.** 1.25
c. 3	**i.** 0.125
d. 4	**j.** 0.75
e. 5	**k.** 0.33
f. 1.5	**l.** 0.95

23. What steps should be followed to simplify multiplication to addition using logarithms?

24. Apply the four-step procedure from Problem 23 to the following:

  a. $2 \times 2 = ?$
  b. $244 \times 30 = ?$
  c. $1.25 \times 150 = ?$
  d. $0.5 \times 0.33 = ?$

25. What steps should be followed to simplify division to subtraction using logarithms?

26. Apply the four-step procedure from Problem 25 to the following:

  a. $12 \div 2 = ?$
  b. $244 \div 2 = ?$
  c. $250 \div 0.25 = ?$
  d. $1255 \div 0.5 = ?$

27. Identify the types of graphs shown in Figure 7-8.

28. Refer to Figure 7-8(a), and answer the following questions:

  a. Which music type has remained the most popular?
  b. Which music type is the least popular?
  c. Which music type became less popular as country music had an increase in popularity in 1992 and 1993?
  d. What was the approximate market share for each of the music types in 1995?

29. Refer to Figure 7-8(b), and answer the following questions:

  a. What is the main reason for the average person to connect to the Internet?
  b. In order, what are the top five reasons why people want to go online?

30. Figure 7-9 shows an electronic device called a *potentiometer*. This device has two terminals (*A* and *B*), a circular resistive track, and a wiper that can be placed at any point on the resistive track by turning a control shaft. (Potentiometers are often used as volume controls for music systems and televisions.) There are two basic types of potentiometers: those with linear resistive tracks, and those with tapered or nonlinear resistive tracks. Figure 7-9 lists the resistance values that were measured between terminals *A* and *B* for both a linear and a tapered potentiometer. When the control shaft was fully counterclockwise (CCW), the resistance between terminals *A* and *B* for both the linear and tapered potentiometer was 0 ohms (0 Ω). When the control shaft was turned one-quarter clockwise (¼ CW), the resistance between terminals *A* and *B* for the linear potentiometer was 250 Ω, and for the tapered potentiometer it was 350 Ω. When the control shaft was turned one-half clockwise (½ CW), the resistance between terminals *A* and *B* for the linear potentiometer was 500 Ω, and for the tapered potentiometer it was 625 Ω. When the control shaft was turned three-quarters clockwise (¾ CW), the resistance between terminals *A* and *B* for the linear potentiometer was 750 Ω, and for the tapered potentiometer it was 900 Ω. Finally, when the control shaft was turned fully clockwise (CW), the resistance between terminals *A* and *B* for the linear potentiometer was 1000 Ω (1 kΩ), and for the tapered potentiometer it was also 1000 Ω. Sketch a "resistance versus control shaft position" line graph, and then plot a linear line and tapered line.

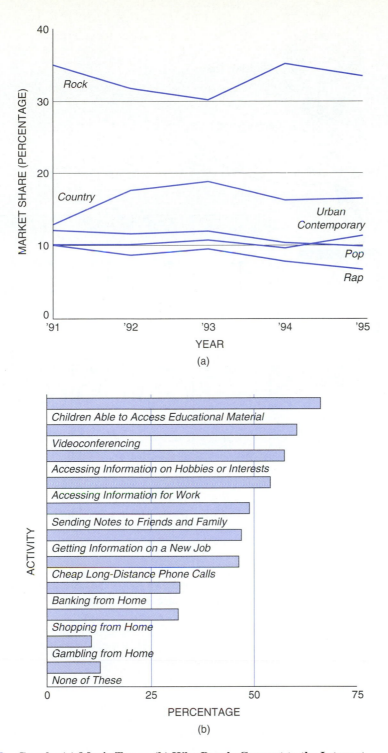

FIGURE 7-8 Graphs (a) Music Types. (b) Why People Connect to the Internet.

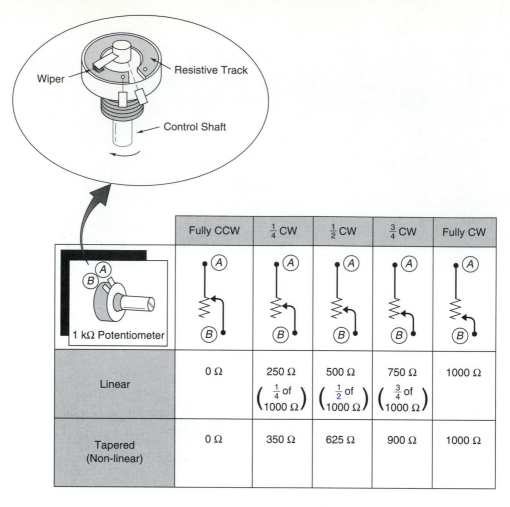

FIGURE 7-9 **A Linear and Tapered Resistive Track Within a Potentiometer.**

Web Site Questions

Go to the web site http://www.prenhall.com/cook, select the textbook *Mathematics for Electronics and Computers,* select this chapter, and then follow the instructions when answering the multiple-choice practice problems.

Current and Voltage

Problem Solver

Charles Proteus Steinmetz (1865–1923) was an outstanding electrical genius who specialized in mathematics, electrical engineering, and chemistry. His three greatest electrical contributions were his investigation and discovery of the law of hysteresis; his investigations in lightning, which resulted in his theory on traveling waves; and his discovery that complex numbers could be used to solve ac circuit problems. Solving problems was in fact his specialty, and on one occasion he was commissioned to troubleshoot a failure on a large company system that no one else had been able to repair. After studying the symptoms and schematics for a short time, he chalked an X on one of the metal cabinets, saying that this was where they would find the problem, and left. He was right, and the problem was remedied to the relief of the company executives; however, they were not pleased when they received a bill for $1000. When they demanded that Steinmetz itemize the charges, he replied—$1 for making the mark and $999 for knowing where to make the mark.

The strong message this vignette conveys is that you will get $1 for physical labor and $999 for mental labor, and this is a good example as to why you should continue in your pursuit of education.

8

Outline and Objectives

VIGNETTE: PROBLEM SOLVER

INTRODUCTION

8-1 CURRENT

Objective 1: Define *electrical current.*

8-1-1 Coulombs per Second

Objective 2: Describe the ampere in relation to coulombs per second.

8-1-2 The Ampere

8-1-3 Units of Current

Objective 3: List the different current units and values.

8-1-4 Conventional versus Electron Flow

Objective 4: Describe the difference between conventional current flow and electron flow.

8-1-5 Measuring Current

Objective 5: List the rules to apply when measuring current.

8-2 VOLTAGE

Objective 6: Define *electrical voltage.*

8-2-1 Symbols

8-2-2 Units of Voltage

Objective 7: List the various voltage units and values.

8-2-3 Measuring Voltage

Objective 8: List the rules to apply when measuring voltage.

8-3 CONDUCTORS

Objective 9: Explain the difference between:
   a. A conductor
   b. An insulator

8-3-1 Conductance

Objective 10: Describe conductance.

8-4 INSULATORS

Objective 11: Explain what makes a good:
   a. Conductor
   b. Insulator

MULTIPLE CHOICE QUESTIONS

COMMUNICATION SKILL QUESTIONS

PRACTICE PROBLEMS

WEB SITE QUESTIONS

Introduction

Now that you have completed the basic math section of this text, we will concentrate on seeing how math is employed in electronics. Direct current, alternating current, and electronic devices and circuits will all be introduced in the next 10 chapters along with all associated mathematics. This foundation in applied mathematics will fully prepare you for your upcoming course in electronics.

8-1   CURRENT

Electron

Smallest subatomic particle of negative charge that orbits the nucleus of the atom.

Positive Ion

Atom that has lost one or more of its electrons and therefore has more protons than electrons, resulting in a net positive charge.

Negative Ion

Atom that has more than the normal neutral amount of electrons.

Positive Charge

The charge that exists in a body that has fewer electrons than normal.

Negative Charge

An electric charge that has more electrons than protons.

The movement of **electrons** from one point to another is known as *electrical current.* Energy in the form of heat or light can cause an outer shell electron to be released from the valence shell of an atom. Once an electron is released, the atom is no longer electrically neutral and is called a **positive ion,** as it now has a net positive charge (more protons than electrons). The released electron tends to jump into a nearby atom, which will then have more electrons than protons and is referred to as a **negative ion.**

Let us now take an example and see how electrons move from one point to another. Figure 8-1 shows a broken metal conductor between two charged objects. The metal conductor can be gold, silver, or copper, but whichever it is, one common trait can be noted: the valence electrons in the outermost shell or orbit of the atom are very loosely bound and can easily be pulled from their parent atom.

In Figure 8-2, the conductor between the two charges has been joined so that a path now exists for current flow. The negative ions on the right in Figure 8-2 have more electrons than protons, whereas the positive ions on the left in Figure 8-2 have fewer electrons than protons and so display a **positive charge.** The metal joining the two charges has its own atoms, which begin in the neutral condition.

Let us now concentrate on one of the negative ions. In Figure 8-2(a), the extra electrons in the outer shells of the negative ions on the right side will feel the attraction of the positive ions on the left side and the repulsion of the outer negative ions, or **negative charge.** This will cause an electron in a negative ion to jump away from its parent atom's orbit and land in an adjacent atom to the left within the metal wire conductor, as shown in Figure 8-2(b). This adjacent atom now has an extra electron and is called a negative ion, and

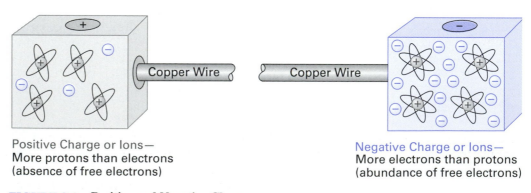

Positive Charge or Ions—
More protons than electrons
(absence of free electrons)

Negative Charge or Ions—
More electrons than protons
(abundance of free electrons)

FIGURE 8-1   Positive and Negative Charges.

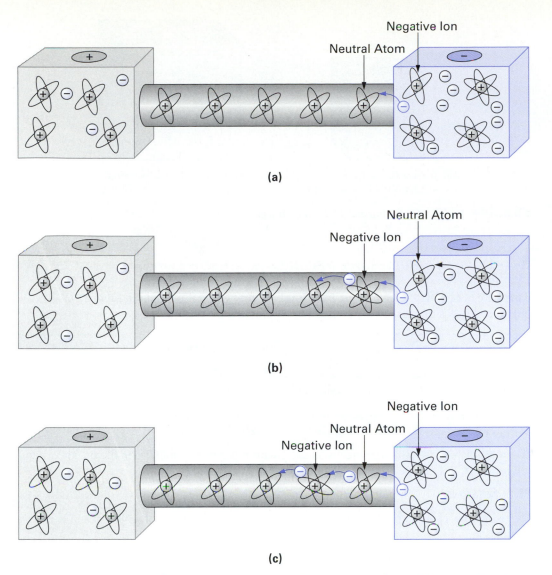

(a)

(b)

(c)

FIGURE 8-2 **Electron Migration Due to Forces of Positive Attraction and Negative Repulsion on Electrons.**

TIME LINE

Ironically, Coulomb's law was not first discovered by Charles A. de Coulomb (1736–1806) but by Henry Cavendish, a wealthy scientist and philosopher. Cavendish did not publish his discovery, which he had made several years before Coulomb discovered the law independently. James Clerk Maxwell published the scientific notebooks of Cavendish in 1879, describing his experiments and conclusions. However, about 100 years had passed, and Coulomb's name was firmly associated with the law. Many scientists demanded that the law be called Cavendish's law, while other scientists refused to change, stating that Coulomb was the discoverer because he made the law known promptly to the scientific community.

the initial parent negative ion becomes a neutral atom, which will now receive an electron from one of the other negative ions, because their electrons are also feeling the attraction of the positive ions on the left side and the repulsion of the surrounding negative ions.

The electrons of the negative ion within the metal conductor feel the attraction of the positive ions, and eventually one of its electrons jumps to the left and into the adjacent atom, as shown in Figure 8-2(c). This continual movement to the left will produce a stream of electrons flowing from right to left. Millions upon millions of atoms within the conductor pass a continuous movement of billions upon billions of electrons from right to left. This electron flow is known as **electric current.**

To summarize, we could say that as long as a force or pressure, produced by the positive charge and negative charge, exists it will cause electrons to flow from the negative to the positive terminal. The positive side has a deficiency of electrons and the negative side has an abundance, and so a continuous flow or migration of electrons takes place between the negative and positive terminal through our metal conducting wire. This electric current or electron flow is a measurable quantity, as will now be explained.

Electric Current (I)

Measured in amperes or amps, it is the flow of electrons through a conductor.

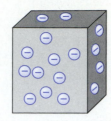 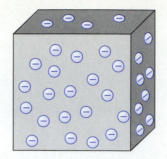

(a) 6.24×10^{18} Electrons
= 1 Coulomb of Charge

(b) 12.48×10^{18} Electrons
= 2 Coulombs of Charge

FIGURE 8-3 **(a) 1 C of Charge; (b) 2 C of Charge.**

8-1-1 *Coulombs per Second*

Coulomb of Charge
Unit of electric charge.
One coulomb equals 6.24
$\times 10^{18}$ electrons.

There are 6.24×10^{18} electrons in 1 **coulomb of charge,** as illustrated in Figure 8-3. To calculate coulombs of charge (designated Q), we can use the formula

$$\text{charge, } Q = \frac{\text{total number of electrons } (n)}{6.24 \times 10^{18}}$$

where Q is the electric charge in coulombs.

■ **INTEGRATED MATH APPLICATION:**

If a total of 3.75×10^{19} free electrons exist within a piece of metal conductor, how many coulombs (C) of charge are within this conductor?

■ *Solution:*

By using the charge (Q) formula we can calculate the number of coulombs (C) in the conductor.

$$\begin{aligned} Q &= \frac{n}{6.24 \times 10^{18}} \\ &= \frac{3.75 \times 10^{19}}{6.24 \times 10^{18}} \\ &= 6 \text{ C} \end{aligned}$$

A total of 6 C of charge exists within the conductor.

CALCULATOR SEQUENCE

Step	Keypad Entry	Display Response
1.	③ . ⑦ ⑤ E (Exponent) ① ⑨	3.75E19
2.	÷	
3.	⑥ . ② ④ E ① ⑧	6.24E18
4.	=	6.0096

In the calculator sequence you will see how the exponent key (E, EE, or EXP) on your calculator can be used.

8-1-2 The Ampere

A coulomb is a **static** or stationary amount of electric charge. In electronics, we are more interested in electrons in motion. Coulombs and time are therefore combined to describe the number of electrons and the rate at which they flow. This relationship is called *current* (I) flow and has the unit of **amperes (A).** By definition, 1 ampere of current is said to be flowing if 6.24×10^{18} electrons (1 C) are drifting past a specific point on a conductor in 1 second of time. Stated as a formula:

$$\text{Current } (I) = \frac{\text{coulombs } (Q)}{\text{time } (t)}$$

$$1 \text{ ampere } = 1 \text{ coulomb per 1 second}$$
$$1 \text{ A} = \frac{1 \text{ C}}{1 \text{ s}}$$

In summary, 1 ampere equals a flow rate of 1 coulomb per second, and current is measured in amperes.

TIME LINE

The unit of electrical current is the ampere, named in honor of André Ampère (1775–1835), a French physicist who pioneered in the study of electromagnetism. After hearing of Hans Oerstad's discoveries, he conducted further experiments and discovered that two current-carrying conductors would attract and repel one another, just like two magnets.

■ INTEGRATED MATH APPLICATION:

If 5×10^{19} electrons pass a point in a conductor in 4 s, what is the amount of current flow in amperes?

■ *Solution:*

Current (I) is equal to Q/t. We must first convert electrons to coulombs.

$$\begin{aligned} Q &= \frac{n}{6.24 \times 10^{18}} \\ &= \frac{5 \times 10^{19}}{6.24 \times 10^{18}} \\ &= 8 \text{ C} \end{aligned}$$

Static

Crackling noise heard on radio receivers caused by electric storms or electric devices in the vicinity.

Ampere (A)

Unit of electric current.

Now, to calculate the amount of current, we use the formula

$$\begin{aligned} I &= \frac{Q}{t} \\ &= \frac{8 \text{ C}}{4 \text{ s}} \\ &= 2 \text{ A} \end{aligned}$$

This means that 2 A or 1.248×10^{19} electrons (2 C) pass a specific point in the conductor every second.

CALCULATOR SEQUENCE

Step	Keypad Entry	Display Response
1.	⑤ Ⓔ (exponent) ① ⑨	5E19
2.	÷	
3.	⑥ . ② ④ Ⓔ ① ⑧	6.24E18
4.	=	8.012
5.	÷	
6.	④	2.003
7.	=	2.003

8-1-3 Units of Current

Current within electronic equipment is normally a value in milliamperes or microamperes and very rarely exceeds 1 ampere. Table 8-1 lists all the prefixes related to current. For example, 1 milliampere is one-thousandth of an ampere, which means that if 1 ampere were divided into 1000 parts, 1 part of the 1000 parts would be flowing through the circuit.

TABLE 8-1   Current Units

NAME	SYMBOL	VALUE
Picoampere	pA	$10^{-12} = \dfrac{1}{1,000,000,000,000}$
Nanoampere	nA	$10^{-9} = \dfrac{1}{1,000,000,000}$
Microampere	μA	$10^{-6} = \dfrac{1}{1,000,000}$
Milliampere	mA	$10^{-3} = \dfrac{1}{1000}$
Ampere	A	$10^{0} = 1$
Kiloampere	kA	$10^{3} = 1000$
Megaampere	MA	$10^{6} = 1,000,000$
Gigaampere	GA	$10^{9} = 1,000,000,000$
Teraampere	TA	$10^{12} = 1,000,000,000,000$

■ **INTEGRATED MATH APPLICATION:**

Convert the following:

    a. 0.0075 A = _____ mA (milliamperes)

    b. 0.04 mA = _____ μA (microamperes)

    c. 7333 mA = _____ A (amperes)

    d. 1275 μA = _____ mA (milliamperes)

■ *Solution:*

    a. 0.0075 A = _____ mA. In this example, 0.0075 A has to be converted so that it is represented in milliamperes (10^{-3} or $^1/_{1000}$ of an ampere). The basic algebraic rule to be remembered is that the expressions on both sides of the equals sign must be equal.

LEFT		RIGHT	
Base	Multiplier	Base	Multiplier
0.0075×10^{0}	=	_____	$\times 10^{-3}$

The multiplier on the right in this example is going to be decreased 1000 times (10^{0} to 10^{-3}), so for the statement to balance, the number on the right will have to be increased 1000 times; that is, the decimal point will have to be moved to the right three places (0.0075 or 7.5). Therefore,

$$0.0075 \times 10^{0} = 7.5 \times 10^{-3}$$

or

$$0.0075 \text{ A} = 7.5 \times 10^{-3} \text{ A or } 7.5 \text{ mA}$$

    b. 0.04 mA = _____ μA. In this example the unit is going from milliamperes to microamperes (10^{-3} to 10^{-6}) or 1000 times smaller, so the number must be made 1000 times greater.

$$0.040 \text{ or } 40.0$$

Therefore, 0.04 mA = 40 μA.

c. 7333 mA = _____ A. The unit is going from milliamperes to amperes, increasing 1000 times, so the number must decrease 1000 times.

$$7\overset{\frown}{3}\overset{\frown}{3}\overset{\frown}{3}. \text{ or } 7.333$$

Therefore, 7333 mA = 7.333 A.

d. 1275 μA = _____ mA. The unit is changing from microamperes to milliamperes, an increase of 1000 times, so the number must decrease by the same factor.

$$1\overset{\frown}{2}\overset{\frown}{7}\overset{\frown}{5}.0 \text{ or } 1.275$$

Therefore, 1275 μA = 1.275 mA.

8-1-4   *Conventional versus Electron Flow*

Electrons drift from a negative to a positive charge, as illustrated in Figure 8-4. As already stated, this current is known as **electron flow.**

In the eighteenth and nineteenth centuries, when very little was known about the atom, researchers believed that current was a flow of positive charges. Although this has now been proved incorrect, many texts still use **conventional current flow,** which is shown in Figure 8-5.

Whether conventional flow or electron flow is used, the same answers to problems, measurements, and designs are obtained. The key point to remember is that direction is not important, but the amount of current flow is.

Throughout this book we will be using electron flow so that we can relate back to the atom when necessary. If you wish to use conventional flow, just reverse the direction of the arrows. To avoid confusion, be consistent with your choice of flow.

8-1-5   *Measuring Current*

Ammeters (ampere meters) are used to measure the current flow within a circuit. Stepping through the sequence detailed in Figure 8-6, you will see how an ammeter is used to measure the value of current within a circuit.

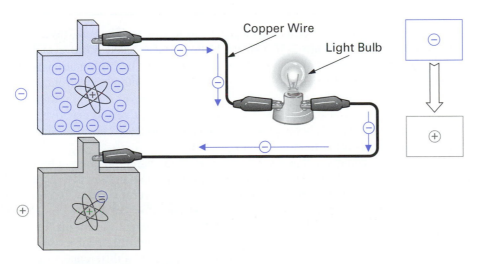

FIGURE 8-4   Electron Current Flow.

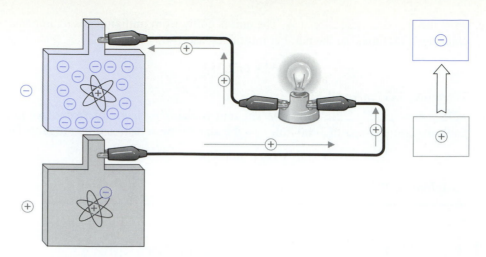

FIGURE 8-5 **Conventional Current Flow.**

In the simple circuit shown in Figure 8-6, an ON/OFF switch is being used to turn ON or OFF a small lightbulb. One of the key points to remember with ammeters is that if you wish to measure the value of current flowing within a wire, the current path must be opened and the ammeter placed in the path so that it can sense and display the value.

SELF-TEST EVALUATION POINT FOR SECTION 8-1

Now that you have completed this section, you should be able to:

■ **Objective 1.** *Define electrical current.*

■ **Objective 2.** *Describe the ampere in relation to coulombs per second.*

■ **Objective 3.** *List the different current units and values.*

■ **Objective 4.** *Describe the difference between conventional current flow and electron flow.*

■ **Objective 5.** *List the rules to apply when measuring current.*

Use the following questions to test your understanding of Section 8-1:

1. What is the unit of current?
2. Define current in relation to coulombs and time.
3. What test instrument is used to measure current?

8-2 VOLTAGE

Electromotive Force (emf)

Force that causes the motion of electrons due to a potential difference between two points.

Potential Difference (PD)

Voltage difference between two points, which will cause current to flow in a closed circuit.

Voltage is the force or pressure exerted on electrons. Referring to Figure 8-7(a) and (b), you will notice two situations. Figure 8-7(a) shows highly concentrated positive and negative charges or potentials connected to each other by a copper wire. In this situation, a large potential difference or voltage is being applied across the copper atom's electrons. This force or voltage causes a large amount of copper atom electrons to move from right to left. On the other hand, Figure 8-7(b) illustrates a low concentration of positive and negative potentials, so a small voltage or pressure is being applied across the conductor, causing a small amount of force, and therefore current, to move from right to left.

In summary, we could say that a highly concentrated charge produces a high voltage, whereas a low concentrated charge produces a low voltage. Voltage is also appropriately known as the "electron moving force" or **electromotive force (emf),** and since two opposite potentials exist (one negative and one positive), the strength of the voltage can also be referred to as the amount of **potential difference (PD)** applied across the circuit. To compare, we can say that a large voltage, electromotive force, or potential difference exists across the copper conductor in Figure 8-7(a), while a small voltage, potential difference, or electromotive force is exerted across the conductor in Figure 8-7(b).

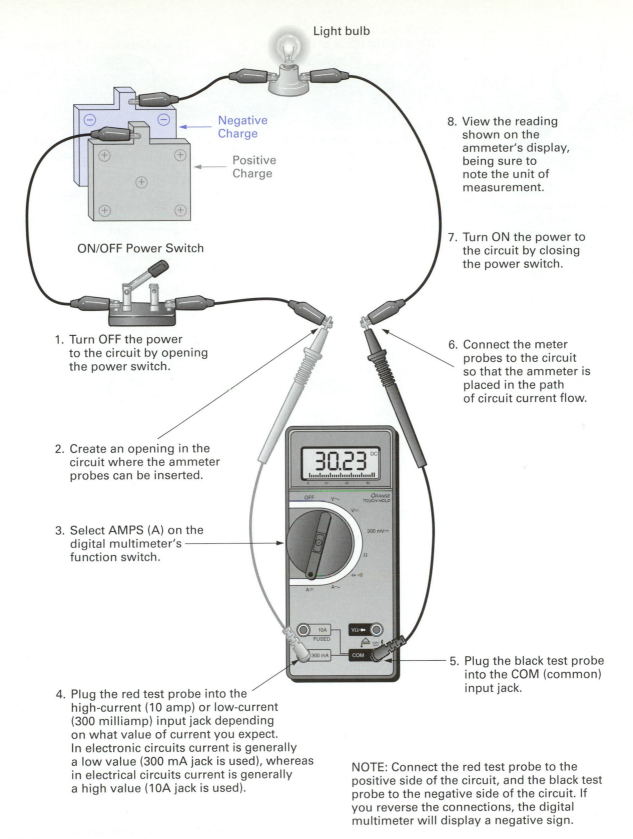

Light bulb

Negative Charge

Positive Charge

ON/OFF Power Switch

8. View the reading shown on the ammeter's display, being sure to note the unit of measurement.

7. Turn ON the power to the circuit by closing the power switch.

1. Turn OFF the power to the circuit by opening the power switch.

6. Connect the meter probes to the circuit so that the ammeter is placed in the path of circuit current flow.

2. Create an opening in the circuit where the ammeter probes can be inserted.

3. Select AMPS (A) on the digital multimeter's function switch.

30.23 DC

OFF   V~   ORANGE TOUCH HOLD

V⎓

300 mV⎓

Ω

⸫ ⸨)

A⎓   A~

10A FUSED   VΩ⸫⸢

300 mA   COM

5. Plug the black test probe into the COM (common) input jack.

4. Plug the red test probe into the high-current (10 amp) or low-current (300 milliamp) input jack depending on what value of current you expect. In electronic circuits current is generally a low value (300 mA jack is used), whereas in electrical circuits current is generally a high value (10A jack is used).

NOTE: Connect the red test probe to the positive side of the circuit, and the black test probe to the negative side of the circuit. If you reverse the connections, the digital multimeter will display a negative sign.

FIGURE 8-6   **Measuring Current with an Ammeter.**

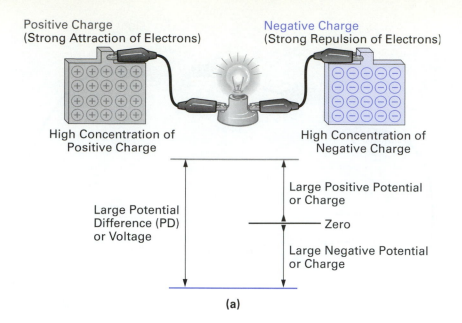

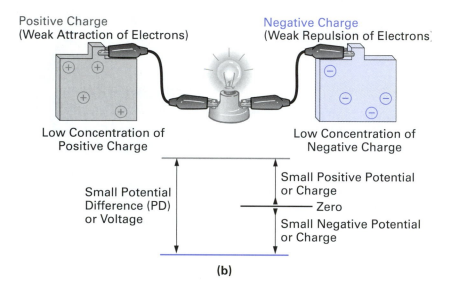

FIGURE 8-7 **(a) Large Potential Difference or Voltage. (b) Small Potential Difference or Voltage.**

Voltage is the force, pressure, potential difference (PD), or electromotive force (emf) that causes electron flow or current and is symbolized by italic uppercase *V*. The unit for voltage is the volt, symbolized by roman uppercase V. This can become a bit confusing. For example, when the voltage applied to a circuit equals 5 volts, the circuit notation would appear as

$$V = 5 \text{ V}$$

You know the first V represents "voltage," not "volt," because 1 volt cannot equal 5 volts. To avoid confusion, some texts and circuits use *E,* symbolizing electromotive force, to represent voltage; for example,

$$E = 5 \text{ V}$$

In this text, we will maintain the original designation for voltage (*V*).

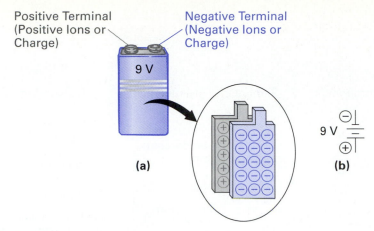

FIGURE 8-8   **The Battery—A Source of Voltage. (a) Physical Appearance. (b) Schematic Symbol.**

8-2-1   *Symbols*

A **battery,** like the one shown in Figure 8-8(a), converts chemical energy into electrical energy. At the positive terminal of the battery, positive charges or ions (atoms with more protons than electrons) are present, and at the negative terminal, negative charges or ions (atoms with more electrons than protons) are available to supply electrons for current flow within a circuit. A battery, therefore, chemically generates negative and positive ions at its respective terminals. The symbol for the battery is shown in Figure 8-8(b).

In Figure 8-9(a) you can see the schematic symbols for many of the devices discussed so far, along with their physical appearance.

In the circuit shown in Figure 8-9(b), a 9 volt battery chemically generates positive and negative ions. The negative ions at the negative terminal force away the negative electrons, which are attracted by the positive charge or absence of electrons at the positive terminal. As the electrons proceed through the copper conductor wire, jumping from one atom to the next, they eventually reach the bulb. As they pass through the bulb, they cause it to glow. When emerging from the lightbulb, the electrons travel through another connector cable and finally reach the positive terminal of the battery.

Studying Figure 8-9(b), you will notice two reasons why the circuit is drawn using symbols rather than illustrating the physical appearance:

1. A circuit with symbols can be drawn faster and more easily.
2. A circuit with symbols has less detail and clutter and is therefore more easily comprehended, since it has fewer distracting elements.

Battery
DC voltage source containing two or more cells that converts chemical energy into electrical energy.

8-2-2   *Units of Voltage*

The unit for voltage is the volt (V). Voltage within electronic equipment is normally measured in volts, whereas heavy-duty industrial equipment normally requires high voltages that are generally measured in kilovolts (kV). Table 8-2 lists all the prefixes and values related to volts.

■ INTEGRATED MATH APPLICATION:

Convert the following:

a. 3000 V = _____ kV (kilovolts)

b. 0.14 V = _____ mV (millivolts)

c. 1500 kV = _____ MV (megavolts)

Component	Symbol	Name	Description
		Incandescent lamp	Incandescence: release of visible radiation (light) by a heated object
		Connecting wire with end alligator clips	Used to connect different components
9 V		Battery	Source of voltage and current
	OPEN / CLOSED	Switch	A device used to open or close a current path
30.23	AM	Ammeter	Used to measure the current flow within a circuit

(a)

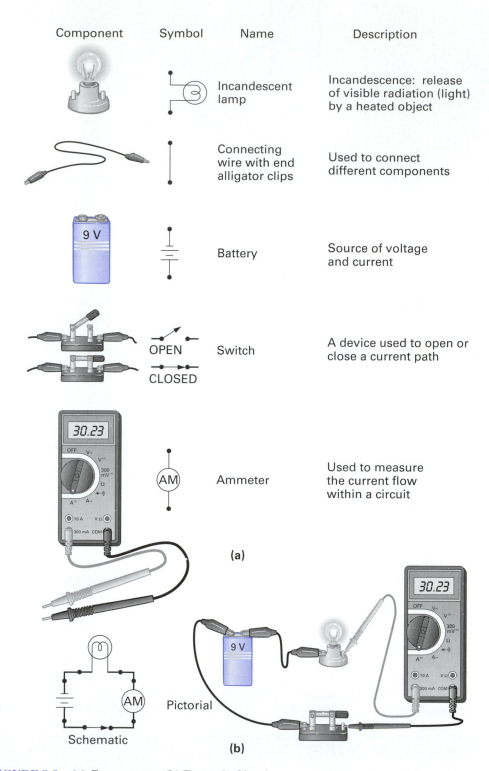

Schematic

Pictorial

(b)

FIGURE 8-9 (a) Components. (b) Example Circuit.

TABLE 8-2 Voltage Units

NAME	SYMBOL	VALUE		
Picovolt	pV	$10^{-12} = \dfrac{1}{1{,}000{,}000{,}000{,}000}$		
Nanovolt	nV	$10^{-9} = \dfrac{1}{1{,}000{,}000{,}000}$		
Microvolt	μV	$10^{-6} = \dfrac{1}{1{,}000{,}000}$		
Millivolt	mV	$10^{-3} = \dfrac{1}{1000}$		
Volt	V	$10^{0} = 1$		
Kilovolt	kV	$10^{3} = 1000$		
Megavolt	MV	$10^{6} = 1{,}000{,}000$		
Gigavolt	GV	$10^{9} = 1{,}000{,}000{,}000$		
Teravolt	TV	$10^{12} = 1{,}000{,}000{,}000{,}000$		

■ *Solution:*

a. 3000 V = 3 kV or 3×10^{3} volts (multiplier ↑ 1000, number ↓ 1000)

b. 0.14 V = 140 mV or 140×10^{-3} volt (multiplier ↓ 1000, number ↑ 1000)

c. 1500 kV = 1.5 MV or 1.5×10^{6} volts (multiplier ↑ 1000, number ↓ 1000)

8-2-3   *Measuring Voltage*

Voltmeter

Instrument designed to measure the voltage or potential difference. Its scale can be graduated in kilovolts, volts, or millivolts.

Voltmeters (voltage meters) are used to measure electrical pressure or voltage. Stepping through the sequence detailed in Figure 8-10, you will see how a voltmeter can be used to measure a battery's voltage.

SELF-TEST EVALUATION POINT FOR SECTION 8-2

Now that you have read this section, you should be able to:

■ *Objective 6.* *Define electrical voltage.*

■ *Objective 7.* *List the various voltage units and values.*

■ *Objective 8.* *List the rules to apply when measuring voltage.*

Use the following questions to test your understanding of Section 8-2:

1. What is the unit of voltage?
2. Convert 3 MV to kilovolts.
3. Which meter is used to measure voltage?

8-3   CONDUCTORS

A lightning bolt that sets fire to a tree and the operation of your calculator are both electrical results achieved by the flow of electrons. The only difference is that your calculator's circuits control the flow of electrons, whereas the lightning bolt is the uncontrolled flow of electrons. In electronics, a **conductor** is used to channel or control the path in which electrons flow.

Any material that passes current easily is called a conductor. These materials are said to have a "low resistance," which means that they will offer very little opposition to current flow.

Conductor

Length of wire whose properties are such that it will carry an electric current.

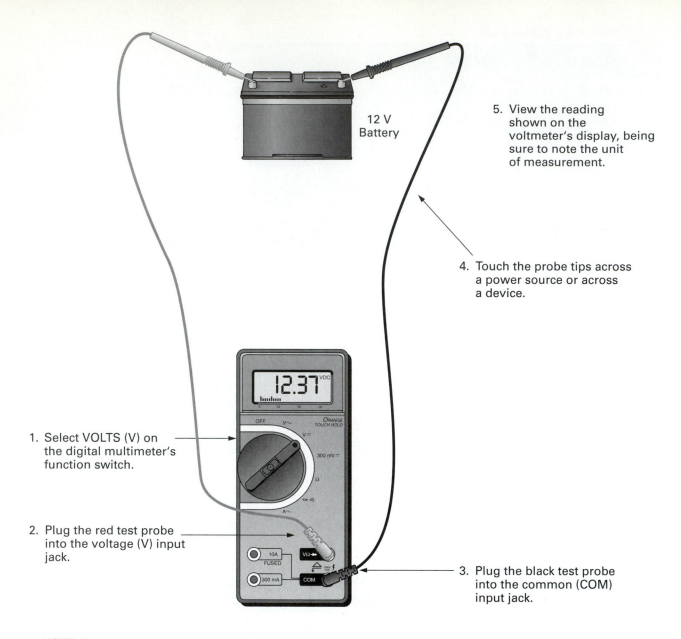

5. View the reading shown on the voltmeter's display, being sure to note the unit of measurement.

4. Touch the probe tips across a power source or across a device.

1. Select VOLTS (V) on the digital multimeter's function switch.

2. Plug the red test probe into the voltage (V) input jack.

3. Plug the black test probe into the common (COM) input jack.

12 V Battery

NOTE: If test leads are reversed, a negative sign will show in the display.

(a)

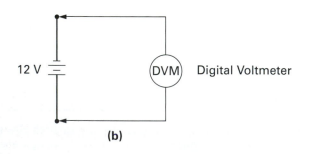

12 V   (DVM)   Digital Voltmeter

(b)

FIGURE 8-10   Using a Voltmeter to Measure Voltage.

8-3-1   *Conductance*

Conductance is the measure of how good a conductor is at carrying current. Conductance (symbolized by G) is equal to the reciprocal of resistance (or opposition) and is measured in the unit siemens (S):

$$\text{Conductance } (G) = \frac{1}{\text{resistance } (R)}$$

Conductance (G) values are measured in siemens (S),
and resistance (R) in ohms (Ω).

This means that conductance is inversely proportional to resistance. For example, if the opposition to current flow (resistance) is low, the conductance is high and the material is said to have a good conductance.

$$\text{High conductance } G \uparrow = \frac{1}{R \downarrow \text{ (low resistance)}}$$

On the other hand, if the resistance of a conducting wire is high ($R \uparrow$), its conductance value is low ($G \downarrow$) and it is called a poor conductor. A good conductor therefore has a high conductance value and a very small resistance to current flow.

■   **INTEGRATED MATH APPLICATION:**

A household electric blanket offers 25 ohms of resistance to current flow. Calculate the conductance of the electric blanket's heating element.

■   *Solution:*

$$\text{Conductance} = \frac{1}{\text{resistance}}$$

$$G = \frac{1}{R} = \frac{1}{25 \text{ ohms}} = 0.04 \text{ siemens}$$

$$\text{or } 40 \text{ mS (millisiemens)}$$

To express how good or poor a conductor is, we must have some sort of reference point. The reference point we use is the best conductor, silver, which is assigned a relative conductivity value of 1.0. Table 8-3 lists other conductors and their relative conductivity values with respect to the best, silver. The formula for calculating relative conductivity is

$$\text{Relative conductivity} = \frac{\text{conductor's relative conductivity}}{\text{reference conductor's relative conductivity}}$$

TABLE 8-3   Relative Conductivity of Conductors

MATERIAL	CONDUCTIVITY (RELATIVE)
Silver	1.000
Copper	0.945
Gold	0.652
Aluminum	0.575
Tungsen	0.297
Nichrome	0.015

TIME LINE

Stephen Gray (1693–1736), an Englishman, discovered that certain substances would conduct electricity.

Conductance (G)

Measure of how well a circuit or path conducts or passes current. It is equal to the reciprocal of resistance.

TIME LINE

Stephen Gray's lead was picked up in 1730 by Charles du Fay, a French experimenter, who believed that there were two types of electricity, which he called *vitreous* and *resinous* electricity.

What is the relative conductivity of tungsten if copper is used as the reference conductor?

■ *Solution:*

CALCULATOR SEQUENCE

Step	Keypad Entry	Display Response
1.	0 . 2 9 7	0.297
2.	÷	
3.	0 . 9 4 5	0.945
4.	=	0.314

$$\text{Tungsten} = 0.297$$
$$\text{Copper} = 0.945$$
$$\text{Relative conductivity} = \frac{\text{conductivity of conductor}}{\text{conductivity of reference}}$$
$$= \frac{0.297}{0.945}$$
$$= 0.314$$

■ INTEGRATED MATH APPLICATION:

What is the relative conductivity of silver if copper is used as the reference?

■ *Solution:*

$$\text{Relative conductivity} = \frac{\text{silver}}{\text{copper}}$$
$$= \frac{1.000}{0.945} = 1.058$$

SELF-TEST EVALUATION POINT FOR SECTION 8-3

Now that you have read this section, you should be able to:

■ **Objective 10.** *Describe conductance.*

Use the following questions to test your understanding of Section 8-3.

1. True or false: A conductor is a material used to block the flow of current.

2. What is the most commonly used conductor in the field of electronics?

3. Calculate the conductance of a 35 ohm heater element.

8-4   INSULATORS

Insulator

A material that has few electrons per atom and those electrons are close to the nucleus and cannot be easily removed.

Breakdown Voltage

The voltage at which breakdown of an insulator occurs.

Dielectric Strength

The maximum potential a material can withstand without rupture.

Any material that offers a high resistance or opposition to current flow is called an **insulator.** Conductors permit the easy flow of current and so have good conductivity, whereas insulators allow a small to almost no amount of free electrons to flow. Insulators can, with sufficient pressure or voltage applied across them, "break down" and conduct current. This **breakdown voltage** must be great enough to dislodge the electrons from their close orbital shells (K, L shells) and release them as free electrons.

Insulators are also called "dielectrics," and the best insulator or dielectric should have the maximum possible resistance and conduct no current at all. The **dielectric strength** of an insulator indicates how good or bad an insulator is by indicating the voltage that will cause the insulating material to break down and conduct a large current. Table 8-4 lists some of the more popular insulators and the value of kilovolts that will cause a centimeter of insulator to break down. As an example, if 1 centimeter of paper is connected to a variable voltage source, a voltage of 500 kilovolts is needed to break down the paper and cause current to flow.

TABLE 8-4 Breakdown Voltages of Certain Insulators

MATERIAL	BREAKDOWN STRENGTH (kV/cm)
Mica	2000
Glass	900
Teflon	600
Paper	500
Rubber	275
Bakelite	151
Oil	145
Porcelain	70
Air	30

The following formula can be used to calculate the dielectric thickness needed to withstand a certain voltage.

$$\text{Dielectric thickness} = \frac{\text{voltage to insulate}}{\text{insulator's breakdown voltage}}$$

■ **INTEGRATED MATH APPLICATION:**

What thickness of mica would be needed to withstand 16,000 V?

■ *Solution:*

$$\text{Mica strength} = 2000 \text{ kV/cm}$$
$$\text{Dielectric thickness} = \frac{16,000 \text{ V}}{2000 \text{ kV/cm}} = 0.008 \text{ cm}$$

CALCULATOR SEQUENCE

Step	Keypad Entry	Display Response
1.	[1] [6] [E] (exponent) [3]	16E3
2.	[÷]	
3.	[2] [E] [6] (or 2000 E3)	2E6
4.	[=]	0.008

■ **INTEGRATED MATH APPLICATION:**

What maximum voltage could 1 mm of air withstand?

■ *Solution:*

There are 10 mm in 1 cm. If air can withstand 30,000 V/cm, it can withstand 3000 V/mm.

SELF-TEST EVALUATION POINT FOR SECTION 8-4

Now that you have completed this section, you should be able to:

■ *Objective 11. Explain what makes a good:*
  a. Conductor
  b. Insulator

Use the following questions to test your understanding of Section 8-4:

1. True or false: An insulator is a material used to block the flow of current.

2. What is considered to be the best insulator material?

3. Define *breakdown voltage*.

4. Would the conductance figure of a good insulator be large or small?

Multiple Choice Questions

1. One coulomb of charge is equal to:
   a. 6.24×10^{18} electrons
   b. $10^{18} \times 10^{12}$ electrons
   c. 6.24×10^{8} electrons
   d. 6.24×10^{81} electrons

2. If 14 C of charge passes by a point in 7 seconds, the current flow is said to be:
   a. 2 A
   b. 98 A
   c. 21 A
   d. 7 A

3. How many electrons are there within 16 C of charge?
   a. 9.98×10^{19}
   b. 14
   c. 16
   d. 10.73×10^{19}

4. Current is measured in:
   a. Volts
   b. Coulombs/second
   c. Ohms
   d. Siemens

5. Voltage is measured in units of:
   a. Amperes
   b. Ohms
   c. Siemens
   d. Volts

6. Another word used to describe voltage is:
   a. Potential difference
   b. Pressure
   c. Electromotive force (emf)
   d. All the above

7. Conductors offer a _____ resistance to current flow.
   a. High
   b. Low
   c. Medium
   d. Maximum

8. Conductance is the measure of how good a conductor is at passing current, and is measured in:
   a. Siemens
   b. Volts
   c. Current
   d. Ohms

Communication Skill Questions

9. Why is current directly proportional to voltage? (Ch. 5)

10. List the four most commonly used current units and their values in terms of the basic unit. (8-1-3)

11. List the four most commonly used voltage units and their values in terms of the basic unit. (8-2-2)

12. Give the unit and symbol for the following:
    a. Voltage (V) is measured in _____ (__). (8-2)
    b. Current (I) is measured in _____ (__). (8-1-2)
    c. Conductance (G) is measured in _____ (__). (8-3-1)
    d. Resistance (R) is measured in _____ (__). (8-3-1)

13. Describe how to use a voltmeter to measure voltage and how to use an ammeter to measure current.

Practice Problems

14. What is the value of conductance in siemens for a 100 ohm resistor?

15. Calculate the total number of electrons in 6.5 C of charge.

16. Calculate the amount of current in amperes passing through a conductor if 3 C of charge passes by a point in 4 s.

17. Convert the following:
    a. 0.014 A = _____ mA
    b. 1374 A = _____ kA
    c. 0.776 μA = _____ nA
    d. 0.91 mA = _____ μA

18. Convert the following:
    a. 1473 mV = _____ V
    b. 7143 V = _____ kV
    c. 0.139 kV = _____ V
    d. 0.390 MV = _____ kV

19. What is the relative conductivity of copper if silver is used as the reference conductor?

20. What minimum thickness of porcelain will withstand 24,000 V?

21. To insulate a circuit from 10 V, what insulator thickness would be needed if the insulator is rated at 750 kV/cm?

22. Convert the following:

    a. 2000 kV/cm to _____ meters

    b. 250 kV/cm to _____ mm

23. What maximum voltage may be placed across 35 mm of mica without its breaking down?

Web Site Questions

Go to the Web site http://www.prenhall.com/cook, select the textbook *Mathematics for Electronics and Computers,* select this chapter, and then follow the instructions when answering the multiple-choice practice problems.

Resistance and Power

Genius of Chippewa Falls

In 1960, Seymour R. Cray, a young vice-president of engineering for Control Data Corporation, informed president William Norris that in order to build the world's most powerful computer he would need a small research lab built near his home. Norris would have shown any other employee the door, but Cray was his greatest asset, so in 1962 Cray moved into his lab, staffed by 34 people and nestled in the woods near his home overlooking the Chippewa River in Wisconsin. Eighteen months later the press was invited to view the 14- by 6-foot 6600 supercomputer that could execute 3 million instructions per second and contained 80 miles of circuitry and 350,000 transistors, which were so densely packed that a refrigeration cooling unit was needed due to the lack of airflow.

Cray left Control Data in 1972 and founded his own company, Cray Research. Four years later the $8.8 million Cray-1 scientific supercomputer outstripped the competition. It included some revolutionary design features, one of which is that since electronic signals cannot travel faster than the speed of light (1 foot per billionth of a second) the wire length should be kept as short as possible, because the longer the wire the longer it takes for a message to travel from one end to the other. With this in mind, Cray made sure that none of the supercomputer's conducting wires exceeded 4 feet in length.

In the summer of 1985, the Cray-2, Seymour Cray's latest design, was installed at Lawrence Livermore Laboratory. The Cray-2 is 12 times faster than the Cray-1, and its densely packed circuits are encased in clear Plexiglas and submerged in a bath of liquid coolant.

Outline and Objectives

VIGNETTE: GENIUS OF CHIPPEWA FALLS

INTRODUCTION

9-1   RESISTANCE

Objective 1: Define *resistance* and *ohm*.

9-1-1 Ohm's Law

Objective 2: Explain Ohm's law and its application.

9-1-2 Ohm's Law Triangle

9-2   RESISTORS

Objective 3: Describe the difference between a fixed- and a variable-value resistor.

9-2-1 Fixed-Value Resistors

Objective 4: Identify the different resistor wattage ratings, and their value and tolerance labeling methods.

9-2-2 Variable-Value Resistors
Mechanically (User) Adjustable Variable Resistor
Thermally (Heat) Adjustable Variable Resistor
Optically (Light) Adjustable Variable Resistor
Filament Resistor

Objective 5: Explain the following types of variable resistors:
    a. Mechanically adjustable: potentiometer
    b. Thermally adjustable: RTD, TFD, and thermistor
    c. Optically adjustable: photoresistor
    d. Filament resistor

9-3   MEASURING RESISTANCE

Objective 6: List the rules to remember when measuring resistance with an ohmmeter.

9-4   RESISTOR CODING

Objective 7: Explain how a resistor's value and tolerance are printed on the body of the component either:
    a. By color coding, or
    b. Typographically

9-4-1 The Color Code
General-Purpose Resistor Code
Precision Resistor Code

Objective 8: Describe the difference between a general-purpose and a precision resistor.

9-4-2 Alphanumeric Coding

9-5   PROTOBOARDS

9-6   ENERGY, WORK, AND POWER

Objective 9: Describe and define the terms *energy, work,* and *power.*

9-6-1 Power

9-6-2 Calculating Energy

9-6-3 Calculating Power

9-6-4 Measuring Power

Objective 10: Describe how to measure power.

9-6-5 The Kilowatt-Hour

Objective 11: Explain what is meant by the term *kilowatt-hour.*

MULTIPLE-CHOICE QUESTIONS

COMMUNICATION SKILL QUESTIONS

PRACTICE PROBLEMS

WEB SITE QUESTIONS

Introduction

Voltage, current, resistance, and power will be the four basic properties of prime importance in your study of electronics. In Chapter 8, voltage and current were introduced, and in this chapter we will examine resistance and power.

9-1   RESISTANCE

Ohm

Unit of resistance, symbolized by the Greek capital letter omega (Ω).

As discussed in the previous chapter, current is measured in amperes and voltage is measured in volts. Resistance is measured in **ohms,** in honor of Georg Simon Ohm, who was the first to formulate the relationship among current, voltage, and resistance.

The larger the resistance, the larger the value of ohms and the more the resistor will oppose current flow. The ohm is given the symbol Ω, which is the Greek capital letter omega. By definition, 1 ohm is the value of resistance that will allow 1 ampere of current to flow through a circuit when a voltage of 1 volt is applied, as shown in Figure 9-1(a), where the resistor is drawn as a zigzag. In some schematics or circuit diagrams the resistor is drawn as a rectangular block, as shown in Figure 9-1(b). The pictorial of this circuit is shown in Figure 9-1(c).

The circuit in Figure 9-2 reinforces our understanding of the ohm. In this circuit, a 1 volt battery is connected across a resistor, whose resistance can be either increased or decreased. As the resistance in the circuit is increased the current will decrease and, conversely, as the resistance of the resistor is decreased the circuit current will increase. So how much is 1 ohm of resistance? The answer to this can be explained by adjusting the resistance of the variable resistor. If the resistor is adjusted until exactly 1 amp of current is flowing around the circuit, the value of resistance offered by the resistor is referred to as 1 ohm (Ω).

9-1-1   *Ohm's Law*

Current flows in a circuit due to the electrical force or voltage applied. The amount of current flow in a circuit, however, is limited by the amount of resistance in the circuit. It can be said,

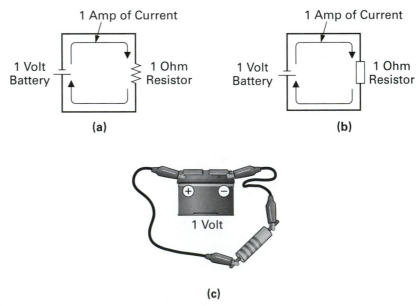

FIGURE 9-1   **One Ohm. (a) New and (b) Old Resistor Symbols. (c) Pictorial.**

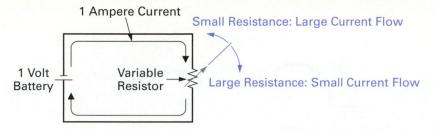

FIGURE 9-2   **1 Ohm Allows 1 Amp to Flow When 1 Volt Is Applied.**

therefore, that the amount of current flow around a circuit is dependent on both voltage and resistance. This relationship among the three electrical properties of current, voltage, and resistance was discovered by Georg Simon Ohm, a German physicist, in 1827. Published originally in 1826, **Ohm's law** states that the current flow in a circuit is directly proportional ($\propto$) to the source voltage applied and inversely proportional ($1/\propto$) to the resistance of the circuit.

Stated in mathematical form, Ohm arrived at this formula:

Ohm's Law
Relationship among the three electrical properties of voltage, current, and resistance. Ohm's law states that the current flow within a circuit is directly proportional to the voltage applied across the circuit and inversely proportional to its resistance.

$$\text{Current } (I) = \frac{\text{voltage } (V)}{\text{resistance } (R)}$$

$$\text{Current } (I) \propto \text{voltage } (V)$$
$$\text{Current } (I) \frac{1}{\propto} \text{resistance } (R)$$

■ INTEGRATED MATH APPLICATION:

How can you calculate voltage ($V = ?$) if I and R are known, and how do you calculate resistance ($R = ?$) if V and I are known?

■ *Solution:*

The answer is, transpose the original formula, as follows:

$$\text{Current } (I) = \frac{\text{voltage } (V)}{\text{resistance } (R)}$$

$$\boxed{I = \frac{V}{R}}$$

Solve for V

$$I = \frac{V}{R}$$

$$\underline{R \times I} = \frac{V \times R}{R} \left(\begin{array}{c}\text{multiply both}\\\text{sides by } R\end{array}\right)$$

$$\underline{R \times I} = \frac{V \times \cancel{R}}{\cancel{R}} \quad (R \div R \text{ cancels})$$

$$\underline{R \times I} = V$$

$$\boxed{V = I \times R}$$

Solve for R

$$I = \frac{V}{R}$$

$$\underline{R \times I} = \frac{V \times R}{R} \left(\begin{array}{c}\text{multiply both sides}\\\text{by } R \text{ to bring } R\\\text{above the line}\end{array}\right)$$

$$\underline{R \times I} = \frac{V \times \cancel{R}}{\cancel{R}} \quad (R \div R \text{ cancels})$$

$$\frac{R \times I}{I} = \frac{V}{I} \left(\begin{array}{c}\text{divide both sides}\\\text{by } I \text{ to isolate } R\end{array}\right)$$

$$\frac{R \times \cancel{I}}{\cancel{I}} = \frac{V}{I} \quad (I \div I \text{ cancels})$$

$$\boxed{R = \frac{V}{I}}$$

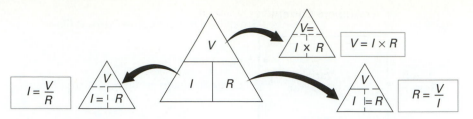

FIGURE 9-3 Ohm's Law Triangle.

To test if all these three formulas are correct, we can insert into each the values of an example, as follows:

$$I = \frac{V}{R} \qquad\qquad V = I \times R \qquad\qquad R = \frac{V}{I}$$

$$2A = \frac{6\,V}{3\,\Omega} \qquad\qquad 6\,V = 2\,A \times 3\,\Omega \qquad\qquad 3\,\Omega = \frac{6\,V}{2\,A}$$

Since the values on both sides of the equals sign are equal in all three equations, the transposition has been successful.

9-1-2   *Ohm's Law Triangle*

Figure 9-3 shows how the three Ohm's law formulas can be placed within a triangle for easy memory recall.

SELF-TEST EVALUATION POINT FOR SECTION 9–1

Now that you have completed this section, you should be able to:

■ **Objective 1.**   *Define* resistance *and* ohm.

■ **Objective 2.**   *Explain Ohm's law and its application.*

Use the following questions to test your understanding of Section 9-1:

1. Define 1 ohm in relation to current and voltage.
2. Calculate I if $V = 24$ V and $R = 6\ \Omega$.
3. What is the Ohm's law triangle?
4. What is the relationship between: (a) current and voltage? (b) current and resistance?
5. Calculate V if $I = 25$ mA and $R = 1$ kΩ.
6. Calculate R if $V = 12$ V and $I = 100\ \mu$A.

9-2   RESISTORS

Conductors are used to connect one device to another, and although they offer a small amount of resistance, this resistance is not normally enough. In electronic circuits, additional resistance is normally needed to control the amount of current flow, and the component used to supply this additional resistance is called a **resistor.**

There are two basic types of resistors: fixed-value and variable-value. Fixed-value resistors, like the examples seen in Figure 9-4, have a resistance value that cannot be changed. On the other hand, variable-value resistors, like the examples seen in Figure 9-5, have a range of values that can be selected by adjusting a control.

Resistor

Component made of a material that opposes the flow of current and therefore has some value of resistance.

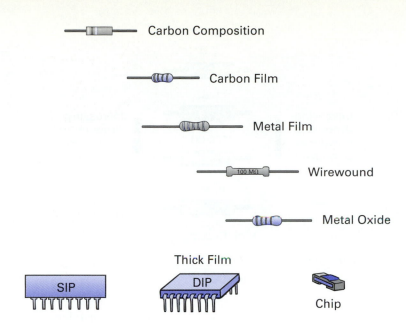

FIGURE 9-4 **Fixed-Value Resistors.**

9-2-1   *Fixed-Value Resistors*

Figure 9-6 illustrates four fixed-value resistors, ranging from 2 Ω to 10 MΩ. The 2 Ω resistor is exactly the same size as the 10 MΩ (10 million Ω) resistor. This is achieved by having more powdered conductor and less powdered insulator in the 2 Ω and less powdered conductor and more powdered insulator in the 10 MΩ. The color-coded rings or bands on the resistors in Figure 9-6 are a means of determining the value of resistance. This and other coding systems will be discussed later.

The physical size of the resistors lets the user know how much power in the form of heat can be **dissipated,** as shown in Figure 9-7. As you already know, resistance is the opposition to current flow, and this opposition causes heat to be generated whenever current passes through a resistor. The amount of heat dissipated each second is measured in watts, and each resistor has its own **wattage rating.** For example, a 2 watt size resistor can dissipate up to 2 joules of heat per second, whereas a ⅛ watt size resistor can dissipate only up to ⅛ joule of heat per second. The key point to remember is that resistors in high-current circuits should have a surface area that is large enough to dissipate the heat faster than it is being generated. If the current passing through a resistor generates heat faster than the resistor can dissipate it, the resistor will burn up and no longer perform its function.

Fixed-Value Resistor

A resistor whose value cannot be changed.

Dissipation

Release of electrical energy in the form of heat.

Wattage Rating

Maximum power a device can safely handle continuously.

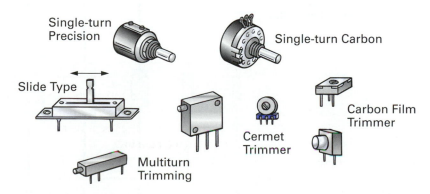

FIGURE 9-5 **Variable-Value Resistors.**

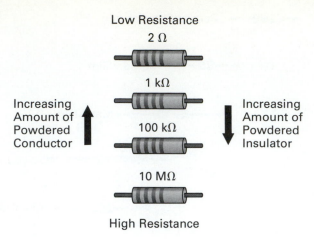

Low Resistance
2 Ω

1 kΩ

Increasing
Amount of
Powdered
Conductor

100 kΩ

Increasing
Amount of
Powdered
Insulator

10 MΩ

High Resistance

FIGURE 9-6   Resistor Ratios.

Tolerance

Permissible deviation from
a specified value, normally
expressed as a
percentage.

Another factor to consider when discussing resistors is their **tolerance.** Tolerance is the amount of deviation or error from the specified value. For example, a 1000 Ω (1 kΩ) resistor with a ±10% (plus and minus 10%) tolerance when manufactured could have a resistance of anywhere between 900 and 1100 Ω.

$$\pm 10\% \text{ of } 1000 = 100$$
$$10\% - \boxed{1000} + 10\%$$
$$\downarrow \qquad\qquad \downarrow$$
$$900 \qquad\qquad 1100$$

This means that two identically marked resistors when measured could be from 900 to 1100 Ω, a difference of 200 Ω. In some applications, this may be acceptable. In other applications, where high precision is required, this deviation could be too large and so a more expensive, smaller tolerance resistor would have to be used.

■ **INTEGRATED MATH APPLICATION:**

Calculate the amount of deviation of the following resistors:

   a. 2.2 kΩ ± 10%

   b. 5 MΩ ± 2%

   c. 3 Ω ± 1%

FIGURE 9-7   **Resistor Wattage Rating Guide. All Resistor Silhouettes Are Drawn to Scale.**

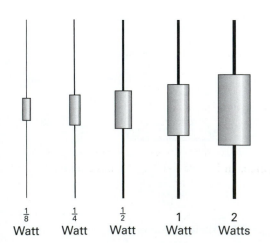

$\frac{1}{8}$
Watt

$\frac{1}{4}$
Watt

$\frac{1}{2}$
Watt

1
Watt

2
Watts

Solution:

a. 10% of 2.2 kΩ = 220 Ω. For + 10%, the value is

$$2200 + 220\ \Omega = 2420\ \Omega$$
$$= 2.42\ k\Omega$$

For −10%, the value is

$$2200 - 220\ \Omega = 1980\ \Omega$$
$$= 1.98\ k\Omega$$

Step	Keypad Entry	Display Response
1.	$\boxed{1}\ \boxed{0}$	10
2.	$\boxed{\%}$	10
3.	$\boxed{\times}$	0.10
4.	$\boxed{2}\ \boxed{.}\ \boxed{2}\ \boxed{E}\ \boxed{3}$	2.2E3
5.	$\boxed{=}$	220

The resistor will measure anywhere from 1.98 kΩ to 2.42 kΩ.

b. 2% of 5 MΩ = 100 kΩ

$$5\ M\Omega + 100\ k\Omega = 5.1\ M\Omega$$
$$5\ M\Omega - 100\ k\Omega = 4.9\ M\Omega$$

Deviation = 4.9 MΩ to 5.1 MΩ

c. 1% of 3 Ω = 0.03 Ω or 30 milliohms (mΩ).

$$3\ \Omega + 0.03\ \Omega = 3.03\ \Omega$$
$$3\ \Omega - 0.03\ \Omega = 2.97\ \Omega$$

Deviation = 2.97 Ω to 3.03 Ω

9-2-2   *Variable-Value Resistors*

A **variable resistor** can have its resistance varied or changed while it is connected in a circuit. In certain applications, the ability to adjust the resistance of a resistor is needed. For example, the volume control on your television set makes use of a variable resistor to vary the amount of current passing to the speakers, and so change the volume of the sound.

Variable Resistor
A resistor whose value can be changed.

Mechanically (User) Adjustable Variable Resistor

Figure 9-8(a) illustrates the physical appearance of a variety of **potentiometers,** also called pots (slang), and Figure 9-8(b) shows the potentiometer's schematic symbol. In Figure 9-8(c), you can see that resistance can actually be measured across three separate combinations: between A and B (X), between B and C (Y), and between C and A (Z). Figure 9-8(d) shows a single-turn potentiometer and a multiturn potentiometer in which a contact arm slides along a shaft, and the resistive track is formed into a helix of 2 to 10 coils.

Potentiometer
Three-lead variable resistor that through mechanical turning of a shaft can be used to produce a variable voltage or potential.

To explain the operation of a potentiometer, Figure 9-9(a) shows a 10 kΩ potentiometer. Looking at Figure 9-9(b), you can see that the resistance measured between terminals A and C will always be the same (10 kΩ) no matter where the user puts the wiper, because current will still have to travel through the complete resistive track between A and C.

The resistance between A and B (X) and B and C (Y) will vary as the wiper's position is moved, as illustrated in Figure 9-9(c). If the user physically turns the shaft in a clockwise direction, the resistance between A and B increases, and the resistance between B and C decreases. Similarly, if the user mechanically turns the shaft counterclockwise, a decrease occurs between A and B, and there is a resulting increase in resistance between B and C. This point is summarized in Figure 9-9(d).

The resistive track can be classified as having either a **linear** or a **tapered** (nonlinear) resistance. In Figure 9-10(a) we have taken a 1 kΩ potentiometer and illustrated the resistance value changes between A and B for a linear and a tapered one-turn potentiometer. The definition of linear is having an output that varies in direct proportion to the input. The input in this case is the user turning the shaft, and the output is the linearly increasing resistance between A and B.

Linear
Relationship between input and output in which the output varies in direct proportion to the input.

Tapered
Nonuniform distribution of resistance per unit length throughout the element.

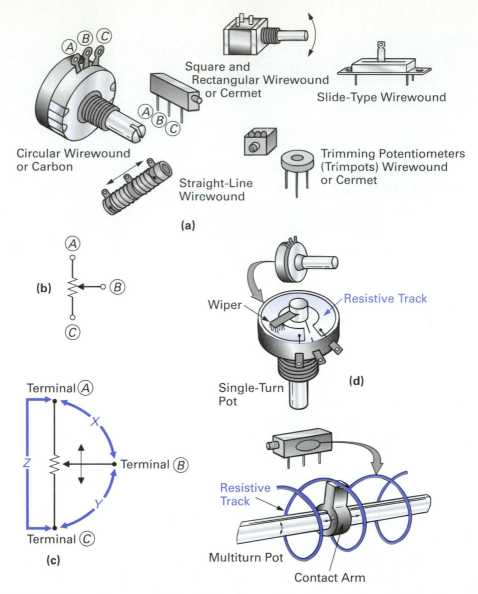

FIGURE 9-8 Potentiometer. (a) Physical Appearance. (b) Schematic Symbol. (c) Operation. (d) Construction.

■ **INTEGRATED MATH APPLICATION:**

Graph the data in Figure 9-10(a) to show the difference between a linear and tapered potentiometer's resistance.

■ *Solution:*

With a tapered potentiometer, the resistance varies nonuniformly along its resistor element, sometimes being greater or less for equal shaft movement at various points along the resistance element, as shown in the table in Figure 9-10(a). Figure 9-10(b) plots the position of the variable resistor's wiper against the resistance between the two output terminals, showing the difference between a linear increase and a nonlinear or tapered increase.

Thermally (Heat) Adjustable Variable Resistor

When we first discussed variable-value resistors, we talked about the potentiometer, which needs a mechanical input (the user turning the shaft) to produce a change in resistance. A

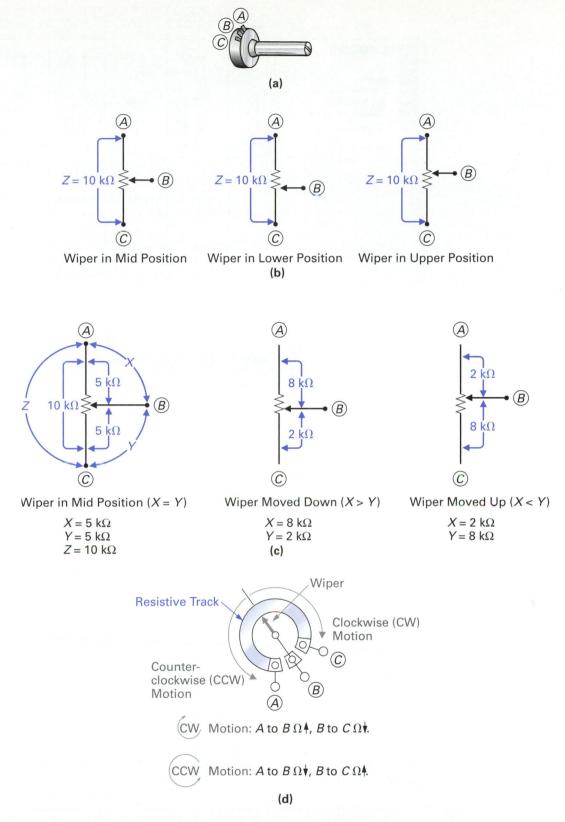

FIGURE 9-9 The Resistance Changes between the Terminals of a Potentiometer.

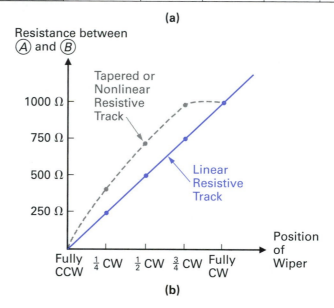

	Fully CCW	$\frac{1}{4}$ CW	$\frac{1}{2}$ CW	$\frac{3}{4}$ CW	Fully CW
1 kΩ Potentiometer					
Linear	0 Ω	250 Ω $\left(\frac{1}{4} \text{ of } 1000\ \Omega\right)$	500 Ω $\left(\frac{1}{2} \text{ of } 1000\ \Omega\right)$	750 Ω $\left(\frac{3}{4} \text{ of } 1000\ \Omega\right)$	1000 Ω
Tapered	0 Ω	350 Ω	625 Ω	900 Ω	1000 Ω

These values have been arbitrarily chosen to illustrate a nonlinear change.

(a)

(b)

FIGURE 9-10   Linear versus Tapered Resistive Track.

Bolometer

Device whose resistance changes when heated.

Thermometry

Relating to the measurement of temperature.

bolometer is a device that, instead of changing its resistance when mechanical energy is applied, changes its resistance when heat energy is applied. Before discussing the three basic types of bolometers, let's first consider the four units of temperature measurement detailed in Table 9-1.

The measurement of temperature (**thermometry**) is probably the most common type of measurement used in industry today. Commercially, temperature is normally expressed in degrees Celsius (°C) or degrees Fahrenheit (°F); however, although not commonly known, kelvin (K) and degrees Rankine (°R) are often used in industry, the kelvin being the international unit of temperature.

■   **INTEGRATED MATH APPLICATION:**

Convert the following.

    a. 74 °F = _____ °C

    b. 45 °C = _____ °F

TABLE 9-1 Four Temperature Scales and Conversion Formulas

	Fahrenheit	Celsius	Kelvin	Rankine
Absolute zero	−459.69 °F	−273.16 °C	0 K	0 °R
Melting point of ice (x)	32 °F	0 °C	273.16 K	491.69 °R
(Difference between x and y)	(180 °F)	(100 °C)	(100 K)	(180 °R)
Boiling Point of water (y)	212 °F	100 °C	373.16 K	671.69 °R

$$F = (\tfrac{9}{5} \times C) + 32$$
$$C = \tfrac{5}{9} \times (F - 32)$$
$$R = F + 460$$
$$F = R - 460$$
$$K = C + 273$$
$$C = K - 273$$

c. 25 °C = _____ K

d. 10 °F = _____ °R

■ *Solution:*

a. $C = \tfrac{5}{9} \times (F - 32)$
   $= \tfrac{5}{9} \times (74 - 32)$
   $= 0.555 \times 42 = 23.3$ °C

b. $F = (\tfrac{9}{5} \times C) + 32$
   $= (\tfrac{9}{5} \times 45) + 32$
   $= (1.8 \times 45) + 32 = 113$ °F

c. $K = C + 273$
   $= 25 + 273 = 298$ K

d. $R = F + 460$
   $= 10 + 460 = 470$ °R

Step	Keypad Entry	Display Response
1.	7 4	74
2.	−	
3.	3 2	32
4.	=	42
5.	STO (store result in memory)	
6.	C/CE (cancel display)	0
7.	5	5
8.	÷	
9.	9	9
10.	=	0.55555
11.	×	
12.	RCL (Recall value from memory	42
13.	=	23.3

Bolometers are normally used as temperature sensors or detectors. The **resistive temperature detector** (RTD) and **thin-film detector** (TFD) shown in Figure 9-11(a) and (b) are both temperature sensors that make use of a copper, nickel, or platinum conducting element and therefore have a positive temperature coefficient (resistance increases as temperature increases).

The thin-film detector (TFD) shown in Figure 9-11(b) is used for very precise temperature readings and is constructed by placing a thin layer of platinum on a ceramic substrate. Because of its small size, the TFD responds rapidly to temperature change and is ideally suited for surface temperature sensing.

Unlike the RTD and TFD, the **thermistor** contains a semiconductor material that has a negative temperature coefficient, which means that its resistance decreases as temperature increases. Semiconductor materials have characteristics that are midway between those of conductors and insulators. In addition to thermistors, semiconductor materials are also used to construct other electronic devices such as diodes and transistors, which will be discussed later.

Optically (Light) Adjustable Variable Resistor

Photo means illumination, and the **photoresistor** is a resistor that is photoconductive. This means that as the material is exposed to light it becomes more conductive and less resistive.

Resistive Temperature Detector (RTD)

A temperature detector consisting of a fine coil of conducting wire (such as platinum) that produces a relatively linear increase in resistance as temperature increases.

Thin-Film Detector (TFD)

A temperature detector containing a thin layer of platinum and used for very precise temperature readings.

Thermistor

Temperature-sensitive semiconductor that has a negative temperature coefficient of resistance (as temperature increases, resistance decreases).

Photoresistor

Also known as a photoconductive cell or light-dependent resistor, it is a device whose resistance varies with the illumination of the cell.

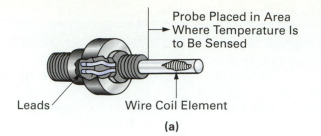

Probe Placed in Area Where Temperature Is to Be Sensed

Leads   Wire Coil Element

(a)

Platinum Winding			
°C	Ohms	°C	Ohms
−200	18.53	+200	175.84
−150	39.65	+250	194.08
−100	60.20	+300	212.03
−50	80.25	+350	229.69
±0	100.0	+400	247.06
+50	119.40	+450	264.14
+100	138.50	+500	280.93
+150	157.32	+550	297.16

(Conductors have a positive temperature coefficient of resistance — temp.↑, R↑)

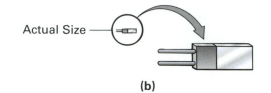

Actual Size

(b)

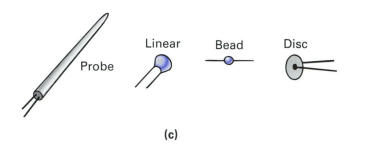

Probe   Linear   Bead   Disc

(c)

°C	Ohms
−50	100,000
0	7,500
+50	7,400
+100	100
+150	50
+200	27
+250	10
+300	7.5

(Semiconductors have a negative temperature coefficient of resistance — T↑, R↓)

FIGURE 9-11   **Temperature Sensors. (a) Resistive Temperature Detector (RTD). (b) Thin-Film Detector (TFD). (c) Thermistor.**

Figure 9-12 shows the construction, schematic symbol, and physical appearance of a photoresistor, which is also called a light-dependent resistor (LDR).

Filament Resistor

Filament Resistor

The resistor in a lightbulb or electron tube.

Figure 9-13(a) illustrates the **filament resistor** within a glass bulb. This component is more commonly known as the household light bulb. The filament resistor is just a coil of wire that glows white hot when current is passed through it and, in so doing, dissipates both heat and light energy.

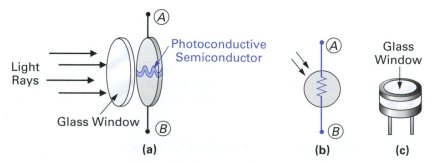

Light Rays   Photoconductive Semiconductor

Glass Window

(a)   **(b)**   **(c)**

FIGURE 9-12   **Photoresistor. (a) Construction. (b) Schematic Symbol. (c) Physical Appearance.**

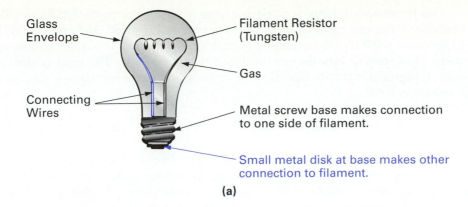

Glass Envelope

Filament Resistor (Tungsten)

Gas

Connecting Wires

Metal screw base makes connection to one side of filament.

Small metal disk at base makes other connection to filament.

(a)

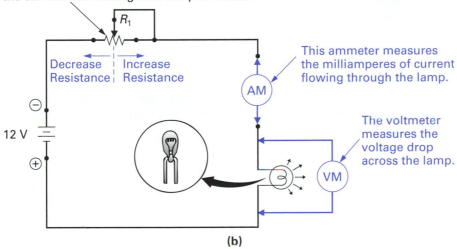

By changing the resistance of the potentiometer, the current flow through the lamp is varied.

R_1

Decrease Resistance | Increase Resistance

12 V

This ammeter measures the milliamperes of current flowing through the lamp.

AM

The voltmeter measures the voltage drop across the lamp.

VM

(b)

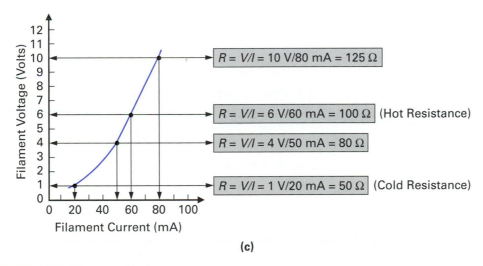

$R = V/I = 10 \text{ V}/80 \text{ mA} = 125 \ \Omega$

$R = V/I = 6 \text{ V}/60 \text{ mA} = 100 \ \Omega$ (Hot Resistance)

$R = V/I = 4 \text{ V}/50 \text{ mA} = 80 \ \Omega$

$R = V/I = 1 \text{ V}/20 \text{ mA} = 50 \ \Omega$ (Cold Resistance)

(c)

FIGURE 9-13 **Filament Resistor.**

This **incandescent lamp** is an electric lamp in which electric current flowing through a filament of resistive material heats the filament until it glows and emits light. Figure 9-13(b) shows a test circuit for varying current through a small incandescent lamp. Potentiometer R_1 is used to vary the current flow through the lamp. The lamp is rated at 6 V, 60 mA. Using Ohm's law we can calculate the lamp's filament resistance at this rated voltage and current:

$$\text{Filament resistance, } R = \frac{V}{1} = \frac{6\text{ V}}{60\text{ mA}} = 100\ \Omega$$

The filament material (tungsten, for example) is like all other conductors in that it has a positive temperature coefficient of resistance. Therefore, as the current through the filament increases, so does the temperature and so does the filament's resistance ($I\uparrow$, temperature $\uparrow$, $R\uparrow$). Consequently, when R_1's wiper is moved to the right so that it produces a high resistance ($R_1\uparrow$), the circuit current will be small ($I\downarrow$) and the lamp will glow dimly. Since the circuit current is small, the filament temperature will be small and so will the lamp's resistance ($I\downarrow$, temperature $\downarrow$, lamp resistance $\downarrow$). This small value of resistance is called the lamp's **cold resistance.** On the other hand, when R_1's wiper is moved to the left so that it produces a small resistance ($R\downarrow$), the circuit current will be large ($I\uparrow$), and the lamp will glow brightly. With the circuit current high, the filament temperature will be high and so will the lamp's resistance ($I\uparrow$, temperature $\uparrow$, lamp resistance $\uparrow$). This large value of resistance is called the lamp's **hot resistance.**

Figure 9-13(c) plots the filament voltage, which is being measured by the voltmeter, against the filament current, which is being measured by the ammeter. As you can see, an increase in current causes a corresponding increase in filament resistance. Studying this graph, you may have also noticed that the lamp has been operated beyond its rated value of 6 V, 60 mA. Although the lamp can be operated beyond its rated value (for example, 10 V, 80 mA), its life expectancy will be decreased dramatically from several hundred hours to only a few hours.

SELF-TEST EVALUATION POINT FOR SECTION 9-2

Now that you have completed this section, you should be able to:

- **Objective 3.** *Describe the difference between a fixed- and a variable-value resistor.*

- **Objective 4.** *Identify the different resistor wattage ratings, and their value and tolerance labeling methods.*

- **Objective 5.** *Explain the following types of variable resistors:*
  a. Mechanically adjustable: potentiometer
  b. Thermally adjustable: RTD, TFD, and thermistor
  c. Optically adjustable: photoresistor
  d. Filament resistor

Use the following questions to test your understanding of Section 9-2:

1. What is the difference between a fixed-value and variable-value resistor?

2. Describe a linear and a tapered potentiometer.

3. True or false: A thermistor has a negative temperature coefficient of resistance.

4. What name is given to an optically adjustable resistor?

5. What is a filament resistor?

9-3   MEASURING RESISTANCE

Resistance is measured with an **ohmmeter.** Stepping through the procedure described in Figure 9-14, you will see how the ohmmeter is used to measure the resistance of a resistor.

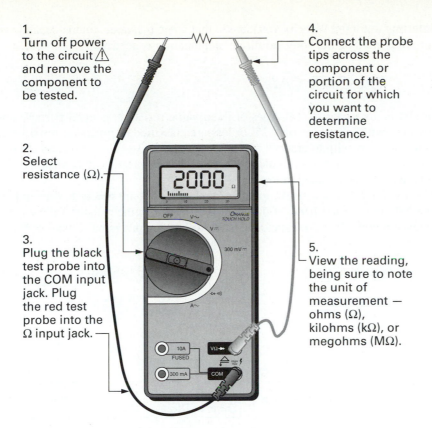

1.
Turn off power to the circuit ⚠ and remove the component to be tested.

2.
Select resistance (Ω).

3.
Plug the black test probe into the COM input jack. Plug the red test probe into the Ω input jack.

4.
Connect the probe tips across the component or portion of the circuit for which you want to determine resistance.

5.
View the reading, being sure to note the unit of measurement — ohms (Ω), kilohms (kΩ), or megohms (MΩ).

FIGURE 9-14   **Measuring Resistance with an Ohmmeter.**

SELF-TEST EVALUATION POINT FOR SECTION 9-3

Now that you have completed this section, you should be able to:

■   **Objective 6.**   *List the rules to remember when measuring resistance with an ohmmeter.*

Use the following questions to test your understanding of Section 9-3:

1.   Name the instrument used to measure resistance.

2.   Describe the five-step procedure that should be used to measure the resistance.

9-4   RESISTOR CODING

Manufacturers indicate the value and tolerance of resistors on the body of the component using either a color code (colored rings or bands) or printed alphanumerics (alphabet and numerals).

Resistors that are smaller in size tend to have the value and tolerance of the resistor encoded using colored rings, as shown in Figure 9-15(a). To determine the specifications of a color-coded resistor, therefore, you will have to decode the rings.

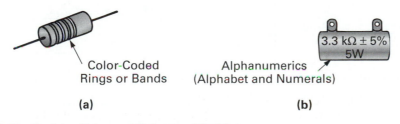

Color-Coded
Rings or Bands

Alphanumerics
(Alphabet and Numerals)

3.3 kΩ ± 5%
5W

(a)

(b)

FIGURE 9-15   **Resistor Value and Tolerance Markings.**

Referring to Figure 9-15(b), you can see that when the value, tolerance, and wattage of a resistor are printed on the body, no further explanation is needed.

9-4-1   *The Color Code*

There are basically two different types of fixed-value resistors: general purpose and precision. Resistors with tolerances of ±2% or less are classified as *precision resistors* and have five bands. Resistors with tolerances of ±5% or greater have four bands and are referred to as *general-purpose resistors*. The color code and differences between precision and general-purpose resistors are explained in Figure 9-16.

When you pick up a resistor, look for the bands that are nearer to one end. This end should be held in your left hand. If there are four bands on the resistor, follow the general-purpose resistor code. If five bands are present, follow the precision resistor code.

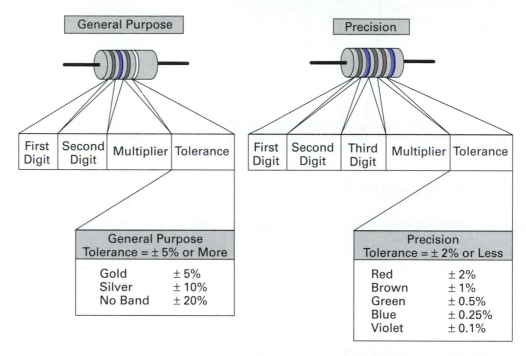

	Color	Digit Value	Multiplier	
Big	Black	0	1 One	1
Beautiful	Brown	1	10 One Zero	10
Roses	Red	2	100 Two Zeros	100
Occupy	Orange	3	1000 Three Zeros	1 k
Your	Yellow	4	10000 Four Zeros	10 k
Garden	Green	5	100000 Five Zeros	100 k
But	Blue	6	1000000 Six Zeros	1 M
Violets	Violet	7	10000000 Seven Zeros	10 M
Grow	Gray	8	–	
Wild	White	9	–	
So	Silver	–	10^{-2} or 0.01	1/100
Get some	Gold	–	10^{-1} or 0.1	1/10
Now	None	–		

FIGURE 9-16   **General-Purpose and Precision Resistor Color Code.**

General-Purpose Resistor Code

1. The first band on either a general-purpose or precision resistor can never be black, and it is the first digit of the number.

2. The second band indicates the second digit of the number.

3. The third band specifies the multiplier to be applied to the number, which ranges from $\times\frac{1}{100}$ to $\times 10,000,000$.

4. The fourth band describes the tolerance or deviation from the specified resistance, which is $\pm 5\%$ or greater.

Precision Resistor Code

1. The first band, like on the general-purpose resistor, is never black and is the first digit of the three-digit number.

2. The second band provides the second digit.

3. The third band indicates the third and final digit of the number.

4. The fourth band specifies the multiplier to be applied to the number.

5. The fifth and final band indicates the tolerance figure of the precision resistor, which is always less than $\pm 2\%$, which is why precision resistors are more expensive than general-purpose resistors.

■ INTEGRATED MATH APPLICATION:

Figure 9-17 illustrates a $\frac{1}{2}$ W resistor.

a. Is this a general-purpose or a precision resistor?

b. What is the resistor's value of resistance?

c. What tolerance does this resistor have, and what plus and minus deviation could this value have?

■ *Solution:*

a. General purpose (four bands)

b. green   blue $\times$ brown

    5   6 $\times$ 10 $= 560\ \Omega$

c. Tolerance band is gold, which is $\pm 5\%$.

$$\text{Deviation} = 5\% \text{ of } 560 = 28$$
$$560 + 28 = 588\ \Omega$$
$$560 - 28 = 532\ \Omega$$

The resistor's value, when measured, could be anywhere from 532 to 588 Ω.

Band 1 = Green
Band 2 = Blue
Band 3 = Brown
Band 4 = Gold

FIGURE 9-17

■ INTEGRATED MATH APPLICATION:

State the resistor's value and tolerance and whether it is general purpose or precision for the examples shown in Figure 9-18.

Orange/Green/Black/Silver

(a)

Green/Blue/Red/None

(b)

Red/Red/Green/Gold/Blue

(c)

FIGURE 9-18

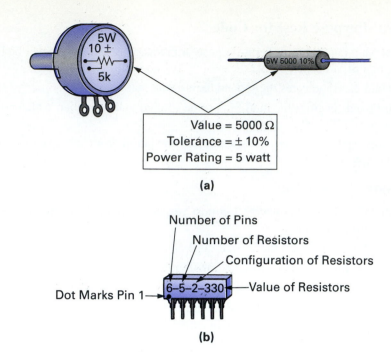

Value = 5000 Ω
Tolerance = ± 10%
Power Rating = 5 watt

(a)

Number of Pins
Number of Resistors
Configuration of Resistors
6–5–2–330 — Value of Resistors
Dot Marks Pin 1

(b)

FIGURE 9-19 **Typographical Value Indication.**

■ *Solution:*

a. orange	green	black	silver	(four bands = general purpose)
3	5 ×	1	10%	= 35 Ω ± 10%
b. green	blue	red	none	(four bands = general purpose)
5	6 ×	100	20%	= 5.6 kΩ ± 20%
c. red	red	green	gold	blue (five bands = precision)
2	2	5 ×	0.1	0.25% = 22.5 Ω ± 0.25%

9-4-2 *Alphanumeric Coding*

If a resistor color code is not found, some other form of marking will always be made on the resistor. The larger resistors and nearly all variable resistors normally have the resistance value, tolerance, and wattage printed on the resistor, as shown in Figure 9-19(a).

SELF-TEST EVALUATION POINT FOR SECTION 9-4

Now that you have completed this section, you should be able to:

■ **Objective 7.** *Explain how a resistor's value and tolerance are printed on the body of the component either:*
a. By color coding, or
b. Typographically

■ **Objective 8.** *Describe the difference between a general-purpose and a precision resistor.*

Use the following questions to test your understanding of Section 9-4:

1. What tolerance differences differentiate a general-purpose from a precision resistor?
2. A red/red/red/gold/blue resistor has what value and tolerance figure?
3. What color code would appear on a 4.7 kΩ, ±10% tolerance resistor?
4. Would a three-band resistor be considered a general-purpose type?

9-5 PROTOBOARDS

The solderless prototyping board (**protoboard**) or breadboard is designed to accommodate the many experiments described in this textbook's associated lab manual. This protoboard will hold and interconnect resistors, capacitors, inductors, and many other components, as well as provide electrical power. Figure 9-20(a) shows an experimental circuit wired up on a protoboard.

Figure 9-20(b) shows the top view of a basic protoboard. As you can see by the cross section on the right side, electrical connector strips are within the protoboard. These conductive strips make a connection between the five hole groups. The bus strips have an electrical connector strip running from end to end, as shown in the cross section. They are usually connected to a power supply, as seen in the example circuit in Figure 9-20(a). In this circuit, you can see that the positive supply voltage is connected to the upper bus strip, while the negative supply (ground) is connected to the lower bus strip. These power supply "rails" can then be connected to a circuit formed on the protoboard with hookup wire, as shown in Figure 9-21(a). In this example, three resistors are connected end-to-end, as shown in the schematic diagram in Figure 9-21(b).

Power in the laboratory can be obtained from a battery, but since the wall-outlet voltage is so accessible, dc power supplies are used. The power supply has advantages over the battery as a source of dc voltage in that it can quickly provide an accurate voltage that can easily be varied by a control on the front panel, and it never runs down.

Protoboard

An experimental arrangement of a circuit on a board. Also called *breadboard*.

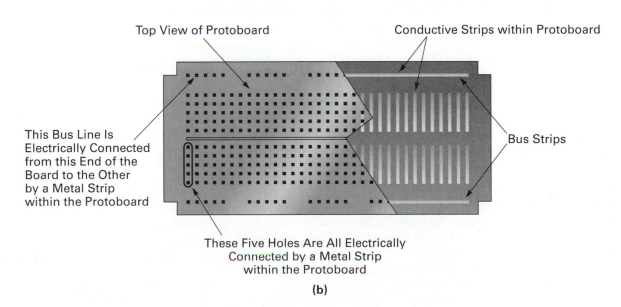

(a)

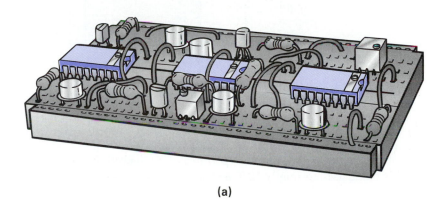

Top View of Protoboard

Conductive Strips within Protoboard

This Bus Line Is Electrically Connected from this End of the Board to the Other by a Metal Strip within the Protoboard

Bus Strips

These Five Holes Are All Electrically Connected by a Metal Strip within the Protoboard

(b)

FIGURE 9-20 **Experimenting with the Protoboard.**

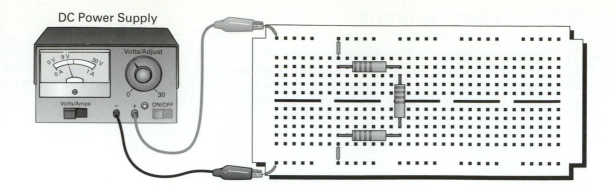

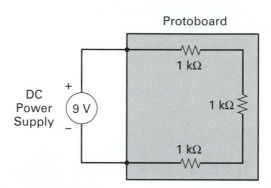

FIGURE 9-21 **Constructing Circuits on the Protoboard.**

During your electronic studies, you will use a power supply in the lab to provide different voltages for various experiments. In Figure 9-21 you can see how a dc power supply is being used to supply 12 V across three 1 kΩ resistors.

Use the following questions to test your understanding of Section 9-5:

1. What is the purpose of the protoboard?
2. What is generally connected to the bus strips?

9-6   ENERGY, WORK, AND POWER

Energy

Capacity to do work.

Work

Work is done anytime energy is transformed from one type to another, and the amount of work done is dependent on the amount of energy transformed.

Joule

The unit of work and energy.

The sun provides us with a consistent supply of energy in the form of light. Coal and oil are fossilized vegetation that grew, among other things, due to the sun, and are examples of **energy** that the earth has stored for millions of years. It can be said, then, that all energy begins from the sun. On the earth, energy is not created or destroyed; it is merely transformed from one form to another. The transforming of energy from one form to another is called **work.** The greater the energy transformed, the more work that is done.

The six basic forms of energy are light, heat, magnetic, chemical, electrical, and mechanical energy. The unit for energy is the **joule** (J). "Potential" (position) and "kinetic" (motion) are two terms used when describing energy. A cart on top of a hill has **potential** (position) **energy,** while a cart rolling down a hill has **kinetic** (motion) **energy.**

Looking at Figure 9-22, let us try to summarize our discussion so far on energy and work. Chemical energy within the battery is converted to electrical energy when the elec-

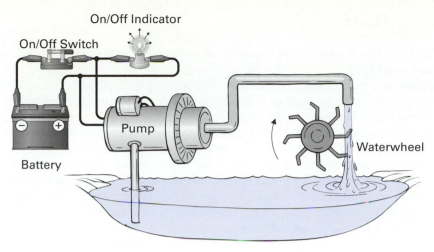

FIGURE 9-22 **Energy Transfer.**

TIME LINE

James P. Joule (1818–1889), an English physicist and self-taught scientist, conducted extensive research into the relationships between electrical, chemical, and mechanical effects, which led him to the discovery that one energy form can be converted into another. For this achievement, his name was given to the unit of energy, the joule.

Potential Energy

Energy that has the potential to do work because of its position.

Kinetic Energy

Energy associated with motion.

Transducer

Any device that converts energy from one form to another.

trons are attracted and repelled and set in motion. To use the two terms, we could say that the battery has the potential energy needed to set electrons in motion, and these moving electrons are said to possess kinetic energy. This electrical energy drives two devices:

1. The lightbulb, which converts electrical energy into light and heat energy
2. The pump, which converts electrical energy into mechanical energy

The pump has the potential energy to cause water flow or kinetic energy, just as the battery has the potential energy to cause kinetic energy in the form of electron flow. The kinetic energy within the water flow is finally used to turn the waterwheel, just as the kinetic energy within the electron flow is used to generate light from the bulb.

Work is done every time one form of energy is transformed to another. A device that converts one form of energy to another is called a **transducer.** In Figure 9-21, the battery, lightbulb, and pump were all examples of transducers doing work at each stage of conversion.

The amount of work done is equal to the amount of energy transformed, and in both cases an equal amount of energy was transformed, so an equal amount of work was done. As a result, energy and work have the same symbol (W), the same formula, and the same unit (the joule). Energy is merely the capacity, potential, or ability to do work, and work is done when a transformation of the potential, capacity, or ability takes place.

■ **INTEGRATED MATH APPLICATION:**

One person walks around a track and takes 5 minutes, while another person runs around the track and takes 50 seconds. Both were full of energy before they walked or ran around the track, and during their travels around the track they converted the chemical energy within their bodies into the mechanical energy of movement.

a. Who exerted the most energy?
b. Who did the most work?

■ *Solution:*

Both exerted the same amount of energy. The runner exerted all her energy (for example, 100 J) in the short time of 50 seconds, while the walker spaced his energy (100 J) over 5 minutes. Since they both did the same amount of work, the only difference between the runner and the walker is time, or the rate at which their energy was transformed.

TIME LINE

The unit of power, the watt, was name after Scottish engineer and inventor James Watt (1736–1819) in honor of his advances in the field of science.

Power

Amount of energy converted by a component or circuit in a unit of time, normally seconds. It is measured in units of watts (joules/second).

Watt (W)

Unit of electric power required to do work at a rate of 1 joule/second. One watt of power is expended when 1 ampere of direct current flows through a resistance of 1 ohm.

9-6-1 *Power*

Power (*P*) is the rate at which work is performed and is given in the unit **watt** (W), which is joules per second (J/s).

Returning to the example involving the two persons walking and running around the track, we could say that the number of joules of energy exerted in 1 second by the runner was far greater than the number of joules of energy exerted in 1 second by the walker, although the total energy exerted by both persons around the entire track was equal and therefore the same amount of work was done. Even though the same amount of energy was used, and therefore the same amount of work was done by the runner and the walker, the output power of each was different. The runner exerted a large value of joules/second or watts (high power output) in a short space of time, whereas the walker exerted only a small value of joules/second or watts (low power output) over a longer period of time.

Whether we are discussing a runner, walker, electric motor, heater, refrigerator, lightbulb, or compact disk player—power is power. The output power, or power ratings, of electrical, electronic, or mechanical devices can be expressed in watts and describes the number of joules of energy converted every second. The output power of rotating machines is given in the unit *horsepower* (hp), the output power of heaters is given in the unit British thermal unit per hour (Btu/h), and the output power of cooling units is given in the unit *ton of refrigeration*. Despite the different names, they can all be expressed quite simply in the unit watt. The conversions are as follows:

$$1 \text{ horsepower (hp)} = 746 \text{ W}$$
$$1 \text{ British thermal unit per hour (Btu/h)} = 0.293 \text{ W}$$
$$1 \text{ ton of refrigeration} = 3.52 \text{ kW (3520 W)}$$

Now that we have an understanding of power, work, and energy, let's reinforce our knowledge by introducing the energy formula and try some examples relating to electronics.

9-6-2 *Calculating Energy*

The amount of energy stored (*W*) is dependent on the coulombs of charge stored (*Q*) and the voltage (*V*).

$$W = Q \times V$$

where *W* = energy stored, in joules (J)
    Q = coulombs of charge (1 coulomb = 6.24×10^{18} electrons)
    V = voltage, in volts (V)

If you consider a battery as an example, you can probably better understand this formula. The battery's energy stored is dependent on how many coulombs of electrons it holds (current) and how much electrical pressure it is able to apply to these electrons (voltage).

■ **INTEGRATED MATH APPLICATION:**

If a 1 V battery can store 6.24×10^{18} electrons, how much energy is the battery said to have?

■ *Solution:*

$$W = Q \times V$$
$$= 1 \text{ C} \times 1 \text{ V}$$
$$= 1 \text{ J of energy}$$

■ INTEGRATED MATH APPLICATION:

How many coulombs of electrons would a 9 V battery have to store to have 63 J of energy?

■ *Solution:*

If $W = Q \times V$, then by transposition:

$$Q = \frac{W}{V}$$

Coulombs of electrons $(Q) = \dfrac{\text{energy in joules } (W)}{\text{battery voltage } (V)}$

$$= \frac{63 \text{ J}}{9 \text{ V}}$$

$$= 7 \text{ C of electrons}$$

or

$$7 \times 6.24 \times 10^{18} = 4.36 \times 10^{19} \text{ electrons}$$

CALCULATOR SEQUENCE

Step	Keypad Entry	Display Response
1.	[6] [3]	63
2.	[÷]	
3.	[9]	9
4.	[=]	7
5.	[×]	7
6.	[6] [.] [2] [4] [E] [1] [8]	6.24E18
7.	[=]	4.36E19

9-6-3 *Calculating Power*

Power, in connection with electricity and electronics, is the rate (t) at which electric energy (W) is converted into some other form. A power formula can therefore be derived as follows:

$$P = \frac{W}{t}$$

where P = power, in watts
W = energy, in joules
t = time, in seconds

Since $W = Q \times V$, we can replace W in the preceding formula with $Q \times V$ to obtain

$$P = \frac{W}{t} = \frac{Q \times V}{t}$$

where Q = coulombs of charge
V = voltage, in volts

Since coulombs of charge $Q = I \times t$, we can replace Q in the preceding formula with $I \times t$ to obtain

$$P = \frac{Q \times V}{t} = \frac{(I \times t) \times V}{t}$$

By canceling t, we arrive at a final formula for power:

$$P = I \times V$$

where P = power, in watts (W)
I = current, in amperes (A)
V = voltage, in volts (V)

This formula states that the amount of power delivered to a device is dependent on the electrical pressure or voltage applied across the device and the current flowing through the device.

■ INTEGRATED MATH APPLICATION:

With regard to electrical and electronic circuits, power is the rate at which electric energy is converted into some other form. In the example in Figure 9-23, electric energy will be transformed into light and heat energy by the lightbulb. Power is expressed in the unit watt, which is the number of joules of energy transformed per second (J/s). If 27 joules of electric energy is being transformed into light and heat per second, how many watts of power does the lightbulb convert?

FIGURE 9-23 **Calculating Power.**

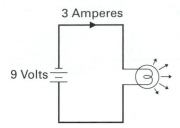

■ *Solution:*

$$\text{Power} = \frac{\text{joules}}{\text{second}}$$
$$= \frac{27 \text{ J}}{1 \text{ s}}$$
$$= 27 \text{ W}$$

The power output in the previous example could easily have been calculated by merely multiplying current by voltage to arrive at the same result.

$$\text{Power} = I \times V = 3 \text{ A} \times 9 \text{ V}$$
$$= 27 \text{ W}$$

We could say, therefore, that the lightbulb dissipates 27 watts of power, or 27 joules of energy per second.

Studying the power formula $P = I \times V$, we can also say that 1 watt of power is expended when 1 ampere of current flows through a circuit that has 1 volt applied.

Like Ohm's law, we can transpose the power formula as follows:

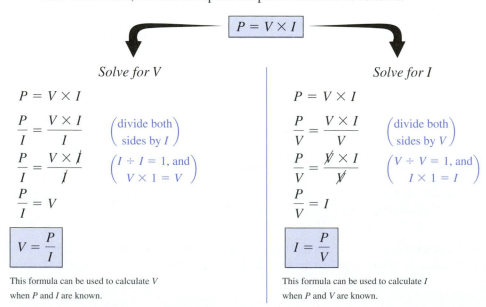

In the previous section we stated that $P = V \times I$, and by transposition we arrived at $V = P/I$ and $I = P/V$. When trying to calculate power we may not always have the values of V and I available. For example, we may know only I and R, or V and R. Substitution enables us to obtain alternative power formulas by replacing one mathematical term with an equivalent mathematical term. For example, since $I = V/R$, we could substitute V/R for I in any formula. Similarly, since $V = I \times R$, we could substitute $I \times R$ for V in any formula. The following shows how we can substitute terms in the $P = V \times I$ formula to arrive at alternative power formulas for wattage calculations.

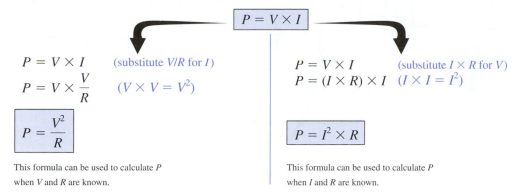

$P = V \times I$ (substitute V/R for I)

$P = V \times \dfrac{V}{R}$ ($V \times V = V^2$)

$$P = \dfrac{V^2}{R}$$

This formula can be used to calculate P when V and R are known.

$P = V \times I$ (substitute $I \times R$ for V)

$P = (I \times R) \times I$ ($I \times I = I^2$)

$$P = I^2 \times R$$

This formula can be used to calculate P when I and R are known.

■ **INTEGRATED MATH APPLICATION:**

In Figure 9-24 a 12 V battery is connected across a 36 Ω resistor. How much power does the resistor dissipate?

■ *Solution:*

Since V and R are known, the V^2/R power formula should be used.

$$\text{Power} = \frac{\text{voltage}^2}{\text{resistance}}$$
$$= \frac{(12\text{ V})^2}{36\ \Omega} = \frac{144}{36}$$
$$= 4\text{ W}$$

Four joules of heat energy are being dissipated every second.

12 V ⎓   36 Ω

FIGURE 9-24

C‍ALCULATOR S‍EQUENCE

Step	Keypad Entry	Display Response
1.	⬚1 ⬚2	12
2.	⬚x²	144
3.	⬚÷	
4.	⬚3 ⬚6	36
5.	⬚=	4

9-6-4   *Measuring Power*

A multimeter can be used to measure power by following the procedure described in Figure 9-25. In this example, the power consumed by a car's music system is being determined. After the current measurement and voltage measurement have been taken, the product of the two values will have to be calculated, since power = current × voltage ($P = I \times V$).

■ **INTEGRATED MATH APPLICATION:**

Calculate the power consumed by the car music system shown in Figure 9-25.

■ *Solution:*

Current drawn from battery by music system = 970 mA. Voltage applied to music system power input = 13.6 V.

$$P = I \times V = 970\text{ mA} \times 13.6\text{ V} = 13.2\text{ W}$$

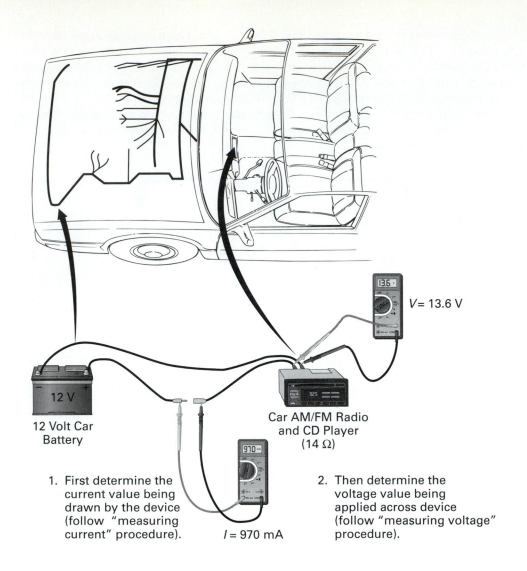

$V = 13.6$ V

12 V

12 Volt Car
Battery

Car AM/FM Radio
and CD Player
(14 Ω)

1. First determine the
   current value being
   drawn by the device
   (follow "measuring
   current" procedure).

$I = 970$ mA

2. Then determine the
   voltage value being
   applied across device
   (follow "measuring voltage"
   procedure).

3. Multiply voltage value by
   current value to obtain
   power value ($P = V \times I$).

FIGURE 9-25   **Measuring Power with a Multimeter.**

With the connection of a special current probe, as shown in Figure 9-26, some multimeters can be used to make power measurements. In this configuration, the current probe senses current, the standard meter probes measure voltage, and the multimeter performs the multiplication process and displays the power reading.

9-6-5   *The Kilowatt-Hour*

Kilowatt-Hour
1000 watts for 1 hour.

Kilowat-Hour Meter
A meter used by electric companies to measure a customer's electric power use in kilowatt-hours.

You and I pay for our electric energy in a unit called the **kilowatt-hour** (kWh). The **kilowatt-hour meter,** shown in Figure 9-27, measures how many kilowatt-hours are consumed, and the electric company then charges accordingly.

So what is a kilowatt-hour? Well, since power is the rate at which energy is used, if we multiply power and time, we can calculate how much energy has been consumed.

Energy consumed (W) = power (P) × time (t)

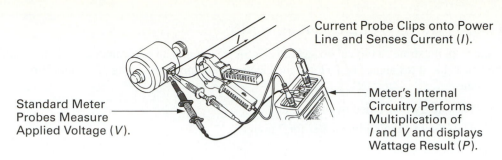

Current Probe Clips onto Power Line and Senses Current (*I*).

Standard Meter Probes Measure Applied Voltage (*V*).

Meter's Internal Circuitry Performs Multiplication of *I* and *V* and displays Wattage Result (*P*).

FIGURE 9-26   **Configuring a Multimeter to Measure Power.**

This formula uses the product of power (in watts) and time (in seconds or hours) and so we can use one of three units: the watt-second (Ws), watt-hour (Wh), or kilowatt-hour (kWh). The kilowatt-hour is most commonly used by electric companies, and by definition, a kilowatt-hour of energy is consumed when 1000 watts of power are used for 1 hour (1 kW for 1 h).

$$\text{Energy consumed (kWh)} = \text{power (kW)} \times \text{time (h)}$$

To see how this would apply, let us look at a couple of examples.

■ **INTEGRATED MATH APPLICATION:**

If we leave a 100 W light bulb on for 10 hours, how many kilowatt-hours will we be charged for?

■ *Solution:*

$$\text{Power consumed (kWh)} = \text{power (kW)} \times \text{time (hours)}$$
$$= 0.1 \text{ kW} \times 10 \text{ hours } (100 \text{ W} = 0.1 \text{ kW})$$
$$= 1 \text{ kWh}$$

CALCULATOR SEQUENCE

Step	Keypad Entry	Display Response
1.	[0] [.] [1] [E] [3]	0.1E3
2.	[×]	
3.	[1] [0]	10
4.	[=]	1E3

FIGURE 9-27   **Digital Kilowatt-Hour Meter.**

■ INTEGRATED MATH APPLICATION:

Figure 9-28 illustrates a typical household electric heater and an equivalent electrical circuit. The heater has a resistance of 7 Ω and the electric company is charging 12 cents/kWh. Calculate:

a. The energy consumed by the heater

b. The cost of running the heater for 7 hours

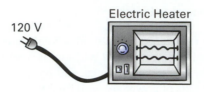
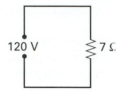

FIGURE 9-28 **An Electric Heater.**

■ *Solution:*

a. Power $(P) = \dfrac{V^2}{R} = \dfrac{120}{7}$

$= \dfrac{14400 \text{ V}}{7} = \dfrac{14.4 \text{ kV}}{7}$

$= 2057 \text{ W (approximately 2 kilowatts or 2 kW)}$

b. Energy consumed = power (kW) × time (hours)
$= 2.057 × 7 \text{ h}$
$= 14.4 \text{ kWh}$

Cost = kWh × rate = 14.4 × 12 cents
$= \$1.73$

SELF-TEST EVALUATION POINT FOR SECTION 9-6

Now that you have completed this section, you should be able to:

Use the following questions to test your understanding of Section 9-6:

■ ***Objective 9.*** *Describe and define the terms* energy, work, *and* power.

■ ***Objective 10.*** *Describe how to measure power.*

■ ***Objective 11.*** *Explain what is meant by the term* kilowatt-hour.

1. List the six basic forms of energy.

2. What is the difference between energy, work, and power?

3. List the formulas for calculating energy and power.

4. What is 1 kilowatt-hour of energy?

REVIEW QUESTIONS

Multiple Choice Questions

1. Resistance is measured in:

   a. Ohms   **c.** Amperes
   b. Volts   **d.** Siemens

2. Current is proportional to:

   a. Resistance   **c.** Both (a) and (b)
   b. Voltage   **d.** None of the above

3. Current is inversely proportional to:

   a. Resistance   **c.** Both (a) and (b)
   b. Voltage   **d.** None of the above

4. If the applied voltage is equal to 15 V and the circuit resistance equals 5 Ω, the total circuit current would be equal to:

   a. 4 A   **c.** 3 A
   b. 5 A   **d.** 75 A

5. Calculate the applied voltage if 3 mA flows through a circuit resistance of 25 kΩ.

    a. 63 mV   **c.** 77 μV

    b. 25 V   **d.** 75 V

6. Energy is measured in:

    a. Volts   **c.** Amperes

    b. Joules   **d.** Watts

7. Chemical energy within a battery is converted into:

    a. Electrical energy

    b. Mechanical energy

    c. Magnetic energy

    d. Heat energy

8. The water pump has the potential to cause water flow, just as the battery has the potential to cause:

    a. Voltage   **c.** Current

    b. Electron flow   **d.** Both (b) and (c)

9. Work is measured in:

    a. Joules   **c.** Amperes

    b. Volts   **d.** Watts

10. The device that converts one energy form to another is called a:

    a. Transformer   **c.** Transit

    b. Transducer   **d.** Transistor

11. Power is the rate at which energy is transformed and is measured in:

    a. Joules   **c.** Volts

    b. Watts   **d.** Amperes

12. Power is measured by using a (an):

    a. Ammeter   **c.** Ohmmeter

    b. Voltmeter   **d.** Wattmeter

13. What would be the power dissipated by a 2 kΩ carbon composition resistor when a current of 20 mA is flowing through it?

    a. 40 W   **c.** 1.25 W

    b. 0.8 W   **d.** None of the above

14. Which of the following is the unit of temperature in the International System of Units?

    a. Celsius   **c.** Kelvin

    b. Fahrenheit   **d.** Rankine

15. Which of the following temperature detectors has a positive temperature coefficient?

    a. RTD

    b. Thermistor

    c. TFD

    d. Both (a) and (b)

    e. Both (a) and (c)

16. Photoresistors are also called:

    a. TFDs   **c.** RTDs

    b. LDRs   **d.** TGFs

17. Photoresistors have:

    a. A positive temperature coefficient

    b. A negative temperature coefficient

    c. Neither (a) nor (b)

18. Resistance is measured with a (an):

    a. Wattmeter   **c.** Ohmmeter

    b. Milliammeter   **d.** A 100 meter

19. Which of the following is *not* true?

    a. Precision resistors have a tolerance of > 5%.

    b. General-purpose resistors can be recognized because they have either three or four bands.

    c. The fifth band indicates the tolerance of a precision resistor.

    d. The third band of a general-purpose resistor specifies the multiplier.

20. The first band on either a general-purpose or a precision resistor can never be:

    a. Brown

    b. Red or black stripe

    c. Black

    d. Red

Communication Skill Questions

21. What is resistance? (9-1)

22. Briefly describe why:

    a. Current is proportional to voltage. (9-1-2)

    b. Current is inversely proportional to resistance. (9-1-2)

23. State Ohm's law. (9-1-1)

24. List the three forms of Ohm's law. (9-1-2)

25. List the six basic forms of energy. (9-6)

26. Briefly describe the following terms:

    a. Potential energy (9-6)

    b. Kinetic energy (9-6)

27. What is a transducer? (9-6)

28. Define power. (9-6-1)

29. Give three formulas for electric power. (9-6-1)

30. Give the units for each of the following:

    a. Energy   **e.** Work

    b. Power   **f.** Current

    c. Voltage   **g.** Charge

    d. Resistance   **h.** Conductance

31. Explain why a resistor's size determines its wattage rating. (9-6-1)

32. Why are resistors given a tolerance figure, and which is best, a small or a large tolerance? (9-4)

33. Define linear and tapered resistive tracks. (9-2)

34. List the rules that should be applied when using an ohmmeter to measure resistance. (9-3)

35. Give the color code for the following resistor values: (9-4)

    a. 1.2 MΩ, ±10%   **c.** 27 kΩ, ±20%

    b. 10 Ω, ±5%   **d.** 273 kΩ, ±0.5%

Practice Problems

36. An electric heater with a resistance of 6 Ω is connected across a 120 V wall outlet.

   a. Calculate the current flow.
   b. Draw the schematic diagram.

37. What source voltage would be needed to produce a current flow of 8 mA through a 16 kΩ resistor?

38. If an electric toaster draws 10 A when connected to a power outlet of 120 V, what is its resistance?

39. Calculate the current flowing through the following light-bulbs when they are connected across 120 V:

   a. 300 W   **c.** 60 W
   b. 100 W   **d.** 25 W

40. Indicate which of the following unit pairs is larger:

   a. Millivolts or volts
   b. Microamperes or milliamperes
   c. Kilowatts or watts
   d. Kilohms or megohms

41. Calculate the unknown resistance in a circuit when an ammeter indicates that a current of 12 mA is flowing and a voltmeter indicates 12 V.

42. What battery voltage would use 1000 J of energy to move 40 C of charge through a circuit?

43. Calculate the resistance of a light bulb that passes 500 mA of current when 120 V is applied. What is the bulb's wattage?

44. Which of the following circuits has the largest resistance and which has the smallest?

   a. $V = 120$ V, $I = 20$ mA
   b. $V = 12$ V, $I = 2$ A
   c. $V = 9$ V, $I = 100\ \mu$A
   d. $V = 1.5$ V, $I = 4$ mA

45. Convert the following:

   a. 1000 W = _____ kW
   b. 0.345 W = _____ mW
   c. 1250×10^3 W = _____ MW
   d. 0.00125 W = _____ μW

46. What is the value of the resistor when a current of 4 A is causing 100 W to be dissipated?

47. How many kilowatt-hours of energy are consumed in each of the following:

   a. 7500 W for 1 hour
   b. 25 W for 6 hours
   c. 127,000 W for half an hour

48. What is the maximum output power of a 12 V, 300 mA power supply?

49. If a 5.6 kΩ resistor has a tolerance of $\pm$ 10%, what would be the allowable deviation in resistance above and below 5.6 kΩ?

50. (a) If a current of 50 mA is flowing through a 10 kΩ, 25 W resistor, how much power is the resistor dissipating? (b) Can the current be increased, and if so by how much?

Web Site Questions

Go to the Web site http://www.prenhall.com/cook, select the textbook *Mathematics for Electronics and Computers,* select this chapter, and then follow the instructions when answering the multiple-choice practice problems.

Series DC Circuits

The First Computer Bug

Mathematician Grace Murray Hopper, an extremely independent U.S. naval officer, was assigned to the Bureau of Ordnance Project at Harvard during World War II. As Hopper recalled, "We were not programmers in those days, the word had not yet come over from England. We were 'coders'," and with her colleagues she was assigned to compute ballistic firing tables on the Harvard Mark 1 computer. In carrying out this task, Hopper developed programming method fundamentals that are still in use.

Hopper is also credited, on a less important note, with creating a term frequently used today falling under the category of computer jargon. During the hot summer of 1945, the computer developed a mysterious problem. Upon investigating, Hopper discovered that a moth had somehow strayed into the computer and prevented the operation of one of the thousands of electromechanical relay switches. In her usual meticulous manner, Hopper removed the remains, then taped and entered them into the logbook. In her own words, "From then on, when an officer came in to ask if we were accomplishing anything, we told him we were 'debugging' the computer," a term that is still used to describe the process of finding problems in a computer program.

Outline and Objectives

VIGNETTE: THE FIRST COMPUTER BUG

INTRODUCTION

10-1 COMPONENTS IN SERIES

Objective 1: Describe a series circuit.

Objective 2: Identify series circuits.

Objective 3: Connect components so that they are in series with one another.

10-2 CURRENT IN A SERIES CIRCUIT

Objective 4: Describe why current remains the same throughout a series circuit.

10-3 RESISTANCE IN A SERIES CIRCUIT

Objective 5: Explain how to calculate total resistance in a series circuit.

Objective 6: Explain how Ohm's law can be applied to calculate current, voltage, and resistance.

10-4 VOLTAGE IN A SERIES CIRCUIT

Objective 7: Describe why the series circuit is known as a voltage divider.

Objective 8: Describe a fixed and a variable voltage divider.

10-4-1 A Voltage Source's Internal Resistance

10-4-2 Fixed Voltage Divider

10-4-3 Variable Voltage Divider

10-5 POWER IN A SERIES CIRCUIT

Objective 9: Explain how to calculate power in a series circuit.

Objective 10: Explain the maximum power transfer theorem.

10-5-1 Maximum Power Transfer

MULTIPLE CHOICE QUESTIONS

COMMUNICATION SKILL QUESTIONS

PRACTICE PROBLEMS

WEB SITE QUESTIONS

Introduction

A series circuit, by definition, is the connecting of components end to end in a circuit to provide a single path for the current. This is true not only for resistors but also for other components that can be connected in series. In all cases, however, the components are connected in succession or strung together one after another so that only one path for current exists between the negative $(-)$ and positive $(+)$ terminals of the supply.

10-1   COMPONENTS IN SERIES

Figure 10-1 illustrates five examples of **series** resistive **circuits**. In all five examples, you will notice that the resistors are connected "in-line" with one another so that the current through the first resistor must pass through the second resistor, and the current through the second resistor must pass through the third, and so on.

■ INTEGRATED MATH APPLICATION:

In Figure 10-2(a), seven resistors are laid out on a table top. Using a protoboard, connect all the resistors in series, starting at R_1, and proceeding in numerical order through the resistors until reaching R_7. After completing the circuit, connect the series circuit to a dc power supply.

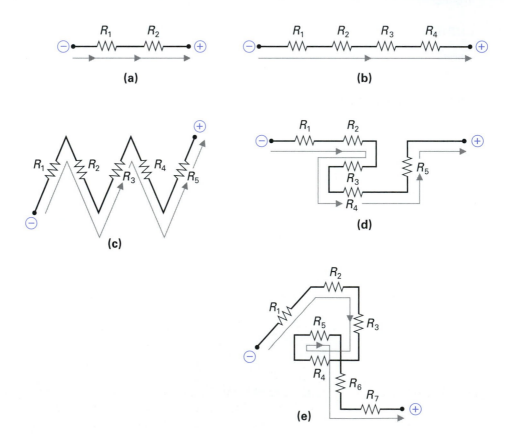

FIGURE 10-1   Five Series Resistive Circuits.

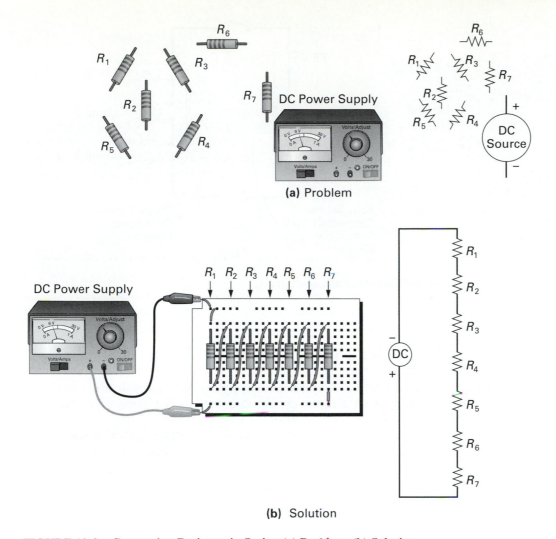

FIGURE 10-2 **Connecting Resistors in Series. (a) Problem. (b) Solution.**

■ *Solution:*

In Figure 10-2(b), you can see that all the resistors are now connected in series (end to end), and the current has only one path to follow from negative to positive.

■ **INTEGRATED MATH APPLICATION:**

Figure 10-3(a) shows four 1.5 V cells and three lamps. Using wires, connect all the cells in series to create a 6 V battery source. Then, connect the three lamps in series with one another, and finally, connect the 6 V battery source across the three-series-connected-lamp load.

■ *Solution:*

In Figure 10-3(b) you can see the final circuit containing a source, made up of four series-connected 1.5 V cells, and a load, consisting of three series-connected lamps. When cells are connected in series, the total voltage (V_T) will be equal to the sum of all the cell voltages:

$$V_{T} = V_1 + V_2 + V_3 + V_4 = 1.5\,\text{V} + 1.5\,\text{V} + 1.5\,\text{V} + 1.5\,\text{V} = 6\,\text{V}$$

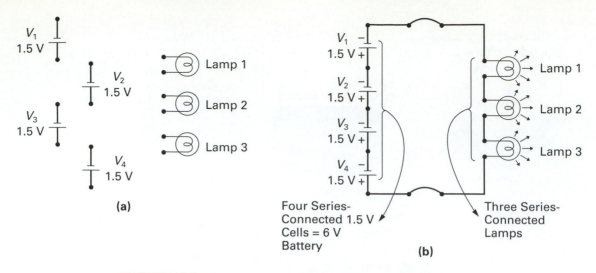

(a)

Four Series-Connected 1.5 V Cells = 6 V Battery

Three Series-Connected Lamps

(b)

FIGURE 10-3 **Series-Connected Cells and Lamps.**

10-2   CURRENT IN A SERIES CIRCUIT

The current in a series circuit has only one path to follow and cannot divert in any other direction. The current through a series circuit, therefore, is the same throughout that circuit.

Returning once again to the water analogy, you can see in Figure 10-4(a) that if 2 gallons of water per second are being supplied by the pump, 2 gallons per second must be pulled into the pump. If the rate at which water is leaving and arriving at the pump is the same, 2 gallons of water per second must be flowing throughout the circuit. It can be said,

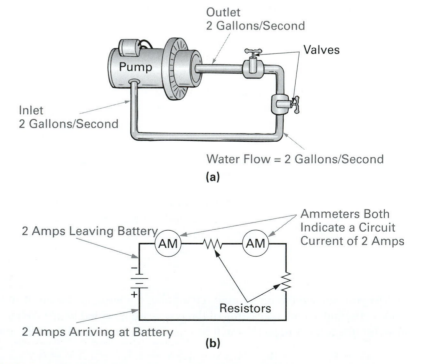

FIGURE 10-4 **Series Circuit Current. (a) Fluid System. (b) Electric System.**

therefore, that the same value of water flow exists throughout a series-connected fluid system. This rule will always remain true, for if the valves were adjusted to double the opposition to flow, then half the flow, or 1 gallon of water per second, would be leaving the pump and flowing throughout the system.

Similarly, with the electronic series circuit shown in Figure 10-4(b), there is a total of 2 A leaving and 2 A arriving at the battery, and so the same value of current exists throughout the series-connected electronic circuit. If the circuit's resistance was changed, a new value of series circuit current would be present throughout the circuit. For example, if the resistance of the circuit was doubled, then half the current, or 1 A, would leave the battery, but that same value of 1 A would flow throughout the entire circuit. This series circuit current characteristic can be stated mathematically as

$$I_T = I_1 = I_2 = I_3 = \cdots$$

Total current = current through R_1 = current through R_2 = current through R_3, and so on.

FIGURE 10-5 **Total Current Example. (a) Schematic. (b) Protoboard Circuit. (c) Circuit Analysis Table.**

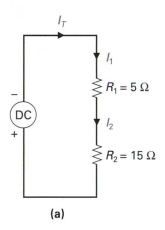

(a)

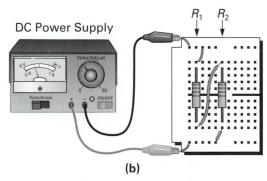

(b)

CIRCUIT ANALYSIS TABLE			
Resistance $R = V/I$	Current $I = V/R$	Voltage $V = I \times R$	Power $P = V \times I$
$R_1 = 5\ \Omega$	$I_1 = 1\ A$		
$R_2 = 15\ \Omega$	$I_2 = 1\ A$		
	$I_T = 1\ A$		

(c)

In Figure 10-5(a) and (b), a total current (I_T) of 1 A is flowing out of the negative terminal of the dc power supply, through two end-to-end resistors R_1 and R_2, and returning to the positive terminal of the dc power supply. Calculate:

    a. The current through R_1 (I_1)

    b. The current through R_2 (I_2)

■ *Solution:*

Since R_1 and R_2 are connected in series, the current through both will be the same as the circuit current, which is equal to 1 A.

$$I_T = I_1 = I_2$$
$$1\text{ A} = 1\text{ A} = 1\text{ A}$$

The details of this circuit are summarized in the analysis table shown in Figure 10-5(c).

SELF-TEST EVALUATION POINT FOR SECTIONS 10-1 AND 10-2

Now that you have completed these sections, you should be able to:

■ **Objective 1.** *Describe a series circuit.*

■ **Objective 2.** *Identify series circuits.*

■ **Objective 3.** *Connect components so that they are in series with one another.*

■ **Objective 4.** *Describe why current remains the same throughout a series circuit.*

Use the following questions to test your understanding of Sections 10-1 and 10-2.

1. What is a series circuit?

2. What is the current flow through each of eight series-connected 8 Ω resistors if 8 A total current is flowing out of a battery?

10-3   RESISTANCE IN A SERIES CIRCUIT

Resistance is the opposition to current flow, and in a series circuit every resistor in series offers opposition to the current flow. In the water analogy of Figure 10-4, the total resistance or opposition to water flow is the sum of the two individual valve opposition values. Like the battery, the pump senses the total opposition in the circuit offered by all the valves or resistors, and the amount of current that flows is dependent on this resistance or opposition.

    The total resistance in a series-connected electronic resistive circuit is thus equal to the sum of all the individual resistances, as shown in Figure 10-6(a) through (d). An equivalent circuit can be drawn for each of the circuits in Figure 10-6(b), (c), and (d) with one resistor of a value equal to the sum of all the series resistance values.

    No matter how many resistors are connected in series, the total resistance or opposition to current flow is always equal to the sum of all the resistor values. This formula can be stated mathematically as

$$R_T = R_1 + R_2 + R_3 + \cdots$$

Total resistance = value of R_1 + value of R_2 + value of R_3 and so on.

Equivalent Resistance (R_{eq})

Total resistance of all the individual resistances in a circuit.

    Total resistance (R_T) is the only opposition a source can sense. It does not see the individual separate resistors but one **equivalent resistance.** Based on the source's voltage and the circuit's total resistance, a value of current will be produced to flow through the circuit (Ohm's law, $I = V/R$).

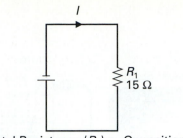

Total Resistance (R_T) or Opposition to
Current Flow =
$R_T = R_1$
$\quad = 15\ \Omega$

(a)

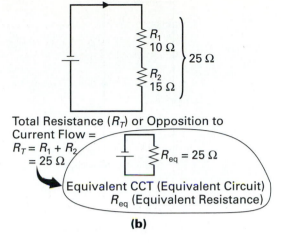

Total Resistance (R_T) or Opposition to
Current Flow =
$R_T = R_1 + R_2$
$\quad = 25\ \Omega$

$R_{eq} = 25\ \Omega$

Equivalent CCT (Equivalent Circuit)
R_{eq} (Equivalent Resistance)

(b)

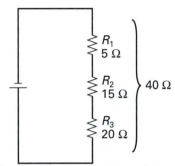

Total Resistance or Opposition to
Current Flow =
$R_T = R_1 + R_2 + R_3$
$\quad = 5 + 15 + 20$
$\quad = 40\ \Omega$

$R_{eq} = 40\ \Omega$

Equivalent CCT

(c)

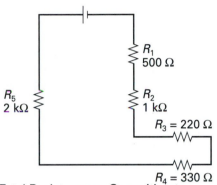

Total Resistance or Opposition to
Current Flow =
$R_T = R_1 + R_2 + R_3 + R_4 + R_5$
$\quad = 500 + 1000 + 220 + 330 + 2000$
$\quad = 4.05\ \text{k}\Omega$

$R_{eq} = 4.05\ \text{k}\Omega$

Equivalent CCT

(d)

FIGURE 10-6 **Total or Equivalent Resistance.**

■ **INTEGRATED MATH APPLICATION:**

Referring to Figure 10-7(a) and (b), calculate

   a. The circuit's total resistance

   b. The current flowing through R_2

■ *Solution:*

   a. $R_T = R_1 + R_2 + R_3 + R_4$
       $= 25\ \Omega + 20\ \Omega + 33\ \Omega + 10\ \Omega$
       $= 88\ \Omega$

   b. $I_T = I_1 = I_2 = I_3 = I_4$. Therefore, $I_2 = I_T = 3\ \text{A}$.

The details of this circuit are summarized in the analysis table shown in Figure 10-7(c).

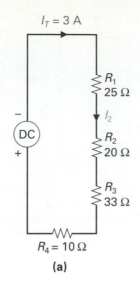

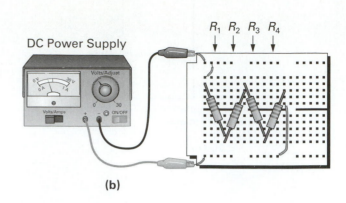

(b)

(a)

CIRCUIT ANALYSIS TABLE			
Resistance $R = V/I$	Current $I = V/R$	Voltage $V = I \times R$	Power $P = V \times I$
$R_1 = 25\ \Omega$	$I_1 = 3\ A$		
$R_2 = 20\ \Omega$	$I_2 = 3\ A$		
$R_3 = 33\ \Omega$	$I_3 = 3\ A$		
$R_4 = 10\ \Omega$	$I_4 = 3\ A$		
$R_T = 88\ \Omega$	$I_T = 3\ A$		

(c)

FIGURE 10-7   **Total Resistance Example. (a) Schematic. (b) Protoboard Circuit. (c) Circuit Analysis Table.**

■ INTEGRATED MATH APPLICATION:

Figure 10-8(a) and (b) shows how a single-pole three-position switch is being used to provide three different lamp brightness levels. In position ① R_1 is placed in series with the lamp, in position ② R_2 is placed in series with the lamp, and in position ③ R_3 is placed in series with the lamp. If the lamp has a resistance of 75 Ω, calculate the three values of current for each switch position.

■ *Solution:*

$$\text{Position 1: } R_T = R_1 + R_{\text{lamp}}$$
$$= 25\ \Omega + 75\ \Omega = 100\ \Omega$$
$$I_T = \frac{V_T}{R_T} = \frac{12\ V}{100\ \Omega} = 120\ \text{mA}$$

$$\text{Position 2: } R_T = R_2 + R_{\text{lamp}} = 50\ \Omega + 75\ \Omega = 125\ \Omega$$
$$I_T = \frac{V_T}{R_T} = \frac{12\ V}{125\ \Omega} = 96\ \text{mA}$$

$$\text{Position 3: } R_T = R_3 + R_{\text{lamp}} = 75\ \Omega + 75\ \Omega = 150\ \Omega$$
$$I_T = \frac{V_T}{R_T} = \frac{12\ V}{150\ \Omega} = 80\ \text{mA}$$

The details of this circuit are summarized in the analysis table shown in Figure 10-8(c).

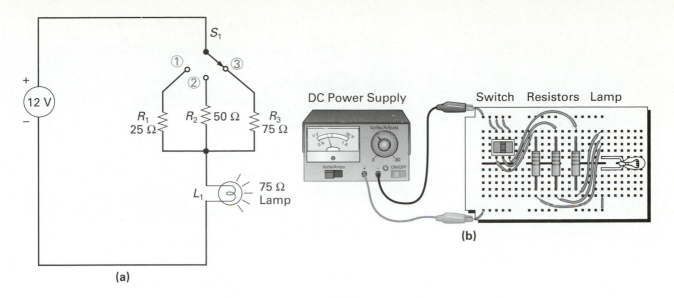

FIGURE 10-8 **Three-Position Switch Controlling Lamp Brightness. (a) Schematic. (b) Proto-board Circuit. (c) Circuit Analysis Table.**

CIRCUIT ANALYSIS TABLE

	Resistance $R = V/I$	Current $I = V/R$	Voltage $V = I \times R$	Power $P = V \times I$
$S_1 = ①$	$R_1 = 25\ \Omega$	$I_1 = 120\ \text{mA}$		
$S_1 = ②$	$R_2 = 50\ \Omega$	$I_2 = 96\ \text{mA}$		
$S_1 = ③$	$R_3 = 75\ \Omega$	$I_3 = 80\ \text{mA}$		

$R_{L1} = 75\ \Omega$

(c)

SELF-TEST EVALUATION POINT FOR SECTION 10-3

Now that you have completed this section, you should be able to:

■ *Objective 5.* *Explain how to calculate total resistance in a series circuit.*

■ *Objective 6.* *Explain how Ohm's law can be applied to calculate current, voltage, and resistance.*

Use the following questions to test your understanding of Section 10-3.

1. State the total resistance formula for a series circuit.
2. Calculate R_T if $R_1 = 2\ \text{k}\Omega$, $R_2 = 3\ \text{k}\Omega$, and $R_3 = 4700\ \Omega$.

10-4   VOLTAGE IN A SERIES CIRCUIT

A potential difference or voltage drop will occur across each resistor in a series circuit when current is flowing. The amount of voltage drop is dependent on the value of the resistor and the amount of current flow. This idea of potential difference or voltage drop is best explained by returning to the water analogy. In Figure 10-9(a), you can see that the high pressure from the pump's outlet is present on the left side of the valve. On the right side of the valve, however, the high pressure is no longer present. The high potential that exists on the left of the valve is not present on the right, so a potential or pressure difference is said to exist across the valve.

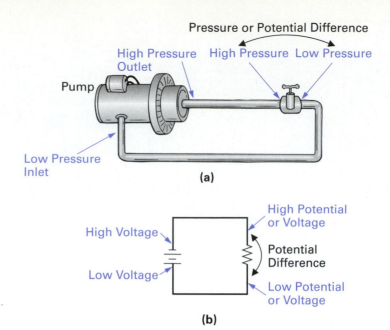

Pressure or Potential Difference

High Pressure Outlet

High Pressure Low Pressure

Pump

Low Pressure Inlet

(a)

High Potential or Voltage

High Voltage

Potential Difference

Low Voltage

Low Potential or Voltage

(b)

FIGURE 10-9 **Series Circuit Voltage. (a) Fluid Analogy of Potential Difference. (b) Electrical Potential Difference.**

Similarly, with the electronic circuit shown in Figure 10-9(b), the battery produces a high voltage or potential that is present at the top of the resistor. The high voltage that exists at the top of the resistor, however, is not present at the bottom. Therefore, a potential difference or voltage drop is said to occur across the resistor. This voltage drop that exists across resistors can be found by utilizing Ohm's law: $V = I \times R$.

■ INTEGRATED MATH APPLICATION:

Referring to Figure 10-10, calculate:

a. Total resistance (R_T)

b. Amount of series current flowing throughout the circuit (I_T)

c. Voltage drop across R_1

d. Voltage drop across R_2

e. Voltage drop across R_3

■ *Solution:*

a. Total resistance $(R_T) = R_1 + R_2 + R_3$
$$= 20\ \Omega + 30\ \Omega + 50\ \Omega$$
$$= 100\ \Omega$$

b. Total current $(I_T) = \dfrac{V_{source}}{R_T}$
$$= \dfrac{100\ V}{100\ \Omega}$$
$$= 1\ A$$

The same current will flow through the complete series circuit, so the current through R_1 will equal 1 A, the current through R_2 will equal 1 A, and the current through R_3 will equal 1 A.

c. Voltage across R_1 $(V_{R1}) = I_1 \times R_1$
$$= 1\ A \times 20\ \Omega$$
$$= 20\ V$$

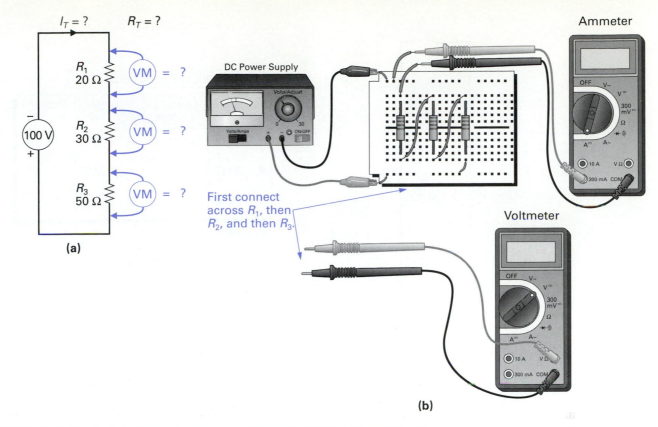

FIGURE 10-10 Series Circuit Example. (a) Schematic. (b) Protoboard Circuit.

d. Voltage across R_2 $(V_{R2}) = I_2 \times R_2$
$$= 1\,A \times 30\,\Omega$$
$$= 30\,V$$

e. Voltage across R_3 $(V_{R3}) = I_3 \times R_3$
$$= 1\,A \times 50\,\Omega$$
$$= 50\,V$$

Figure 10-11(a) shows the schematic, and Figure 10-11(b) shows the analysis table for this example, with all of the calculated data inserted. As you can see, the 20 Ω resistor drops 20 V, the 30 Ω resistor has 30 V across it, and the 50 Ω resistor drops 50 V. From this example, you will notice that the larger the resistor value, the larger the voltage drop. Resistance and voltage drops are consequently proportional to one another.

> The voltage drop across a device is proportional to its resistance.

$$V_{\text{drop}} \uparrow = I(\text{constant}) \times R \uparrow$$
$$V_{\text{drop}} \downarrow = I(\text{constant}) \times R \downarrow$$

Another interesting point you may have noticed from Figure 10-11 is that if you were to add up all the voltage drops around a series circuit, they would equal the source (V_S) applied:

> Total voltage applied (V_S or V_T) $= V_{R1} + V_{R2} + V_{R3} + \cdots$

In the example in Figure 10-11, you can see that this is true, since

$$100\,V = 20\,V + 30\,V + 50\,V$$
$$100\,V = 100\,V$$

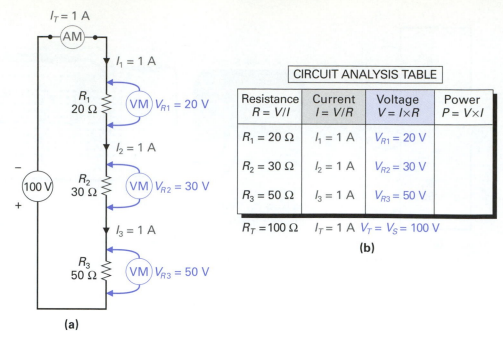

FIGURE 10-11   Voltage Drop and Resistance. (a) Schematic. (b) Circuit Analysis Table.

TIME LINE

German physicist Gustav Robert Kirchhoff (1824–1879) extended Ohm's law and developed two important laws of his own, known as Kirchhoff's voltage law and Kirchhoff's current law.

Kirchhoff's Voltage Law

The algebraic sum of the voltage drops in a closed path circuit is equal to the algebraic sum of the source voltage applied.

The series circuit has in fact divided up the applied voltage, and it appears proportionally across all the individual resistors. This characteristic was first observed by Gustav Kirchhoff in 1847. In honor of his discovery, this effect is known as **Kirchhoff's voltage law,** which states: The sum of the voltage drops in a series circuit is equal to the total voltage applied.

To summarize the effects of current, resistance, and voltage in a series circuit so far, we can say that:

1. The current in a series circuit has only one path to follow.
2. The value of current in a series circuit is the same throughout the entire circuit.
3. The total resistance in a series circuit is equal to the sum of all the resistances.
4. Resistance and voltage drops in a series circuit are proportional to one another, so a large resistance will have a large voltage drop, and a small resistance will have a small voltage drop.
5. The sum of the voltage drops in a series circuit is equal to the total voltage applied.

■   **INTEGRATED MATH APPLICATION:**

First, calculate the voltage drop across the resistor R_1 in the circuit in Figure 10-12(a) for a resistance of 4 Ω. Then, change the 4 Ω resistor to a 2 Ω resistor and recalculate the voltage drop across the new resistance value. Use a constant source of 4 V.

■   *Solution:*

Referring to Figure 10-12(b), you can see that when $R_1 = 4\ \Omega$, the voltage across R_1 can be calculated by using Ohm's law and is equal to

$$V_{R1} = I_1 \times R_1$$
$$V_{R1} = 1\,\text{A} \times 4\,\Omega$$
$$= 4\,\text{V}$$

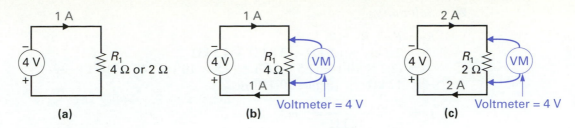

FIGURE 10-12   Single Resistor Circuit.

If the resistance is now changed to 2 Ω, as shown in Figure 10-12(c), the current flow within the circuit will be equal to

$$I = \frac{V_S}{R} = \frac{4\ V}{2\ \Omega} = 2\ A$$

The voltage dropped across the 2 Ω resistor will still be equal to

$$V_{R1} = I_1 \times R_1$$
$$= 2\ A \times 2\ \Omega$$
$$= 4\ V$$

As you can see from this example, if only one resistor is connected in a series circuit, the entire applied voltage appears across this resistor. The value of this single resistor will determine the amount of current flow through the circuit, and this value of circuit current will remain the same throughout.

■   **INTEGRATED MATH APPLICATION:**

Referring to Figure 10-13(a), calculate the following, and then draw the circuit schematic again with all of the new values inserted.

   a. Total circuit resistance
   b. Value of circuit current (I_T)
   c. Voltage drop across each resistor

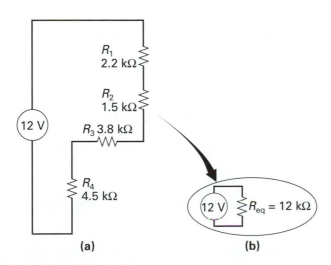

FIGURE 10-13   Series Circuit Example.

Solution:

a. $R_T = R_1 + R_2 + R_3 + R_4$
   $= 2.2 \text{ k}\Omega + 1.5 \text{ k}\Omega + 3.8 \text{ k}\Omega + 4.5 \text{ k}\Omega$
   $= (2.2 \times 10^3) + (1.5 \times 10^3) + (3.8 \times 10^3) + (4.5 \times 10^3)$
   $= 12 \text{ k}\Omega$ [Figure 10-13(b)]

b. $I_T = \dfrac{V_S}{R_T} = \dfrac{12 \text{ V}}{12 \text{ k}\Omega} = 1 \text{ mA}$

c. Voltage drop across each resistor:

$$V_{R1} = I_T \times R_1$$
$$= 1 \text{ mA} \times 2.2 \text{ k}\Omega$$
$$= 2.2 \text{ V}$$

$$V_{R2} = I_T \times R_2$$
$$= 1 \text{ mA} \times 1.5 \text{ k}\Omega$$
$$= 1.5 \text{ V}$$

$$V_{R3} = I_T \times R_3$$
$$= 1 \text{ mA} \times 3.8 \text{ k}\Omega$$
$$= 3.8 \text{ V}$$

$$V_{R4} = I_T \times R_4$$
$$= 1 \text{ mA} \times 4.5 \text{ k}\Omega$$
$$= 4.5 \text{ V}$$

Figure 10-14(a) shows the schematic diagram for this example with all of the values inserted, and Figure 10-14(b) shows this circuit's analysis table.

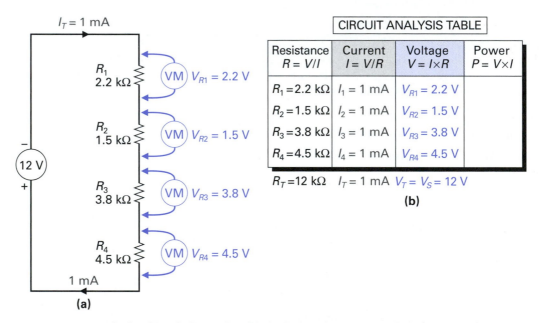

FIGURE 10-14   **Series Circuit Example with All Values Inserted. (a) Schematic. (b) Circuit Analysis Table.**

10-4-1   *A Voltage Source's Internal Resistance*

Figure 10-15(a) illustrates a battery on the left and its symbol on the right. When a circuit or system is connected across a battery, as shown in Figure 10-15(a), the circuit or system can be represented by its equivalent value of resistance, called the *load resistance* (R_L). In Figure

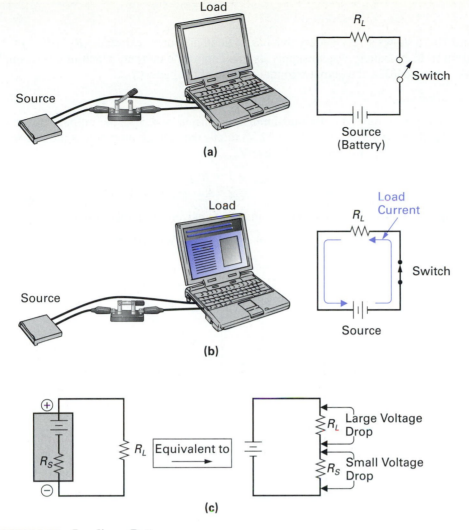

FIGURE 10-15   Loading a Battery.

10-15(a) the switch is open, so the battery is not loaded and no load current flows. In Figure 10-15(b) the switch has been closed and so the battery has a completed current path. As a result, a load current will flow, the value of which depends on the load resistance and battery voltage of 12 V.

The battery just discussed is known as an *ideal voltage source* because its voltage did not change from a *no-load* to *full-load* condition. In reality, there is no such thing as an ideal voltage source. Batteries or any other type of voltage source are not 100% efficient. If a voltage source were 100% efficient, it would be able to generate electrical energy at its output without the accompanying heat. A voltage source's inefficiency is represented by an internal resistor connected in series with the battery symbol, as seen in Figure 10-15(c). As you can see in the schematic circuit illustrated in this figure, the load resistance R_L and the source resistance R_S (pronounced "R sub L", and "R sub S") together form the total resistive load.

Very little voltage will be dropped across R_S, as it is normally small compared with R_L. Internal inefficiencies of batteries and all other voltage sources must always be kept small because a large R_S would drop a greater amount of the supply voltage, resulting in less output voltage to the load and therefore a waste of electrical power. To give an example, a lead–acid cell will typically have an internal resistance of 0.01 Ω (10 mΩ).

■ **INTEGRATED MATH APPLICATION:**

Figure 10-16 shows a 12 V battery that has an internal source resistance (R_S or R_{int}) of 0.5 Ω (500 mΩ). If the battery was to supply its maximum safe current, which in this example is 2.5 A, what would be the output terminal voltage of the battery?

■ *Solution:*

If the battery was supplying its maximum safe current of 2.5 A, the output voltage (V_{out}) would equal the source voltage ($V_S = 12$ V) minus the voltage drop across the internal battery resistance (V_{Rint}). First let us calculate V_{Rint}.

$$V_{Rint} = 1 \times R_{int}$$
$$= 2.5 \text{ A} \times 0.5 \text{ Ω} = 1.25 \text{ V}$$

The output voltage under full load (maximum safe current) will therefore be

$$V_{out} = V_S - V_{Rint} = 12 \text{ V} - 1.25 \text{ V} = 10.75 \text{ V}$$

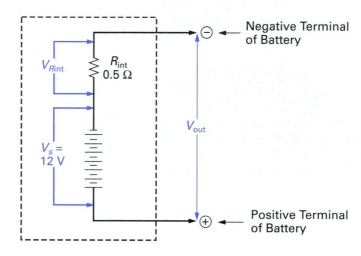

FIGURE 10-16 Voltage Drop across a Battery's Internal Series Resistance.

10-4-2 *Fixed Voltage Divider*

A series-connected circuit is often referred to as a *voltage-divider circuit,* because the total voltage applied (V_T) or source voltage (V_S) is divided and dropped proportionally across all the resistors in a series circuit. The amount of voltage dropped across a resistor is proportional to the value of resistance, and so a larger resistance will have a larger voltage drop, while a smaller resistance will have a smaller voltage drop.

The voltage dropped across a resistor is normally a factor that needs to be calculated. The voltage-divider formula allows you to calculate the voltage drop across any resistor without having to first calculate the value of circuit current. This formula is stated as

$$V_X = \frac{R_X}{R_T} \times V_S$$

where V_X = voltage dropped across selected resistor
  R_X = selected resistor's value
  R_T = total series circuit resistance
  V_S = source or applied voltage

Figure 10-17 illustrates a circuit from a previous example. To calculate the voltage drop across each resistor, we would normally have to:

1. Calculate the total resistance by adding up all the resistance values.
2. Once we have the total resistance and source voltage (V_S), we could then calculate current.
3. Having calculated the current flowing through each resistor, we could then use the current value to calculate the voltage dropped across any one of the four resistors merely by multiplying current by the individual resistance value.

The voltage-divider formula allows us to bypass the last two steps in this procedure. If we know total resistance, supply voltage, and the individual resistance values, we can calculate the voltage drop across the resistor without having to calculate steps 2 and 3. For example, what would be the voltage dropped across R_2 and R_4 for the circuit shown in Figure 10-17? The voltage dropped across R_2 is

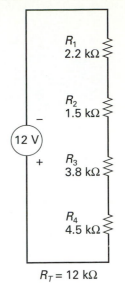

$$R_T = 12 \text{ k}\Omega$$

FIGURE 10-17 Series Circuit Example.

CALCULATOR SEQUENCE

Step	Keypad Entry	Display Response
1.	$\boxed{1}\boxed{.}\boxed{5}\boxed{E}\boxed{3}$	1.5E3
2.	$\boxed{+}$	
3.	$\boxed{1}\boxed{2}\boxed{E}\boxed{3}$	12E3
4.	$\boxed{\times}$	0.125
5.	$\boxed{1}\boxed{2}$	12
6.	$\boxed{=}$	1.5

$$V_{R2} = \frac{R_2}{R_T} \times V_S$$
$$= \frac{1.5 \text{ k}\Omega}{1.2 \text{ k}\Omega} \times 12 \text{ V}$$
$$= 1.5 \text{ V}$$

The voltage dropped across R_4 is

$$V_{R4} = \frac{R_4}{R_T} \times V_S$$
$$= \frac{4.5 \text{ k}\Omega}{12 \text{ k}\Omega} \times 12 \text{ V}$$
$$= 4.5 \text{ V}$$

The voltage-divider formula could also be used to find the voltage drop across two or more series-connected resistors. For example, referring again to the example circuit shown in Figure 10-17, what would be the voltage dropped across R_2 and R_3 combined? The voltage across $R_2 + R_3$ can be calculated using the voltage-divider formula as follows:

$$V_{R2} \text{ and } V_{R3} = \frac{R_2 + R_3}{R_T} \times V_S$$
$$= \frac{5.3 \text{ k}\Omega}{12 \text{ k}\Omega} \times 12 \text{ V}$$
$$= 5.3 \text{ V}$$

As can be seen in Figure 10-18, the voltage drop across R_2 and R_3 is 5.3 V.

FIGURE 10-18 Series Circuit Example.

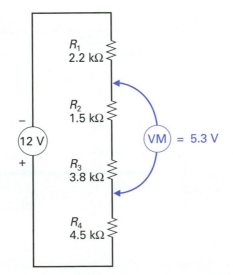

■ INTEGRATED MATH APPLICATION:

Referring to Figure 10-19(a), calculate the voltage drop across:

   a. R_1, R_2, and R_3 separately

   b. R_2 and R_3 combined

   c. R_1, R_2, and R_3 combined

■ *Solution:*

   a. The voltage drop across a resistor is proportional to the resistance value. The total resistance (R_T) in this circuit is 100 Ω or 100% of R_T. R_1 is 20% of the total resistance, so 20% of the source voltage will appear across R_1. R_2 is 30% of the total resistance, so 30% of the source voltage will appear across R_2. R_3 is 50% of the total resistance, so 50% of the source voltage will appear across R_3. This was a very simple problem in which the figures worked out very neatly. The voltage-divider formula achieves the very same thing by calculating the ratio of the resistance value to the total resistance. This percentage is then multiplied by the source voltage in order to find the desired resistor's voltage drop:

$$
\begin{aligned}
V_{R1} &= \frac{R_1}{R_T} \times V_S \\
&= \frac{20\ \Omega}{100\ \Omega} \times V_S \\
&= 0.2 \times 100\ \text{V} \qquad (20\%\ \text{of}\ 100\ \text{V}) \\
&= 20\ \text{V}
\end{aligned}
$$

$$
\begin{aligned}
V_{R2} &= \frac{R_2}{R_T} \times V_S \\
&= \frac{30\ \Omega}{100\ \Omega} \times V_S \\
&= 0.3 \times 100\ \text{V} \qquad (30\%\ \text{of}\ 100\ \text{V}) \\
&= 30\ \text{V}
\end{aligned}
$$

$$
\begin{aligned}
V_{R3} &= \frac{R_3}{R_T} \times V_S \\
&= \frac{50\ \Omega}{100\ \Omega} \times V_S \\
&= 0.5 \times 100\ \text{V} \qquad (50\%\ \text{of}\ 100\ \text{V}) \\
&= 50\ \text{V}
\end{aligned}
$$

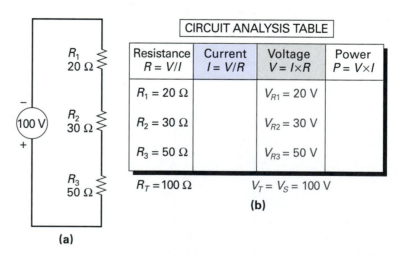

CIRCUIT ANALYSIS TABLE

Resistance $R = V/I$	Current $I = V/R$	Voltage $V = I \times R$	Power $P = V \times I$
$R_1 = 20\ \Omega$		$V_{R1} = 20\ \text{V}$	
$R_2 = 30\ \Omega$		$V_{R2} = 30\ \text{V}$	
$R_3 = 50\ \Omega$		$V_{R3} = 50\ \text{V}$	
$R_T = 100\ \Omega$		$V_T = V_S = 100\ \text{V}$	

(b)

(a)

FIGURE 10-19   **Series Circuit Example. (a) Schematic. (b) Circuit Analysis Table.**

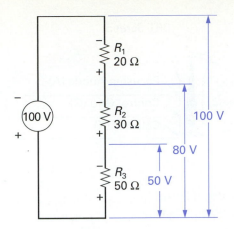

FIGURE 10-20 Series Circuit Example.

b. Voltage dropped across R_2 and R_3 = 30 + 50 = 80 V.

c. Voltage dropped across R_1, R_2, and R_3 = 20 + 30 + 50 = 100 V.

The details of this circuit are summarized in the analysis table shown in Figure 10-19(b).

To summarize the voltage-divider formula, we can say: The voltage drop across a resistor or group of resistors in a series circuit is equal to the ratio of that resistance (R_X) to the total resistance (R_T), multiplied by the source voltage (V_S).

To show an application of a voltage divider, let us imagine that three voltages of 50, 80, and 100 V were required by an electronic system in order to make it operate. To meet this need, we could use three individual power sources, which would be very expensive, or use one 100 V voltage source connected across three resistors, as shown in Figure 10-20, to divide up the 100 V.

■ INTEGRATED MATH APPLICATION:

Figure 10-21(a) shows a simplified circuit of an oscilloscope's cathode ray tube (CRT) or picture tube. In this example, a three-resistor series circuit and a −600 V supply voltage are being used to produce the needed three supply voltages for the CRT's three electrodes, called the focusing anode, the control grid, and the cathode. The heated cathode emits electrons that are collected and concentrated into a beam by the combined electrostatic effect of the control grid and focusing anode. This beam of electrons passes through the apertures of the control grid and focusing anode and strikes the inner surface of the CRT screen. This inner surface is coated with a phosphorescent material that emits light when it is struck by electrons. The voltage between the cathode (K) and grid (G) of the CRT (V_{KG}) determines the intensity of the electron beam and therefore the brightness of the trace seen on the screen. The voltage between the grid (G) and anode (A) of the CRT (V_{GA}) determines the sharpness of the electron beam and therefore the focus of the trace seen on the screen. For the resistance values given, calculate the voltages on each of the three CRT electrodes.

■ *Solution:*

Referring to the illustration and calculations in Figure 10-21(b), you can see how the voltage-divider formula can be used to calculate the voltage drop across R_1 (V_{R1} = 300 V), R_2 (V_{R2} = 255 V), and R_3 (V_{R3} = 45 V). Since the grid of the CRT is connected directly to the −600 V supply, the grid voltage (V_G) will be

$$V_G = V_T = -600V$$

The cathode voltage (V_K) will be equal to the total supply voltage (V_T) minus the voltage drop across R_3 (V_{R3}), so

$$V_K = V_T - V_{R3} = (-600) - (-45) = -555 \text{ V}$$

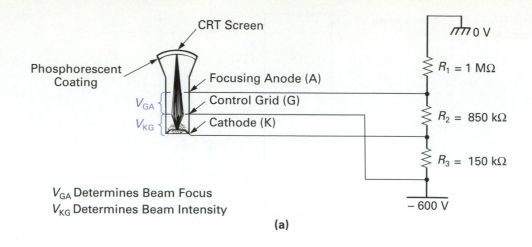

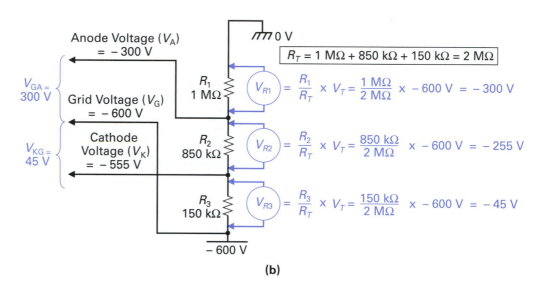

FIGURE 10-21   **Fixed Voltage-Divider Circuit for Supplying Voltages to the Electrodes of a CRT.**

The anode voltage (V_A) will be equal to the total supply voltage (V_T) minus the voltage drops across R_3 (V_{R3}) and R_2 (V_{R2}), and therefore

$$V_A = V_T - (V_{R3} + V_{R2}) = (-600) - (-45 + -225)$$
$$= -600 + 300 \text{ V} = -300 \text{ V}$$

The V_A, V_K, and V_G voltages are all negative with respect to 0 V. The voltages V_{GA} and V_{KG}, shown on the left of Figure 10-21(b), will be the potential difference between the two electrodes. Therefore, V_{KG} is equal to the difference between V_K and V_G (the difference between -600 V and -555 V), and is equal to 45 V ($V_{KG} = 45$ V $= V_{R3}$). The voltage between the CRT's grid and anode (V_{GA}) is equal to the difference between V_G and V_A (the difference between -600 V and -300 V), which will be 300 V ($V_{GA} = 300$ V $= V_{R1}$).

■   **INTEGRATED MATH APPLICATION:**

Figure 10-22(a) shows a 24 V voltage source driving a 10 Ω resistor that is located 1000 ft from the battery. If two 1000-ft lengths of wire are used to connect the source to the load, what will be the voltage applied across the load?

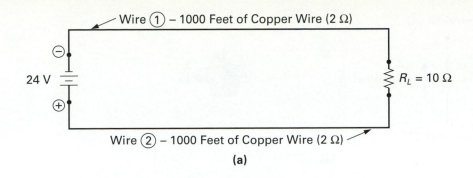

Wire ① – 1000 Feet of Copper Wire (2 Ω)

24 V

$R_L = 10 \, \Omega$

Wire ② – 1000 Feet of Copper Wire (2 Ω)

(a)

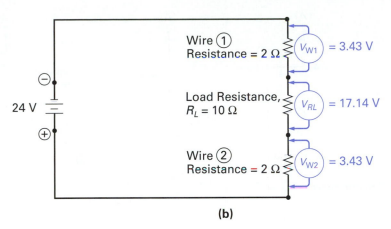

24 V

Wire ①
Resistance = 2 Ω V_{W1} = 3.43 V

Load Resistance,
$R_L = 10 \, \Omega$ V_{RL} = 17.14 V

Wire ②
Resistance = 2 Ω V_{W2} = 3.43 V

(b)

FIGURE 10-22   Series Wire Resistance.

■ *Solution:*

The copper cable has a resistance of 2 Ω for every 1000 ft. To be more accurate, this means that our circuit should be redrawn as shown in Figure 10-22(b) to show the series resistances of wire 1 and wire 2. Using the voltage-divider formula, we can calculate the voltage drop across wire 1 and wire 2.

$$V_{W1} = \frac{R_{W1}}{R_T} \times V_T$$

$$= \frac{2 \, \Omega}{14 \, \Omega} \times 24 \text{ V} = 3.34 \text{ V}$$

Since the voltage drop across wire 2 will also be 3.43 V, the total voltage drop across both wires will be 6.86 V. The remainder, 17.14 V (24 V − 6.86 V = 17.14 V), will appear across the load resistor, R_L.

10-4-3   *Variable Voltage Divider*

When discussing variable-value resistors in Chapter 9, we talked about a potentiometer, or variable voltage divider, which consists of a fixed value of resistance between two terminals and a wiper that can be adjusted to vary resistance between its terminal and one of the other two.

To review, Figure 10-23(a) shows the potentiometer's schematic symbol and physical appearance. If the wiper of the potentiometer is moved down, as seen in Figure 10-23(b), the resistance between terminals *A* and *B* will increase, and the resistance between *B* and *C* will decrease. On the other hand, if the wiper is moved up, as seen in Figure 10-23(c), the resistance between terminals *A* and *B* will decrease, and the resistance between *B* and *C* will

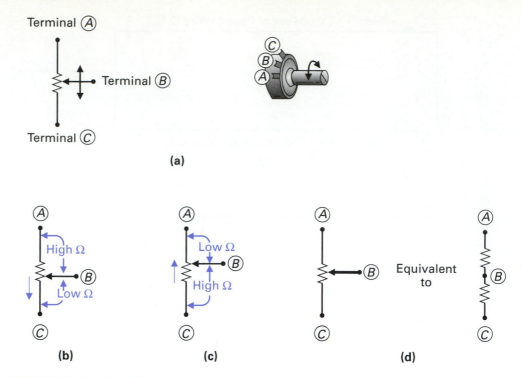

(a)

(b)   **(c)**   **(d)**

FIGURE 10-23   **Potentiometer.**

increase. The resistances between *A* and *B* and *B* and *C* are therefore inversely proportional to one another in that if one increases the other decreases, and vice versa. The resistance values between terminals *A* and *B* and between *B* and *C* can be thought of as two separate resistors, as seen in Figure 10-23(d). No matter what position the wiper is put in, the total resistance between *A* and *B* (R_{AB}) and *B* and *C* (R_{BC}) will always be equal to the rated value of the potentiometer, and equal to the resistance between *A* and *C* (R_{AC}).

$$R_{AB} + R_{BC} = R_{AC}$$

As an example, Figure 10-24(a) illustrates a 10 kΩ potentiometer that has been hooked up across a 10 V dc source with a voltmeter between terminals *B* and *C*. If the wiper terminal is positioned midway between *A* and *C*, the voltmeter should read 5 V, and the potentiometer will be equivalent to two 5 kΩ resistors in series, as shown in Figure 10-24(b). Kirchhoff's voltage law states that the entire source voltage will be dropped across the resistances in the circuit, and since the resistance values are equal, each will drop half of the source voltage, that is, 5 V.

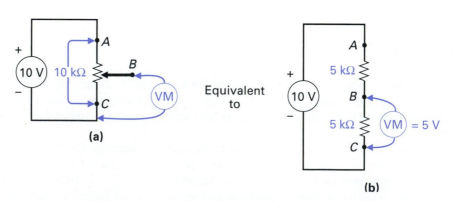

FIGURE 10-24   **Potentiometer Wiper in Midposition.**

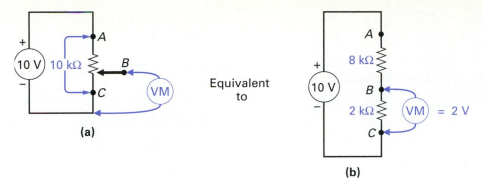

FIGURE 10-25 Potentiometer Wiper in Lower Position.

In Figure 10-25(a), the wiper has been moved down so that the resistance between A and B is equal to 8 kΩ, and the resistance between B and C equals 2 kΩ. This will produce 2 V on the voltmeter, as shown in Figure 10-25(b). The amount of voltage drop is proportional to the resistance, so a larger voltage will be dropped across the larger resistance. Using the voltage-divider formula, you can calculate that the 8 kΩ resistance is 80% of the total resistance and therefore will drop 80% of the voltage:

$$V_{AB} = \frac{R_{AB}}{R_{\text{total}}} \times V_S = \frac{8 \text{ k}\Omega}{10 \text{ k}\Omega} \times 10 \text{ V} = 8 \text{ V}$$

The 2 kΩ resistance between B and C is 20% of the total resistance and consequently will drop 20% of the total voltage:

$$V_{BC} = \frac{R_{BC}}{R_{\text{total}}} \times V_S = \frac{2 \text{ k}\Omega}{10 \text{ k}\Omega} \times 10 \text{ V} = 2 \text{ V}$$

In Figure 10-26(a), the wiper has been moved up, and now 2 kΩ exists between A and B, and 8 kΩ is present between B and C. In this situation, 2 V will be dropped across the 2 kΩ between A and B, and 8 V will be dropped across the 8 kΩ between B and C, as shown in Figure 10-26(b).

From this discussion, you can see that the potentiometer can be adjusted to supply different voltages on the wiper. This voltage can be decreased by moving the wiper down to supply a minimum of 0 V as shown in Figure 10-27(a), or the wiper can be moved up to supply a maximum of 10 V as shown in Figure 10-27(b). By adjusting the wiper position, the potentiometer can be made to deliver any voltage between its maximum and minimum value, which is why the potentiometer is known as a variable voltage divider.

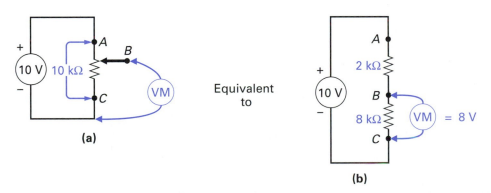

FIGURE 10-26 Potentiometer Wiper in Upper Position.

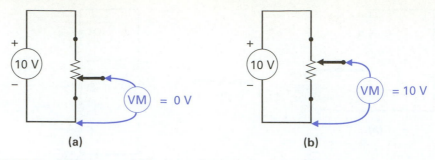

(a)   **(b)**

FIGURE 10-27 **Minimum and Maximum Settings of a Potentiometer.**

■ INTEGRATED MATH APPLICATION:

Figure 10-28 illustrates how a potentiometer can be used to control the output volume of an amplifier that is being driven by a compact disk (CD) player. The preamplifier is producing an output of 2 V, which is developed across a 50 kΩ potentiometer. If the wiper of the potentiometer is in its upper position, the full 2 V from the preamp will be applied into the input of the power amplifier. The power amplifier has a fixed voltage gain (A_V) of 12, and therefore the power amplifier's output is always 12 times larger than the input voltage. An input of 2 V (V_{in}) will therefore produce an output voltage (V_{out}) of 24 V ($V_{out} = V_{in} \times A_V = 2\,V \times 12 = 24\,V$).

As the wiper is moved down, less of the 2 V from the preamp will be applied to the power amplifier, and therefore the output of the power amplifier and volume of the music heard will decrease. If the wiper of the potentiometer is adjusted so that a resistance of 20 kΩ exists between the wiper and the lower end of the potentiometer, what will be the input voltage to the power amplifier and output voltage to the speaker?

■ *Solution:*

By using the voltage-divider formula, we can determine the voltage developed across the potentiometer with 20 kΩ of resistance between the wiper (B) and lower end (C).

$$V_{in} = \frac{R_{AB}}{R_{AC}} \times V_{preamp}$$

$$= \frac{20\,k\Omega}{50\,k\Omega} \times 2\,V = 0.8\,V\ (800\,mV)$$

The voltage at the input of the power amplifier (V_{in}) is applied to the input of the power amplifier, which has an amplification factor or gain of 12, and therefore the output voltage will be

$$V_{out} = V_{in} \times A_V$$
$$= 0.8\,V \times 12 = 9.6\,V$$

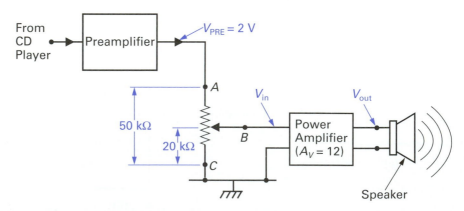

FIGURE 10-28 **The Potentiometer as a Volume Control.**

Now that you have completed this section, you should be able to:

■ **Objective 7.** *Describe why the series circuit is known as a voltage divider.*

■ **Objective 8.** *Describe a fixed and a variable voltage divider.*

Use the following questions to test your understanding of Section 10-4.

1. True or false: A series circuit is also known as a voltage-divider circuit.

2. True or false: The voltage drop across a series resistor is proportional to the value of the resistor.

3. If 6 Ω and 12 Ω resistors are connected across a dc 18 V supply, calculate I_T and the voltage drop across each.

4. State the voltage-divider formula.

5. Which component can be used as a variable voltage divider?

6. Could a rheostat be used in place of a potentiometer?

10-5 POWER IN A SERIES CIRCUIT

As discussed earlier in Chapter 9, power is the rate at which work is done, and work is said to have been done when energy is converted from one energy form to another. Resistors convert electrical energy into heat energy, and the rate at which they dissipate energy is called *power* and is measured in *watts* (joules per second). Resistors all have a resistive value, a tolerance, and a wattage rating. The wattage of a resistor is the amount of heat energy a resistor can safely dissipate per second, and this wattage is directly proportional to the resistor's size. Figure 10-29 reviews the size versus wattage rating of commercially available resistors. As you know, any of the power formulas can be used to calculate wattage. Resistors are manufactured in several different physical sizes, and if, for example, it is calculated that for a certain value of current and voltage a 5 W resistor is needed, and a ½ W resistor is put in its place, the ½ W resistor will burn out, because it is generating heat (5 W) faster than it can dissipate heat (½ W). A 10 W or 25 W resistor or greater could be used to replace a 5 W resistor but anything less than a 5 W resistor will burn out.

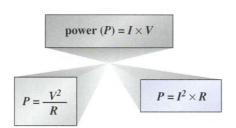

It is therefore necessary that we have some way of calculating which wattage rating is necessary for each specific application.

A question you may be asking is why not just use large-wattage resistors everywhere. The two disadvantages of doing this are:

1. The larger the wattage, the greater the cost.

2. The larger the wattage, the greater the size and area the resistor occupies within the equipment.

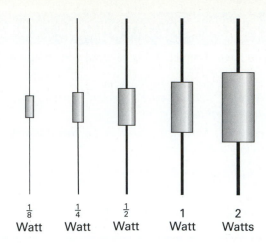

FIGURE 10-29 **Resistor Wattage Ratings.**

$\frac{1}{8}$ Watt $\frac{1}{4}$ Watt $\frac{1}{2}$ Watt   1 Watt   2 Watts

---

■ INTEGRATED MATH APPLICATION:

Figure 10-30 shows a 20 V battery driving a 12 V/1 A television set. R_1 is in series with the TV set and is being used to drop 8 V of the 20 V supply, so 12 V will be developed across the television.

   a. What is the wattage rating for R_1?

   b. What is the series load resistance of the TV set?

   c. What is the amount of power being consumed by the TV set?

   d. Compile a circuit analysis table for this circuit.

FIGURE 10-30 **Series Circuit Example.**

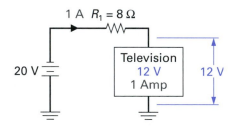

1 A $R_1 = 8\ \Omega$

20 V

Television
12 V
1 Amp

12 V

■ *Solution:*

   a. Everything is known about R_1. Its resistance is 8 Ω, it has 1 A of current flowing through it, and 8 V is being dropped across it. As a result, any one of the three power formulas can be used to calculate the wattage rating of R_1.

$$\text{Power } (P) = I \times V = 1\ \text{A} \times 8\ \text{V} = 8\ \text{W}$$

or

$$P = I^2 \times R = 1^2 \times 8 = 8\ \text{W}$$

or

$$P = \frac{V^2}{R} = \frac{8^2}{8} = 8\ \text{W}$$

The nearest commercially available device would be a 10 W resistor. If size is not a consideration, it is ideal to double the wattage needed and use a 16 W resistor.

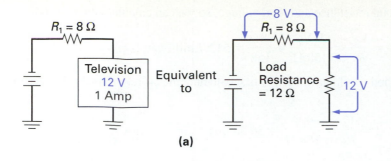

CIRCUIT ANALYSIS TABLE			
Resistance $R = V/I$	Current $I = V/R$	Voltage $V = I \times R$	Power $P = V \times I$
$R_1 = 8\,\Omega$	$I_1 = 1$ A	$V_{R1} = 8$ V	$P_1 = 8$ W
$R_{TV} = 12\,\Omega$	$I_2 = 1$ A	$V_{RTV} = 12$ V	$P_{TV} = 12$ W
$R_T = 20\,\Omega$	$I_T = 1$ A	$V_T = V_S = 20$ V	$P_T = 20$ W

(b)

FIGURE 10-31 Series Circuit Example with Values Inserted. (a) Schematic. (b) Circuit Analysis Table.

b. You may recall that any piece of equipment is equivalent to a load resistance. The TV set has 12 V across it and is pulling 1 A of current. Its load resistance can be calculated simply by using Ohm's law and deriving an equivalent circuit, as shown in Figure 5-31(a).

$$R_L \text{ (load resistance)} = \frac{V}{I}$$
$$= \frac{12 \text{ V}}{1 \text{ A}}$$
$$= 12\,\Omega$$

c. The amount of power being consumed by the TV set is

$$P = V \times I = 12 \text{ V} \times 1 \text{ A} = 12 \text{ W}$$

d. Figure 10-31(b) shows the circuit analysis table for this example.

◼ INTEGRATED MATH APPLICATION:

Calculate the total amount of power dissipated in the series circuit in Figure 10-32, and insert any calculated values in a circuit analysis table.

◼ *Solution:*

The total power dissipated in a series circuit is equal to the sum of all the power dissipated by all the resistors. The easiest way to calculate the total power is to simplify the circuit to one resistance, as shown in Figure 10-32.

$$R_T = R_1 + R_2 + R_3 + R_4$$
$$= 5\,\Omega + 33\,\Omega + 45\,\Omega + 75\,\Omega$$
$$= 158\,\Omega$$

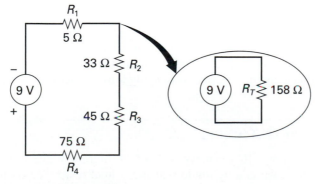

FIGURE 10-32 Series Circuit Example.

We now have total resistance and total voltage, so we can calculate the total power:

$$P_T = \frac{V_S^{\,2}}{R_T} = \frac{(9\ \text{V})^2}{158\ \Omega} = \frac{81}{158} = 512.7\ \text{milliwatts (mW)}$$

The longer method would have been to first calculate the current through the series circuit:

$$I = \frac{V_S}{R_T} = \frac{9\ \text{V}}{158\ \Omega} = 56.96\ \text{mA} \quad \text{or} \quad 57\ \text{mA}$$

We could then calculate the power dissipated by each separate resistor and add up all the individual values to gain a total power figure. This is illustrated in Figure 10-33(a).

$$P_T = P_1 + P_2 + P_3 + P_4 + \cdots$$

Total power = addition of all the individual power losses

$$P_T = 16\ \text{mW} + 107\ \text{mW} + 146\ \text{mW} + 243\ \text{mW}$$
$$= 512\ \text{mW}$$

The calculated values for this example are shown in the circuit analysis table in Figure 10-33(b).

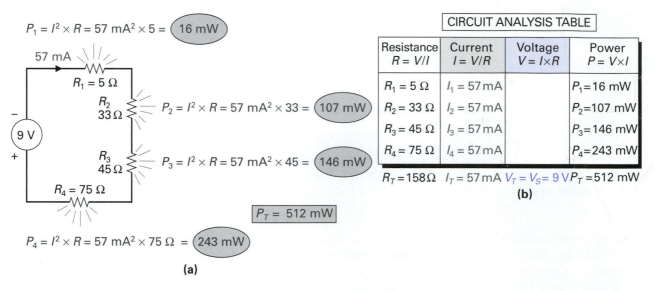

FIGURE 10-33 Series Circuit Example with Values Inserted. (a) Schematic. (b) Circuit Analysis Table.

10-5-1 *Maximum Power Transfer*

The **maximum power transfer theorem** states that maximum power will be delivered to a load when the resistance of the load (R_L) is equal to the resistance of the source (R_S). The best way to see if this theorem is correct is to apply it to a series of examples and then make a comparison.

■ INTEGRATED MATH APPLICATION:

Figure 10-34(a) illustrates a 10 V battery with a 5 Ω internal resistance connected across a load. The load in this case is a lightbulb that has a load resistance of 1 Ω. Calculate the power delivered to this lightbulb or load when $R_L = 1\ \Omega$.

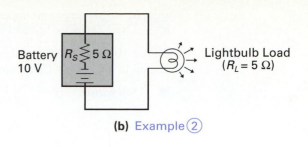

(b) Example ②

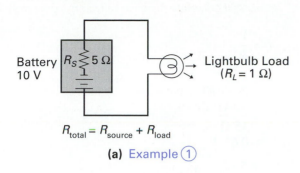

(a) Example ①

$R_{total} = R_{source} + R_{load}$

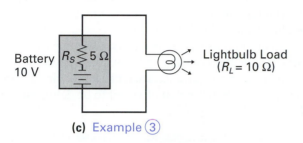

(c) Example ③

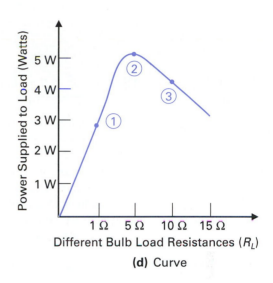

(d) Curve

FIGURE 10-34 **Maximum Power Transfer.**

■ *Solution:*

$$\text{Circuit current, } I = \frac{V}{R}$$

$$= \frac{10 \text{ V}}{R_S + R_L}$$

$$= \frac{10 \text{ V}}{6 \text{ }\Omega}$$

$$= 1.66 \text{ A}$$

The power supplied to the load is $P = I^2 \times R = (1.66 \text{ A})^2 \times 1 \text{ }\Omega = 2.8 \text{ W}$.

■ **INTEGRATED MATH APPLICATION:**

Figure 10-34(b) illustrates the same battery and R_S, but in this case connected across a 5 Ω lightbulb. Calculate the power delivered to this lightbulb now that $R_L = 5 \text{ }\Omega$.

■ *Solution:*

$$I = \frac{V}{R}$$

$$= \frac{10 \text{ V}}{R_S + R_L}$$

$$= \frac{10 \text{ V}}{10 \text{ }\Omega}$$

$$= 1 \text{ A}$$

Power supplied is $P = I^2 \times R = (1 \text{ A})^2 \times 5 \text{ }\Omega = 5 \text{ W}$.

Figure 10-34(c) illustrates the same battery again, but in this case a 10 Ω lightbulb is connected in the circuit. Calculate the power delivered to this lightbulb now that $R_L = 10\ \Omega$.

■ *Solution:*

$$I = \frac{V}{R}$$
$$= \frac{10\ V}{R_S + R_L}$$
$$= \frac{10\ V}{15\ \Omega}$$
$$= 0.67\ A$$

Thus, $P = I^2 \times R = (0.67\ A)^2 \times 10\ \Omega = 4.5\ W$.

Optimum Power Transfer

Since the ideal maximum power transfer conditions cannot always be achieved, most designers try to achieve optimum power transfer and have the source resistance and load resistance as close in value as possible.

As can be seen from the graph in Figure 10-34(d), which plots the power supplied to the load against load resistance, maximum power is delivered to the load (5 W) when the load resistance is equal to the source resistance.

The maximum power transfer condition is used only in special cases such as the automobile starter, where the load resistance remains constant and maximum power is needed. In most other cases, where load resistance can vary over a range of values, circuits are designed for a load resistance that will cause the best amount of power to be delivered. This is known as **optimum power transfer.**

SELF-TEST EVALUATION POINT FOR SECTION 10-5

Now that you have completed this section, you should be able to:

■ **Objective 9.** *Explain how to calculate power in a series circuit.*

■ **Objective 10.** *Explain the maximum power transfer theorem.*

Use the following questions to test your understanding of Section 10-5.

1. State the power formula.
2. Calculate the power dissipated by a 12 Ω resistor connected across a 12 V supply.
3. What fixed resistor type should probably be used for Question 2, and what would be a safe wattage rating?
4. What would be the total power dissipated if R_1 dissipates 25 W, and R_2 dissipates 3800 mW?

REVIEW QUESTIONS

Multiple Choice Questions

1. A series circuit:
   a. Is the connecting of components end to end
   b. Provides a single path for current
   c. Functions as a voltage divider
   d. All the above

2. The total current in a series circuit is equal to:
   a. $I_1 + I_2 + I_3 + \ldots$ c. $I_1 = I_2 = I_3 = \ldots$
   b. $I_1 - I_2$ d. All the above

3. If R_1 and R_2 are connected in series with a total current of 2 A, what will be the current flowing through R_1 and R_2, respectively?
   a. 1 A, 1 A
   b. 2 A, 1 A
   c. 2 A, 2 A
   d. All the above could be true on some occasions.

4. The total resistance in a series circuit is equal to:
   a. The total voltage divided by the total current
   b. The sum of all the individual resistor values
   c. $R_1 + R_2 + R_3 + \ldots$
   d. All the above
   e. None of the above is even remotely true.

5. Which of Kirchhoff's laws applies to series circuits:

   a. His voltage law
   b. His current law
   c. His power law
   d. None of them apply to series circuits, only parallel.

6. The amount of voltage dropped across a resistor is proportional to:

   a. The value of the resistor
   b. The current flow in the circuit
   c. Both (a) and (b)
   d. None of the above

7. If three resistors of 6 kΩ, 4.7 kΩ, and 330 Ω are connected in series with one another, what total resistance will the battery sense?

   a. 11.03 MΩ **c.** 6 kΩ
   b. 11.03 Ω **d.** 11.03 kΩ

8. The voltage-divider formula states that the voltage drop across a resistor or multiple resistors in a series circuit is equal to the ratio of that _____ to the _____ multiplied by the _____.

   a. Resistance, source voltage, total resistance
   b. Resistance, total resistance, source voltage
   c. Total current, resistance, total voltage
   d. Total voltage, total current, resistance

9. The _____ can be used as a variable voltage divider.

   a. Potentiometer   **c.** SPDT switch
   b. Fixed resistor   **d.** None of the above

10. A resistor of larger physical size will be able to dissipate _____ heat than a small resistor.

   a. More   **c.** About the same
   b. Less   **d.** None of the above

11. The _____ is the most useful tool when checking series circuits.

   a. Ammeter   **c.** Voltmeter
   b. Wattmeter   **d.** Both (a) and (b)

12. When an open component occurs in a series circuit, it can be noticed because:

   a. Zero volts appears across it
   b. The supply voltage appears across it
   c. 1.3 V appears across it
   d. None of the above

13. Power can be calculated by:

   a. The addition of all the individual power figures
   b. The product of the total current and the total voltage
   c. The square of the total voltage divided by the total resistance
   d. All the above

14. A series circuit is known as a:

   a. Current divider   **c.** Current subtractor
   b. Voltage divider   **d.** All the above

15. In a series circuit only _____ path(s) exists for current flow, whereas the voltage applied is distributed across all the individual resistors.

   a. Three   **c.** Four
   b. Several   **d.** One

Communication Skill Questions

16. Describe a series-connected circuit. (10-1)

17. Describe and state mathematically what happens to current flow in a series circuit. (10-2)

18. Describe how total resistance can be calculated in a series circuit. (10-3)

19. Describe why voltage is dropped across a series circuit and how each voltage drop can be calculated. (10-4)

20. Briefly describe why resistance and voltage drops are proportional to one another. (10-4)

21. Describe a fixed and a variable voltage divider. (10-4-2 and 10-4-3)

Practice Problems

26. If three resistors of 1.7 kΩ, 3.3 kΩ, and 14.4 kΩ are connected in series with one another across a 24 V source as shown in Figure 10-35, calculate:

   a. Total resistance (R_T)
   b. Circuit current
   c. Individual voltage drops
   d. Individual and total power dissipated

27. If 40 Ω and 35 Ω resistors are connected across a 24 V source, what will be the current flow through the resistors, and what resistance will cause half the current to flow?

22. How can individual and total power be calculated in a series circuit? (10-5)

23. Is a series circuit a voltage divider? (10-4)

24. What is a variable voltage divider? (10-4)

25. State Kirchhoff's voltage (series circuit) law. (10-4)

FIGURE 10-35

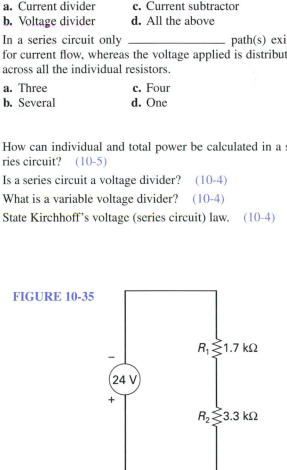

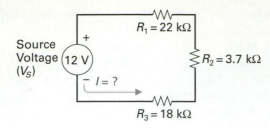

FIGURE 10-36

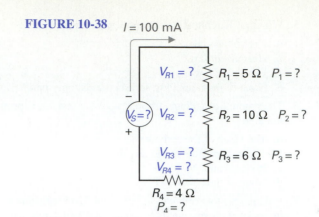

FIGURE 10-38

28. Calculate the total resistance (R_T) of the following series-connected resistors: 2.7 kΩ, 3.4 MΩ, 370 Ω, and 4.6 MΩ.

29. Calculate the value of resistors needed to divide up a 90 V source to produce 45 V and 60 V outputs, with a divider-circuit current of 1 A.

30. If R_1 = 4.7 kΩ and R_2 = 6.4 kΩ and both are connected across a 9 V source, how much voltage will be dropped across R_2?

31. What current would flow through R_1 if it was one-third the ohmic value of R_2 and R_3, and all were connected in series with a total current of 6.5 mA flowing out of V_S?

32. Draw a circuit showing R_1 = 2.7 kΩ, R_2 = 3.3 kΩ, and R_3 = 0.027 MΩ in series with one another across a 20 V source. Calculate:

    a. I_T **d.** P_2 **g.** V_{R2}
    b. P_T **e.** P_3 **h.** V_{R3}
    c. P_1 **f.** V_{R1} **i.** I_1

33. Calculate the current flowing through three lightbulbs that are dissipating 120 W, 60 W, and 200 W when they are connected in series across a 120 V source. How is the voltage divided around the series circuit?

34. If three equal-value resistors are series connected across a dc power supply adjusted to supply 10 V, what percentage of the source voltage will appear across R_1?

35. Refer to the following figures and calculate:

    a. I (Figure 10-36)
    b. R_T and P_T (Figure 10-37)
    c. V_S, V_{R1}, V_{R2}, V_{R3}, V_{R4}, P_1, P_2, P_3, and P_4 (Figure 10-38)
    d. P_T, I, R_1, R_2, R_3, and R_4 (Figure 10-39)

FIGURE 10-37

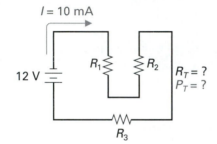

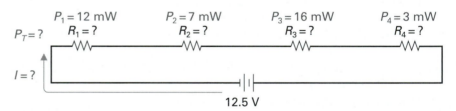

FIGURE 10-39

Web Site Questions

Go to the Web site http://www.prenhall.com/cook, select the textbook *Mathematics for Electronics and Computers,* select this chapter, and then follow the instructions when answering the multiple-choice practice problems.

Parallel DC Circuits

Let's Toss for It!

David Packard

In 1938, Bill Hewlett and Dave Packard, close friends and engineering graduates at Stanford University, set up shop in the one-car garage behind the Packards' rented home in Palo Alto, California. In the garage, the two worked on what was to be the first product of their lifetime business together, an electronic oscillating instrument that represented a breakthrough in technology and was specifically designed to test sound equipment.

Late in the year, the oscillator (designated the 200A "because the number sounded big") was presented at a West Coast meeting of the Institute of Radio Engineers (now the Institute of Electrical and Electronics Engineers—IEEE) and the orders began to roll in. Along with the first orders for the 200A was a letter from Walt Disney Studios asking the two to build a similar oscillator covering a different frequency range. The Model 200B was born shortly thereafter, and Disney purchased eight to help develop the unique sound system for the classic film *Fantasia*. By 1940, the young company had out-grown the garage and moved into a small rented building nearby.

Over the years, the company continued a steady growth, expanding its product line to more than 10,000, including computer systems and peripheral products, test and measur-ing instruments, hand-held calculators, medical electronic equipment, and systems for chemical analysis. Employees have increased from 2 to almost 82,000, and the company is one of the largest industrial corporations in America, with a net revenue of $8.1 billion. What name should go first was decided by the toss of a coin on January 1, 1939, and the outcome was Hewlett-Packard or HP.

Outline and Objectives

VIGNETTE: LET'S TOSS FOR IT!

INTRODUCTION

11-1 COMPONENTS IN PARALLEL

Objective 1: Describe the difference between a series and a parallel circuit.

Objective 2: Be able to recognize and determine whether circuit components are connected in series or parallel.

11-2 VOLTAGE IN A PARALLEL CIRCUIT

Objective 3: Explain why voltage measures the same across parallel-connected components.

11-3 CURRENT IN A PARALLEL CIRCUIT

Objective 4: State Kirchhoff's current law.

Objective 5: Describe why branch current and resistance are inversely proportional to each other.

11-4 RESISTANCE IN A PARALLEL CIRCUIT

Objective 6: Determine the total resistance of any parallel-connected resistive circuit.

11-4-1 Two Resistors in Parallel

11-4-2 Equal-Value Resistors in Parallel

11-5 POWER IN A PARALLEL CIRCUIT

Objective 7: Describe and be able to apply all formulas associated with the calculation of voltage, current, resistance, and power in a parallel circuit.

MULTIPLE CHOICE QUESTIONS

COMMUNICATION SKILL QUESTIONS

PRACTICE PROBLEMS

WEB SITE QUESTIONS

Introduction

By tracing the path of current, we can determine whether a circuit has series-connected or parallel-connected components. In a series circuit, there is only one path for current, whereas in parallel circuits the current has two or more paths. These paths are known as *branches*. A parallel circuit, by definition, is when two or more components are connected to the same voltage source so that the current can branch out over two or more paths. In a parallel resistive circuit, two or more resistors are connected to the same source, so current splits to travel through each separate branch resistance.

11-1   COMPONENTS IN PARALLEL

Parallel Circuit

Also called shunt; circuit having two or more paths for current flow.

Many components, other than resistors, can be connected in parallel, and a **parallel circuit** can easily be identified because current is split into two or more paths. Being able to identify a parallel connection requires some practice, because they can come in many different shapes and sizes. The means for recognizing series circuits is that if you can place your pencil at the negative terminal of the voltage source (battery) and follow the wire connections through components to the positive side of the battery and only have one path to follow, the circuit is connected in series. If, however, you can place your pencil at the negative terminal of the voltage source and follow the wire and at some point have a choice of two or more routes, the circuit is connected with two or more parallel branches. The number of routes determines the number of parallel branches. Figure 11-1 illustrates five examples of parallel resistive circuits.

■   **INTEGRATED MATH APPLICATION:**

Figure 11-2(a) illustrates four resistors laid out on a table top.

    a. With wire leads, connect all four resistors in parallel on a protoboard, and then connect the circuit to a dc power supply.

    b. Draw the schematic diagram of the parallel-connected circuit.

■   *Solution:*

Figure 11-2(b) shows how to connect the resistors in parallel and the circuit's schematic.

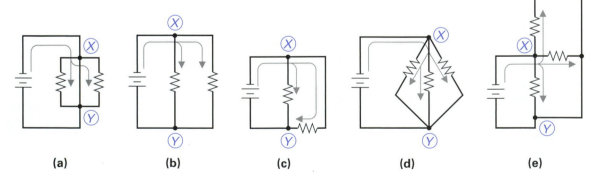

   (a)   **(b)**   **(c)**   **(d)**   **(e)**

FIGURE 11-1   **Parallel Circuits.**

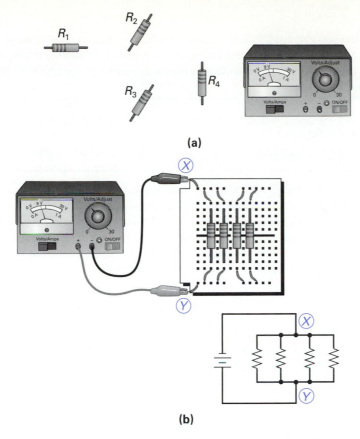

(a)

(b)

FIGURE 11-2   **Connecting Resistors in Parallel. (a) Problem. (b) Solution.**

■   **INTEGRATED MATH APPLICATION:**

Figure 11-3(a) shows four 1.5 V cells and three lamps. Using wires, connect all the cells in parallel to create a 1.5 V source. Then, connect the three lamps in parallel with one another, and finally, connect the 1.5 V source across the three-parallel-connected-lamp load.

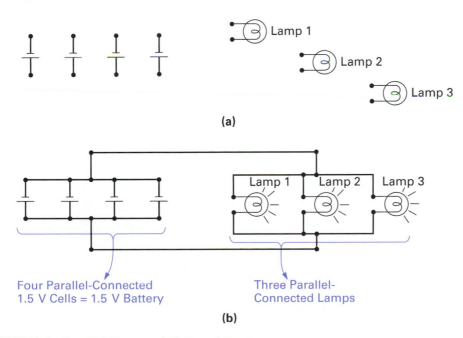

FIGURE 11-3   **Parallel-Connected Cells and Lamps.**

■ *Solution:*

In Figure 11-3(b) you can see the final circuit containing a source consisting of four parallel-connected 1.5 V cells, and a load consisting of three parallel-connected lamps. When cells are connected in parallel, the total voltage remains the same as for one cell, but the current demand can now be shared.

11-2   VOLTAGE IN A PARALLEL CIRCUIT

Figure 11-4(a) shows a simple circuit with four resistors connected in parallel across the voltage source of a 9 V battery. The current from the negative side of the battery will split between the four different paths or branches, yet the voltage drop across each branch of a parallel circuit is equal to the voltage drop across all the other branches in parallel. This means that if the voltmeter were to measure the voltage across *A* and *B* or *C* and *D* or *E* and *F* or *G* and *H,* the voltages would all be the same or, in this example, would all drop 9 V.

It is quite easy to imagine why there will be the same voltage drop across all the resistors, seeing that points *A, C, E,* and *G* are all one connection and points *B, D, F,* and *H* are all one connection. Measuring the voltage drop with the voltmeter across any of the resistors is the same as measuring the voltage across the battery, as shown in Figure 11-4(b). As long as the voltage source remains constant, the voltage drop will always be common (9 V) across the parallel resistors, no matter what value or how many resistors are connected in parallel. The voltmeter is therefore measuring the voltage between two common points that

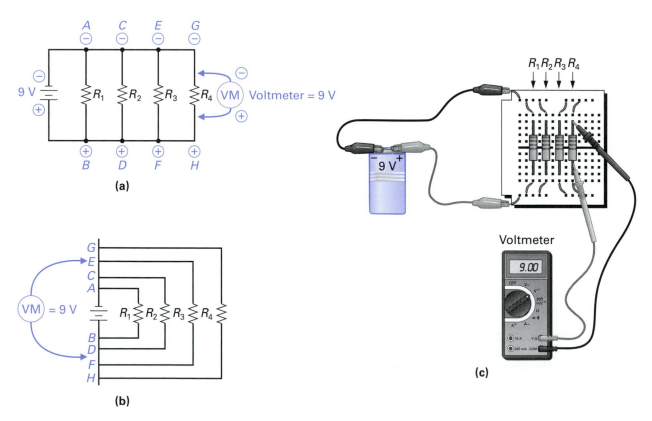

FIGURE 11-4   Voltage in a Parallel Circuit.

CHAPTER 11 / PARALLEL DC CIRCUITS

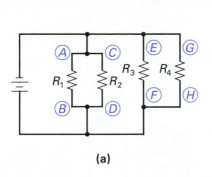

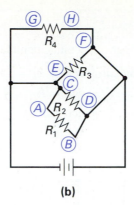

FIGURE 11-5   Parallel-Circuit Voltage Drop.

are directly connected to the battery, and the voltage dropped across all these parallel resistors will be equal to the source voltage.

In Figure 6-5(a) and (b), the same circuit is shown in two different ways, so you can see how the same circuit can look completely different. In both examples, the voltage drop across any of the resistors will always be the same and, as long as the voltage source is not heavily loaded, equal to the source voltage. Just as you can trace the positive side of the battery to all four resistors, you can also trace the negative side to all four resistors.

Mathematically stated, we can say that in a parallel circuit

$$V_{B1} = V_{B2} = V_{B3} = V_{B4} = V_S$$

Voltage drop across branch 1 = voltage drop across branch 2 = voltage drop across branch 3 (etc.) = source voltage.

To reinforce the concept, let us examine this parallel circuit characteristic using a water analogy, as seen in Figure 11-6. The pressure across valves A and B will always be the same, even if one offers more opposition than the other. This is because the pressure measured across either valve will be the same as the pressure difference between piping X and Y. Since the piping X and Y runs directly back to the pump, the pressure across A and B is the same as the pressure difference across the pump.

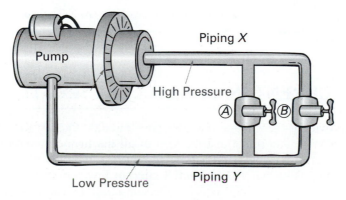

FIGURE 11-6   Fluid Analogy of Parallel-Circuit Pressure.

■ **INTEGRATED MATH APPLICATION:**

Refer to Figure 11-7 and calculate:

   a. Voltage drop across R_1

   b. Voltage drop across R_2

   c. Voltage drop across R_3

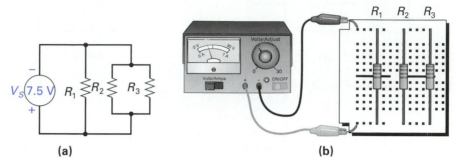

FIGURE 11-7 **A Parallel Circuit Example. (a) Schematic. (b) Protoboard Circuit.**

■ *Solution:*

Because all these resistors are connected in parallel, the voltage across every branch will be the same and equal to the source voltage applied. Therefore,

$$V_{R1} = V_{R2} = V_{R3} = V_S$$
$$7.5 \text{ V} = 7.5 \text{ V} = 7.5 \text{ V} = 7.5 \text{ V}$$

SELF-TEST EVALUATION POINT FOR SECTIONS 11-1 AND 11-2

Now that you have completed these sections, you should be able to:

■ ***Objective 1.*** *Describe the difference between a series and a parallel circuit.*

■ ***Objective 2.*** *Be able to recognize and determine whether circuit components are connected in series or parallel.*

■ ***Objective 3.*** *Explain why voltage measures the same across parallel-connected components.*

Use the following questions to test your understanding of Sections 11-1 and 11-2.

1. Describe a parallel circuit.

2. True or false: A parallel circuit is also known as a voltage-divider circuit.

3. What will be the voltage drop across R_1 if $V_S = 12$ V and R_1 and R_2 are in parallel with one another and equal to 24 Ω each?

4. Can Kirchhoff's voltage law be applied to parallel circuits?

Branch Current

A portion of the total current that is present in one path of a parallel circuit.

Kirchhoff's Current Law

The sum of the currents flowing into a point in a circuit is equal to the sum of the currents flowing out of that same point.

11-3   CURRENT IN A PARALLEL CIRCUIT

In addition to providing the voltage law for series circuits, Gustav Kirchhoff (in 1847) was the first to observe and prove that the sum of all the **branch currents** in a parallel circuit ($I_1 + I_2 + I_3$, etc.) is equal to the total current (I_T). In honor of his second discovery, this phenomenon is known as **Kirchhoff's current law,** which states that the sum of all the currents entering a junction is equal to the sum of all the currents leaving that same junction.

Figure 11-8(a) and (b) illustrate two examples of how this law applies. In both examples, the sum of the currents entering a junction is equal to the sum of the currents leaving

280   CHAPTER 11 / PARALLEL DC CIRCUITS

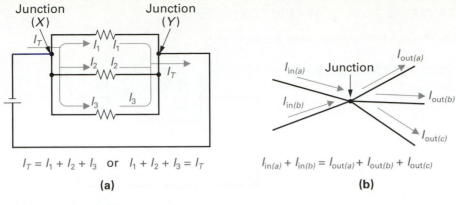

$I_T = I_1 + I_2 + I_3$ or $I_1 + I_2 + I_3 = I_T$

(a)

$I_{in(a)} + I_{in(b)} = I_{out(a)} + I_{out(b)} + I_{out(c)}$

(b)

FIGURE 11-8 Kirchhoff's Current Law.

that same junction. In Figure 11-8(a) the total current arrives at junction X and splits to produce three branch currents, I_1, I_2, and I_3, which cumulatively equal the total current (I_T) that arrived at the junction X. The same three branch currents combine at junction Y, and the total current (I_T) leaving that junction is equal to the sum of the three branch currents arriving at junction Y.

$$I_T = I_1 + I_2 + I_3 + I_4 = \ldots$$

As another example, in Figure 11-8(b) you can see that there are two branch currents entering a junction [$I_{in(a)}$ and $I_{in(b)}$] and three branch currents leaving that same junction [$I_{out(a)}$, $I_{out(b)}$, and $I_{out(c)}$]. As stated below the illustration, the sum of the input currents will equal the sum of the output currents: $I_{in(a)} + I_{in(b)} = I_{out(a)} + I_{out(b)} + I_{out(c)}$.

■ **INTEGRATED MATH APPLICATION:**

Refer to Figure 11-9 and calculate the value of I_T.

■ *Solution:*

By Kirchhoff's current law,

$$I_T = I_1 + I_2 + I_3 + I_4$$
$$? = 2 \text{ mA} + 17 \text{ mA} + 7 \text{ mA} + 37 \text{ mA}$$
$$I_T = 63 \text{ mA}$$

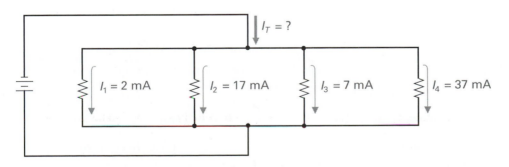

FIGURE 11-9 Calculating Total Current.

INTEGRATED MATH APPLICATION:

Refer to Figure 11-10 and calculate the value of I_1.

FIGURE 11-10 Calculating Branch Current.

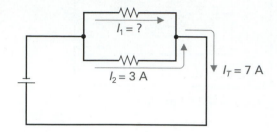

Solution:

By transposing Kirchhoff's current law, we can determine the unknown value (I_1):

$$I_T = I_1 + I_2 \qquad I_T = I_1 + I_2 \qquad (-I_2)$$
$$7\,A = ? + 3\,A \quad \text{or} \quad I_T - I_2 = I_1$$
$$I_1 = 4\,A \qquad I_1 = I_T - I_2 = 7\,A - 3\,A = 4\,A$$

As with series circuits, to find out how much current will flow through a parallel circuit, we need to find out how much opposition or resistance is being connected across the voltage source.

$$I_T = \frac{V_S}{R_T}$$

Total current equals source voltage divided by total resistance.

When we connect resistors in parallel, the total resistance in the circuit will actually decrease. In fact, the total resistance in a parallel circuit will always be less than the value of the smallest resistor in the circuit.

To prove this point, Figure 11-11 shows how two sets of identical resistors (R_1, R_2, and R_3) were used to build both a series and a parallel circuit. The total current flow in the paral-

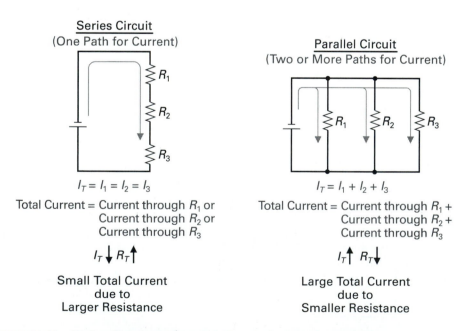

FIGURE 11-11 Series-Circuit and Parallel-Circuit Current Comparison.

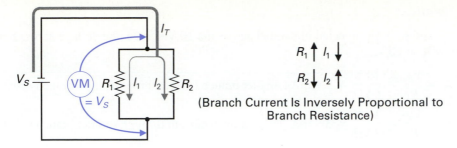

FIGURE 11-12 The Parallel-Circuit Current Divider.

lel circuit will be larger than the total current in the series circuit, because the parallel circuit has two or more paths for current to flow, whereas the series circuit only has one.

To explain why the total current will be larger in a parallel circuit, let us use the analogy of a freeway with only one path for traffic to flow. A single-lane freeway is equivalent to a series circuit, and only a small amount of traffic is allowed to flow along this freeway. If the freeway is expanded to accommodate two lanes, a greater amount of traffic can flow along the freeway in the same amount of time. Having more lanes permits a greater total amount of traffic flow. In parallel circuits, more branches will allow a greater total amount of current flow because there is less resistance in more paths than there is with only one path. This concept is summarized in Figure 11-11.

Just as a series circuit is often referred to as a voltage-divider circuit, a parallel circuit is often referred to as a **current-divider** circuit, because the total current arriving at a junction will divide or split into branch currents (Kirchhoff's law), as shown in Figure 11-12.

The current division is inversely proportional to the resistance in the branch, assuming that the voltage across both resistors is constant and equal to the source voltage (V_S). This means that a large branch resistance will cause a small branch current ($I{\downarrow} = V/R{\uparrow}$), and a small branch resistance will cause a large branch current ($I{\uparrow} = V/R{\downarrow}$).

Current Divider

A parallel network designed to proportionally divide the circuit's total current.

■ INTEGRATED MATH APPLICATION:

Calculate the following for Figure 11-13(a), and then insert the values in a circuit analysis table.

   a. I_1

   b. I_2

   c. I_T

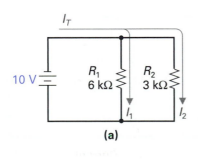

	CIRCUIT ANALYSIS TABLE		
Resistance $R = V/I$	Voltage $V = I \times R$	Current $I = V/R$	Power $P = V \times I$
$R_1 = 6\ k\Omega$	$V_{R1} = 10\ V$	$I_1 = 1.6\ mA$	
$R_2 = 3\ k\Omega$	$V_{R2} = 10\ V$	$I_2 = 3.3\ mA$	

$V_T = V_S = 10\ V$ $I_T = 4.9\ mA$

(b)

FIGURE 11-13 Parallel Circuit Example. (a) Schematic. (b) Circuit Analysis Table.

■ *Solution:*

Since R_1 and R_2 are connected in parallel across the 10 V source, the voltage across both resistors will be 10 V.

a. $I_1 = \dfrac{V_{R1}}{R_1} = \dfrac{10\ \text{V}}{6\ \text{k}\Omega} = 1.6\ \text{mA}$ (smaller branch current through larger branch resistance)

b. $I_2 = \dfrac{V_{R2}}{R_2} = \dfrac{10\ \text{V}}{3\ \text{k}\Omega} = 3.3\ \text{mA}$ (larger branch current through smaller branch resistance)

c. By Kirchhoff's current law,

$$I_T = I_1 + I_2$$
$$= 1.6\ \text{mA} + 3.3\ \text{mA}$$
$$= 4.9\ \text{mA}$$

The circuit analysis table for this example is shown in Figure 11-13(b).

By rearranging Ohm's law, we can arrive at another formula, which is called the *current-divider formula* and can be used to calculate the current through any branch of a multiple-branch parallel circuit.

$$I_x = \dfrac{R_T}{R_x} \times I_T$$

where I_x = branch current desired
R_T = total resistance
R_x = resistance in branch
I_T = total current

■ **INTEGRATED MATH APPLICATION:**

Refer to Figure 11-14 and calculate the following if the total circuit resistance (R_T) is equal to 1 kΩ:

a. $I_1 =$

b. $I_2 =$

c. $I_3 =$

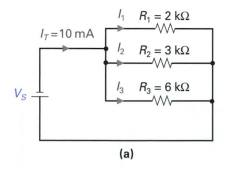

CIRCUIT ANALYSIS TABLE			
Resistance $R = V/I$	Voltage $V = I \times R$	Current $I = V/R$	Power $P = V \times I$
$R_1 = 2\ \text{k}\Omega$		$I_1 = 5\ \text{mA}$	
$R_2 = 3\ \text{k}\Omega$		$I_2 = 3.33\ \text{mA}$	
$R_3 = 6\ \text{k}\Omega$		$I_3 = 1.67\ \text{mA}$	
$R_T = 1\ \text{k}\Omega$		$I_T = 10\ \text{mA}$	

(b)

FIGURE 11-14 **Parallel Circuit Example. (a) Schematic. (b) Circuit Analysis Table.**

■ **Solution:**

Since the source and therefore the voltage across each branch resistor are not known, we will use the current-divider formula to calculate I_1, I_2, and I_3.

a. $I_1 = \dfrac{R_T}{R_1} \times I_T = \dfrac{1\text{ k}\Omega}{2\text{ k}\Omega} \times 10\text{ mA} = 5\text{ mA}$

(smallest branch resistance has largest branch current)

b. $I_2 = \dfrac{R_T}{R_2} \times I_T = \dfrac{1\text{ k}\Omega}{3\text{ k}\Omega} \times 10\text{ mA}$

$= 3.33\text{ mA}$

c. $I_3 = \dfrac{R_T}{R_3} \times I_T = \dfrac{1\text{ k}\Omega}{6\text{ k}\Omega} \times 10\text{ mA}$

$= 1.67\text{ mA}$

(largest branch resistance has smallest branch current)

To double-check that the values for I_1, I_2, and I_3 are correct, you can apply Kirchhoff's current law, which is

$$I_T = I_1 + I_2 + I_3$$
$$10\text{ mA} = 5\text{ mA} + 3.33\text{ mA} + 1.67\text{ mA}$$
$$= 10\text{ mA}$$

CALCULATOR SEQUENCE

Step	Keypad Entry	Display Response
1.	[1][E][3]	1.3
2.	[÷]	
3.	[2][E][3]	2.3
4.	[×]	0.5
5.	[1][0][E][3][+/−]	10 −3
6.	[=]	5.−03

■ **INTEGRATED MATH APPLICATION:**

A common use of parallel circuits is in the residential electrical system. All the household lights and appliances are wired in parallel, as seen in the typical room wiring circuit in Figure 11-15(a). If it is a cold winter morning, and lamps 1 and 2 are switched ON, together with the space heater and hair dryer, what will the individual branch currents be, and what will be the total current drawn from the source?

■ **Solution:**

Figure 11-15(b) shows the schematic of the pictorial in Figure 11-15(a). Since all resistances are connected in parallel across a 120 V source, the voltage across all devices will be 120 V. Using Ohm's law we can calculate the four branch currents:

$$I_1 = \frac{V_{\text{lamp1}}}{R_{\text{lamp1}}} = \frac{120\text{ V}}{125\text{ }\Omega} = 960\text{ mA}$$

$$I_2 = \frac{V_{\text{lamp2}}}{R_{\text{lamp2}}} = \frac{120\text{ V}}{125\text{ }\Omega} = 960\text{ mA}$$

$$I_3 = \frac{V_{\text{hairdryer}}}{R_{\text{hairdryer}}} = \frac{120\text{ V}}{40\text{ }\Omega} = 3\text{ A}$$

$$I_4 = \frac{V_{\text{heater}}}{R_{\text{heater}}} = \frac{120\text{ V}}{12\text{ }\Omega} = 10\text{ A}$$

By Kirchhoff's current law,

$$I_T = I_1 + I_2 + I_3 + I_4$$
$$= 960\text{ mA} + 960\text{ mA} + 3\text{ A} + 10\text{ A}$$
$$= 14.92\text{ A}$$

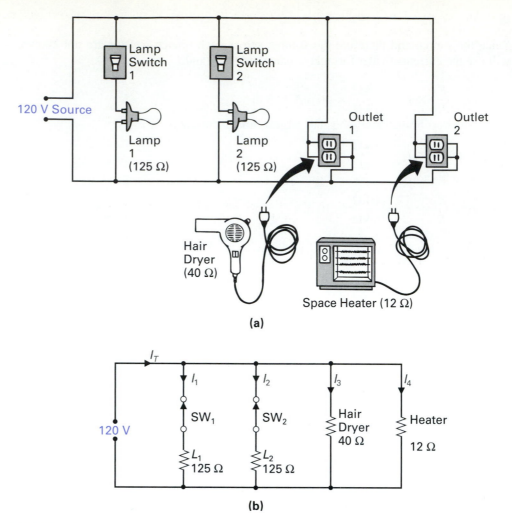

(a)

(b)

FIGURE 11-15 **Parallel Home Electrical System.**

Now that you have completed this section, you should be able to:

- **Objective 4.** *State Kirchhoff's current law.*
- **Objective 5.** *Describe why branch current and resistance are inversely proportional to each other.*

Use the following questions to test your understanding of Section 11-3.

1. State Kirchhoff's current law.
2. If $I_T = 4$ A and $I_1 = 2.7$ A in a two-resistor parallel circuit, what will be the value of I_2?
3. State the current-divider formula.
4. Calculate I_1 if $R_T = 1$ kΩ, $R_1 = 2$ kΩ, and $V_T = 12$ V.

11-4   RESISTANCE IN A PARALLEL CIRCUIT

We now know that parallel circuits will have a larger current flow than a series circuit containing the same resistors due to the smaller total resistance. To calculate exactly how much total current will flow, we need to be able to calculate the total resistance that the parallel circuit presents to the source.

The ability of a circuit to conduct current is a measure of that circuit's conductance, and you will remember from Chapter 8 that conductance (G) is equal to the reciprocal of resistance and is measured in siemens.

$$G = \frac{1}{R} \text{ (siemens)}$$

Every resistor in a parallel circuit will have a conductance figure that is equal to the reciprocal of its resistance, and the total conductance (G_T) of the circuit will be equal to the sum of all the individual resistor conductances. Therefore,

$$G_T = G_{R1} + G_{R2} + G_{R3} + \cdots$$

Total conductance is equal to the conductance of R_1 + the conductance of R_2 + the conductance of R_3 + . . .

Once you have calculated total conductance, the reciprocal of this figure will give you total resistance. If, for example, we have two resistors in parallel, as shown in Figure 11-16, the conductance for R_1 will equal

$$G_{R1} = \frac{1}{R_1} = \frac{1}{20 \; \Omega} = 0.05 \text{ S}$$

The conductance for R_2 will equal

$$G_{R2} = \frac{1}{R_2} = \frac{1}{40 \; \Omega} = 0.025 \text{ S}$$

The total conductance will therefore equal

$$\begin{aligned} G_{\text{total}} &= G_{R1} + G_{R2} \\ &= 0.05 + 0.025 \\ &= 0.075 \text{ S} \end{aligned}$$

Since total resistance is equal to the reciprocal of total conductance, total resistance for the parallel circuit in Figure 11-16 will be

$$R_{\text{total}} = \frac{1}{G_{\text{total}}} = \frac{1}{0.075 \text{ S}} = 13.3 \; \Omega$$

Combining these three steps (first calculate individual conductances, total conductance and then total resistance) we can arrive at the following *reciprocal formula:*

$$R_{\text{total}} = \frac{1}{(1/R_1) + (1/R_2)}$$

This formula states that the conductance of R_1 (G_{R1})
+ conductance of R_2 (G_{R2})
= total conductance (G_T),
and the reciprocal of total conductance is equal to total resistance.

In the example for Figure 11-16, this combined general formula for total resistance can be verified by plugging in the example values.

FIGURE 11-16 **Parallel Circuit Conductance and Resistance.**

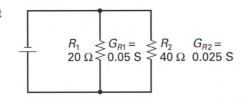

$$R_T = \frac{1}{(1/R_1) + (1/R_2)}$$

$$= \frac{1}{(1/20) + (1/40)}$$

$$= \frac{1}{0.05 + 0.025}$$

$$= \frac{1}{0.075}$$

$$= 13.3 \ \Omega$$

The *reciprocal formula* for calculating total parallel circuit resistance for any number of resistors is

$$R_T = \frac{1}{(1/R_1) + (1/R_2) + (1/R_3) + (1/R_4) + \ldots}$$

■ INTEGRATED MATH APPLICATION:

Referring to Figure 11-17(a), calculate:

a. Total resistance

b. Voltage drop across R_2

c. Voltage drop across R_3

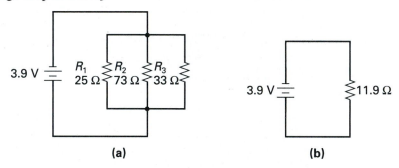

(a)   **(b)**

FIGURE 11-17   **Parallel Circuit Example. (a) Schematic. (b) Equivalent Circuit.**

■ *Solution:*

Total resistance can be calculated using the reciprocal formula:

$$R_T = \frac{1}{(1/R_1) + (1/R_2) + (1/R_3)}$$

$$= \frac{1}{(1/25 \ \Omega) + (1/73 \ \Omega) + (1/33 \ \Omega)}$$

$$= \frac{1}{0.04 + 0.014 + 0.03}$$

$$= 11.9 \ \Omega$$

With parallel resistance circuits, the total resistance is always smaller than the smallest branch resistance. In this example the total opposition of this circuit is equivalent to 11.9 Ω, as shown in Figure 11-17(b).

With parallel resistive circuits, the voltage drop across any branch is equal to the voltage drop across each of the other branches and is equal to the source voltage, in this example 3.9 V.

11-4-1   *Two Resistors in Parallel*

If only two resistors are connected in parallel, a quick and easy formula called the *product-over-sum formula* can be used to calculate total resistance.

$$R_T = \frac{R_1 \times R_2}{R_1 + R_2}$$

$$\text{Total resistance} = \frac{\text{product of both resistance values}}{\text{sum of both resistance values}}$$

Using the example shown in Figure 11-18, let us compare the *product-over-sum* formula with the *reciprocal* formula.

(a) PRODUCT-OVER-SUM FORMULA

$$R_T = \frac{R_1 \times R_2}{R_1 + R_2}$$
$$= \frac{3.7 \text{ k}\Omega \times 2.2. \text{ k}\Omega}{3.7 \text{ k}\Omega + 2.2 \text{ k}\Omega}$$
$$= \frac{8.14 \text{ k}\Omega^2}{5.9 \text{ k}\Omega}$$
$$= 1.38 \text{ k}\Omega$$

(b) RECIPROCAL FORMULA

$$R_T = \frac{1}{(1/R_1) + (1/R_2)}$$
$$= \frac{1}{(1/3.7 \text{ k}\Omega) + (1/2.2 \text{ k}\Omega)}$$
$$= \frac{1}{(270.2 \times 10^{-6}) + (454.5 \times 10^{-6})}$$
$$= \frac{1}{724.7 \times 10^{-6}}$$
$$= 1.38 \text{ k}\Omega$$

As you can see from this example, the advantage of the product-over-sum parallel resistance formula (a) is its ease of use. Its disadvantage is that it can be used only for two resistors in parallel. The rule to adopt, therefore, is that if a circuit has two resistors in parallel use the *product-over-sum* formula, and in circuits containing more than two resistors, use the *reciprocal* formula.

11-4-2   *Equal-Value Resistors in Parallel*

If resistors of equal value are connected in parallel, a special case *equal-value formula* can be used to calculate the total resistance.

$$R_T = \frac{\text{value of one resistor } (R)}{\text{number of parallel resistors } (n)}$$

■   **INTEGRATED MATH APPLICATION:**

Figure 11-19(a) shows how a stereo music amplifier is connected to drive two 8 Ω speakers, which are connected in parallel with each other. What is the total resistance connected across the amplifier's output terminals?

FIGURE 11-18   **Two Resistors in Parallel.**

6.5 V R_1 3.7 kΩ R_2 2.2 kΩ

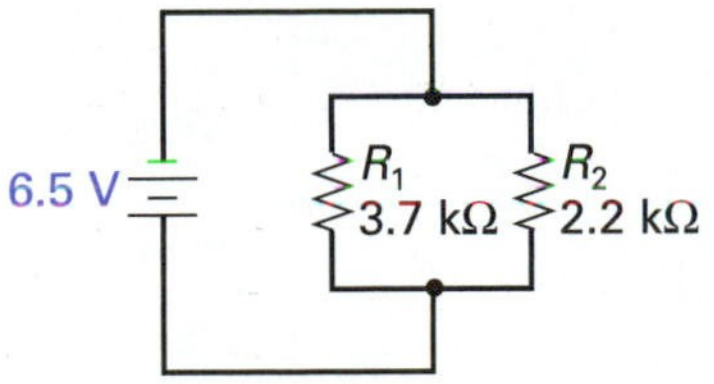

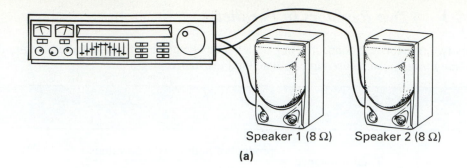

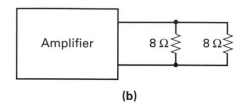

(a)

(b)

FIGURE 11-19 **Parallel-Connected Speakers.**

■ *Solution:*

Referring to the schematic in Figure 11-19(b), you can see that since both parallel-connected speakers have the same resistance, the total resistance is most easily calculated by using the equal-value formula:

$$R_T = \frac{R}{n} = \frac{8 \; \Omega}{2} = 4 \; \Omega$$

■ **INTEGRATED MATH APPLICATION:**

Refer to Figure 11-20 and calculate:

   a. Total resistance in part (a)

   b. Total resistance in part (b)

■ *Solution:*

CALCULATOR SEQUENCE FOR (A)

Step	Keypad Entry	Display Response
1.	[4][.][5][E][6]	4.5E6
2.	[+]	
3.	[3][.][2][E][6]	3.2E6
4.	[=]	7.7E6
5.	[STO] (store in memory)	
6.	[C/CE]	0.
7.	[4][.][5][E][6]	4.5E6
8.	[×]	
9.	[3][.][2][E][6]	3.2E6
10.	[=]	1.44E13
11.	[÷]	
12.	[RM] (Recall memory)	7.7E6
13.	[=]	1.87E6

a. Figure 11-20(a) has only two resistors in parallel, and therefore the product-over-sum resistor formula can be used.

$$R_T = \frac{R_1 \times R_2}{R_1 + R_2}$$
$$= \frac{4.5 \; \text{M}\Omega \times 3.2 \; \text{M}\Omega}{4.5 \; \text{M}\Omega + 3.2 \; \text{M}\Omega}$$
$$= \frac{14.4 \; \text{M}\Omega^2}{7.7 \; \text{M}\Omega}$$
$$= 1.9 \; \text{M}\Omega$$

The equivalent circuit is seen in Figure 11-20(c).

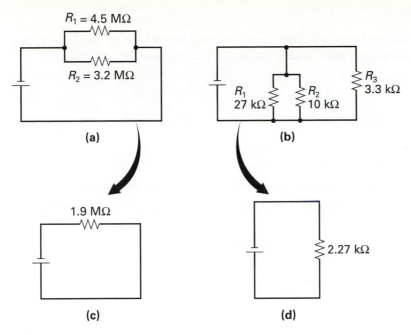

FIGURE 11-20   Parallel Circuit Examples.

b. Figure 11-20(b) has more than two resistors in parallel, and therefore the reciprocal formula must be used.

$$R_T = \frac{1}{(1/R_1) + (1/R_2) + (1/R_3)}$$

$$= \frac{1}{(1/27 \text{ k}\Omega) + (1/10 \text{ k}\Omega) + (1/3.3 \text{ k}\Omega)}$$

$$= \frac{1}{440.0 \times 10^{-6}}$$

$$= 2.27 \text{ k}\Omega$$

The equivalent circuit can be seen in Figure 11-20(d).

■ **INTEGRATED MATH APPLICATION:**

Find the total resistance of the parallel circuit in Figure 11-21.

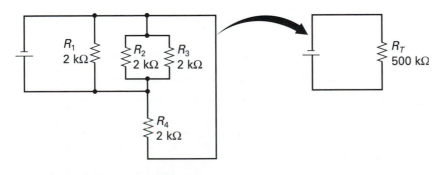

FIGURE 11-21   Parallel Circuit Example.

■ *Solution:*

Since all four resistors are connected in parallel and are all of the same value, the equal-value resistors in parallel formula can be used to calculate total resistance.

$$R_T = \frac{R}{n} = \frac{2\text{ k}\Omega}{4} = 500 \text{ }\Omega$$

To summarize what we have learned so far about parallel circuits, we can say that:

1. Components are said to be connected in parallel when the current has to travel two or more paths between the negative and positive sides of the voltage source.
2. The voltage across all the parallel branches is always the same.
3. The total current from the source is equal to the sum of all the branch currents (Kirchhoff's current law).
4. The amount of current flowing through each branch is inversely proportional to the resistance value in that branch.
5. The total resistance of a parallel circuit is always less than the value of the smallest branch resistance.

SELF-TEST EVALUATION POINT FOR SECTION 11-4

Now that you have completed this section, you should be able to:

■ *Objective 6.* *Determine the total resistance of any parallel-connected resistive circuit.*

Use the following questions to test your understanding of Section 11-4.

State which parallel resistance formula should be used in questions 1 through 3 for calculating R_T.

1. Two resistors in parallel.
2. More than two resistors in parallel.
3. Equal-value resistors in parallel.
4. Calculate the total parallel resistance when $R_1 = 2.7$ kΩ, $R_2 = 24$ kΩ, and $R_3 = 1$ MΩ.

11-5   POWER IN A PARALLEL CIRCUIT

As with series circuits, the total power in a parallel resistive circuit is equal to the sum of all the power losses for each of the resistors in parallel.

$$P_T = P_1 + P_2 + P_3 + P_4 + \cdots$$

Total power = addition of all the power losses

The formulas for calculating the amount of power dissipated are

$$P = I \times V$$
$$P = \frac{V^2}{R}$$
$$P = I^2 \times R$$

■ **INTEGRATED MATH APPLICATION:**

Calculate the total amount of power dissipated in Figure 11-22.

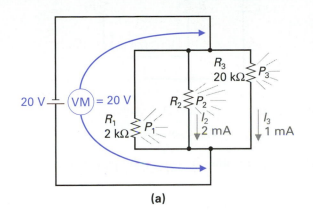

CIRCUIT ANALYSIS TABLE			
Resistance $R = V/I$	Voltage $V = I \times R$	Current $I = V/R$	Power $P = V \times I$
$R_1 = 2\ k\Omega$	$V_{R1} = 20\ V$	$I_1 =$	$P_1 = 200\ mW$
$R_2 =$	$V_{R2} = 20\ V$	$I_2 = 2\ mA$	$P_2 = 40\ mW$
$R_3 = 20\ k\Omega$	$V_{R3} = 20\ V$	$I_3 = 1\ mA$	$P_3 = 20\ mW$
	$V_T = V_S = 20\ V$		$P_T = 260\ mW$

(a)   (b)

FIGURE 11-22 **Parallel Circuit Example. (a) Schematic. (b) Circuit Analysis Table.**

■ *Solution:*

The total power dissipated in a parallel circuit is equal to the sum of all the power dissipated by all the resistors. With P_1, we know only voltage and resistance and therefore we can use the formula

$$P_1 = \frac{V_{R1}^2}{R_1} = \frac{20^2}{2\ k\Omega} = 0.2\ W \quad or \quad 200\ mW$$

With P_2, we know only current and voltage, and therefore we can use the formula

$$P_2 = I_2 \times V_{R2} = 2\ mA \times 20\ V = 40\ mW$$

With P_3, we know V, I, and R; however, we will use the third power formula:

$$P_3 = I_3^2 \times R_3 = (1\ mA)^2 \times 20\ k\Omega = 20\ mW$$

Total power (P_T) equals the sum of all the power or wattage losses for each resistor:

$$P_T = P_1 + P_2 + P_3$$
$$= 200\ mW + 40\ mW + 20\ mW$$
$$= 260\ mW$$

CALCULATOR SEQUENCE

Step	Keypad Entry	Display Response
1.	[2][0]	20.
2.	[x²]	400.
3.	[÷]	
4.	[2][E][3]	2.E3
5.	[=]	0.2

CALCULATOR SEQUENCE

Step	Keypad Entry	Display Response
1.	[2][E][3][+/−]	2.E-3
2.	[×]	
3.	[2][0]	20.
4.	[=]	40E-3

CALCULATOR SEQUENCE

Step	Keypad Entry	Display Response
1.	[1][E][3][+/−][x²]	1.E-6
2.	[×]	1.E-6
3.	[2][0][E][3]	20E3
4.	[=]	20.E-3

■ **INTEGRATED MATH APPLICATION:**

In Figure 11-23, there are two $\frac{1}{2}$ W (0.5 W) resistors connected in parallel. Should the wattage rating for each of these resistors be increased or decreased, or can they remain the same?

FIGURE 11-23 **Parallel Circuit Example.**

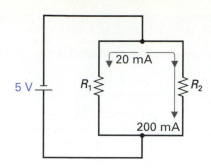

■ *Solution:*

Since current and voltage are known for both branches, the power formula used in both cases can be $P = I \times V$.

$$P_1 = I \times V$$
$$= 20 \text{ mA} \times 5 \text{ V}$$
$$= 0.1 \text{ W}$$

$$P_2 = I \times V$$
$$= 200 \text{ mA} \times 5 \text{ V}$$
$$= 1 \text{ W}$$

R_1 is dissipating 0.1 W, and so a 0.5 W resistor will be fine in this application. On the other hand, R_2 is dissipating 1 W and is designed to dissipate only 0.5 W. R_2 will therefore overheat unless it is replaced with a resistor of the same ohmic value, but with a 1 W or greater rating.

■ **INTEGRATED MATH APPLICATION:**

Figure 11-24(a) shows a simplified diagram of an automobile external light system. A 12 V lead–acid battery is used as a source and is connected across eight parallel-connected lamps. The left and right brake lights are controlled by the brake switch, which is attached to the brake pedal. When the light switch is turned on, both the rear taillights and the low-beam headlights are brought into circuit and turned on. The high-beam set of headlights is activated only if the high-beam switch is closed. For the lamp resistances given, calculate the output power of each lamp, when in use.

■ *Solution:*

Figure 11-24(b) shows the schematic diagram of the pictorial in Figure 11-24(a). Since both V and R are known, we can use the V^2/R power formula.

Brake lights: Left lamp wattage $(P) = \dfrac{V^2}{R} = \dfrac{(12 \text{ V})^2}{4 \, \Omega} = 36 \text{ W}$

Right lamp wattage is the same as left.

Taillights: Left lamp wattage $(P) = \dfrac{V^2}{R} = \dfrac{(12 \text{ V})^2}{6 \, \Omega} = 24 \text{ W}$

Right lamp wattage is the same as left.

Low-beam headlights: $(P) = \dfrac{V^2}{R} = \dfrac{(12 \text{ V})^2}{3 \, \Omega} = 48 \text{ W}$

Each low-beam headlight is a 48 W lamp.

High-beam headlights: $(P) = \dfrac{V^2}{R} = \dfrac{(12 \text{ V})^2}{3 \, \Omega} = 48 \text{ W}$

Each high-beam headlight is a 48 W lamp.

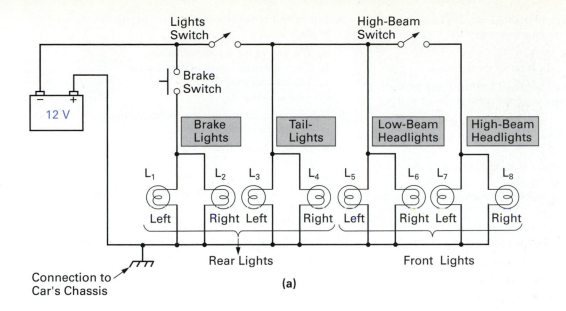

(a)

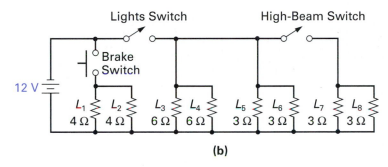

(b)

FIGURE 11-24 **Parallel Automobile External Light System.**

REVIEW QUESTIONS

Multiple Choice Questions

1. A parallel circuit has _____ path(s) for current to flow.

   a. One
   b. Two or more
   c. Only three
   d. None of the above

2. If a source voltage of 12 V is applied across four resistors of equal value in parallel, the voltage drops across each resistor will be equal to:

   a. 12 V
   b. 3 V
   c. 4 V
   d. 48 V

3. What would be the voltage drop across two 25 Ω resistors in parallel if the source voltage was equal to 9 V?

   a. 50 V   c. 12 V
   b. 25 V   d. None of the above

4. If a four-branch parallel circuit has 15 mA flowing through each branch, the total current into the parallel circuit will be equal to:

   a. 15 mA   c. 30 mA
   b. 60 mA   d. 45 mA

5. If the total three-branch parallel circuit current is equal to 500 mA, and 207 mA is flowing through one branch and 153 mA through another, what will be the current flow through the third branch?

   a. 707 mA   c. 140 mA
   b. 653 mA   d. None of the above

6. A large branch resistance will cause a _____ branch current.

   a. Large   c. Medium
   b. Small   d. None of the above are true

7. What would be the conductance of a 1 kΩ resistor?

   a. 10 mS   c. 2 kΩ
   b. 1 mS   d. All the above

8. If only two resistors are connected in parallel, the total resistance equals:

   a. The sum of the resistance values
   b. Three times the value of one resistor
   c. The product over the sum
   d. All the above

9. If resistors of equal value are connected in parallel, the total resistance can be calculated by:

   a. Dividing one resistor value by the number of parallel resistors
   b. Summing the resistor values
   c. Dividing the number of parallel resistors by one resistor value
   d. All the above could be true

10. The total power in a parallel circuit is equal to the:

    a. Product of total current and total voltage
    b. Reciprocal of the individual power losses
    c. Sum of the individual power losses
    d. Both (a) and (b)
    e. Both (a) and (c)

Communication Skill Questions

11. Describe the difference between a series and a parallel circuit. (11-1)

12. Explain and state mathematically the situation regarding voltage in a parallel circuit. (11-2)

13. State Kirchhoff's current law for parallel circuits. (11-3)

14. What is the current-divider formula? (11-3)

15. List the formulas for calculating the following total resistances: (11-4)

    a. Two resistors of different values
    b. More than two resistors of different values
    c. Equal-value resistors

16. Describe the relationship between branch current and branch resistance. (11-3)

17. Briefly describe total and individual power in a parallel resistive circuit. (11-5)

18. Why is a parrallel circuit called a current divider? (11-3)

19. Explain why parallel circuits have a smaller total resistance and larger total current than series circuits. (11-3)

20. Briefly describe why Kirchhoff's voltage law applies to series circuits and why Kirchhoff's current law relates to parallel circuits. (11-3)

Practice Problems

21. Calculate the total resistance of four 30 kΩ resistors in parallel.

22. Find the total resistance for each of the following parallel circuits:

    a. 330 Ω and 560 Ω
    b. 47 kΩ, 33 kΩ, and 22 kΩ
    c. 2.2 MΩ, 3 kΩ, and 220 Ω

23. If 10 V is connected across three 25 Ω resistors in parallel, what will be the total and individual branch currents?

24. If a four-branch parallel circuit has branch currents equal to 25 mA, 37 mA, 220 mA, and 0.2 A, what is the total circuit current?

25. If three resistors of equal value are connected across a 14 V supply and the total resistance is equal to 700 Ω, what is the value of each branch current?

26. If three 75 W lightbulbs are connected in parallel across a 110 V supply, what is the value of each branch current? What is the branch current through the other two lightbulbs if one burns out?

27. If 33 kΩ and 22 kΩ resistors are connected in parallel across a 20 V source, calculate:

    a. Total resistance   d. Total power dissipated
    b. Total current   e. Individual resistor power
    c. Branch currents   dissipated

28. If four parallel-connected resistors are dissipating 75 mW each, what is the total power being dissipated?

29. Calculate the branch currents through the following parallel resistor circuits when they are connected across a 10 V supply:

    a. 22 kΩ and 33 kΩ   b. 220 Ω, 330 Ω, and 470 Ω

30. If 30 Ω and 40 Ω resistors are connected in parallel, which resistor will generate the greatest amount of heat?

FIGURE 11-25   **Connect in Parallel across a 12 Volt Source.**

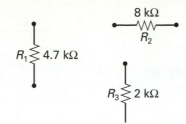

31. Calculate the total conductance and resistance of the following parallel circuits:

   a. Three 5 Ω resistors   **c.** 1 MΩ, 500 MΩ, 3.3 MΩ

   b. Two 200 Ω resistors   **d.** 5 Ω, 3 Ω, 2 Ω

32. Connect the three resistors in Figure 11-25 in parallel across a 12 V battery and then calculate the following:

   a. V_{R1}, V_{R2}, V_{R3} **d.** P_T

   b. I_1, I_2, I_3 **e.** P_1, P_2, P_3

   c. I_T **f.** G_{R1}, G_{R2}, G_{R3}

33. Calculate R_T in Figure 11-26 (a), (b), (c), and (d).

34. Calculate the branch currents through four 60 W bulbs connected in parallel across 110 V. How much is the total current, and what would happen to the total current if one of the bulbs was to burn out? What change would occur in the remaining branch currents?

35. Calculate the following in Figure 11-27:

   a. I_2 **c.** V_S, I_1, I_2

   b. I_T **d.** R_2, I_1, I_2, P_T

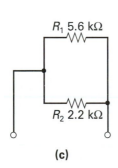

(a)

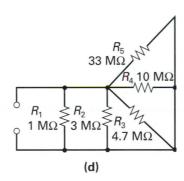

(b)

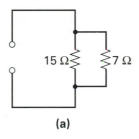

(c)

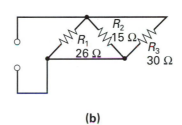

(d)

FIGURE 11-26   **Calculate Total Resistance.**

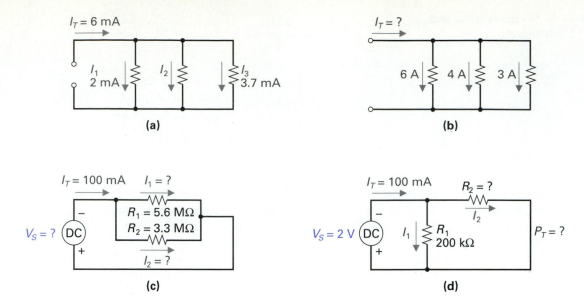

FIGURE 11-27 **Calculate the Unknowns.**

Web Site Questions

Go to the Web site http://www.prenhall.com/cook, select the textbook *Mathematics for Electronics and Computers,* select this chapter, and then follow the instructions when answering the multiple-choice practice problems.

Series–Parallel DC Circuits and Theorems

The Christie Bridge Circuit

Who invented the Wheatstone bridge circuit? It was obviously Sir Charles Wheatstone. Or was it?

The Wheatstone bridge was actually invented by S.H. Christie of the Royal Military Academy at Woolwich, England. He described the circuit in detail in the *Philosophical Transactions* paper dated February 28, 1833. Christie's name, however, was unknown and his invention was ignored.

Ten years later, Sir Charles Wheatstone called attention to Christie's circuit. Sir Charles was very well known, and from that point on, and even to this day, the circuit is known as a Wheatstone bridge. Later, Werner Siemens would modify Christie's circuit and invent the variable-resistance arm bridge circuit, which would also be called a Wheatstone bridge.

No one has given full credit to the real inventors of these bridge circuits, until now!

Charles Wheatstone

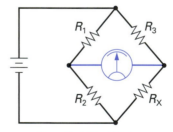

The Christie Bridge

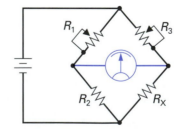

The Siemens Bridge

Outline and Objectives

VIGNETTE: THE CHRISTIE BRIDGE CIRCUIT

INTRODUCTION

12-1   SERIES- AND PARALLEL-CONNECTED COMPONENTS

Objective 1: Identify the difference between a series, a parallel, and a series–parallel circuit.

12-2   TOTAL RESISTANCE IN A SERIES-PARALLEL CIRCUIT

Objective 2: Describe how to use a three-step procedure to determine total resistance.

12-3   VOLTAGE DIVISION IN A SERIES–PARALLEL CIRCUIT

12-4   BRANCH CURRENTS IN A SERIES–PARALLEL CIRCUIT

12-5   POWER IN A SERIES–PARALLEL CIRCUIT

12-6   FIVE-STEP METHOD FOR SERIES–PARALLEL CIRCUIT ANALYSIS

Objective 3: Describe for the series–parallel circuit how to use a five-step procedure to calculate:
   a. Total resistance.
   b. Total current.
   c. Voltage division.
   d. Branch current.
   e. Total power dissipated.

12-7   SERIES–PARALLEL CIRCUITS

12-7-1   Loading of Voltage-Divider Circuits

Objective 4: Explain what loading effect a piece of equipment will have when connected to a voltage divider.

12-7-2   The Wheatstone Bridge
Balanced Bridge
Unbalanced Bridge
Determining Unknown Resistance

Objective 5: Identify and describe the Wheatstone bridge circuit in both the balanced and unbalanced condition.

12-7-3   The $R–2R$ Ladder Circuit

Objective 6: Describe the $R–2R$ ladder circuit used for digital-to-analog conversion.

12-8   THEOREMS FOR DC CIRCUITS

Objective 7: Describe the differences between a voltage and a current source.

Objective 8: Analyze series–parallel networks using:
   a. The superposition theorem.
   b. Thévenin's theorem.
   c. Norton's theorem.

12-8-1   Voltage and Current Sources
Voltage Source
Current Source

12-8-2   Superposition Theorem

12-8-3   Thévenin's Theorem

12-8-4   Norton's Theorem

MULTIPLE CHOICE QUESTIONS

COMMUNICATION SKILL QUESTIONS

PRACTICE PROBLEMS

WEB SITE QUESTIONS

Introduction

Very rarely are we lucky enough to run across straightforward series or parallel circuits. In general, all electronic equipment is composed of many components that are interconnected to form a combination of series and parallel circuits. In this chapter, we will be combining our knowledge of the series and parallel circuits discussed in the previous two chapters.

12-1   SERIES- AND PARALLEL-CONNECTED COMPONENTS

Series–Parallel Circuit

Network or circuit that contains components that are connected in both series and parallel.

Figure 12-1(a) through (f) shows six examples of **series–parallel resistive circuits.** The most important point to learn is how to distinguish between the resistors that are connected in series and the resistors that are connected in parallel, which will take a little practice.

One thing that you may not have noticed when examining Figure 12-1 is that:

Circuit 12-1(a) is equivalent to 12-1(b).

Circuit 12-1(c) is equivalent to 12-1(d).

Circuit 12-1(e) is equivalent to 12-1(f).

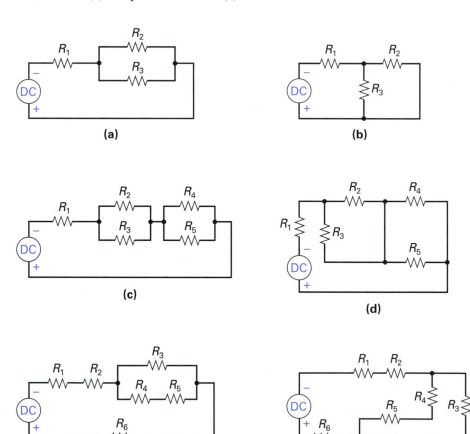

FIGURE 12-1   Series–Parallel Resistive Circuits.

CHAPTER 12 / SERIES–PARALLEL DC CIRCUITS AND THEOREMS

When analyzing these series–parallel circuits, always remember that current flow determines whether the resistor is connected in series or parallel. Begin at the negative side of the battery and apply these two rules:

1. If the total current has only one path to follow through a component, that component is connected in series.
2. If the total current has two or more paths to follow through two or more components, those components are connected in parallel.

Referring again to Figure 12-1, you can see that series or parallel resistor networks are easier to identify in parts (a), (c), and (e) than in parts (b), (d), and (f). Redrawing the circuit so that the components are arranged from left to right or from top to bottom is your first line of attack in your quest to identify series- and parallel-connected components.

■ INTEGRATED MATH APPLICATION:

Refer to Figure 12-2 and identify which resistors are connected in series and which are in parallel.

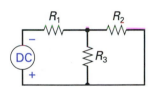

FIGURE 12-2 Series–Parallel Circuit Example.

Solution:

First, let's redraw the circuit so that the components are aligned either from left to right as shown in Figure 12-3(a), or from top to bottom as shown in Figure 12-3(b). Placing your pencil at the negative terminal of the battery on whichever figure you prefer, either Figure 12-3(a) or (b), trace the current paths through the circuit toward the positive side of the battery, as illustrated in Figure 12-4.

The total current arrives first at R_1. There is only one path for current to flow, which is through R_1, and therefore R_1 is connected in series. The total current proceeds on past R_1 and arrives at a junction where current divides and travels through two branches, R_2 and R_3. Since current had to split into two paths, R_2 and R_3 are therefore connected in parallel. After the parallel connection of R_2 and R_3, total current combines and travels to the positive side of the battery.

In this example, therefore, R_1 is in series with the parallel combination of R_2 and R_3.

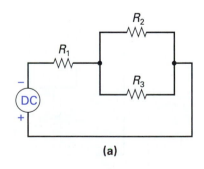

(a)

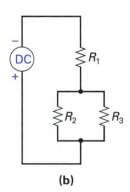

(b)

FIGURE 12-3 Redrawn Series–Parallel Circuit. (a) Left to Right.
(b) Top to Bottom.

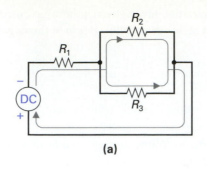

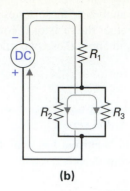

(a)

(b)

FIGURE 12-4   **Tracing Current through a Series–Parallel Circuit.**

■   **INTEGRATED MATH APPLICATION:**

Refer to Figure 12-5 and identify which resistors are connected in series and which are connected in parallel.

Solution:

Figure 12-6 illustrates the simplified, redrawn schematic of Figure 12-5. Total current leaves the negative terminal of the battery, and all of this current has to travel through R_1, which is therefore a series-connected resistor. Total current will split at junction *A*, and consequently, R_3 and R_4 with R_2 make up a parallel combination. The current that flows through R_3 (I_2) will also flow through R_4 and therefore R_3 is in series with R_4. I_1 and I_2 branch currents combine at junction *B* to produce total current, which has only one path to follow through the series resistor R_5, and finally, to the positive side of the battery.

In this example, therefore, R_3 and R_4 are in series with each other and both are in parallel with R_2, and this combination is in series with R_1 and R_5.

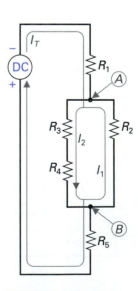

FIGURE 12-6   **Redrawn Series–Parallel Circuit Example.**

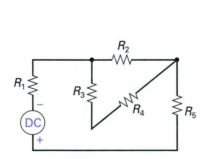

FIGURE 12-5   **Series–Parallel Circuit Example.**

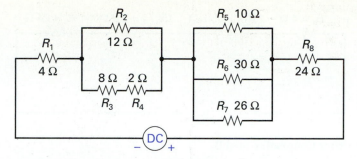

FIGURE 12-7   Total Series–Parallel Circuit Resistance.

12-2   TOTAL RESISTANCE IN A SERIES–PARALLEL CIRCUIT

No matter how complex or involved the series–parallel circuit, there is a simple three-step method to simplify the circuit to a single equivalent total resistance. Figure 12-7 illustrates an example of a series–parallel circuit. Once we have analyzed and determined the series–parallel relationship, we can proceed to solve for total resistance.
    The three-step method is:

Step A: Determine the equivalent resistances of all branch series-connected resistors.
Step B: Determine the equivalent resistances of all parallel-connected combinations.
Step C: Determine the equivalent resistance of the remaining series-connected resistances.

Let's now put this procedure to work with the example circuit in Figure 12-7.

    STEP A   Solve for all branch series-connected resistors. In our example, this applies only to R_3 and R_4, and since this is a series connection, we have to use the series resistance formula.

$$R_{3,4} = R_3 + R_4 = 8 + 2 = 10 \ \Omega \quad \text{(series resistance formula)}$$

With R_3 and R_4 solved, the circuit now appears as indicated in Figure 12-8.

    STEP B   Solve for all parallel combinations. In this example, they are the two parallel combinations of (a) R_2 and $R_{3,4}$ and (b) R_5 and R_6 and R_7. Since these are parallel connections, use the parallel resistance formulas.

$$R_{2,3,4} = \frac{R_2 \times R_{3,4}}{R_2 + R_{3,4}} = \frac{12 \times 10}{12 + 10} = 5.5 \ \Omega \quad \text{(product-over-sum formula)}$$

$$R_{5,6,7} = \frac{1}{(1/R_5) + (1/R_6) + (1/R_7)} = 5.8 \ \Omega \quad \text{(reciprocal formula)}$$

With $R_{2,3,4}$ and $R_{5,6,7}$ solved, the circuit now appears as illustrated in Figure 12-9.

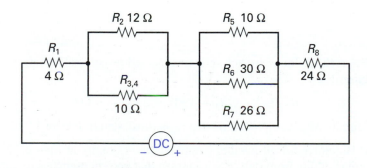

FIGURE 12-8   After Completing Step A.

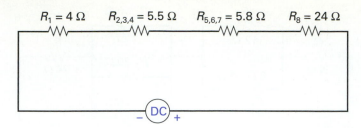

FIGURE 12-9 After Completing Step B.

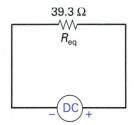

FIGURE 12-10 After Completing Step C.

STEP C   Solve for the remaining series resistances. There are now four remaining series resistances, which can be reduced to one equivalent resistance (R_{eq}) or total resistance (R_T). As seen in Figure 12-10, according to the series resistance formula, the total equivalent resistance for this example circuit will be

$$R_{eq} = R_1 + R_{2,3,4} + R_{5,6,7} + R_8$$
$$= 4\ \Omega + 5.5\ \Omega + 5.8\ \Omega + 24\ \Omega$$
$$= 39.3\ \Omega$$

■   **INTEGRATED MATH APPLICATION:**

Find the total resistance of the circuit in Figure 12-11.

FIGURE 12-11   **Calculate Total Resistance.**

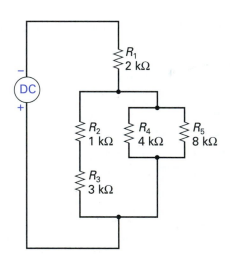

Solution:

STEP A   Solve for all branch series-connected resistors. This applies to R_2 and R_3 (series connection):

$$R_{2,3} = R_2 + R_3 = 1\ k\Omega + 3\ k\Omega = 4\ k\Omega$$

The resulting circuit, after completion of step A, is illustrated in Figure 12-12(a).

STEP B   Solve for all parallel combinations. Looking at Figure 12-12(a), which shows the circuit resulting from step A, you can see that current branches into three paths, so the parallel reciprocal formula must be used for this step.

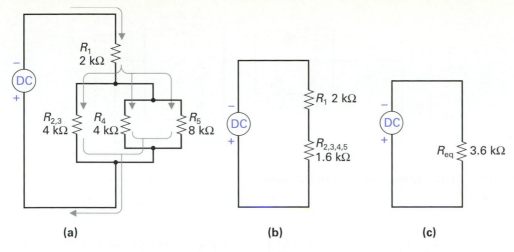

FIGURE 12-12 Calculating Total Resistance. (a) Step A. (b) Step B. (c) Step C.

$$R_{2,3,4,5} = \frac{1}{(1/R_{2,3}) + (1/R_4) + (1/R_5)}$$

$$= \frac{1}{(1/4\ k\Omega) + (1/4\ k\Omega) + (1/8\ k\Omega)} = 1.6\ k\Omega$$

The resulting circuit, after completion of step B, is illustrated in Figure 12-12(b).

STEP C   Solve for the remaining series resistances. Looking at Figure 12-12(b), which shows the circuit resulting from step B, you can see that there are two remaining series resistances. The equivalent resistance (R_{eq}) is therefore equal to

$$R_{eq} = R_1 + R_{2,3,4,5} = 2\ k\Omega + 1.6\ k\Omega = 3.6\ k\Omega$$

The total equivalent resistance, after completion of all three steps, is illustrated in Figure 12-12(c).

SELF-TEST EVALUATION POINT FOR SECTIONS 12-1 AND 12-2

Now that you have completed these sections, you should be able to:

■ **Objective 1.**   *Identify the difference between a series, a parallel, and a series–parallel circuit.*

■ **Objective 2.**   *Describe how to use a three-step procedure to determine total resistance.*

Use the following questions to test your understanding of Sections 12-1 and 12-2.

1. How can we determine which resistors are connected in series and which are connected in parallel in a series–parallel circuit?

2. Calculate the total resistance if two series-connected 12 kΩ resistors are connected in parallel with a 6 kΩ resistor.

3. State the three-step procedure used to determine total resistance in a circuit made up of both series and parallel resistors.

4. Sketch the following series–parallel resistor network made up of three resistors. R_1 and R_2 are in series with each other and are connected in parallel with R_3. If $R_1 = 470\ \Omega$, $R_2 = 330\ \Omega$, and $R_3 = 270\ \Omega$, what is R_T?

12-3   VOLTAGE DIVISION IN A SERIES–PARALLEL CIRCUIT

There is a simple three-step procedure for finding the voltage drop across each part of a series–parallel circuit. Figure 12-13 illustrates an example of a series–parallel circuit to which we will apply the three-step method for determining voltage drop.

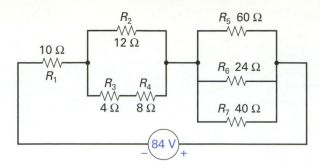

FIGURE 12-13 Series–Parallel Circuit Example.

STEP 1   Determine the circuit's total resistance. This is achieved by following the three-step method used previously for calculating total resistance.

Step A: $R_{3,4} = 4 + 8 = 12 \ \Omega$

Step B: $R_{2,3,4} = \dfrac{1}{(1/R_2) + (1/R_{3,4})} = 6 \ \Omega$

$R_{5,6,7} = \dfrac{1}{(1/R_5) + (1/R_6) + (1/R_7)} = 12 \ \Omega$

Figure 12-14 illustrates the equivalent circuit up to this point. We end up with one series resistor (R_1) and two series equivalent resistors ($R_{2,3,4}$ and $R_{5,6,7}$). R_T is therefore equal to 28 Ω.

STEP 2   Determine the circuit's total current. This step is achieved simply by utilizing Ohm's law.

$$I_T = \frac{V_T}{R_T} = \frac{84 \text{ V}}{28 \ \Omega} = 3 \text{ A}$$

STEP 3   Determine the voltage across each series resistor and each parallel combination (series equivalent resistor) in Figure 12-14. Since these are all in series, the same current (I_T) will flow through all three.

$$V_{R1} = I_T \times R_1 = 3 \text{ A} \times 10 \ \Omega = 30 \text{ V}$$
$$V_{R2,3,4} = I_T \times R_{2,3,4} = 3 \text{ A} \times 6 \ \Omega = 18 \text{ V}$$
$$V_{R5,6,7} = I_T \times R_{5,6,7} = 3 \text{ A} \times 12 \ \Omega = 36 \text{ V}$$

The voltage drops across the series resistor (R_1) and series equivalent resistors ($R_{2,3,4}$ and $R_{5,6,7}$) are illustrated in Figure 12-15.

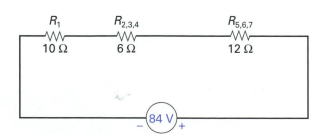

FIGURE 12-14   After Completing Step 1.

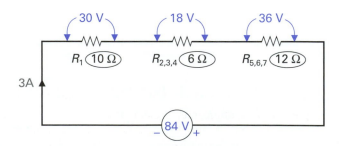

FIGURE 12-15   After Completing Steps 2 and 3.

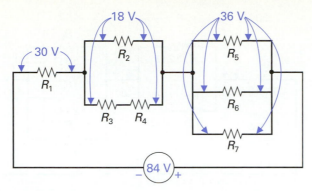

FIGURE 12-16 Detail of Step 3.

Kirchhoff's voltage law states that the sum of all the voltage drops is equal to the source voltage applied. This law can be used to confirm that our calculations are all correct:

$$
\begin{aligned}
V_T &= V_{R1} + V_{R2,3,4} + V_{R5,6,7} \\
&= 30\ \text{V} + 18\ \text{V} + 36\ \text{V} \\
&= 84\ \text{V}
\end{aligned}
$$

To summarize, refer to Figure 12-16, which shows these voltage drops inserted into our original circuit. As you can see from this illustration:

30 V is dropped across R_1.

18 V is dropped across R_2.

18 V is dropped across both R_3 and R_4.

36 V is dropped across R_5.

36 V is dropped across R_6.

36 V is dropped across R_7.

SELF-TEST EVALUATION POINT FOR SECTION 12-3

Use the following questions to test your understanding of Section 12-3.

1. State the three-step procedure used to calculate the voltage drop across each part of a series–parallel circuit.

2. Referring to Figure 12-13, double the values of all the resistors. Would the voltage drops calculated previously change, and if so, what would they be?

12-4   BRANCH CURRENTS IN A SERIES–PARALLEL CIRCUIT

In the preceding example, step 2 calculated the total current flowing in a series–parallel circuit. The next step is to find out exactly how much current is flowing through each parallel branch. This will be called step 4. Figure 12-17 shows the previously calculated data inserted in the appropriate places in our example circuit.

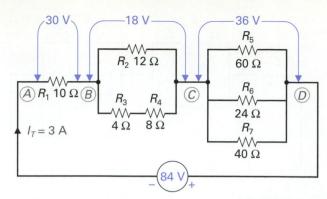

FIGURE 12-17 **Series–Parallel Circuit Example with Previously Calculated Data.**

STEP 4   Total current (I_T) will exist at points A, B, C, and D. Between A and B, current has only one path through which to flow, that is, through R_1. R_1 is therefore a series resistor, so $I_1 = I_T = 3$ A. Between points B and C, current has two paths: through R_2 (12 Ω) and through R_3 and R_4 (12 Ω).

$$I_2 = \frac{V_{R2}}{R_2} = \frac{18 \text{ V}}{12 \text{ Ω}} = 1.5 \text{ A}$$

$$I_{3,4} = \frac{V_{R3,4}}{R_{3,4}} = \frac{18 \text{ V}}{12 \text{ Ω}} = 1.5 \text{ A}$$

Not surprisingly, the total current of 3 A is split equally because both branches have equal resistance.

The two 1.5 A branch currents will combine at point C to produce once again the total current of 3 A. Between points C and D, current has three paths to flow through, R_5, R_6, and R_7.

$$I_5 = \frac{V_{R5}}{R_5} = \frac{36 \text{ V}}{60 \text{ Ω}} = 0.6 \text{ A}$$

$$I_6 = \frac{V_{R6}}{R_6} = \frac{36 \text{ V}}{24 \text{ Ω}} = 1.5 \text{ A}$$

$$I_7 = \frac{V_{R7}}{R_7} = \frac{36 \text{ V}}{40 \text{ Ω}} = 0.9 \text{ A}$$

All three branch currents will combine at point D to produce the total current of 3 A ($I_T = I_5 + I_6 + I_7 = 0.6 + 1.5 + 0.9 = 3$ A), proving Kirchhoff's current law.

12-5   POWER IN A SERIES–PARALLEL CIRCUIT

Whether resistors are connected in series or in parallel, the total power in a series–parallel circuit is equal to the sum of all the individual power losses.

$$P_T = P_1 + P_2 + P_3 + P_4 + \ldots$$

Total power = sum of all power losses

The formulas for calculating the amount of power lost by each resistor are

$$P = \frac{V^2}{R} \qquad\qquad P = I \times V \qquad\qquad P = I^2 \times R$$

Let us calculate the power dissipated by each resistor. This final calculation will be called step 5.

STEP 5   Since resistance, voltage, and current are known, any one of the three formulas for power can be used to determine power.

$$P_1 = \frac{V_{R1}^2}{R_1} = \frac{(30 \text{ V})^2}{10 \text{ }\Omega} = 90 \text{ W}$$

$$P_2 = \frac{V_{R2}^2}{R_2} = \frac{(18 \text{ V})^2}{12 \text{ }\Omega} = 27 \text{ W}$$

$$P_3 = I_{R3}^2 \times R_3 = (1.5 \text{ A})^2 \times 4 \text{ }\Omega = 9 \text{ W}$$

$$P_4 = I_{R4}^2 \times R_4 = (1.5 \text{ A})^2 \times 8 \text{ }\Omega = 18 \text{ W}$$

$$P_5 = \frac{V_{R5}^2}{R_5} = \frac{(36 \text{ V})^2}{60 \text{ }\Omega} = 21.6 \text{ W}$$

$$P_6 = \frac{V_{R6}^2}{R_6} = \frac{(36 \text{ V})^2}{24 \text{ }\Omega} = 54 \text{ W}$$

$$P_7 = \frac{V_{R7}^2}{R_7} = \frac{(36 \text{ V})^2}{40 \text{ }\Omega} = 32.4 \text{ W}$$

$$P_T = P_1 + P_2 + P_3 + P_4 + P_5 + P_6 + P_7$$
$$= 90 + 27 + 9 + 18 + 21.6 + 54 + 32.4$$
$$= 252 \text{ W}$$

or

$$P_T = \frac{V_T^2}{R_T} = \frac{(84 \text{ V})^2}{28 \text{ }\Omega}$$
$$= 252 \text{ W}$$

The total power dissipated in this example circuit is 252 W. All the information can now be inserted in a final diagram for the example, as shown in Figure 12-18.

12-6   FIVE-STEP METHOD FOR SERIES–PARALLEL CIRCUIT ANALYSIS

Let's now combine and summarize all the steps for calculating resistance, voltage, current, and power in a series–parallel circuit by solving another problem. Before we begin, however, let us review the five-step procedure.

Solving for Resistance, Voltage, Current, and Power in a Series–Parallel Circuit

STEP 1   Determine the circuit's total resistance.

Step A   Solve for series-connected resistors in all parallel combinations.
Step B   Solve for all parallel combinations.
Step C   Solve for remaining series resistances.

STEP 2   Determine the circuit's total current.

STEP 3   Determine the voltage across each series resistor and each parallel combination (series equivalent resistor).

STEP 4   Determine the value of current through each parallel resistor in every parallel combination.

STEP 5   Determine the total and individual power dissipated by the circuit.

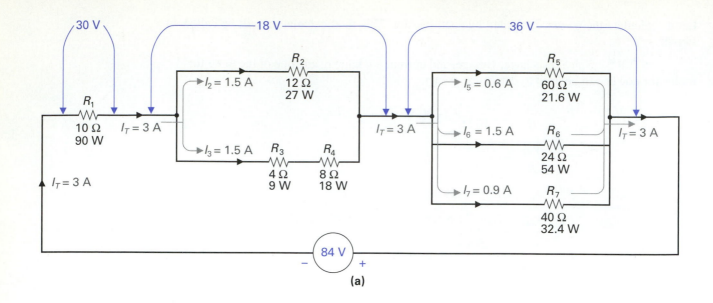

(a)

CIRCUIT ANALYSIS TABLE			
Resistance $R = V/I$	Voltage $V = I \times R$	Current $I = V/R$	Power $P = V \times I$
$R_1 = 10\ \Omega$	$V_{R1} = 30\ V$	$I_1 = 3\ A$	$P_1 = 90\ W$
$R_2 = 12\ \Omega$	$V_{R2} = 18\ V$	$I_2 = 1.5\ A$	$P_2 = 27\ W$
$R_3 = 4\ \Omega$	$V_{R3} = 6\ V$	$I_3 = 1.5\ A$	$P_3 = 9\ W$
$R_4 = 8\ \Omega$	$V_{R4} = 12\ V$	$I_4 = 1.5\ A$	$P_4 = 18\ W$
$R_5 = 60\ \Omega$	$V_{R5} = 36\ V$	$I_5 = 0.6\ A$	$P_5 = 21.6\ W$
$R_6 = 24\ \Omega$	$V_{R6} = 36\ V$	$I_6 = 1.5\ A$	$P_6 = 54\ W$
$R_7 = 40\ \Omega$	$V_{R7} = 36\ V$	$I_7 = 0.9\ A$	$P_7 = 32.4\ W$
$R_T = 28\ \Omega$	$V_T = V_S = 84V$	$I_T = 3\ A$	$P_T = 252\ W$

(b)

FIGURE 12-18 **Series–Parallel Circuit Example with All Information Inserted.**
(a) Schematic. (b) Circuit Analysis Table.

■ **INTEGRATED MATH APPLICATION:**

Referring to Figure 12-19, calculate:

a. Total resistance

b. Total current

c. Voltage drop across all resistors

d. Current through each resistor

e. Total power dissipated by the circuit

Solution:

This problem has asked us to calculate everything about the series–parallel circuit shown in Figure 12-19, and is an ideal application for our five-step series–parallel circuit analysis procedure.

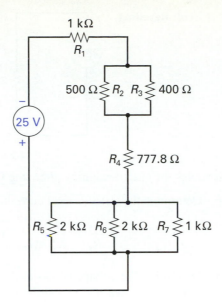

STEP 1   Determine the circuit's total resistance.

Step A: There are no series resistors within parallel combinations.

Step B: There are two-resistor (R_2, R_3) and three-resistor (R_5, R_6, R_7) parallel combinations in this circuit.

$$R_{2,3} = \frac{1}{(1/R_2) + (1/R_3)} = 222.2 \ \Omega$$

$$R_{5,6,7} = \frac{1}{(1/R_5) + (1/R_6) + (1/R_7)} = 500 \ \Omega$$

Figure 12-20 illustrates the circuit resulting after step B.

Step C: Solve for the remaining four resistances to obtain the circuit's total resistance (R_T) or equivalent resistance (R_{eq}).

$$\begin{aligned}
R_{eq} &= R_1 + R_{2,3} + R_4 + R_{5,6,7} \\
&= 1000 \ \Omega + 222.2 \ \Omega + 777.8 \ \Omega + 500 \ \Omega \\
&= 2500 \ \Omega \quad \text{or} \quad 2.5 \ \text{k}\Omega
\end{aligned}$$

FIGURE 12-20   **Circuit Resulting after Step 1B.**

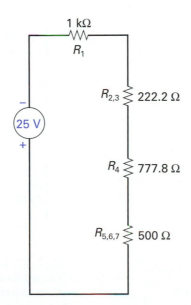

FIGURE 12-21 Circuit Resulting
after Step 1C.

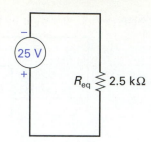

Figure 12-21 illustrates the circuit resulting after step C.

STEP 2 Determine the circuit's total current.

$$I_T = \frac{V_S}{R_T} = \frac{25 \text{ V}}{2.5 \text{ k}\Omega} = 10 \text{ mA}$$

STEP 3 Determine the voltage across each series resistor and each series equivalent resistor. To achieve this, we utilize the diagram obtained after completing step B (Figure 12-20):

$$V_{R1} = I_T \times R_1 = 10 \text{ mA} \times 1 \text{ k}\Omega = 10 \text{ V}$$
$$V_{R2,3} = I_T \times R_{2,3} = 10 \text{ mA} \times 222.2 \text{ }\Omega = 2.222 \text{ V}$$
$$V_{R4} = I_T \times R_4 = 10 \text{ mA} \times 777.8 \text{ }\Omega = 7.778 \text{ V}$$
$$V_{R5,6,7} = I_T \times R_{5,6,7} = 10 \text{ mA} \times 500 \text{ }\Omega = 5 \text{ V}$$

Figure 12-22 illustrates the results after step 3.

STEP 4 Determine the value of current through each parallel resistor (Figure 12-23). R_1 and R_4 are series-connected resistors, and therefore their current will equal 10 mA.

$$I_1 = 10 \text{ mA}$$
$$I_4 = 10 \text{ mA}$$

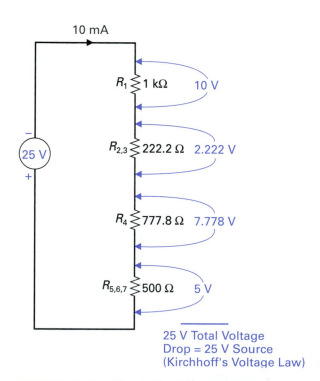

FIGURE 12-22 Circuit Resulting after Step 3.

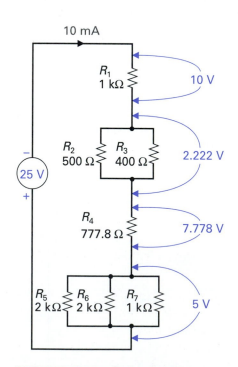

FIGURE 12-23 Series–Parallel Circuit Example with Steps 1, 2, and 3 Data Inserted.

The current through the parallel resistors is calculated by Ohm's law.

$$I_2 = \frac{V_{R2}}{R_2} = \frac{2.222 \text{ V}}{500 \text{ } \Omega} = 4.4 \text{ mA}$$

$$I_3 = \frac{V_{R3}}{R_3} = \frac{2.222 \text{ V}}{400 \text{ } \Omega} = 5.6 \text{ mA}$$

$$\left. \begin{array}{l} I_T = I_2 + I_3 \\ 10 \text{ mA} = 4.4 \text{ mA} + 5.6 \text{ mA} \end{array} \right\} \text{Kirchhoff's current law}$$

$$I_5 = \frac{V_{R5}}{R_5} = \frac{5 \text{ V}}{2 \text{ k}\Omega} = 2.5 \text{ mA}$$

$$I_6 = \frac{V_{R6}}{R_6} = \frac{5 \text{ V}}{2 \text{ k}\Omega} = 2.5 \text{ mA}$$

$$I_7 = \frac{V_{R7}}{R_7} = \frac{5 \text{ V}}{1 \text{ k}\Omega} = 5 \text{ mA}$$

$$\left. \begin{array}{l} I_T = I_5 + I_6 + I_7 \\ 10 \text{ mA} = 2.5 \text{ mA} + 2.5 \text{ mA} + 5 \text{ mA} \end{array} \right\} \text{Kirchhoff's current law}$$

STEP 5   Determine the total power dissipated by the circuit.

$$P_T = P_1 + P_2 + P_3 + P_4 + P_5 + P_6 + P_7$$

or

$$P_T = \frac{V_T^2}{R_T}$$

Each resistor's power figure can be calculated, and the sum would be the total power dissipated by the circuit. Since the problem does not ask for the power dissipated by each individual resistor but for the total power dissipated, it will be easier to use the formula:

$$
\begin{aligned}
P_T &= \frac{V_T^2}{R_T} \\
&= \frac{(25 \text{ V})^2}{2.5 \text{ k}\Omega} \\
&= 0.25 \text{ W}
\end{aligned}
$$

SELF-TEST EVALUATION POINT FOR 12-4, 12-5, AND 12-6

Now that you have completed these sections, you should be able to:

■ *Objective 3.* *Describe for the series–parallel circuit how to use a five-step procedure to calculate:*

  a. *Total resistance.*
  b. *Total current.*
  c. *Voltage division.*
  d. *Branch current.*
  e. *Total power dissipated.*

Use the following questions to test your understanding of Sections 12-4, 12-5, and 12-6.

1. State the five-step method used for series–parallel circuit analysis.

2. Design your own five-resistor series–parallel circuit, assign resistor values and a source voltage, and then apply the five-step analysis method.

Loading

The adding of a load to a source.

12-7-1   *Loading of Voltage-Divider Circuits*

The straightforward voltage divider was discussed in Chapter 10, but at that point we did not explore some changes that will occur if a load resistance is connected to the voltage divider's output. Figure 12-24 shows a voltage divider, and as you can see, the advantage of a voltage-divider circuit is that it can be used to produce several different voltages from one main voltage source by the use of a few chosen resistor values.

In our discussion on load resistance, we discussed how every circuit or piece of equipment offers a certain amount of resistance, and this resistance represents how much a circuit or piece of equipment will load down the source supply.

Figure 12-25 shows an example voltage-divider circuit that is being used to develop a 10 V source from a 20 V dc supply. Figure 12-25(a) illustrates this circuit in the unloaded condition, and by making a few calculations we can analyze this circuit condition.

STEP 1 $R_T = R_1 + R_2 = 1\text{ k}\Omega + 1\text{ k}\Omega = 2\text{ k}\Omega$

STEP 2 $I_T = \dfrac{V_T}{R_T} = \dfrac{20\text{ V}}{2\text{ k}\Omega} = 10\text{ mA}$

The current that flows through a voltage divider, without a load connected, is called the **bleeder current.** In this example the bleeder current is equal to 10 mA. It is called the bleeder current because it is continually drawing or bleeding this current from the voltage source.

Bleeder Current

Current drawn continuously from a voltage source. A bleeder resistor is generally added to lessen the effect of load changes or to provide a voltage drop across a resistor.

STEP 3 $V_{R1} = V_{R2}$ (as resistors are the same value)
$V_{R1} = 10\text{ V}$
$V_{R2} = 10\text{ V}$

In Figure 12-25(b) we have connected a piece of equipment represented as a resistance (R_3) across the 10 V supply. This automatically turns the previous series circuit of R_1 and R_2 into a series–parallel circuit made up of R_1, R_2, and the 100 kΩ load resistance. By making a few more calculations, we can discover the changes that are produced by connecting this load resistance.

STEP 1   Total resistance (R_T)

Step B: $R_{2,3} = \dfrac{R_2 \times R_3}{R_2 + R_3} = \dfrac{1\text{ k}\Omega \times 100\text{ k}\Omega}{1\text{ k}\Omega + 100\text{ k}\Omega} = 990.1\ \Omega$

FIGURE 12-24   Voltage-Divider Circuit.

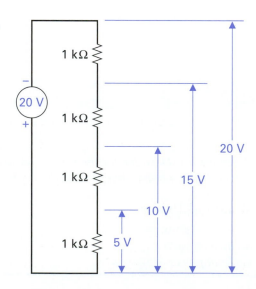

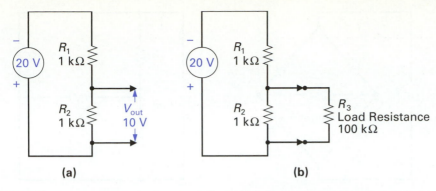

FIGURE 12-25 Voltage-Divider Circuit. (a) Unloaded Output Voltage. (b) Loaded Output Voltage.

Step C:

$$R_{1,2,3} = R_1 + R_{2,3}$$
$$= 1 \text{ k}\Omega + 990.1 \text{ }\Omega$$
$$= 1.99 \text{ k}\Omega$$

STEP 2   Total current (I_T)

$$I_T = \frac{20 \text{ V}}{1.99 \text{ k}\Omega} = 10.05 \text{ mA}$$

STEP 3 $V_{R1} = I_T \times R_1 = 10.05 \text{ mA} \times 1 \text{ k}\Omega = 10.05 \text{ V}$
$V_{R2,3} = I_T \times R_{2,3} = 10.05 \text{ mA} \times 990.1 \text{ }\Omega = 9.95 \text{ V}$

STEP 4 $I_1 = I_T = 10.05 \text{ mA}$

$$I_2 = \frac{V_{R2}}{R_2} = \frac{9.95 \text{ V}}{1 \text{ k}\Omega} = 9.95 \text{ mA}$$

$$I_3 = \frac{V_{R3}}{R_3} = \frac{9.95 \text{ V}}{100 \text{ k}\Omega} = 99.5 \text{ }\mu\text{A}$$

$$\left.\begin{array}{l} I_2 + I_3 = I_T \\ 9.95 \text{ mA} + 99.5 \text{ }\mu\text{A} = 10.05 \text{ mA} \end{array}\right\} \text{Kirchhoff's current law}$$

As you can see, the load resistance is pulling 99.5 μA from the source, and this pulls the voltage down to 9.95 V from the required 10 V that was desired and is normally present in the unloaded condition.

When designing a voltage divider, design engineers need to calculate how much current a particular load will pull and then alter the voltage-divider resistor values to offset the loading effect when the load is connected.

12-7-2   *The Wheatstone Bridge*

In 1850, Charles Wheatstone developed a circuit to measure resistance. This circuit, which is still widely used today, is called the **Wheatstone bridge** and is illustrated in Figure 12-26(a). In Figure 12-26(b), the same circuit has been redrawn so that the series and parallel resistor connections are easier to see.

Wheatstone Bridge

A four-arm, generally resistive, bridge that is used to measure resistance.

Balanced Bridge

Figure 12-27 illustrates an example circuit in which four resistors are connected together to form a series–parallel arrangement. Let us now use the five-step procedure to find out exactly what resistance, current, voltage, and power values exist throughout the circuit.

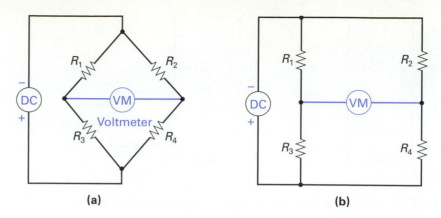

(a)   **(b)**

FIGURE 12-26   **Wheatstone Bridge. (a) Actual Circuit. (b) Redrawn Simplified Circuit.**

STEP 1   Total resistance (R_T)

Step A:

$$R_{1,3} = R_1 + R_3 = 10 + 20 = 30$$
$$R_{2,4} = R_2 + R_4 = 10 + 20 = 30$$

Step B:

$$R_T: (R_{1,2,3,4}) = \frac{R_{1,3} \times R_{2,4}}{R_{1,3} + R_{2,4}}$$

$$= \frac{30 \times 30}{30 + 30} = 15 \ \Omega$$

Total resistance = 15 Ω

STEP 2   Total current (I_T):

$$I_T = \frac{V_T}{R_T} = \frac{30 \text{ V}}{15 \ \Omega} = 2 \text{ A}$$

STEP 3   Since $R_{1,3}$ is in parallel with $R_{2,4}$, 30 V will appear across both $R_{1,3}$ and $R_{2,4}$.

$$V_T = V_{R1,3} = V_{R2,4} = 30 \text{ V}$$

voltage-divider formula $$V_{R1} = \frac{R_1}{R_{1,3}} \times V_T$$

$$= \frac{10}{30} \times 30 = 10 \text{ V}$$

$$V_{R3} = \frac{R_3}{R_{1,3}} \times V_T = 20 \text{ V}$$

FIGURE 12-27   Wheatstone Bridge Circuit Example.

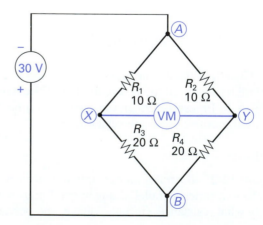

$$V_{R2} = \frac{R_2}{R_{2,4}} \times V_T = 10 \text{ V}$$

$$V_{R4} = \frac{R_4}{R_{2,4}} \times V_T = 20 \text{ V}$$

STEP 4

$$I_{1,3} = \frac{V_{R1,3}}{R_{1,3}} = \frac{30 \text{ V}}{30 \text{ }\Omega} = 1 \text{ A}$$

$$I_{2,4} = \frac{V_{R2,4}}{R_{2,4}} = \frac{30 \text{ V}}{30 \text{ }\Omega} = 1 \text{ A}$$

$$\left. \begin{array}{l} I_{1,3} + I_{2,4} = I_T \\ 1 \text{ A} + 1 \text{ A} = 2 \text{ A} \end{array} \right\} \text{Kirchhoff's current law}$$

STEP 5   Total power dissipated (P_T):

$$\begin{aligned} P_T &= I_T^2 \times R_T \\ &= (2 \text{ A})^2 \times 15 \text{ }\Omega \\ &= 4 \text{ A}^2 \times 15 \text{ }\Omega \\ &= 60 \text{ W} \end{aligned}$$

Figure 12-28 shows all of the step results inserted in the Wheatstone bridge example schematic. The Wheatstone bridge is said to be in the balanced condition when the voltage at point X equals the voltage at point Y ($V_{R3} = V_{R4}$, 20 V = 20 V). This same voltage exists across R_3 and R_4, so the voltmeter, which is measuring the voltage difference between X and Y, will indicate 0 V potential difference, and the circuit is said to be a *balanced bridge.*

Unbalanced Bridge

In Figure 12-29 we have replaced R_3 with a variable resistor and set it to 10 Ω. The R_2 and R_4 resistor combination will not change its voltage drop; however, R_1 and R_3, which are now equal, will split the 30 V supply, producing 15 V across R_3. The voltmeter will indicate the difference in potential (5 V) from the voltage across R_3 at point X (15 V) and across R_4 at point Y (20 V). The voltmeter is actually measuring the imbalance in the circuit, which is why this circuit in this condition is known as an *unbalanced bridge.*

Determining Unknown Resistance

Figure 12-30 shows how a Wheatstone bridge circuit can be used to find the value of an unknown resistor (R_{un}). The variable-value resistor (R_{va}) is a calibrated resistor, which means that its resistance has been checked against a known, accurate resistance and its value can be adjusted and read from a calibrated dial.

FIGURE 12-28   Balanced Wheatstone Bridge.

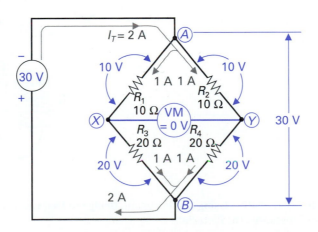

FIGURE 12-29 Unbalanced Wheatstone Bridge.

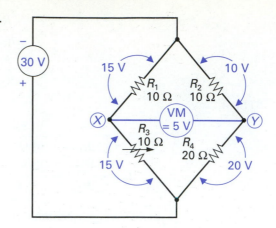

The procedure to follow to find the value of the unknown resistor is as follows:

1. Adjust the variable-value resistor until the voltmeter indicates that the Wheatstone bridge is balanced (voltmeter indicates 0 V).

2. Read the value of the variable-value resistor. As long as $R_1 = R_2$, the variable resistance value will be the same as the unknown resistance value.

$$R_{va} = R_{un}$$

Since R_1 and R_2 are equal to each other, the voltage will be split across the two resistors, producing 10 V at point Y. The variable-value resistor must therefore be adjusted so that it equals the unknown resistance, and therefore the same situation will occur, in that the 20 V source will be split, producing 10 V at point X, indicating a balanced 0 V condition on the voltmeter. For example, if the unknown resistance is equal to 5 Ω, then only when the variable-value resistor is adjusted and equal to 5 Ω will 10 V appear at point X and allow the circuit meter to read 0 V, indicating a balance. The variable-value resistor resistance can be read (5 Ω) and the unknown resistor resistance will be known (5 Ω).

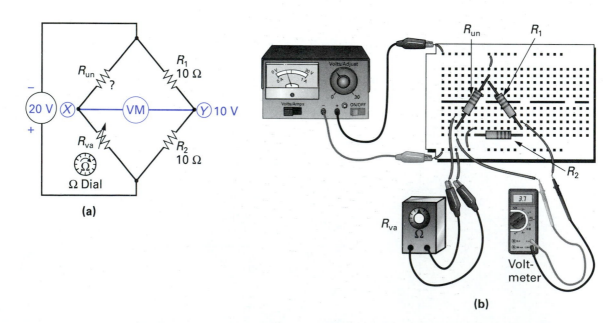

FIGURE 12-30 Using a Wheatstone Bridge to Determine an Unknown Resistance.
(a) Schematic. (b) Pictorial.

INTEGRATED MATH APPLICATION:

What is the unknown resistance in Figure 12-31?

FIGURE 12-31   Wheatstone Bridge Circuit Example.

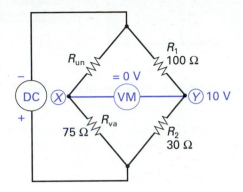

Solution:

The bridge is in a balanced condition, as the voltmeter is reading a 0 V difference between points X and Y. In the previous section we discovered that if $R_1 = R_2$, then:

$$R_{va} = R_{un}$$

In this case, R_1 does not equal R_2, so a variation in the formula must be applied to take into account the ratio of R_1 and R_2.

$$R_{un} = R_{va} \times \frac{R_1}{R_2}$$

$$= 75\ \Omega \times \frac{100}{30}$$

$$= 75\ \Omega \times 3.33 = 250\ \Omega$$

Since R_1 is 3.33 times greater than R_2, then R_{un} must be 3.33 times greater than R_{va} if the Wheatstone bridge is in the balanced condition.

CALCULATOR SEQUENCE

Step	Keypad Entry	Display Response
1.	[1] [0] [0]	100
2.	[÷]	
3.	[3] [0]	30
4.	[=]	3.333
5.	[×]	
6.	[7] [5]	75
7.	[=]	250

12-7-3   *The R–2R Ladder Circuit*

Figure 12-32 illustrates an **R–2R ladder circuit,** which is a series–parallel circuit used within computer systems to convert digital information to analog information. To fully understand this circuit, our first step should be to find out exactly which branches will have which values of current. This can be obtained by finding out what value of resistance the current sees when it arrives at the three junctions *A, B,* and *C.* Let us begin with point *C* first and simplify the circuit. This is illustrated in Figure 12-33(a). No specific resistance value has been chosen, but in all cases 2R resistors (2 × R) are twice the resistance of an R resistor.

R–2R Ladder Circuit

A network or circuit composed of a sequence of L network connected in tandem. This R–2R circuit is used in digital-to-analog converters.

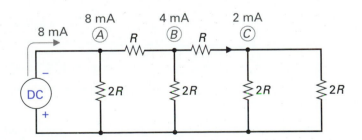

FIGURE 12-32   *R–2R Ladder Circuit.*

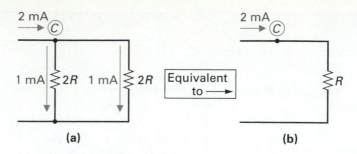

(a)   **(b)**

FIGURE 12-33   *R–2R Equivalent Circuit at Junction C.*

In Figure 12-33(a), if 2 mA of current arrives at point C, it sees $2R$ of resistance in parallel with a $2R$ resistance, and so the 2 mA of current splits, and 1 mA flows through each branch. Two $2R$ resistors in parallel with each other would consequently be equivalent to one R, as seen in Figure 7-33(b).

In Figure 12-34(a), if 4 mA of current arrives at point B, it sees two series resistances to the right, which is equivalent to $2R$, as seen in Figure 12-34(b), and one resistance below it of $2R$. The 4 mA therefore splits, causing 2 mA down one path and 2 mA down the other. The two $2R$ resistors in parallel, as seen in Figure 12-34(b), are equivalent to one R, as shown in Figure 12-34(c).

In Figure 12-35(a), if 8 mA of current arrives at point A, it sees two series resistors to the right, which is equivalent to $2R$, as seen in Figure 12-35(b), and one resistance below it of $2R$. The 8 mA of current therefore splits equally, causing 4 mA down one path and 4 mA down the other. The two $2R$ resistors in parallel, as seen in Figure 12-35(b), are equivalent to one resistance R, as shown in Figure 12-35(c).

Figure 12-36(a) through (f) summarizes the step-by-step simplification of this circuit.

The question you may have at this point is: What is the primary application of this circuit? The answer is as a current divider, as seen in Figure 12-37(a). The 8 mA of reference current is repeatedly divided by 2 as it moves from left to right, producing currents of 4 mA, 2 mA, and 1 mA. The result of this R–$2R$ current division can be used in a circuit known as a *digital-to-analog converter* (DAC), which is illustrated in Figure 12-37(b).

Digital data or information exists within a computer, and this system expresses numbers and letters in two discrete steps; for example, on–off, high–low, open–closed, or 0–1. Only two conditions exist within the computer, and all information is represented by this two-state system.

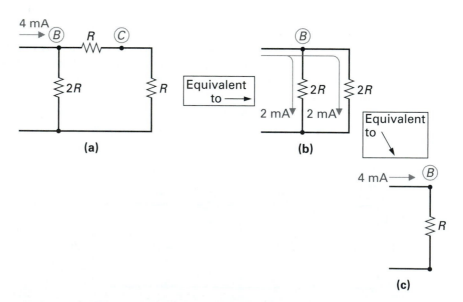

FIGURE 12-34   *R–2R Equivalent Circuit at Junction B.*

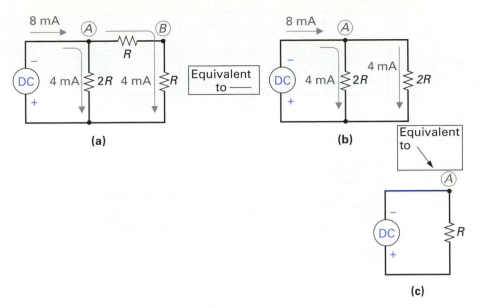

(a)

(b)

(c)

FIGURE 12-35 *R–2R* Equivalent Circuit at Junction *A*.

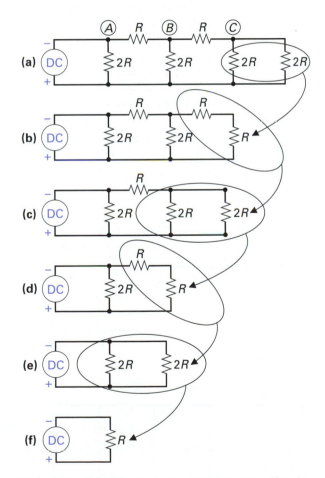

FIGURE 12-36 Step-by-Step Simplification of an *R–2R* Ladder Circuit.

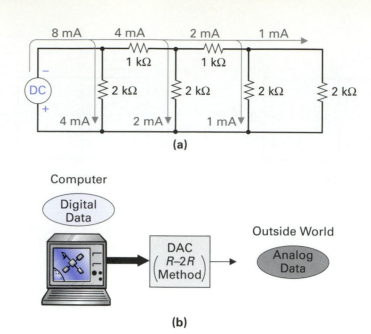

(a)

(b)

FIGURE 12-37 *R–2R* Ladder in a Digital-to-Analog Converter.

Analog data or information exists outside a computer in our environment, and in this system, data are expressed in many different levels, as opposed to just two with digital information. Numbers are expressed in one of 10 different levels (0–9), as opposed to only two (0–1) in digital.

Due to these differences, a device is needed that will interface (convert or link two different elements) the digital information within a computer to the analog information that you and I understand. This device, called a digital-to-analog converter, uses the *R–2R* ladder circuit that we just discussed.

■ INTEGRATED MATH APPLICATION:

Determine the reference current and branch currents for the circuit if Figure 12-38.

Solution:

In our simplification of the ladder circuit, we discussed previously that any *R–2R* ladder circuit can be simplified to one resistor equal to *R*, as seen in Figure 12-39(a). The reference or total current supplied will therefore equal

$$ I_T = \frac{V_T}{R_T} = \frac{24\ \text{V}}{1\ \text{k}\Omega} = 24\ \text{mA} $$

and the current will split through each branch, as shown in Figure 12-39(b).

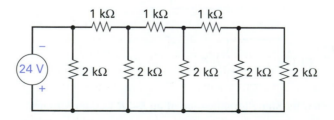

FIGURE 12-38 Calculate *R–2R* Circuit References Current and Branch Currents.

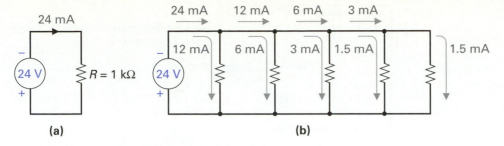

FIGURE 12-39 *R–2R* Circuit with Data Inserted.

SELF-TEST EVALUATION POINT FOR SECTION 12-7

Now that you have completed this section, you should be able to:

■ **Objective 4.** *Explain what loading effect a piece of equipment will have when connected to a voltage divider.*

■ **Objective 5.** *Identify and describe the Wheatstone bridge circuit in both the balanced and unbalanced condition.*

■ **Objective 6.** *Describe the R–2R ladder circuit used for digital-to-analog conversion.*

Use the following questions to test your understanding of Section 12-7.

1. What is meant by loading of a voltage-divider circuit?
2. Sketch a Wheatstone bridge circuit and list an application of this circuit.
3. What value will the total resistance of an *R–2R* ladder circuit always equal?
4. For what application could the *R–2R* ladder be used?

12-8   THEOREMS FOR DC CIRCUITS

Series–parallel circuits can become very complex in some applications, and the more help you have in simplifying and analyzing these networks, the better. The following theorems can be used as powerful analytical tools for evaluating circuits. To begin, let's discuss the differences between voltage and current sources.

12-8-1 *Voltage and Current Sources*

The easiest way to understand a current source is to compare its features with those of a voltage source, so let's begin by discussing voltage sources.

Voltage Source

The battery is an example of a **voltage source** that in the ideal condition will produce a fixed output voltage regardless of what load resistance is connected across its terminals. This means that even if a large load current is drawn from the battery (heavy load, due to a small load resistance) or if a small load current results (light load, due to a large load resistance), the battery will always produce a constant output voltage, as seen in Figure 12-40.

In reality, every voltage source, whether a battery, power supply, or generator, will have some level of inefficiency and not only generate an output electrical dc voltage but

Voltage Source
The circuit or device that supplies voltage to a load circuit.

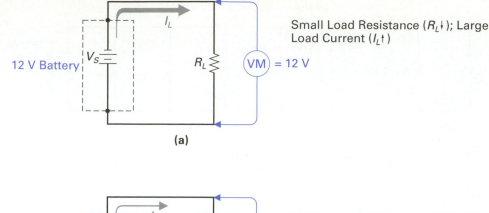

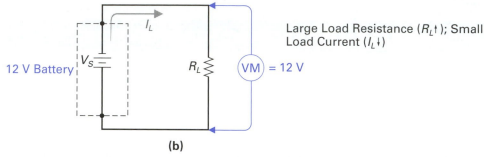

FIGURE 12-40 Ideal Voltage Source. (a) Heavy Load. (b) Light Load.

also generate heat. This inefficiency is represented as an internal resistance, as seen in Figure 12-41(a), and in most cases this internal source resistance (R_{int}) is very low (several ohms) compared to the load resistance (R_L). In Figure 12-41(a), no load has been connected, so the output or open-circuit voltage will be equal to the source voltage, V_S. When a load is connected across the battery, as shown in Figure 12-41(b), R_{int} and R_L form a series circuit, and some of the source voltage appears across R_{int}. As a result, the output or load voltage is always less than V_S. Since R_{int} is normally quite small compared with R_L, the voltage source approaches ideal, as almost all the source voltage (V_S) appears across R_L.

In conclusion, a voltage source should have the smallest possible internal resistance so that the output voltage (V_{out}) will remain constant and approximately equal to V_S independent of whether a light load (large R_L, small I_L) or heavy load (small R_L, large I_L) is connected across its output terminals.

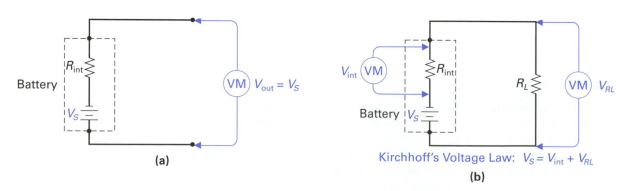

FIGURE 12-41 Realistic Voltage Source. (a) Unloaded. (b) Loaded.

CHAPTER 12 / SERIES–PARALLEL DC CIRCUITS AND THEOREMS

■ INTEGRATED MATH APPLICATION:

Calculate the output voltage in Figure 12-42 if R_L is equal to:

a. $100\ \Omega$

b. $1\ k\Omega$

c. $100\ k\Omega$

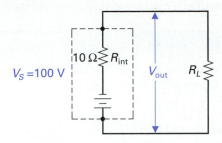

Solution:

(Voltage-divider formula)

a. $R_L = 100\ \Omega$:

$$V_{out} = \frac{R_L}{R_T} \times V_S$$

$$= \frac{100\ \Omega}{110\ \Omega} \times 100\ V = 90.9\ V$$

FIGURE 12-42 Voltage Source
Circuit Example.

b. $R_L = 1\ k\Omega\ (1000\ \Omega)$

$$V_{out} = \frac{1000\ \Omega}{1010\ \Omega} \times 100\ V = 99.0\ V$$

c. $R_L = 100\ k\Omega\ (100{,}000\ \Omega)$

$$V_{out} = \frac{100{,}000\ \Omega}{100{,}010\ \Omega} \times 100\ V = 99.99\ V$$

From this example you can see that the larger the load resistance, the greater the output voltage (V_{out} or V_{RL}). To explain this in a little more detail, we can say that a large R_L is considered a light load for the voltage source, as it only has to produce a small load current ($R_L\uparrow$, $I_L\downarrow$), and consequently, the heat generated by the source is small ($P_{R_{int}}\downarrow = I_2\downarrow \times R$) and the voltage source is more efficient (V_{out} almost equals V_S), approaching ideal ($V_{out} = V_S$). The voltage source in this example, however, produced an almost constant output voltage (within 10% of V_S) despite the very large changes in R_L.

Current Source

Just as a voltage source has a voltage rating, a **current source** has a current rating. An ideal voltage source should deliver a constant output voltage, and similarly an ideal current source should deliver its constant rated current, regardless of what value of load resistance is connected across its output terminals, as seen in Figure 12-43.

A current source can be thought of as a voltage source with an extremely large internal resistance, as seen in Figure 12-44(a) and symbolized in Figure 12-44(b).

Current Source

The circuit or device that supplies current to a load circuit.

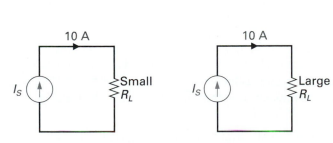

FIGURE 12-43 The Current Source Maintains a Constant Output Current Whether the Load is (a) Heavy (Small R_L) or (b) Light (Large R_L).

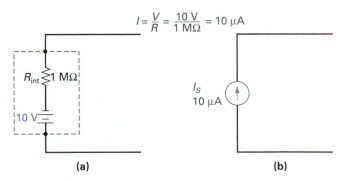

FIGURE 12-44 The Large Internal Resistance of the Current Source.

The current source should have a large internal resistance so that whatever the load resistance connected across the output, it will have very little effect on the total resistance, and the load current will remain constant. The symbol for a constant current source has an arrow within a circle, and this arrow points in the direction of current flow.

■ **INTEGRATED MATH APPLICATION:**

Calculate the load current supplied by the current source in Figure 12-45 if the following values of R_L are connected across the output terminals:

    a. 100 Ω

    b. 1 kΩ

    c. 100 kΩ

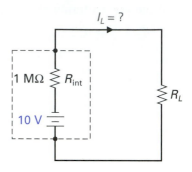

Solution:

    a. $R_L = 100\ \Omega$ ($R_T = R_{int} + R_L = 1,000,000\ \Omega$
          $+\ 100\ \Omega = 1,000,100\ \Omega$)

$$I_L = \frac{V_S}{R_T}$$

$$= \frac{10\ \text{V}}{1,000,100\ \Omega} = 9.999\ \mu\text{A}$$

    b. $R_L = 1\ \text{k}\Omega$ ($R_T = R_{int} + R_L = 1,001,000\ \Omega$)

$$I_L = \frac{10\ \text{V}}{1,001,000\ \Omega} = 9.99\ \mu\text{A}$$

    c. $R_L = 100\ \text{k}\Omega$ ($R_T = R_{int} + R_L = 1,100,000\ \Omega$)

$$I_L = \frac{10\ \text{V}}{1,100,000\ \Omega} = 9.09\ \mu\text{A}$$

FIGURE 12-45 **Current Source Circuit Example.**

From this example you can see that the current source delivered an almost constant output current regardless of the large load resistance change.

12-8-2 *Superposition Theorem*

Superposition Theorem

In a network or circuit containing two or more voltage sources, the current at any point is equal to the algebraic sum of the individual source currents produced by each source acting separately.

The **superposition theorem** is used not only in electronics but also in physics and even economics. It is used to determine the net effect in a circuit that has two or more sources connected. The basic idea behind this theorem is that if two voltage sources are both producing a current through the same circuit, the net current can be determined by first finding the individual currents and then adding them together. Stated formally: *In a network containing two or more voltage sources, the current at any point is equal to the algebraic sum of the individual source currents produced by each source acting separately.*

The best way to fully understand the theorem is to apply it to a few examples. Figure 12-46(a) illustrates a simple series circuit with two resistors and two voltage sources. The 12 V source (V_1) is trying to produce a current in a clockwise direction while the 24 V source (V_2) is trying to force current in a counterclockwise direction. What will be the resulting net current in this circuit?

        STEP 1   To begin, let's consider what current would be produced in this circuit if only V_1 were connected, as shown in Figure 12-46(b).

$$I_1 = \frac{V_1}{R_T} = \frac{12\ \text{V}}{9\ \Omega} = 1.33\ \text{A}$$

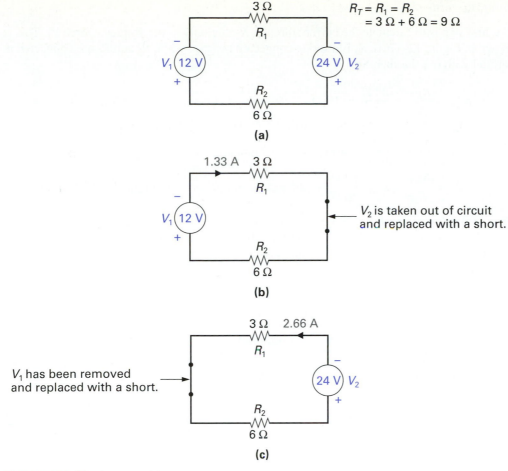

$$R_T = R_1 = R_2$$
$$= 3\,\Omega + 6\,\Omega = 9\,\Omega$$

(a)

(b)

(c)

FIGURE 12-46 Superposition.

STEP 2   The next step is to determine how much current V_2 would produce if V_1 were not connected in the circuit, as shown in Figure 12-46(c).

$$I_2 = \frac{V_2}{R_T} = \frac{24\ \text{V}}{9\ \Omega} = 2.66\ \text{A}$$

V_1 is attempting to produce 1.33 A in the clockwise direction while V_2 is trying to produce 2.66 A in the counterclockwise direction. The net current will consequently be 1.33 A in the counterclockwise direction.

■   INTEGRATED MATH APPLICATION:

Calculate the current through R_2 in Figure 12-47 using the superposition theorem.

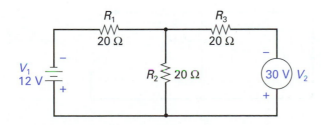

FIGURE 12-47 Superposition Circuit Example.

Solution:

The first step is to calculate the current through R_2 due only to the voltage source V_1. This is shown in Figure 12-47(a). R_1 is a series-connected resistor, while R_2 and R_3 are connected in parallel with one another. So:

$$R_{2,3} = \frac{R_{EV}}{n} = \frac{20\ \Omega}{2} = 10\ \Omega \qquad \text{(equal-value parallel resistor formula)}$$

$$R_T = R_1 + R_{2,3} = 20\ \Omega + 10\ \Omega = 30\ \Omega \qquad \text{(total resistance)}$$

$$I_T = \frac{V_1}{R_T} = \frac{12\ \text{V}}{30\ \Omega} = 400\ \text{mA} \qquad \text{(total current)}$$

$$I_2 = \frac{R_{2,3}}{R_2} \times I_T = \frac{10\ \Omega}{20\ \Omega} \times 400\ \text{mA} = 200\ \text{mA} \qquad \text{(current-divider formula)}$$

Thus 200 mA of current is flowing down through R_2.

The next step is to find the current flow through R_2 due only to the voltage source V_2. This is shown in Figure 12-48(b). In this instance, R_3 is a series-connected resistor, and R_1 and R_2 make up a parallel circuit. So:

$$R_{1,2} = 10\ \Omega$$

$$R_T = R_{1,2} + R_3 = 10\ \Omega + 20\ \Omega = 30\ \Omega$$

$$I_T = \frac{V_2}{R_T} = \frac{30\ \text{V}}{30\ \Omega} = 1\ \text{A} \qquad (1000\ \text{mA})$$

$$I_2 = \frac{R_{1,2}}{R_2} \times I_T = \frac{10\ \Omega}{20\ \Omega} \times 1000\ \text{mA} = 500\ \text{mA}$$

Thus 500 mA of current will flow down through R_2.

Both V_1 and V_2 produce a current flow down through R_2, so I_{R2} and I_{R3} have the same algebraic sign, and the total current through R_2 is equal to the sum of the two currents produced by V_1 and V_2.

$$I_2\ (\text{total}) = I_2\ \text{due to}\ V_1 + I_2\ \text{due to}\ V_2$$
$$= 200\ \text{mA} + 500\ \text{mA}$$
$$= 700\ \text{mA}$$

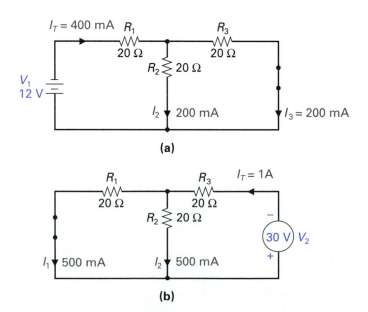

FIGURE 12-48 **Superposition Circuit Solution.**

CHAPTER 12 / SERIES–PARALLEL DC CIRCUITS AND THEOREMS

■ INTEGRATED MATH APPLICATION:

Calculate the current flow through R_1 in Figure 12-49.

FIGURE 12-49   Superposition Circuit Example.

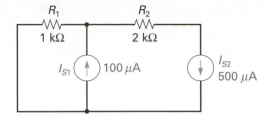

Solution:

With the superposition theorem, current sources are treated differently from voltage sources in that *each current source is removed from the circuit and replaced with an open,* as illustrated in the first step of the solution shown in Figure 12-50(a). In this instance you can see that current has only one path (series circuit), so the current through R_1 is counterclockwise and equal to the I_{S1} source current, 100 μA.

In Figure 12-50(b), the current source I_{S1} has been removed and replaced with an open. R_1 and R_2 form a series circuit, so the total source current from I_{S2} flows through R_1 ($I_1 = 500 \mu$A) in a clockwise direction.

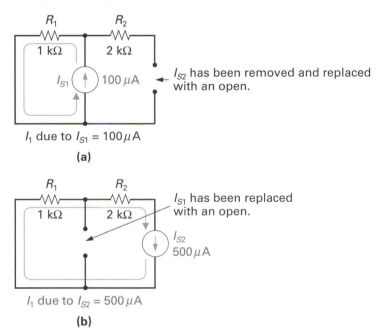

FIGURE 12-50   Superposition Circuit Solution.

Since I_{S1} is producing 100 μA of current through R_1 in a counterclockwise direction, and I_{S2} is producing 500 μA of current in a clockwise direction, the resulting current through R_1 will equal

$$I_1 = I_{S2} \text{ (cw)} - I_{S1} \text{ (ccw)}$$
$$= 500 \ \mu\text{A} - 100 \ \mu\text{A} = 400 \ \mu\text{A clockwise}$$

12-8-3   *Thévenin's Theorem*

Thévenin's theorem allows us to replace the complex networks in Figure 12-51(a) with an equivalent circuit containing just one source voltage (V_{TH}) and one series-connected resistance (R_{TH}), as in Figure 12-51(b). Stated formally: *Any network of voltage sources and*

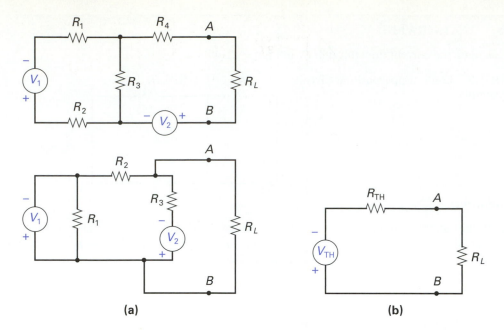

FIGURE 12-51 Thévenin's Theorem. (a) Complex Multiple Resistors and Source Networks Are Replaced by (b) One Source Voltage (V_{TH}) and One Series-Connected Resistance (R_{TH}).

resistors can be replaced by a single equivalent voltage source (V_{TH}) in series with a single equivalent resistance (R_{TH}).

Figure 12-52(a) illustrates an example circuit. As with any theorem, a few rules must be followed to obtain an equivalent V_{TH} and R_{TH}.

STEP 1   The first step is to disconnect the load (R_L) and calculate the voltage that will appear across points *A* and *B*, as in Figure 12-52(b). This open-circuit voltage will be the same as the voltage drop across R_2 (V_{R2}) and is called the *Thévenin equivalent voltage* (V_{TH}). First, let's calculate current:

$$I_T = \frac{V_S}{R_T} = \frac{12 \text{ V}}{9 \text{ }\Omega} = 1.333 \text{ A}$$

Therefore, V_{R2} or V_{TH} will equal

$$V_{R2} = I_T \times R_2 = 1.333 \times 6 \text{ }\Omega = 8 \text{ V}$$

so $V_{TH} = 8 \text{ V}$.

STEP 2   Now that the Thévenin equivalent voltage has been calculated, the next step is to calculate the Thévenin equivalent resistance. In this step, the source voltage is removed and replaced with a short, as seen in Figure 12-52(c), and the Thévenin equivalent resistance is equal to whatever resistance exists between points *A* and *B*. In this example, R_1 and R_2 form a parallel circuit, the total resistance of which is equal to

$$R_T = \frac{R_1 \times R_2}{R_1 + R_2} = \frac{3 \times 6}{3 + 6} = \frac{18}{9} = 2 \text{ }\Omega$$

so $R_{TH} = 2 \text{ }\Omega$. The circuit to be Thévenized in Figure 12-52(a) can be represented by the Thévenin equivalent circuit shown in Figure 12-52(d).

The question you may have at this time is why we would need to simplify such a basic circuit, when Ohm's law could just as easily have been used to analyze the network. Thévenin's theorem has the following advantages:

1. If you had to calculate load current and load voltage (I_{RL} and V_{RL}) for 20 different values of R_L, it would be far easier to use the Thévenin equivalent circuit with the series-

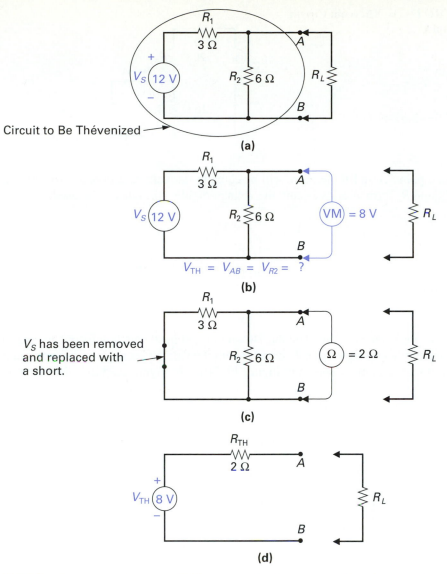

FIGURE 12-52 Thévenin's Theorem. (a) Example Circuit. (b) Obtaining Thévenin Voltage (V_{TH}). (c) Obtaining Thévenin Resistance. (d) Thévenin Equivalent Circuit.

connected resistors R_{TH} and R_L, rather than to apply Ohm's law to the series–parallel circuit made up of R_1, R_2, and R_L.

2. Thévenin's theorem permits you to solve complex circuits that cannot easily be analyzed using Ohm's law.

■ **INTEGRATED MATH APPLICATION:**

Determine V_{TH} and R_{TH} for the circuit in Figure 12-53.

Solution:

The first step is to remove the load resistor R_L and calculate what voltage will appear between points A and B, as shown in Figure 12-54(a). Removing R_L will open the path for current to flow through R_4, which will consequently have no voltage drop across it. The voltage

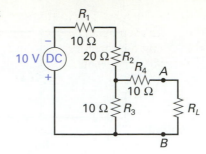

FIGURE 12-53 Thévenin Circuit Example.

between points A and B therefore will be equal to the voltage dropped across R_3, and since R_1, R_2, and R_3 form a series circuit, the voltage-divider formula can be used:

$$V_{R3} = \frac{R_3}{R_T} \times V_S$$

$$= \frac{10\ \Omega}{40\ \Omega} \times 10\ \text{V} = 2.5\ \text{V}$$

Therefore,

$$V_{AB} = V_{R3} = V_{TH} = 2.5\ \text{V}$$

The next step is to calculate the Thévenin resistance, which will equal whatever resistance appears across the terminals A and B with the voltage source having been removed and replaced with a short, as shown in Figure 12-54(b). In Figure 12-54(c), the circuit has been

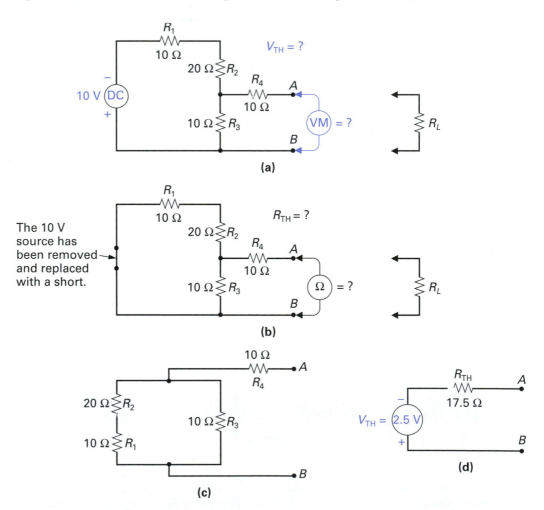

FIGURE 12-54 Thévenin Circuit Solution.

CHAPTER 12 / SERIES–PARALLEL DC CIRCUITS AND THEOREMS

redrawn so that the relationship between the resistors can be seen in more detail. As you can see, R_1 and R_2 are in series with each other, and both are in parallel with R_3, and this combination is in series with R_4. Using our three-step procedure for calculating total resistance in a series–parallel circuit, the following results are obtained:

1. $R_{1,2} = R_1 + R_2 = 10\ \Omega + 20\ \Omega = 30\ \Omega$

2. $R_{1,2,3} = \dfrac{R_{1,2} \times R_3}{R_{1,2} + R_3} = \dfrac{30\ \Omega \times 10\ \Omega}{30\ \Omega + 10\ \Omega} = \dfrac{300\ \Omega^2}{40\ \Omega} = 7.5\ \Omega$

3. $R_T = R_{1,2,3} + R_4 = 7.5\ \Omega + 10\ \Omega = 17.5\ \Omega$

Figure 12-54(d) illustrates the Thévenin equivalent circuit.

12-8-4   *Norton's Theorem*

Norton's theorem, like Thévenin's theorem, is a tool for simplifying a complex circuit into a more manageable one. Figure 12-55 illustrates the difference between a Thévenin equivalent and a Norton equivalent circuit. Thévenin's theorem simplifies a complex network and uses an equivalent voltage source (V_{TH}) and an equivalent series resistance (R_{TH}). Norton's theorem, on the other hand, simplifies a complex circuit and represents it with an equivalent current source (I_N) in parallel with an equivalent Norton resistance (R_N), as shown in Figure 12-55.

As with any theorem, a set of steps has to be carried out to arrive at an equivalent circuit. The example we will use is shown in Figure 12-56(a) and is the same example used in Figure 12-55.

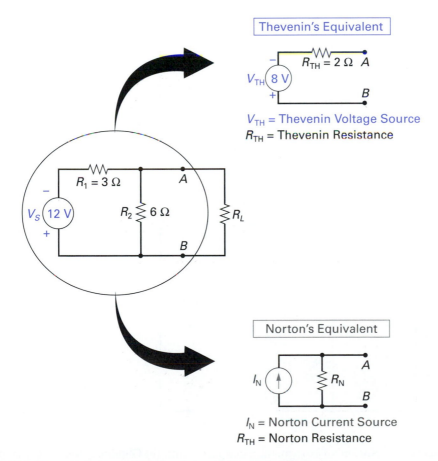

FIGURE 12-55   **Comparison of Thévenin's and Norton's Circuits.**

STEP 1 Calculate the Norton equivalent current source, which will be equal to the current that would flow between terminals A and B if the load resistor were removed and replaced with a short, as seen in Figure 12-56(b). Placing a short between terminals A and B will short out the resistor R_2, so the only resistance in the circuit will be R_1. The Norton equivalent current source in this example will therefore be equal to

$$I_N = \frac{V_S}{R_T} = \frac{12\ V}{3\ \Omega} = 4\ A$$

STEP 2 The next step is to determine the value of the Norton equivalent resistance that will be placed in parallel with the current source, unlike the Thévenin equivalent resistance, which was placed in series. As with Thévenin's theorem, though, the Norton equivalent resistance (R_N) is equal to the resistance between terminals A and B when the voltage source is removed and replaced with a short, as shown in Figure 12-56(c). Since R_1 and R_2 form a parallel circuit, the Norton equivalent resistance will be equal to

$$R_N = \frac{R_1 \times R_2}{R_1 + R_2} = \frac{3\ \Omega \times 6\ \Omega}{3\ \Omega + 6\ \Omega} = \frac{18\ \Omega^2}{9\ \Omega} = 2\ \Omega$$

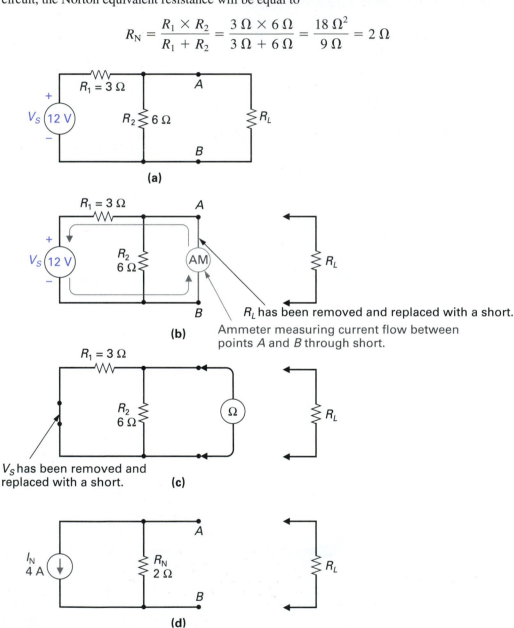

FIGURE 12-56 Norton's Theorem. (a) Example Circuit. (b) Obtaining Norton Current. (c) Obtaining Norton Resistance (R_N). (d) Norton Equivalent Circuit.

The Norton equivalent circuit has been determined simply by carrying out these two steps and is illustrated in Figure 12-56(d).

■ **INTEGRATED MATH APPLICATION:**

Determine I_N and R_N for the circuit in Figure 12-57.

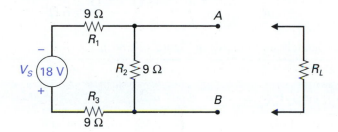

FIGURE 12-57 **Norton Circuit Example.**

Solution:

If a short is placed between terminals A and B, R_2 will be shorted out, so the current between points A and B, and therefore the Norton equivalent current, will be limited by only R_1 and R_3 and will equal [Figure 12-58(a)]

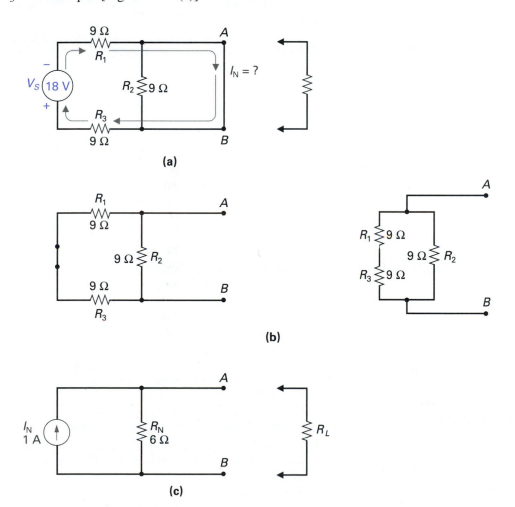

FIGURE 12-58 **Norton Circuit Solution.**

$$I_N = \frac{V_S}{R_T} = \frac{V_S}{R_1 + R_3} = \frac{18 \text{ V}}{9 \ \Omega + 9 \ \Omega} = 1 \text{ A}$$

You can see that replacing V_S with a short means that the Norton equivalent resistance between terminals A and B is made up of R_1 and R_3 in series with each other, and both are in parallel with R_2. R_N will therefore be equal to [Figure 12-58(b)]

$$R_{1,3} = R_1 + R_3 = 9 \ \Omega + 9 \ \Omega = 18 \ \Omega$$
$$R_T = \frac{R_{1,3} \times R_2}{R_{1,3} + R_2} = \frac{18 \ \Omega \times 9 \ \Omega}{18 \ \Omega + 9 \ \Omega} = 6 \ \Omega$$

The Norton equivalent circuit is shown in Figure 12-58(c).

SELF-TEST EVALUATION POINT FOR SECTION 12-8

Now that you have completed this section, you should be able to:

■ **Objective 7.** *Describe the differences between a voltage and a current source.*

■ **Objective 8.** *Analyze series–parallel networks using:*

  a. *The superposition theorem.*
  b. *Thévenin's theorem.*
  c. *Norton's theorem.*

Use the following questions to test your understanding of Section 12-8.

1. A constant voltage source will have a _____ internal resistance, whereas a constant current source will have a _____ internal resistance. (large or small)

2. The superposition theorem is a logical way of analyzing networks with more than one _____.

3. Thévenin's theorem represents a complex two-terminal network as a single _____ source with a series-connected single _____.

4. Norton's theorem also allows you to analyze complex two-terminal networks as a single _____ source in parallel with a single resistor.

REVIEW QUESTIONS

Multiple Choice Questions

1. A series–parallel circuit is a combination of:

   a. Components connected end to end
   b. Series (one current path) circuits
   c. Both series and parallel circuits
   d. Parallel (two or more current paths) circuits

2. Total resistance in a series–parallel circuit is calculated by applying the _____ resistance formula to series-connected resistors and the _____ resistance formula to resistors connected in parallel.

   a. Series, parallel   **c.** Series, series
   b. Parallel, series   **d.** Parallel, parallel

3. Total current in a series–parallel circuit is determined by dividing the total _____ by the total _____.

   a. Power, current   **c.** Current, resistance
   b. Voltage, resistance   **d.** Voltage, power

4. Branch current within series–parallel circuits can be calculated by:

   a. Ohm's law
   b. The current-divider formula
   c. Kirchhoff's current law
   d. All the above

5. All voltages on a circuit diagram are with respect to _____ unless otherwise stated.

   a. The other side of the component
   b. The high-voltage source
   c. Ground
   d. All the above

6. A _____ ground has the negative side of the source voltage connected to ground, whereas a _____ ground has the positive side of the source voltage connected to ground.

   a. Positive, negative   **c.** Positive, earth
   b. Chassis, earth   **d.** Negative, positive

7. The output voltage will always _____ when a load or voltmeter is connected across a voltage divider.

   a. Decrease
   b. Remain the same
   c. Increase
   d. All the above could be considered true.

8. A Wheatstone bridge was originally designed to measure:

    a. An unknown voltage   **c.** An unknown power
    b. An unknown current   **d.** An unknown resistance

9. A balanced bridge has an output voltage:

    a. Equal to the supply voltage   **c.** Of 0 V
    b. Equal to half the supply voltage   **d.** Of 5 V

10. The total resistance of a ladder circuit is best found by starting at the point _____ the source.

    a. Nearest to   **c.** Midway between
    b. Farthest from   **d.** All the above

11. The R–$2R$ ladder circuit finds its main application as a(an):

    a. Analog-to-digital converter
    b. Digital-to-analog converter
    c. Device to determine unknown resistance values
    d. All the above

12. The Norton equivalent resistance (R_N) is always in _____ with the Norton equivalent current source (I_N).

    a. Proportion   **c.** Parallel
    b. Series   **d.** Either series or parallel

13. An ideal current source has _____ internal resistance, while an ideal voltage source has _____ internal resistance.

    a. An infinite, 0 Ω of   **c.** 0 Ω of, an infinite
    b. No, a large amount   **d.** No, an infinite

14. The Thévenin equivalent resistance (R_{TH}) is always in _____ with the Thévenin equivalent voltage (V_{TH}).

    a. Proportion   **c.** Parallel
    b. Series   **d.** Either series or parallel

15. The superposition theorem is useful for analyzing circuits with:

    a. Two or more voltage sources
    b. A single voltage source
    c. Only two voltage sources
    d. A single current source

16. A resistor, when it burns out, will generally:

    a. Decrease slightly in value   **c.** Short
    b. Increase slightly in value   **d.** Open

17. In a series–parallel resistive circuit, an open series-connected resistor will cause _____ current, whereas an open parallel-connected resistor will result in a total current _____.

    a. An increase in, decrease   **c.** Zero, decrease
    b. A decrease in, increase   **d.** None of the above

18. In a series–parallel resistive circuit, a shorted series-connected resistor will cause _____ current, whereas a shorted parallel-connected resistor will result in a total current _____.

    a. An increase in, increase
    b. A decrease in, decrease
    c. An increase in, decrease
    d. A decrease in, increase

19. A resistor's resistance will typically _____ with age, resulting in a total circuit current _____.

    a. Decrease, decrease   **c.** Decrease, increase
    b. Increase, increase   **d.** All the above

20. The maximum power transfer theorem states that maximum power is transferred from source to a load when load resistance is equal to source resistance.

    a. True   **b.** False

Communication Skill Questions

21. State the five-step method for determining a series–parallel circuit's resistance, voltage, current, and power values. (12-6)

22. Illustrate the following series–parallel circuits:

    a. R_1 in series with a parallel combination R_2, R_3, and R_4.
    b. R_1 in series with a two-branch parallel combination consisting of R_2 and R_3 in series and R_4 in parallel.
    c. R_1 in parallel with R_2, which is in series with a three-resistor parallel combination, R_3, R_4, and R_5.

23. Using the example in Question 22(c), apply values of your choice and apply the five-step procedure. (12-6)

24. Describe what is meant by "loading of a voltage-divider circuit." (12-7-1)

25. Illustrate and describe the Wheatstone bridge in the: (12-7-2)

    a. Balanced condition
    b. Unbalanced condition
    c. Application of measuring unknown resistances

26. Describe how the ladder circuit acts as a current divider. (12-7-3)

27. Briefly describe the difference between a voltage source and a current source. (12-8-1)

28. Briefly describe the following theorems: (12-8)

    a. Superposition   **c.** Norton's
    b. Thévenin's   **d.** Maximum power transfer

29. What would be the advantages of Thévenin's and Norton's theorems for obtaining an equivalent circuit? (12-8)

30. Draw the components that would exist in a: (12-8)

    a. Thévenin equivalent circuit
    b. Norton equivalent circuit

31. Describe the steps involved in obtaining a: (12-8)

    a. Thévenin equivalent circuit
    b. Norton equivalent circuit

32. Give the divider formula, Ohm's law, and Kirchhoff's laws used to determine: (12-6)

    a. Branch currents
    b. Voltage drops in a series–parallel circuit (List all six.)

Practice Problems

33. R_3 and R_4 are in series with each other and are both in parallel with R_5. This parallel combination is in series with two series-connected resistors, R_1 and R_2. $R_1 = 2.5$ kΩ, $R_2 = 10$ kΩ, $R_3 = 7.5$ kΩ, $R_4 = 2.5$ kΩ, $R_5 = 2.5$ MΩ, and $V_S = 100$ V. For these values, calculate:

  a. Total resistance
  b. Total current
  c. Voltage across series resistors and parallel combinations
  d. Current through each resistor
  e. Total and individual power figures

34. Referring to the example in Question 33, calculate the voltage at every point of the circuit with respect to ground.

35. A 10 V source is connected across a series–parallel circuit made up of R_1 in parallel with a branch made up of R_2 in series with a parallel combination of R_3 and R_4. $R_1 = 100$ Ω, $R_2 = 100$ Ω, $R_3 = 200$ Ω, and $R_4 = 300$ Ω. For these values, apply the five-step procedure, and also determine the voltage at every point of the circuit with respect to ground.

36. Calculate the output voltage (V_{RL}) in Figure 12-59 if R_L is equal to:

  a. 25 Ω **b.** 2.5 kΩ **c.** 2.5 MΩ

37. What load current will be supplied by the current source in Figure 12-60 if R_L is equal to:

  a. 25 Ω **b.** 2.5 kΩ **c.** 2.5 MΩ

38. Use the superposition theorem to calculate total current:

  a. Through R_2 in Figure 12-61(a)
  b. Through R_3 in Figure 12-61(b)

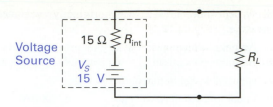

FIGURE 12-59

39. Convert the following voltage sources to equivalent current sources:

  a. $V_S = 10$ V, $R_{int} = 15$ Ω
  b. $V_S = 36$ V, $R_{int} = 18$ Ω
  c. $V_S = 110$ V, $R_{int} = 7$ Ω

40. Use Thévenin's theorem to calculate the current through R_L in Figure 12-62. Sketch the Thévenin and Norton equivalent circuits for Figure 12-62.

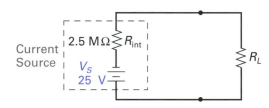

FIGURE 12-60

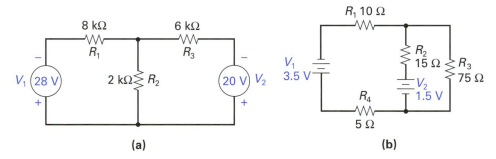

      (a)   **(b)**

FIGURE 12-61

41. Sketch the Thévenin and Norton equivalent circuits for the networks in Figure 12-63.

42. Convert the following current sources to equivalent voltage sources:

   a. $I_S = 5$ mA, $R_{int} = 5$ MΩ
   b. $I_S = 10$ A, $R_{int} = 10$ kΩ
   c. $I_S = 0.0001$ A, $R_{int} = 2.5$ kΩ

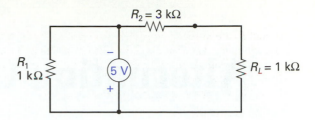

FIGURE 12-62

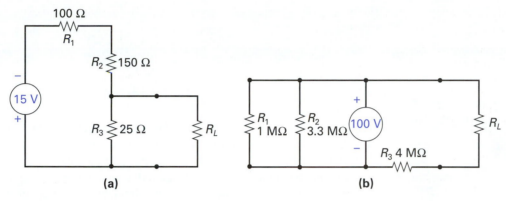

FIGURE 12-63

Web Site Questions

Go to the Web site http://www.prenhall.com/cook, select the textbook *Mathematics for Electronics and Computers,* select this chapter, and then follow the instructions when answering the multiple-choice practice problems.

Alternating Current (AC)

The Laser

Theodore Maiman

In 1898, H. G. Wells's famous book *The War of the Worlds* had Martian invaders with laserlike death rays blasting bricks, incinerating trees, and piercing iron as if it were paper. In 1917, Albert Einstein stated that, under certain conditions, atoms or molecules could absorb light and then be stimulated to release this borrowed energy. In 1954, Charles H. Townes, a professor at Columbia University, conceived and constructed with his students the first "maser" (acronym for "microwave amplification by stimulated emission of radiation"). In 1958, Townes and Arthur L. Shawlow wrote a paper showing how stimulated emission could be used to amplify light waves as well as microwaves, and the race was on to develop the first "laser." In 1960, Theodore H. Maiman, a scientist at Hughes Aircraft Company, directed a beam of light from a flash lamp into a rod of synthetic crystal, which responded with a burst of crimson light so bright that it outshone the sun.

An avalanche of new lasers emerged, some as large as football fields, others no bigger than a pinhead. Lasers can be made to produce invisible infrared or ultraviolet light or any visible color in the rainbow; and the high-power type can vaporize any material a million times faster and more intensely than a nuclear blast, whereas the low-power lasers are safe to use in children's toys.

At present, the laser is being used by the FBI to detect fingerprints that are 40 years old, in defense programs, in compact disk players, in underground fiber optic communication to transmit hundreds of telephone conversations, to weld car bodies, to drill holes in baby-bottle nipples, to create three-dimensional images called holograms, and as a surgeon's scalpel in the operating room. Not a bad beginning for a device that when first developed was called "a solution looking for a problem."

Outline and Objectives

VIGNETTE: THE LASER

INTRODUCTION

13-1   THE DIFFERENCE BETWEEN DC AND AC

Objective 1: Explain the difference between alternating current and direct current.

13-2   WHY ALTERNATING CURRENT?

Objective 2: Describe how ac is used to deliver power and to represent information.

Objective 3: Give the three advantages that ac has over dc from a power point of view.

13-2-1   Power Transfer

Objective 4: Describe basically the ac power distribution system from the electric power plant to the home or industry.

13-2-2   Information Transfer

Objective 5: Describe the three waves used to carry information between two points.

Objective 6: Explain how the three basic information carriers are used to carry many forms of information on different frequencies within the frequency spectrum.

13-2-3   Electrical Equipment or Electronic Equipment?

13-3   AC WAVE SHAPES

Objective 7: Explain the different characteristics of the five basic wave shapes.

Objective 8: Describe fundamental and harmonic frequencies.

13-3-1   The Sine Wave
Amplitude
Peak Value
Peak-to-Peak Value
RMS or Effective Value

Average Value
Frequency and Period
Wavelength
Phase Relationships
The Meaning of Sine

13-3-2   The Square Wave
Duty Cycle
Average
Frequency-Domain Analysis

13-3-3   The Rectangular or Pulse Wave
PRF, PRT, and Pulse Length
Duty Cycle
Average
Frequency-Domain Analysis

13-3-4   The Triangular Wave
Frequency-Domain Analysis

13-3-5   The Sawtooth Wave

13-3-6   Other Waveforms

13-4   MEASURING AC SIGNALS

Objective 9: Describe how a multimeter can be used to measure ac voltage and current.

13-4-1   The AC Meter

13-4-2   The Oscilloscope
Controls
Measurements

13-5   DECIBELS

Objective 10: Explain how gain and loss can be expressed in decibels and how to convert between arithmetic gain and decibel gain, and vice versa.

13-5-1   The Human Ear's Response to Sound

13-5-2   The Bel and the Decibel

13-5-3   Power Gain or Loss in Decibels
Current gain and Voltage/Current Ratio

13-5-4   Voltage Gain or Loss in Decibels

13-5-5   Common Power and Voltage/Current Ratios

In this chapter you will be introduced to alternating current (ac). While direct current (dc) can be used in many instances to deliver power or represent information, there are certain instances in which ac is preferred. For example, ac is easier to generate and is more efficiently transmitted as a source of electrical power for the home or business. Audio (speech and music) and video (picture) information are generally always represented in electronic equipment as an alternating current or alternating voltage signal.

In this chapter, we will begin by describing the difference between dc and ac, and then examine where ac is used. Following this we will discuss all the characteristics of ac waveform shapes, and finally, the differences between electricity and electronics.

13-1   THE DIFFERENCE BETWEEN DC AND AC

Direct Current

Current flow in only one direction.

One of the best ways to describe anything new is to begin by redescribing something known and then to discuss the unknown. The known topic in this case is **direct current.** Direct current (dc) is the flow of electrons in one DIRECTion and one direction only. Dc voltage is nonvarying and normally obtained from a battery or power supply unit, as seen in Figure 13-1(a). The only variation in voltage from a battery occurs due to the battery's discharge, but even then, the current will still flow in only one direction, as seen in Figure 13-1(b). A dc voltage of 9 or 6 V could be illustrated graphically as shown in Figure 13-1(c). Whether 9 or 6 V, the voltage can be seen to be constant or the same at any time.

Some power supplies supply a form of dc known as *pulsating dc,* which varies periodically from zero to a maximum, back to zero, and then repeats. Figure 13-2(a) illustrates the physical appearance and schematic diagram of a battery charger that is connected across two series resistors. The battery charger is generating a waveform, as shown in Figure 13-2(b), known as pulsating dc. At time 1 [Figure 13-2(c)], the power supply is generating 9 V, and direct current is flowing from negative to positive. At time 2 [Figure 13-2(d)], the power supply is producing 0 V, and therefore no current is being produced. In between time 1 and time 2, the voltage out of the power supply will decrease from 9 V to 0 V. No matter what the voltage, whether 8, 7, 6, 5, 4, 3, 2, or 1 V, current will be flowing in only one direction (unidirectional) and is therefore referred to as dc.

Pulsating dc is normally supplied by a battery charger and is used to charge secondary batteries. It is also used to operate motors that convert the pulsating dc electrical energy into a mechanical rotation output. Whether steady or pulsating, direct current is current in only one DIRECTion.

Alternating Current

Electric current that rises from zero to a maximum in one direction, falls to zero, then rises to a maximum in the opposite direction, and then repeats another cycle, the positive and negative alternations being equal.

Alternating current (ac) flows first in one direction and then in the opposite direction. This reversing current is produced by an alternating voltage source, as shown in Figure 13-3(a) which reaches a maximum in one direction (positive), decreases to zero, and then reverses itself and reaches a maximum in the opposite direction (negative). This is graphically illustrated in Figure 13-3(b). During the time of the positive voltage alternation, the polarity of the voltage will be as shown in Figure 13-3(c), so current will flow from negative to positive in a counterclockwise direction. During the time of the negative voltage alternation, the polarity of the voltage will reverse, as shown in Figure 13-3(d), causing current to flow once again from negative to positive, but in this case, in the opposite clockwise direction.

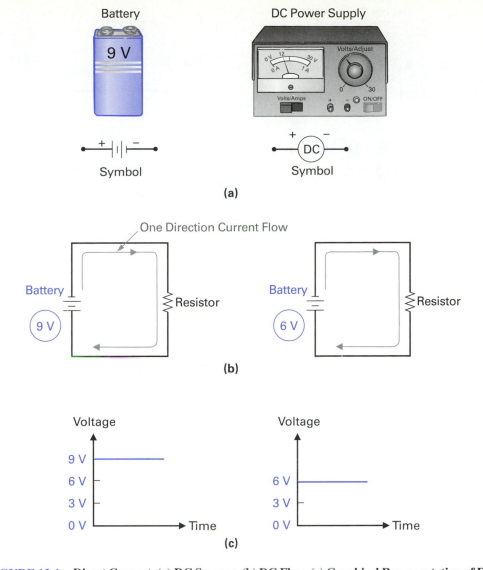

FIGURE 13-1 Direct Current. (a) DC Sources. (b) DC Flow. (c) Graphical Representation of DC.

Now that you have completed this section, you should be able to:

■ *Objective 1.* *Explain the difference between alternating current and direct current.*

Use the following questions to test your understanding of Section 13-1.

1. Give the full names of the following abbreviations: (a) ac; (b) dc.

2. Is the pulsating waveform generated by a battery charger considered to be ac or dc? Why?

3. The polarity of a/an _____ voltage source will continually reverse, and therefore so will the circuit current.

4. The polarity of a/an _____ voltage source will remain constant, and therefore current will flow in only one direction.

5. List the two main applications of ac.

6. State briefly the difference between ac and dc.

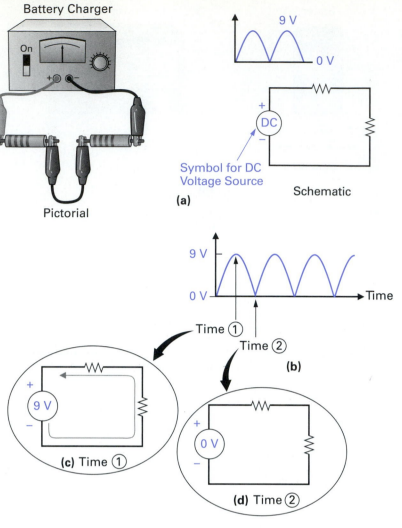

FIGURE 13-2 Pulsating DC.

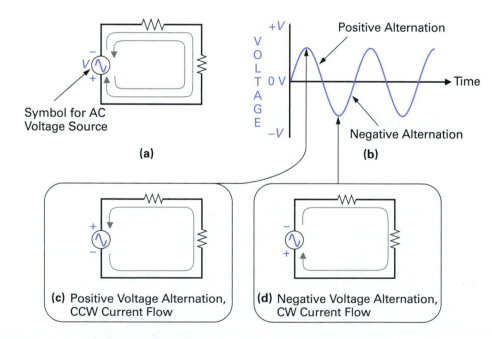

FIGURE 13-3 Alternating Current.

13-2   WHY ALTERNATING CURRENT?

The question that may be troubling you at this point is: If we have been managing fine for the past chapters with dc, why do we need ac?

There are two main applications for ac:

1. *Power transfer:* to supply electrical power for lighting, heating, cooling, appliances, and machinery in both home and industry
2. *Information transfer:* to communicate or carry information, such as radio music and television pictures, between two points

To begin, let us discuss the first of these applications, power transfer.

13-2-1   *Power Transfer*

There are three advantages that ac has over dc from a power point of view, and these are:

1. Flashlights, radios, and portable televisions all use batteries (dc) as a source of power. In these applications in which a small current is required, batteries will last a good length of time before there is a need to recharge or replace them. Many appliances and most industrial equipment need a large supply of current, and in this situation a generator would have to be used to generate this large amount of current. The operation of generators is opposite that of motors, in that a **generator** converts a mechanical rotation input into an electrical output. Generators can be used to generate either dc or ac, but ac generators can be larger, less complex internally, and cheaper to operate, and this is the first reason why we use ac instead of dc for supplying power.

2. From a power point of view, ac is always used by electric companies when transporting power over long distances to supply both the home and industry with electrical energy. Recalling the power formula, you will remember that power is proportional to either current or voltage squared ($P \propto I^2$ or $P \propto V^2$), which means that to supply power to the home or industry, we would supply either a large current or voltage. As you can see in Figure 13-4, between the electric power plant and home or industry are power lines carrying the power. The amount of power lost (heat) in these power lines can be calculated by using the formula $P = I^2 \times R$, where I is the current flowing through the line and R is the resistance of the power lines. This means that the larger the current, the greater the amount of power lost in the lines in the form of heat and therefore the smaller the amount of power supplied to the home or industry. For this reason, power companies transport electric energy at a very high voltage, between 200,000 and 600,000 V. Since the voltage is high, the current can be low and provide the same amount of power to the consumer ($P = V\uparrow \times I\downarrow$). Yet, because the current is low, the amount of heat loss generated in the power lines is minimal.

   Now that we have discovered why it is more efficient over a long distance to transport high voltages than high current, what does this have to do with ac? An ac voltage can easily and efficiently be transformed up or down to a higher or lower voltage by utilizing a device known as a **transformer,** and even though dc voltages can be stepped up and down, the method is inefficient and more complex.

3. Nearly all electronic circuits and equipment are powered by dc voltages, which means that once the ac power arrives at the home or industry, in most cases it will have to be converted into dc power to operate electronic equipment. It is a relatively simple process to convert ac to dc, but conversion from dc to ac is a complex and comparatively inefficient process.

Figure 13-4 illustrates ac power distribution from the electric power plant to the home and industry. The ac power distribution system begins at the electric power plant, which has

Generator
Device used to convert a mechanical energy input into an electrical energy output.

Transformer
Device consisting of two or more coils that are used to couple electric energy from one circuit to another, yet maintain electrical isolation between the two.

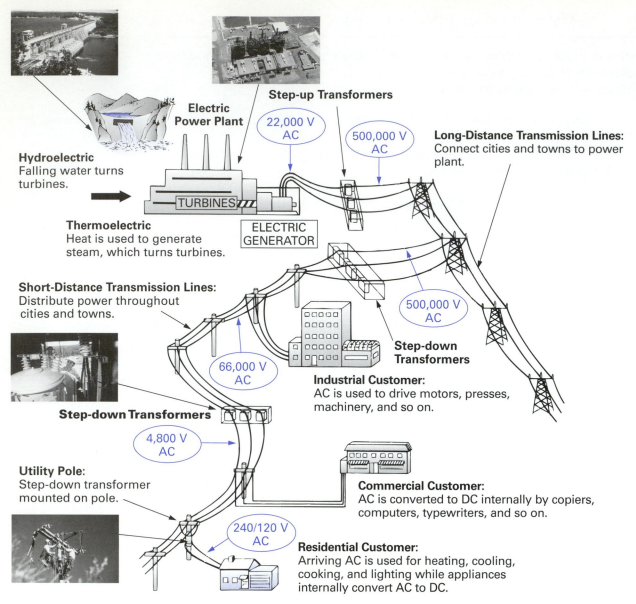

FIGURE 13-4 **AC Power Distribution.**

Text labels within figure:

Step-up Transformers

Electric
Power Plant

22,000 V
AC

500,000 V
AC

Long-Distance Transmission Lines:
Connect cities and towns to power
plant.

Hydroelectric
Falling water turns
turbines.

TURBINES

ELECTRIC
GENERATOR

Thermoelectric
Heat is used to generate
steam, which turns turbines.

Short-Distance Transmission Lines:
Distribute power throughout
cities and towns.

500,000 V
AC

**Step-down
Transformers**

66,000 V
AC

Industrial Customer:
AC is used to drive motors, presses,
machinery, and so on.

Step-down Transformers

4,800 V
AC

Utility Pole:
Step-down transformer
mounted on pole.

Commercial Customer:
AC is converted to DC internally by copiers,
computers, typewriters, and so on.

240/120 V
AC

Residential Customer:
Arriving AC is used for heating, cooling,
cooking, and lighting while appliances
internally convert AC to DC.

powerful large generators driven by turbines to generate large ac voltages. The turbines can be driven either by falling water (hydroelectric) or by steam, which is produced with intense heat by the combustion of coal, gas, or oil or by a nuclear reactor (thermoelectric). The turbine supplies the mechanical energy to the generator, to be transformed into ac electrical energy.

The generator generates an ac voltage of approximately 22,000 V, which is stepped up by transformers to approximately 500,000 V. This voltage is applied to the long-distance transmission lines that connect the power plant to the city or town. At each city or town, the voltage is tapped off the long-distance transmission lines and stepped down to approximately 66,000 V and is distributed to large-scale industrial customers. The 66,000 V is stepped down again to approximately 4800 V and distributed throughout the city or town by short-distance transmission lines. This 4800 V is used by small-scale industrial customers and residential customers who receive the ac power via step-down transformers on utility poles, which step down the 4800 V to 240 V and 120 V.

Much equipment and many devices within industry and the home will run directly on the ac power, such as heating, lighting, and cooling. Some equipment that runs on dc, such as televisions and computers, internally converts the 120 V ac to the dc voltages required.

Now that you have completed this section, you should be able to:

- **Objective 2.** *Describe how ac is used to deliver power and to represent information.*
- **Objective 3.** *Give the three advantages that ac has over dc from a power point of view.*
- **Objective 4.** *Describe basically the ac power distribution system from the electric power plant to the home or industry.*

Use the following questions to test your understanding of Section 13-2-1.

1. In relation to power transfer, what three advantages does ac have over dc?
2. True or false: A generator converts an electrical input into a mechanical output.
3. What formula is used to calculate the amount of power lost in a transmission line?
4. What is a transformer?
5. What voltage is provided to the wall outlet in the home?
6. Most appliances internally convert the _____ input voltage into a _____ voltage.

13-2-2 *Information Transfer*

Information, by definition, is the property of a signal or message that conveys something meaningful to the recipient. **Communication,** which is the transfer of information between two points, began with speech and progressed to handwritten words in letters and printed words in newspapers and books. To achieve communication over greater distances, face-to-face communications evolved into telephone and radio communications.

A simple communication system can be seen in Figure 13-5(a).

The voice information or sound wave produced by the sender is a variation in air pressure, and travels at the speed of sound, as detailed in Figure 13-5. Sound waves or sounds in the case of musical instruments are normally generated by a vibrating reed or plucked string. In this example the sender's vocal cords vibrate backward and forward, producing a rarefaction or decreased air pressure, where few air molecules exist, and a compression or increased air pressure, where many air molecules exist. Like the ripples produced by a stone falling in a pond, the sound waves produced by the sender constantly expand and travel outward.

The microphone is in fact a **transducer** (energy converter), because it converts the sound wave (which is a form of mechanical energy) into electrical energy in the form of voltage and current, which varies in the same manner as the sound wave and therefore contains the sender's message or information.

The **electrical wave,** shown in Figure 13-5(c), is a variation in voltage or current and can exist only in a wire conductor or circuit. This electrical signal travels at the speed of light.

The speaker, like the microphone, is also an electroacoustical transducer that converts the electrical energy input into a mechanical sound-wave output. These sound waves strike the outer eardrum, causing the ear diaphragm to vibrate, and these mechanical vibrations actuate nerve endings in the ear, which convert the mechanical vibrations into electrochemical impulses that are sent to the brain. The brain decodes this information by comparing these impulses with a library of previous sounds and so provides the sensation of hearing.

For communication between two distant points, a wire must be connected between the microphone and speaker; however, if an electrical wave is applied to an antenna, the electrical wave is converted into a radio or electromagnetic wave, as shown in the inset in Figure 13-5(a), and communication is established without the need for a connecting wire—hence the term **wireless communication.** Antennas are designed to radiate and receive electromagnetic waves, which vary in field strength, as shown in Figure 13-5(d), and can exist in either air or space. These radio waves, as they are also known, travel at the speed of light and allow us to achieve communication over great distances.

More specifically, radio waves are composed of two basic components. The electrical voltage applied to the antenna is converted into an electric field, and the electrical current into a magnetic field. This **electromagnetic** (electric–magnetic) **wave** is used to carry a variety of information, such as speech, radio broadcasts and television signals.

Communication

Transmission of information between two points.

Transducer

Any device that converts energy from one form to another.

Electrical Wave

Travelling wave propagated in a conductive medium that is a variation in voltage or current and travels at slightly less than the speed of light.

Wireless Communication

Term describing radio communication that requires no wires between the two communicating points.

Electromagnetic (Radio) Wave

Wave that consists of both an electric and magnetic variation, and travels at the speed of light.

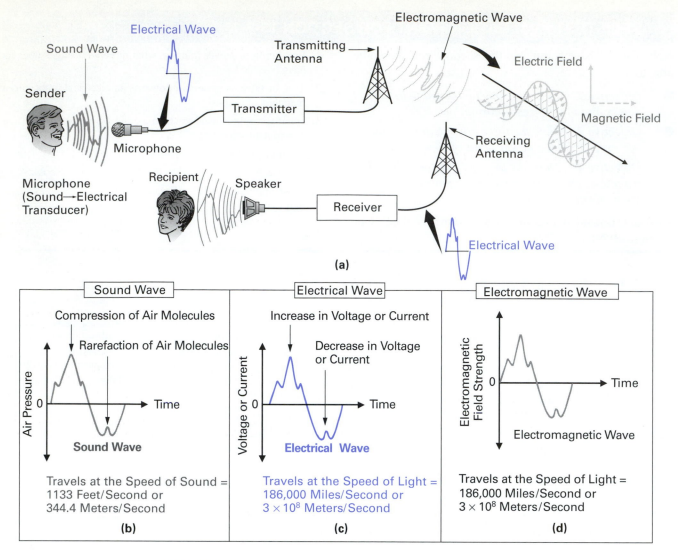

(a)

Sound Wave	Electrical Wave	Electromagnetic Wave
Compression of Air Molecules Rarefaction of Air Molecules **Sound Wave** Travels at the Speed of Sound = 1133 Feet/Second or 344.4 Meters/Second **(b)**	Increase in Voltage or Current Decrease in Voltage or Current **Electrical Wave** Travels at the Speed of Light = 186,000 Miles/Second or 3×10^8 Meters/Second **(c)**	Electromagnetic Wave Travels at the Speed of Light = 186,000 Miles/Second or 3×10^8 Meters/Second **(d)**

FIGURE 13-5 Information Transfer.

In summary, the sound wave is a variation in air pressure, the electrical wave is a variation of voltage or current, and the electromagnetic wave is a variation of electric and magnetic field strength.

■ INTEGRATED MATH APPLICATION:

How long will it take the sound wave produced by a rifle shot to travel 9630.5 feet?

■ *Solution:*

This problem makes use of the following formula:

$$\text{Distance} = \text{velocity} \times \text{time}$$

or

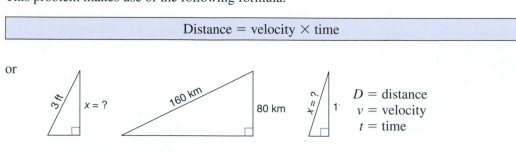

D = distance
v = velocity
t = time

If someone travels at 20 mph for 2 hours, the person will travel 40 miles ($D = v \times t = $ 20 mph $\times$ 2 hours = 40 miles). In this problem, the distance (9630.5 ft) and the sound wave's velocity (1130 ft/sec) are known, and so by rearranging the formula, we can find time:

$$\text{Time} = \frac{\text{distance}}{\text{velocity}} = \frac{9630.5 \text{ ft}}{1130 \text{ ft/s}} = 8.5 \text{ s}$$

■ INTEGRATED MATH APPLICATION:

How long will it take an electromagnetic (radio) wave to reach a receiving antenna that is 2000 miles away from the transmitting antenna?

■ *Solution:*

In this problem, both distance (2000 miles) and velocity (186,000 miles/s) are known, and time has to be calculated:

$$\text{Time} = \frac{\text{distance}}{\text{velocity}} = \frac{2000 \text{ miles}}{186,000 \text{ miles/s}}$$
$$= 1.075 \times 10^{-2} \quad \text{or} \quad 10.8 \text{ ms}$$

13-2-3   *Electrical Equipment or Electronic Equipment?*

At the beginning of this chapter, it was stated that ac is basically used in two applications: (1) power transfer and (2) information transfer. These two uses for ac help define the difference between electricity and electronics. Electronic equipment manages the flow of information, whereas electrical equipment manages the flow of power. In summary:

EQUIPMENT	MANAGES
Electrical	Power (large values of V and I)
Electronic	Information (small values of V and I)

To use an example, we can say that a dc power supply is a piece of electrical equipment, since it is designed to manage the flow of power. A TV set, however, is an electronic system, since its electronic circuits are designed to manage the flow of audio (sound) and video (picture) information.

Because most electronic systems include a dc power supply, we can also say that the electrical circuits manage the flow of power, and this power supply enables the electronic circuits to manage the flow of information.

SELF-TEST EVALUATION POINT FOR 13-2-2 AND 13-2-3

Now that you have completed these sections, you should be able to:

■ **Objective 5.** *Describe the three waves used to carry information between two points.*

■ **Objective 6.** *Explain how the three basic information carriers are used to carry many forms of information on different frequencies within the frequency spectrum.*

Use the following questions to test your understanding of Sections 13-2-2 and 13-2-3.

1. Define *information* and *communication*.
2. The _____ wave is a variation in air pressure, the _____ wave is a variation in field strength, and the _____ wave is a variation of voltage or current.

3. Which of the three waves described in Question 2:
   a. Can exist only in the air?
   b. Can exist in either air or a vacuum?
   c. Exists in a wire conductor?

4. Sound waves travel at the speed of sound, which is _____, while electrical and electromagnetic waves travel at the speed of light, which is _____.

5. A human ear is designed to receive _____ waves, an antenna is designed to transmit or receive _____ waves, and an electronic circuit is designed to pass only _____ waves.

6. Give the names of the following energy converters or transducers:
   a. Sound wave (mechanical energy) to electrical wave
   b. Electrical wave to sound wave
   c. Electrical wave to electromagnetic wave
   d. Sound wave to electrochemical impulses

7. _____ equipment manages the flow of information, and these ac waveforms normally have small values of current and voltage.

8. _____ equipment manages the flow of power, and these ac waveforms normally have large values of current and voltage.

13-3   AC WAVE SHAPES

In all fields of electronics, whether medical, industrial, consumer, or data processing, different types of information are conveyed between two points, and electronic equipment manages the flow of this information.

Let's now discuss the basic types of ac wave shapes. The way in which a wave varies in magnitude with respect to time describes its wave shape. All ac waves can be classified into one of six groups, and these are illustrated in Figure 13-6.

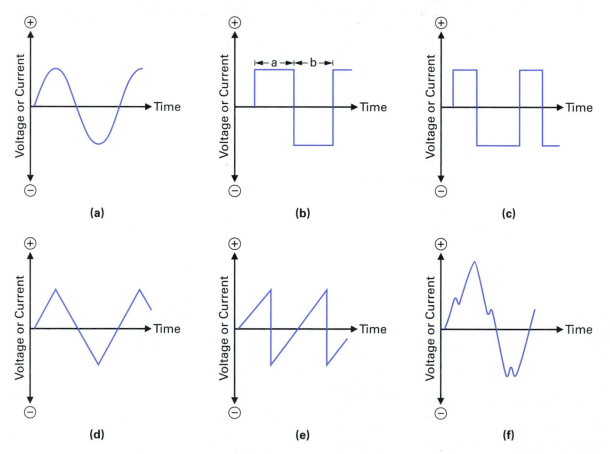

FIGURE 13-6   AC Wave Shapes. (a) Sine Wave. (b) Square Wave. (c) Pulse Wave. (d) Triangular Wave. (e) Sawtooth Wave. (f) Irregular Wave.

13-3-1   *The Sine Wave*

The **sine wave** is the most common type of waveform. It is the natural output of a generator that converts a mechanical input, in the form of a rotating shaft, into an electrical output in the form of a sine wave. In fact, for one cycle of the input shaft, the generator will produce one sinusoidal ac voltage waveform, as shown in Figure 13-7. When the input shaft of the generator is at 0°, the ac output is 0 V. As the shaft is rotated through 360° the ac output voltage will rise to a maximum positive voltage at 90°, fall back to 0 V at 180°, and then reach a maximum negative voltage at 270°, and finally return to 0 V at 360°. If this ac voltage is applied across a closed circuit, it produces a current that continually reverses or alternates in each direction.

Figure 13-8 illustrates the sine wave, with all the characteristic information labeled, which at first glance looks a bit ominous. Let's analyze and discuss each piece of information individually, beginning with the amplitude of the sine wave.

Amplitude

Figure 13-9 plots direction and amplitude against time. The **amplitude** or magnitude of a wave is often represented by a **vector** arrow, also illustrated in Figure 13-9. The vector's length indicates the magnitude of the current or voltage, while the arrow's point is used to show the direction, or polarity.

Peak Value

The peak of an ac wave occurs on both the positive and negative alternation, but is only at the peak (maximum) for an instant. Figure 13-10(a) illustrates an ac current waveform rising to a positive peak of 10 A, falling to zero, and then reaching a negative peak of 10 A in the reverse direction. Figure 13-10(b) shows an ac voltage waveform reaching positive and negative peaks of 9 V.

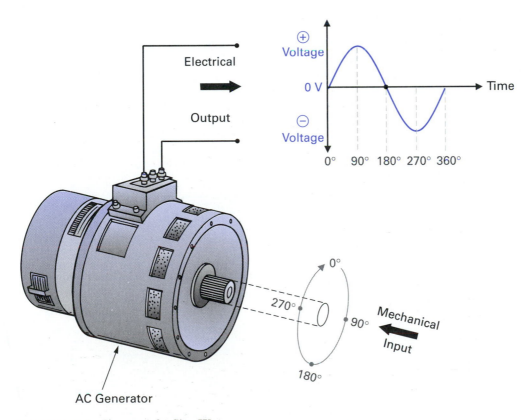

FIGURE 13-7   **Degrees of a Sine Wave.**

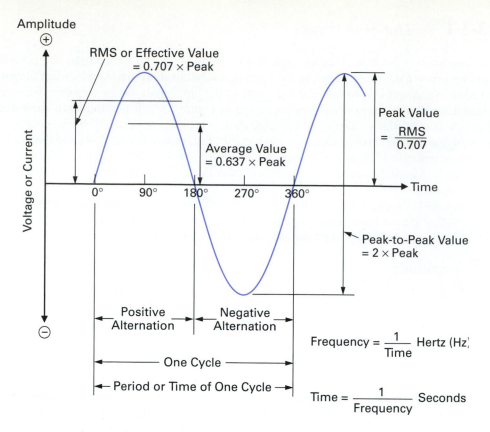

FIGURE 13-8 Sine Wave.

Peak-to-Peak Value

The **peak-to-peak value** of a sine wave is the value of voltage or current between the positive and negative maximum values of a waveform. For example, the peak-to-peak value of the current waveform in Figure 13-10(a) is equal to $I_{p-p} = 2 \times I_p = 20$ A. In Figure 13-10(b), it would be equal to $V_{p-p} = 2 \times V_p = 18$ V.

$$\text{p-p} = 2 \times \text{peak}$$

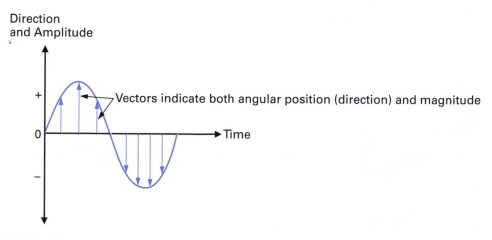

FIGURE 13-9 Sine-Wave Amplitude.

CHAPTER 13 / ALTERNATING CURRENT (AC)

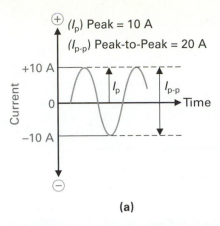

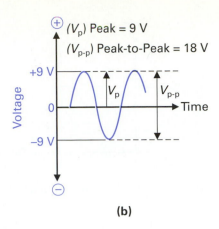

FIGURE 13-10 **Peak and Peak-to-Peak Value of a Sine Wave.**

RMS or Effective Value

Both the positive and negative alternation of a sine wave can accomplish the same amount of work, but the ac waveform is at its maximum value only for an instant in time, spending most of its time between peak currents. Our examples in Figure 13-10(a) and (b), therefore, cannot supply the same amount of power as a dc value of 10 A or 9 V.

The effective value of a sine wave is equal to 0.707 of the peak value. Let's now see how this value was obtained. Power is equal to either $P = I^2 \times R$ or $P = V^2/R$. Said another way, power is proportional to the voltage or current squared. If every instantaneous value of either the positive or negative half-cycle of any voltage or current sinusoidal waveform is squared, as shown in Figure 13-11, and then averaged to obtain the mean value, the square root of this mean value would be equal to 0.707 of the peak.

For example, if the process is carried out on the 10 A current waveform considered previously, the result would equal 7.07 A (0.707 × 10 A = 7.07 A), which is 0.707 of the peak value of 10 A.

$$\text{rms} = 0.707 \times \text{peak}$$

This **root-mean-square (rms) value** of 0.707 can always be used to determine how effective an ac sine wave will be. For example, a 10 A dc source will be 10 A effective because it is continually at its peak value and always delivering power to the circuit to which it is connected, whereas a 10 A ac source will be only 7.07 A effective, as seen in Figure 13-12,

RMS Value

RMS value of an ac voltage, current, or power waveform is equal to 0.707 of the peak value. The rms value is the effective or dc value equivalent of the ac wave.

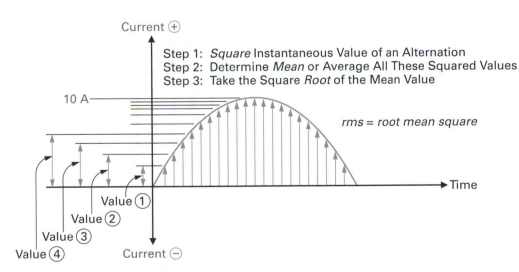

FIGURE 13-11 **Obtaining the RMS Value of 0.707 for a Sine Wave.**

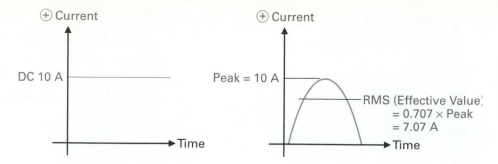

FIGURE 13-12 Effective Equivalent.

because it is at 10 A for only a short period of time. As another example, a 10 V ac sine-wave alternation would be as effective or supply the same amount of power to a circuit as a 7.07 V dc source.

Unless otherwise stated, ac values of voltage or current are always given in rms. The peak value can be calculated by transposing the original rms formula, rms = peak × 0.707, to obtain:

$$\text{Peak} = \frac{\text{rms}}{0.707}$$

Because 1/0.707 = 1.414, the peak can also be calculated from

$$\text{Peak} = \text{rms} \times 1.414$$

Average Value

Average Value

Mean value found when the area of a wave above a line is equal to the area of the wave below the line.

The **average value** of the positive or negative alternation is found by taking either the positive or negative alternation, and listing the amplitude or vector length of current or voltages at 1° intervals, as shown in Figure 13-13(a). The sum of all these values is then divided by the total number of values (averaging), which for all sine waves will calculate out to be 0.637 of the peak voltage or current. For example, the average value of a sine-wave alternation with a peak of 10 V, as seen in Figure 13-13(b), is equal to

$$\text{Average} = 0.637 \times \text{peak}$$

$$= 0.637 \times 10 \text{ V}$$
$$= 6.37 \text{ V}$$

The average of a positive or negative alternation (half-cycle) is equal to 0.637 × peak; however, the average of the complete cycle, including both the positive and negative half-cycles, is mathematically zero, as the amount of voltage or current above the zero line is equal to but the opposite of the amount of voltage or current below the zero line, as shown in Figure 13-14.

■ **INTEGRATED MATH APPLICATION:**

Calculate V_p, $V_\text{p-p}$, V_rms, and V_avg of a 16 V peak sine wave.

■ *Solution:*

$$V_\text{p} = 16 \text{ V}$$
$$V_\text{p-p} = 2 \times V_\text{p} = 2 \times 16 \text{ V} = 32 \text{ V}$$
$$V_\text{rms} = 0.707 \times V_\text{p} = 0.707 \times 16 \text{ V} = 11.3 \text{ V}$$
$$V_\text{avg} = 0.637 \times V_\text{p} = 0.637 \times 16 \text{ V} = 10.2 \text{ V}$$

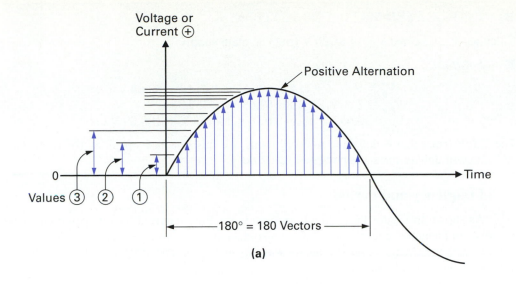

Voltage or
Current ⊕

Positive Alternation

0

Values ③ ② ①

180° = 180 Vectors

Time

(a)

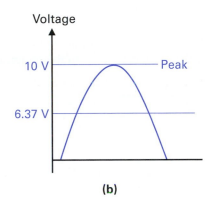

Voltage

10 V — Peak

6.37 V

(b)

FIGURE 13-13   **Average Value of a Sine-Wave Alternation = 0.637 × Peak.**

FIGURE 13-14   **Average Value of a**   Voltage or Current
Complete Sine-Wave Cycle = 0.

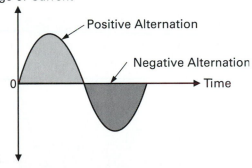

Positive Alternation

Negative Alternation

0

Time

Positive Alternation = Negative Alternation

Area =   Area

and Mathematically Cancel to Equal Zero

■ **INTEGRATED MATH APPLICATION:**

Calculate V_p, $V_{p\text{-}p}$, and V_{avg} of a 120 V (rms) ac main supply.

■ *Solution:*

$$V_p = V_{rms} \times 1.414 = 120\ V \times 1.414 = 169.68\ V$$
$$V_{p\text{-}p} = 2 \times V_p = 2 \times 169.68\ V = 339.36\ V$$
$$V_{avg} = 0.637 \times V_p = 0.637 \times 169.68\ V = 108.09\ V$$

The 120 V (rms) that is delivered to every home and business has a peak of 169.68 V. This ac value will deliver the same power as 120 V dc.

Frequency and Period

Period

Time taken to complete one complete cycle of a periodic or repeating waveform.

Frequency

Rate of recurrences of a periodic wave normally within a unit of one second, measured in hertz (cycles/second).

As shown in Figure 13-15, the **period** (t) is the time required for one complete cycle (positive and negative alternation) of the sinusoidal current or voltage waveform. A *cycle,* by definition, is the change of an alternating wave from zero to a positive peak, to zero, then to a negative peak, and finally, back to zero.

Frequency is the number of repetitions of a periodic wave in a unit of time. It is symbolized by *f* and is given the unit hertz (cycles per second), in honor of a German physicist, Heinrich Hertz.

Sinusoidal waves can take a long or a short amount of time to complete one cycle. This time is related to frequency in that period is equal to the reciprocal of frequency, and vice versa.

$$f(\text{hertz}) = \frac{1}{t}$$

$$t(\text{seconds}) = \frac{1}{f}$$

where t = period
          f = frequency

TIME LINE

Heinrich R. Hertz (1857–1894), a German physicist, was the first to demonstrate the production and reception of electromagnetic (radio) waves. In honor of his work in this field, the unit of frequency is called the hertz.

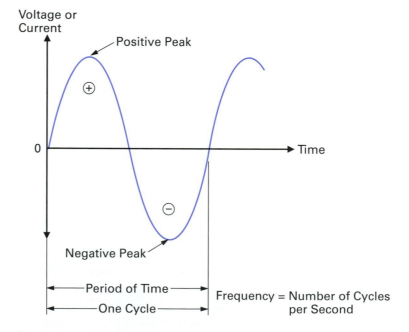

FIGURE 13-15   Frequency and Period.

358   CHAPTER 13 / ALTERNATING CURRENT (AC)

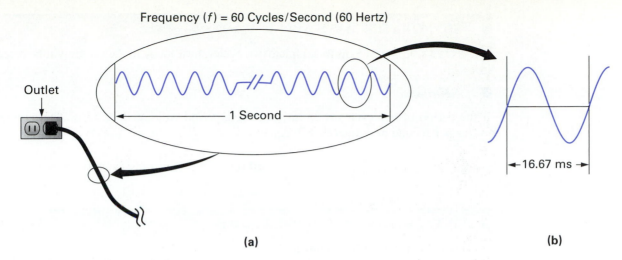

Frequency (*f*) = 60 Cycles/Second (60 Hertz)

Outlet

1 Second

16.67 ms

(a)   (b)

FIGURE 13-16   120 V, 60 Hz AC Supply.

For example, the ac voltage of 120 V (rms) arrives at the household electrical outlet alternating at a frequency of 60 hertz (Hz). This means that 60 cycles arrive at the household electrical outlet in 1 second. If 60 cycles occur in 1 second, as seen in Figure 13-16(a), it is actually taking $\frac{1}{60}$ of a second for one of the 60 cycles to complete its cycle, which calculates out to be

$$\frac{1}{60} \text{ of 1 second} = \frac{1}{60 \text{ cycles}} \times 1 \text{ second} = 16.67 \text{ milliseconds (ms)}$$

So the time or period of one cycle can be calculated by using the formula period (t) = $1/f$ = 1/60 Hz = 16.67 ms, as shown in Figure 13-16(b).

If the period or time of a cycle is known, the frequency can be calculated. For example:

$$\text{Frequency } (f) = \frac{1}{\text{period}} = 1/16.67 \text{ ms} = 60 \text{ Hz}$$

As illustrated in Figure 13-4, all homes in the United States receive at their wall outlets an ac voltage of 120 V rms at a frequency of 60 Hz. This frequency was chosen for convenience, as a lower frequency would require larger transformers, and if the frequency were too low, the slow switching (alternating) current through lightbulbs would cause the lights to flicker. A higher frequency than 60 Hz was found to cause an increase in the amount of heat generated in the core of all power distribution transformers due to eddy currents and hysteresis losses. A frequency of 60 Hz was chosen in the United States; however, other countries, such as England, and most of Europe use an ac power line frequency of 50 Hz (240 V).

■   **INTEGRATED MATH APPLICATION:**

If a sine wave has a period of 400 μs, what is its frequency?

■   *Solution:*

$$\text{Frequency } (f) = \frac{1}{\text{time}(t)} = \frac{1}{400 \ \mu\text{s}} = 2.5 \text{ kHz or 2500 cycles per second}$$

■ **INTEGRATED MATH APPLICATION:**

If it takes a sine wave 25 ms to complete two cycles, how many of the cycles will be received in 1 s?

■ *Solution:*

If the period of two cycles is 25 ms, one cycle period will equal 12.5 ms. The number of cycles per second or frequency will equal

$$f = \frac{1}{t} = \frac{1}{12.5 \text{ ms}} = 80 \text{ Hz} \quad \text{or} \quad 80 \text{ cycles/second}$$

■ **INTEGRATED MATH APPLICATION:**

Calculate the period of the following:

  a. 100 MHz
  b. 40 cycles every 5 seconds
  c. 4.2 kilocycles/second
  d. 500 kHz

■ *Solution:*

$$f = \frac{1}{t} \quad \text{therefore, } t = \frac{1}{f}$$

a. $t = \dfrac{1}{100 \text{ MHz}} = 10 \text{ nanoseconds (ns)}$

b. 40 cycles/5 s = 8 cycles/second (8 Hz)

$$t = \frac{1}{8 \text{ Hz}} = 125 \text{ ms}$$

c. $t = \dfrac{1}{4.2 \text{ kHz}} = 238 \text{ } \mu\text{s}$

d. $t = \dfrac{1}{500 \text{ kHz}} = 2 \text{ } \mu\text{s}$

Wavelength

Wavelength

Distance between two points of corresponding phase and equal to waveform velocity or speed divided by frequency.

Wavelength, as its name states, is the physical length of one complete cycle and is generally measured in meters. The wavelength (λ, lambda) of a complete cycle is dependent on the frequency and velocity of the transmission:

$$\lambda = \frac{\text{velocity}}{\text{frequency}}$$

Electromagnetic Waves Radio waves travel at the speed of light in air or a vacuum, which is 3×10^8 meters/second or 3×10^{10} centimeters/second.

$$\lambda \text{ (m)} = \frac{3 \times 10^8 \text{ m/s}}{f \text{ (Hz)}} \qquad \text{or} \qquad \lambda \text{ (cm)} = \frac{3 \times 10^{10} \text{ cm/s}}{\text{frequency (Hz)}}$$

(There are 100 centimeters [cm] in 1 meter [m]; therefore, cm = 10^{-2}, and m = 10^0 or 1.) Consequently, the higher the frequency, the shorter the wavelength, which is why a short-wave radio receiver is designed to receive high frequencies ($\lambda \downarrow = 3 \times 10^8/f \uparrow$).

CHAPTER 13 / ALTERNATING CURRENT (AC)

INTEGRATED MATH APPLICATION:

Calculate the wavelength of the electromagnetic waves illustrated in Figure 13-17.

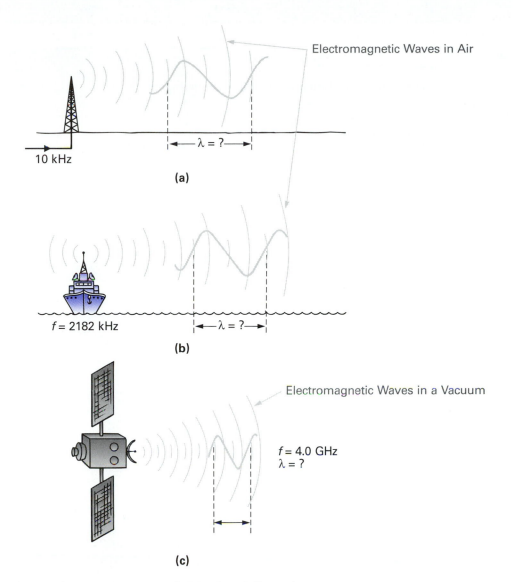

Electromagnetic Waves in Air

λ = ?

10 kHz

(a)

f = 2182 kHz   λ = ?

(b)

Electromagnetic Waves in a Vacuum

f = 4.0 GHz
λ = ?

(c)

FIGURE 13-17   **Electromagnetic Wavelength Examples.**

TIME LINE

During World War II, there was a need for microwave-frequency vacuum tubes. British inventor Henry Boot developed the magnatron in 1939.

TIME LINE

In 1939, a U.S. brother duo, Russel and Sigurd Varian, invented the klystron. In 1943, the travelling wave tube amplifier was invented by Rudolf Komphner, and up to this day these three microwave tubes are still used extensively.

■ *Solution:*

a. $\lambda = \dfrac{3 \times 10^8}{f\,(\text{Hz})}\ \text{m/s} = \dfrac{3 \times 10^8}{10\ \text{kHz}} = 30{,}000\ \text{m}$ or $30\ \text{km}$

b. $\lambda = \dfrac{3 \times 10^{10}}{f\,(\text{Hz})}\ \text{cm/s} = \dfrac{3 \times 10^{10}}{2182\ \text{kHz}} = 13{,}748.9\ \text{cm}$ or $137.489\ \text{m}$

c. $\lambda = \dfrac{3 \times 10^{10}}{f\,(\text{Hz})}\ \text{cm/s} = \dfrac{3 \times 10^{10}}{4.0\ \text{GHz}} = \dfrac{3 \times 10^{10}}{4 \times 10^9} = 7.5\ \text{cm}$ or $0.075\ \text{m}$

Sound Waves. Sound waves travel at a slower speed than electromagnetic waves, as their mechanical vibrations depend on air molecules, which offer resistance to the traveling wave. For sound waves, the wavelength formula is

$$\lambda \text{ (m)} = \frac{344.4 \text{ m/s}}{f \text{ (Hz)}}$$

■ **INTEGRATED MATH APPLICATION:**

Calculate the wavelength of the sound waves illustrated in Figure 13-18.

■ *Solution:*

a. $\lambda \text{ (m)} = \dfrac{344.4 \text{ m/s}}{f \text{ (Hz)}} = \dfrac{344.4 \text{ m/s}}{35 \text{ kHz}} = 9.8 \times 10^{-3} \text{ m} = 9.8 \text{ mm}$

b. 300 Hz: $\lambda \text{ (m)} = \dfrac{344.4 \text{ m/s}}{300 \text{ Hz}} = 1.15 \text{ m}$

   3000 Hz: $\lambda \text{ (m)} = \dfrac{344.4 \text{ m/s}}{3000 \text{ Hz}} = 0.115 \text{ m}$ or   11.5 cm

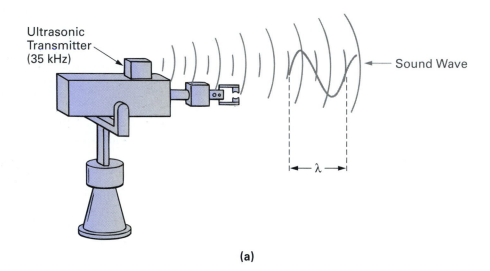

(a)

Frequency Range = 300 Hz to 3 kHz
Wavelength Range = ? to ?

(b)

FIGURE 13-18   **Sound Wavelength Examples.**

Phase Relationships

The **phase** of a sine wave is always relative to another sine wave of the same frequency. Figure 13-19(a) illustrates two sine waves that are in phase with one another, while

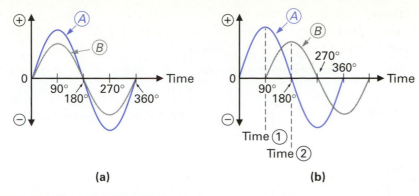

(a)   **(b)**

FIGURE 13-19   **Phase Relationship. (a) In Phase. (b) Out of Phase.**

Figure 13-19(b) shows two sine waves that are out of phase with each other. Sine wave *A* is the reference, since the positive-going zero crossing is at 0°, its positive peak is at 90°, its negative-going zero crossing is at 180°, its negative peak is at 270°, and the cycle completes at 360°. In Figure 13-19(a), sine wave *B* is in phase with *A*, since its peaks and zero crossings occur at the same time as sine wave *A*'s.

In Figure 13-19(b), sine wave *B* has been shifted to the right by 90° with respect to the reference sine wave *A*. This **phase shift** or **phase angle** of 90° means that sine wave *A leads B* by 90°, or *B lags A* by 90°. Sine wave *A* is said to lead *B*, as its positive peak, for example, occurs first at time 1, while the positive peak of *B* occurs later at time 2.

■ **INTEGRATED MATH APPLICATION:**

What are the phase relationships between the two waveforms illustrated in Figure 19-20(a) and (b)?

■ *Solution:*

a. The phase shift or angle is 90°. Sine wave *B* leads sine wave *A* by 90°, or *A* lags *B* by 90°.

b. The phase shift or angle is 45°. Sine wave *A* leads sine wave *B* by 45°, or *B* lags *A* by 45°.

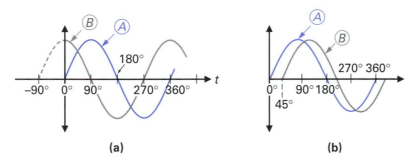

(a)   **(b)**

FIGURE 13-20   **Phase Relationship Examples.**

The Meaning of Sine

The square, rectangular, triangular, and sawtooth waveform shapes are given their names because of their waveform shapes. The sine wave is the most common type of waveform shape, and why the name sine was given to this wave needs to be explained further.

Figure 13-21(a) shows the correlation between the 360° of a circle and the 360° of a sine wave. Within the circle a triangle has been drawn to represent the right-angle tri-

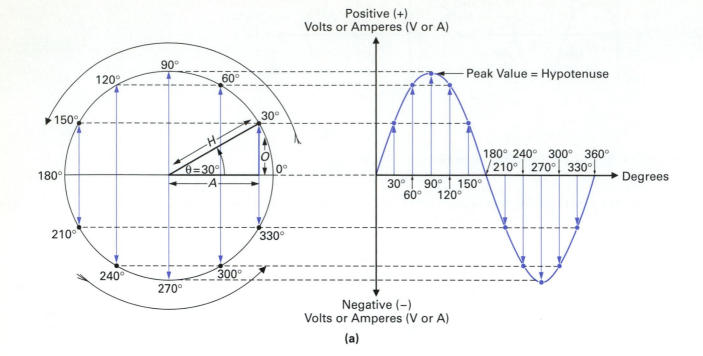

Peak Value = Hypotenuse

(a)

Angle	Sine
0	0.000
15	0.259
30	0.500
45	0.707
60	0.866
75	0.966
90	1.000

(b)

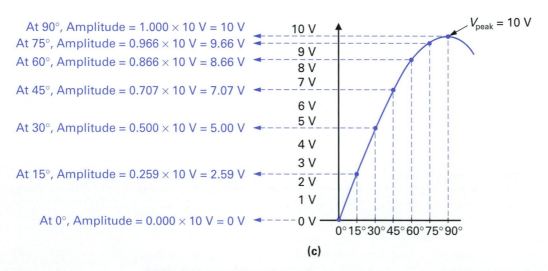

At 90°, Amplitude = 1.000 × 10 V = 10 V
At 75°, Amplitude = 0.966 × 10 V = 9.66 V
At 60°, Amplitude = 0.866 × 10 V = 8.66 V
At 45°, Amplitude = 0.707 × 10 V = 7.07 V
At 30°, Amplitude = 0.500 × 10 V = 5.00 V
At 15°, Amplitude = 0.259 × 10 V = 2.59 V
At 0°, Amplitude = 0.000 × 10 V = 0 V

V_{peak} = 10 V

(c)

FIGURE 13-21 Meaning of Sine.

angle formed at 30°. The hypotenuse side will always remain at the same length throughout 360° and will equal the peak of the sine wave. The opposite side of the triangle is equal to the amplitude vector of the sine wave at 30°. To calculate the amplitude of the opposite side, and therefore the amplitude of the sine wave at 30°, we use the sine of theta formula.

$$\sin \theta = \frac{\text{opposite}}{\text{hypotenuse}}$$

$$\sin 30° = \frac{O}{H} \left(\begin{array}{c} \textit{Calculator sequence:} \\ \boxed{3}\ \boxed{0}\ \boxed{\text{SIN}} \end{array} \right)$$

$$0.500 = \frac{O}{H}$$

This tells us that at 30°, the opposite side is 0.500 or half the size of the hypotenuse. At 30°, therefore, the amplitude of the sine wave will be 0.5 (50%) of the peak value.

Figure 13-21(b) lists the sine values at 15° increments. Figure 13-21(c) shows an example of a 10 V–peak sine wave. At 15°, a sine wave will always be at 0.259 (sine of 15°) of the peak value, which for a 10 V sine wave will be 2.59 V (0.259 × 10 V = 2.59 V). At 30°, a sine wave will have increased to 0.500 (sine of 30°) of the peak value. At 45°, a sine wave will be at 0.707 of the peak, and so on. The sine wave is called a sine wave, therefore, because it changes in amplitude at the same rate as the sine trigonometric function.

CALCULATOR KEYS

Name: Graphing a Sine Wave

Function and Example:

1. Press $\boxed{\text{Y=}}$ to display the Y= editor. Press $\boxed{\text{SIN}}$ $\boxed{\text{X,T,}\Theta,n}$ $\boxed{)}$ to store sin(X) in Y1.

2. Press $\boxed{\text{ZOOM}}$ 7 to select 7:ZTrig, which graphs the equation in the Zoom Trig window.

3. Press the TABLE key to view the data for every value of *X*.

13-3-2   *The Square Wave*

The **square wave** is a periodic (repeating) wave that alternates from a positive peak value to a negative peak value, and vice versa, for equal lengths of time.

In Figure 13-22 you can see an example of a square wave that is at a frequency of 1 kHz and has a peak of 10 V. If the frequency of a wave is known, its period or time of one cycle can be calculated by using the formula $t = 1/f = 1/1\text{ kHz} = 1\text{ ms}$ or $\frac{1}{1000}$ of a second. One complete cycle will take 1 ms to complete, so the positive and negative alternations will last for 0.5 ms each.

If the peak of the square wave is equal to 10 V, the peak-to-peak value of this square wave will equal $V_{\text{p-p}} = 2 \times V_{\text{p}} = 20$ V.

To summarize this example, the square wave alternates from a positive peak value of +10 V to a negative peak value of −10 V for equal time lengths (half-cycles) of 0.5 ms.

Square Wave

Wave that alternates between two fixed values for an equal amount of time.

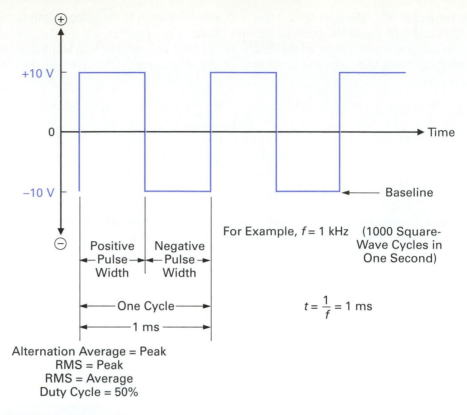

FIGURE 13-22 Square Wave.

Duty Cycle

Duty cycle is an important relationship, which has to be considered when discussing square waveforms. The **duty cycle** is the ratio of a pulse width (positive or negative pulse or cycle) to the overall period or time of the wave and is normally given as a percentage.

Duty Cycle

A term used to describe the amount of ON time versus OFF time. ON time is usually expressed as a percentage.

$$\text{Duty cycle (\%)} = \frac{\text{pulse width } (P_w)}{\text{period } (t)} \times 100\%$$

■ INTEGRATED MATH APPLICATION:

What is the duty cycle of the waveform shown in Figure 13-22?

■ *Solution:*

The duty cycle of the example square wave in Figure 13-22 will equal

$$\text{Duty cycle (\%)} = \frac{\text{pulse width } (P_w)}{\text{period } (t)} \times 100\%$$

$$= \frac{0.5 \text{ ms}}{1 \text{ ms}} \times 100\%$$

$$= 50\%$$

Since a square wave always has a positive and a negative alternation that are equal in time, the duty cycle of all square waves is equal to 50%, which actually means that the positive cycle lasts for 50% of the time of one cycle.

Average

The average or mean value of a square wave can be calculated by using the formula

$$V \text{ or } I \text{ average} = \text{baseline} + (\text{duty cycle} \times \text{peak to peak})$$

The average of the complete square-wave cycle in Figure 13-22 should calculate out to be zero, as the amount above the line equals the amount below. If we apply the formula to this example, you can see that

$$
\begin{aligned}
V_{avg} &= \text{baseline} + (\text{duty cycle} \times \text{peak to peak}) \\
&= -10 \text{ V} + (0.5 \times 20 \text{ V}) \\
&= -10 \text{ V} + 10 \text{ V} \\
&= 0 \text{ V}
\end{aligned}
$$

However, a square wave does not always alternate about 0. For example, Figure 13-23 illustrates a 16 $V_{p\text{-}p}$ square wave that rests on a baseline of 2 V. The average value of this square wave is equal to

$$
\begin{aligned}
V_{avg} &= \text{baseline} + (\text{duty cycle} \times \text{peak to peak}) \\
&= 2 \text{ V} + (0.5 \times 16 \text{ V}) \\
&= (+2 \text{ V}) + (+8 \text{ V}) \\
&= 10 \text{ V}
\end{aligned}
$$

■ **INTEGRATED MATH APPLICATION:**

Calculate the duty cycle and V_{avg} of a square wave of 0 to 5 V.

■ *Solution:*

The duty cycle of a square wave is always 0.5 or 50%. The average is:

$$
\begin{aligned}
V_{avg} &= \text{baseline} + (\text{duty cycle} \times V_{p\text{-}p}) \\
&= 0 \text{ V} + (0.5 \times 5 \text{ V}) \\
&= 0 \text{ V} + 2.5 \text{ V} = 2.5 \text{ V}
\end{aligned}
$$

Up to this point, we have seen the ideal square wave, which has instantaneous transition from the negative to the positive values, and vice versa, as shown in Figure 13-24(a). In fact, the transitions from negative to positive (positive or leading edge) and from positive to negative (negative or trailing edge) are not as ideal as shown here. It takes a small amount of time for the wave to increase to its positive value (the **rise time**), and an equal amount of time for a wave to decrease to its negative value (the **fall time**). Rise time (T_R), by definition, is the time it takes for an edge to rise from 10% to 90% of its full amplitude, while fall time (T_F) is the time it takes for an edge to fall from 90% to 10% of its full amplitude, as shown in Figure 13-24(b).

With a waveform such as that in Figure 13-24(b), it is difficult, unless a standard is used, to know exactly what points to use when measuring the width of either the positive or

Rise Time

Time it takes a positive edge of a pulse to rise from 10% to 90% of its peak value.

Fall Time

Time it takes a negative edge of a pulse to fall from 90% to 10% of its peak value.

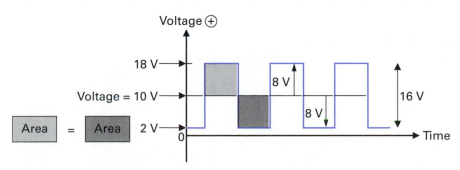

FIGURE 13-23   **2 to 18 V Square Wave.**

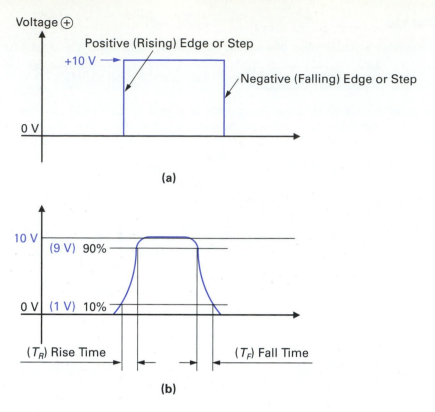

(a)

(b)

FIGURE 13-24   Square Wave's Rise and Fall Times. (a) Ideal. (b) Actual.

negative alternation. The standard width is always measured between the two 50% amplitude points, as shown in Figure 13-25.

Frequency-Domain Analysis

A *periodic wave* is a wave that repeats the same wave shape from one cycle to the next. Figure 13-26(a) is a **time-domain** representation of a periodic sine wave, which is the same way it would appear on an oscilloscope display as it plots the sine wave's amplitude against time. Figure 13-26(b) is a **frequency-domain** representation of the same periodic sine wave, and this graph, which shows the wave as it would appear on a spectrum analyzer, plots the sine wave's amplitude against frequency instead of time. This graph shows all the frequency components contained within a wave, and since, in this example, the sine wave has a period of 1 ms and therefore a frequency of 1 kHz, there is one bar at the 1 kHz point of the graph, and its size represents the sine wave's amplitude. Pure sine waves have no other frequency components.

Other periodic wave shapes, such as square, pulse, triangular, sawtooth, or irregular, are actually made up of a number of sine waves having a particular frequency, amplitude,

Time-Domain Analysis

A method of representing a waveform by plotting its amplitude versus time.

Frequency-Domain Analysis

A method of representing a waveform by plotting its amplitude versus frequency.

FIGURE 13-25   **Pulse Width of a Square Wave.**

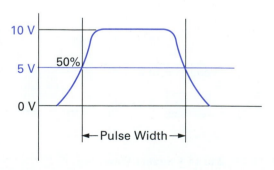

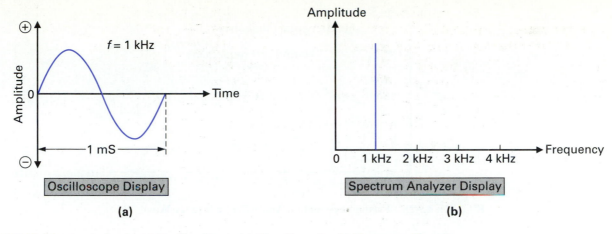

FIGURE 13-26   Analysis of a 1 kHz Sine Wave. (a) Time Domain. (b) Frequency Domain.

and phase. To produce a square wave, for instance, you would start with a sine wave, as shown in Figure 13-27(a), whose frequency is equal to the resulting square-wave frequency. This sine wave is called the **fundamental frequency,** and all the other sine waves that will be added to this fundamental are called **harmonics** and will always be lower in amplitude and higher in frequency. These harmonics or multiples are harmonically related to the fundamental, in that the second harmonic is twice the fundamental frequency, the third harmonic is three times the fundamental frequency, and so on.

Fundamental Frequency

This sine wave is always the lowest-frequency and largest-amplitude component of any waveform shape and is used as a reference.

Harmonic Frequency

Sine wave that is smaller in amplitude and is some multiple of the fundamental frequency.

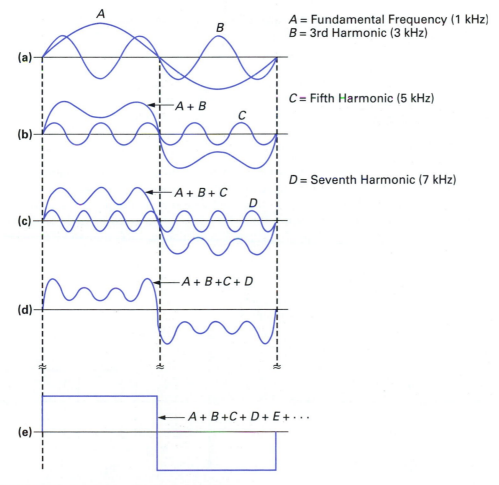

A = Fundamental Frequency (1 kHz)
B = 3rd Harmonic (3 kHz)

C = Fifth Harmonic (5 kHz)

D = Seventh Harmonic (7 kHz)

FIGURE 13-27   Square-Wave Composition.

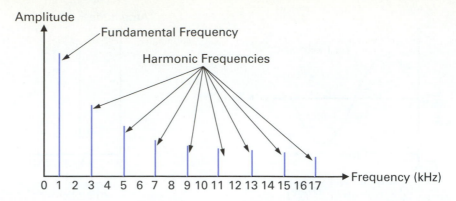

FIGURE 13-28 **Frequency-Domain Analysis of a Square Wave.**

Square waves are composed of a fundamental frequency and an infinite number of odd harmonics (third, fifth, seventh, and so on). If you look at the progression in Figure 13-27(a) through (d), you see that by continually adding these odd harmonics the waveform comes closer to a perfect square wave, as shown in Figure 13-27(e).

Figure 13-28 plots the frequency domain of a square wave, with the bars representing the odd harmonics of decreasing amplitude.

13-3-3   *The Rectangular or Pulse Wave*

The **rectangular wave** is similar to the square wave in many respects, in that it is a periodic wave that alternately switches between one of two fixed values. The difference in the rectangular wave is that it does not remain at the two peak values for equal lengths of time, as shown in the examples in Figure 13-29(a) and (b).

In Figure 13-29(a), the rectangular wave remains at its negative level for a longer period than its positive, while the rectangular wave in Figure 13-29(b) stays at its positive value for the longer period of time and is only momentarily at its negative value.

PRF, PRT, and Pulse Length

In the discussion of a rectangular wave, a few terms change. Instead of stating the cycles per second as frequency, we call it **pulse repetition frequency** (PRF), which is far more descriptive. The reciprocal of frequency is time, and with rectangular pulse waveforms the reciprocal of the PRF is **pulse repetition time** (PRT), as summarized in Table 13-1. With rectangular or pulse waves, therefore, frequency is equivalent to PRF, time to PRT, and the only difference is the name.

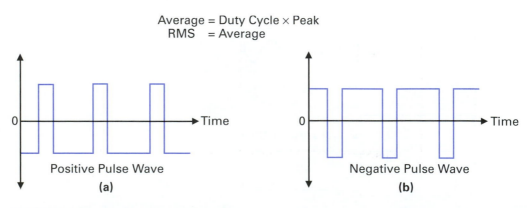

FIGURE 13-29 **Rectangular or Pulse Wave.**

TABLE 13-1

SQUARE AND SINE WAVE		RECTANGULAR WAVE	
Frequency $(f) = \dfrac{1}{\text{time } (t)}$		Pulse repetition frequency (PRF) $= \dfrac{1}{\text{pulse repetition time (PRT)}}$	
	Equivalent to		
Time $(t) = \dfrac{1}{\text{frequency } (f)}$		Pulse repetition time (PRT) $= \dfrac{1}{\text{pulse repetition frequency (PRF)}}$	

Let us look at the example in Figure 13-30 of a 5 V rectangular wave at a frequency of 1 kHz and a pulse width of 1 μs, and practice with these new terms. With a pulse repetition frequency of 1 kHz, the time between the leading edges of pulses (PRT) will be 1/1 kHz = 1 ms. **Pulse width** (P_w), **pulse duration** (P_d), or **pulse length** (P_l) are all terms that describe the length of time for which the pulse lasts, and in this example it is equal to 1 μs, which means that 999 μs exists between the end of one pulse and the beginning of the next.

Duty Cycle

The duty cycle is calculated in exactly the same way as for the square wave and is a ratio of the pulse width to the overall time (PRT). In our example in Figure 13-30 the duty cycle will be equal to

$$\text{Duty cycle } (\%) = \frac{\text{pulse width } (P_w)}{\text{PRT}} \times 100\%$$

$$= \frac{1\ \mu s}{1000\ \mu s} \times 100\%$$

$$= \text{duty cycle figure of } 0.001 \times 100\%$$

$$= 0.001 \times 100\%$$

$$= 0.1\%$$

The result tells us that the positive pulse lasts for 0.1% of the total time (PRT).

Average

The average or mean value of this waveform is calculated by using the same square-wave formula. The average of the pulse wave in Figure 13-30 will be

$$V \text{ or } I \text{ average} = \text{baseline} + (\text{duty cycle} \times \text{peak to peak})$$
$$V_{\text{avg}} = 0\text{ V} + (0.001 \times 5\text{ V})$$
$$= 0\text{ V} + (5\text{ mV})$$
$$= 5\text{ mV}$$

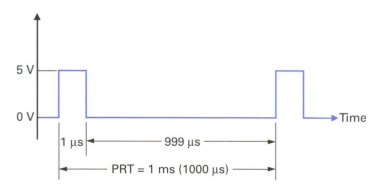

FIGURE 13-30 PRF and PRT of a Pulse Wave.

Pulse Width, Pulse Length, or Pulse Duration

The time interval between the leading edge and trailing edge of a pulse at which the amplitude reaches 50% of the peak pulse amplitude.

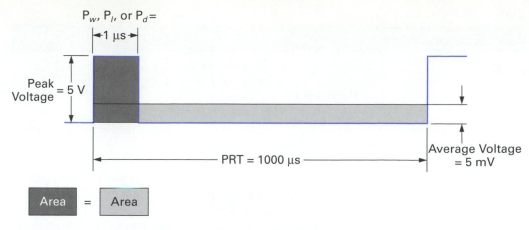

$$P_w, P_l, \text{ or } P_d =$$

|←1 μs→|

Peak Voltage = 5 V

PRT = 1000 μs

Average Voltage = 5 mV

Area = Area

FIGURE 13-31 **Average of a Pulse Wave.**

Figure 13-31 illustrates the average value of this rectangular waveform. If the voltage and width of the positive pulse are taken and spread out over the entire PRT, they will have a mean level equal, in this example, to 5 mV.

■ INTEGRATED MATH APPLICATION:

Calculate the duty cycle and average voltage of the following radar pulse waveform:

Peak voltage, V_p = 20 kV
Pulse length, P_l = 1 μs
Baseline voltage = 0 V
PRF = 3300 pulses per second (pps)

■ *Solution:*

$$\text{Duty cycle} = \frac{\text{pulse length } (P_l)}{\text{PRT}} \times 100\%$$

$$= \frac{1 \text{ μs}}{303 \text{ μs}} \times 100\% \left(\text{PRT} = \frac{1}{\text{PRF}} = \frac{1}{3300} = 303 \text{ μs} \right)$$

$$= (3.3 \times 10^{-3}) \times 100\%$$

$$= 0.33\%$$

$$V_{avg} = \text{baseline} + (\text{duty cycle} \times V_{p\text{-}p})$$
$$= 0 \text{ V} + [(3.3 \times 10^{-3}) \times 20 \times 10^3]$$
$$= 66 \text{ V}$$

Frequency-Domain Analysis

The pulse or rectangular wave is closely related to the square wave, as shown in Figure 13-32, but some changes occur in its harmonic content. One is that even-numbered harmonics are present, and their amplitudes do not fall off as quickly as do those of the square wave. The amplitude and phase of these sine-wave harmonics are determined by the pulse width and pulse repetition frequency, and the narrower the pulse, the greater the number of harmonics present.

Triangular Wave

A repeating wave that has equal positive and negative ramps that have linear rates of change with time.

13-3-4 *The Triangular Wave*

A **triangular wave** consists of a positive and negative ramp of equal values, as shown in Figure 13-33. Both the positive and negative ramps have a linear increase and decrease, respectively. Linear, by definition, is the relationship between two quantities that exists when

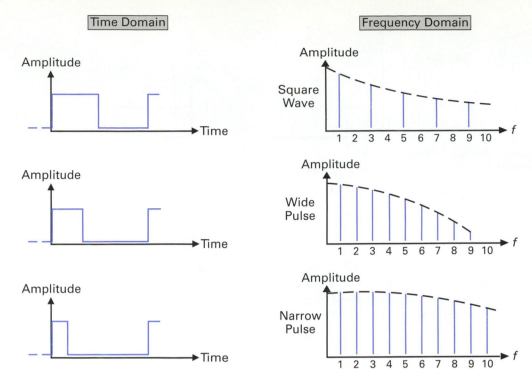

FIGURE 13-32 Time- and Frequency-Domain Analysis of a Pulse Waveform.

a change in a second quantity is directly proportional to a change in the first quantity. The two quantities in this case are voltage or current and time. As shown in Figure 13-33, if the increment of change of voltage ΔV (pronounced "delta vee") is changing at the same rate as the increment of time Δt ("delta tee"), the ramp is said to be **linear**.
Δt ("delta tee"), the ramp is said to be **linear**.

    With Figure 13-34(a), the voltage has risen 1 V in 1 second (time 1) and maintains that rise through to time 2 and, consequently, is known as a linearly rising slope. In Fig-

Linear

Relationship between input and output in which the output varies in direct proportion to the input.

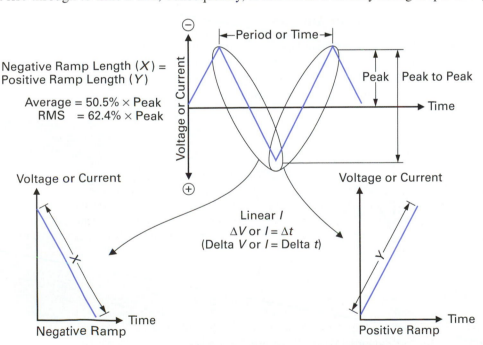

FIGURE 13-33 Triangular Wave.

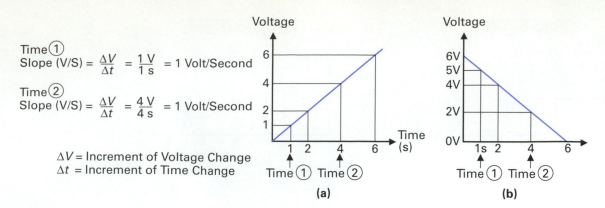

FIGURE 13-34 Linear Triangular Wave Rise and Fall.

ure 13-34(b), the voltage is falling first from 6 to 5 V, which is a 1 V drop in 1 second, and in time 2 from 6 V to 2 V, which is a 4 V drop in 4 s. The rate of fall still remains the same, so the waveform is referred to as a linearly falling slope.

This formula for slope will also apply to a current waveform, and is

$$
\text{Slope (A/s)} = \frac{\Delta I}{\Delta t}
$$

where ΔI = increment of current change
         Δt = increment of time change

With triangular waves, frequency and time apply as usual, as seen in Figure 13-35, with

$$
\text{frequency} = \frac{1}{\text{time}} \text{ (Hz)}
$$

$$
\text{time} = \frac{1}{\text{frequency}} \text{ (s)}
$$

Frequency-Domain Analysis

The time domain of a triangular wave is shown in Figure 13-36(a), while the frequency domain of a triangular wave is shown in Figure 13-36(b). Frequency-domain analysis is often used to test electronic circuits, as it tends to highlight problems that would normally not show up when a sine-wave signal is applied.

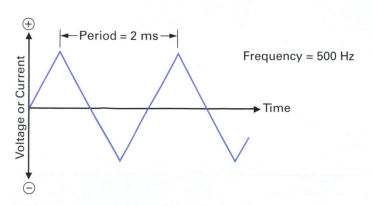

FIGURE 13-35 Triangular Wave Period and Frequency.

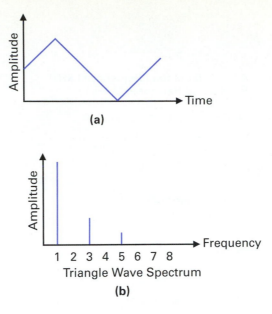

FIGURE 13-36 Analysis of a Triangular Wave. (a) Time Domain. (b) Frequency Domain.

(a)

(b)

Triangle Wave Spectrum

13-3-5 *The Sawtooth Wave*

On an oscilloscope display (time-domain presentation), the **sawtooth wave** is very similar to a triangular wave, in that a sawtooth wave has a linear ramp; however, unlike the triangular wave, which reverses and has an equal but opposite ramp back to its starting level, the sawtooth "flies" back to its starting point immediately and then repeats the previous ramp, as seen in Figure 13-37, which shows both a positive and a negative ramp sawtooth.

Figure 13-38 shows the odd and even harmonics contained in the frequency-domain analysis of a negative-going ramp.

13-3-6 *Other Waveforms*

The waveforms discussed so far are some of the more common types; however, since every waveform shape (except a pure sine wave) is composed of a large number of sine waves combined in an infinite number of ways, any waveform shape is possible. Figure 13-39 illustrates a variety of waveforms that can be found in all fields of electronics, and the appendix contains a detailed frequency spectrum chart detailing the applications of many electromagnetic frequencies.

Sawtooth Wave
Repeating waveform that rises from zero to a maximum value linearly and then falls to zero and repeats.

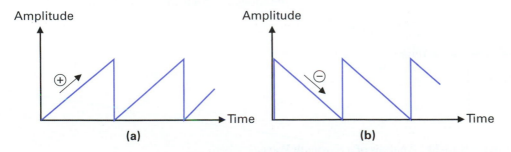

(a)

(b)

FIGURE 13-37 Sawtooth Wave. (a) Positive Ramp. (b) Negative Ramp.

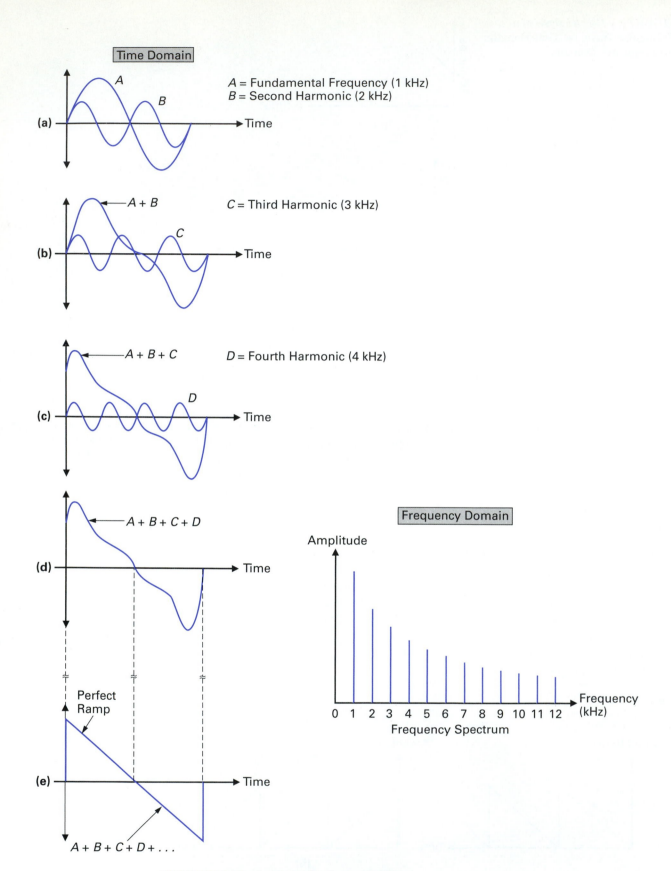

FIGURE 13-38 **Analysis of a Sawtooth Wave.**

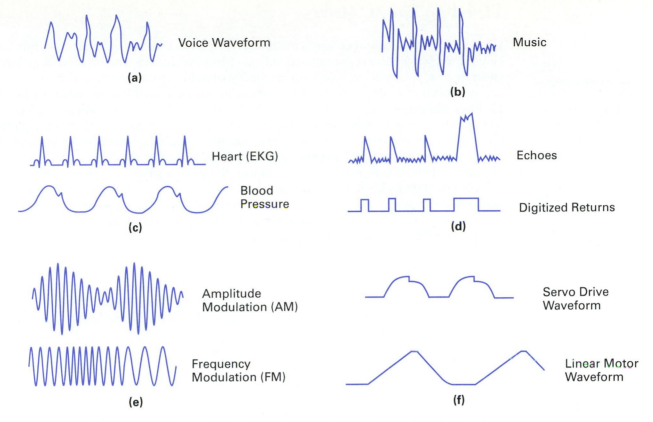

FIGURE 13-39 Waveforms. **(a)** Telephone Communications. **(b)** Radio Broadcast. **(c)** Medical. **(d)** Radar/Sonar. **(e)** Communications. **(f)** Industrial.

SELF-TEST EVALUATION POINT FOR SECTION 13-3

Now that you have completed this section, you should be able to:

- ■ **Objective 7.** *Explain the different characteristics of the five basic wave shapes.*

- ■ **Objective 8.** *Describe fundamental and harmonic frequencies.*

Use the following questions to test your understanding of Section 13-3.

1. Sketch the following waveforms:

    a. Sine wave
    b. Square wave
    c. Rectangular wave
    d. Triangular wave
    e. Sawtooth wave

2. An oscilloscope gives a _____ -domain representation of a periodic wave, whereas a spectrum analyzer gives a _____ -domain representation.

3. What are the wavelength formulas for sound waves and electromagnetic waves, and why are they different?

4. A square wave is composed of an infinite number of _____ harmonics.

13-4   MEASURING AC SIGNALS

As a technician or engineer, you are going to be required to diagnose and repair a problem in the shortest amount of time possible. To make this fault-finding and repair process more efficient, you can make use of certain pieces of test equipment.

13-4-1   *The AC Meter*

Rectifier

A device that converts alternating current into a unidirectional or dc current.

Up to this point we have seen how a multimeter, like the one shown in Figure 13-40(a), can be used to measure direct current and voltage. Most multimeters can be used to measure either dc or ac. When the technician or engineer wishes to measure ac, the ac current or voltage is converted to dc internally by a circuit known as a **rectifier,** as shown in Figure 13-40(b), before passing on to the meter. The dc produced by the rectification process is in fact pulsating, as shown in the waveforms in Figure 13-40(b), so the current through the meter is a series of pulses rising from zero to maximum (peak) and from maximum back to zero. Frequencies below 10 Hz (lower frequency limit) will cause the digits on a digital multimeter to continually increase in value and then decrease in value as the meter follows the pulsating dc. This makes it difficult to read the meter. This effect will also occur with an analog multimeter, which, from 10 Hz to approximately 2 kHz, will not be able to follow the fluctuation, causing the needle to remain in a position equal to the average value of the pulsating dc from the rectifier (0.637 of peak). Most meters are normally calibrated internally to indicate rms values (0.707 of peak) rather than average values, because this effective value is most commonly used when expressing ac voltage or current. The upper frequency limit of the ac meter is approximately 2 to 8 kHz. Beyond this limit the meter becomes progressively inaccurate due to the reactance of the capacitance in the rectifier. This reactance, which will be discussed later, will result in inaccurate indications due to the change in opposition at different ac input frequencies.

13-4-2   *The Oscilloscope*

Oscilloscope

Instrument used to view signal amplitude, frequency, and shape at different points throughout a circuit.

Figure 13–41 illustrates a typical **oscilloscope** (sometimes abbreviated to *scope*), which is used primarily to display the shape and spacing of electrical signals. The oscilloscope displays the actual sine, square, rectangular, triangular, or sawtooth wave shape that is occurring at any point in a circuit. This display is made on a cathode-ray tube (CRT), which is also used in television sets and computers to display video information. From the display on the CRT, we can measure or calculate time, frequency, and amplitude characteristics such as rms, average, peak, and peak-to-peak.

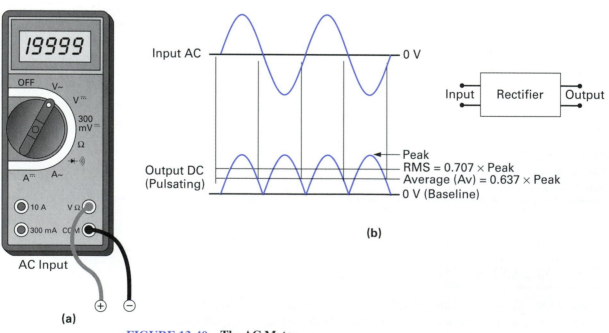

FIGURE 13-40   The AC Meter.

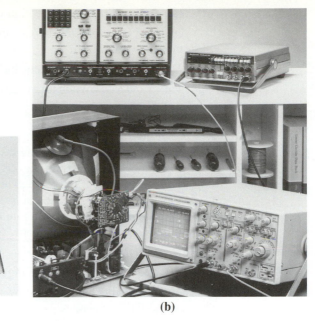

(a)　　　　　　　　(b)

FIGURE 13-41　Typical Oscilloscope. (a) Oscilloscope. (b) Localizing the Faulty Block or Board.

Controls

Oscilloscopes come with a wide variety of features and functions, but the basic operational features are almost identical. Figure 13-42 illustrates the front panel of a typical oscilloscope. Some of these controls are difficult to understand without practice and experience, so practical experimentation is essential if you hope to gain a clear understanding of how to operate an oscilloscope.

TIME LINE

The father of television, Vladymir Zworykin (1889–1982), developed the first television picture tube, called the kinescope, in 1920.

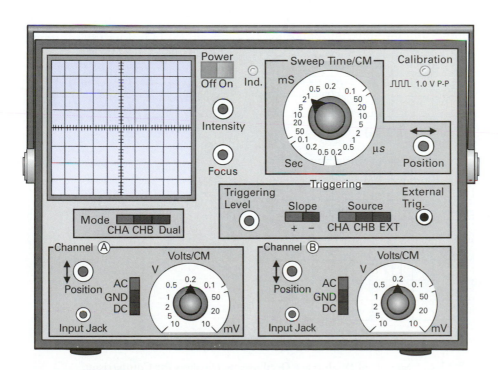

FIGURE 13-42　Oscilloscope Controls.

(a) *General controls* (see Figure 13-42)

Intensity control: Controls the brightness of the trace, which is the pattern produced on the screen of a CRT.

Focus control: Used to focus the trace.

Power OFF/ON: Switch turns on oscilloscope, and indicator shows when oscilloscope is turned on.

Some oscilloscopes have the ability to display more than one pattern or trace on the CRT screen, as seen by the examples in Figure 13-43. A dual-trace oscilloscope can produce two traces or patterns on the CRT screen at the same time, whereas a single-trace oscilloscope can trace out only one pattern on the screen. The dual-trace oscilloscope is very useful, as it allows us to make comparisons between the phase, amplitude, shape, and timing of two signals from two separate test points. One signal or waveform is applied to the channel A input of the oscilloscope, while the other waveform is applied to the channel *B* input.

(b) *Channel selection* (see Figure 13-42)

Mode switch: This switch allows us to select which channel input should be displayed on the CRT screen.

CHA: The input arriving at channel A's jack is displayed on the screen as a single trace.

CHB: The input arriving at channel B's jack is displayed on the screen as a single trace.

Dual: The inputs arriving at jacks A and B are both displayed on the screen, as a dual trace.

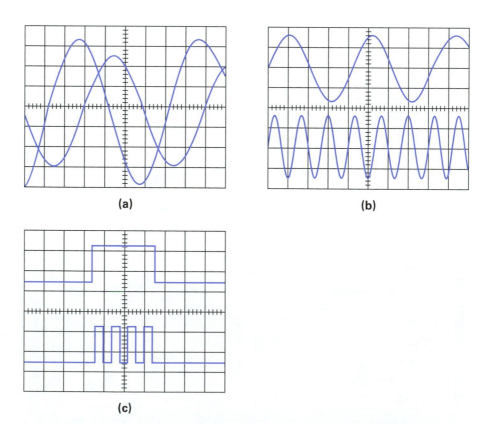

(a)

(b)

(c)

FIGURE 13-43 **Sample of Dual-Trace Oscilloscope Displays for Comparison.**

(c) *Calibration output* This output connection provides a point where a fixed 1 V peak-to-peak square-wave signal can be obtained at a frequency of 1 kHz. This signal is normally fed into either channel A or B's input to test probes and the oscilloscope operation.

(d) *Channel **A** and **B** horizontal controls*

↔*Position control:* This control moves the position of the one (single-trace) or two (dual-trace) waveforms horizontally (left or right) on the CRT screen.

Sweep Time/CM switch: The oscilloscope contains circuits that produce a beam of light that is swept continuously from the left to the right of the CRT screen. When no input signal is applied, this sweep produces a straight horizontal line in the center of the screen. When an input signal is present, this horizontal sweep is influenced by the input signal, which moves it up and down to produce a pattern on the CRT screen that is the same as the input pattern (sine, square, sawtooth, and so on). This sweep time/cm switch selects the speed of the sweep from left to right, and it can be either fast (0.2 microseconds per centimeter; 0.2 μs/cm) or slow (0.5 seconds per centimeter; 0.5 s/cm). A low-frequency input signal (long cycle time or period) will require a long time setting (0.5 s/cm) so that the sweep can capture and display one or more cycles of the input. A number of settings are available, with lower time settings displaying fewer cycles, and higher time settings showing more cycles of an input.

(e) *Triggering controls* These provide the internal timing control between the sweep across the screen and the input waveform.

Triggering Level control: This determines the point where the sweep starts.

Slope switch (+): Sweep is triggered on positive-going slope.

(−): Sweep is triggered on negative-going slope.

Source switch, CHA: The input arriving into channel A jack triggers the sweep.

CHB: The input arriving into channel B jack triggers the sweep.

EXT: The signal arriving at the external trigger jack is used to trigger the sweep.

(f) *Channel **A** and **B** vertical controls* The A and B channel controls are identical.

Volts/CM switch: This switch sets the number of volts to be displayed by each major division on the vertical scale of the screen.

↕*Position control:* Moves the trace up or down for easy measurement or viewing.

AC-GND-DC *switch:* In the AC position, a capacitor on the input will pass the ac component entering the input jack but will block any dc components.

In the GND position, the input is grounded (0 V) so that the operator can establish a reference.

In the DC position, both AC and DC components are allowed to pass on to and be displayed on the screen.

Measurements

The oscilloscope is probably the most versatile of test equipment, as it can be used to test:

DC voltage

AC voltage

Waveform duration

Waveform frequency

Waveform shape

(a) Voltage Measurement The screen is divided into 8 vertical and 10 horizontal divisions, as shown in Figure 13-44. This 8 × 10 cm grid is called the *graticule.* Every

Calibration

To determine, by measurement or comparison with a standard, the correct value of each scale reading.

Triggering

Initiation of an action in a circuit which then functions for a predetermined time, for example, the duration of one sweep in a cathode-ray tube.

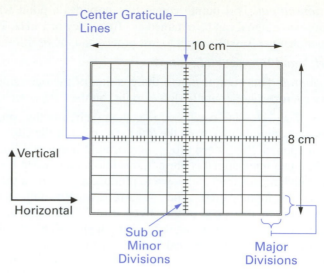

FIGURE 13-44 **Oscilloscope Grid.**

vertical division has a value depending on the setting of the volts/cm control. For example, if the volts/cm control is set to 5 V/div or 5 V/cm, the waveform shown in Figure 13-45(a), which rises up four major divisions, will have a peak positive alternation value of 20 V (4 div × 5 V/div = 20 V).

As another example, look at the positive alternation in Figure 13-45(b). The positive alternation rises up three major divisions and then extends another three subdivisions, which are equal to 1 V each because five subdivisions exist within one major division, and one major division is, in this example, equal to 5 V. The positive alternation shown in Figure 13-45(b) therefore has a peak of three major divisions (3 cm × 5 V/cm = 15 V), plus three subdivisions (3 × 1 V = 3 V), which equals 18 V peak.

In Figure 13-45(c), we have selected the 10 volt/cm position, which means that each major division is equal to 10 V, and each subdivision is equal to 2 V. In this example, the waveform peak will be equal to two major divisions (2 × 10 V = 20 V), plus four subdivisions (4 × 2 V = 8 V), which is equal to 28 V. Once the peak value of a sine wave is known, the peak to peak, average, and rms can be calculated mathematically.

In measurements of a dc voltage with the oscilloscope, the volts/cm is applied in the same way, as shown in Figure 13-46. A positive dc voltage in this situation will cause deflection toward the top of the screen, whereas a negative voltage will cause deflection toward the bottom of the screen.

To determine the dc voltage, count the number of major divisions, and to this add the number of minor divisions. In Figure 13-46, a major division equals 1 V/cm, and therefore, a minor division equals 0.2 V/cm, so the dc voltage being measured is interpreted as +2.6 V.

(b) Time and Frequency Measurement The frequency of an alternating wave, such as that seen in Figure 13-47(a), is inversely proportional to the amount of time it takes to complete one cycle ($f = 1/t$). Consequently, if time can be measured, frequency can be determined.

The time/cm control relates to the horizontal line on the oscilloscope graticule and is used to determine the period of a cycle so that frequency can be calculated. For example, in Figure 8-62(b), a cycle lasts five major horizontal divisions, and since the 20 μs/division setting has been selected, the period of the cycle will equal 5 × 20 μs/division = 100 μs. If the period is equal to 100 μs, the frequency of the waveform will be equal to $f = 1/t = 1/100$ μs = 10 kHz.

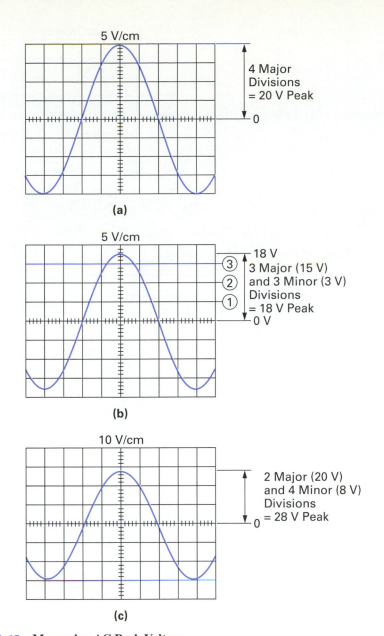

(a)

5 V/cm

4 Major Divisions = 20 V Peak

0

(b)

5 V/cm

18 V

③ 3 Major (15 V)
② and 3 Minor (3 V)
① Divisions = 18 V Peak

0 V

(c)

10 V/cm

2 Major (20 V) and 4 Minor (8 V) Divisions = 28 V Peak

0

FIGURE 13-45   **Measuring AC Peak Voltage.**

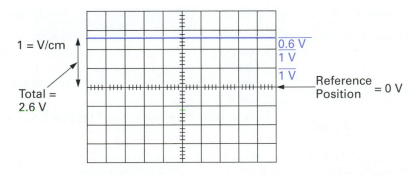

1 = V/cm

Total = 2.6 V

0.6 V
1 V
1 V

Reference Position = 0 V

FIGURE 13-46   **Measuring DC Voltage.**

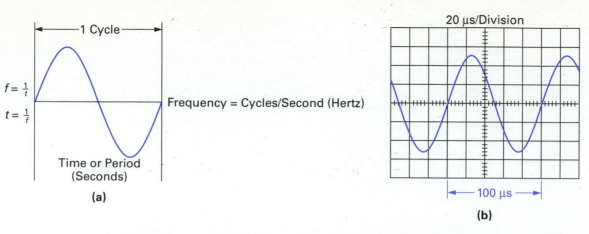

FIGURE 13-47 Time and Frequency Measurement.

■ **INTEGRATED MATH APPLICATION:**

A complete sine-wave cycle occupies four horizontal divisions and four vertical divisions from peak to peak. If the oscilloscope is set on the 20 ms/cm and 500 mV/cm, calculate:

  a. $V_{\text{p-p}}$

  b. t

  c. f

  d. V_{p}

  e. V_{avg} of peak

  f. V_{rms}

■ *Solution:*

> 4 horizontal divisions × 20 ms/div = 80 ms
> 4 vertical divisions × 500 mV/div = 2 V

  a. $V_{\text{p-p}} = 2\text{ V}$

  b. $t = 80\text{ ms}$

  c. $f = \dfrac{1}{t} = 12.5\text{ Hz}$

  d. $V_{\text{p}} = 0.5 \times V_{\text{p-p}} = 1\text{ V}$

  e. $V_{\text{avg}} = 0.637 \times V_{\text{p}} = 0.637\text{ V}$

  f. $V_{\text{rms}} = 0.707 \times V_{\text{p}} = 0.707\text{ V}$

SELF-TEST EVALUATION POINT FOR SECTION 13-4

Use the following questions to test your understanding of Section 13-4.

1. The _____ is used primarily to display the shape and spacing of electrical signals.

2. Name the device within the oscilloscope on which the waveforms can be seen.

3. On the 20 μs/div time setting, a full cycle occupies four major divisions. What is the waveform's frequency and period?

4. On the 2 V/cm voltage setting, the waveform swings up and down two major divisions (a total of 4 cm of vertical swing). Calculate the waveform's peak and peak-to-peak voltage.

5. What is/are the advantage(s) of the dual-trace oscilloscope?

13-5   DECIBELS

In Chapter 7 we discussed how any number can be expressed as a power of 10 and that the exponent of this power of 10 is called the common logarithm. Once a number has been converted to its power-of-10 equivalent, certain mathematical operations, such as multiplication and division, are simplified. Today multiplication, division, and other mathematical operations such as raising a number to some power or extracting square roots, can be accomplished more easily with a calculator. The use of logarithms to simplify problem solving has dramatically decreased. Logarithms, however, are still used in a variety of applications in most of the sciences. In electronics, one of the primary uses of logarithms is in power-to-decibel conversions, a topic that will be discussed in the following section.

13-5-1   *The Human Ear's Response to Sound*

It has been proven through experimentation that the human ear responds logarithmically to changes in sound level. To explain this in more detail, refer to the illustration in Figure 13-48(a), which shows a sine-wave generator connected to a speaker. As the volume control of the generator is increased the power output of the speaker is gradually increased from 1 W to 100 W, and the listener is asked to indicate the points at which a change in volume is detected. These points are listed in the table in Figure 13-48(b). Studying this table, you can see that the power has to be increased from its start point of 1 W to approximately 1.26 W before the ear can detect the increase in volume. As the power is further increased a change in volume is detected at 1.58 W, 2 W, 2.5 W, 3.17 W, and so on, as shown in the table. This table demonstrates an important point: *The human ear is not sensitive to the amount of change that occurs but rather to the ratio of change that occurs.*

In other words, at a low-power level the amount of power needed to sense a change in sound is small (between 1 W and 1.26 W there is an increase of 0.26 W), whereas at a high-power level, the amount of power needed to sense a change in sound is high (between 79.4 W and 100 W there is an increase of 20.6 W). Although the amount of change between any two points is not the same, the ratio of change between any two powers is the same:

$$\frac{1.26}{1} = 1.26 \quad \frac{100}{79.4} = 1.26$$

13-5-2   *The Bel and the Decibel*

Studying the table in Figure 13-48(b), you can see that the human ear is rather slow to respond to increased stimulation. For example, you can see that as the sound is increased in intensity to twice its original value (1 W to 2 W), it is not heard by the listener as twice as loud. In fact, the listener seems to hear only a few small changes in loudness. To make the sound twice as loud, the power must be increased by 10 times (1 W to 10 W), and to make the sound three times as loud, the power must be increased by 100 times (1 W to 100 W). To connect this point to something we all understand, let us imagine that you are about to purchase an audio amplifier for your home music system, and you cannot decide whether to buy a 50 W amplifier or spend twice as much and buy a 100 W amplifier. Although the 100 W amplifier delivers twice the output power, your ears will perceive this as only a small difference in volume. For your ears to detect a significant difference in volume, you would need an amplifier that could deliver 10 times the power. Only a 500 W amplifier will sound as though you are getting twice your money's worth, compared with a 50 W amplifier. This relationship between detected sound and power output is logarithmic because the sound detected goes up in multiples of 10. Because the human ear responds to sound intensity in this way, it is convenient for us to measure sound intensity in a

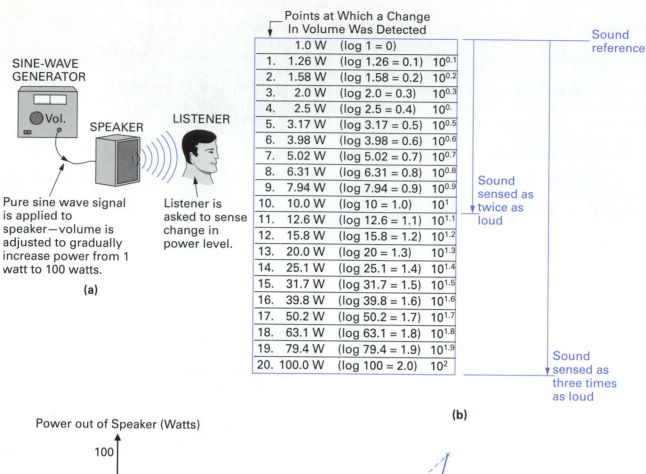

(a)

Pure sine wave signal is applied to speaker—volume is adjusted to gradually increase power from 1 watt to 100 watts.

SINE-WAVE GENERATOR

Vol.

SPEAKER

LISTENER

Listener is asked to sense change in power level.

Points at Which a Change In Volume Was Detected

	1.0 W	(log 1 = 0)	
1.	1.26 W	(log 1.26 = 0.1)	$10^{0.1}$
2.	1.58 W	(log 1.58 = 0.2)	$10^{0.2}$
3.	2.0 W	(log 2.0 = 0.3)	$10^{0.3}$
4.	2.5 W	(log 2.5 = 0.4)	$10^{0.}$
5.	3.17 W	(log 3.17 = 0.5)	$10^{0.5}$
6.	3.98 W	(log 3.98 = 0.6)	$10^{0.6}$
7.	5.02 W	(log 5.02 = 0.7)	$10^{0.7}$
8.	6.31 W	(log 6.31 = 0.8)	$10^{0.8}$
9.	7.94 W	(log 7.94 = 0.9)	$10^{0.9}$
10.	10.0 W	(log 10 = 1.0)	10^{1}
11.	12.6 W	(log 12.6 = 1.1)	$10^{1.1}$
12.	15.8 W	(log 15.8 = 1.2)	$10^{1.2}$
13.	20.0 W	(log 20 = 1.3)	$10^{1.3}$
14.	25.1 W	(log 25.1 = 1.4)	$10^{1.4}$
15.	31.7 W	(log 31.7 = 1.5)	$10^{1.5}$
16.	39.8 W	(log 39.8 = 1.6)	$10^{1.6}$
17.	50.2 W	(log 50.2 = 1.7)	$10^{1.7}$
18.	63.1 W	(log 63.1 = 1.8)	$10^{1.8}$
19.	79.4 W	(log 79.4 = 1.9)	$10^{1.9}$
20.	100.0 W	(log 100 = 2.0)	10^{2}

Sound reference

Sound sensed as twice as loud

Sound sensed as three times as loud

(b)

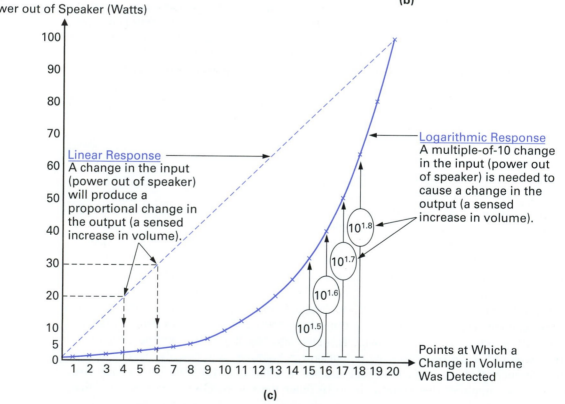

Power out of Speaker (Watts)

Linear Response A change in the input (power out of speaker) will produce a proportional change in the output (a sensed increase in volume).

Logarithmic Response A multiple-of-10 change in the input (power out of speaker) is needed to cause a change in the output (a sensed increase in volume).

$10^{1.5}$ $10^{1.6}$ $10^{1.7}$ $10^{1.8}$

Points at Which a Change in Volume Was Detected

(c)

FIGURE 13-48 The Human Ear's Logarithmic Response to Changes in Sound Level.

logarithmic unit, which is a unit that goes up in equal steps for every 10-fold increase in intensity. This unit is called the **bel**, in honor of the great American scientist Alexander Graham Bell. The bel was originally used to indicate the amount of signal loss in 2 miles of telephone wire. When dealing with small electronic circuits like amplifiers, we need to use a smaller unit of measure. A **decibel (dB)** is equal to $^1/_{10}$th of a bel (1×10^{-1} bel). To understand the relationship between bel and decibel, consider

$$0.1 \text{ amp} = 1 \text{ deciamp } (1 \times 10^{-1} \text{ A, or } 1 \text{ dA}) = 10 \text{ centiamps}$$
$$(10 \times 10^{-2} \text{ A}) = 100 \text{ milliamps } (100 \times 10^{-3} \text{ A})$$

The same is true for bels.

$$0.1 \text{ bel} = 1 \text{ decibel } (1 \times 10^{-1} \text{ B, or } 1 \text{ dB})$$

Referring again to the table in Figure 13-48(b), you can see that it requires a 1 dB (0.1 bel) change in sound level before you are aware that a change in sound intensity has taken place. Figure 13-48(c) shows the logarithmic response curve of the human ear as a plot of the information in the table in Figure 13-48(b). Comparing this logarithmic response with the linear response, which is also shown in Figure 13-48(c), you can see that a linear response increases by 1, 2, 3, 4, and so on, whereas a logarithmic response increases by 0.1, 1, 10, 100, and so on.

To show how the decibel is used to indicate the level of sound, here are some typical decibel ratings.

Whisper	0 dB
Normal conversation	45 dB
Industrial equipment	90 dB
Very loud music	110 dB
Threshold of pain	130 dB

13-5-3 *Power Gain or Loss in Decibels*

By definition the bel is the unit of sound level and is equal to the logarithm (base 10) of the ratio of output power to input power. Stated mathematically:

$$10^{\text{bel}} = \frac{P_{\text{out}}}{P_{\text{in}}} \qquad \text{bel} = \log_{10} \frac{P_{\text{out}}}{P_{\text{in}}}$$

In exponential form   In logarithmic form

Power gain in bels is expressed as the logarithm of the ratio of power levels and is equal to

$$A_P(\text{power gain in bels}) = \log_{10} \frac{P_{\text{out}}}{P_{\text{in}}}$$

However, as stated earlier, the bel is too large a unit for our use, and so we will generally use the decibel, which is $^1/_{10}$th of a bel. Therefore, power gain in decibels is equal to

$$A_{P(\text{dB})}(\text{power gain in decibels}) = 10 \times \log_{10} \frac{P_{\text{out}}}{P_{\text{in}}}$$

This formula can be used to convert arithmetic power ratios to logarithmic decibel units.

Watts to Decibels   Let us now apply this formula to a few examples to see how the decibel can give us an indication of power gain.

Logarithmic Unit

A unit that goes up in equal steps for every 10-fold increase in intensity.

Bel

The logarithmic unit of sound.

Decibel (dB)

One tenth of a bel (1×10^{-1} bel). The logarithmic unit for the difference in power at the output of an amplifier compared with the input.

Calculate the power gain in decibels of an amplifier that has an input of 10 mW and an output of 1 W.

■ *Solution:*

$$A_{P(\text{dB})} = 10 \times \log \frac{P_{\text{out}}}{P_{\text{in}}} = 10 \times \log \frac{1000 \text{ mW}}{10 \text{ mW}} = 10 \times \log 100 = 10 \times 2 = +20 \text{ dB}$$

Remember that a power gain of 20 dB does not mean that the output of the amplifier is 20 times greater than the input. From the values given, we can see that 1 W (or 1000 mW) is 100 times greater than 10 mW. To convert the decibel power gain value ($A_{P(\text{dB})}$) to a normal power gain value (A_P), we simply reorganize the original power formula as follows:

$$A_{P(\text{dB})} = 10 \times \log \frac{P_{\text{out}}}{P_{\text{in}}} = 10 \times \log A_P$$

To solve for A_P, divide both sides by 10

$$\frac{A_{P(\text{dB})}}{10} = \frac{10 \times \log A_P}{10}$$

and take the antilog of both sides.

$$\text{antilog} \frac{A_{P(\text{dB})}}{10} = A_P$$

$$\boxed{A_P = \text{antilog} \frac{A_{P(\text{dB})}}{10}}$$

For the previous example,

$$A_P = \text{antilog} \frac{A_{P(\text{dB})}}{10} = \text{antilog} \frac{+20 \text{ dB}}{10} = \text{antilog } 2 = 100$$

■ INTEGRATED MATH APPLICATION:

A cable has an input of 1.5 kW and an output of 1.2 kW. Calculate the power loss in decibels.

■ *Solution:*

$$A_{P(\text{dB})} = 10 \times \log \frac{P_{\text{out}}}{P_{\text{in}}} = 10 \times \log \frac{1.2 \text{ kW}}{1.5 \text{ kW}} = 10 \times \log 0.8 = 10 \times 0.0969 = -1 \text{ dB}$$

Once again, a -1 dB loss does not mean that the output from the cable is equal to the input minus 1. To convert this loss in dB to a normal gain figure, simply reverse the decibel formula as before.

$$A_P = \text{antilog} \frac{A_{P(\text{dB})}}{10} = \text{antilog} \frac{-1 \text{ dB}}{10} = \text{antilog } (-0.1) = 0.79$$

The output from the cable is equal to 0.79, or 79%, of the input.

INTEGRATED MATH APPLICATION:

Using the following table, calculate the total decibel loss if 38 ft of RG 58 coaxial cable is used to connect a factory's computer controller to a manufacturing robot.

RG/U TYPE COAXIAL CABLE	IMPEDANCE (Ω)	DIAMETER (IN.)	dB/100 FT
RG8A/U	52	0.405	9.0
RG58C/U	50	0.195	17.5
RG108A/U	78	0.235	26.2
RG179B/U	75	0.100	25.0
RG188A/U	50	0.110	30.0
RG196A/U	50	0.08	45.0
RG213/U	50	0.405	9.0

■ *Solution:*

The RG 58 coaxial cable has a loss of 17.5 dB per 100 ft of cable. For 38 ft, the loss will be

$$\text{Cable loss} = \frac{38}{100} \times 17.5 = 38\% \text{ of } 17.5 \text{ dB} = 6.65 \text{ dB}$$

The previous examples illustrate an important point regarding decibels. If there is a power gain (the power out is greater than the power in), the decibel value is positive. On the other hand, if there is a power loss (the power out is less than the power in), the decibel value is negative. Table 13-2 lists some of the more common decibel power gain and loss ratios.

Quickly Converting between Power and Decibels   By applying the following two relationships, you can quickly convert power ratios to decibels and decibel values to power ratios.

1. An arithmetic power ratio of 2:1 is equivalent to +3 dB, which means if power is doubled, the ratio increases by +3 dB. Using a 2:1 power ratio and +3 dB as a base, we can say that a 4:1 power ratio (2 × 2) is equal to +6 dB (3 dB + 3 dB), an 8:1 ratio (2 × 2 × 2) is equal to +9 dB (3 dB + 3 dB + 3 dB), and so on.

   Similarly, a power ratio of 1:2 is equivalent to −3 dB, which means if power is halved, the ratio is decreased by −3 dB. Using a 1:2 power ratio and −3 dB as a base, we can say that a 6 dB loss (−6 dB) is equivalent to one-fourth ($\frac{1}{4}$) the power, a 9 dB loss (−9 dB) is equal to one-eighth the power, and so on.

TABLE 13-2   Common Decibel Power Gain and Loss Ratios

GAIN (+dB)	LOSS (−dB)
+3 dB = × 2	−3 dB = ÷ 2 or 1/2
+6 dB = × 4	−6 dB = ÷ 4 or 1/4
+9 dB = × 8	−9 dB = ÷ 8
+10 dB = × 10	−10 dB = ÷ 10
+12 dB = × 16	−12 dB = ÷ 16
+15 dB = × 32	−15 dB = ÷ 32
+20 dB = × 100	−20 dB = ÷ 100
+30 dB = × 1,000	−30 dB = ÷ 1,000
+40 dB = × 10,000	−40 dB = ÷ 10,000
+50 dB = × 100,000	−50 dB = ÷ 100,000

2. A power ratio of 10:1 is equivalent to $+10$ dB, which means if power is increased by 10, the ratio increases by 10 dB. Using a 10:1 ratio and 10 dB as a base, we can say that a 100:1 power ratio (10×10) is equal to $+20$ dB (10 dB + 10 dB), a 1000:1 power ratio ($10 \times 10 \times 10$) is equal to $+30$ dB (10 dB + 10 dB + 10 dB), and so on.

Similarly, a power ratio of 1:10 is equivalent to -10 dB, which means if power is reduced to one-tenth, the ratio is decreased by -10 dB. Using a 1:10 power ratio and -10 dB as a base, we can say that a 20 dB loss (-20 dB) is equivalent to one-hundredth the power, a 30 dB loss (-30 dB) is equal to one-thousandth ($\frac{1}{1000}$) the power, and so on.

Multiple-Stage Gain   Until now we have been dealing with only a single device. How would we calculate the total gain or loss of a system that contains several stages? If the gain of several stages is given as an arithmetic power ratio, as seen in Figure 13-49(a), the total gain is equal to the product of all of the stages. Therefore, the total gain in this system will be

$$A_{\text{total}} = A_1 \times A_2 \times A_3 \times \cdots$$

■   **INTEGRATED MATH APPLICATION:**

Calculate the total gain of the amplifier in Figure 13-49(a) and its final output power.

■   *Solution:*

$$A_{\text{total}} = A_1 \times A_2 \times A_3 \times A_4 = 5 \times 10 \times 10 \times 10 = 5000$$

Once we know the multistage gain of the amplifier, we can calculate the output from this four-stage amplifier because we know that the output power is 5000 times larger than the input power. An input of 10 mW, therefore, would produce an output of

$$P_{\text{out}} = A_P \times P_{\text{in}} = 5000 \times 10 \text{ mW} = 50 \text{ W}$$

The next question is: How do we deal with a multiple-stage gain or loss when the ratio of change for each stage is given in decibels? To answer this question, let us use an example.

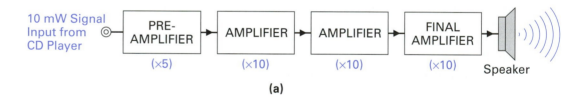

(a)

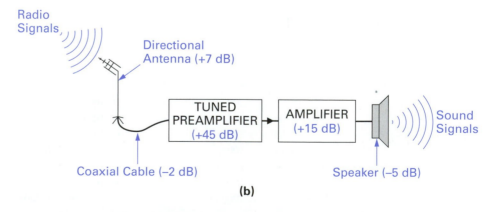

(b)

FIGURE 13-49   **Calculating Multiple Stage Gain.**

■ **INTEGRATED MATH APPLICATION:**

Calculate the gain of the radio receiver shown in Figure 13-49(b).

■ *Solution:*

Because decibels are logarithmic, the total gain or loss of the circuit shown in Figure 13-49(b) can be calculated by simply adding all the individual dB values. The directional antenna captures a signal of +7 dB, there is a −2 dB loss in the coaxial cable between the antenna and the tuned preamplifier, the preamplifier has a gain of +45 dB, the amplifier a gain of 15 dB, and the speaker has a 5 dB loss. The radio receiver gain will be

$$A_{\text{total}} = (+7 \text{ dB}) + (-2 \text{ dB}) + (+45 \text{ dB}) + (+15 \text{ dB}) + (-5 \text{ dB}) = +60 \text{ dB}$$

When the gain of a multistage circuit or system is measured in decibels the total gain or loss of the amplifier can be calculated by simply adding all the individual decibel values.

$$A_{\text{total}} = A_{1(\text{dB})} + A_{2(\text{dB})} + A_{3(\text{dB})} + \cdots$$

Scaling Power in dBm In the preceding examples, both the input power and the output power were given. In some applications, the power at a circuit test point is given with respect to a reference level of power. For example, cable and telecommunication companies have established a standard of 1 mW as a reference power. With this **decibels relative to 1 milliwatt** or **dBm** system, power levels below 1 mW are indicated as −dBm values, and power levels above 1 mW are shown as +dBm values. The formula for calculating dBm is

$$A_{P(\text{dBm})} = \log \frac{P_{\text{out}}}{1 \text{ mW}}$$

dBm
Decibels relative to 1 milliwatt.

■ **INTEGRATED MATH APPLICATION:**

An amplifier has an output of 1.5 W; what is this value in dBm?

■ *Solution:*

$$A_{P(\text{dBm})} = \log \frac{P_{\text{out}}}{1 \text{ mW}} = \log \frac{1.5 \text{ W}}{1 \text{ mW}} = \log 1500 = 3.18 \text{ dBm}$$

Decibels to Watts Transposing the original decibel power formula, we can derive formulas for calculating power out and power in.

$$A_{P(\text{dB})} = 10 \times \log \frac{P_{\text{out}}}{P_{\text{in}}} \quad \text{(original dB formula)}$$

$$P_{\text{out}} = P_{\text{in}} \times \text{antilog} \frac{\text{dB}}{10}$$

$$P_{\text{in}} = \frac{P_{\text{out}}}{\text{antilog} \dfrac{\text{dB}}{10}}$$

■ **INTEGRATED MATH APPLICATION:**

Calculate the power out of a 35 dB amplifier for an input of 4 mW.

■ *Solution:*

$$P_{out} = P_{in} \times antilog \frac{dB}{10} = 4 \text{ mW} \times antilog \frac{35}{10} = 12.65 \text{ W}$$

■ **INTEGRATED MATH APPLICATION:**

Calculate the input power needed for a 30 dB amplifier to produce a 2 W output.

■ *Solution:*

$$P_{in} = \frac{P_{out}}{antilog \dfrac{dB}{10}} = \frac{2 \text{ W}}{antilog \dfrac{30}{10}} = \frac{2 \text{ W}}{antilog\ 3} = 2 \text{ mW}$$

13-5-4   *Voltage Gain or Loss in Decibels*

A voltage gain or loss can also be expressed in decibels by substituting V^2/R for power in the decibel power formula.

$$A_{P(dB)} = 10 \times log \frac{P_{out}}{P_{in}}$$

Since

$$P_{out} = \frac{V_{out}^2}{R_{out}} \quad \text{and} \quad P_{in} = \frac{V_{in}^2}{R_{in}}$$

we can write the formula for voltage gain in decibels:

$$A_{V(dB)} = 10 \times log \left(\frac{\dfrac{V_{out}^2}{R_{out}}}{\dfrac{V_{in}^2}{R_{in}}} \right)$$

$$= 10 \times log \left(\frac{V_{out}^2}{V_{in}^2} \times \frac{R_{in}}{R_{out}} \right)$$

$$= 10 \times log \frac{V_{out}^2}{V_{in}^2} + 10 \times log \frac{R_{in}}{R_{out}}$$

If $R_{out} = R_{in}$,

$$= 10 \times log \left(\frac{V_{out}}{V_{in}} \right)^2 + (10 \times log\ 1)$$

$$= 20 \times log \frac{V_{out}}{V_{in}} + 0 \quad \text{therefore,} \quad A_{V(dB)} = 20 \times log \frac{V_{out}}{V_{in}}$$

$$A_{V(dB)} = 20 \times log \frac{V_{out}}{V_{in}}$$

CHAPTER 13 / ALTERNATING CURRENT (AC)

3. The advantage(s) of ac over dc from a power distribution point of view is/are:
   a. Generators can supply more power than batteries
   b. AC can be transformed to a high or low voltage easily, minimizing power loss
   c. AC can easily be converted into dc
   d. All the above
   e. Only (a) and (c)

4. The approximate voltage appearing on long-distance transmission lines in the ac distribution system is:
   a. 250 V
   b. 2500 V
   c. 500,000 V
   d. 250,000 V

5. The most common type of alternating wave shape is the:
   a. Square wave
   b. Sine wave
   c. Rectangular wave
   d. Triangular wave

6. _____ equipment manages the flow of information.
   a. Electronic
   b. Electrical
   c. Discrete
   d. Integrated

7. _____ equipment manages the flow of power.
   a. Electronic
   b. Electrical
   c. Discrete
   d. Integrated

8. The peak-to-peak value of a sine wave is equal to:
   a. Twice the rms value
   b. 0.707 times the rms value
   c. Twice the peak value
   d. 1.14 × the average value

9. The rms value of a sine wave is also known as the:
   a. Effective value
   b. Average value
   c. Peak value
   d. All the above

10. The peak value of a 115 V (rms) sine wave is:
    a. 115 V
    b. 230 V
    c. 162.7 V
    d. Two of the above could be true.

11. The mathematical average value of a sine wave cycle is:
    a. 0.637 × peak
    b. 0.707 × peak
    c. 1.414 × rms
    d. Zero

12. The frequency of a sine wave is equal to the reciprocal of _____.
    a. The period
    b. One cycle time
    c. One alternation
    d. Both (a) and (b)
    e. None of the above

13. What is the period of a 1 MHz sine wave?
    a. 1 ms
    b. One millionth of a second
    c. 10 ms
    d. 100 μs

14. The sine of 90° is:
    a. 0
    b. 0.5
    c. 1
    d. Any of the above

15. What is the frequency of a sine wave that has a cycle time of 1 ms?
    a. 1 MHz
    b. 1 kHz
    c. 200 m
    d. 10 kHz

16. The pulse width (P_w) is the time between the _____ points on the positive and negative edges of a pulse.
    a. 10%
    b. 90%
    c. 50%
    d. All the above

17. The duty cycle is the ratio of _____ to period.
    a. Peak
    b. Average power
    c. Pulse length
    d. Both (a) and (c)

18. With a pulse waveform, PRF can be calculated by taking the reciprocal of:
    a. The duty cycle
    b. PRT
    c. P_d
    d. P_l

19. A sound wave exists in _____ and travels at approximately _____.
    a. Space, 1130 ft/s
    b. Wires, 186,282.397 miles/s
    c. Air, 3×10^6 m/s
    d. None of the above

20. Electrical and electromagnetic waves travel at a speed of:
    a. 186,000 miles/s
    b. 3×10^8 meters/s
    c. 162,000 nautical miles/s
    d. All the above

21. If an amplifier has an input of 25 mW and an output of 25 W, its gain is dB will be
    a. +30 dB
    b. 1000 dB
    c. −20 dB
    d. +10 dB

22. If a cable has a loss of 4 dB, the output from the cable will equal _____ % of the input.
    a. 60.2
    b. 39.8
    c. 70.7
    d. 63.2

23. If the first stage of a multistage amplifier has a voltage gain of 10 and the second stage has a voltage gain of 2, what will be the overall voltage gain?
    a. 12
    b. 6
    c. 20
    d. 8

24. If an amplifier has a gain of 35 dB and the speaker it is driving has an 8 dB loss, what will be the overall gain?
    a. +43 dB
    b. 4.38 dB
    c. 35 dB
    d. +27 dB

25. A common-emitter power amplifier has a gain of 30 dB and an input of 20 mW. What will be the power out of this amplifier?
    a. 20 W
    b. 10 W
    c. 75 W
    d. 100 W

Communication Skill Questions

26. Describe the three advantages that ac has over dc from a power point of view. (13-2-1)

27. Describe briefly the ac power distribution system. (13-2-1)

28. What is the difference between electricity and electronics? (13-4)

29. What are the six basic ac information wave shapes? (13-3)

30. Describe briefly the following terms as they relate to the sine wave: (13-3-1)

a. RMS   d. Average   g. Period
b. Peak   e. The name sine   h. Wavelength
c. Peak to peak   f. Frequency   i. Phase

31. Describe briefly the following terms as they relate to the square wave: (13-3-1)

a. Duty cycle   b. Average

32. Describe briefly the following terms as they relate to the rectangular wave: (13-3-3)

a. PRT   c. Duty cycle
b. PRF   d. Average

33. Briefly describe the meaning of the terms *fundamental frequency* and *harmonics*. (13-3-2)

34. List and describe all the pertinent information relating to the following information carriers: (13-2-2)

a. Sound wave   c. Electromagnetic wave
b. Electrical wave

35. Describe the difference between frequency- and time-domain analysis. (13-3-2)

Practice Problems

36. Calculate the periods of the following sine-wave frequencies:

a. 27 kHz   d. 365 Hz
b. 3.4 MHz   e. 60 Hz
c. 25 Hz   f. 200 kHz

37. Calculate the frequency for each of the following values of time:

a. 16 ms   d. 0.05 s
b. 1 s   e. 200 μs
c. 15 μs   f. 350 ms

38. A 22 V peak sine wave will have the following values:

a. Rms voltage =   c. Peak-to-peak voltage =
b. Average voltage =

39. A 40 mA rms sine wave will have the following values:

a. Peak current =   c. Average current =
b. Peak-to-peak current =

40. How long would it take an electromagnetic wave to travel 60 miles?

41. An 11 kHz rectangular pulse, with a pulse width of 10 μs, will have a duty cycle of _____ %.

42. Calculate the PRT of a 400 kHz pulse waveform.

43. Calculate the average current of the pulse waveform in Question 42 if its peak current is equal to 15 A.

44. What is the duty cycle of a 10 V peak square wave at a frequency of 1 kHz?

45. Considering a fundamental frequency of 1 kHz, calculate the frequency of its:

a. Third harmonic   c. Seventh harmonic
b. Second harmonic

46. If one cycle of a sine wave occupies 4 cm on the oscilloscope horizontal grid and 5 cm from peak to peak on the vertical grid, calculate frequency, period, rms, average, and peak for the following control settings:

a. 0.5 V/cm, 20 μs/cm   c. 50 mV/cm, 0.2 μs/cm
b. 10 V/cm, 10 ms/cm

47. If the volts/cm switch is positioned to 10 V/cm and the waveform extends 3.5 divisions from peak to peak, what is the peak-to-peak value of this wave?

48. If a square wave occupies 5.5 horizontal cm on the 1 μs/cm position, what is its frequency?

49. A 12-ft coaxial cable has an input of 25 W and an output of 23.7 W. What is the cable loss in decibels?

50. Calculate the arithmetic gain or loss of the following:

a. Power gain = 20 dB   c. Current gain = 6 dB
b. Voltage loss = −20 dB   d. Power gain = 3 dB

Web Site Questions

Go to the web site http://www.prenhall.com/cook, select the textbook *Mathematics for Electronics and Computers,* select this chapter, and then follow the instructions when answering the multiple-choice practice problems.

Capacitors

The Turing Enigma

Alan Turing

During the Second World War, the Germans developed a cipher-generating apparatus called "Enigma." This electromechanical teleprinter would scramble messages with several randomly spinning rotors that could be set to a predetermined pattern by the sender. This key and plug pattern was changed three times a day by the Germans and cracking the secrets of Enigma became of the utmost importance to British Intelligence. With this objective in mind, every brilliant professor and eccentric researcher was gathered at a Victorian estate near London called Bletchley Park. They specialized in everything from engineering to literature and were collectively called the Backroom Boys.

By far the strangest and definitely most gifted of the group was an unconventional theoretician from Cambridge University named Alan Turing. He wore rumpled clothes and had a shrill stammer and crowing laugh that aggravated even his closest friends. He had other legendary idiosyncrasies that included setting his watch by sighting on a certain star from a specific spot and then mentally calculating the time of day. He also insisted on wearing his gas mask whenever he was out, not for fear of a gas attack, but simply because it helped his hay fever.

Turing's eccentricities may have been strange but his genius was indisputable. At the age of twenty-six he wrote a paper outlining his "universal machine" that could solve any mathematical or logical problem. The data or, in this case, the intercepted enemy messages could be entered into the machine on paper tape and then compared with known Enigma codes until a match was found.

In 1943 Turing's ideas took shape as the Backroom Boys began developing a machine that used 2,000 vacuum tubes and incorporated five photoelectric readers that could process 25,000 characters per second. It was named "Colossus," and it incorporated the stored program and other ideas from Turing's paper written seven years earlier.

Turing could have gone on to accomplish much more. However, his idiosyncrasies kept getting in the way. He became totally preoccupied with abstract questions concerning machine intelligence. His unconventional personal lifestyle led to his arrest in 1952 and, after a sentence of psychoanalysis, his suicide two years later.

Before joining the Backroom Boys at Bletchley Park, Turing's genus was clearly apparent at Cambridge. How much of a role he played in the development of Colossus is still unknown and remains a secret guarded by the British Official Secrets Act. Turing was never fully recognized for his important role in the development of this innovative machine, except by one of his Bletchley Park colleagues at his funeral who said, "I won't say what Turing did made us win the war, but I daresay we might have lost it without him."

Outline and Objectives

VIGNETTE: THE TURING ENIGMA

14-1 THE UNIT OF CAPACITANCE

Objective 1: State the unit of capacitance and explain how it relates to charge and voltage.

14-2 FACTORS DETERMINING CAPACITANCE

Objective 2: List and explain the factors determining capacitance.

14-2-1 Plate Area (A)

14-2-2 Distance between the Plates (d)

14-2-3 Dielectric Constant

14-2-4 The Capacitance Formula

14-3 CAPACITORS IN COMBINATION

Objective 3: Calculate total capacitance in parallel and series capacitive circuits.

14-3-1 Capacitors in Parallel

14-3-2 Capacitors in Series

14-4 CAPACITIVE TIME CONSTANT

Objective 4: Explain the capacitor time constant as it relates to dc charging and discharging.

14-4-1 DC Charging

14-4-2 DC Discharging

14-5 CAPACITIVE REACTANCE

Objective 5: Define and explain capacitive reactance.

14-6 SERIES *RC* CIRCUIT

Objective 6: Describe impedance, phase angle, power, and power factor as they relate to a series and parallel *RC* circuit.

14-6-1 Vector Diagram

14-6-2 Voltage

14-6-3 Impedance

14-6-4 Phase Angle or Shift (θ)

14-6-5 Power
Purely Resistive Circuit
Purely Capacitive Circuit
Resistive and Capacitive Circuit
Power Factor

14-7 PARALLEL *RC* CIRCUIT

14-7-1 Voltage

14-7-2 Current

14-7-3 Phase Angle

14-7-4 Impedance

14-7-5 Power

MULTIPLE CHOICE QUESTIONS

COMMUNICATION SKILL QUESTIONS

PRACTICE PROBLEMS

WEB SITE QUESTIONS

Capacitor

Device that stores electric energy in the form of an electric field that exists within a dielectric (insulator) between two conducting plates, each of which is connected to a lead. This device was originally called a condenser.

Farad

Unit of capacitance.

14-1 THE UNIT OF CAPACITANCE

Capacitance is the ability of a capacitor to store an electrical charge, and the unit of capacitance is the **farad** (F), named in honor of Michael Faraday's work in 1831 in the field of capacitance. A capacitor with the capacity of 1 farad (1 F) can store 1 coulomb of electrical charge (6.24×10^{18} electrons) if 1 volt is applied across the capacitor's plates, as seen in Figure 14-1.

A 1 F capacitor is a very large value and not frequently found in electronic equipment. Most values of capacitance found in electronic equipment are in the units between the microfarad ($\mu F = 10^{-6}$) and picofarad ($pF = 10^{-12}$). A microfarad is 1 millionth of a farad (10^{-6}). So if a 1 F capacitor can store 6.24×10^{18} electrons with 1 V applied, a 1 μF capacitor, which has one-millionth the capacity of a 1 F capacitor, can store only one-millionth of a coulomb, or $(6.24 \times 10^{18}) \times (1 \times 10^{-6}) = 6.24 \times 10^{12}$ electrons when 1 V is applied, as shown in Figure 14-2.

■ **INTEGRATED MATH APPLICATION:**

Convert the following to either microfarads or picofarads (whichever is more appropriate):

   a. 0.00002 F
   b. 0.00000076 F
   c. 0.00047×10^{-7} F

■ *Solution:*

   a. 20 μF
   b. 0.76 μF
   c. 47 pF

Since there is a direct relationship between capacitance, charge, and voltage, there must be a way of expressing this relationship in a formula.

$$\text{Capacitance, } C \text{ (farads)} = \frac{\text{charge, } Q \text{ (coulombs)}}{\text{voltage, } V \text{ (volts)}}$$

where C = capacitance
       Q = charge, in coulombs
       V = voltage, in volts

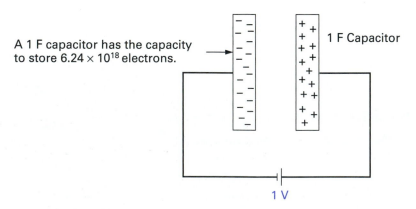

A 1 F capacitor has the capacity to store 6.24×10^{18} electrons.

1 F Capacitor

1 V

FIGURE 14-1   One Farad of Capacitance.

CHAPTER 14 / CAPACITORS

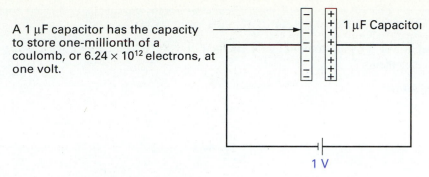

A 1 μF capacitor has the capacity to store one-millionth of a coulomb, or 6.24×10^{12} electrons, at one volt.

1 μF Capacitor

1 V

FIGURE 14-2   **One-Millionth of a Farad.**

By transposition of the formula, we arrive at the following combinations for the same formula:

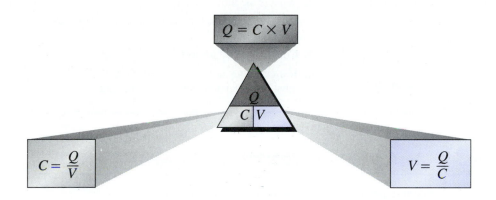

$$Q = C \times V$$

$$C = \frac{Q}{V}$$

$$V = \frac{Q}{C}$$

■ INTEGRATED MATH APPLICATION:

If a capacitor has the capacity to hold 36 C ($36 \times 6.24 \times 10^{18} = 2.25 \times 10^{20}$ electrons) when 12 V is applied across its plates, what is the capacitance of the capacitor?

■ *Solution:*

$$\begin{aligned} C &= \frac{Q}{V} \\ &= \frac{36 \text{ C}}{12 \text{ V}} \\ &= 3 \text{ F} \end{aligned}$$

■ INTEGRATED MATH APPLICATION:

How many electrons can a 3 μF capacitor store when 5 V is applied across it?

■ *Solution:*

$$\begin{aligned} Q &= C \times V \\ &= 3 \text{ } \mu F \times 5 \text{ V} \\ &= 15 \text{ } \mu C \end{aligned}$$

(15 microcoulombs is 15 millionths of a coulomb.) Since 1 C = 6.24×10^{18} electrons, 15 μC = $(15 \times 10^{-6}) \times 6.24 \times 10^{18} = 9.36 \times 10^{13}$ electrons.

If a capacitor of 2 F has stored 42 C of charge (2.63×10^{20} electrons), what is the voltage across the capacitor?

■ *Solution:*

$$V = \frac{Q}{C}$$
$$= \frac{42 \text{ C}}{2 \text{ F}}$$
$$= 21 \text{ V}$$

SELF-TEST EVALUATION POINT FOR SECTION 14-1

Now that you have completed this section, you should be able to:

■ ***Objective 1.*** *State the unit of capacitance and explain how it relates to charge and voltage.*

Use the following questions to test your understanding of Section 14-1.

1. What is the unit of capacitance?
2. State the formula for capacitance in relation to charge and voltage.
3. Convert 30,000 μF to farads.
4. If a capacitor holds 17.5 C of charge when 9 V is applied, what is the capacitance of the capacitor?

14-2   FACTORS DETERMINING CAPACITANCE

The capacitance of a capacitor is determined by three factors:

1. The plate area of the capacitor
2. The distance between the plates
3. The type of dielectric used

Let's now discuss these three factors in more detail, beginning with the plate area.

14-2-1   *Plate Area* (A)

The capacitance of a capacitor is directly proportional to the plate area. This area in square centimeters is the area of only one plate and is calculated by multiplying length by width. This is illustrated in Figure 14-3(a) and (b). In these two examples, the (b) capacitor plate is twice as large as the (a) capacitor plate, and since capacitance is proportional to plate area ($C \propto A$), the capacitor in example (b) will have double the capacity or capacitance of the capacitor in example (a). Since the energy of a charged capacitor is in the electric field between the plates and the plates of the capacitor (b) are double those of (a), there is twice as much area for the electric field to exist, and this doubles the capacitor's capacitance.

14-2-2   *Distance between the Plates* (d)

The distance or separation between the plates is dependent on the thickness of the dielectric used. The capacitance of a capacitor is inversely proportional to this distance between the plates, in that an increase in the distance ($d\uparrow$) causes a decrease in the capacitor's capacitance ($C\downarrow$). In Figure 14-4(a), a large distance between the capacitor plates results in a small capacitance, whereas in Figure 14-4(b) the dielectric thickness and the plate separation are

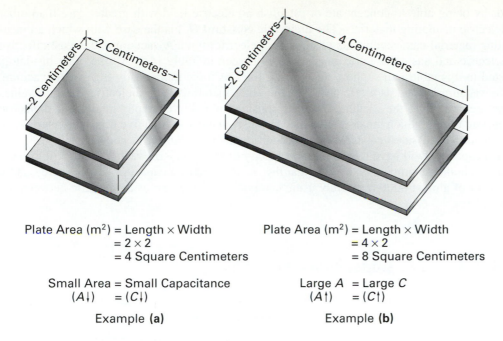

Plate Area (m²) = Length × Width
          = 2 × 2
          = 4 Square Centimeters

Plate Area (m²) = Length × Width
          = 4 × 2
          = 8 Square Centimeters

Small Area = Small Capacitance
  $(A\downarrow)$ = $(C\downarrow)$

Large A = Large C
  $(A\uparrow)$ = $(C\uparrow)$

Example **(a)**

Example **(b)**

FIGURE 14-3 **Capacitance Is Proportional to Plate Area (A).**

half that of capacitor (a). This illustrates also how the capacitance of a capacitor can be doubled, in this case by halving the space between the plates. The gap across which the electric lines of force exist is halved in capacitor (b), and this doubles the strength of the electric field, which consequently doubles capacitance. Simply stated, an electric line of force (Z) in Figure 14-4(a) can be used to produce two electric lines of force (X and Y) in Figure 14-4(b) if the distance is half.

14-2-3   *Dielectric Constant*

The insulating dielectric of a capacitor concentrates the electric lines of force between the two plates. As a result, different dielectric materials can change the capacitance of a capac-

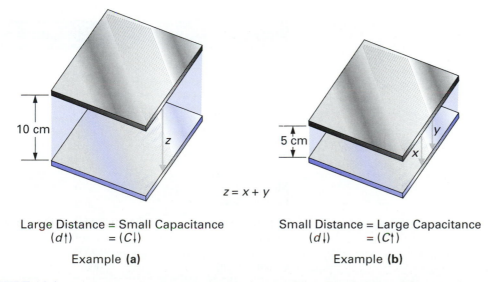

10 cm

z

5 cm

y

x

$z = x + y$

Large Distance = Small Capacitance
  $(d\uparrow)$ = $(C\downarrow)$

Small Distance = Large Capacitance
  $(d\downarrow)$ = $(C\uparrow)$

Example **(a)**

Example **(b)**

FIGURE 14-4 **Capacitance Is Inversely Proportional to Plate Separation or Distance (d).**

itor by being able to concentrate or establish an electric field with greater ease than other dielectric insulating materials. The **dielectric constant** (*K*) is the ease with which an insulating material can establish an electrostatic (electric) field. A vacuum is the least effective dielectric and has a dielectric constant of 1, as seen in Table 14-1. All the other insulators listed in this table will support electrostatic lines of force more easily than will a vacuum. The vacuum is used as a reference. All the other materials have dielectric constant values that are relative to the vacuum dielectric constant of 1. For example, mica has a dielectric constant of 5.0, which means that mica can cause an electric field five times the intensity of a vacuum; and, since capacitance is proportional to the dielectric constant ($C \propto K$), the mica capacitor will have five times the capacity of the same-size vacuum dielectric capacitor. In another example, we can see that the capacitance of a capacitor can be increased by a factor of almost 8000 by merely using ceramic rather than air as a dielectric between the two plates.

14-2-4   *The Capacitance Formula*

Thus plate area, separation, and the dielectric used are the three factors that change the capacitance of a capacitor. The formula that combines these three factors is

$$ C = \frac{(8.85 \times 10^{-12}) \times K \times A}{d} $$

where C = capacitance, in farads (F)
8.85×10^{-12} is a constant
K = dielectric constant
A = plate area, in square meters (m^2)
d = distance between the plates, in meters (m)

This formula summarizes what has been said in relation to capacitance. The capacitance of a capacitor is directly proportional to the dielectric constant (*K*) and the area of the plates (*A*) and is inversely proportional to the dielectric thickness or distance between the plates (*d*).

TABLE 14-1   Dielectric Constants

MATERIAL	DIELECTRIC CONSTANT (*K*)[a]
Vacuum	1.0
Air	1.0006
Teflon	2.0
Wax	2.25
Paper	2.5
Amber	2.65
Rubber	3.0
Oil	4.0
Mica	5.0
Ceramic (low)	6.0
Bakelite	7.0
Glass	7.5
Water	78.0
Ceramic (high)	8000.

[a]Different material compositions can cause different values of *K*.

What is the capacitance of a ceramic capacitor with a 0.3 m² plate area and a dielectric thickness of 0.0003 m?

■ *Solution:*

$$C = \frac{(8.85 \times 10^{-12}) \times K \times A}{d}$$

$$= \frac{(8.85 \times 10^{-12}) \times 6 \times 0.3 \text{ m}^2}{0.0003 \text{ m}}$$

$$= 5.31 \times 10^{-8}$$

$$= 0.0531 \ \mu\text{F}$$

Step	Keypad Entry	Display Response
1.	8 . 8 5 E 1 2 +/−	8.85E-12
2.	×	
3.	6	6
4.	×	5.31E-11
5.	0 . 3	
6.	÷	1.59E-11
7.	0 . 0 0 0 3	
8.	=	5.31E-8

SELF-TEST EVALUATION POINT FOR SECTION 14-2

Now that you have completed this section, you should be able to:

■ *Objective 2.* *List and explain the factors determining capacitance.*

Use the following questions to test your understanding of Section 14-2.

1. List the three variable factors that determine the capacitance of a capacitor.

2. State the formula for capacitance.

3. If a capacitor's plate area is doubled, the capacitance will _____.

4. If a capacitor's dielectric thickness is halved, the capacitance will _____.

14-3 CAPACITORS IN COMBINATION

Like resistors, capacitors can be connected in either series or parallel. As you will see in this section, the rules for determining total capacitance for parallel- and series-connected capacitors are the opposite of those for series- and parallel-connected resistors.

14-3-1 *Capacitors in Parallel*

In Figure 14-5(a), you can see a 2 μF and 4 μF capacitor connected in parallel with each other. Because the top plate of capacitor *A* is connected to the top plate of capacitor *B* with a wire, and a similar situation occurs with the bottom plates, you can see that this is the same as if the top and bottom plates were touching one another, as shown in Figure 14-5(b). When drawn so that the respective plates are touching, the dielectric constant and plate separation is the same as shown in Figure 14-5(a), but now we can easily see that the plate area is actually increased. Consequently, if capacitors are connected in parallel, the effective plate area is increased; and since capacitance is proportional to plate area $[C\uparrow = (8.85 \times 10^{-12}) \times K \times A\uparrow/d\,]$, the capacitance will also increase. Total capacitance is actually calculated by adding the plate areas, so total capacitance is equal to the sum of all the individual capacitances in parallel.

$$C_T = C_1 + C_2 + C_3 + C_4 + \cdots$$

FIGURE 14-5   Capacitors in Parallel.

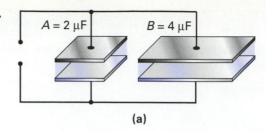

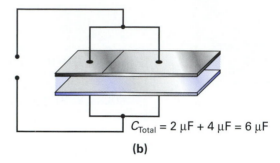

$C_{\text{Total}} = 2\ \mu F + 4\ \mu F = 6\ \mu F$

(b)

■ INTEGRATED MATH APPLICATION:

Determine the total capacitance of the circuit in Figure 14-6(a). What will be the voltage drop across each capacitor?

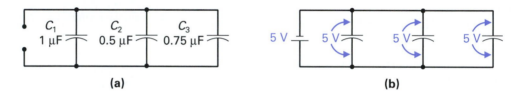

(a)   **(b)**

FIGURE 14-6   Example of Parallel-Connected Capacitors.

■ *Solution:*

$$C_T = C_1 + C_2 + C_3$$
$$= 1\ \mu F + 0.5\ \mu F + 0.75\ \mu F$$
$$= 2.25\ \mu F$$

As with any parallel-connected circuit, the source voltage appears across all the components. If, for example, 5 V is connected to the circuit of Figure 14-6(b), all the capacitors will charge to the same voltage of 5 V because the same voltage always exists across each section of a parallel circuit.

14-3-2   *Capacitors in Series*

In Figure 14-7(a), we have taken the two capacitors of 2 μF and 4 μF and connected them in series. Since the bottom plate of the *A* capacitor is connected to the top plate of the *B* capacitor, they can be redrawn so that they are touching, as shown in Figure 14-7(b).

The top plate of the *A* capacitor is connected to a wire into the circuit, and the bottom plate of *B* is connected to a wire into the circuit. This connection creates two center plates that are isolated from the circuit and can therefore be disregarded, as shown in Figure 14-7(c). The first thing you will notice in this illustration is that the dielectric thickness ($d\uparrow$) has increased, causing a greater separation between the plates. The effec-

FIGURE 14-7   Capacitors in Series.

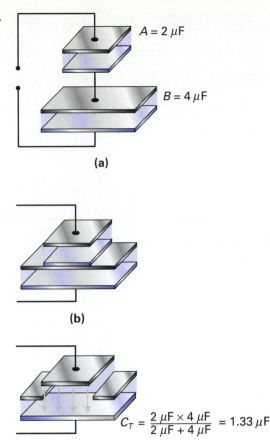

$A = 2\ \mu\text{F}$

$B = 4\ \mu\text{F}$

(a)

(b)

$$C_T = \frac{2\ \mu\text{F} \times 4\ \mu\text{F}}{2\ \mu\text{F} + 4\ \mu\text{F}} = 1.33\ \mu\text{F}$$

(c)

tive plate area of this capacitor has decreased, as it is just the area of the top plate only. Even though the bottom plate extends out farther, the electric field can exist only between the two plates, so the surplus metal of the bottom plate has no metal plate opposite for the electric field to exist.

Consequently, when capacitors are connected in series the effective plate area is decreased ($A\downarrow$) and the dielectric thickness increased ($d\uparrow$), and both these effects result in an overall capacitance decrease ($C\downarrow\downarrow = (8.85 \times 10^{-12}) \times K \times A\downarrow/d\uparrow$).

The plate area is actually decreased to that of the smallest individual capacitance connected in series, which in this example is the plate area of A. If the plate area were the only factor, then capacitance would always equal the smallest capacitor value; however, the dielectric thickness is always equal to the sum of all the capacitor dielectrics, and this factor always causes the total capacitance (C_T) to be less than the smallest individual capacitance when capacitors are connected in series.

The total capacitance of two or more capacitors in series therefore is calculated by using the following formulas: For two capacitors in series,

$$C_T = \frac{C_1 \times C_2}{C_1 + C_2} \quad \text{(product-over-sum formula)}$$

For more than two capacitors in series,

$$C_T = \frac{1}{(1/C_1) + (1/C_2) + (1/C_3) + \cdots} \quad \text{(reciprocal formula)}$$

Determine the total capacitance of the circuit in Figure 14-8.

■ *Solution:*

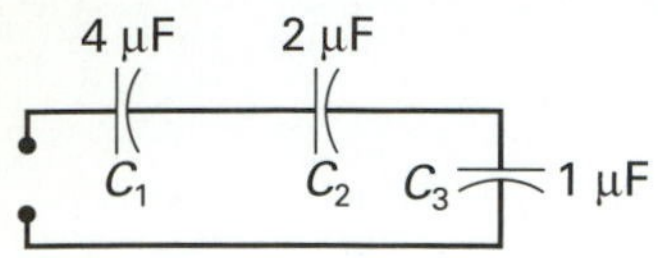

FIGURE 14-8 **Example of Series-Connected Capacitors.**

$$C_T = \frac{1}{(1/C_1) + (1/C_2) + (1/C_3)}$$

$$= \frac{1}{(1/4\ \mu F) + (1/2\ \mu F) + (1/1\ \mu F)}$$

$$= \frac{1}{1.75 \times 10^6} = 5.7143 \times 10^{-7}$$

$$= 0.5714\ \mu F \quad \text{or} \quad 0.6\ \mu F$$

The total capacitance for capacitors in series is calculated in the same way as total resistance when resistors are in parallel.

As with series-connected resistors, the sum of all of the voltage drops across the series-connected capacitors will equal the voltage applied (Kirchhoff's voltage law). With capacitors connected in series, the charged capacitors act as a voltage divider, and therefore the voltage-divider formula can be applied to capacitors in series.

$$V_{cx} = \frac{C_T}{C_x} \times V_T$$

where V_{cx} = voltage across desired capacitor
C_T = total capacitance
C_x = desired capacitor's value
V_T = total supplied voltage

■ **INTEGRATED MATH APPLICATION:**

Using the voltage-divider formula, calculate the voltage dropped across each of the capacitors in Figure 14-8 if $V_T = 24$ V.

■ *Solution:*

$$V_{C1} = \frac{C_T}{C_1} \times V_T = \frac{0.5714\ \mu F}{4\ \mu F} \times 24\ V = 3.4\ V$$

$$V_{C2} = \frac{C_T}{C_2} \times V_T = \frac{0.5714\ \mu F}{2\ \mu F} \times 24\ V = 6.9\ V$$

$$V_{C3} = \frac{C_T}{C_3} \times V_T = \frac{0.5714\ \mu F}{1\ \mu F} \times 24\ V = 13.7\ V$$

$$V_T = V_{C1} + V_{C2} + V_{C3} = 3.4 + 6.9 + 13.7 = 24\ V$$
(Kirchhoff's voltage law)

If the capacitor values are the same, as seen in Figure 14-9(a), the voltage divides equally across each capacitor, as each capacitor has an equal amount of charge and therefore has half of the applied voltage (in this example, 3 V across each capacitor).

When the capacitor values are different, the smaller value of capacitor will actually charge to a higher voltage than the larger capacitor. In the example in Figure 14-9(b), the smaller capacitor is actually half the size of the other capacitor, and it has charged to twice the voltage. Since Kirchhoff's voltage law has to apply to this and every series circuit, you can easily calculate that the voltage across C_1 will equal 4 V and is twice that of C_2, which is 2 V. To understand this fully, we must first understand that although the capacitance is different, both capacitors have an equal value of coulomb charge held within them, which in this example is 8 μC.

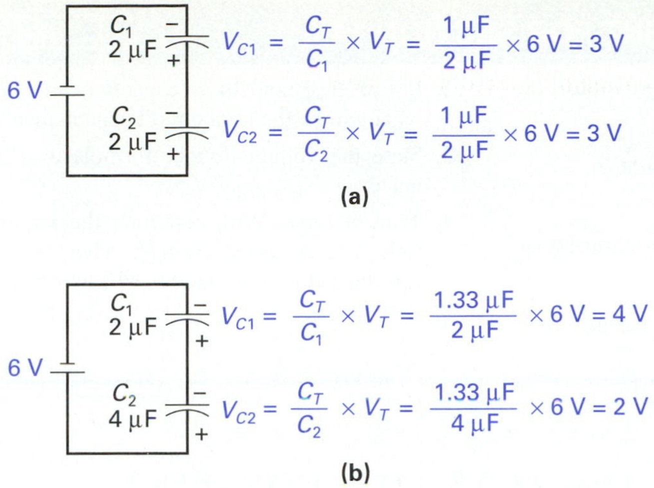

$V_{C1} = \dfrac{C_T}{C_1} \times V_T = \dfrac{1\,\mu F}{2\,\mu F} \times 6\,V = 3\,V$

$V_{C2} = \dfrac{C_T}{C_2} \times V_T = \dfrac{1\,\mu F}{2\,\mu F} \times 6\,V = 3\,V$

(a)

$V_{C1} = \dfrac{C_T}{C_1} \times V_T = \dfrac{1.33\,\mu F}{2\,\mu F} \times 6\,V = 4\,V$

$V_{C2} = \dfrac{C_T}{C_2} \times V_T = \dfrac{1.33\,\mu F}{4\,\mu F} \times 6\,V = 2\,V$

(b)

FIGURE 14-9   Voltage Drops across Series-Connected Capacitors.

$$
\begin{aligned}
Q_1 &= C_1 \times V_1 \\
&= 2\,\mu F \times 4\,V = 8\,\mu C \\
Q_2 &= C_2 \times V_2 \\
&= 4\,\mu F \times 2\,V = 8\,\mu C
\end{aligned}
$$

This equal charge occurs because the same amount of current flow exists throughout a series circuit, so both capacitors are being supplied with the same number or quantity of electrons. The charge held by C_1 is large with respect to its small capacitance, whereas the same charge held by C_2 is small with respect to its larger capacitance.

If the charge remains the same (Q is constant) and the capacitance is small, the voltage drop across the capacitor will be large, because the charge is large with respect to the capacitance:

$$
V\uparrow = \frac{Q}{C\downarrow}
$$

On the other hand, for a constant charge, a large capacitance will have a small charge voltage because the charge is small with respect to the capacitance:

$$
V\downarrow = \frac{Q}{C\uparrow}
$$

We can apply the water analogy once more and imagine two series-connected buckets, one of which is twice the size of the other. Both are being supplied by the same series pipe, which has an equal flow of water throughout, and are consequently each holding an equal amount of water, for example, 1 gallon. The 1 gallon of water in the small bucket is large with respect to the size of the bucket, and a large amount of pressure exists within that bucket. The 1 gallon of water in the large bucket is small with respect to the size of the bucket, so a small amount of pressure exists within this bucket. The pressure within a bucket is similar to the voltage across a capacitor, and therefore a small bucket or capacitor will have a greater pressure or voltage associated with it, whereas a large bucket or capacitor will develop a small pressure or voltage.

To summarize capacitors in series, all the series-connected components will have the same charging current throughout the circuit, and because of this, two or more capacitors in series will always have equal amounts of coulomb charge. If the charge (Q) is equal, the voltage across the capacitor is determined by the value of the capacitor. A small capacitance will charge to a larger voltage ($V\uparrow = Q/C\downarrow$), whereas a large value of capacitance will charge to a smaller voltage ($V\downarrow = Q/C\uparrow$).

Now that you have completed this section, you should be able to:

■ **Objective 3.** *Calculate total capacitance in parallel and series capacitive circuits.*

Use the following questions to test your understanding of Section 14-3.

1. If 2 μF, 3 μF, and 5 μF capacitors are connected in series, what will be the total circuit capacitance?

2. If 7 pF, 2 pF, and 14 pF capacitors are connected in parallel, what will be the total circuit capacitance?

3. State the voltage-divider formula as it applies to capacitance.

4. True or false: With resistors, the larger value of resistor will drop a larger voltage, whereas with capacitors the smaller value of capacitor will actually charge to a higher voltage.

14-4 CAPACITIVE TIME CONSTANT

When a capacitor is connected across a dc voltage source, it will charge to a value equal to the voltage applied. If the charged capacitor is then connected across a load, the capacitor will then discharge through the load. The time it takes a capacitor to charge or discharge can be calculated if the circuit's resistance and capacitance are known. Let us now see how we can calculate a capacitor's charge time and discharge time.

14-4-1 *DC Charging*

When a capacitor is connected across a dc voltage source, such as a battery or power supply, current will flow and the capacitor will charge up to a value equal to the dc source voltage, as shown in Figure 14-10. When the charge switch is first closed, as seen in Figure 14-10(a), there is no voltage across the capacitor at that instant and therefore a potential difference exists between the battery and capacitor. This causes current to flow and to begin charging the capacitor.

Once the capacitor begins to charge, the voltage across the capacitor does not instantaneously rise to 100 V. It takes a certain amount of time before the capacitor voltage is equal to the battery voltage. When the capacitor is fully charged no potential difference exists between the voltage source and the capacitor. Consequently, no more current flows in the circuit as the capacitor has reached its full charge, as seen in Figure 14-10(b). The amount of time it takes for a capacitor to charge to the supplied voltage (in this example, 100 V) is dependent on the circuit's resistance and capacitance value. If the circuit's resistance is increased, the opposition to current flow will be increased, and it will take the capacitor a longer period of time to obtain the same amount of charge because the circuit current available to charge the capacitor is less.

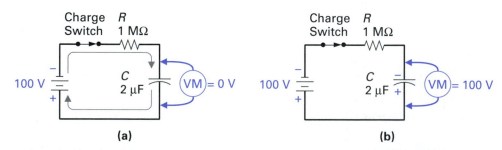

(a)   **(b)**

FIGURE 14-10 Capacitor Charging. (a) Switch Is Closed and Capacitor Begins to Charge. (b) Capacitor Charged.

If the value of capacitance is increased, it again takes a longer time to charge to 100 V because a greater amount of charge is required to build up the voltage across the capacitor to 100 V.

The circuit's resistance (R) and capacitance (C) are the two factors that determine the charge time (τ). Mathematically, this can be stated as

$$\tau = R \times C$$

where τ = **time constant** (s)
        R = resistance (Ω)
        C = capacitance (F)

In this example, we are using a resistance of 1 MΩ and a capacitance of 2 μF, which means that the time constant is equal to

$$\tau = R \times C$$
$$= 2\ \mu F \times 1\ M\Omega$$
$$= (2 \times 10^{-6}) \times (1 \times 10^{6})$$
$$= 2\ s$$

Two seconds is the time, so what is the constant? The constant value that should be remembered throughout this discussion is "**63.2**."

Figure 14-11 illustrates the rise in voltage across the capacitor from 0 to a maximum of 100 V in five time constants (5 $\times$ 2 s = 10 s). So where does 63.2 come into all this?

First time constant: In 1RC seconds (1 $\times$ R $\times$ C = 2 s), the capacitor will charge to 63.2% of the applied voltage (63.2% $\times$ 100 V = 63.2 V).

Time Constant

Time needed for either a voltage or current to rise to 63.2% of the maximum or fall to 36.8% of the initial value. The time constant of an RC circuit is equal to the product of R and C.

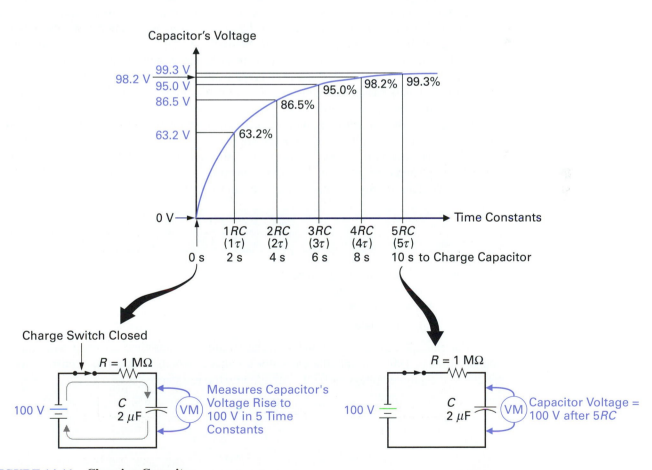

FIGURE 14-11   Charging Capacitor.

Second time constant: In $2RC$ seconds ($2 \times R \times C = 4$ s), the capacitor will charge to 63.2% of the remaining voltage. In the example, the capacitor will be charged to 63.2 V in one time constant, and therefore the voltage remaining is equal to 100 V − 63.2 V = 36.8 V. At the end of the second time constant, therefore, the capacitor will have charged to 63.2% of the remaining voltage (63.2% × 36.8 V = 23.3 V), which means that it will have reached 86.5 V (63.2 V + 23.3 V = 86.5 V) or 86.5% of the applied voltage.

Third time constant: In $3RC$ seconds (6 s), the capacitor will charge to 63.2% of the remaining voltage:

$$\text{Remaining voltage} = 100 \text{ V} - 86.5 \text{ V}$$
$$= 13.5 \text{ V}$$
$$63.2\% \text{ of } 13.5 \text{ V} = 8.532 \text{ V}$$

At the end of the third time constant, therefore, the capacitor will have charged to 86.5 V + 8.532 V = 95 V, or 95% of the applied voltage.

Fourth time constant: In $4RC$ seconds (8 s), the capacitor will charge to 63.2% of the remaining voltage (100 V − 95 V = 5 V); therefore, 63.2% of 5 V = 3.2 V. So the capacitor will have charged to 95 V + 3.2 V = 98.2 V, or 98.2% of the applied voltage.

Fifth time constant: In $5RC$ seconds (10 s), the capacitor is considered to be fully charged since the capacitor will have reached 63.2% of the remaining voltage (100 V − 98.2 V = 1.8 V); therefore, 63.2% of 1.8 V = 1.1 V. So the capacitor will have charged to 98.2 V + 1.1 V = 99.3 V, or 99.3% of the applied voltage.

The voltage waveform produced by the capacitor acquiring a charge is known as an *exponential* waveform, and the voltage across the capacitor is said to rise exponentially. An exponential rise is also referred to as a *natural increase.* There are many factors that exponentially rise and fall. For example, we grow exponentially, in that there is quite a dramatic change in our height in the early years and then this increase levels off and height reaches a maximum.

Before the switch is closed and even at the instant the switch is closed, the capacitor is not charged, which means that there is no capacitor voltage to oppose the supply voltage and, therefore, a maximum current of V/R, 100 V/1 MΩ = 100 μA, flows. This current begins to charge the capacitor, a potential difference begins to build up across the plates of the capacitor, and this voltage opposes the supply voltage, causing a decrease in charging current. As the capacitor begins to charge, less of a potential difference exists between the supply voltage and capacitor voltage and so the current begins to decrease.

To calculate the current at any time, we can use the formula

$$i = \frac{V_S - V_C}{R}$$

where i = instantaneous current
V_S = source voltage
V_C = capacitor voltage
R = resistance

For example, the current flowing in the circuit after one time constant will equal the source voltage, 100 V, minus the capacitor's voltage, which in one time constant will be 63.2% of the source voltage or 63.2 V, divided by the resistance.

$$i = \frac{V_S - V_C}{R}$$
$$= \frac{100 \text{ V} - 63.2 \text{ V}}{1 \text{ M}\Omega}$$
$$= 36.8 \text{ }\mu\text{A}$$

As the charging continues, the potential difference across the plates exponentially rises to equal the supply voltage, as seen in Figure 14-12(a), while the current exponentially falls to zero, as shown in Figure 14-12(b). The constant of 63.2 can be applied to the exponential fall of current from 100 μA to 0 μA in 5RC seconds.

When the switch was closed to start the charging of the capacitor, there was no charge on the capacitor; therefore, a maximum potential difference existed between the battery and capacitor, causing a maximum current flow of 100 μA ($I = V/R$).

First time constant: In 1RC seconds, the current will exponentially decrease 63.2% (63.2% of 100 μA = 63.2 μA) to a value of 36.8 μA (100 μA − 63.2 μA). In the example of 2 μF and 1 MΩ, this occurs in 2 s.

Second time constant: In 2RC seconds (2 $\times$ R $\times$ C = 4 s), the current will decrease 63.2% of the remaining current, which is

$$63.2\% \text{ of } 36.8 \ \mu A = 23.26 \ \mu A$$

The current will drop 23.26 μA from 36.8 μA and reach 13.5 μA or 13.5%.

Third time constant: In 3RC seconds (6 s), the capacitor's charge current will decrease 63.2% of the remaining current (13.5 μA) to 5 μA or 5%.

Fourth time constant: In 4RC seconds (8 s), the current will have decreased to 1.8 μA or 1.8%.

Fifth time constant: After 5RC seconds (10 s), the charge current is now 0.7 μA or 0.7%. At this time, the charge current is assumed to be zero and the capacitor is now charged to a voltage equal to the applied voltage.

Studying the exponential rise of the voltage and the exponential decay of current in a capacitive circuit, you will notice an interesting relationship. In a pure resistive circuit, the current flow through a resistor will be in step with the voltage across that same resistor, in that an increased current will cause a corresponding increase in voltage drop across the resistor. Voltage and current are consequently said to be *in step* or *in phase* with each other. With the capacitive circuit, the current flow in the circuit and voltage across the capacitor are not in step or in phase with each other. When the switch is closed to charge the capacitor, the current is maximum (100 μA), while the voltage across the capacitor is zero. After five time constants (10 s), the capacitor's voltage is now maximum (100 V) and the circuit current is zero, as seen in Figure 14-13. The circuit current flow is out of phase with the capacitor voltage, and this difference is referred to as a *phase shift.* In any circuit containing capacitance, current will lead voltage.

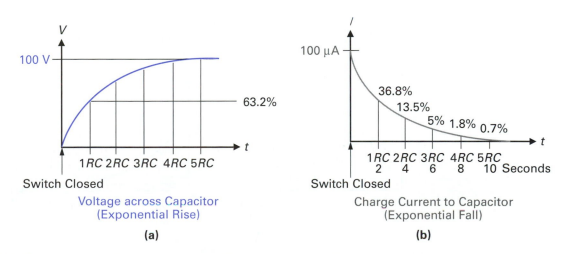

FIGURE 14-12 **Exponential Rise in Voltage and Fall in Current in a Charging Capacitive Circuit.**

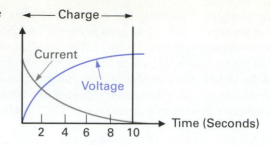

14-4-2 *DC Discharging*

Figure 14-14 illustrates the circuit, voltage, and current waveforms that occur when a charged capacitor is discharged from 100 V to 0 V. The 2 μF capacitor, which was charged to 100 V in 10 s (5RC), is discharged from 100 to 0 V in the same amount of time.

Looking at the voltage curve, you can see that the voltage across the capacitor decreases exponentially, dropping 63.2% to 36.8 V in 1RC seconds, another 63.2% to 13.5 V in 2RC seconds, another 63.2% to 5 V in 3RC seconds, and so on, until zero.

The current flow within the circuit is dependent on the voltage in the circuit, which is across the 2 μF capacitor. As the voltage decreases, the current will also decrease by the same amount ($I\downarrow = V\downarrow/R$).

$$\text{Discharge switch closed: } I = \frac{V}{R} = \frac{100 \text{ V}}{1 \text{ M}\Omega} = 100 \ \mu\text{A} \quad \text{maximum}$$

$$1RC \ (2) \text{ seconds: } I = \frac{V}{R} = \frac{36.8 \text{ V}}{1 \text{ M}\Omega} = 36.8 \ \mu\text{A}$$

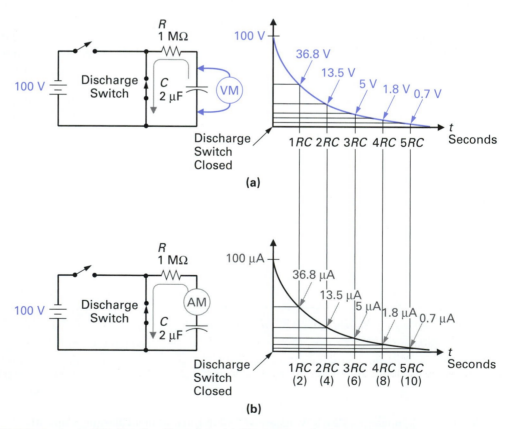

FIGURE 14-14   Discharging Capacitor. (a) Voltage Waveform. (b) Current Waveform.

$$2RC \text{ (4) seconds: } I = \frac{V}{R} = \frac{13.5 \text{ V}}{1 \text{ M}\Omega} = 13.5 \ \mu\text{A}$$

$$3RC \text{ (6) seconds: } I = \frac{V}{R} = \frac{5 \text{ V}}{1 \text{ M}\Omega} = 5.0 \ \mu\text{A}$$

$$4RC \text{ (8) seconds: } I = \frac{V}{R} = \frac{1.8 \text{ V}}{1 \text{ M}\Omega} = 1.8 \ \mu\text{A}$$

$$5RC \text{ (10) seconds: } I = \frac{V}{R} = \frac{0.7 \text{ V}}{1 \text{ M}\Omega} = 0.7 \ \mu\text{A} \quad \text{zero}$$

SELF-TEST EVALUATION POINT FOR SECTION 14-4

Now that you have completed this section, you should be able to:

■ *Objective 4.* *Explain the capacitor time constant as it relates to dc charging and discharging.*

Use the following questions to test your understanding of Section 14-4.

1. What is the capacitor time constant?
2. In one time constant, a capacitor will have charged to what percentage of the applied voltage?
3. In one time constant, a capacitor will have discharged to what percentage of its full charge?
4. True or false: The charge or discharge of a capacitor follows a linear rate of change.

14-5 CAPACITIVE REACTANCE

Resistance (R), by definition, is the opposition to current flow with the dissipation of energy and is measured in ohms. Capacitors oppose current flow like a resistor, but a resistor dissipates energy, whereas a capacitor stores energy (when it charges) and then gives back its energy into the circuit (when it discharges). Because of this difference, a new term had to be used to describe the opposition offered by a capacitor. **Capacitive reactance (X_C)**, by definition, is the opposition to current without the dissipation of energy and is also measured in ohms.

If capacitive reactance is basically opposition, it is inversely proportional to the amount of current flow. If a large current is within a circuit, the opposition or reactance must be low ($I\uparrow, X_C\downarrow$). Conversely, a small circuit current will be the result of a large opposition or reactance ($I\downarrow, X_C\uparrow$).

When a dc source is connected across a capacitor, current will flow only for a short period of time ($5RC$ seconds) to charge the capacitor. After this time, there is no further current flow. Consequently, the capacitive reactance or opposition offered by a capacitor to dc is infinite (maximum).

Alternating current continuously reverses in polarity, resulting in the continuous charging and discharging of the capacitor. This means that charge and discharge currents are always flowing around the circuit, and if we have a certain value of current, we must also have a certain value of reactance or opposition.

Initially, when the capacitor's plates are uncharged, they will not oppose or react against the charging current, and therefore maximum current will flow ($I\uparrow$), and the reactance will be very low ($X_C\downarrow$). As the capacitor charges, it will oppose or react against the charge current, which will decrease ($I\downarrow$), so the reactance will increase ($X_C\uparrow$). The discharge current is also highest at the start of discharge ($I\uparrow, X_C\downarrow$) as the voltage of the charged capacitor is also high; but as the capacitor discharges, its voltage decreases and the discharge current will also decrease ($I\downarrow, X_C\uparrow$).

To summarize, at the start of a capacitor charge or discharge, the current is maximum, so the reactance is low. This value of current then begins to fall to zero, so the reactance increases.

Capacitive Reactance (X_C)
Measured in ohms, it is the ability of a capacitor to oppose current flow without the dissipation of energy.

If the applied alternating current is at a high frequency, as shown in Figure 14-15(a), it is switching polarity more rapidly than a lower frequency and there is very little time between the start of charge and discharge. As the charge and discharge currents are largest at the beginning of the charge and discharge of the capacitor, the reactance has very little time to build up and oppose the current, which is why the current is a high value and the capacitive reactance is small at higher frequencies. With lower frequencies, as shown in Figure 14-15(b), the applied alternating current is switching at a slower rate, and therefore the reactance, which is low at the beginning, has more time to build up and oppose the current.

Capacitive reactance is therefore inversely proportional to frequency:

$$\text{Capacitive reactance } (X_C) \propto \frac{1}{f \text{ (frequency)}}$$

Frequency, however, is not the only factor that determines capacitive reactance. Capacitive reactance is also inversely proportional to the value of capacitance. If a larger capacitor value is used, a longer time is required to charge the capacitor ($\tau\uparrow = C\uparrow R$), which means that current will be present for a longer period of time, so the overall current will be large ($I\uparrow$); consequently, the reactance must be small ($X_C\downarrow$). On the other hand, a small capacitance value will charge in a small amount of time ($\tau\downarrow = C\downarrow R$) and the current will be present for only a short period of time. The overall current will therefore be small ($I\downarrow$), indicating a large reactance ($X_C\uparrow$).

$$\text{Capacitive reactance } (X_C) \propto \frac{1}{C \text{ (capacitance)}}$$

FIGURE 14-15 Capacitive Reactance Is Inversely Proportional to Frequency.

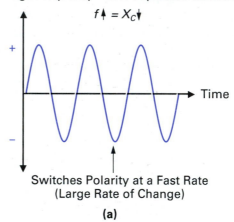

High Frequency = Low Capacitive Reactance

$f\uparrow = X_C\downarrow$

Time

Switches Polarity at a Fast Rate
(Large Rate of Change)

(a)

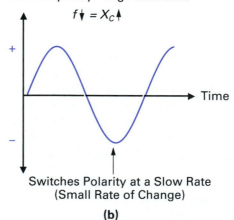

Low Frequency = High Reactance

$f\downarrow = X_C\uparrow$

Time

Switches Polarity at a Slow Rate
(Small Rate of Change)

(b)

Capacitive reactance (X_C) is therefore inversely proportional to both frequency and capacitance and can be calculated by using the formula

$$X_C = \frac{1}{2\pi fC}$$

where X_C = capacitive reactance, in ohms
2π = constant
f = frequency, in hertz
C = capacitance, in farads

■ **INTEGRATED MATH APPLICATION:**

Calculate the reactance of a 2 μF capacitor when a 10 kHz sine wave is applied.

■ *Solution:*

CALCULATOR SEQUENCE

$$X_C = \frac{1}{2\pi fC}$$

$$= \frac{1}{2 \times \pi \times 10\ \text{kHz} \times 2\ \mu\text{F}} = 8\ \Omega$$

Step	Keypad Entry	Display Response
1.	2	2.0
2.	×	
3.	π	3.1415927
4.	×	6.283185
5.	1 0 EE 3	10E3
6.	×	6.2831.3
7.	2 EE 6 +/−	2.−06
8.	=	0.1256637
9.	1/x	7.9577

SELF-TEST EVALUATION POINT FOR SECTION 14-5

Now that you have completed this section, you should be able to:

■ *Objective 5.* *Define and explain capacitive reactance.*

Use the following questions to test your understanding of Section 14-5.

1. Define *capacitive reactance*.
2. State the formula for capacitive reactance.
3. Why is capacitive reactance inversely proportional to frequency and capacitance?
4. If $C = 4\ \mu$F and $f = 4$ kHz, calculate X_C.

14-6 SERIES *RC* CIRCUIT

In a purely resistive circuit, as shown in Figure 14-16(a), the current flowing within the circuit and the voltage across the resistor are in phase with one another. In a purely capacitive circuit, as shown in Figure 14-16(b), the current flowing in the circuit leads the voltage across the capacitor by 90°.

Purely resistive: 0° phase shift (*I* is in phase with *V*)
Purely capacitive: 90° phase shift (*I* leads *V* by 90°)

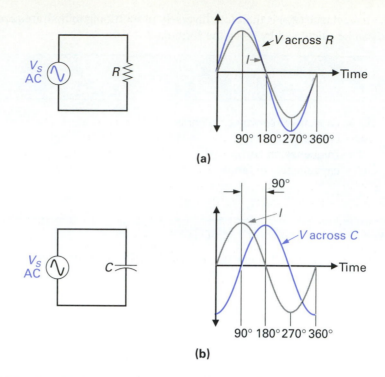

(a)

(b)

FIGURE 14-16 Phase Relationships between *V* and *I*. (a) Resistive Circuit: Current and Voltage Are in Phase. (b) Capacitive Circuit: Current Leads Voltage by 90°.

If we connect a resistor and capacitor in series, as shown in Figure 14-17(a), we have probably the most commonly used electronic circuit, which has many applications. The voltage across the resistor (V_R) is always in phase with the circuit current (I), as can be seen in Figure 14-17(b), because maximum points and zero crossover points occur at the same time. The voltage across the capacitor (V_C) lags the circuit current by 90°.

Since the capacitor and resistor are in series, the same current is supplied to both components; Kirchhoff's voltage law can be applied, which states that the sum of the voltage drops around a series circuit is equal to the voltage applied (V_S). The voltage drop across the resistor (V_R) and the voltage drop across the capacitor (V_C) are out of phase with each other, which means that their peaks occur at different times. The signal for the applied voltage (V_S) is therefore obtained by adding the values of V_C and V_R at each instant in time, plotting the results, and then connecting the points with a line; this is represented in Figure 14-17(b) by the shaded waveform.

Although the waveforms in Figure 14-17(b) indicate the phase relationship between *I*, V_S, V_R, and V_C, it seems difficult to understand clearly the relationship among all four be-

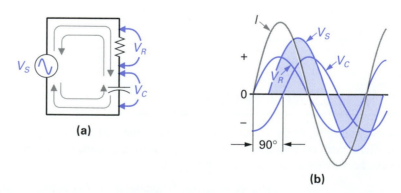

(a)

(b)

FIGURE 14-17 *RC* Series Circuit. (a) Circuit. (b) Waveforms.

cause of the crisscrossing of waveforms. An easier method of representation is to return to the circle and vectors that were introduced in Chapter 6.

14-6-1   *Vector Diagram*

A vector (or phasor) is a quantity that has both magnitude and direction and is represented by a line terminated by an arrowhead, as seen in Figure 14-18(a). Vectors are generally always used to represent a physical quantity that has two properties. For example, if you were traveling 60 miles per hour in a southeast direction, the size or magnitude of the vector would represent 60 mph, and the vector would point southeast. In an ac circuit containing a reactive component such as a capacitor, the vector is used to represent a voltage or current. The magnitude of the vector represents the value of voltage or current, and the direction of the vector represents the phase of the voltage or current.

A **vector diagram** is an arrangement of vectors to illustrate the magnitude and phase relationships between two or more quantities of the same frequency within an ac circuit. Figure 14-18(b) illustrates the basic parts of a vector diagram. As an example, the current (I) vector is at the 0° position, and the size of the arrow represents the peak value of alternating current.

14-6-2   *Voltage*

Figure 14-19(a), (b), and (c) repeat our previous *RC* series circuit with waveforms and a vector diagram. In Figure 14-19(b), the current peak flowing in the series *RC* circuit occurs at 0° and will be used as a reference; therefore, the vector of current in Figure 14-19(c) is to the right in the 0° position. The voltage across the resistor (V_R) is in phase or coincident with the current (I), as shown in Figure 14-19(b), and the vector that represents the voltage across the resistor (V_R) overlaps or coincides with the I vector, at 0°.

The voltage across the capacitor (V_C) is, as shown in Figure 14-19(b), 90° out of phase (lagging) with the circuit's current, so the V_C vector in Figure 14-19(c) is drawn at $-90°$ (minus sign indicates lag) to the current vector, and the length of this vector represents the magnitude of the voltage across the capacitor. Because the ohmic values of the resistor (R) and the capacitor (X_C) are equal, the voltage drop across both components is the same. The V_R and the V_C vectors are subsequently equal in length.

The source voltage (V_S) is, by Kirchhoff's voltage law, equal to the sum of the series voltage drops (V_C and V_R). However, since these voltages are not in phase with each other, we cannot simply add the two together. The source voltage (V_S) will be the sum of both V_C

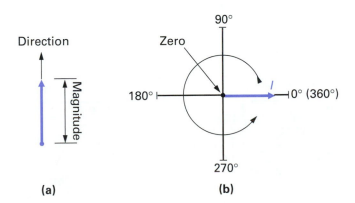

(a)   **(b)**

FIGURE 14-18   **Vectors. (a) Vector. (b) Vector Diagram.**

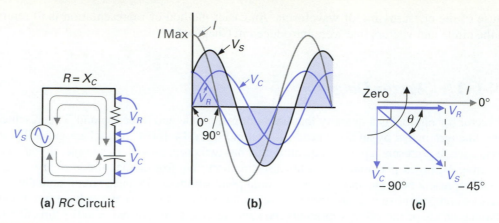

(a) *RC* Circuit   **(b)**   **(c)**

FIGURE 14-19 *RC* **Series Circuit Analysis.**

and V_R at a particular time. Studying the waveforms in Figure 14-19(b), you will notice that peak source voltage occurs at 45°. By vectorially adding the two voltages V_R and V_C in Figure 14-19(c), we obtain a resultant V_S vector that has both magnitude and phase. The angle theta (θ) formed between circuit current (I) and source voltage (V_S) will always be less than 90° and in this example is equal to $-45°$, since the voltage drops across R and C are equal because R and X_C are of the same ohmic value.

If V_C and V_R are drawn to scale, the peak source voltage (V_S) can be calculated by using the same scale, and a mathematical rather than graphical method can be used to save the drafting time.

In Figure 14-20(a), we have taken the three voltages (V_R, V_C, and V_S) and formed a right-angle triangle, as shown in Figure 14-20(b). The Pythagorean theorem for right-angle triangles states that if you take the square of a (V_R) and add it to the square of b (V_C), the square root of the result will equal c (the source voltage, V_S).

$$V_S = \sqrt{V_R^2 + V_C^2}$$

By transposing the formula according to the rules of algebra we can calculate any unknown if we know two variables.

$$V_C = \sqrt{V_S^2 - V_R^2}$$
$$V_R = \sqrt{V_S^2 - V_C^2}$$

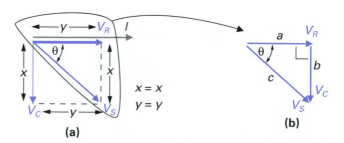

(a)   **(b)**

FIGURE 14-20 **Voltage and Current Vector Diagram of a Series *RC* Circuit.**

■ INTEGRATED MATH APPLICATION:

Calculate the source voltage applied across an *RC* series circuit if $V_R = 12$ V and $V_C = 8$ V.

■ *Solution:*

$$V_S = \sqrt{V_R^2 + V_C^2}$$
$$= \sqrt{(12 \text{ V})^2 + (8 \text{ V})^2}$$
$$= \sqrt{144 \text{ V}^2 + 64 \text{ V}^2}$$
$$= 14.42 \text{ V}$$

CALCULATOR SEQUENCE

Step	Keypad Entry	Display Response
1.	1 2	12.0
2.	x²	144.0
3.	+	
4.	8	8.0
5.	χ²	64.0
6.	=	208.0
7.	√x	14.42220

14-6-3 *Impedance*

Since resistance is the opposition to current with the dissipation of heat, and reactance is the opposition to current without the dissipation of heat, a new term is needed to describe the total resistive and reactive opposition to current. **Impedance** (designated *Z*) is also measured in ohms and is the total circuit opposition to current flow. It is a combination of resistance (*R*) and reactance (X_C); however, in our capacitive and resistive circuit, a phase shift or difference exists, and just as V_C and V_R cannot be merely added together to obtain V_S, *R* and X_C cannot simply be summed to obtain *Z*.

> **Impedance (*Z*)**
> Measured in ohms, it is the total opposition a circuit offers to current flow (reactive and resistive).

If the current within a series circuit is constant (the same throughout the circuit), the resistance of a resistor (*R*) or reactance of a capacitor (X_C) will be directly proportional to the voltage across the resistor (V_R) or the capacitor (V_C).

$$V_R] = I \times R], \qquad V_C] = I \times X_C]$$

A vector diagram similar to the voltage vector diagram can be drawn to illustrate opposition, as shown in Figure 14-21(a). The current is used as a reference (0°); the resistance vector (*R*) is in phase with the current vector (*I*), since V_R is always in phase with *I*. The capacitive reactance (X_C) vector is at −90° to the resistance vector, due to the 90° phase shift between a resistor and capacitor. The lengths of the resistance vector (*R*) and capacitive reactance vector (X_C) are equal in this example. By vectorially adding *R* and X_C, we have a resulting impedance (*Z*) vector.

By using the three variables, which have again formed a right-angle triangle [Figure 14-21(b)], we can apply the Pythagorean theorem to calculate the total opposition or impedance (*Z*) to current flow, taking into account both *R* and X_C.

$$Z = \sqrt{R^2 + X_C^2}$$

FIGURE 14-21 Resistance, Reactance, and Impedance Vector Diagram of a Series *RC* Circuit.

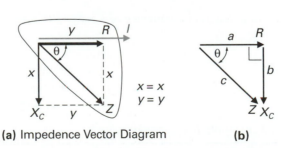

(a) Impedence Vector Diagram   **(b)**

$$R = \sqrt{Z^2 - X_C^2}$$
$$X_C = \sqrt{Z^2 - R^2}$$

■ **INTEGRATED MATH APPLICATION:**

Calculate the total impedance of a series RC circuit if $R = 27\ \Omega$, $C = 0.005\ \mu F$, and the source frequency $= 1$ kHz.

■ *Solution:*

The total opposition (Z) or impedance is equal to

$$Z = \sqrt{R^2 + X_C^2}$$

R is known, but X_C will need to be calculated.

$$X_C = \frac{1}{2\pi f C}$$
$$= \frac{1}{2 \times \pi \times 1\ \text{kHz} \times 0.005\ \mu F}$$
$$= 31.8\ \text{k}\Omega$$

Since $R = 27\ \Omega$ and $X_C = 31.8$ kΩ, then

$$Z = \sqrt{R^2 + X_C^2}$$
$$= \sqrt{(27\ \Omega)^2 + (31.8\ \text{k}\Omega)^2}$$
$$= \sqrt{729\ \Omega^2 + 1 \times 10^9\ \Omega^2}$$
$$= 31.8\ \text{k}\Omega$$

As you can see in this example, the small resistance of 27 Ω has very little effect on the circuit's total opposition or impedance, due to the relatively large capacitive reactance of 31,800 Ω.

■ **INTEGRATED MATH APPLICATION:**

Calculate the total impedance of a series RC circuit if $R = 45$ kΩ and $X_C = 45\ \Omega$.

■ *Solution:*

$$Z = \sqrt{R^2 + X_C^2}$$
$$= \sqrt{(45\ \text{k}\Omega)^2 + (45\ \Omega)^2}$$
$$= 45\ \text{k}\Omega$$

In this example, the relatively small value of X_C had very little effect on the circuit's opposition or impedance, due to the large circuit resistance.

■ **INTEGRATED MATH APPLICATION:**

Calculate the total impedance of a series RC circuit if $X_C = 100\ \Omega$ and $R = 100\ \Omega$.

■ *Solution:*

$$Z = \sqrt{R^2 + X_C^2}$$
$$= \sqrt{100^2 + 100^2}$$
$$= 141.4\ \Omega$$

In this example R was equal to X_C.

We can define the total opposition or impedance in terms of Ohm's law, in the same way as we defined resistance.

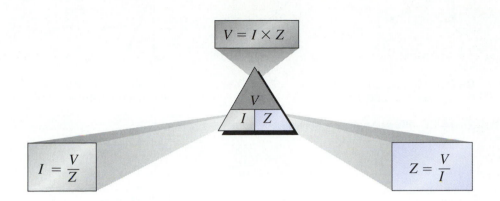

$$V = I \times Z$$

$$I = \frac{V}{Z}$$

$$Z = \frac{V}{I}$$

By transposition, we can arrive at the usual combinations of Ohm's law.

14-6-4   *Phase Angle or Shift (θ)*

In a purely resistive circuit, the total opposition (Z) is equal to the resistance of the resistor, so the phase shift (θ) is equal to $0°$ [Figure 14-22(a)].

In a purely capacitive circuit, the total opposition (Z) is equal to the capacitive reactance (X_C) of the capacitor, so the phase shift (θ) is equal to $-90°$ [Figure 14-22(b)].

When a circuit contains both resistance and capacitive reactance, the total opposition or impedance has a phase shift that is between $0°$ and $90°$. Referring back to the impedance vector diagram in Figure 14-21 and the preceding example, you can see that we have used a simple example where R has equaled X_C, so the phase shift (θ) has always been $-45°$. If the resistance and reactance are different from each other, as was the case in the other two examples, the phase shift (θ) will change, as shown in Figure 14-23. As can be seen in this illustration, the phase shift (θ) is dependent on the ratio of capacitive reactance to resistance (X_C/R). A more resistive circuit will have a phase shift between $0°$ and $45°$, whereas a more reactive circuit will have a phase shift between $45°$ and $90°$. By the use of trigonometry (the science of triangles), we can derive a formula to calculate the degree of phase shift, since two quantities X_C and R are known.

The phase angle, θ, is equal to

$$\theta = \text{invtan}\, \frac{X_C}{R}$$

FIGURE 14-22   **Phase Angles.**
(a) Purely Resistive Circuit. (b) Purely Capacitive Circuit.

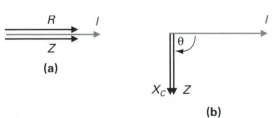

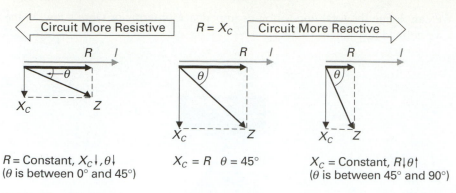

FIGURE 14-23 Phase Angle of a Series *RC* Circuit.

This formula will determine by what angle Z leads R. Since X_C/R is equal to V_C/V_R, the phase angle can also be calculated if V_R and V_C are known.

$$\theta = \text{invtan}\,\frac{V_C}{V_R}$$

This formula will determine by what angle V_R leads V_S.

■ **INTEGRATED MATH APPLICATION:**

Calculate the phase shift or angle in two different series *RC* circuits if:
   a. $V_R = 12$ V, $V_C = 8$ V
   b. $R = 27\ \Omega$, $X_C = 31.8$ kΩ

■ *Solution:*

a. $\theta = \text{invtan}\,\dfrac{V_C}{V_R}$

$= \text{invtan}\,\dfrac{8\text{ V}}{12\text{ V}}$

$= 33.7°\ (V_R \text{ leads } V_S \text{ by } 33.7°)$

b. $\theta = \text{invtan}\,\dfrac{X_C}{R}$

$= \text{invtan}\,\dfrac{31.8\text{ k}\Omega}{27\ \Omega}$

$= 89.95°\ (R \text{ leads } Z \text{ by } 89.95°)$

14-6-5   *Power*

In this section we examine power in a series ac circuit. Let us begin with a simple resistive circuit and review the power formulas used previously.

Purely Resistive Circuit

In Figure 14-24 you can see the current, voltage, and power waveforms generated by applying an ac voltage across a purely resistive circuit. The applied voltage causes current to

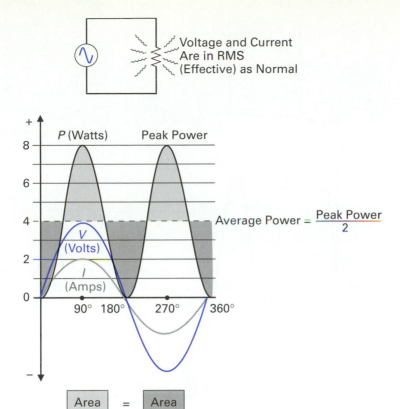

FIGURE 14-24 **Power in a Purely Resistive Circuit.**

flow around the circuit, and the electrical energy is converted into heat energy. This heat or power is dissipated and lost and can be calculated by using the power formula.

$$P = V \times I$$
$$P = I^2 \times R$$
$$P = \frac{V^2}{R}$$

Voltage and current are in phase with each other in a resistive circuit, and instantaneous power is calculated by multiplying voltage by current at every instant through 360° ($P = V \times I$). The sinusoidal power waveform is totally positive, because a positive voltage multiplied by a positive current gives a positive value of power, and a negative voltage multiplied by a negative current will also produce a positive value of power. For these reasons, a resistor is said to generate a positive power waveform, which you may have noticed is twice the frequency of the voltage and current waveforms; two power cycles occur in the same time as one voltage and current cycle.

The power waveform has been split in half, and this line that exists between the maximum point (8 W) and zero point (0 W) is the average value of power (4 W) that is being dissipated by the resistor.

Purely Capacitive Circuit

In Figure 14-25, you can see the current, voltage, and power waveforms generated by applying an ac voltage source across a purely capacitive circuit. As expected, the current leads the voltage by 90°, and the power wave is calculated by multiplying voltage by current, as before, at every instant through 360°. The resulting power curve is both positive and negative. During the positive alternation of the power curve, the capacitor is taking power as the capacitor charges. When the power alternation is negative, the capacitor is giving back the power it took as it discharges back into the circuit.

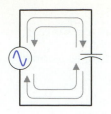

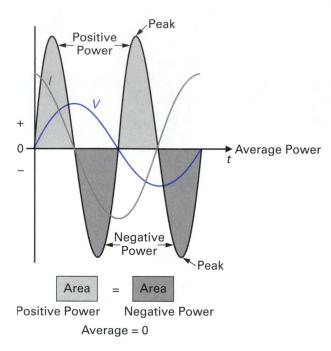

FIGURE 14-25 **Power in a Purely Capacitive Circuit.**

The average power dissipated is once again the value that exists between the maximum positive and maximum negative points and causes the area above this line to equal the area below. This average power level calculates out to be zero, which means that no power is dissipated in a purely capacitive circuit.

Resistive and Capacitive Circuit

In Figure 14-26 you can see the current, voltage, and power waveforms generated by applying an ac voltage source across a series-connected *RC* circuit. The current leads the voltage by some phase angle less than 90°, and the power waveform is once again determined by the product of voltage and current. The negative alternation of the power cycle indicates that the capacitor is discharging and giving back the power that it consumed during the charge.

The positive alternation of the power cycle is much larger than the negative alternation because it is the combination of both the capacitor taking power during charge and the resistor consuming and dissipating power in the form of heat. The average power being dissipated will be some positive value, due to the heat being generated by the resistor.

Power Factor

In a purely resistive circuit, all the energy supplied to the resistor from the source is dissipated in the form of heat. This form of power is referred to as **resistive power** (P_R) or **true power,** and is calculated with the formula

$$P_R = I^2 \times R$$

Resistive Power or True Power

The average power consumed by a circuit during one complete cycle of alternating current.

CHAPTER 14 / CAPACITORS

FIGURE 14-26   **Power in a Resistive and Capacitive Circuit.**

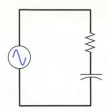

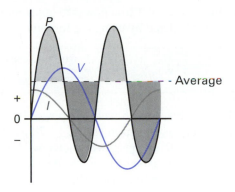

In a purely capacitive circuit, all the energy supplied to the capacitor is stored from the source and then returned to the source, without energy loss. This form of power is referred to as **reactive power** (P_X) or **imaginary power.**

$$P_{X = I}{}^{2} \times X_C$$

When a circuit contains both capacitance and resistance, some of the supply is stored and returned by the capacitor and some of the energy is dissipated and lost by the resistor.

Figure 14-27(a) illustrates another vector diagram. Just as the voltage across a resistor is 90° out of phase with the voltage across a capacitor, and resistance is 90° out of phase with reactance, resistive power will be 90° out of phase with reactive power.

If we take the three variables from Figure 14-27(b) to form a right-angle triangle as in Figure 14-27(c), we can vectorially add true power and imaginary power to produce a resultant **apparent power** vector. Apparent power is the power that appears to be supplied to the load and includes both the true power dissipated by the resistance and the imaginary power delivered to the capacitor.

Reactive Power or Imaginary Power

Also called wattless power, it is the power value obtained by multiplying the effective value of current by the effective value of voltage and the sine of the angular phase difference between current and voltage.

Apparent Power

The power value obtained in an ac circuit by multiplying together effective values of voltage and current, which reach their peaks at different times.

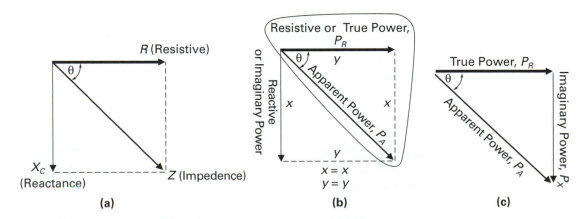

FIGURE 14-27   **Apparent Power.**

Applying the Pythagorean theorem once again, we can calculate apparent power by

$$P_A = \sqrt{P_R^2 + P_X^2}$$

where P_A = apparent power, in volt-amperes (VA)
P_R = true power, in watts (W)
P_X = reactive power, in volt-amperes reactive (VAR)

The **power factor** is a ratio of true power to apparent power and is a measure of the loss in a circuit. It can be calculated by using the formula

$$PF = \frac{\text{true power } (P_R)}{\text{apparent power } (P_A)}$$

Figure 14-28 helps explain what the power factor of a circuit actually indicates. In a purely resistive circuit, as shown in Figure 14-28(a), the apparent power will equal the true power ($P_A = \sqrt{P_R^2 + 0^2}$). The power factor will therefore equal 1 (PF = P_R/P_A). In a purely capacitive circuit, as shown in Figure 14-28(b), the true power will be zero, and therefore the power factor will equal 0 (PF = $0/P_A$). A power factor of 1 therefore indicates a maximum power loss (circuit is resistive), whereas a power factor of 0 indicates no power loss (circuit is capacitive). With circuits that contain both resistance and reactance, the power factor will be somewhere between zero (0 = reactive) and one (1 = resistive).

Since true power is determined by resistance, and apparent power is dependent on impedance, as shown in Figure 14-27(a), the power factor can also be calculated by using the formula

$$PF = \frac{R}{Z}$$

As the ratio of true power (adjacent) to apparent power (hypotenuse) determines the angle θ, the power factor can also be determined by the cosine of angle θ.

$$PF = \cos \theta$$

Purely Resistive	Purely Reactive
In a resistive circuit the reactive power is zero, and therefore the true power = apparent power.	In a reactive circuit the resistive power (P_R) is zero.

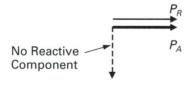

The power factor is therefore equal to:

$$PF = \frac{P_R}{P_A} = 1 \quad \text{(maximum value)}$$

The power factor is therefore equal to:

$$PF = \frac{P_R}{P_A} = 0 \quad \text{(minimum value)}$$

(a)   (b)

FIGURE 14-28 **Circuit's Power Factor. (a) Purely Resistive, PF = 1. (b) Purely Reactive, PF = 0.**

■ **INTEGRATED MATH APPLICATION:**

Calculate the following for a series RC circuit if $R = 2.2$ kΩ, $X_C = 3.3$ kΩ, and $V_S = 5$ V.

  a. Z
  b. I
  c. θ
  d. P_R
  e. P_X
  f. P_A
  g. PF

■ *Solution:*

a. $Z = \sqrt{R^2 + X_C^2}$
$\quad = \sqrt{(2.2 \text{ k}\Omega)^2 + (3.3 \text{ k}\Omega)^2}$
$\quad = 3.96 \text{ k}\Omega$

b. $I = \dfrac{V_S}{Z} = \dfrac{5 \text{ V}}{3.96 \text{ k}\Omega} = 1.26 \text{ mA}$

c. $\theta = \text{invtan} \dfrac{X_C}{R} = \text{invtan} \dfrac{3.3 \text{ k}\Omega}{2.2 \text{ k}\Omega}$
$\quad = \text{invtan } 1.5 = 56.3°$

d. True power $= I^2 \times R$
$\qquad\qquad\quad = (1.26 \text{ mA})^2 \times 2.2 \text{ k}\Omega$
$\qquad\qquad\quad = 3.49 \text{ mW}$

e. Reactive power $= I^2 \times X_C$
$\qquad\qquad\qquad\quad = (1.26 \text{ mA})^2 \times 3.3. \text{ k}\Omega$
$\qquad\qquad\qquad\quad = 5.24 \times 10^{-3} \text{ or } 5.24 \text{ mVAR}$

f. Apparent power $= \sqrt{P_R^2 + P_X^2}$
$\qquad\qquad\qquad\quad = \sqrt{(3.49 \text{ mW})^2 + (5.24 \text{ mW})^2}$
$\qquad\qquad\qquad\quad = 6.29 \times 10^{-3} \text{ or } 6.29 \text{ mVA}$

g. Power factor $= \dfrac{R}{Z} = \dfrac{2.2 \text{ k}\Omega}{3.96 \text{ k}\Omega} = 0.55$

or

$\qquad\qquad\quad = \dfrac{P_R}{P_A} = \dfrac{3.49 \text{ mW}}{6.29 \text{ mW}} = 0.55$

or

$\qquad\qquad\quad = \cos \theta = \cos 56.3° = 0.55$

CALCULATOR SEQUENCE

Step	Keypad Entry	Display Response
1.	[3] [.] [3] [E] [3]	3.3E3
2.	[÷]	
3.	[2] [.] [2] [E] [3]	2.22E3
4.	[=]	1.5
5.	[inv] [tan]	56.309932

SELF-TEST EVALUATION POINT FOR SECTION 14-6

Now that you have completed this section, you should be able to:

■ *Objective 6. Describe impedance, phase angle, power, and power factor as they relate to a series and parallel RC circuit.*

Use the following questions to test your understanding of Section 14-6.

1. What is the phase relationship between current and voltage in a series RC circuit?

2. What is a phasor diagram?

3. Define and state the formula for impedance.

4. What is the phase angle or shift in:
   a. A purely resistive circuit?
   b. A purely capacitive circuit?
   c. A series circuit consisting of R and C?

14-7 PARALLEL *RC* CIRCUIT

Now that we have analyzed the characteristics of a series *RC* circuit, let us connect a resistor and capacitor in parallel.

14-7-1 *Voltage*

As with any parallel circuit, the voltage across all components in parallel is equal to the source voltage; therefore,

$$V_R = V_C = V_S$$

14-7-2 *Current*

In Figure 14-29(a), you will see a parallel circuit containing a resistor and a capacitor. The current through the resistor and capacitor is simply calculated by applying Ohm's law.

$$\text{(Resistor current) } I_R = \frac{V_S}{R}$$

$$\text{(Capacitor current) } I_C = \frac{V_S}{X_C}$$

Total current (I_T), however, is not as simply calculated. As expected, resistor current (I_R) is in phase with the applied voltage (V_S), as shown in the vector diagram in Figure 14-29(b). Capacitor current will always lead the applied voltage by 90°, and as the applied voltage is being used as our reference at 0° on the vector diagram, the capacitor current will have to be drawn at +90° in order to lead the applied voltage by 90°, since vector diagrams rotate in a counterclockwise direction.

Total current is therefore the vector sum of both the resistor and capacitor currents. Total current can be calculated by using the Pythagorean theorem:

$$I_T = \sqrt{I_R^2 + I_C^2}$$

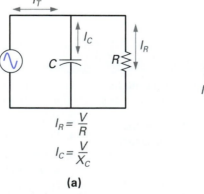

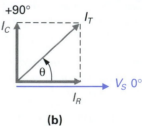

(a)

(b)

FIGURE 14-29 **Parallel *RC* Circuit.**

14-7-3　*Phase Angle*

The angle by which the total current (I_T) leads the source voltage (V_S) can be determined with either of the following formulas:

$$\theta = \text{invtan}\frac{I_C\ \text{(opposite)}}{I_R\ \text{(adjacent)}} \qquad \theta = \text{invtan}\frac{R}{X_C}$$

14-7-4　*Impedance*

Since the circuit is both capacitive and resistive, the total opposition or impedance of the parallel *RC* circuit can be calculated by

$$Z = \frac{V_S}{I_T}$$

The impedance of a parallel *RC* circuit is equal to the total voltage divided by the total current. Using basic algebra, we can rearrange this basic formula to express impedance in terms of reactance and resistance.

$$Z = \frac{R \times X_C}{\sqrt{R^2 \times X_C^2}}$$

14-7-5　*Power*

With respect to power, there is no difference between a series circuit and a parallel circuit. The true power or resistive power (P_R) dissipated by an R_C circuit is calculated with the formula

$$P_R = I_R^2 \times R$$

The imaginary power or reactive power (P_X) of the circuit can be calculated with the formula

$$P_X = I_C^2 \times X_C$$

The apparent power is equal to the vector sum of the true power and the reactive power.

$$P_A = \sqrt{P_R^2 + P_X^2}$$

As with series *RC* circuits the power factor is calculated as

$$PF = \frac{P_R\ \text{(resistive power)}}{P_A\ \text{(apparent power)}}$$

A power factor of 1 indicates a purely resistive circuit, whereas a power factor of 0 indicates a purely reactive circuit.

Calculate the following for a parallel RC circuit in which $R = 24\,\Omega$, $X_C = 14\,\Omega$, and $V_S = 10\,\text{V}$.

a. I_R

b. I_C

c. I_T

d. Z

e. θ

■ *Solution:*

a. $I_R = \dfrac{V_S}{R} = \dfrac{10\,\text{V}}{24\,\Omega} = 416.66\,\text{mA}$

b. $I_C = \dfrac{V_S}{X_C} = \dfrac{10\,\text{V}}{14\,\Omega} = 714.28\,\text{mA}$

c. $I_T = \sqrt{I_R^2 + I_C^2}$
$\quad = \sqrt{(416.66\,\text{mA})^2 + (714.28\,\text{mA})^2}$
$\quad = \sqrt{0.173 + 0.510}$
$\quad = 826.5\,\text{mA}$

d. $Z = \dfrac{V_S}{I_T} = \dfrac{10\,\text{V}}{826.5\,\text{mA}} = 12\,\Omega$

or

$\quad = \dfrac{R \times X_C}{\sqrt{R^2 + X_C^2}} = \dfrac{24 \times 14}{\sqrt{24^2 + 14^2}}$
$\quad = 12\,\Omega$

e. $\theta = \arctan \dfrac{I_C}{I_R} = \arctan \dfrac{714.28\,\text{mA}}{416.66\,\text{mA}}$
$\quad = \arctan 1.714 = 59.7°$

or

$\quad \theta = \arctan \dfrac{R}{X_C} = \arctan \dfrac{24\,\Omega}{14\,\Omega}$
$\quad = \arctan 1.714 = 59.7°$

CALCULATOR SEQUENCE

Step	Keypad Entry	Display Response
1.	7 1 4 . 2 8 E 3 +/−	714.28E-3
2.	÷	
3.	4 1 6 . 6 6 E 3 +/−	416.66E-3
4.	=	1.7142994
5.	inv tan	59.743762

SELF-TEST EVALUATION POINT FOR SECTION 14-7

Use the following questions to test your understanding of Section 14-7.

1. What is the phase relationship between current and voltage in a parallel RC circuit?

2. Could a parallel RC circuit be called a voltage divider?

3. State the formula used to calculate:

   a. I_T

   b. Z

4. Will capacitor current lead or lag resistor current in a parallel RC circuit?

REVIEW QUESTIONS

Multiple Choice Questions

1. When a capacitor charges:

   a. The voltage across the plates rises exponentially
   b. The circuit current falls exponentially
   c. The capacitor charges to the source voltage in $5RC$ seconds
   d. All the above

2. What is the capacitance of a capacitor if it can store 24 C of charge when 6 V is applied across the plates?

   a. $2\ \mu F$
   b. $3\ \mu F$
   c. $4.7\ \mu F$
   d. None of the above

3. The capacitance of a capacitor is directly proportional to:

   a. The plate area
   b. The distance between the plates
   c. The constant of the dielectric used
   d. Both (a) and (c)
   e. Both (a) and (b)

4. The capacitance of a capacitor is inversely proportional to:

   a. The plate area
   b. The distance between the plates
   c. The dielectric used
   d. Both (a) and (c)
   e. Both (a) and (b)

5. Total series capacitance of two capacitors is calculated by:

   a. Using the product-over-sum formula
   b. Using the voltage-divider formula
   c. Using the series resistance formula on the capacitors
   d. Adding all the individual values

6. Total parallel capacitance is calculated by:

   a. Using the product-over-sum formula
   b. Using the voltage-divider formula
   c. Using the parallel resistance formula on capacitors
   d. Adding all the individual values

7. In one time constant, a capacitor will charge to _____ of the source voltage.

   a. 86.5%
   b. 63.2%
   c. 99.3%
   d. 98.2%

8. A 47 μF capacitor charged to 6.3 V will have a stored charge of:

   a. 296.1 μC
   b. 2.96×10^{-4} C
   c. 0.296 mC
   d. All of the above
   e. Both (a) and (b)

9. Capacitive reactance is inversely proportional to:

   a. Capacitance and resistance
   b. Frequency and capacitance
   c. Capacitance and impedance
   d. Both (a) and (c)

10. The impedance of an RC series circuit is equal to:

    a. The sum of R and X_C
    b. The square root of the sum of R^2 and $X_C{}^2$
    c. The square of the sum of R and X_C
    d. The sum of the square root of R and X_C

11. In a purely resistive circuit:

    a. The current flowing in the circuit leads the voltage across the capacitor by 90°.
    b. The circuit current and resistor voltage are in phase with each other.
    c. The current leads the voltage by 45°.
    d. The current leads the voltage by a phase angle between 0° and 90°.

12. In a purely capacitive circuit:

    a. The current flowing in the circuit leads the voltage across the capacitor by 90°.
    b. The circuit current and resistor voltage are in phase with each other.
    c. The current leads the voltage by 45°.
    d. The current leads the voltage by a phase angle between 0° and 90°.

13. In a series circuit containing both capacitance and resistance:

    a. The current flowing in the circuit leads the voltage across the capacitor by 90°.
    b. The circuit current and resistor voltage are in phase with each other.
    c. The current leads the voltage by 45°.
    d. Both (a) and (b).

14. In a series RC circuit, the source voltage is equal to:

    a. The sum of V_R and V_C
    b. The difference between V_R and V_C
    c. The vectoral sum of V_R and V_C
    d. The sum of V_R and V_C squared

15. As the source frequency is increased, the capacitive reactance will:

    a. Increase
    b. Decrease
    c. Be unaffected
    d. Increase, depending on harmonic content

16. The phase angle of a series RC circuit indicates by what angle V_S _____ V_R.

    a. Lags   c. Leads or lags
    b. Leads   d. None of the above

17. In a series RC circuit, the vector combination of R and X_C is the circuit's _____.

    a. Phase angle   c. Source voltage
    b. Apparent power   d. Impedance

18. In a parallel RC circuit, the total current is equal to:

    a. The sum of I_R and I_C
    b. The difference between I_R and I_C
    c. The vectoral sum of I_R and I_C
    d. The sum of I_R and I_C squared

19. _____ is the opposition offered by a capacitor to current flow without the dissipation of energy.

   a. Capacitive reactance   **d.** Phase angle
   b. Resistance   **e.** The power factor
   c. Impedance

20. _____ is the total reactive and resistive circuit opposition to current flow.

   a. Capacitive reactance   **d.** Phase angle
   b. Resistance   **e.** The power factor
   c. Impedance

Communication Skill Questions

21. What happens to a capacitor during:

   a. Charge (14-4-1)
   b. Discharge (14-4-2)

22. Briefly explain the relationship between capacitance, charge, and voltage. (14-1-4)

23. Describe the three factors affecting the capacitance of a capacitor. (14-2)

24. List the formula(s) used to calculate total capacitance when capacitors are connected in:

   a. Parallel (14-3-1)
   b. Series (14-3-2)

25. Explain how the constant 63.2 is used in relation to the charge and discharge of a capacitor. (14-4)

26. Give the formula and define the term _capacitive reactance._ (14-5).

Practice Problems

30. If a 10 μF capacitor is charged to 10 V, how many coulombs of charge has it stored?

31. If a 0.006 μF capacitor has stored 125×10^{-6} C of charge, what potential difference will appear across the plates?

32. Calculate the capacitance of the capacitor that has the following parameter values: $A = 0.008$ m^2; $d = 0.00095$ m; the dielectric used is paper.

33. Calculate the total capacitance if the following are connected in:

   a. Parallel: 1.7 μF, 2.6 μF, 0.03 μF, 1200 pF
   b. Series: 1.6 μF, 1.4 μF, 4 μF

34. If three capacitors of 0.025 μF, 0.04 μF, and 0.037 μF are connected in series across a 12 V source, as shown in Figure 14-30, what will be the voltage drop across each?

27. In a series RC circuit, give the formulas for calculating: (14-6)

   a. V_S, when V_R and V_C are known
   b. Z, when R and X_C are known
   c. Z, when I and V are known
   d. θ, when X_C and R are known
   e. θ, when V_C and V_R are known
   f. Power factor, when R and Z are known
   g. Power factor, when P_R and P_A are known

28. In a parallel RC circuit, give the formulas for calculating: (14-7)

   a. I_R, when V and R are known
   b. I_C, when V and X_C are known
   c. I, when I_R and I_C are known
   d. Z, when V and I are known
   e. Z, when R and X_C are known

29. What is meant by _long_ or _short time constant,_ and do large or small values of RC produce a long or a short time constant? (14–4)

35. Calculate the capacitive reactance of the capacitor circuits with the following parameters:

   a. $f = 1$ kHz, $C = 2$ μF
   b. $f = 100$ Hz, $C = 0.01$ μF
   c. $f = 17.3$ MHz, $C = 47$ μF

36. In a series RC circuit, the voltage across the capacitor is 12 V and the voltage across the resistor is 6 V. Calculate the source voltage.

37. Calculate the impedance for the following series RC circuits:

   a. 2.7 MΩ, 3.7 μF, 20 kHz
   b. 350 Ω, 0.005 μF, 3 MHz
   c. $R = 8.6$ kΩ, $X_C = 2.4$ Ω
   d. $R = 4700$ Ω, $X_C = 2$ kΩ

38. In a parallel RC circuit with parameters of $V_S = 12$ V, $R = 4$ MΩ, and $X_C = 1.3$ kΩ, calculate:

   a. I_R **d.** Z
   b. I_C **e.** θ
   c. I_T

39. Calculate the total reactance in:

   a. A series circuit where $X_{C1} = 200$ Ω, $X_{C2} = 300$ Ω, $X_{C3} = 400$ Ω
   b. A parallel circuit where $X_{C1} = 3.3$ kΩ, $X_{C2} = 2.7$ kΩ

40. Calculate the capacitance needed to produce 10 kΩ of reactance at 20 kHz.

41. At what frequency will a 4.7 μF capacitor have a reactance of 2000 Ω?

FIGURE 14-30

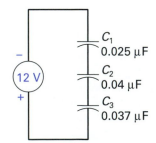

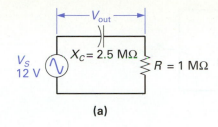

(a)

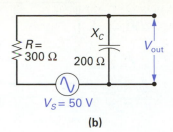

(b)

FIGURE 14-31

42. A series *RC* circuit contains a resistance of 40 Ω and a capacitive reactance of 33 Ω across a 24 V source.

    a. Sketch the schematic diagram.

    b. Calculate Z, I, V_R, V_C, I_R, I_C, and θ.

43. A parallel *RC* circuit contains a resistance of 10 kΩ and a capacitive reactance of 5 kΩ across a 100 V source.

    a. Sketch the schematic diagram.

    b. Calculate I_R, I_C, I_T, Z, V_R, V_C, and θ.

44. Calculate V_R and V_C for the circuits shown in Figure 14-31(a) and (b).

45. Calculate the impedance of the four circuits shown in Figure 14-32.

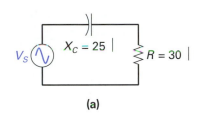

(a)

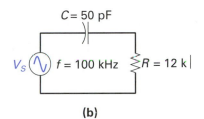

(b)

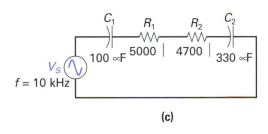

(c)

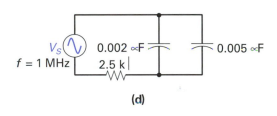

(d)

FIGURE 14-32

46. In Figure 14-33, the output voltage, since it is taken across the capacitor, will ——————— the voltage across the resistor by ——————— degrees.

47. If the positions of the capacitor and resistor in Figure 14-33 are reversed, the output voltage, since it is now taken across the resistor, will ——————— the voltage across the capacitor by ——————— degrees.

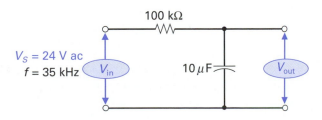

FIGURE 14-33

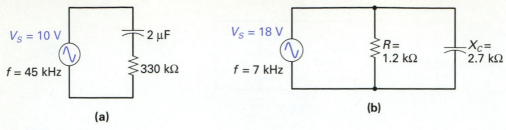

(a)

(b)

FIGURE 14-34

48. Calculate the resistive power, reactive power, apparent power, and power factor for the circuit seen in Figure 14-33; $V_{in} = 24$ V and $f = 35$ kHz.

49. Refer to Figure 14-34 and calculate the following:
   a. [Figure 14-34(a)] $X_C, I, Z, I_R, \theta, V_R, V_C$
   b. [Figure 14-34(b)] $V_R, V_C, I_R, I_C, I_T, Z, \theta$

Web Site Questions

Go to the Web site http://www.prenhall.com/cook, select the textbook *Mathematics for Electronics and Computers,* select this chapter, and then follow the instructions when answering the multiple-choice practice problems.

Inductors and Transformers

The Wizard of Menlo Park

Thomas Alva Edison was born to Samuel and Nancy Edison on February 11, 1847. As a young boy he had a keen and inquisitive mind, yet he did not do well at school, so his mother, a former school teacher, withdrew him from school and tutored him at home. In later life he said that his mother taught him to read well and instilled a love for books that lasted the rest of his life. In fact, the inventor's personal library of more than ten thousand volumes is preserved at the Edison Laboratory National Monument in West Orange, New Jersey.

At the age of twenty-nine, after several successful inventions, Edison put into effect what is probably his greatest idea—the first industrial research laboratory. Choosing Menlo Park in New Jersey, which was then a small rural village, Edison had a small building converted into a laboratory for his 15-member staff, and a house built for his wife and two small daughters. When asked to explain the point of this lab, Edison boldly stated that it would produce "a minor invention every ten days and a big thing every six months or so." At the time, most of the scientific community viewed his prediction as preposterous; however, in the next 10 years Edison would be granted 420 patents, including those for the electric lightbulb, the motion picture, the phonograph, the universal electric motor, the fluorescent lamp, and the medical fluoroscope.

Over one thousand patents were granted to Edison during his lifetime; his achievements at what he called his "invention factory" earned him the nickname "the wizard of Menlo Park." When asked about his genius he said, "Genius is two percent inspiration and ninety-eight percent perspiration."

15

Outline and Objectives

VIGNETTE: THE WIZARD OF MENLO PARK

INTRODUCTION

15-1 SELF-INDUCTION

Objective 1: Describe self-induction.

15-2 THE INDUCTOR

15-3 FACTORS DETERMINING INDUCTANCE

Objective 2: List and explain the factors affecting inductance.

Objective 3: Give the formula for inductance.

15-3-1 Number of Turns (N)

15-3-2 Area of Coil (A)

15-3-3 Length of Coil (l)

15-3-4 Core Material (μ)

15-3-5 Formula for Inductance

15-4 INDUCTORS IN COMBINATION

Objective 4: Identify inductors in series and parallel and understand how to calculate total inductance when inductors are in combination.

15-4-1 Inductors in Series

15-4-2 Inductors in Parallel

15-5 INDUCTIVE TIME CONSTANT

Objective 5: Explain the inductive time constant.

15-5-1 DC Current Rise

15-5-2 DC Current Fall

15-5-3 AC Rise and Fall

15-6 INDUCTIVE REACTANCE

Objective 6: Give the formula for inductive reactance.

15-7 SERIES *RL* CIRCUIT

Objective 7: Describe all aspects relating to a series *RL* circuit.

15-7-1 Voltage

15-7-2 Impedance (Z)

15-7-3 Phase Shift

15-7-4 Power
Purely Resistive Circuit
Purely Inductive Circuit
Resistive and Inductive Circuit
Power Factor

15-8 PARALLEL *RL* CIRCUIT

Objective 8: Describe all aspects relating to a parallel *RL* circuit.

15-8-1 Current

15-8-2 Phase Angle

15-8-3 Impedance

15-8-4 Power

15-9 MUTUAL INDUCTANCE

Objective 9: Define mutual inductance and how it relates to transformers.

15-10 BASIC TRANSFORMER

Objective 10: Describe the basic operation of a transformer.

15-11 TRANSFORMER LOADING

Objective 11: Explain the differences between a loaded and unloaded transformer.

15-12 **TRANSFORMER RATIOS AND APPLICATIONS**

Objective 12: List the three basic applications of transformers.

Objective 13: Describe how a transformer's turns ratio can be used to step up or step down voltage or current, or to match impedances.

15-12-1 Turns Ratio

15-12-2 Voltage Ratio
Step Up
Step Down

15-12-3 Power and Current Ratio

15-12-4 Impedance Ratio

15-13 **TRANSFORMER RATINGS**

Introduction

There is no physical difference between an inductor and an electromagnet, since they are both coils. The two devices are given different names because they are used in different applications even though their construction and principle of operation are the same. An electromagnet is used to generate a magnetic field in response to current, while an inductor is used to oppose any change in current.

When alternating current was introduced in Chapter 13, it was mentioned that an ac voltage could be stepped up to a larger voltage or stepped down to a smaller voltage by a device called a **transformer.** The transformer is an electrical device that makes use of electromagnetic induction to transfer alternating current from one circuit to another. The transformer consists of two inductors that are placed in very close proximity to each other. When an alternating current flows through the first coil or **primary winding** the inductor sets up a magnetic field. The expanding and contracting magnetic field produced by the primary cuts across the windings of the second inductor or **secondary winding** and induces a voltage in this coil.

By changing the ratio between the number of turns in the secondary winding to the number of turns in the primary winding, some characteristics of the ac signal can be changed or transformed as it passes from primary to secondary. For example, a low ac voltage can be stepped up to a higher ac voltage, or a high ac voltage can be stepped down to a lower ac voltage.

15-1 SELF-INDUCTION

In Figure 15-1 an external voltage source has been connected across a coil, forcing a current through the coil. This current will generate a magnetic field that will increase in field strength in a short time from zero to maximum, expanding from the center of the conductor (electromagnetism). The expanding magnetic lines of force have relative motion with respect to the stationary conductor, so an induced voltage results (electromagnetic induction). The blooming magnetic field generated by the conductor actually causes a voltage to be induced in the conductor that is generating the magnetic field. This effect of a current-carrying coil of conductor inducing a voltage within itself is known as **self-inductance.** This phenomenon was first discovered by Heinrich Lenz, who observed that the induced voltage causes an induced (bucking) current to flow in the coil, which opposes the source current producing it.

Figure 15-2(a) shows an inductor connected across a dc source. When the switch is closed, a circuit current will exist through the inductor and the resistor. As the current rises toward its maximum value the magnetic field expands, and throughout this time of relative

Self-Inductance

The property that causes a counterelectromotive force to be produced in a conductor when the magnetic field expands or collapses with a change in current.

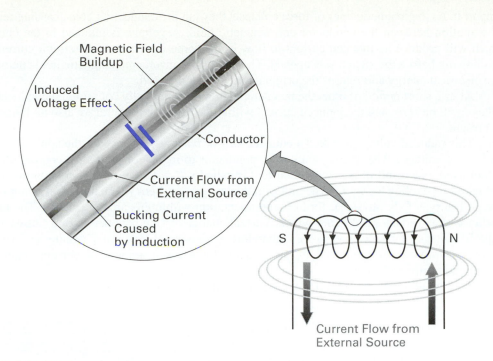

FIGURE 15-1 Self-Induction of a Coil.

motion between field and conductor, an induced voltage will be present. This induced voltage will produce an induced current to oppose the change in the circuit current.

When the current reaches its maximum, the magnetic field, which is dependent on current, will also reach a maximum value and then no longer expand but remain stationary. When the current remains constant, no change will occur in the magnetic field and therefore no relative motion will exist between the conductor and magnetic field, resulting in no induced voltage or current to oppose circuit current, as shown in Figure 15-2(b). The coil has accepted electrical energy and is storing it in the form of a magnetic energy field, just as the capacitor stored electrical energy in the form of an electric field.

If the switch is put in position *B*, as shown in Figure 15-2(c), the current from the battery will be zero, and the magnetic field will collapse, as it no longer has circuit current to

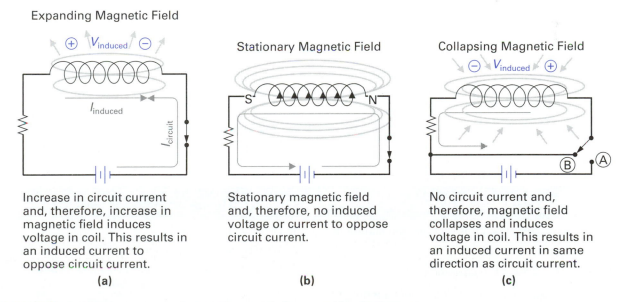

Expanding Magnetic Field

Increase in circuit current and, therefore, increase in magnetic field induces voltage in coil. This results in an induced current to oppose circuit current.

(a)

Stationary Magnetic Field

Stationary magnetic field and, therefore, no induced voltage or current to oppose circuit current.

(b)

Collapsing Magnetic Field

No circuit current and, therefore, magnetic field collapses and induces voltage in coil. This results in an induced current in same direction as circuit current.

(c)

FIGURE 15-2 Self-Inductance. (a) Switch Closed. (b) Constant Circuit Current. (c) Switch Opened.

Counter emf (Counter Electromotive Force)

Abbreviated "counter emf," or "back emf," it is the voltage generated in an inductor due to an alternating or pulsating current and is always of opposite polarity to that of the applied voltage.

Inductance

Property of a circuit or component to oppose any change in current as the magnetic field produced by the change in current causes an induced countercurrent to oppose the original change.

Henry

Unit of inductance.

TIME LINE

Joseph Henry (1797–1878), a U.S. physicist, conducted extensive studies into electromagnetism. Henry was the first to insulate the magnetic coil of wire and developed coils for telegraphy and motors. In recognition of his discovery of self-induction in 1832, the unit of inductance is called the henry.

support it. As the magnetic lines of force collapse, they cut the conducting coils, causing relative motion between the conductor and magnetic field. A voltage is induced in the coil, which will produce an induced current to flow in the same direction as the circuit current was flowing before the switch was opened. The coil now converts the magnetic field energy into electrical energy and returns the original energy that it stored.

After a short period of time, the magnetic field will have totally collapsed, the induced voltage will be zero, and the induced current within the circuit will therefore also no longer be present.

This induced voltage is called a **counter emf** or *back emf*. It opposes the applied emf (or battery voltage). The ability of a coil or conductor to induce or produce a counter emf within itself as a result of a change in current is called *self-inductance,* or more commonly **inductance** (symbolized by L). The unit of inductance is the **henry** (H), named in honor of Joseph Henry, a U.S. physicist, for his experimentation within this area of science. The inductance of an inductor is 1 henry when a current change of 1 ampere per second causes an induced voltage of 1 volt. Inductance is therefore a measure of how much counter emf (induced voltage) can be generated by an inductor for a given amount of current change through that same inductor.

This counter emf or induced voltage can be calculated by the formula

$$V_{ind} = L \times \frac{\Delta I}{\Delta t}$$

where L = inductance, in henrys (H)
   ΔI = increment of change of current (I)
   Δt = increment of change with respect to time (t)

A larger inductance ($L\uparrow$) will create a larger induced voltage ($V_{ind}\uparrow$), and if the rate of change of current with respect to time is increased ($\Delta I/\Delta t\uparrow$), the induced voltage or counter emf will also increase ($V_{ind}\uparrow$).

■ INTEGRATED MATH APPLICATION:

What voltage is induced across an inductor of 4 H when the current is changing at a rate of:

a. 1 A/s?

b. 4 A/s?

■ *Solution:*

a. $V_{ind} = L \times \dfrac{\Delta I}{\Delta t} = 4\ \text{H} \times 1\ \text{A/s} = 4\ \text{V}$

b. $V_{ind} = L \times \dfrac{\Delta I}{\Delta t} = 4\ \text{H} \times 4\ \text{A/s} = 16\ \text{V}$

The faster the coil current changes, the larger the induced voltage.

SELF-TEST EVALUATION POINT FOR SECTION 15-1

Now that you have completed this section, you should be able to:

■ *Objective 1. Describe self-induction.*

Use the following questions to test your understanding of Section 15-1.

1. Define self-induction.

2. What is counter emf, and how can it be calculated?

3. Calculate the voltage induced in a 2 mH inductor if the current is increasing at a rate of 4 kA/s.

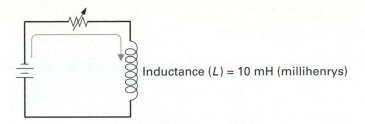

Inductance (L) = 10 mH (millihenrys)

FIGURE 15-3 Inductor's Ability to Oppose Current Change.

15-2   THE INDUCTOR

An **inductor** is basically an electromagnet, as its construction and principle of operation are the same. We use the two different names because they have different applications. The purpose of the electromagnet or solenoid is to generate a magnetic field, whereas the purpose of an inductor or coil is to oppose any change of circuit current.

In Figure 15-3 a steady value of direct current is present within the circuit, and the inductor is creating a steady or stationary magnetic field. If the current in the circuit is suddenly increased (by lowering the circuit resistance), the change in the expanding magnetic field will induce a counter emf within the inductor. This induced voltage will oppose the source voltage from the battery and attempt to hold current at its previous low level.

The counter emf cannot completely oppose the current increase, for if it did, the lack of current change would reduce the counter emf to zero. Current therefore incrementally increases up to a new maximum, which is determined by the applied voltage and the circuit resistance ($I = V/R$). Once the new higher level of current has been reached and remains constant, there will no longer be a change. This lack of relative motion between field and conductor will no longer generate a counter emf, so the current will remain at its new higher constant value.

This effect also happens in the opposite respect. If current decreases (by increasing circuit resistance), the magnetic lines of force will collapse because of the reduction of current and induce a voltage in the inductor, which will produce an induced current in the same direction as the circuit current. These two combine and tend to maintain the current at the higher previous constant level. Circuit current will fall, however, as the induced voltage and current are present only during the change (in this case the decrease from the higher current level to the lower); and once the new lower level of current has been reached and remains constant, the lack of change will no longer induce a voltage or current, so the current will then remain at its new lower constant value.

The inductor is therefore an electronic component that will oppose any changes in circuit current, and this ability or behavior is referred to as *inductance*. Since the current change is opposed by a counter emf, inductance may also be defined as the ability of a device to induce a counter emf within itself for a change in current.

SELF-TEST EVALUATION POINT FOR SECTION 15-2

Use the following questions to test your understanding of Section 15-2.

1. What is the difference between an electromagnet and an inductor?

2. True or false: The inductor will oppose any changes in circuit current.

15-3   FACTORS DETERMINING INDUCTANCE

The inductance of an inductor is determined by four factors:

1. Number of turns
2. Area of the coil

3. Length of the coil

4. Core material used within the coil

Let's now discuss how these four factors can affect inductance, beginning with the number of turns.

15-3-1 *Number of Turns (N) (Figure 15-4)*

If an inductor has a greater number of turns, the magnetic field produced by passing current through the coil will have more magnetic force than an inductor with fewer turns. A greater magnetic field will cause a larger counter emf, because more magnetic lines of flux will cut more coils of the conductor, producing a larger inductance value. Inductance (L) is therefore proportional to the number of turns (N):

$$L \propto N$$

15-3-2 *Area of Coil (A) (Figure 15-5)*

If the area of the coil is increased for a given number of turns, more magnetic lines of force will be produced, and if the magnetic field is increased, the inductance will also increase. Inductance (L) is therefore proportional to the cross-sectional area of the coil (A):

$$L \propto A$$

15-3-3 *Length of Coil (l) (Figure 15-6)*

If, for example, four turns are spaced out (long-length coil), the summation that occurs between all the individual coil magnetic fields will be small. On the other hand, if four turns are wound close to one another (short-length coil), all the individual coil magnetic

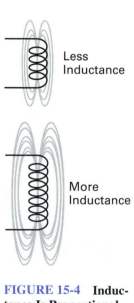

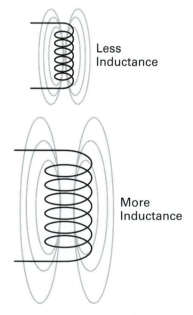

FIGURE 15-4   Inductance Is Proportional to the Number of Turns ($L \propto N$) in a Coil.

FIGURE 15-5   Inductance Is Proportional to the Area ($L \propto A$) in a Coil.

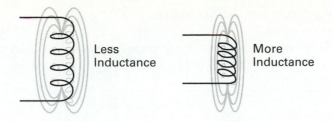

FIGURE 15-6   Inductance Is Inversely Proportional to the Length ($L \propto 1/l$) of a Coil.

fields will easily interact and add together to produce a larger magnetic field and, therefore, greater inductance. Inductance is therefore inversely proportional to the length of the coil, in that a longer coil, for a given number of turns, produces a smaller inductance, and vice versa.

$$L \propto \frac{1}{l}$$

15-3-4   *Core Material (μ) (Figure 15-7)*

Most inductors have core materials such as nickel, cobalt, iron, steel, ferrite, or an alloy. These cores have magnetic properties that concentrate or intensify the magnetic field. Permeability is another factor that is proportional to inductance, and the values for various materials are shown in Table 15-1. The greater the permeability of the core material, the greater the inductance.

$$L \propto \mu$$

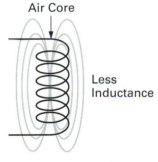

Air Core

Less Inductance

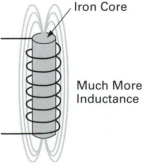

Iron Core

Much More Inductance

FIGURE 15-7   Inductance Is Proportional to the Permeability of the Coil Material ($L \propto \mu$).

TABLE 15-1   Permeabilities of Various Materials

MATERIAL	PERMEABILITY(μ)
Air or vacuum	1.26×10^{-6}
Nickel	6.28×10^{-5}
Cobalt	7.56×10^{-5}
Cast iron	1.1×10^{-4}
Machine steel	5.65×10^{-4}
Transformer iron	6.9×10^{-3}
Silicon iron	8.8×10^{-3}
Permalloy	0.126
Superalloy	1.26

15-3-5   *Formula for Inductance*

All four factors described can be placed in a formula for calculating inductance.

$$L = \frac{N^2 \times A \times \mu}{l}$$

where L = inductance, in henrys (H)
        N = number of turns
        A = cross-sectional area, in square meters (m²)
        μ = permeability
        l = length of core, in meters (m)

■   INTEGRATED MATH APPLICATION:

Refer to Figure 15-8(a) and (b) and calculate the inductance of each.

■   *Solution:*

a. $L = \dfrac{5^2 \times 0.01 \times (6.28 \times 10^{-5})}{0.001} = 15.7 \text{ mH}$

b. $L = \dfrac{10^2 \times 0.1 \times (1.1 \times 10^{-4})}{0.1} = 11 \text{ mH}$

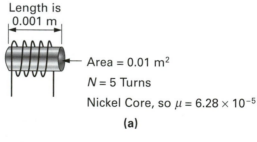

Length is
0.001 m

Area = 0.01 m²

N = 5 Turns

Nickel Core, so $\mu = 6.28 \times 10^{-5}$

(a)

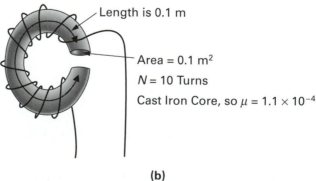

Length is 0.1 m

Area = 0.1 m²

N = 10 Turns

Cast Iron Core, so $\mu = 1.1 \times 10^{-4}$

(b)

FIGURE 15-8   **Inductor Examples.**

Now that you have completed this section, you should be able to:

■ **Objective 2.** *List and explain the factors affecting inductance.*

■ **Objective 3.** *Give the formula for inductance.*
Use the following questions to test your understanding of Section 15-3.

1. List the four factors that determine the inductance of an inductor.

15-4 INDUCTORS IN COMBINATION

Inductors oppose the change of current in a circuit and so are treated in a manner similar to that used for resistors connected in combination. Two or more inductors in series merely extend the coil length and increase inductance. Inductors in parallel are treated in a manner similar to that used for resistors, with the total inductance being less than the value of the smallest inductor.

15-4-1 *Inductors in Series*

When inductors are connected in series with one another, the total inductance is calculated by summing all the individual inductances.

$$L_T = L_1 + L_2 + L_3 + \cdots$$

■ **INTEGRATED MATH APPLICATION:**

Calculate the total inductance of the circuit shown in Figure 15-9.

■ *Solution:*

$$L_T = L_1 + L_2 + L_3$$
$$= 5 \text{ mH} + 7 \text{ mH} + 10 \text{ mH}$$
$$= 22 \text{ mH}$$

FIGURE 15-9 Inductors in Series.

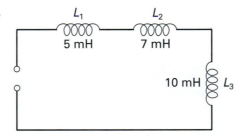

15-4-2 *Inductors in Parallel*

When inductors are connected in parallel with one another, the reciprocal (two or more inductors) or product-over-sum (two inductors) formula can be used to find total inductance, which will always be less than the value of the smallest inductor.

$$L_T = \frac{1}{(1/L_1) + (1/L_2) + (1/L_3) + \ldots}$$

$$L_T = \frac{L_1 \times L_2}{L_1 + L_2}$$

■ **INTEGRATED MATH APPLICATION:**

Determine L_T for the circuits in Figure 15-10(a) and (b).

■ *Solution:*

a. Reciprocal formula:

$$L_T = \frac{1}{(1/L_1) + (1/L_2) + (1/L_3)}$$

$$= \frac{1}{(1/10 \text{ mH}) + (1/5 \text{ mH}) + (1/20 \text{ mH})}$$

$$= 2.9 \text{ mH}$$

b. Product over sum:

$$L_T = \frac{L_1 \times L_2}{L_1 + L_2}$$

$$= \frac{10 \ \mu\text{H} \times 2 \ \mu\text{H}}{10 \ \mu\text{H} + 2 \ \mu\text{H}}$$

$$= \frac{20 \times 10^{-12} \text{H}}{12 \ \mu\text{H}} = 1.67 \ \mu\text{H}$$

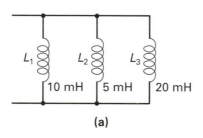

(a)

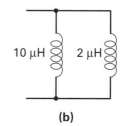

(b)

FIGURE 15-10　**Inductors in Parallel.**

SELF-TEST EVALUATION POINT FOR SECTION 15-4

Now that you have completed this section, you should be able to:

■ *Objective 4.　Identify inductors in series and parallel and understand how to calculate total inductance when inductors are in combination.*

Use the following questions to test your understanding of Section 15-4.

1. True or false: In calculating total inductance, inductors can be treated in the same manner as capacitors.

2. State the formula for calculating total inductance in:
   a. A series circuit
   b. A parallel circuit

3. Calculate the total circuit inductance if 4 mH and 2 mH are connected:
   a. In series
   b. In parallel

15-5 INDUCTIVE TIME CONSTANT

Inductors will not have any effect on a steady value of direct current (dc) from a dc voltage source. If, however, the dc is changing (pulsating), the inductor will oppose the change whether it is an increase or decrease in direct current, because a change in current causes the magnetic field to expand or contract, and in so doing it cuts the coil of the inductor and induces a voltage that counters the applied emf.

15-5-1 *DC Current Rise*

Figure 15-11(a) illustrates an inductor (L) connected across a dc source (battery) through a switch and a series-connected resistor. When the switch is closed, current will flow and the magnetic field will begin to expand around the inductor. This field cuts the coils of the inductor and induces a counter emf to oppose the rise in current. Current in an inductive circuit therefore cannot rise instantly to its maximum value, which is determined by Ohm's law ($I = V/R$). Current will in fact take a time to rise to maximum, as graphed in Figure 15-11(b), due to the inductor's ability to oppose change.

It will actually take five time constants (5τ) for the current in an inductive circuit to reach maximum value. This time can be calculated by using the formula

$$\tau = \frac{L}{R} \quad \text{seconds}$$

The constant to remember is the same as before: 63.2%. In one time constant ($1 \times L/R$) the current in the RL circuit reaches 63.2% of its maximum value. In two time constants ($2 \times L/R$), the current increases by 63.2% of the remaining current, and so on, through five time constants.

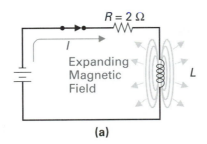

(a)

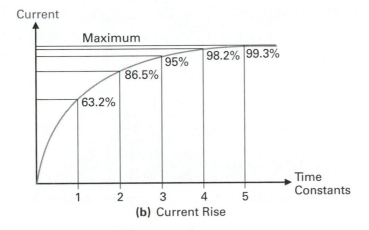

(b) Current Rise

FIGURE 15-11 **DC Inductor Current Rise.**

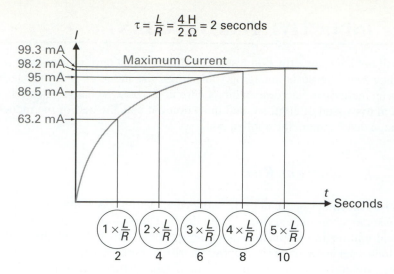

$$\tau = \frac{L}{R} = \frac{4\,H}{2\,\Omega} = 2 \text{ seconds}$$

FIGURE 15-12   **Exponential Current Rise.**

For example, if the maximum possible circuit current is 100 mA and an inductor of 4 H is connected in series with a resistor of 2 Ω, the current will increase as shown in Figure 15-12.

Referring to the $\tau = L/R$ formula, you will notice that how quickly an inductor will allow the current to rise to its maximum value is proportional to the inductance and inversely proportional to the resistance. A larger inductance increases the strength of the magnetic field, so the opposition or counter emf increases, and the longer it takes for current to rise to a maximum ($\tau\uparrow = L\uparrow/R$). If the circuit resistance is increased, the maximum current will be smaller, and a smaller maximum is reached more quickly than a higher ($\tau\downarrow = L/R\uparrow$).

15-5-2   *DC Current Fall*

When the inductor's dc source of current is removed, as shown in Figure 15-13(a) by placing the switch in position *B,* the magnetic field will collapse and cut the coils of the inductor, inducing a voltage and causing a current to flow in the same direction as the original source current. This current will exponentially decay, or fall from the maximum to zero level, in five time constants ($5 \times L/R = 5 \times 4/2 = 10$ seconds), as shown in Figure 15-13(b).

■   **INTEGRATED MATH APPLICATION:**

Calculate the circuit current at each of the five time constants if a 12 V dc source is connected across a series *RL* circuit, $R = 60\ \Omega$ and $L = 24$ mH. Plot the results on a graph showing current against time.

■   *Solution:*

$$\text{Maximum current, } I_{max} = \frac{V_S}{R} = \frac{12\,V}{60\,\Omega} = 200 \text{ mA}$$

$$\text{Time constant, } \tau = \frac{L}{R} = \frac{24\,mH}{60\,\Omega} = 400\ \mu s$$

At one time constant (400 μs after source voltage is applied), the current will be

$$I = 63.2\% \text{ of } I_{max}$$
$$= 0.632 \times 200 \text{ mA} = 126.4 \text{ mA}$$

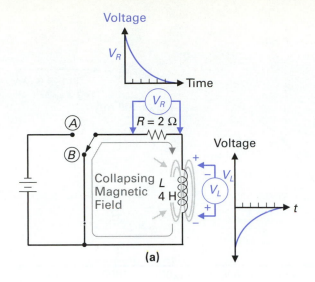

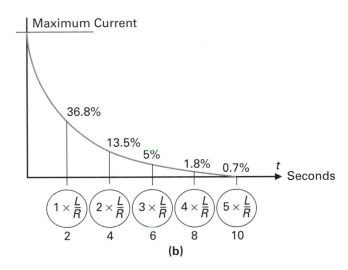

FIGURE 15-13 **Exponential Current Fall.**

At two time constants (800 μs after source voltage is applied):

$$I = 86.5\% \text{ of } I_{max}$$
$$= 0.865 \times 200 \text{ mA} = 173 \text{ mA}$$

At three time constants (1200 μs or 1.2 ms):

$$I = 95\% \text{ of } I_{max}$$
$$= 0.95 \times 200 \text{ mA} = 190 \text{ mA}$$

At four time constants (1.6 ms):

$$I = 98.2\% \text{ of } I_{max}$$
$$= 0.982 \times 200 \text{ mA} = 196.4 \text{ mA}$$

At five time constants (2 ms):

$$I = 99.3\% \text{ of } I_{max}$$
$$= 0.993 \times 200 \text{ mA} = 198.6 \text{ mA, approximately maximum (200 mA)}$$

See Figure 15-14.

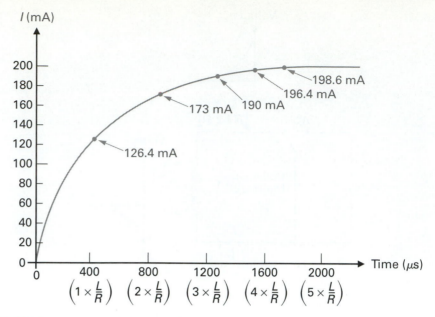

FIGURE 15-14 Exponential Current Rise Example.

15-5-3 *AC Rise and Fall*

If an alternating (ac) voltage is applied across an inductor, as shown in Figure 15-15(a), the inductor will continuously oppose the alternating current because it is always changing. Figure 15-15(b) shows the phase relationship between the voltage across an inductor or counter emf and the circuit current.

The current in the circuit causes the magnetic field to expand and collapse and cut the conducting coils, resulting in an induced counter emf. At points X and Y, the steepness of the current waveform indicates that the current will be changing at its maximum rate, and therefore the opposition or counter emf will also be maximum. When the current is at its maximum positive or negative value, it has a very small or no rate of change (flat peaks). Therefore, the opposition or counter emf should be very small or zero, as can be seen by the waveforms. The counter emf is, therefore, said to be 90° out of phase with the circuit current.

To summarize, we can say that the voltage across the inductor (V_L) or counter emf leads the circuit current (I) by 90°.

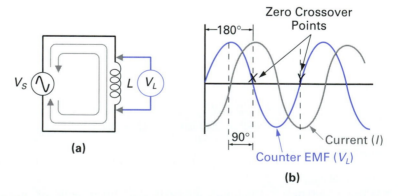

FIGURE 15-15 The Voltage across an Inductor Leads the Circuit Current by 90° in an Inductive Circuit.

Now that you have completed this section, you should be able to:

■ **Objective 5.** *Explain the inductive time constant.*

Use the following questions to test your understanding of Section 15-5.

1. How does the inductive time constant relate to the capacitive time constant?

2. True or false: The greater the value of the inductor, the longer it would take for current to rise to a maximum.

3. True or false: A constant dc level is opposed continuously by an inductor.

4. What reaction does an inductor have to ac?

15-6 INDUCTIVE REACTANCE

Reactance is the opposition to current flow without the dissipation of energy, as opposed to resistance, which is the opposition to current flow with the dissipation of energy.

Inductive reactance (X_L) is the opposition to current flow offered by an inductor without the dissipation of energy. It is measured in ohms and can be calculated by using the formula:

$$X_L = 2\pi \times f \times L$$

> **Inductive Reactance**
> Measured in ohms, it is the opposition to alternating or pulsating current flow without the dissipation of energy.

where X_L = inductive reactance, in ohms (Ω)
  2π = 2π radians, 360° or 1 cycle
  f = frequency, in hertz (Hz)
  L = inductance, in henrys (H)

Inductive reactance is proportional to frequency ($X_L \propto f$) because a higher frequency (fast-switching current) will cause a greater amount of current change, and a greater change will generate a larger counter emf, which is an opposition or reactance against current flow. When 0 Hz is applied to a coil (dc), no change exists, so the inductive reactance of an inductor to dc is zero ($X_L = 2\pi \times 0 \times L = 0$).

Inductive reactance is also proportional to inductance because a larger inductance will generate a greater magnetic field and subsequent counter emf, which is the opposition to current flow.

Ohm's law can be applied to inductive circuits just as it can be applied to resistive and capacitive circuits. The current flow in an inductive circuit (I) is proportional to the voltage applied (V), and inversely proportional to the inductive reactance (X_L). Expressed mathematically,

$$I = \frac{V}{X_L}$$

■ **INTEGRATED MATH APPLICATION:**

Calculate the current flowing in the circuit illustrated in Figure 15-16.

■ *Solution:*

The current can be calculated by Ohm's law and is a function of the voltage and opposition, which in this case is inductive reactance.

$$I = \frac{V}{X_L}$$

FIGURE 15-16

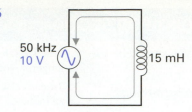

However, we must first calculate X_L:

$$X_L = 2\pi \times f \times L$$
$$= 6.28 \times 50 \text{ kHz} \times 15 \text{ mH}$$
$$= 4710 \ \Omega \text{ or } 4.71 \text{ k}\Omega$$

Current is therefore equal to

$$I = \frac{V}{X_L} = \frac{10 \text{ V}}{4.71 \text{ k}\Omega} = 2.12 \text{ mA}$$

■ INTEGRATED MATH APPLICATION:

What opposition or inductive reactance will a motor winding or coil offer if $V = 12$ V and $I = 4.5$ A?

■ *Solution:*

$$X_L = \frac{V}{I} = \frac{12 \text{ V}}{4.5 \text{ A}} = 2.66 \ \Omega$$

SELF-TEST EVALUATION POINT FOR SECTION 15-6

Now that you have completed this section, you should be able to:

■ *Objective 6.* *Give the formula for inductive reactance.*

Use the following questions to test your understanding of Section 15-6.

1. Define and state the formula for inductive reactance.
2. Why is inductive reactance proportional to frequency and the inductance value?
3. How does inductive reactance relate to Ohm's law?
4. True or false: Inductive reactance is measured in henrys.

15-7   SERIES *RL* CIRCUIT

In a purely resistive circuit, as seen in Figure 15-17, the current flowing within the circuit and the voltage across the resistor are in phase with each other. In a purely inductive circuit, as shown in Figure 15-18, the current will lag the applied voltage by 90°. If we connect a resistor and inductor in series, as shown in Figure 15-19(a), we will have the most common combination of R and L used in electronic equipment.

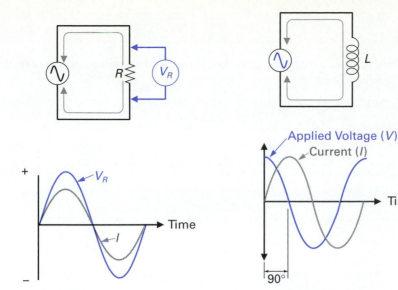

FIGURE 15-17 **Purely Resistive Circuit: Current and Voltage Are in Phase.**

FIGURE 15-18 **Purely Inductive Circuit: Current Lags Applied Voltage by 90°.**

15-7-1 *Voltage*

The voltage across the resistor and inductor shown in Figure 15-19(a) can be calculated by using Ohm's law:

$$V_R = I \times R$$

$$V_L = I \times X_L$$

The vector diagram in Figure 15-19(b) illustrates current (I) as the reference at 0°, and as expected, the voltage across the resistor (V_R) is in phase with the circuit current. The voltage across the inductor (V_L) leads the circuit current and the voltage across the resistor (V_R) by 90°, or the circuit current vector lags the voltage across the inductor by 90°.

As with any series circuit, we have to apply Kirchhoff's voltage law when calculating the value of applied or source voltage (V_S), which, due to the phase difference between V_R and V_L, is the vector sum of all the voltage drops. By creating a right triangle from the three quantities, as shown in Figure 15-19(c), and applying the Pythagorean theorem, we arrive at a formula for source voltage.

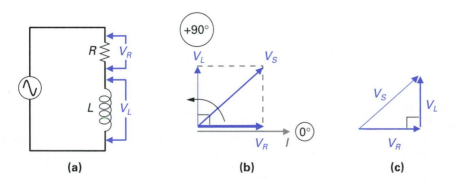

FIGURE 15-19 **Series *RL* Circuit.**

$$V_S = \sqrt{V_R^2 + V_L^2}$$

As with any formula with three quantities, if two are known we can calculate the other by simply rearranging the formula to

$$V_S = \sqrt{V_R^2 + V_L^2}$$

$$V_R = \sqrt{V_S^2 - V_L^2}$$

$$V_L = \sqrt{V_S^2 - V_R^2}$$

■ **INTEGRATED MATH APPLICATION:**

Calculate V_R, V_L, and V_S for the circuit shown in Figure 15-20

■ *Solution:*

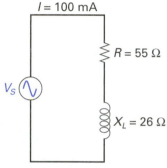

I = 100 mA

$R = 55\ \Omega$

V_S

$X_L = 26\ \Omega$

FIGURE 15-20 **Voltage in a Series *RL* Circuit.**

$$V_R = I \times R$$
$$= 100\ \text{mA} \times 55\ \Omega$$
$$= 5.5\ \text{V}$$
$$V_L = I \times X_L$$
$$= 100\ \text{mA} \times 26\ \Omega$$
$$= 2.6\ \text{V}$$

$$V_S = \sqrt{V_R^2 + V_L^2}$$

$$= \sqrt{(5.5\ \text{V})^2 + (2.6\ \text{V})^2}$$
$$= 6\ \text{V}$$

15-7-2   *Impedance (Z)*

Impedance is the total opposition to current flow offered by a circuit with both resistance and reactance. It is measured in ohms and can be calculated by using Ohm's law:

$$Z = \frac{V}{I}$$

Just as a phase shift or difference exists between V_R and V_L and they cannot be added to find applied voltage, the same phase difference exists between R and X_L, so impedance or total opposition cannot simply be the sum of the two, as shown in Figure 15-21

FIGURE 15-21   **Impedance in a Series *RL* Circuit.**

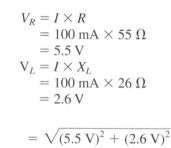

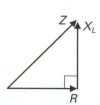

The impedance of a series R_L circuit is equal to the square root of the sum of the squares of resistance and reactance, and by rearrangement, X_L and R can also be calculated if the other two values are known:

$$Z = \sqrt{R^2 + X_L^2}$$

$$R = \sqrt{Z^2 - X_L^2}$$

$$X_L = \sqrt{Z^2 - R^2}$$

■ **INTEGRATED MATH APPLICATION:**

Referring to Figure 15-20, calculate Z.

■ *Solution:*

$$\begin{aligned}
Z &= \sqrt{R^2 + X_L^2} \\
&= \sqrt{55^2 + 26^2} \\
&= 60.8 \ \Omega
\end{aligned}$$

15-7-3 *Phase Shift*

If a circuit is purely resistive, the phase shift (θ) between the source voltage and circuit current is zero. If a circuit is purely inductive, voltage leads current by 90°; therefore, the phase shift is +90°. If the resistance and inductive reactance are equal, the phase shift will equal +45°, as shown in Figure 15-22.

The phase shift in an inductive and resistive circuit is the degrees of lead between the source voltage (V_S) and current (I), and by looking at the examples in Figure 15-22, you can see that the phase angle is proportional to reactance and inversely proportional to resistance. Mathematically, the phase shift can be expressed as

$$\theta = \arctan \frac{X_L}{R}$$

Because the current is the same in both the inductor and resistor in a series circuit, the voltage drops across the inductor and resistor are directly proportional to reactance and resistance:

$$V_R \updownarrow = I \text{ (constant)} \times R\updownarrow, \quad V_L\updownarrow = I \text{ (constant)} \times X_L\updownarrow$$

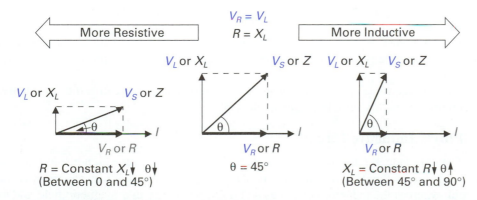

FIGURE 15-22 **Phase Shift (θ) in a Series RL Circuit.**

The phase shift can also be calculated by using the voltage drop across the inductor and resistor:

$$\theta = \arctan \frac{V_L}{V_R}$$

■ **INTEGRATED MATH APPLICATION:**

Referring to Figure 15-20, calculate the phase shift between source voltage and circuit current.

■ *Solution:*

Since the ratio of X_L/R and V_L/V_R are known for this example circuit, either of the phase shift formulas can be used.

$$\begin{aligned}
\theta &= \arctan \frac{X_L}{R} \\
&= \arctan \frac{26\ \Omega}{55\ \Omega} \\
&= \arctan 0.4727 \\
&= 25.3°
\end{aligned}$$

or

$$\begin{aligned}
\theta &= \arctan \frac{V_L}{V_R} \\
&= \arctan \frac{2.6\ \text{V}}{5.5\ \text{V}} \\
&= \arctan 0.4727 \\
&= 25.3°
\end{aligned}$$

The source voltage in this example circuit leads the circuit current by 25.3°.

15-7-4 *Power*

Purely Resistive Circuit

Figure 15-23 illustrates the current, voltage, and power waveforms produced when an ac voltage is applied across a purely resistive circuit. Voltage and current are in phase, and true power (P_R) in watts can be calculated by multiplying current by voltage ($P = V \times I$).

The sinusoidal power waveform is totally positive, as a positive voltage multiplied by a positive current produces a positive value of power, and a negative voltage multiplied by a negative current also produces a positive value of power. For this reason, the resistor is said to develop a positive power waveform that is twice the frequency of the voltage or current waveform.

The average value of power dissipated by a purely resistive circuit is the value halfway between maximum and zero, in this example 4 W.

Purely Inductive Circuit

The pure inductor, like the capacitor, is a reactive component, which means that it will consume power without the dissipation of energy. The capacitor holds its energy in an electric field, whereas the inductor consumes and holds its energy in a magnetic field and then releases it back into the circuit.

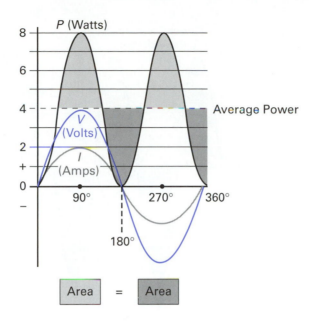

FIGURE 15-23 Purely Resistive Circuit.

Voltage and Current Are in RMS (Effective) as Normal

P (Watts)

Average Power

V (Volts)

I (Amps)

90°   270°   360°

180°

Area = Area

The power curve alternates equally above and below the zero line, as seen in Figure 15-24. During the first positive power half-cycle, the circuit current is on the increase, to maximum (point *A*), the magnetic field is building up, and the inductor is storing electrical energy. When the circuit current is on the decline between *A* and *B*, the magnetic field begins to collapse and self-induction occurs and returns electrical energy back into the circuit. The power alternation is both positive when the inductor is consuming power and negative when the inductor is returning the power back into the circuit. Because the positive and negative power alternations are equal but opposite, the average power dissipated is zero.

Resistive and Inductive Circuit

An inductor is different from a capacitor in that it has a small amount of resistance no matter how pure the inductor. For this reason, inductors will never have an average power figure of zero, because even the best inductor will have some value of inductance and resistance within it, as seen in Figure 15-25. The reason an inductor has resistance is that it is simply a piece of wire, and any piece of wire has a certain value of resistance, as well as inductance. This coil resistance should be, and normally is, very small and can usually be ignored; however, in some applications even this small resistance can prevent the correct operation of a circuit, so a value or term had to be created to specify the differences in the quality of inductor available.

The **quality factor** (*Q*) of an inductor is the ratio of the energy stored in the coil by its inductance to the energy dissipated in the coil by the resistance; therefore, the higher the *Q*, the better the coil is at storing energy rather than dissipating it:

Quality Factor

Quality factor of an inductor or capacitor is the ratio of a component's reactance (energy stored) to its effective series resistance (energy dissipated).

$$\text{Quality } (Q) = \frac{\text{energy stored}}{\text{energy dissipated}}$$

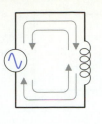

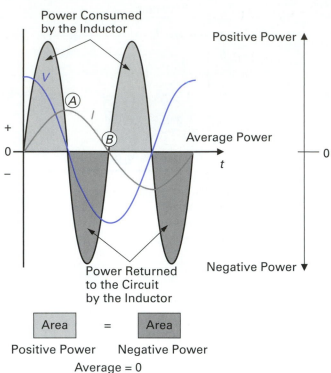

Power Consumed
by the Inductor

Positive Power ▲

Average Power

0

t

Power Returned
to the Circuit
by the Inductor

Negative Power ▼

| Area | = | Area |
| Positive Power | | Negative Power |

Average = 0

FIGURE 15-24 **Purely Inductive Circuit.**

The energy stored is dependent on the inductive reactance (X_L) of the coil, and the energy dissipated is dependent on the resistance (R) of the coil. The quality factor of a coil or inductor can therefore also be calculated by using the formula

$$Q = \frac{X_L}{R}$$

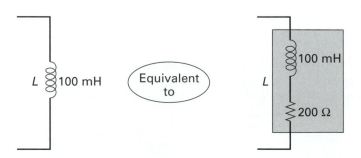

FIGURE 15-25 **Resistance within an Inductor.**

CHAPTER 15 / INDUCTORS AND TRANSFORMERS

Calculate the quality factor Q of a 22 mH coil connected across a 2 kHz, 10 V source if its internal coil resistance is 27 Ω.

■ *Solution:*

$$Q = \frac{X_L}{R}$$

The reactance of the coil is not known but can be calculated by the formula

$$X_L = 2\pi fL$$
$$= 2\pi \times 2\ \text{kHz} \times 22\ \text{mH} = 276.5\ \Omega$$

Therefore,

$$Q = \frac{X_L}{R} = \frac{276.5\ \Omega}{27\ \Omega} = 10.24$$

Inductors, therefore, will never appear as pure inductance but rather as an inductive and resistive (RL) circuit, and the resistance within the inductor will dissipate *true power.*

Figure 15-26 illustrates a circuit containing R and L and the power waveforms produced when $R = X_L$; the phase shift (θ) is equal to 45°.

The positive power alternation, which is above the zero line, is the combination of the power dissipated by the resistor and the power consumed by the inductor while circuit current was on the rise. The negative power alternation is the power that was given back to the circuit by the inductor while the inductor's magnetic field was collapsing and returning the energy that was consumed.

Power Factor

When a circuit contains both inductance and resistance, some of the energy is consumed and then returned by the inductor (reactive or imaginary power), and some of the energy is dissipated and lost by the resistor (resistive or true power).

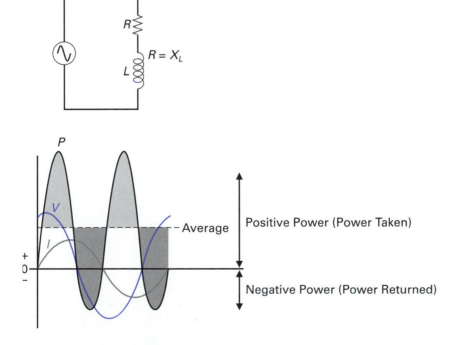

FIGURE 15-26   Power in a Resistive and Inductive Circuit.

Apparent power is the power that appears to be supplied to the load and is the vector sum of both the reactive and true power; it can be calculated by using the formula

$$P_A = \sqrt{P_R^2 + P_X^2}$$

where P_A = apparent power, in volt-amperes (VA)
P_R = true power, in watts (W)
P_X = reactive power, in volt-amperes reactive (VAR)

The power factor is a ratio of the true power to the apparent power and is therefore a measure of the loss in a circuit.

$$PF = \frac{\text{true power } (P_R)}{\text{apparent power } (P_A)}$$

or

$$PF = \frac{R}{Z}$$

or

$$PF = \cos\theta$$

■ **INTEGRATED MATH APPLICATION:**

Calculate the following for a series RL circuit if $R = 40$ kΩ, $L = 450$ mH, $f = 20$ kHz, and $V_S = 6$ V:

a. X_L

b. Z

c. I

d. θ

e. Apparent power

f. PF

■ *Solution:*

a. $X_L = 2\pi fL = 2\pi \times 20 \text{ kHz} \times 450 \text{ mH}$
   $= 56.5$ kΩ

b. $Z = \sqrt{R^2 + X_L^2}$
   $= \sqrt{(40 \text{ k}\Omega)^2 + (56.5 \text{ k}\Omega)^2} = 69.23$ kΩ

c. $I = \dfrac{V_S}{Z} = \dfrac{6 \text{ V}}{69.23 \text{ k}\Omega} = 86.6$ μA

d. $\theta = \arctan \dfrac{X_L}{R} = \arctan \dfrac{56.5 \text{ k}\Omega}{40 \text{ k}\Omega} = 54.7°$

e. Apparent power $= \sqrt{(\text{true power})^2 + (\text{reactive power})^2}$
   $P_R = I^2 \times R = (86.6 \text{ }\mu A)^2 \times 40 \text{ k}\Omega = 300 \text{ }\mu W$
   $P_X = I^2 \times X_L = (86.6 \text{ }\mu A)^2 \times 56.5 \text{ k}\Omega = 423.7 \text{ }\mu VAR$
   $P_A = \sqrt{P_R^2 + P_X^2}$
   $= \sqrt{(300 \text{ }\mu W)^2 + (423.7 \text{ }\mu W)^2} = 519.2 \text{ }\mu VA$

f. $PF = \dfrac{P_R}{P_A} = \dfrac{300\ \mu W}{519.2\ \mu W} = 0.57$

$= \dfrac{R}{Z} = \dfrac{40\ k\Omega}{69.23\ k\Omega} = 0.57$

$= \cos\theta = \cos 54.7° = 0.57$

SELF-TEST EVALUATION POINT FOR SECTION 15-7

Now that you have completed this section, you should be able to:

■ **Objective 7.** *Describe all aspects relating to a series RL circuit.*

Use the following questions to test your understanding of Section 15-7.

1. True or false: In a purely inductive circuit, the current will lead the applied voltage by 90°.

2. Calculate the applied source voltage V_S in an *RL* circuit where $V_R = 4$ V and $V_L = 2$ V.

3. Define and state the formula for impedance when R and X_L are known.

4. If $R = X_L$, the phase shift will equal _____.

5. What is positive power?

6. Define Q and state the formula when X_L and R are known.

7. What is the difference between true and reactive power?

8. State the power factor formula.

15-8 PARALLEL *RL* CIRCUIT

Now that we have observed the behavior of resistors and inductors in series, let us analyze the parallel *RL* circuit.

15-8-1 *Current*

In Figure 15-27(a) you will see a parallel combination of a resistor and inductor. The voltage across both components is equal because of the parallel connection, and the current through each branch is calculated by applying Ohm's law.

$$\text{(Resistor current) } I_R = \dfrac{V_S}{R}$$

$$\text{(Inductor current) } I_L = \dfrac{V_S}{X_L}$$

Total current (I_T) is equal to the vector combination of the resistor current and inductor current, as shown in Figure 15-27(b). Figure 15-27(c) illustrates the current waveforms.

$$I_T = \sqrt{I_R^2 + I_L^2}$$

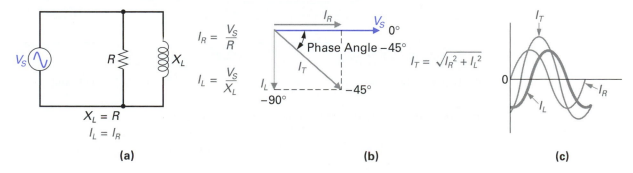

FIGURE 15-27 Parallel *RL* Circuit.

15-8-2　*Phase Angle*

The angle by which the total current (I_T) leads the source voltage (V_S) can be determined with either of the following formulas:

$$\theta = \arctan \frac{I_L}{I_R}$$

or

$$\theta = \arctan \frac{R}{X_L}$$

15-8-3　*Impedance*

The total opposition or impedance of a parallel *RL* circuit can be calculated by

$$Z = \frac{V_S}{I_T}$$

Using basic algebra, we can rearrange this formula to express impedance in terms of reactance and resistance.

$$Z = \frac{R \times X_L}{\sqrt{R^2 + X_L^2}}$$

15-8-4　*Power*

As with series *RL* circuits, resistive power and reactive power can be calculated by

$$P_R = I_R^2 \times R$$

$$P_X = I_L^2 \times X_L$$

The apparent power of the circuit is calculated by

$$P_A = \sqrt{P_R^2 + P_X^2}$$

and, finally, the power factor is equal to

$$PF = \cos \theta = \frac{P_R}{P_A}$$

■ INTEGRATED MATH APPLICATION:

Calculate the following for a parallel *RL* circuit if $R = 45\ \Omega$, $X_L = 1100\ \Omega$, and $V_S = 24$ V:

   a. I_R

   b. I_L

   c. I_T

   d. Z

   e. θ

Solution:

a. $I_R = \dfrac{V_S}{R} = \dfrac{24 \text{ V}}{45 \text{ }\Omega} = 533.3 \text{ mA}$

b. $I_L = \dfrac{V_S}{X_L} = \dfrac{24 \text{ V}}{1100 \text{ }\Omega} = 21.8 \text{ mA}$

c. $I_T = \sqrt{I_R^2 + I_L^2}$
   $= \sqrt{(533.3 \text{ mA})^2 + (21.8 \text{ mA})^2} = 533.7 \text{ mA}$

d. $Z = \dfrac{R \times X_L}{\sqrt{R^2 + X_L^2}} = \dfrac{45 \text{ }\Omega \times 1100 \text{ }\Omega}{\sqrt{(45 \text{ }\Omega)^2 + (1100 \text{ }\Omega)^2}}$
   $= \dfrac{49.5 \text{ k}\Omega}{1100.9 \text{ }\Omega} = 44.96 \text{ }\Omega$

e. $\theta = \arctan \dfrac{R}{X_L} = \arctan \dfrac{45 \text{ }\Omega}{1100 \text{ }\Omega} = 2.34°$

Therefore, I_T lags V_S by 2.34°.

15-9 MUTUAL INDUCTANCE

As was discussed earlier in the chapter, self-inductance is the process by which a coil induces a voltage within itself. The principle on which a transformer is based is an inductive effect known as **mutual inductance,** which is the process by which an inductor induces a voltage in another inductor.

Figure 15-28 illustrates two inductors that are magnetically linked yet electrically isolated from each other. As the alternating current continually rises, falls, and then rises in the opposite direction a magnetic field will build up, collapse, and then build up in the opposite direction.

If a second inductor or secondary coil (L_2) is in close proximity with the first inductor or primary coil (L_1), which is producing the alternating magnetic field, a voltage will be induced into the nearby inductor, which causes current to flow in the secondary circuit through the load resistor. This phenomenon is known as mutual inductance or transformer action.

Mutual Inductance
Ability of one inductor's magnetic lines of force to link with another inductor.

FIGURE 15-28 **Mutual Inductance.**

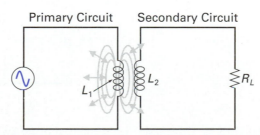

Magnetic Field Alternately Builds up and Collapses

As with self-inductance, mutual inductance is dependent on change. Direct current (dc) is a constant current and produces a constant or stationary magnetic field that does not change, as reviewed in Figure 15-29(a). Alternating current, however, varies continuously and because the polarity of the magnetic field is dependent on the direction of current flow (left-hand rule), the magnetic field will also alternate in polarity, as seen in Figure 15-29(b); it is this continual building up and collapsing of the magnetic field that cuts the adjacent inductor's conducting coils and induces a voltage in the secondary circuit. Mutual induction is possible only with ac and cannot be achieved with dc due to the lack of change.

Self-induction is a measure of how much voltage an inductor can induce within itself. Mutual inductance is a measure of how much voltage is induced in the secondary coil due to the change in current in the primary coil.

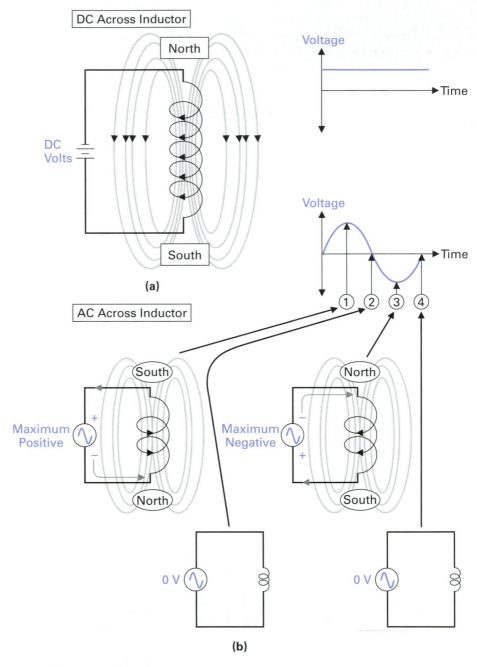

FIGURE 15-29 **Magnetic Field Produced by (a) DC across an Inductor and (b) AC across an Inductor.**

Now that you have completed this section, you should be able to:

■ **Objective 9.** *Define mutual inductance and how it relates to transformers.*

Use the following questions to test your understanding of Section 15-9.

1. Define mutual inductance and explain how it differs from self-inductance.

2. True or false: Mutual induction is possible only with direct current flow.

15-10   BASIC TRANSFORMER

Figure 15-30(a) illustrates a basic transformer, which consists of two coils within close proximity of each other, to ensure that the second coil will be cut by the magnetic flux lines produced by the first coil, and thereby ensure mutual inductance. The ac voltage source is electrically connected (through wires) to the primary coil or winding, and the load (R_L) is electrically connected to the secondary coil or winding.

In Figure 15-30(b), the ac voltage source has produced current flow in the primary circuit, as illustrated. This current flow produces a north pole at the top of the primary winding, and as the ac voltage input swings more negative, the current increase causes the magnetic field being developed by the primary winding to increase. This expanding magnetic field cuts the coils of the secondary winding and induces a voltage, and a subsequent current flows in the secondary circuit, which travels up through the load resistor. The ac voltage follows a sinusoidal pattern and moves from a maximum negative to zero and then begins to build up toward a maximum positive.

In Figure 15-30(c), the current flow in the primary circuit is in the opposite direction due to the ac voltage increase in the positive direction. As voltage increases, current increases, and the magnetic field expands and cuts the secondary winding, inducing a voltage and causing current to flow in the reverse direction down through the load resistor.

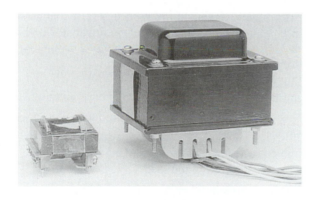

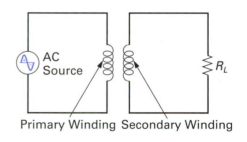

(a)

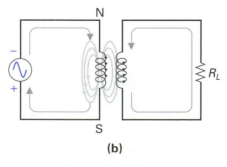

(b)

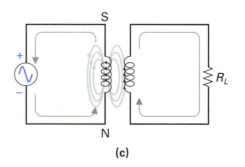

(c)

FIGURE 15-30   Transformer Action.

You may have noticed a few interesting points about the basic transformer just discussed.

1. As primary current increases, secondary current increases, and as primary current decreases, secondary current also decreases. It can therefore be said that the frequency of the alternating current in the secondary is the same as the frequency of the alternating current in the primary.
2. Although the two coils are electrically isolated from each other, energy can be transferred from primary to secondary, because the primary converts electrical energy into magnetic energy, and the secondary converts magnetic energy back into electrical energy.

15-11 TRANSFORMER LOADING

Let's now carry our discussion of the basic transformer a little further and see what occurs when the transformer is not connected to a load, as shown in Figure 15-31. Primary circuit current is determined by $I = V/Z$, where Z is the impedance of the primary coil (both its inductive reactance and resistance), and V is the applied voltage. Since no current can flow in the secondary, because an open in the circuit exists, the primary acts as a simple inductor, and the primary current is small due to the inductance of the primary winding. This small primary current lags the applied voltage due to the counter emf by approximately 90° because the coil is mainly inductive and has very little resistance.

When a load is connected across the secondary, as shown in Figure 15-32, a change in conditions occurs, and the transformer acts differently. The important point that will be observed is that as we go from a no-load to a load condition the primary current will increase due to mutual inductance. Let's follow the steps one by one.

1. The ac applied voltage sets up an alternating magnetic field in the primary winding.
2. The continuously changing flux of this primary field induces and produces a counter emf into the primary to oppose the applied voltage.
3. The primary's magnetic field also induces a voltage in the secondary winding, which causes current to flow in the secondary circuit through the load.
4. The current in the secondary winding produces another magnetic field opposite the field being produced by the primary.

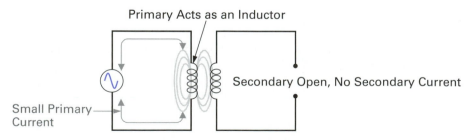

FIGURE 15-31 **Unloaded Transformer.**

FIGURE 15-32   Loaded Transformer.

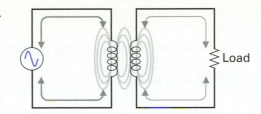

5. This secondary magnetic field feeds back to the primary and induces a voltage that tends to cancel or weaken the counter emf that was set up in the primary by the primary current.

6. The primary's counter emf is therefore reduced, so primary current can now increase.

7. This increase in primary current is caused by the secondary's magnetic field; consequently, the greater the secondary current, the stronger the secondary magnetic field, which causes a reduction in the primary's counter emf, and therefore a primary current increase.

In summary, an increase in secondary current $(I_s\uparrow)$ causes an increase in primary current $(I_p\uparrow)$. This effect, in which the primary induces a voltage in the secondary (V_s), and the secondary induces a voltage into the primary (V_p), is known as *mutual inductance*.

SELF-TEST EVALUATION POINT FOR SECTION 15-11

Now that you have completed this section, you should be able to:

■ *Objective 11. Explain the differences between a loaded and an unloaded transformer.*

Use the following questions to test your understanding of Section 15-11.

1. True or false: An increase in secondary current causes an increase in primary current.

2. True or false: The greater the secondary current, the greater the primary's counter emf.

15-12   TRANSFORMER RATIOS AND APPLICATIONS

Basically, transformers are used for one of three applications:

1. To step up (increase) or step down (decrease) voltage

2. To step up (increase) or step down (decrease) current

3. To match impedances

In all three cases, any of the applications can be achieved by changing the ratio of the number of turns in the primary winding compared to the number of turns in the secondary winding. This ratio is appropriately called the *turns ratio*.

15-12-1   *Turns Ratio*

The **turns ratio** is the ratio between the number of turns in the secondary winding (N_s) and the number of turns in the primary winding (N_p).

$$\text{Turns ratio} = \frac{N_s}{N_p}$$

Let us use a few examples to see how the turns ratio can be calculated.

Turns Ratio
Ratio of the number of turns in the secondary winding to the number of turns in the primary winding of a transformer.

■ **INTEGRATED MATH APPLICATION:**

If the primary has 200 turns and the secondary has 600, what is the turns ratio (Figure 15-33)?

FIGURE 15-33 **Step-Up Transformer Example.**

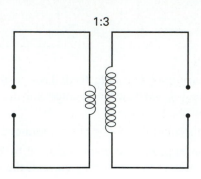

1:3

■ *Solution:*

$$
\begin{aligned}
\text{Turns ratio} &= \frac{N_s}{N_p} \\
&= \frac{600}{200} \\
&= \frac{3(\text{secondary})}{1(\text{primary})} \\
&= 3
\end{aligned}
$$

This simply means that there are three windings in the secondary to every one winding in the primary. Moving from a small number (1) to a larger number (3) means that we *stepped up* in value. Stepping up always results in a turns ratio figure greater than 1, in this case, 3.

■ **INTEGRATED MATH APPLICATION:**

If the primary has 120 turns and the secondary has 30 turns, what is the turns ratio (Figure 15-34)?

FIGURE 15-34 **Step-Down Transformer Example.**

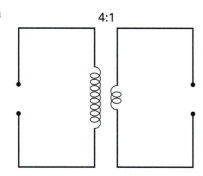

4:1

■ *Solution:*

$$
\begin{aligned}
\text{Turns ratio} &= \frac{N_s}{N_p} \\
&= \frac{30}{120} \\
&= \frac{1\ (\text{secondary})}{4\ (\text{primary})} \\
&= 0.25
\end{aligned}
$$

Said simply, there are four primary windings to every one secondary winding. Moving from a larger number (4) to a smaller number (1) means that we *stepped down* in value. Stepping down always results in a turns ratio figure of less than 1, in this case 0.25.

15-12-2 *Voltage Ratio*

Transformers are used within the power supply unit of almost every piece of electronic equipment to step up or step down the 115 V ac from the outlet. Some electronic circuits require lower-power supply voltages, whereas other devices may require higher-power supply voltages. The transformer is used in both instances to convert the 115 V ac to the required value of voltage.

Step Up

If the secondary voltage (V_s) is greater than the primary voltage (V_p), the transformer is called a **step-up transformer** ($V_s > V_p$), as shown in Figure 15-35. The voltage is stepped up or increased in much the same way as a generator voltage can be increased by increasing the number of turns.

If the ac primary voltage is 100 V and the turns ratio is a 1:5 step-up, the secondary voltage will be five times that of the primary voltage, or 500 V, because the magnetic flux established by the primary cuts more turns in the secondary and therefore induces a larger voltage.

In this example, you can see that the ratio of the secondary voltage to the primary voltage is equal to the turns ratio; in other words,

$$\frac{V_s}{V_p} = \frac{N_s}{N_p}$$

or

$$\frac{500}{100} = \frac{500}{100}$$

To calculate V_s, therefore, we can rearrange the formula and arrive at

$$V_s = \frac{N_s}{N_p} \times V_p$$

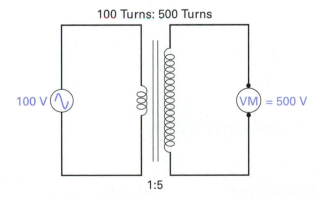

FIGURE 15-35 **Step-Up Transformer.**

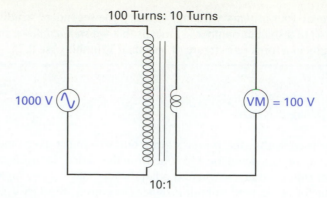

100 Turns: 10 Turns

1000 V

VM = 100 V

10:1

FIGURE 15-36 Step-Down Transformer.

In our example, this is

$$V_s = \frac{500}{100} \times 100 \text{ V}$$
$$= 500 \text{ V}$$

Step Down

If the secondary voltage (V_s) is smaller than the primary voltage (V_p), the transformer is called a **step-down transformer** ($V_s < V_p$), as shown in Figure 15-36. The secondary voltage will be equal to

$$V_s = \frac{N_s}{N_p} \times V_p$$
$$= \frac{10}{100} \times 1000 \text{ V} = 100 \text{ V}$$

■ **INTEGRATED MATH APPLICATION:**

Calculate the secondary voltage (V_s) if a 1:6 step-up transformer has 24 V ac applied to the primary.

■ *Solution:*

$$V_s = \frac{N_s}{N_p} \times V_p$$
$$= \frac{6}{1} \times 24 \text{ V} = 144 \text{ V}$$

The coupling coefficient (k) in this formula is always assumed to be 1, which for most iron-core transformers is almost always the case. This means that all the primary magnetic flux is linking the secondary, and the secondary voltage is dependent on the number of secondary turns that are being cut by the primary magnetic flux.

The transformer can be used to transform the primary ac voltage into any other voltage, either up or down, merely by changing the transformer's turns ratio.

15-12-3 *Power and Current Ratio*

The power in the secondary of the transformer is equal to the power in the primary ($P_p = P_s$). Power, as we know, is equal to $P = V \times I$, and if voltage is stepped up or down, the current automatically is stepped down or up, respectively, in the direction opposite the voltage to maintain the power constant.

For example, if the secondary voltage is stepped up ($V_s\uparrow$), the secondary current is stepped down ($I_s\downarrow$), so the output power is the same as the input power.

$$P_s = V_s \uparrow \times I_s \downarrow$$

This is an equal but opposite change. Therefore, $P_s = P_p$; you cannot get more power out than you put in. The current ratio is therefore inversely proportional to the voltage ratio:

$$\frac{V_s}{V_p} = \frac{I_p}{I_s}$$

If the secondary voltage is stepped up, the secondary current goes down:

$$\frac{V_s\uparrow}{V_p} = \frac{I_p}{I_s\downarrow}$$

If the secondary voltage is stepped down, the secondary current goes up:

$$\frac{V_s\downarrow}{V_p} = \frac{I_p}{I_s\uparrow}$$

If the current ratio is inversely proportional to the voltage ratio, it is also inversely proportional to the turns ratio:

$$\frac{I_p}{I_s} = \frac{V_s}{V_p} = \frac{N_s}{N_p}$$

By rearranging the current and turns ratio, we can arrive at a formula for secondary current, which is

$$I_s = \frac{N_p}{N_s} \times I_p$$

■ INTEGRATED MATH APPLICATION:

The step-up transformer in Figure 15-37 has a turns ratio of 1 to 5. Calculate:

a. Secondary voltage (V_s)
b. Secondary current (I_s)
c. Primary power (P_p)
d. Secondary power (P_s)

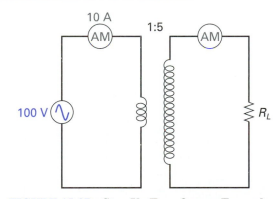

FIGURE 15-37 **Step-Up Transformer Example.**

■ *Solution:*

The secondary has five times as many windings as the primary; consequently, the voltage will be stepped up by a factor of 5 between primary and secondary. If the secondary voltage is going to be five times that of the primary, the secondary current is going to decrease to one-fifth of the primary current.

a. $V_s = \dfrac{N_s}{N_p} \times V_p$

$ = \dfrac{5}{1} \times 100\text{ V}$

$ = 500\text{ V}$

b. $I_s = \dfrac{N_p}{N_s} \times I_p$

$ = \dfrac{1}{5} \times 10\text{ A}$

$ = 2\text{ A}$

c. $P_p = V_p \times I_p = 100\text{ V} \times 10\text{ A} = 1000\text{ VA}$

d. $P_s = V_s \times I_s = 500\text{ V} \times 2\text{ A} = 1000\text{ VA}$

Therefore, $P_p = P_s$.

15-12-4 *Impedance Ratio*

Maximum Power Transfer Theorem

The maximum power will be absorbed by the load from the source when the impedance of the load is equal to the impedance of the source.

The **maximum power transfer theorem,** which was discussed previously and is summarized in Figure 15-38, states that maximum power is transferred from source (ac generator) to load (equipment) when the impedance of the load is equal to the internal impedance of the source. If these impedances are different, a large amount of power can be wasted.

In most cases it is required to transfer maximum power from a source that has an internal impedance (Z_s) that is not equal to the load impedance (Z_L). In this situation, a transformer can be inserted between the source and the load to make the load impedance appear to equal the source's internal impedance.

For example, let's imagine that your car stereo system (source) has an internal impedance of 100 Ω and is driving a speaker (load) of 4 Ω impedance, as seen in Figure 15-39.

By choosing the correct turns ratio, you can make the 4 Ω speaker appear as a 100 Ω load impedance, which will match the 100 Ω internal source impedance of the stereo system, resulting in maximum power transfer.

The turns ratio can be calculated by using the formula

$$\text{Turns ratio} = \sqrt{\dfrac{Z_L}{Z_s}}$$

where

Z_L = load impedance in ohms
Z_s = source impedance in ohms

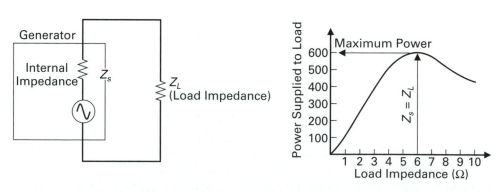

FIGURE 15-38 **Maximum Power Transfer Theorem.**

CHAPTER 15 / INDUCTORS AND TRANSFORMERS

FIGURE 15-39   **Impedance**
Matching.

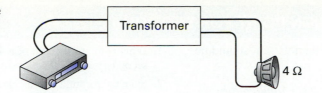

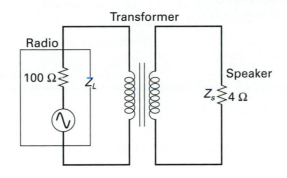

In our example, this will calculate out to be

$$\text{Turns ratio} = \sqrt{\frac{Z_L}{Z_s}}$$

$$= \sqrt{\frac{4}{100}} = \frac{\sqrt{4}}{\sqrt{100}}$$

$$= \frac{2}{10} = \frac{1}{5}$$

$$= 0.2$$

If the turns ratio is less than 1, a step-down transformer is required. A turns ratio of 0.2 means a step-down transformer is needed with a turns ratio of 5:1 ($\frac{1}{5} = 0.2$).

■ INTEGRATED MATH APPLICATION:

Calculate the turns ratio needed to match the 22.2 Ω output impedance of an amplifier to two 16 Ω speakers connected in parallel.

■ *Solution:*

The total impedance of two 16 Ω speakers in parallel will be

$$Z_L = \frac{\text{product}}{\text{sum}} = \frac{16 \times 16}{16 + 16} = 8\ \Omega$$

The turns ratio will be

$$\text{Turns ratio} = \sqrt{\frac{Z_L}{Z_s}}$$

$$= \sqrt{\frac{8\ \Omega}{22.2\ \Omega}}$$

$$= \sqrt{0.36} = 0.6$$

Therefore, a step-down transformer is needed with a turns ratio of 1.67:1.

Now that you have completed this section, you should be able to:

■ **Objective 12.** *List the three basic applications of transformers.*

■ **Objective 13.** *Describe how a transformer's turns ratio can be used to step up or step down voltage or current, or to match impedances.*

Use the following questions to test your understanding of Section 15-12.

1. What is the turns ratio of a 402 turn primary and 1608 turn secondary, and is this transformer step up or step down?
2. State the formula for calculating the secondary voltage (V_s).
3. True or false: A transformer can, by an adjustment of the turns ratio, be made to step up both current and voltage between primary and secondary.
4. State the formula for calculating the secondary current (I_s).
5. What turns ratio is needed to match a 25 Ω source to a 75 Ω load?
6. Calculate V_s if $N_s = 200$, $N_p = 112$, and $V_p = 115$ V.

15-13   TRANSFORMER RATINGS

A typical transformer rating could read 1 kVA, 500/100, 60 Hz. The 500 normally specifies the maximum primary voltage, the 100 normally specifies the maximum secondary voltage, and the 1 kVA is the apparent power rating. In this example, the maximum load current will equal

$$I_s = \frac{\text{apparent power } (P_A)}{\text{secondary voltage } (V_s)} \qquad \left(\begin{array}{l} P = V \times I; \text{ therefore,} \\ \quad I = \dfrac{P}{V} \end{array} \right)$$

$$= \frac{1 \text{ kVA}}{100 \text{ V}} = 10 \text{ A}$$

With this secondary voltage at 100 V and a maximum current of 10 A, the smallest load resistor that can be connected across the output of the secondary is

$$R_L = \frac{V_s}{I_s}$$

$$= \frac{100 \text{ V}}{10 \text{ A}}$$

$$= 10 \text{ } \Omega$$

Exceeding the rating of the transformer will cause overheating and even burning out of the windings.

■ **INTEGRATED MATH APPLICATION:**

Calculate the smallest value of load resistance that can be connected across a 3 kVA, 600/200, 60 Hz step-down transformer.

■ *Solution:*

$$I = \frac{\text{apparent power } (P_A)}{\text{secondary voltage } (V_s)}$$

$$\frac{3 \text{ kVA}}{200 \text{ V}} = 15 \text{ A}$$

$$R_L = \frac{V_s}{I_s} = \frac{200 \text{ V}}{15 \text{ A}} = 13.3 \text{ } \Omega$$

Use the following questions to test your understanding of Section 15-13.

1. What does each of the values mean when a transformer is rated as a 10 kVA, 200/100, 60 Hz?

2. If a 100 Ω resistor is connected across the secondary of a transformer that is to supply 1 kV and is rated at a maximum current of 8 A, will the transformer overheat and possibly burn out?

REVIEW QUESTIONS

Multiple Choice Questions

1. Self-inductance is a process by which a coil will induce a voltage within _____.
   a. Another inductor
   b. Two or more inductors in close proximity
   c. Itself
   d. Both (a) and (b)

2. Inductance is a process by which a coil will induce a voltage within _____.
   a. Another inductor
   b. Two or more inductors in close proximity
   c. Itself
   d. Both (a) and (b)

3. The inductor is basically:
   a. An electromagnet
   b. A coil of wire
   c. A coil of conductor formed around a core material
   d. All the above
   e. None of the above

4. The inductance of an inductor is proportional to _____ and inversely proportional to _____.
   a. $N, A, \mu; l$
   b. $A, \mu, l; N$
   c. $\mu, l, N; A$
   d. $N, A, l; \mu$

5. The total inductance of a series circuit is:
   a. Less than the value of the smallest inductor
   b. Equal to the sum of all the inductance values
   c. Equal to the product over sum
   d. All the above

6. The total inductance of a parallel circuit can be calculated by:
   a. Using the product-over-sum formula
   b. Using L divided by N for equal-value inductors
   c. Using the reciprocal resistance formula
   d. All the above

7. It will actually take _____ time constants for the current in an inductive circuit to reach a maximum value.
   a. 63.2
   b. 1
   c. 1.414
   d. 5

8. The time constant for a series inductive/resistive circuit is equal to:
   a. $L \times R$
   b. L/R
   c. V/R
   d. $2\pi \times f \times L$

9. Inductive reactance (X_L) is proportional to:
   a. Time or period of the ac applied
   b. Frequency of the ac applied
   c. The stray capacitance that occurs due to the air acting as a dielectric between two turns of a coil
   d. The value of inductance
   e. Two of the above are true.

10. In a series RL circuit, the source voltage (V_S) is equal to:
    a. The square root of the sum of V_R^2 and V_L^2
    b. The vector sum of V_R and V_L
    c. $I \times Z$
    d. Two of the above are partially true.
    e. Answers (a), (b), and (c) are correct.

11. In a purely resistive circuit, the phase shift is equal to _____, whereas in a purely inductive or capacitive circuit, the phase shift is _____ degrees.
    a. 45, 0
    b. 90, 0
    c. 45, 90
    d. None of the above

12. Inductive reactance:
    a. Increases with frequency
    b. Is proportional to inductance
    c. Reduces the amplitude of alternating current
    d. All the above

13. The current through an inductor _____ the voltage across the same inductor by _____.
    a. Lags, 90°
    b. Lags, 45°
    c. Leads, 90°
    d. Leads, 45°

14. The phasor combination of X_L and R is the circuit's:
    a. Reactance
    b. Total resistance
    c. Power factor
    d. Impedance

15. In a series RL circuit, where $V_R = 200$ mV and $V_L = 0.2$V, $\theta = $ _____.
    a. 45°
    b. 90°
    c. 0°
    d. 1°

16. Transformer action is based on:

   a. Self-inductance   **c.** Mutual capacitance
   b. Air between the coils   **d.** Mutual inductance

17. An increase in transformer secondary current will cause a/an _____ in primary current.

   a. Decrease   **b.** Increase

18. A step-up transformer will always have a turns ratio _____, and a step-down transformer has a turns ratio _____.

   a. $< 1, > 1$ **c.** $> 1, < 1$
   b. $> 1, > 1$ **d.** $< 1, < 1$

19. Transformers can only be used with alternating current because:

   a. It produces an alternating magnetic field.
   b. It produces a fixed magnetic field.
   c. Its magnetic field is greater than that of dc.
   d. Its rms is 0.707 of the peak.

20. Assuming 100% efficiency, the output power, P_s, is always equal to:

   a. P_p **d.** Both (a) and (c)
   b. $V_s \times I_s$ **e.** Both (a) and (b)
   c. $0.5 \times P_p$

Communication Skill Questions

21. What is self-induction, and how does it relate to an inductor? (15-1)

22. Give the formula for inductance, and explain the four factors that determine inductance. (15-3)

23. List all the formulas for calculating total inductance when inductors are connected in: (15-4)

   a. Series   **b.** Parallel

24. Define the following terms:

   a. Inductive reactance (15-6)   **d.** Q factor (15-7-4)
   b. Impedance (15-7-2)   **e.** Power factor (15-7-4)
   c. Phase shift (15-7-3)

25. Explain briefly why an inductor acts like an open to an instantaneous change. Why does an inductor act like a short to dc? (15-1)

26. What is a transformer? (Introduction)

27. Describe mutual inductance and how it relates to transformers. (15-9)

28. Why does loading the transformer's secondary circuit affect primary transformer current? (15-11)

29. Could a transformer be considered a dc block? (15-9)

30. Would a step-up voltage transformer step up or step down current? What would happen to secondary power? (15-12)

Practice Problems

31. Convert the following:

   a. 0.037 H to mH   **c.** 862 mH to H
   b. 1760 μH to mH   **d.** 0.256 mH to μH

32. Calculate the impedance (Z) of the following series RL combinations:

   a. 22 MΩ, 25 μH, $f = 1$ MHz
   b. 4 kΩ, 125 mH, $f = 100$ kHz
   c. 60 Ω, 0.05 H, $f = 1$ MHz

33. Calculate the voltage across a coil if: ($d = \Delta$ or delta)

   a. $d_i/d_t = 120$ mA/ms and $L = 2$ μH
   b. $d_i/d_t = 62$ μA/μs and $L = 463$ mH
   c. $d_i/d_t = 4$ A/s and $L = 25$ mH

34. Calculate the total inductance of the following series circuits:

   a. 75 μH, 61 μH, 50 mH
   b. 8 mH, 4 mH, 22 mH

35. Calculate the total inductance of the following parallel circuits:

   a. 12 mH, 8 mH
   b. 75 μH, 34 μH, 27 μH

36. Calculate the total inductance of the following series–parallel circuits:

   a. 12 mH in series with 4 mH, and both in parallel with 6 mH
   b. A two-branch parallel arrangement made up of 6 μH and 2 μH in series with one another, and 8 μH and 4 μH in series with one another
   c. Two parallel arrangements in series with each other, made up of 1 μH and 2 μH in parallel and 4 μH and 15 μH in parallel

37. In a series RL circuit, if $V_L = 12$ V and $V_R = 6$ V, calculate:

   a. V_S **d.** Q
   b. I if $Z = 14$ kΩ **e.** Power factor
   c. Phase angle

38. What value of inductance is needed to produce 3.3 kΩ of reactance at 15 kHz?

39. At what frequency will a 330 μH inductor have a reactance of 27 kΩ?

40. Calculate the impedance of the circuits seen in Figure 15-40.

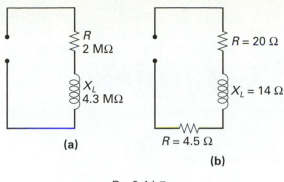

(a)

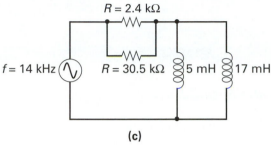

(b)

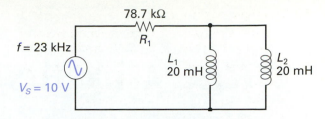

FIGURE 15-42 *RL* Circuit.

(c)

FIGURE 15-40 **Calculating Impedance.**

41. Referring to Figure 15-41, calculate the voltage across the inductor for all five time constants after the switch has been closed.

42. Referring to Figure 15-42, calculate:

a. L_T
b. X_L
c. Z
d. I_{R1}, I_{L1}, and I_{L2}
e. θ
f. True power, reactive power, and apparent power
g. Power factor

43. Referring to Figure 15-43, calculate:

a. R_T **f.** I_R, I_L
b. L_T **g.** I_T
c. X_L **h.** θ
d. Z **i.** Apparent power
e. V_R, V_L **j.** PF

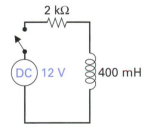

FIGURE 15-41
Inductive Time
Constant Example.

44. Calculate the turns ratio of the following transformers and state whether they are step up or step down:

a. P = 12 T, S = 24 T   **c.** P = 24 T,S = 5 T
b. P = 3 T, S = 250 T   **d.** P = 240 T,S = 120 T

45. Calculate the secondary ac voltage for all the examples in Question 44 if the primary voltage equals 100 V.

46. What turns ratio would be needed to match a source impedance of 24 Ω to a load impedance of 8 Ω?

47. What turns ratio would be needed to step:

a. 120 V to 240 V   **c.** 30 V to 14 V
b. 240 V to 720 V   **d.** 24 V to 6 V

48. For a 24 V, 12-turn primary with a 16-, 2-, 1-, and 4-turn multiple secondary transformer, calculate each of the secondary voltages.

49. If a 2:1 step-down transformer has a primary input voltage of 120 V, 60 Hz, and a 2 kVA rating, calculate the maximum secondary current and smallest load resistor that can be connected across the output.

50. If a transformer is rated at 500 VA, 60 Hz, the primary voltage is 240 V ac, and the secondary voltage is 600 V ac, calculate:

a. Maximum load current   **b.** Smallest value of R_L

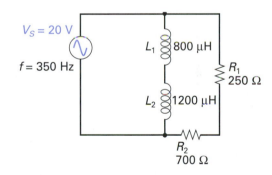

FIGURE 15-43 **Parallel *RL* Circuit.**

Web Site Questions

Go to the Web site http://www.prenhall.com/cook, select the textbook *Mathematics for Electronics and Computers,* select this chapter, and then follow the instructions when answering the multiple-choice practice problems.

RLC Circuits and Complex Numbers

The Fairchildren

William Shockley, John Bardeen, and Walter Brattain (left to right)

On December 23, 1947, John Bardeen, Walter Brattain, and William Shockley first demonstrated how a semiconductor device, named the *transistor,* could be made to amplify; however, the device had mysterious problems and was very unpredictable. Shockley continued his investigations, and in 1951 he presented the world with the first reliable junction transistor. In 1956, the three shared the Nobel Prize in physics for their discovery, and much later, in 1972, Bardeen would win a rare second Nobel Prize for his research at the University of Illinois in the field of superconductivity.

Shockley left Bell Labs in 1955 to start his own semiconductor company near his home in Palo Alto and began recruiting personnel. He was, however, very selective, hiring only those who were bright, young, and talented. The company was a success, although many of the employees could not tolerate Shockley's eccentricities, such as posting everyone's salary and requiring that the employees rate one another.

Two years later, eight of Shockley's most talented defected. The "traitorous eight," as Shockley called them, started their own company only a dozen blocks away, named Fairchild Semiconductor.

More than 50 companies would be founded by former Fairchild employees. One of the largest was started by Robert Noyce and two other colleagues from the group of eight Shockley defectors; they named their company Intel, which was short for "intelligence."

16

Outline and Objectives

VIGNETTE: THE FAIRCHILDREN

INTRODUCTION

16-1 SERIES *RLC* CIRCUIT

Objective 1: Identify the difference between a series and a parallel *RLC* circuit.

Objective 2: Explain the following as they relate to series *RLC* circuits:
   a. Impedance
   b. Current
   c. Voltage
   d. Phase angle
   e. Power

16-1-1 Impedance

16-1-2 Current

16-1-3 Voltage

16-1-4 Phase Angle

16-1-5 Power

16-2 PARALLEL *RLC* CIRCUIT

Objective 3: Explain the following as they relate to parallel *RLC* circuits:
   a. Voltage
   b. Current
   c. Impedance
   d. Phase angle
   e. Power

16-2-1 Voltage

16-2-2 Current

16-2-3 Phase Angle

16-2-4 Impedance

16-2-5 Power

16-3 RESONANCE

Objective 4: Define resonance, and explain the characteristics of:
   a. Series resonance
   b. Parallel resonance

16-3-1 Series Resonance
   Quality Factor
   Bandwidth

16-3-2 Parallel Resonance
   Flywheel Action
   The Reality of Tanks
   Quality Factor
   Bandwidth
   Selectivity

16-4 APPLICATIONS OF *RLC* CIRCUITS

Objective 5: Identify and explain the following *RLC* circuit applications:
   a. Low-pass filter
   b. High-pass filter
   c. Bandpass filter
   d. Band-stop filter

16-4-1 Low-Pass Filter

16-4-2 High-Pass Filter

16-4-3 Bandpass Filter

16-4-4 Band-Stop Filter

16-5 COMPLEX NUMBERS

Objective 6: Describe complex numbers in both rectangular and polar form.

Objective 7: Perform complex number arithmetic.

Objective 8: Describe how complex numbers apply to ac circuits containing series–parallel *RLC* components.

16-5-1 The Real Number Line

16-5-2 The Imaginary Number Line

16-5-3 The Complex Plane

16-5-4 Polar Complex Numbers

16-5-5 Rectangular/Polar Conversions
Polar-to-Rectangular Conversion
Rectangular-to-Polar Conversion

16-5-6 Complex Number Arithmetic
Addition
Subtraction
Multiplication
Division

16-5-7 How Complex Numbers Apply to AC Circuits
Series AC Circuits
Series–Parallel AC Circuits

Introduction

In this chapter we will combine resistors (R), inductors (L), and capacitors (C) into series and parallel ac circuits. Resistors, as we have discovered, operate and react to voltage and current in a very straightforward way, in that the voltage across a resistor is in phase with the resistor current.

Inductors and capacitors operate in essentially the same way, in that they both store energy and then return it back to the circuit; however, they have completely opposite reactions to voltage and current. To help you remember the phase relationships between voltage and current for capacitors and inductors, you may wish to use the following memory phrase:

ELI the ICE man

This phrase states that voltage (symbolized E) leads current (I) in an inductive (L) circuit (abbreviated by the word "ELI"), while current (I) leads voltage (E) in a capacitive (C) circuit (abbreviated by the word "ICE").

In this chapter we will study the relationships among voltage, current, impedance, and power in both series and parallel RLC circuits. We will also examine the important RLC circuit characteristic called resonance, and see how RLC circuits can be made to operate as filters. In the final section we will discuss how complex numbers can be used to analyze series and parallel ac circuits containing resistors, inductors, and capacitors.

16-1 SERIES *RLC* CIRCUIT

Figure 16-1 begins our analysis of series RLC circuits by illustrating the current and voltage relationships. The circuit current is always the same throughout a series circuit and can therefore be used as a reference. Studying the waveforms and vector diagrams shown alongside the components, you can see that the voltage across a resistor is always in phase with the current, whereas the voltage across the inductor leads the current by 90° and the voltage across the capacitor lags the current by 90°.

Now let's analyze the impedance, current, voltage, and power distribution of this circuit in a little more detail.

16-1-1 *Impedance*

Impedance is the total opposition to current flow and is a combination of both reactance (X_L, X_C) and resistance (R). An example circuit is illustrated in Figure 16-2(a).

FIGURE 16-1 Series *RLC* Circuit. (a) *RLC* Series Circuit Current: Current Flow Is Always the Same in All Parts of a Series Circuit. (b) *RLC* Series Circuit Voltages: *I* Is in Phase with V_R, *I* Lags V_L by 90°, and *I* Leads V_C by 90°.

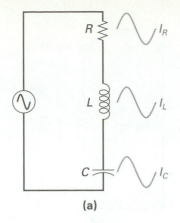

(a)

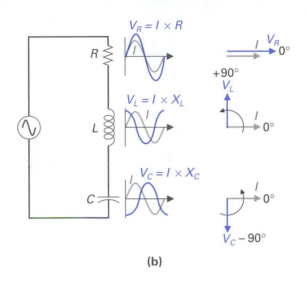

(b)

Capacitive reactance can be calculated by using the formula

$$X_C = \frac{1}{2\pi f C}$$

In the example,

$$X_C = \frac{1}{2\pi \times 60 \text{ Hz} \times 10 \ \mu\text{F}} = 265.3 \ \Omega$$

Inductive reactance is calculated by using the formula

$$X_L = 2\pi f L$$

In the example,

$$X_L = 2\pi \times 60 \text{ Hz} \times 20 \text{ mH} = 7.5 \ \Omega$$

Resistance in the example is equal to $R = 33 \ \Omega$. Figure 16-2(b) illustrates these values of resistance and reactance in a vector diagram. In this vector diagram you can see that X_L is drawn 90° ahead of R, and X_C is drawn 90° behind R. The capacitive and inductive reactances are 180° out of phase with each other and counteract to produce the vector diagram shown in Figure 16-2(c). The difference between X_L and X_C is equal to 257.8, and since X_C is greater than X_L, the resultant reactive vector is capacitive. Reactance, however, is not in

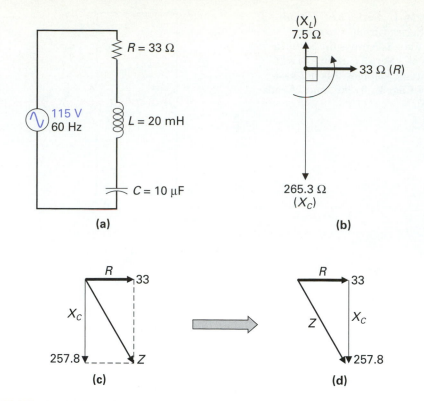

(a)

(b)

(c)

(d)

FIGURE 16-2 Series Circuit Impedance.

phase with resistance, and impedance is the vector sum of the reactive (X) and resistive (R) vectors. The formula, based on the Pythagorean theorem, as illustrated in Figure 16-2(d), is

$$Z = \sqrt{R^2 + X^2}$$

In this example, therefore, the circuit impedance will be equal to

$$Z = \sqrt{R^2 + X^2} = \sqrt{33^2 + 257.8^2} = 260 \ \Omega$$

Since reactance (X) is equal to the difference between (symbolized by $\sim$) X_L and X_C $(X_L \sim X_C)$, the impedance formula can be modified slightly to incorporate the calculation to determine the difference between X_L and X_C.

$$Z = \sqrt{R^2 + (X_L \sim X_C)^2}$$

Using our example with this new formula, we arrive at the same value of impedance, and since the difference between X_L and X_C resulted in a capacitive vector, the circuit is said to act capacitively.

$$
\begin{aligned}
Z &= \sqrt{R^2 + (X_L \sim X_C)^2} \\
&= \sqrt{33^2 + (7.5 \sim 265.3)^2} \\
&= \sqrt{33^2 + 257.8^2} \\
&= 260 \ \Omega
\end{aligned}
$$

If, on the other hand, the component values were such that the difference was an inductive vector, then the circuit would be said to act inductively.

16-1-2   *Current*

The current in a series circuit is the same at all points throughout the circuit, and therefore

$$I = I_R = I_L = I_C$$

Once the total impedance of the circuit is known, Ohm's law can be applied to calculate the circuit current:

$$I = \frac{V_S}{Z}$$

In the example, circuit current is equal to

$$
\begin{aligned}
I &= \frac{V_S}{Z} \\
&= \frac{115 \text{ V}}{260 \ \Omega} \\
&= 0.44 \text{ A} \quad \text{or} \quad 440 \text{ mA}
\end{aligned}
$$

16-1-3   *Voltage*

Now that we know the value of current flowing in the series circuit, we can calculate the voltage drops across each component, as shown in Figure 16-3(a).

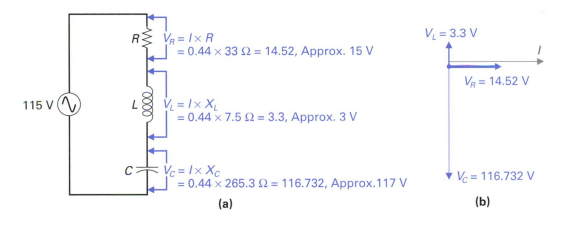

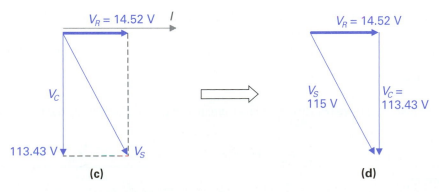

FIGURE 16-3   Series Voltage Drops.

$$V_R = I \times R$$

$$V_L = I \times X_L$$

$$V_C = I \times X_C$$

Since none of these voltages are in phase with one another, as shown in Figure 16-3(b), they must be added vectorially to obtain the applied voltage.

The following formula, based on the Pythagorean theorem and illustrated in Figure 16-3(c) and (d), can be used to calculate V_S.

$$V_S = \sqrt{V_R{}^2 + (V_L \sim V_C)^2}$$

In the example circuit, the applied voltage is, as we already know, 115 V.

$$
\begin{aligned}
V_S &= \sqrt{15^2 + (3 \sim 117)^2} \\
&= \sqrt{225 + 12{,}996} \\
&= \sqrt{13{,}221} \\
&= 115 \text{ V}
\end{aligned}
$$

16-1-4   *Phase Angle*

As can be seen in Figure 16-3(c), there is a phase difference between the source voltage (V_S) and the circuit current (I). This phase difference can be calculated with either of the following formulas.

$$\theta = \arctan \frac{V_L \sim V_C}{V_R}$$

$$\theta = \arctan \frac{X_L \sim X_C}{R}$$

In the example circuit, θ is

$$
\begin{aligned}
\theta &= \arctan \frac{3.3 \text{ V} \sim 116.732 \text{ V}}{14.52 \text{ V}} \\
&= \arctan 7.812 \\
&= 82.7°
\end{aligned}
$$

Because the example circuit is capacitive (ICE), the phase angle will be $-82.7°$, since V_S lags I in a circuit that acts capacitively.

16-1-5   *Power*

The true power or resistive power (P_R) dissipated by a circuit can be calculated using the formula

$$P_R = I^2 \times R$$

which in our example is

$$P_R = (0.44 \text{ A})^2 \times 33 \ \Omega = 6.4 \text{ W}$$

The apparent power (P_A) consumed by the circuit is calculated by

$$P_A = V_S \times I$$

which in our example is

$$P_A = 115 \text{ V} \times 0.44 \text{ A} = 50.6 \text{ volt-amperes (VA)}$$

The true or actual power dissipated by the resistor is, as expected, smaller than the apparent power that appears to be being used.

The power factor can be calculated, as usual, by

$$\text{PF} = \cos\theta = \frac{R}{Z} = \frac{P_R}{P_A}$$

PF of 0 = reactive circuit
PF of 1 = resistive circuit

In the example circuit, PF = 0.126, indicating that the circuit is mainly reactive.

■ INTEGRATED MATH APPLICATION:

For a series circuit where $R = 10 \ \Omega$, $L = 5$ mH, $C = 0.05 \ \mu$F, and $V_S = 100$ V/2 kHz, calculate:

a. X_C f. Apparent power
b. X_L g. True power
c. Z h. Power factor
d. I i. Phase angle
e. V_R, V_C, and V_L

■ *Solution:*

a. $X_C = \dfrac{1}{2\pi f C} = 1.6 \text{ k}\Omega$

b. $X_L = 2\pi f L = 62.8 \ \Omega$

c. $Z = \sqrt{R^2 + (X_L \sim X_C)^2}$
$= \sqrt{(10 \ \Omega)^2 + (1.6 \text{ k}\Omega \sim 62.8 \ \Omega)^2}$
$= \sqrt{10 \ \Omega^2 + (1.54 \text{ k}\Omega)^2}$
$= 1.54 \text{ k}\Omega$ (capacitive circuit due to high X_C)

d. $I = \dfrac{V_S}{Z} = \dfrac{100 \text{ V}}{1.54 \text{ k}\Omega} = 64.9 \text{ mA}$

e. $V_R = I \times R = 64.9 \text{ mA} \times 10 \ \Omega = 0.65 \text{ V}$
$V_C = I \times X_C = 64.9 \text{ mA} \times 1.6 \text{ k}\Omega = 103.9 \text{ V}$
$V_L = I \times X_L = 64.9 \text{ mA} \times 62.8 \ \Omega = 4.1 \text{ V}$

f. Apparent power $= V_S \times I = 100 \text{ V} \times 64.9 \text{ mA} = 6.49 \text{ W}$

g. True power $= I^2 \times R = (64.9 \text{ mA})^2 \times 10 \ \Omega = 42.17 \text{ mW}$

h. $\text{PF} = \dfrac{R}{Z} = \dfrac{10 \ \Omega}{1.5 \text{ k}\Omega} = 0.006$ (reactive circuit)

i. $\theta = \arctan \dfrac{V_L \sim V_C}{V_R}$

$\quad = \arctan \dfrac{4.1\ \text{V} \sim 103.9\ \text{V}}{0.65\ \text{V}}$

$\quad = \arctan 153.54$

$\quad = 89.63°$

Capacitive circuit (ICE); therefore, V_S lags I by $-89.63°$.

SELF-TEST EVALUATION POINT FOR SECTION 16-1

Now that you have completed this section, you should be able to:

■ **Objective 1.** *Identify the difference between a series and a parallel RLC circuit.*

■ **Objective 2.** *Explain the following as they relate to series RLC circuits:*

a. *Impedance*
b. *Current*
c. *Voltage*
d. *Phase angle*
e. *Power*

Use the following questions to test your understanding of Section 16-1.

1. List in order the procedure that should be followed to fully analyze a series *RLC* circuit.

2. State the formulas for calculating the following in relation to a series *RLC* circuit.

a. Impedance   f. V_R
b. Current   g. V_L
c. Apparent power   h. V_C
d. V_S i. Phase angle (θ)
e. True power   j. Power factor (PF)

16-2   PARALLEL *RLC* CIRCUIT

Now that the characteristics of a series circuit are understood, let us connect a resistor, inductor, and capacitor in parallel with one another. Figure 16-4(a) and (b) show the current and voltage relationships of a parallel *RLC* circuit.

16-2-1   *Voltage*

As can be seen in Figure 16-4(a), the voltage across any parallel circuit will be equal and in phase. Therefore,

$$V_R = V_L = V_C = V_S$$

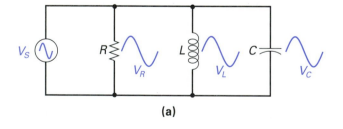
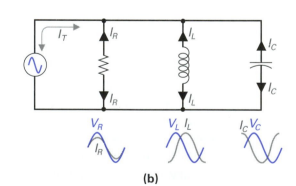

(a)

(b)

FIGURE 16-4   **Parallel *RLC* Circuit. (a) *RLC* Parallel Circuit Voltage: Voltages across Each Component Are All Equal and in Phase with One Another in a Parallel Circuit. (b) *RLC* Parallel Circuit Currents: I_R Is in Phase with V_R, I_L Lags V_L by 90°, and I_C Leads V_C by 90°.**

16-2-2 *Current*

With current, we must first calculate the individual branch currents (I_R, I_L, and I_C) and then calculate the total circuit current (I_T). An example circuit is illustrated in Figure 16-5(a), and the branch currents can be calculated by using the formulas

$$I_R = \frac{V}{R}$$

$$I_L = \frac{V}{X_L}$$

$$I_C = \frac{V}{X_C}$$

Figure 16-5(b) illustrates these branch currents vectorially, with I_R in phase with V_S, I_L lagging by 90°, and I_C leading I_R by 90°. The 180° phase difference between I_C and I_L results in a cancellation, as shown in Figure 16-5(c).

The total current (I_T) can be calculated by using the Pythagorean theorem on the right triangle, as illustrated in Figure 16-5(d).

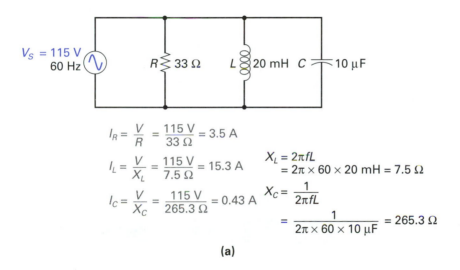

(a)

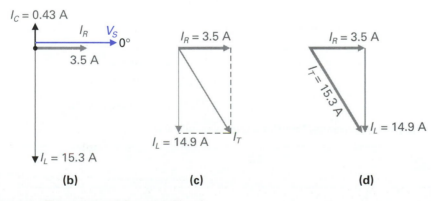

(b)   (c)   (d)

FIGURE 16-5 Example *RLC* Parallel Circuit.

$$I_T = \sqrt{I_R^2 + I_X^2}$$

$$(I_X = I_L \sim I_C)$$
$$= \sqrt{3.5^2 + 14.9^2}$$
$$= 15.3 \text{ A}$$

16-2-3   *Phase Angle*

As shown in Figure 16-5(b) and (c), there is a phase difference between the source voltage (V_S) and the circuit current (I_T). This phase difference can be calculated using the formula

$$\theta = \arctan \frac{I_L \sim I_C}{I_R}$$

$$= \arctan \frac{15.3 \text{ A} - 0.43 \text{ A}}{3.5 \text{ A}}$$
$$= \arctan 4.25$$
$$= 76.7°$$

Since this is an inductive circuit (ELI), the total current (I_T) will lag the source voltage (V_S) by $-76.7°$.

16-2-4   *Impedance*

With the total current (I_T) known, the impedance of all three components in parallel can be calculated by the formula

$$Z = \frac{V}{I_T}$$

$$= \frac{115 \text{ V}}{15.3 \text{ A}}$$
$$= 7.5 \ \Omega$$

16-2-5   *Power*

The true power dissipated can be calculated using

$$P_R = I_R^2 \times R$$

$$= (3.5 \text{ A})^2 \times 33 \ \Omega$$
$$= 404.3 \text{ W}$$

The apparent power consumed by the circuit is calculated by

$$P_A = V_S \times I_T$$

$$= 115 \text{ V} \times 15.3 \text{ A}$$
$$= 1759.5 \text{ volt-amperes (VA)}$$

Finally, the power factor can be calculated, as usual, with

$$PF = \cos \theta = \frac{P_R}{P_A}$$

In the example circuit, PF $= 0.23$.

CHAPTER 16 / *RLC* CIRCUITS AND COMPLEX NUMBERS

Now that you have completed this section, you should be able to:

■ **Objective 3.** *Explain the following as they relate to parallel RLC circuits:*

   a. *Voltage*
   b. *Current*
   c. *Impedance*
   d. *Phase angle*
   e. *Power*

Use the following questions to test your understanding of Section 16-2.

1. State the formulas for calculating the following in relation to a parallel *RLC* circuit:

   a. I_R **d.** I_L
   b. I_T **e.** θ
   c. I_C **f.** Z

2. State the formulas for:

   a. P_R **c.** P_A
   b. P_X **d.** PF

16-3   RESONANCE

Resonance is a circuit condition that occurs when the inductive reactance (X_L) and the capacitive reactance (X_C) have been balanced. Figure 16-6 illustrates a parallel- and a series-connected *LC* circuit. If a dc voltage is applied to the input of either circuit, the capacitor will act as an open (X_C = infinite Ω), and the inductor will act as a short ($X_L = 0\ \Omega$).

    If a low-frequency ac is now applied to the input, X_C will decrease from maximum, and X_L will increase from zero. As the ac frequency is increased further the capacitive reactance will continue to fall ($X_C\!\downarrow \propto 1/f\!\uparrow$) and the inductive reactance to rise ($X_L\!\uparrow \propto f\!\uparrow$), as shown in Figure 16-7.

    As the input ac frequency is increased further a point will be reached at which X_L will equal X_C, and this condition is known as *resonance*. The frequency at which $X_L = X_C$ in either a parallel or a series *LC* circuit is known as the *resonant frequency* (f_0) and can be calculated by the following formula, which has been derived from the capacitive and inductive reactance formulas:

Resonance
Circuit condition that occurs when the inductive reactance (X_L) is equal to the capacitive reactance (X_C).

$$f_0 = \frac{1}{2\pi\sqrt{LC}}$$

where f_0 = resonant frequency, in hertz (Hz)
       L = inductance, in henrys (H)
       C = capacitance, in farads (F)

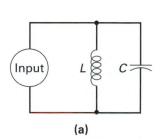
  (a)

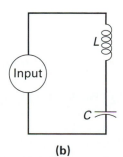
  (b)

FIGURE 16-6   **Resonance. (a) Parallel *LC* Circuit. (b) Series *LC* Circuit.**

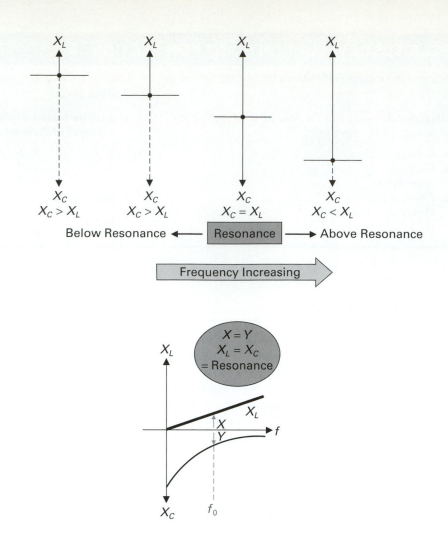

FIGURE 16-7 **Frequency versus Reactance.**

■ **INTEGRATED MATH APPLICATION:**

Calculate the resonant frequency (f_0) of a series LC circuit if $L = 750$ mH and $C = 47$ μF.

■ *Solution:*

$$f_0 = \frac{1}{2\pi\sqrt{L \times C}}$$

$$= \frac{1}{2\pi\sqrt{(750 \times 10^{-3}) \times (47 \times 10^{-6})}}$$

$$= 26.8 \text{ Hz}$$

16-3-1 *Series Resonance*

Series Resonant Circuit

A resonant circuit in which the capacitor and coil are in series with the applied ac voltage.

Figure 16-8(a) illustrates a series RLC circuit at resonance ($X_L = X_C$), or **series resonant circuit.** The ac input voltage causes current to flow around the circuit, and since all the components are connected in series, the same value of current (I_S) will flow through all the components. Since R, X_L, and X_C are all equal to 100 Ω and the current flow is the same throughout, the voltage dropped across each component will be equal, as illustrated vectorially in Figure 16-8(b).

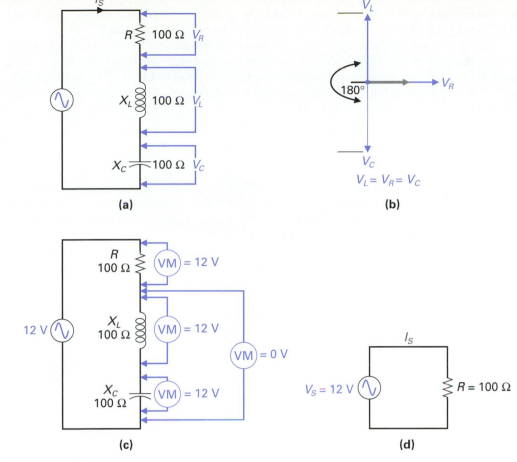

FIGURE 16-8 Series Resonant Circuit.

The voltage across the resistor is in phase with the series circuit current (I_S); however, since the voltage across the inductor (V_L) is 180° out of phase with the voltage across the capacitor (V_C), and both are equal to each other, V_L cancels V_C, when both are measured in series.

Three unusual characteristics occur when a circuit is at resonance that do not occur at any other frequency.

(1) The first is that if V_L and V_C cancel, the voltage across L and C will measure 0 V on a voltmeter. Since there is effectively no voltage being dropped across these two components, all the voltage must be across the resistor ($V_R = 12$ V). This is true; however, since the same current flows throughout the series circuit, a voltmeter will measure 12 V across C, 12 V across L, and 12 V across R, as shown in Figure 16-8(c). It now appears that the voltage drops around the series circuit (36 V) do not equal the voltage applied (12 V). This is not true, as V_L and V_C cancel, because they are out of phase with each other, so Kirchhoff's voltage law is still valid.

(2) The second unusual characteristic of resonance is that because the total opposition or impedance (Z) is equal to

$$Z = \sqrt{R^2 + (X_L \sim X_C)^2}$$

and the difference between X_L and X_C is 0 ($Z = \sqrt{R^2 + 0}$, the impedance of a series circuit at resonance is equal to the resistance value R ($Z = \sqrt{R^2} = R$). As a result, the applied ac voltage of 12 V is forcing current to flow through this series RLC circuit. Since current is equal to $I_S = V/Z$ and $Z = R$, the circuit current at resonance is dependent only on the value of resistance. The capacitor and inductor are invisible and are seen by the source as simply a piece of conducting wire with no resistance, as illustrated in Figure 16-8(d). Since only resistance exists in the circuit, current (I_S) and voltage (V_S) are in phase with each other, and as expected for a purely resistive circuit, the power factor will be equal to 1.

(3) To emphasize the third strange characteristic of series resonance, we will take another example, shown in Figure 16-9. In this example, R is made smaller (10 Ω) than X_L and X_C (100 Ω each).The circuit current in this example is equal to $I = V/R = $ 12 V/10 Ω = 1.2 A, as $Z = R$ at resonance. Since the same current flows throughout a series circuit, the voltage across each component can be calculated.

$$V_R = I \times R = 1.2\,\text{A} \times 10\,\Omega = 12\,\text{V}$$
$$V_L = I \times X_L = 1.2\,\text{A} \times 100\,\Omega = 120\,\text{V}$$
$$V_C = I \times X_C = 1.2\,\text{A} \times 100\,\Omega = 120\,\text{V}$$

Because V_L is 180° out of phase with V_C, the 120 V across the capacitor cancels with the 120 V across the inductor, resulting in 0 V across L and C combined, as shown in Figure 16-9(b). Since L and C have the ability to store energy, the voltage across them individually will appear larger than the applied voltage.

If the resistance in the circuit is removed completely, as shown in Figure 16-9(c), the circuit current, which is determined by the resistance only, will increase to a maximum ($I\uparrow = V/R\downarrow$) and, consequently, cause an infinitely high voltage across the inductor and capacitor ($V\uparrow = I\uparrow \times R$). In reality, the ac source will have some value of internal resistance, and the inductor, which is a long length of thin wire ($R\uparrow$), will have some value of resistance, as shown in Figure 16-9(d), which limits the series resonant circuit current.

In summary, we can say that in a series resonant circuit:

1. The inductor and capacitor electrically disappear due to their equal but opposite effect, resulting in a 0 V drop across the series combination, and the circuit seems purely resistive.

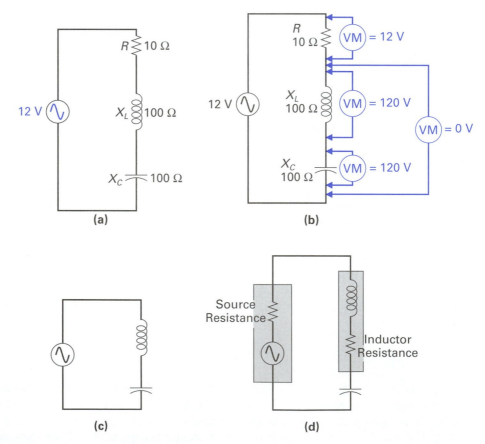

FIGURE 16-9 **Circuit Effects at Resonance.**

CHAPTER 16 / *RLC* CIRCUITS AND COMPLEX NUMBERS

2. The current flow is large because the impedance of the circuit is low and equal to the series resistance (R), which has the source voltage developed across it.

3. The individual voltage drops across the inductor or capacitor can be larger than the source voltage if R is smaller than X_L and X_C.

Quality Factor

As discussed previously in Chapter 15, the Q factor is a ratio of inductive reactance to resistance and is used to express how efficiently an inductor will store rather than dissipate energy. In a series resonant circuit, the Q factor indicates the quality of the series resonant circuit, or is the ratio of the reactance to the resistance.

$$Q = \frac{X_L}{R}$$

or, since $X_L = X_C$,

$$Q = \frac{X_C}{R}$$

Another way to calculate the Q of a series resonant circuit is by using the formula

$$Q = \frac{V_L}{V_R} = \frac{V_C}{V_R}$$

(at resonance only)

or, since $V_R = V_S$,

$$Q = \frac{V_L}{V_S} = \frac{V_C}{V_S}$$

(at resonance only)

If the Q and source voltage are known, the voltage across the inductor or capacitor can be found by transposition of the formula, as can be seen in the example in Figure 16-10.

FIGURE 16-10 Quality Factor at Resonance.

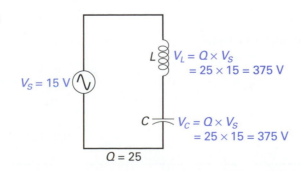

$V_S = 15 \text{ V}$

$V_L = Q \times V_S$
$= 25 \times 15 = 375 \text{ V}$

$V_C = Q \times V_S$
$= 25 \times 15 = 375 \text{ V}$

$Q = 25$

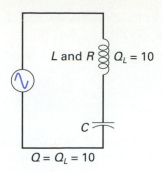

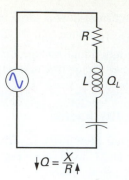

$Q = Q_L = 10$

$\uparrow Q = \dfrac{X}{R} \uparrow$

FIGURE 16-11   Resistance within Inductor.

The Q of a resonant circuit is almost entirely dependent on the inductor's coil resistance, because capacitors tend to have almost no resistance at all, only reactance, which makes them very efficient.

The inductor has a Q value of its own, and if only L and C are connected in series with each other, the Q of the series resonant circuit will be equal to the Q of the inductor, as shown in Figure 16-11. If the resistance is added in with L and C, the Q of the series resonant circuit will be less than that of the inductor's Q.

■   **INTEGRATED MATH APPLICATION:**

Calculate the resistance of the series resonant circuit illustrated in Figure 16-12.

**FIGURE 16-12   Series Resonant
Circuit Example.**

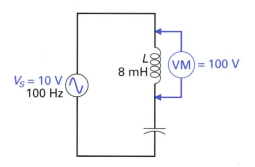

■   *Solution:*

$$Q = \frac{V_L}{V_S} = \frac{100 \text{ V}}{10 \text{ V}} = 10$$

Since $Q = X_L/R$ and $R = X_L/Q$, if the inductive reactance can be found, the R can be determined.

$$X_L = 2\pi \times f \times L$$
$$= 2\pi \times 100 \text{ Hz} \times 8 \text{ mH}$$
$$= 5 \ \Omega$$

R will therefore equal

$$R = \frac{X_L}{Q}$$
$$= \frac{5 \ \Omega}{10}$$
$$= 0.5 \ \Omega$$

Bandwidth

A series resonant circuit is selective in that frequencies at resonance or slightly above or below will cause a larger current than frequencies well above or below the circuit's resonant frequency. The group or band of frequencies that causes the larger current is called the circuit's **bandwidth.**

Figure 16-13 illustrates a series resonant circuit and its bandwidth. The × marks on the curve illustrate where different frequencies were applied to the circuit and the resulting value of current measured in the circuit. The resulting curve produced is called a **frequency response curve,** as it illustrates the circuit's response to different frequencies. At resonance, $X_L = X_C$ and the two cancel, which is why maximum current was present in the circuit (100 mA) when the resonant frequency (100 Hz) was applied.

The bandwidth includes the group or band of frequencies that cause 70.7% or more of the maximum current to flow within the series resonant circuit. In this example, frequencies from 90 to 110 Hz cause 70.7 mA or more, which is 70.7% of maximum (100 mA), to flow. The bandwidth in this example is equal to

$$BW = 110 - 90 = 20 \text{ Hz}$$

(110 Hz and 90 Hz are known as **cutoff frequencies.**)

Referring to the bandwidth curve in Figure 16-13(b), you may notice that 70.7% is also called the **half-power point,** although it does not exist halfway between 0 and maximum. This value of 70.7% is not the half-current point but the half-power point, as we can prove with a simple example.

<hr>

■ **INTEGRATED MATH APPLICATION:**

$R = 2 \text{ k}\Omega$ and $I = 100$ mA; therefore, power $= I^2 \times R = (100 \text{ mA})^2 \times 2 \text{ k}\Omega = 20$ W. If the current is now reduced so that it is 70.7% of its original value, calculate the power dissipated.

■ **Solution:**

$$P = I^2 \times R = (70.7 \text{ mA})^2 \times 2 \text{ k}\Omega = 10 \text{ W}$$

In summary, the 70.7% current points are equal to the 50% or half-power points. A circuit's bandwidth is the band of frequencies that exist between the 70.7% current points or half-power points.

<hr>

Bandwidth

Width of the group or band of frequencies between the half-power points.

Frequency Response Curve

A graph indicating a circuit's response to different frequencies.

Cutoff Frequency

Frequency at which the gain of the circuit falls below 0.707 of the maximum current or half-power (−3 dB).

Half-Power Point

A point at which power is 50%. This half-power point corresponds to 70.7% of the total current.

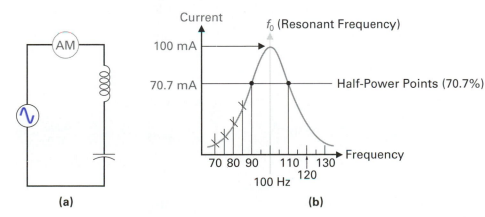

FIGURE 16-13 Series Resonant Circuit Bandwidth. (a) Circuit. (b) Frequency Response Curve.

The bandwidth of a series resonant circuit can also be calculated by use of the formula

$$BW = \frac{f_0}{Q_{f_0}}$$

where f_0 = resonant frequency
         Q_{f_0} = quality factor at resonance

This formula states that the BW is proportional to the resonant frequency of the circuit and inversely proportional to the Q of the circuit.

Figure 16-14 illustrates three example response curves. In these three examples, the value of R is changed from 100 Ω to 200 Ω to 400 Ω. This does not vary the resonant frequency but simply alters the Q and therefore the BW. The resistance value determines the Q of the circuit, and since Q is inversely proportional to resistance, Q is proportional to current; consequently, a high value of Q will cause a high value of current.

In summary, the bandwidth of a series resonant circuit will increase as the Q of the circuit decreases (BW↑ = f_0/Q↓), and vice versa.

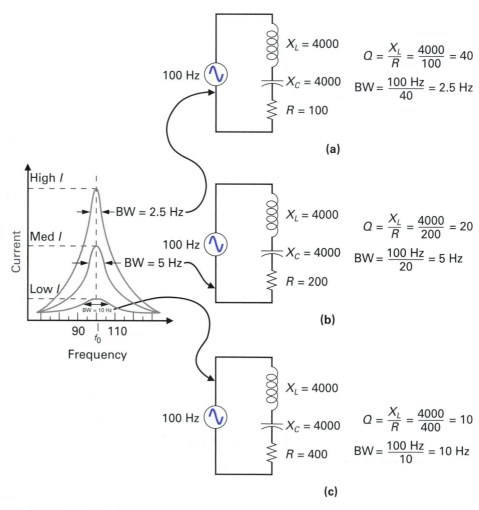

FIGURE 16-14   **Bandwidth of a Series Resonant Circuit.**

Use the following questions to test your understanding of Section 16-3-1.

1. Define *resonance*.

2. What is series resonance?

3. In a series resonant circuit, what are the three rather unusual circuit phenomena that take place?

4. How does Q relate to series resonance?

5. Define *bandwidth*.

6. Calculate BW if $f_0 = 12$ kHz and $Q = 1000$.

16-3-2 *Parallel Resonance*

The **parallel resonant circuit** acts differently than the series resonant circuit, and these different characteristics need to be analyzed and discussed. Figure 16-15 illustrates a parallel resonant circuit. The inductive current could be calculated by using the formula

$$I_L = \frac{V_L}{X_L}$$
$$= \frac{10 \text{ V}}{1 \text{ k}\Omega}$$
$$= 10 \text{ mA}$$

The capacitive current could be calculated by using the formula

$$I_C = \frac{V_C}{X_C}$$
$$= \frac{10 \text{ V}}{1 \text{ k}\Omega}$$
$$= 10 \text{ mA}$$

Parallel Resonant Circuit

Circuit having an inductor and capacitor in parallel with each other, offering a high impedance at the frequency of resonance.

Looking at the vector diagram in Figure 16-15(b), you can see that I_C leads the source voltage by 90° (ICE), and I_L lags the source voltage by 90° (ELI), creating a 180° phase difference between I_C and I_L. This means that when 10 mA of current flows up through the inductor, 10 mA of current will flow in the opposite direction down through the capacitor, as shown in Figure 16-16(a). During the opposite alternation, 10 mA will flow down through the inductor and 10 mA will travel up through the capacitor, as shown in Figure 16-16(b).

If 10 mA arrives into point *X,* and 10 mA of current leaves point *X,* no current can be flowing from the source (V_S) to the parallel *LC* circuit; the current is simply swinging or oscillating back and forth between the capacitor and inductor.

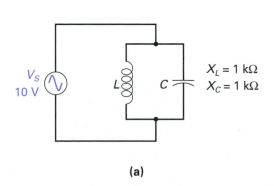

(a)

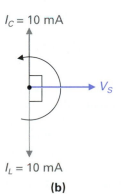

(b)

FIGURE 16-15 **Parallel Resonant Circuit. (a) Circuit. (b) Vector Diagram.**

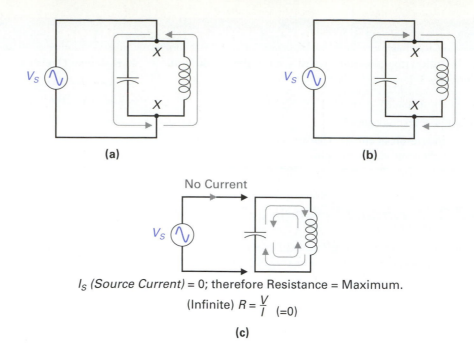

I_S *(Source Current)* = 0; therefore Resistance = Maximum.

(Infinite) $R = \dfrac{V}{I}$ (=0)

(c)

FIGURE 16-16   Current in a Parallel Resonant Circuit.

The source voltage (V_S) is needed initially to supply power to the *LC* circuit and start the oscillations; but once the oscillating process is in progress (assuming the ideal case), current flows back and forth only between inductor and capacitor, and no current flows from the source. So the *LC* circuit appears as an infinite impedance and the source can be disconnected, as shown in Figure 16-16(c).

Flywheel Action

Let's discuss this oscillating effect, called **flywheel action,** in a little more detail. The name is derived from the fact that it resembles a mechanical flywheel, which, once started, will keep going continuously until friction reduces the magnitude of the rotations to zero.

The electronic equivalent of the mechanical flywheel is a resonant parallel-connected *LC* circuit. Figure 16-17(a) through (h) illustrates the continuous energy transfer between capacitor and inductor, and vice versa. The direction of the circulating current reverses each half-cycle at the frequency of resonance. Energy is stored in the capacitor in the form of an electric field between the plates on one half-cycle, and then the capacitor discharges, supplying current to build up a magnetic field on the other half-cycle. The inductor stores its energy in the form of a magnetic field, which will collapse, supplying a current to charge the capacitor, which will then discharge, supplying a current back to the inductor, and so on. Due to the "storing action" of this circuit, it is sometimes related to the fluid analogy and referred to as a **tank circuit.**

The Reality of Tanks

Under ideal conditions a tank circuit should oscillate indefinitely if no losses occur within the circuit. In reality, the resistance of the coil reduces that 100% efficiency, as does friction with the mechanical flywheel. This coil resistance is illustrated in Figure 16-18(a), and, unlike reactance, resistance is the opposition to current flow, with the dissipation of energy in the form of heat. Because a small part of the energy is dissipated with each cycle, the oscillations will be reduced in size and eventually fall to zero, as shown in Figure 16-18(b).

If the ac source is reconnected to the tank, as shown in Figure 16-18(c), a small amount of current will flow from the source to the tank to top up the tank or replace the dissipated power. The higher the coil resistance is, the higher the loss and the larger the current flow from source to tank to replace the loss.

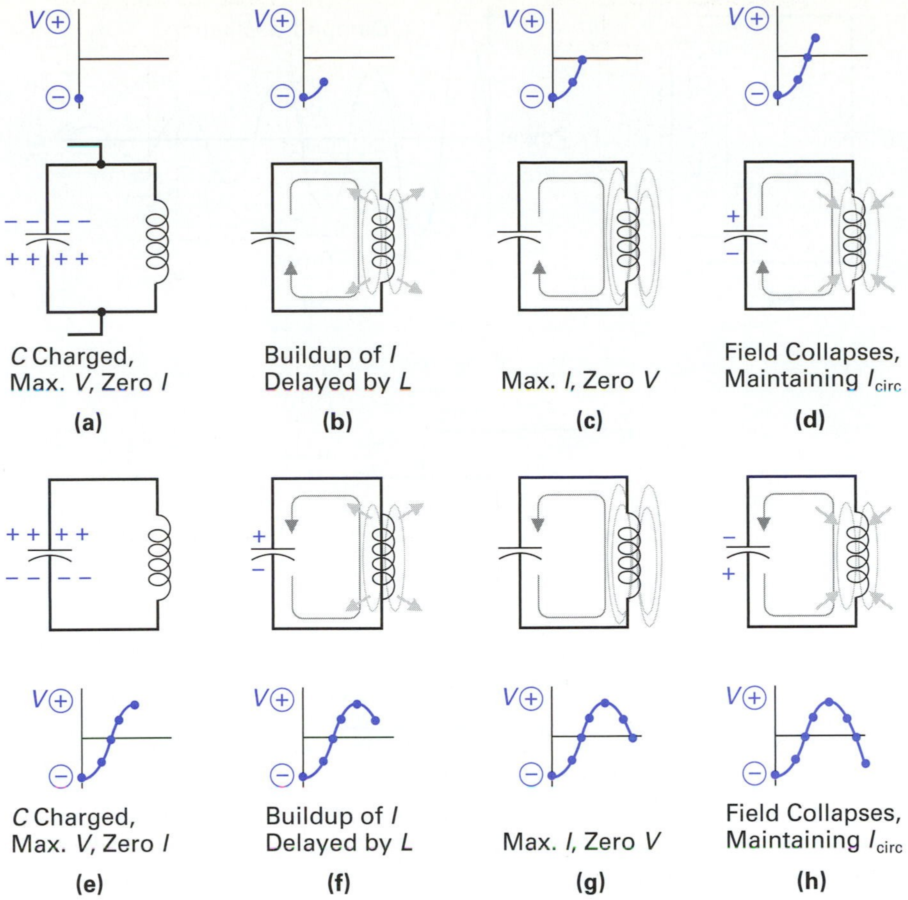
V(+)
(−)
C Charged,
Max. V, Zero I
(a)
Buildup of I
Delayed by L
(b)
Max. I, Zero V
(c)
Field Collapses,
Maintaining I circ
(d)
C Charged,
Max. V, Zero I
(e)
Buildup of I
Delayed by L
(f)
Max. I, Zero V
(g)
Field Collapses,
Maintaining I circ
(h)

FIGURE 16-17 Energy and Current in an *LC* Parallel Circuit at Resonance.

Quality Factor

In the series resonant circuit, we were concerned with voltage drops, since current remains the same throughout a series circuit, so

$$Q = \frac{V_C \text{ or } V_L}{V_S} \quad \text{(at resonance only)}$$

In a parallel resonant circuit, we are concerned with circuit currents rather than voltage, so

$$Q = \frac{I_{\text{tank}}}{I_S}$$
(at resonance only)

The quality factor, Q, can also be expressed as the ratio between reactance and resistance:

$$Q = \frac{X_L}{R} \quad \text{(at any frequency)}$$

Another formula, which is the most frequently used when discussing and using parallel resonant circuits, is

$$Q = \frac{Z_{\text{tank}}}{X_L}$$
(at resonance only)

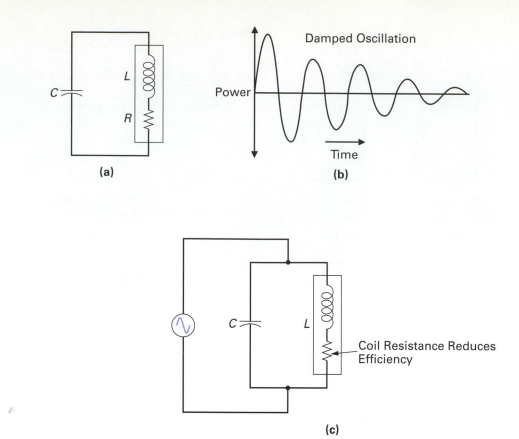

(a)

(b)

(c)

FIGURE 16-18 Losses in Tanks.

This formula states that the Q of the tank is proportional to the tank impedance. A higher tank impedance results in a smaller current flow from source to tank. This assures that less power is dissipated, and that means a higher-quality tank.

Of all the three Q formulas for parallel resonant circuits, $Q = I_{tank}/I_S$, $Q = X_L/R$, and $Q = Z_{tank}/X_L$, the latter is the easiest to use, as both X_L and the tank impedance can easily be determined in most cases where C, L, and R internal for the inductor are known.

Bandwidth

Figure 16-19 illustrates a parallel resonant circuit and two typical response curves. These response curves summarize what we have described previously, in that a parallel resonant circuit has maximum impedance [Figure 16-19(b)] and minimum current [Figure 16-19(c)] at

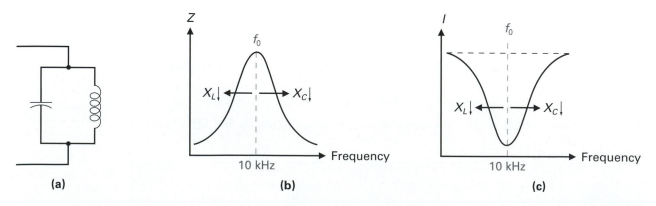

(a)

(b)

(c)

FIGURE 16-19 Parallel Resonant Circuit Bandwidth.

resonance. The current versus frequency response curve shown in Figure 16-19(c) is the complete opposite of the series resonant response curve. At frequencies below resonance (< 10 kHz), X_L is low and X_C is high, and the inductor offers a low reactance, producing a high current path and low impedance. On the other hand, at frequencies above resonance (> 10 kHz), the capacitor displays a low reactance, producing a high current path and low impedance. The parallel resonant circuit is like the series resonant circuit in that it responds to a band of frequencies close to its resonant frequency.

The bandwidth (BW) can be calculated by use of the formula

$$BW = \frac{f_0}{Q_{f_0}}$$

where f_0 = resonant frequency
 Q_{f_0} = quality factor at resonance

■ INTEGRATED MATH APPLICATION:

Calculate the bandwidth of the circuit illustrated in Figure 16-20.

FIGURE 16-20 Bandwidth Example.

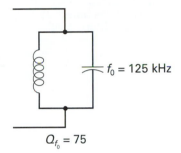

$f_0 = 125$ kHz

$Q_{f_0} = 75$

■ *Solution:*

$$BW = \frac{f_0}{Q_{f_0}}$$
$$= \frac{125 \text{ kHz}}{75}$$
$$= 1.7 \text{ kHz}$$
$$\frac{1.7 \text{ kHz}}{2} = 0.85 \text{ kHz}$$

Therefore, the bandwidth extends from

$$f_0 + 0.85 \text{ kHz} = 125.85 \text{ kHz}$$
$$f_0 - 0.85 \text{ kHz} = 124.15 \text{ kHz}$$
$$BW = 124.15 \text{ kHz to } 125.85 \text{ kHz}$$

Selectivity

Circuits containing inductance and capacitance are often referred to as **tuned circuits,** since they can be adjusted to make the circuit responsive to a particular frequency (the resonant frequency). **Selectivity,** by definition, is the ability of a tuned circuit to respond to a desired frequency and to ignore all others. Parallel resonant *LC* circuits are sometimes too selective, as the *Q* is too large, producing too narrow a bandwidth, as shown in Figure 16-21(a) (BW↓ = f_0/Q↑).

In this situation, because of the very narrow response curve, a high resistance value can be placed in parallel with the *LC* circuit to provide an alternative path for source current. This process is known as *loading* or *damping* the tank and will cause an increase in source current and decrease in *Q* (Q↓ = I_{tank}/I_{source}↑). The decrease in *Q* will cause a corresponding increase

Tuned Circuit

Circuit that can have its components' values varied so that the circuit responds to one selected frequency yet heavily attenuates all other frequencies.

Selectivity

Characteristic of a circuit to discriminate between the wanted signal and the unwanted signal.

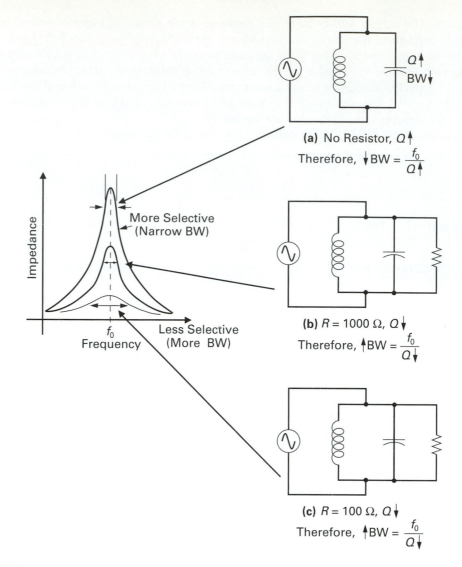

(a) No Resistor, $Q\uparrow$
Therefore, $\downarrow$BW $= \dfrac{f_0}{Q\uparrow}$

(b) $R = 1000\ \Omega,\ Q\downarrow$
Therefore, $\uparrow$BW $= \dfrac{f_0}{Q\downarrow}$

(c) $R = 100\ \Omega,\ Q\downarrow$
Therefore, $\uparrow$BW $= \dfrac{f_0}{Q\downarrow}$

FIGURE 16-21 **Varying Bandwidth by Loading the Tank Circuit.**

in BW (BW$\uparrow = f_0/Q\downarrow$), as shown by the examples in Figure 16-21, which illustrates a 1000 Ω loading resistor [Figure 16-21(b)] and a 100 Ω loading resistor [Figure 16-21(c)].

In summary, a parallel resonant circuit can be made less selective with a broader bandwidth if a resistor is added in parallel, providing an increase in current and a decrease in impedance, which widens the bandwidth.

SELF-TEST EVALUATION POINT FOR SECTION 16-3-2

Now that you have completed this section, you should be able to:

■ *Objective 4.* *Define resonance, and explain the characteristics of:*

 a. Series resonance
 b. Parallel resonance

Use the following questions to test your understanding of Section 16-3-2.

1. What are the differences between a series and a parallel resonant circuit?
2. Describe flywheel action.
3. Calculate the value of Q of a tank if $X_L = 50\ \Omega$ and $R = 25\ \Omega$.
4. When calculating bandwidth for a parallel resonant circuit, can you use the series resonant bandwidth formula?
5. What is selectivity?

Filter circuits are used to pass some frequencies and block others. There are basically four types of filters:

1. Low-pass filter, which passes frequencies below a cutoff frequency
2. High-pass filter, which passes frequencies above a cutoff frequency
3. Bandpass filter, which passes a band of frequencies
4. Band-stop filter, which stops a band of frequencies

16-4-1   *Low-Pass Filter*

Figure 16-22(a) illustrates how an inductor and capacitor can be connected to act as a low-pass filter. At low frequencies, X_L has a small value compared with the load resistor (R_L), so nearly all the low-frequency input is developed and appears at the output across R_L. Since X_C is high at low frequencies, nearly all the current passes through R_L rather than C.

At high frequencies, X_L increases and drops more of the applied input across the inductor rather than the load. The capacitive reactance, X_C, aids this low output–at–high frequency effect by decreasing its reactance and providing an alternative path for current to flow.

Since the inductor basically blocks alternating current and the capacitor shunts alternating current, the net result is that high-frequency signals are prevented from reaching the load. The way in which this low-pass filter responds to frequencies is graphically illustrated in Figure 16-22(b).

16-4-2   *High-Pass Filter*

Figure 16-23(a) illustrates how an inductor and capacitor can be connected to act as a high-pass filter. At high frequencies, the reactance of the capacitor (X_C) is low, and the reactance of the inductor (X_L) is high, so all the high frequencies are easily passed by the capacitor and blocked by the inductor, so they all are routed through to the output and load.

At low frequencies, the reverse condition exists, resulting in a low X_L and a high X_C. The capacitor drops nearly all the input, and the inductor shunts the signal current away from the output load.

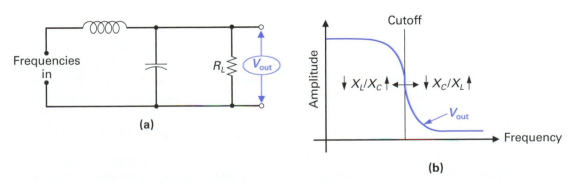

FIGURE 16-22   Low-Pass Filter. (a) Circuit. (b) Frequency Response.

FIGURE 16-23 High-Pass Filter.
(a) Circuit. (b) Frequency Response.

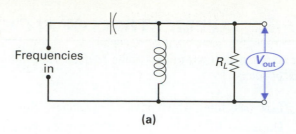

(a)

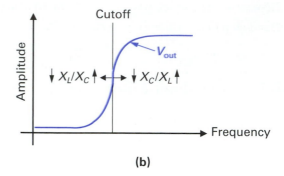

(b)

16-4-3 *Bandpass Filter*

Bandpass Filter

Filter circuit that passes a group or band of frequencies between a lower and an upper cutoff frequency while heavily attenuating any other frequency outside this band.

Figure 16-24(a) illustrates a series resonant **bandpass filter,** and Figure 16-24(b) shows a parallel resonant bandpass filter. Figure 16-24(c) shows the frequency response curve produced by the bandpass filter. At resonance, the series resonant LC circuit has a very low impedance and will consequently pass the resonant frequency to the load with very little drop across the L and C components.

Below resonance, X_C is high, and the capacitor drops a large amount of the input signal; above resonance, X_L is high, and the inductor drops most of the input frequency voltage. This circuit will therefore pass a band of frequencies centered around the resonant frequency of the series LC circuit and block all other frequencies above and below this resonant frequency.

Figure 16-24(b) illustrates how a parallel resonant LC circuit can be used to provide a bandpass response. The series resonant circuit was placed in series with the output, whereas the parallel resonant circuit will have to be placed in parallel with the output to provide the same results. At resonance, the parallel resonant circuit or tank has a high impedance, so very little current will be shunted away from the output; it will be passed on to the output, and almost all the input will appear at the output across the load.

Above resonance, X_C is small, so most of the input is shunted away from the output by the capacitor; below resonance, X_L is small, and the shunting action occurs again, but this time through the inductor.

Figure 16-25 illustrates how a transformer can be used to replace the inductor to produce a bandpass filter. At resonance, maximum flywheel current flows within the parallel circuit made up of the capacitor and the primary of the transformer (L), which is known as a *tuned transformer.* With maximum flywheel current, there will be a maximum magnetic field, which means that there will be maximum power transfer between primary and secondary. Thus, nearly all the input will be coupled to the output (coupling coefficient $k = 1$) and appear across the load at and around a small band of frequencies centered on resonance.

Above and below resonance, current within the parallel resonant circuit will be smaller, so the power transfer ability will be less, effectively keeping the frequencies outside the passband from appearing at the output.

FIGURE 16-24 Bandpass Filter.
(a) Series Resonant Bandpass Filter.
(b) Parallel Resonant Bandpass Filter.
(c) Frequency Response.

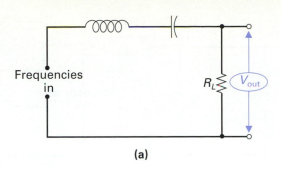

(a)

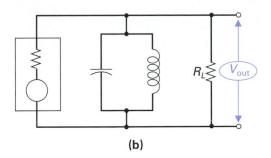

(b)

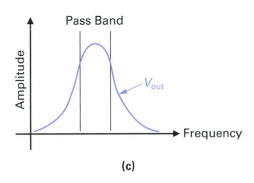

(c)

16-4-4   *Band-Stop Filter*

Figure 16-26(a) illustrates a series resonant and Figure 16-26(b) a parallel resonant **band-stop filter.** Figure 16-26(c) shows the frequency response curve produced by a band-stop filter. The band-stop filter operates exactly the opposite to a bandpass filter in that it blocks or attenuates a band of frequencies centered on the resonant frequency of the *LC* circuit.

Band-Stop Filter

A filter that attenuates alternating currents whose frequencies are between given upper and lower cutoff values while passing frequencies above and below this band.

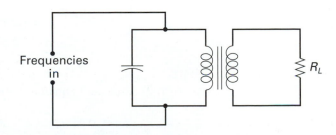

FIGURE 16-25   Parallel Resonant Bandpass Circuit Using a Transformer.

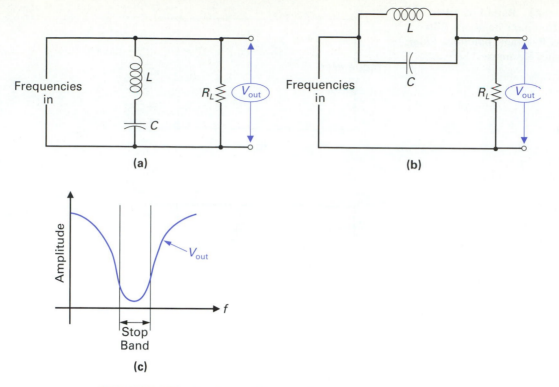

(a)

(b)

(c)

FIGURE 16-26 **Band-Stop Filter. (a) Series Resonant Band-Stop Filter. (b) Parallel Resonant Band-Stop Filter. (c) Frequency Response.**

In the series resonant circuit in Figure 16-26(a), the *LC* impedance is very low at and around resonance, so these frequencies are rejected or shunted away from the output. Above and below resonance, the series circuit has a very high impedance, which results in almost no shunting of the signal away from the output.

In the parallel resonant circuit in Figure 16-26(b), the *LC* circuit is in series with the load and output. At resonance, the impedance of a parallel resonant circuit is very high, and the band of frequencies centered around resonance is blocked. Above and below resonance, the impedance of the tank is very low, so nearly all the input is developed across the output.

Filters are necessary in applications such as television or radio where it is necessary to tune in (select or pass) one frequency that contains the information desired yet to block all the millions of other frequencies that are also carrying information, as shown in Figure 16-27.

FIGURE 16-27 **Tuning in of Station by Use of a Bandpass Filter.**

Now that you have completed this section, you should be able to:

■ **Objective 5.** *Identify and explain the following RLC circuit applications:*

a. *Low-pass filter*
b. *High-pass filter*
c. *Bandpass filter*
d. *Band-stop filter*

Use the following questions to test your understanding of Section 16-4.

1. Of the four types of filters, which:
   a. Would utilize the inductor as a shunt?
   b. Would utilize the capacitor as a shunt?
   c. Would use a series resonant circuit as a shunt?
   d. Would use a parallel resonant circuit as a shunt?

2. In what applications can filters be found?

16-5   COMPLEX NUMBERS

After reading this section you will realize that there is really nothing complex about **complex numbers.** The complex number system allows us to determine the *magnitude* and *phase angle* of electrical quantities by adding, subtracting, multiplying, and dividing phasor quantities and is an invaluable tool in ac circuit analysis.

> **Complex Numbers**
> Numbers composed of a real number part and an imaginary number part.

16-5-1   *The Real Number Line*

Real numbers can be represented on a horizontal line, known as the real number line, as in Figure 16-28. Referring to this line, you can see that positive numbers exist to the right of the center point corresponding to zero, and negative numbers exist to the left. This representation satisfied most mathematicians for a short time, as they could indicate numbers such as 2 or 5 as a point on the line. Numbers corresponding to $\sqrt{9}$ could also be represented, as three points to the right of zero ($\sqrt{9} = +3$); however, a problem was reached if they wished to indicate a point corresponding to $\sqrt{-9}$. The $\sqrt{-9}$ is not $+3$ [since $(+3) \times (+3) = +9$], and it is not -3 [since $(-3) \times (-3) = +9$]. So it was eventually realized that the square root of a negative number could not be indicated on the real number line, as it is not a real number.

> **Real Numbers**
> Numbers that have no imaginary parts.

16-5-2   *The Imaginary Number Line*

Mathematicians decided to call the square root of a negative number, such as $\sqrt{-4}$ or $\sqrt{-9}$, an **imaginary number,** which is not fictitious or imaginary but simply a particular type of number.

Just as real numbers can be represented on a real number line, imaginary numbers can be represented on an imaginary number line, as shown in Figure 16-29. The imaginary number line is vertical, to distinguish it from the real number line, and in calculations with electrical quantities a $\pm j$ prefix, known as the *j* **operator,** is used for values that appear on the imaginary number line.

> **Imaginary Number**
> A complex number whose imaginary part is not zero.

> **j Operator**
> A prefix used to indicate an imaginary number.

16-5-3   *The Complex Plane*

A complex number is the combination of a real and an imaginary number and is represented on a two-dimensional plane called the **complex plane,** shown in Figure 16-30. Generally,

> **Complex Plane**
> A plane whose points are identified by means of complex numbers.

FIGURE 16-28   Real-Number Line.

FIGURE 16-29 **Imaginary Number Line.**

Positive Imaginary Numbers

Negative Imaginary Numbers

the real number appears first, followed by the imaginary number. Here are some examples of complex numbers.

REAL NUMBERS	IMAGINARY NUMBERS
3	$+j4$
-2	$+j4$
-3	$-j2$

Complex numbers, therefore, are merely terms that need to be added as phasors, and all you have to do basically is draw a vector representing the real number and then draw another vector representing the imaginary number.

■ **INTEGRATED MATH APPLICATION:**

Find the points in the complex plane in Figure 16-31 that correspond to the following complex numbers.

$$W = 3 + j4$$
$$X = 5 - j7$$
$$Y = -4 + j6$$
$$Z = -3 - j5$$

■ *Solution:*

By first locating the point corresponding to the real number on the horizontal line and then plotting it against the imaginary number on the vertical line, the points can be determined as shown in Figure 16-31.

A number like $3 + j4$ specifies two phasors in **rectangular coordinates,** so this system is the *rectangular representation of a complex number.* There are several other ways to describe a complex number, one of which is the polar representation of a complex number, using **polar coordinates,** which will be discussed next.

Rectangular Coordinates

A Cartesian coordinate system whose straight-line axes or coordinate planes are perpendicular.

Polar Coordinates

Either of two numbers that locate a point in a plane by its distance from a fixed point on a line and the angle this line makes with a fixed line.

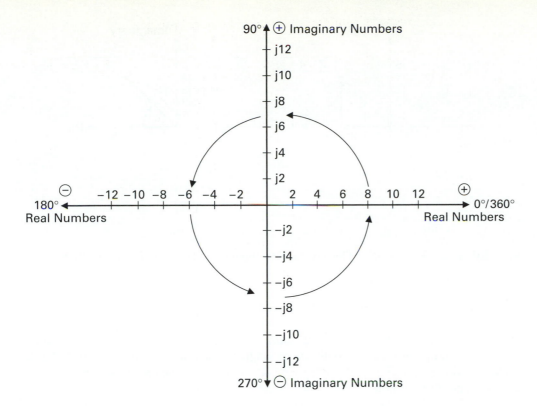

FIGURE 16-30 Complex Plane.

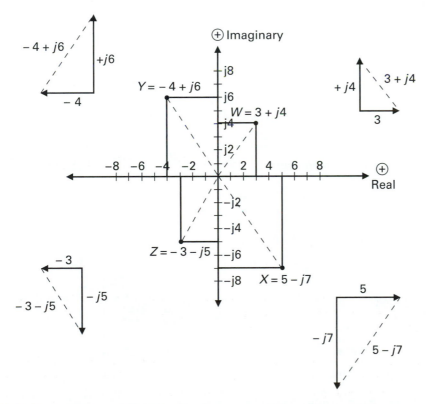

FIGURE 16-31 Examples of Complex Numbers.

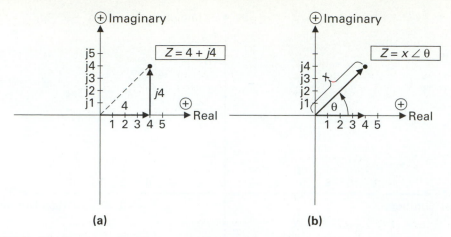

(a)   **(b)**

FIGURE 16-32   Representing Phasors. (a) Rectangular Notation. (b) Polar Notation.

16-5-4   *Polar Complex Numbers*

Phasors can also be expressed in polar form, as shown in Figure 16-32, which compares rectangular and polar notation. With the rectangular notation in Figure 16-32(a), the horizontal coordinate is the real part and the vertical coordinate is the imaginary part of the complex number. With the polar notation shown in Figure 16-32(b), the magnitude of the phasor (x, or size) and the angle ($\angle \theta$, meaning "angle theta") relative to the positive real axis (measured in a counterclockwise direction) are stated.

■   **INTEGRATED MATH APPLICATION:**

Sketch the following polar numbers:

  a. $5 \angle 60°$

  b. $3 \angle 220°$

■   *Solution:*

As you can see in Figure 16-33, an equivalent negative angle, which is calculated by subtracting the given positive angle from 360°, can also be used.

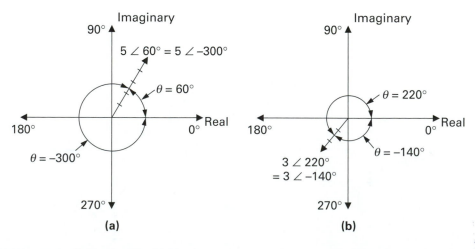

(a)   **(b)**

FIGURE 16-33   Polar Number Examples.

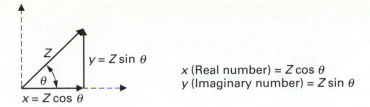

FIGURE 16-34 Polar-to-Rectangular Conversion.

16-5-5 *Rectangular/Polar Conversions*

Many scientific calculators have a feature that allows you to convert rectangular numbers to polar numbers, and vice versa. These conversions are based on the Pythagorean theorem and trigonometric functions, discussed previously in Chapter 6.

Polar-to-Rectangular Conversion

The polar notation states the magnitude and angle, as shown in Figure 16-34. The following examples show how this conversion can be achieved.

■ **INTEGRATED MATH APPLICATION:**

Convert the following polar numbers to rectangular form:

a. $5 \angle 30°$　　b. $18 \angle -35°$　　c. $44 \angle 220°$

■ *Solution:*

a. Real number $= 5 \cos 30° = 4.33$
   Imaginary number $= 5 \sin 30° = j2.5$
   Polar number, $5 \angle 30° =$ rectangular number, $4.33 + j2.5$

b. Real number $= 18 \cos (-35°) = 14.74$
   Imaginary number $= 18 \sin (-35°) = -j10.32$
   Polar number, $18 \angle -35° =$ rectangular number, $14.74 - j10.32$

c. Real number $= 44 \cos 220° = -33.7$
   Imaginary number $= 44 \sin 220° = -j28.3$
   Polar number, $44 \angle 220° =$ rectangular number, $-33.7 - j28.3$

Rectangular-to-Polar Conversion

The rectangular notation states the horizontal (real) and vertical (imaginary) sides of a triangle, as shown in Figure 16-35. The following examples show how the conversion can be achieved.

FIGURE 16-35 Rectangular-to-Polar Conversion.

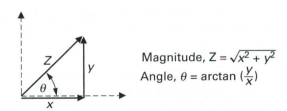

Magnitude, $Z = \sqrt{x^2 + y^2}$
Angle, $\theta = \arctan \left(\frac{y}{x} \right)$

■ **INTEGRATED MATH APPLICATION:**

Convert the following rectangular numbers to polar form:

 a. $4 + j3$ b. $16 - j14$

■ *Solution:*

 a. Magnitude $= \sqrt{4^2 + 3^2} = 5$
   Angle $= \arctan (3/4) = 36.9°$
   Rectangular number, $4 + j3 =$ polar number, $5 \angle 36.9°$

 b. Magnitude $= \sqrt{16^2 + (-14)^2} = 21.3$
   Angle $= \arctan (-14/16) = -41.2°$
   Rectangular number, $16 - j14 =$ polar number, $21.3 \angle -41.2°$

16-5-6   *Complex Number Arithmetic*

Since a phase difference exists between real and imaginary (j) numbers, certain rules should be applied when adding, subtracting, multiplying, or dividing complex numbers.

Addition

The sum of two complex numbers is equal to the sum of their real and imaginary parts.

■ **INTEGRATED MATH APPLICATION:**

Add the following complex numbers:

 a. $(3 + j4) + (2 + j5)$
 b. $(4 + j5) + (2 - j3)$

■ *Solution:*

 a. $(3 + j4) + (2 + j5) = (3 + 2) + (j4 + j5) = 5 + j9$
 b. $(4 + j5) + (2 - j3) = (4 + 2) + (j5 - j3) = 6 + j2$

Subtraction

The difference of two complex numbers is equal to the difference between the separate real and imaginary parts.

■ **INTEGRATED MATH APPLICATION:**

Subtract the following complex numbers:

 a. $(4 + j3) - (2 + j2)$
 b. $(12 + j6) - (6 - j3)$

■ *Solution:*

 a. $(4 + j3) - (2 + j2) = (4 - 2) + j(3 - 2) = 2 + j1$
 b. $(12 + j6) - (6 - j3) = (12 - 6) + j[6 - (-3)] = 6 + j9$

Multiplication

Multiplication of two complex numbers is achieved more easily if they are in polar form. The simple rule to remember is to multiply the magnitudes and then add the angles algebraically.

INTEGRATED MATH APPLICATION:

Multiply the following complex numbers:

   a. $5 \angle 35° \times 7 \angle 70°$

   b. $4 \angle 53° \times 12 \angle -44°$

■ *Solution:*

   a. Multiply the magnitudes: $5 \times 7 = 35$.
      Algebraically add the angles: $\angle (35° + 70°) = \angle 105° = 35 \angle 105°$.

   b. Multiply the magnitudes: $4 \times 12 = 48$.
      Algebraically add the angles: $\angle [53° + (-44°)] = \angle 9° = 48 \angle 9°$.

Division

Division is also more easily carried out in polar form. The rule to remember is to divide the magnitudes, and then subtract the denominator angle from the numerator angle.

■ **INTEGRATED MATH APPLICATION:**

Divide the following complex numbers:

   a. $60 \angle 30°$ by $30 \angle 15°$ b. $100 \angle 20°$ by $5 \angle -7°$

■ *Solution:*

   a. Divide the magnitudes: $60/30 = 2$.
      Subtract the denominator angle from the numerator angle: $\angle (30° - 15°) = \angle 15° = 2 \angle 15°$.

   b. Divide the magnitudes: $100/5 = 20$.
      Subtract the angles: $\angle [20° - (-7°)] = \angle 27° = 20 \angle 27°$.

16-5-7   *How Complex Numbers Apply to AC Circuits*

Complex numbers find an excellent application in ac circuits due to all the phase differences that occur between different electrical quantities, such as X_L, X_C, R, and Z, as shown in Figure 16-36. The positive real number line, at an angle of 0°, is used for resistance, which in this example is 3 Ω, as shown in Figure 16-36(a) and (b).

     On the positive imaginary number line, at an angle of 90° ($+j$), inductive reactance (X_L) is represented, which in this example is $j4$ Ω ($X_L = 4$ Ω). The voltage drop across an inductor (V_L) is proportional to its inductive reactance (X_L), and both are represented on the $+j$ imaginary number line, since the voltage drop across an inductor will always lead the current (which in a series circuit is always in phase with the resistance) by 90°.

     On the negative imaginary number line, at an angle of $-90°$ or 270° ($-j$), capacitive reactance (X_C) is represented, which in this example is $-j2$ ($X_C = 2$ Ω). The voltage drop across a capacitor (V_C) is proportional to its capacitive reactance (X_C), and both are represented on the $-j$ imaginary number line, since the voltage drop across a capacitor always lags current (charge and discharge) by $-90°$.

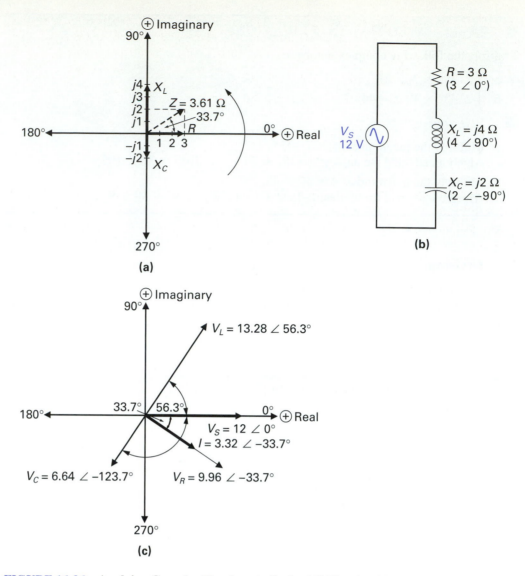

FIGURE 16-36 Applying Complex Numbers to Series AC Circuits. (a) Impedance Phasors. (b) Series Circuit. (c) Voltage and Current Phasors.

Series AC Circuits

Referring to Figure 16-36(a) and (b) once again, we can calculate total impedance simply by adding the phasors.

Z_T **(Rectangular)**   Total series impedance is equal to the sum of all the resistances and reactances:

$$Z = R + (jX_L \sim jX_C)$$
$$= 3 + (+j4 \sim -j2)$$
$$= 3 + j2$$

Z_T **(Polar)**   The total series impedance can be converted from rectangular to polar form:

$$\text{Magnitude} = \sqrt{3^2 + 2^2} = 3.61 \ \Omega$$
$$\text{Angle} = \arctan \frac{2}{3} = 33.7°$$
$$= 3.61 \angle 33.7°$$

Current   Once the magnitude of Z_T is known (3.61 Ω), I can be calculated. The source voltage of 12 V is a real positive number (0°) and is therefore represented as $12 \angle 0°$. Current is equal to

$$I = \frac{V_S}{Z_T} = \frac{12 \angle 0°}{3.61 \angle 33.7°}$$

Polar division   Divide the magnitudes: $\dfrac{12}{3.61} = 3.32$ A

Subtract the angles: $0° - 33.7° = -33.7°$

$$I = 3.32 \angle -33.7° \text{ A}$$

Phase Angle   The circuit current has an angle of $-33.7°$, which means that it lags V_T (inductive circuit, therefore ELI). This negative phase angle is expected, since this series circuit is inductive ($X_L > X_C$), in which case current should lag voltage by some phase angle. This phase angle is less than 45° because the net reactance is less than the circuit resistance and is shown in Figure 16-36(c).

Voltage Drops   The component voltage drops are calculated with the following formulas and shown in Figure 16-36(c).

$V_R = I \times R = (3.32 \angle -33.7°) \times (3 \angle 0°)$

Multiply the magnitudes: $3.32 \times 3 = 9.96$ V

Algebraically add the angles: $\angle (-33.7° + 0°) = -33.7°$

$V_R = 9.96 \angle -33.7°$ V

$V_L = I \times X_L = (3.32 \angle -33.7°) \times (4 \angle 90°)$

Multiply the magnitudes: $3.32 \times 4 = 13.28$ V

Algebraically add the angles: $\angle (-33.7° + 90°) = 56.3°$

$V_L = 13.28 \angle 56.3°$ V

$V_C = I \times X_C = (3.32 \angle -33.7°) \times (2 \angle -90°)$

Multiply the magnitudes: $3.32 \times 2 = 6.64$ V

Algebraically add the angles: $\angle (-33.7° + -90°) = -123.7°$

$V_C = 6.64 \angle -123.7°$ V

Phase Relationships   As shown in Figure 16-36(c), the voltage across an inductor leads the circuit current by $+90°$, whereas the voltage across a capacitor lags the circuit current by $-90°$. The source voltage acts as the zero reference phase and leads the circuit current and the voltage across the resistor (V_R) by 33.7°.

Source Voltage   Although the source voltage is known, it can be checked to verify all the previous calculations, since the sum of all the individual voltage drops should equal the source voltage.

$$\text{POLAR} \longrightarrow \text{RECTANGULAR}$$

$V_L = 13.28 \angle 56.3°$	$= 7.37 + j11.05$
$V_R = 9.96 \angle -33.7°$	$= 8.29 - j5.53$
$V_C = 6.64 \angle -123.7°$	$= -3.68 - j5.52$
	$\overline{11.98 + j0}$

Series–Parallel AC Circuits

Figure 16-37 illustrates a series–parallel ac circuit containing an *RL* branch, an *RC* branch, and an *RLC* branch.

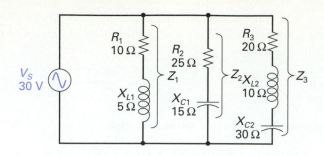

FIGURE 16-37 Applying Complex Numbers to Series–Parallel AC.

Impedance of Each Branch   Each of the three branches will have a value of impedance that will be equal to:

$$\text{RECTANGULAR} \longrightarrow \text{POLAR}$$

$$Z_1 = 10 + j5 = 11.2 \angle 26.6° \ \Omega$$
$$Z_2 = 25 - j15 = 29.2 \angle -31.0° \ \Omega$$

The third branch is capacitive, since the difference between $-j30$ (X_{C2}) and $+j10$ (X_{L2}) is $-j20$.

$$Z_3 = 20 - j20 = 28.3 \angle -45° \ \Omega$$

Branch Currents   The three branch currents, I_1, I_2, and I_3, are calculated by dividing the source voltage (V_S) by the individual branch impedances.

$$I_1 = \frac{V_S}{Z_1} = \frac{30 \angle 0°}{11.2 \angle 26.6°}$$

Divide magnitudes: $30 \div 11.2 = 2.68$

Subtract angles: $\angle (0° - 26.6°) = -26.6°$

$$I_1 = 2.68 \angle -26.6° = 2.4 - j1.2 \ \text{A}$$
$$I_2 = \frac{V_S}{Z_2} = \frac{30 \angle 0°}{29.2 \angle -31°} = 1.03 \angle +31° = 0.88 + j0.5 \ \text{A}$$
$$I_3 = \frac{V_S}{Z_3} = \frac{30 \angle 0°}{28.3 \angle -45°} = 1.06 \angle +45° = 0.75 + j0.7 \ \text{A}$$

Total Current

$$
\begin{aligned}
I_T &= I_1 + I_2 + I_3 \\
&= (2.4 - j1.2) + (0.88 + j0.5) + (0.75 + j0.7) \\
&= (2.4 + 0.88 + 0.75) + [-j1.2 + (+j0.5) + (+j0.7)] \\
&= 4.03 \ \text{A}
\end{aligned}
$$

In polar form, this result equals $4.03 \angle 0° \ \text{A}$.

Total Impedance

$$Z_T = \frac{V_S}{I_T} = \frac{30 \angle 0°}{4.30 \angle 0°} = 7.44 \angle 0° \ \Omega$$

$$\text{POLAR} \longrightarrow \text{RECTANGULAR}$$

$$7.44 \angle 0° = 7.44 + j0$$

The complex ac circuit seen in Figure 16-37 is therefore equivalent to a 7.44 Ω resistor in series with no reactance.

SELF-TEST EVALUATION POINT FOR SECTION 16-5

Now that you have completed this section, you should be able to:

■ **Objective 6.** *Describe complex numbers in both rectangular and polar form.*

■ **Objective 7.** *Perform complex number arithmetic.*

■ **Objective 8.** *Describe how complex numbers apply to ac circuits containing series–parallel RLC components.*

Use the following questions to test your understanding of Section 16-5.

1. In complex numbers, resistance is a/an _____ term and reactance is a/an _____ term. (imaginary/real)

2. Convert the following rectangular number to polar form: 5 + *j*6.

3. Convert the following polar number to rectangular form: 33 ∠ 25°.

4. What is a complex number?

REVIEW QUESTIONS

Multiple Choice Questions

1. Capacitive reactance is _____ to frequency and capacitance, whereas inductive reactance is _____ to frequency and inductance.

   a. Proportional, inversely proportional
   b. Inversely proportional, proportional
   c. Proportional, proportional
   d. Inversely proportional, inversely proportional

2. Resonance is a circuit condition that occurs when:

   a. V_L equals V_C d. Both (a) and (c)
   b. X_L equals X_C e. Both (a) and (b)
   c. L equals C

3. As frequency is increased X_L will _____, and X_C will _____.

   a. Decrease, increase   c. Remain the same, decrease
   b. Increase, decrease   d. Increase, remain the same

4. In an *RLC* series resonant circuit, with $R = 500\ \Omega$ and $X_L = 250\ \Omega$, what would be the value of X_C?

   a. $2\ \Omega$ c. $250\ \Omega$
   b. $125\ \Omega$ d. $500\ \Omega$

5. At resonance, the voltage drop across both a series-connected inductor and capacitor will equal:

   a. 70.7 V   c. 10 V
   b. 50% of the source   d. Zero

6. In a series resonant circuit the current flow is _____, as the impedance is _____ and equal to _____.

   a. Large, small, R c. Large, small, X
   b. Small, large, X d. Small, large, R

7. A circuit's bandwidth includes a group or band of frequencies that cause _____ or more of the maximum current, or more than _____ of the maximum power to appear at the output.

   a. 110%, 90%   c. 70.7%, 50%
   b. 50%, 70.7%   d. Both (a) and (c)

8. The bandwidth of a circuit is proportional to the:

   a. Frequency of resonance   c. Tank current
   b. Q of the tank   d. Two of the above

9. Series or parallel resonant circuits can be used to create:

   a. Low-pass filters
   b. Low-pass and high-pass filters
   c. Bandpass and band-stop filters
   d. All the above

10. Flywheel action occurs in:

    a. A tank circuit   d. Two of the above
    b. A parallel *LC* circuit   e. None of the above
    c. A series *LC* circuit

11. $25 \angle 39°$ is an example of a complex number in:

    a. Polar form   c. Algebraic form
    b. Rectangular form   d. None of the above

12. $3 + j10$ is an example of a complex number in:

    a. Polar form   c. Algebraic form
    b. Rectangular form   d. None of the above

13. Which complex number form is usually more convenient for addition and subtraction?

    a. Rectangular   b. Polar

14. Which complex number form is usually more convenient for multiplication and division?

    a. Rectangular   b. Polar

15. In complex numbers, *resistance* is a real term, and *reactance* is a/an _____.

    a. j term
    b. Imaginary term
    c. Value appearing on the vertical axis
    d. All the above

Communication Skill Questions

16. Illustrate with phasors and describe the current and voltage relationships in a series *RLC* circuit. (16-1)

17. Describe the procedure for the analysis of a series *RLC* circuit. (16-1)

18. Define resonance and give the formula for calculating the frequency of resonance. (16-3)

19. Describe the three unusual characteristics of a circuit that is at resonance. (16-3)

20. Define the following: (16-3)

    a. Flywheel action   c. Bandwidth
    b. Quality factor   d. Selectivity

21. Describe a frequency response curve. (16-3)

22. Illustrate with phasors and describe the current and voltage relationships in a parallel *RLC* circuit. (16-2)

23. Describe the differences between a series and a parallel resonant circuit. (16-3)

24. Explain how loading a tank affects bandwidth and selectivity. (16-3-2)

25. Illustrate the circuit and explain the operation of the following, with their corresponding response curves. (16-4)

    a. Low-pass filter   c. Bandpass filter
    b. High-pass filter   d. Band-stop filter

26. Describe why capacitive reactance is written as $-jX_C$ and inductive reactance is written as jX_L. (16-5)

27. How are capacitive and inductive reactances written in polar form? (16-5-7)

28. List the rules used to perform complex number: (16-5-6)

    a. Addition (rectangular)   c. Multiplication (polar)
    b. Subtraction (rectangular) d. Division (polar)

29. Describe briefly how the real number and imaginary number lines are used for ac circuit analysis and what electrical phasors are represented at 0°, 90°, and −90°. (16-5-7)

30. Referring to Figure 16-36, describe why the series circuit current (I) is not in phase with the source voltage (V_S). (16-5-7)

Practice Problems

31. Calculate the values of capacitive or inductive reactance for the following when connected across a 60 Hz source:

a. 0.02 μF   e. 4 mH
b. 18 μF   f. 8.18 H
c. 360 pF   g. 150 mH
d. 2700 nF   h. 2 H

32. If a 1.2 kΩ resistor, a 4 mH inductor, and an 8 μF capacitor are connected in series across a 120 V/60 Hz source, calculate:

a. X_C g. V_C
b. X_L h. Apparent power
c. Z i. True power
d. I j. Resonant frequency
e. V_R k. Circuit quality factor
f. V_L l. Bandwidth

33. If a 270 Ω resistor, a 150 mH inductor, and a 20 μF capacitor are all connected in parallel with one another across a 120 V/60 Hz source, calculate:

a. X_L f. I_T
b. X_C g. Z
c. I_R h. Resonant frequency
d. I_L i. Q factor
e. I_C j. Bandwidth

34. Calculate the impedance of a series circuit if $R = 750$ Ω, $X_L = 25$ Ω, and $X_C = 160$ Ω.

35. Calculate the impedance of a parallel circuit with the same values as those of Question 34 when a 1 V source voltage is applied.

36. State the following series circuit impedances in rectangular and polar form:

a. $R = 33$ Ω, $X_C = 24$ Ω
b. $R = 47$ Ω, $X_L = 17$ Ω

37. Convert the following impedances to rectangular form:

a. $25 \angle 37°$ c. $114 \angle -114°$
b. $19 \angle -20°$ d. $59 \angle 99°$

38. Convert the following impedances to polar form:

a. $-14 + j14$
b. $27 + j17$
c. $-33 - j18$
d. $7 + j4$

39. Add the following complex numbers:

a. $(4 + j3) + (3 + j2)$
b. $(100 - j50) + (12 + j9)$

40. Perform the following mathematical operations:

a. $(35 \angle -24°) \times (13 \angle 50°)$
b. $(100 - j25) - (25 + j5)$
c. $(98 \angle 80°) \div (40 \angle 17°)$

41. State the impedances of the circuits seen in Figure 16-38 in rectangular and polar form. What is Z_T in ohms, and what is its phase angle?

42. Calculate in polar form the impedances of both circuits shown in Figure 16-39. Then, combine the two impedances as if the circuits were parallel connected, using the product-over-sum method. Express the combined impedance in polar form.

43. Referring to Figure 16-40, calculate:

a. Z_T (rectangular and polar)
b. Circuit current and phase angle
c. Voltage drops
d. V_C, V_L, and V_R phase relationships

44. Sketch an impedance and voltage phasor diagram for the circuits in Figure 16-38.

45. Referring to Figure 16-41, calculate:

a. Impedance of the two branches
b. Branch currents
c. Total current
d. Total impedance

46. Sketch an impedance and current phasor diagram for the circuit shown in Figure 16-39.

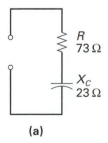

(a)

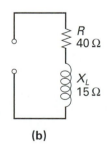

(b)

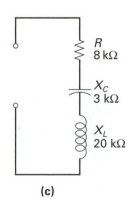

(c)

FIGURE 16-38

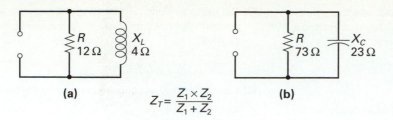

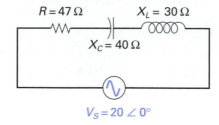

$$Z_T = \frac{Z_1 \times Z_2}{Z_1 + Z_2}$$

(a) **(b)**

FIGURE 16-39

FIGURE 16-40

$R = 47\,\Omega$ $X_L = 30\,\Omega$
$X_C = 40\,\Omega$
$V_S = 20 \angle 0°$

FIGURE 16-41

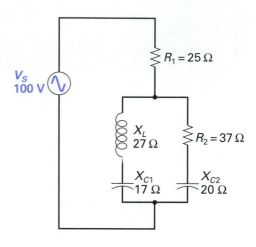

47. Referring to Figure 16-38, verify that the sum of all the individual voltage drops is equal to the total voltage.

48. Referring to Figure 16-39, verify that the sum of all the branch currents is equal to the total current.

49. Determine whether Questions 43 and 45 would be easier to answer with or without the use of complex numbers.

50. Is the circuit in Figure 16-39 more inductive or capacitive?

Web Site Questions

Go to the web site http://www.prenhall.com/cook, select the textbook *Mathematics for Electronics and Comput-* *ers,* select this chapter, and then follow the instructions when answering the multiple-choice practice problems.

Diodes and Transistors

Spitting Lightning Bolts

Nikola Tesla was born in Yugoslavia in 1856. He studied mathematics and physics in Prague, and in 1884 he emigrated to the United States.

In New York he met Thomas Edison, the self-educated inventor who is best known for his development of the phonograph and the incandescent light bulb. Both men were gifted and eccentric and, due to their common interest in "invention," they got along famously. Because Tesla was unemployed, Edison offered him a job. In his lifetime, Edison would go on to take out 1,033 patents and become one of the most prolific inventors of all time. Tesla would go on to invent many different types of motors, generators, and transformers, one of which is named the "Tesla coil" and produces five-foot lightning bolts. With this coil, Tesla investigated "wireless power transmission," the only one of his theories that has not come into being.

Both Tesla and Edison had very strong views on different aspects of electricity and, as time passed, the two men began to engage in very long, loud, and angry arguments. One such discussion concerned whether power should be distributed as alternating current or direct current. Eventually, the world would side with Tesla and choose ac. At the time, this topic, like many others, would cause a hatred to develop between the two men. Eventually Tesla left Edison and started his own company. However, the anger remained, and on one occasion when they were both asked to attend a party for their friend Mark Twain, both refused to come because the other had been invited.

In 1912, Tesla and Edison were both nominated for the Nobel prize in physics, but because neither one would have anything to do with the other, the prize went to a third party—proving that bitterness really will cause a person to cut off his nose to spite his face.

When angry count up to four; when very angry, swear.
Mark Twain

Outline and Objectives

VIGNETTE: SPITTING LIGHTNING BOLTS

INTRODUCTION

17-1   THE JUNCTION DIODE

Objective 1: Name and identify the terminals of a junction diode package and its schematic symbol.

Objective 2: Determine whether a diode is forward or reverse biased by observing the applied bias voltage's polarity.

Objective 3: Explain the forward and reverse characteristics of junction diodes.

Objective 4: Interpret the graphically plotted voltage-current (*V-I*) characteristic curve for a typical junction diode.

17-1-1   Diode Operation

17-1-2   Basic Diode Application

17-1-3   A Junction Diode's Characteristic Curve

17-2   THE ZENER DIODE

Objective 5: Sketch the schematic symbol and describe the different package types for the zener diode.

Objective 6: Describe the forward and reverse voltage-current characteristics of a typical zener diode and the key parameters that govern its operation.

Objective 7: Explain the basic operating principles of the zener diode.

17-2-1   Zener Diode Voltage Current (*V–I*) Characteristics

17-3   THE LIGHT-EMITTING DIODE

Objective 8: Sketch the schematic symbol, and describe the different package types for the light emitting diode or LED.

Objective 9: Describe the forward and reverse voltage-current characteristics of a typical LED and the key parameters that govern its operation.

Objective 10: Explain the basic operating principles of the LED.

17-3-1   LED Characteristics

17-4   INTRODUCTION TO THE TRANSISTOR

Objective 11: Name the three terminals of the bipolar junction transistor.

Objective 12: Describe the difference between the construction and schematic symbol of an NPN and a PNP bipolar transistor.

Objective 13: Identify the base, collector, and emitter terminals of typical low-power and high-power transistor packages.

Objective 14: Describe the two basic actions of a bipolar transistor:
  a. ON/OFF switching action
  b. Variable-resistor action

Objective 15: Define the terms *transistor* and *transistance*.

Objective 16: Explain how the transistor can be used in the following basic applications:
  a. A digital (two-state) circuit, such as a logic gate
  b. An analog (linear) circuit, such as an amplifier

17-4-1   Transistor Types (NPN and PNP)

17-4-2   Transistor Construction and Packaging

17-4-3   Transistor Operation
          The Transistor's ON/OFF Switching Action
          The Transistor's Variable-Resistor Action

17-4-4   Transistor Applications
          Digital Logic Gate Circuit
          Analog Amplifier Circuit

17-5 DETAILED TRANSISTOR OPERATION

17-5-1 Basic Bipolar Transistor Action

Objective 17: Describe in more detail the bipolar junction transistor action.

Objective 18: Explain in more detail how the transistor can be used as an amplifier.

> A Correctly Biased NPN Transistor Circuit
> The Current-Controlled Transistor
> Operating a Transistor in the Active Region
> Operating the Transistor in Cutoff and Saturation
> Biasing PNP Bipolar Transistors

17-5-2 Bipolar Transistor Circuit Configurations and Characteristics.

Objective 19: Identify and list the characteristics of the three transistor circuit configurations, namely:
  a. Common base
  b. Common emitter
  c. Common collector

Objective 20: Explain the meaning of the following:
  a. Transistor voltage and current abbreviations
  b. DC alpha
  c. DC beta
  d. Collector characteristic curve

  e. AC beta
  f. Input resistance or impedance
  g. Output resistance or impedance

Objective 21: Calculate a transistor circuit's
  a. DC current gain
  b. AC current gain
  c. Voltage gain
  d. Power gain

> Common-Emitter Circuits
> Common-Base Circuits
> Common-Collector Circuits

17-5-3 Bipolar Transistor Biasing Circuits
> Base Biasing
> Voltage-Divider Biasing

Objective 22: Calculate the different values of circuit voltage and current for the following transistor biasing methods:
  a. Base biasing
  b. Voltage-divider biasing

Objective 23: Define the following terms:
  a. DC load line
  b. Cutoff point
  c. Saturation point
  d. Quiescent point

Introduction

The first diode was accidentally created by Edison in 1883 when he was experimenting with his lightbulb. At this time he did not place any importance on the device and its effect, as he could not see any practical application for it. The word *diode* is derived from the fact that the device has two (*di*) electrodes (*ode*).

Once the importance of diodes was realized, construction of the device began. The first diodes were vacuum-tube devices having a hot-filament negative cathode that released free electrons that were collected by a positive plate called the anode. Today's diode is made of a p-n semiconductor junction but still operates on the same principle. The N-type region (cathode) is used to supply free electrons, which are then collected by the P-type region (anode). The operation of both the vacuum tube and semiconductor diode is identical in that the device will pass current in only one direction. That is, it will act as a conductor and pass current easily in one direction when the bias voltage across it is of one polarity, yet it will block current and imitate an insulator when the bias voltage applied is of the opposite polarity.

17-1 THE JUNCTION DIODE

The two electrodes or terminals of the diode are called the anode and cathode, as seen in Figure 17-1(a) which shows the schematic symbol of a diode. To help you remember which terminal is the anode and which is the cathode, and which terminal is positive and which is

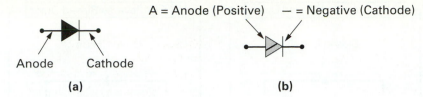

A = Anode (Positive)   — = Negative (Cathode)

(a)   **(b)**

FIGURE 17-1 **Schematic Symbol of a Diode.**

negative, Figure 17-1(b) shows how a line drawn through the triangle section of the symbol will make the letter *A* and indicate the "anode" terminal. Similarly, if the vertical flat side of the diode symbol is aligned horizontally "—", as in Figure 17-1(b), it becomes the "negative" symbol. This memory system helps us remember that the anode terminal of the diode is next to the triangle part of the symbol and is positive, and the cathode terminal of a diode is next to the vertical line of the symbol and is negative.

The diode is generally mounted in one of the three basic packages shown in Figure 17-2. These packages are designed to protect the diode from mechanical stresses and the environment. The difference in the size of the packages is due to the different current rating of the diode. A black band or stripe is generally placed on the package closest to the cathode terminal for identification purposes, as seen in Figure 17-2(a) and (b). Larger diode packages, like the one seen in Figure 17-2(c), usually have the diode symbol stamped on the package to indicate anode/cathode terminals.

17-1-1   *Diode Operation*

As far as operation is concerned, the diode operates like a switch. If you give the diode what it wants, that is, make the anode terminal positive with respect to the cathode terminal, as seen in Figure 17-3(a), the device is equivalent to a closed switch as seen in Figure 17-3(b). In this condition, the diode is said to be ON or *forward biased*.

On the other hand, if you do not give the diode what it wants, that is, if you make the anode terminal negative with respect to the cathode, as seen in Figure 17-3(c), the device is equivalent to an open switch, as seen in Figure 17-3(d). In this condition the diode is said to be OFF or *reverse biased*.

17-1-2   *Basic Diode Application*

As an application, Figure 17-4 shows how the diode can be used as a switch within an **encoder circuit.** The pull-up resistors R_1, R_2, and R_3 ensure that lines *A*, *B*, and *C* are normally

Encoder Circuit

A circuit that produces different output voltage codes, depending on the position of a rotary switch.

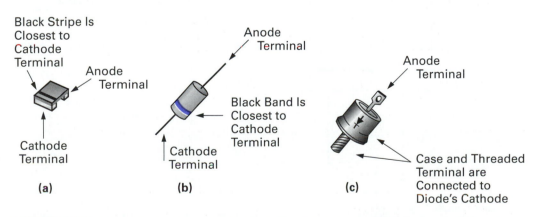

(a)   **(b)**   **(c)**

FIGURE 17-2 **Diode Packaging. (a) Chip Package—1/4 A. (b) Small Current Package—Less than 3 A. (c) Large Current Package—Greater than 3 A.**

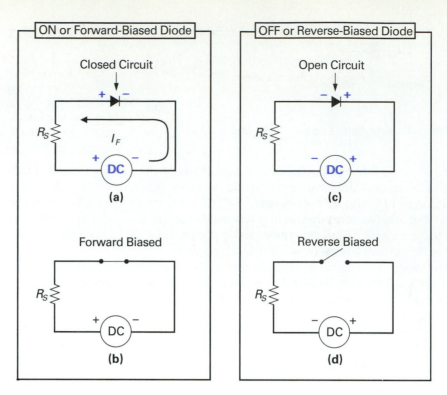

FIGURE 17-3   Diode Operation. (a)(b) Forward-Biased (ON) Diode. (c)(d) Reverse-Biased (OFF) Diode.

all at +5 V. This is the output voltage on each line when the rotary switch is in position 2, as seen in the table in Figure 17-4.

When the rotary switch is turned to position 1, D_1 is connected in circuit, and because its anode is made positive via R_2, and its cathode is at 0 V, the diode D_1 will turn ON and be equivalent to a closed switch. The 0 V on the cathode of D_1 will be switched through to line B (all 5 V will be dropped across R_2) producing an output voltage code of $A = +5$ V, $B = 0$ V, $C = +5$ V, as seen in the table in Figure 17-4.

When the rotary switch is turned to position 3, D_2 and D_3 are connected in circuit, and because both anodes are made positive via R_1 and R_3, and both diode cathodes are at 0 V,

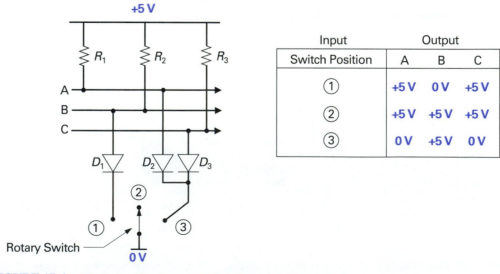

Input	Output		
Switch Position	A	B	C
①	+5 V	0 V	+5 V
②	+5 V	+5 V	+5 V
③	0 V	+5 V	0 V

FIGURE 17-4   Diode Application: A Switch Encoder Circuit.

D_2 and D_3 will turn ON. These forward-biased diodes will switch 0 V through to lines A and C, producing an output voltage code of $A = 0$ V, $B = +5$ V, $C = 0$ V, as seen in the table in Figure 17-4.

This *code generator* or *encoder circuit* will produce three different output voltage codes for each of the three positions of the rotary switch. These codes could then be used to initiate one of three different operations based on the operator setting of the rotary control switch.

17-1-3 *A Junction Diode's Characteristic Curve*

Semiconductor devices such as diodes and transistors are constructed using *p-n* junctions. A diode, for example, has only one *p-n* junction and is created by doping a single piece of pure semiconductor to produce an N-type and a P-type region. A bipolar junction transistor, on the other hand, has two *p-n* junctions and is created by doping a single piece of pure semiconductor with three alternate regions (*npn* or *pnp*). The point at which these two oppositely doped materials come in contact with each other is called a junction, which is why these devices are called **junction diodes** and *bipolar junction transistors.*

These *p-n* junctions need voltages of a certain amplitude and polarity to control their operation. These voltages, which incline or cause the diode to operate in a certain manner, are known as **bias voltages.** Bias voltages control the resistance of the junction and, therefore, the amount of current that can pass through the *p-n* junction diode.

The upper right quadrant of the four sections in Figure 17-5 shows what forward current will pass through the diode when a forward bias voltage is applied. As you can see from the inset, the diode is forward biased by applying a positive potential to its anode and a neg-

Junction Diode

A semiconductor diode whose ON/OFF characteristics occur at a junction between the *N*-type and *P*-type semiconductor materials.

Bias Voltage

Voltage that inclines or causes the diode to operate in a certain manner.

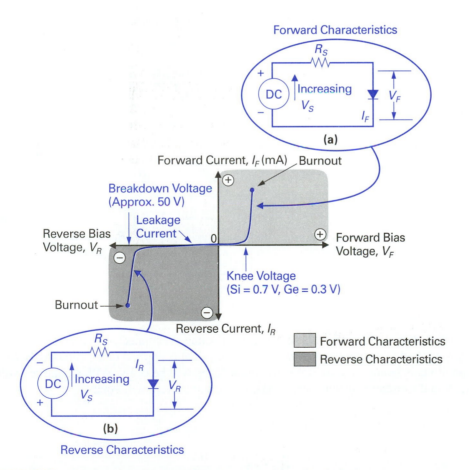

FIGURE 17-5 **The Junction Diode Voltage–Current Characteristic Curve.**

ative potential to its cathode. In this instance the diode is said to be ON and is equivalent to a closed switch. Beginning at the graph origin and following the curve into the forward quadrant, you can see that the forward current through a diode is extremely small until the forward bias voltage exceeds the diode's internal barrier voltage, which for silicon is 0.7 V and for germanium is 0.3 V.

Referring to the linearly increasing current portion of the forward curve in Figure 17-5, you will notice that although there is a large change in forward current, the forward voltage drop across the diode remains almost constant between 0.7 V and 0.75 V.

The amount of heat produced in the diode is proportional to the value of current through the diode ($P\uparrow = I^2\uparrow \times R$). For example, an IN4001 diode, which is a commonly used low-power silicon diode, has a manufacturer's maximum forward (I_F max.) rating of 1 A. If this value of current is exceeded, the diode will begin generating more heat than it can dissipate and burn out. A series current-limiting resistor (R_S) is generally always included to limit the forward current, as shown in the inset in Figure 17-5. Although the series resistor will limit forward current, it cannot prevent a damaging forward current if enough pressure or forward voltage is applied ($V\uparrow = I\uparrow \times R$). The value of forward current is equal to

$$I_F = \frac{V_s - V_{\text{diode}}}{R_s}$$

■ **INTEGRATED MATH APPLICATION:**

Calculate the value of current for the circuit shown in Figure 17-6.

FIGURE 17-6 A P-N Junction Diode Circuit.

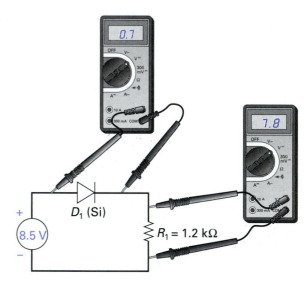

■ *Solution:*

The diode is forward biased because the applied voltage is connected so that its positive terminal is applied to the anode, and the negative terminal is applied to the cathode. Because a silicon diode is being used, the forward voltage drop will be 0.7 V. With an applied voltage of 8.5 V and a circuit resistance of 1.2 kΩ, the circuit current will equal

$$I_F = \frac{V_S - V_{\text{diode}}}{R_s}$$

$$I_F = \frac{8.5 \text{ V} - 0.7 \text{ V}}{1.2 \text{ k}\Omega}$$

$$I_F = \frac{7.8 \text{ V}}{1.2 \text{ k}\Omega}$$

$$I_F = 6.5 \text{ mA}$$

The lower left quadrant of the four sections in Figure 17-5 shows what reverse current will pass through the diode when a reverse bias voltage is applied. As you can see from the inset, a diode is reverse biased by applying a negative potential to its anode and a positive potential to its cathode. In this instance, current is effectively reduced to zero, and the diode is said to be OFF and equivalent to an open switch.

These characteristics can be seen in the reverse curve in Figure 17-5. Beginning at the graph origin and following the curve into the reverse quadrant, you can see that the reverse current through the diode increases only slightly (approximately 100 μA). Throughout this part of the curve the diode is said to be blocking current because the leakage current is generally so small it is ignored for most practical applications. If the reverse voltage (V_R) is further increased, a point will be reached at which the diode will break down, resulting in a sudden increase in current. The point on the reverse voltage scale at which the diode breaks down and there is a sudden increase in reverse current is called the **breakdown voltage.** Referring to the reverse curve in Figure 17-5, you can see that most silicon diodes break down as the reverse bias voltage approaches 50 V. For example, the IN4001 low-power silicon diode has a reverse breakdown voltage (which is sometimes referred to as the **peak inverse voltage** or **PIV**) of 50 V listed on its manufacturer's data sheet. If this reverse bias voltage is exceeded, an avalanche of continuously rising current will eventually generate more heat than can be dissipated, resulting in the destruction of the diode.

Semiconductor materials, and therefore diodes, have a negative temperature coefficient of resistance. This means that as temperature increases ($T\uparrow$) their resistance decreases ($R\downarrow$).

Breakdown Voltage or Peak Inverse Voltage (PIV)

The point on the reverse voltage scale at which the diode breaks down and there is a sudden increase in the reverse current.

SELF-TEST EVALUATION POINT FOR SECTION 17-1

Now that you have completed this section, you should be able to:

■ *Objective 1.* *Name and identify the terminals of a junction diode package and its schematic symbol.*

■ *Objective 2.* *Determine whether a diode is forward or reverse biased by observing the applied bias voltage's polarity.*

■ *Objective 3.* *Explain the forward and reverse characteristics of junction diodes.*

■ *Objective 4.* *Interpret the graphically plotted voltage–current (V–I) characteristics curve for a typical junction diode.*

Use the following questions to test your understanding of Section 17-1.

1. What is the typical forward drop across a silicon diode?
2. What value of barrier voltage has to be overcome in order to forward bias a silicon diode?
3. How many *p-n* junctions are within a junction diode?

17-2 THE ZENER DIODE

Figure 17-7(a) shows the two schematic symbols used to represent the **zener diode.** As you can see, the zener diode symbol resembles the basic *p-n* junction diode symbol in appearance; however, the zener diode symbol has a zigzag bar instead of the straight bar. This zigzag bar at the cathode terminal is included as a memory aid since it is Z-shaped and will always remind us of zener.

Figure 17-7(b) shows two typical low-power zener diode packages, and one high-power zener diode package. The surface mount low-power zener package has two metal

Zener Diode

A diode constructed to operate at voltages that are equal to or greater than the reverse breakdown voltage rating.

FIGURE 17-7 The Zener Diode.
(a) Zener Diode Schematic Symbols.
(b) Packages.

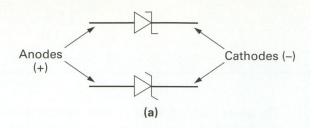

Anodes
(+)
Cathodes (−)

(a)

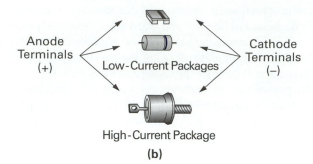

Anode
Terminals
(+)
Low-Current Packages
Cathode
Terminals
(−)

High-Current Package

(b)

pads for direct mounting to the surface of a circuit board, whereas the axial lead low-power zener package has the zener mounted in a glass or epoxy case. The high-power zener package is generally stud mounted and contained in a metal case. These packages are identical to the basic *p-n* junction diode low-power and high-power packages. Once again, a band or stripe is used to identify the cathode end of the zener diode in the low-power packages, whereas the threaded terminal of a high-power package is generally always the cathode.

17-2-1   *Zener Diode Voltage–Current (V–I) Characteristics*

Figure 17-8 shows the *V–I* (voltage–current) characteristic curve of a typical zener diode. This characteristic curve is almost identical to the basic *p-n* junction diode's characteristic curve. For example, when forward biased at or beyond 0.7 V, the zener diode will turn ON and be equivalent to a closed switch; whereas, when reverse biased the zener diode will turn OFF and be equivalent to an open switch. The main difference, however, is that the zener diode has been specifically designed to operate in the reverse breakdown region of the curve. This is achieved, as can be seen in the inset in Figure 17-8, by making sure that the external bias voltage applied to a zener diode will not only reverse bias the zener diode (+ → cathode, − → anode) but also be large enough to drive the zener diode into its reverse breakdown region.

As the reverse voltage across the zener diode is increased from the graph origin (which represents 0 V) the value of **reverse leakage current** (I_R) begins to increase. Comparing *the* voltage developed across the zener (V_Z) with the value of current through the zener (I_Z), you may have noticed that *the voltage drop across a zener diode (V_Z) remains almost constant when it is operated in the reverse zener breakdown region, even though current through the zener (I_Z) can vary considerably. This ability of the zener diode to maintain a relatively constant voltage regardless of variations in zener current is the key characteristic of the zener diode.*

Generally, manufacturers rate zener diodes based on their **zener voltage** (V_Z) rather than their breakdown voltage (V_{Br}). A wide variety of zener diode voltage ratings are available, ranging from 1.8 V to several hundred volts. For example, *many of the frequently used low-voltage zener diodes have ratings of 3.3 V, 4.7 V, 5.1 V, 5.6 V, 6.2 V, and 9.1 V.*

Reverse Leakage Current (I_R)

The undesirable flow of current through a device in the reverse direction.

Zener Voltage (V_Z)

The voltage drop across the zener when it is being operated in the reverse zener breakdown region.

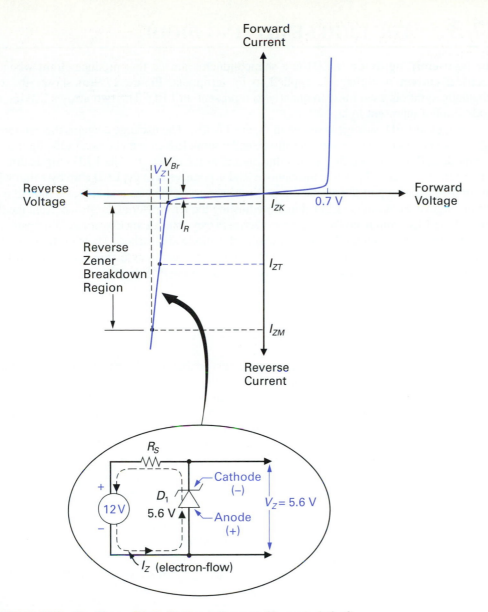

FIGURE 17-8 The Zener Diode Voltage–Current Characteristic Curve.

Now that you have completed this section, you should be able to:

■ **Objective 5.** *Sketch the schematic symbol and describe the different package types for the zener diode.*

■ **Objective 6.** *Describe the forward and reverse voltage–current characterisitics of a typical zener diode and the key parameters that govern its operation.*

■ **Objective 7.** *Explain the basic operating principles of the zener diode.*

Use the following questions to test your understanding of Section 17-2.

1. True or false: The zener diode is designed specifically to operate at voltages exceeding breakdown.

2. In most applications, a zener diode is _____ biased.

3. What is the difference between a zener diode's schematic symbol and a basic *p-n* junction's symbol?

The **light-emitting diode (LED)** is a semiconductor device that produces light when an electrical current or voltage is applied to its terminals. Figure 17-9(a) shows the two schematic symbols used most frequently to represent an LED. The two arrows leaving the diode symbol represent light.

A typical LED package is shown in Figure 17-9(b). The package contains the two terminals for connection to the anode and cathode and a semiclear case that contains the light-emitting diode and a lens. Looking at this illustration, you can see that the LED chip is directly connected to the anode lead, and the cathode lead is connected to the LED chip by a thin wire. The dome-shaped top of the plastic (epoxy) case serves as the lens and acts as a magnifier to conduct light away from the LED chip. By adjusting the lens material, lens shape, and the distance of the LED chip from the lens, manufacturers can obtain a variety of radiation patterns.

There are three methods used to identify the anode and cathode leads of the LED, and these are shown in Figure 17-9(c). In all three cases, the cathode lead is distinguished from the anode lead by having its lead shorter or its lead flattened, or being nearer to the flat side of the case.

17-3-1 *LED Characteristics*

Light-emitting diodes have V–I characteristic curves that are almost identical to the basic *p-n* junction diode, as shown in Figure 17-10.

Studying the forward characteristics, you can see that LEDs have a high forward voltage rating (V_F is normally between $+1$ V and $+3$ V) and a low maximum forward current rating (I_F is normally between 20 mA and 50 mA). In most circuit applications, the LED will have a forward voltage drop of 2 V, and a forward current of 20 mA. The LED will usually always need to be protected from excessive forward current damage by a series current-limiting resistor, which will limit the forward current so that it does not exceed the LED's *maximum forward current* (I_{FM} or I_{FMAX}) rating, as seen in the inset in Figure 17-10. The circuit in the inset shows how to forward bias an LED. In this example circuit, the source voltage of 5 V is being applied to a 130 Ω series resistor (R_S) and an LED with an I_{FM} rating of 50 mA and a VF rating of 1.8 V. To calculate the value of current in this circuit, we first deduct the 1.8 V developed across the LED (V_{LED}) from the source voltage (V_S) to determine the voltage drop across the resistor (V_{RS}). Then using Ohm's law, we divide V_{RS} by R_S to determine current.

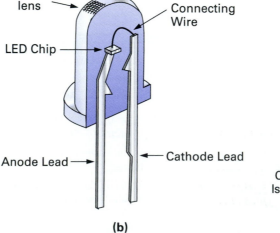

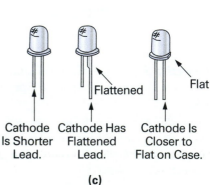

(a) (b) (c)

FIGURE 17-9 **The Light-Emitting Diode (LED). (a) Schematic Symbol. (b) Construction. (c) Lead Identification.**

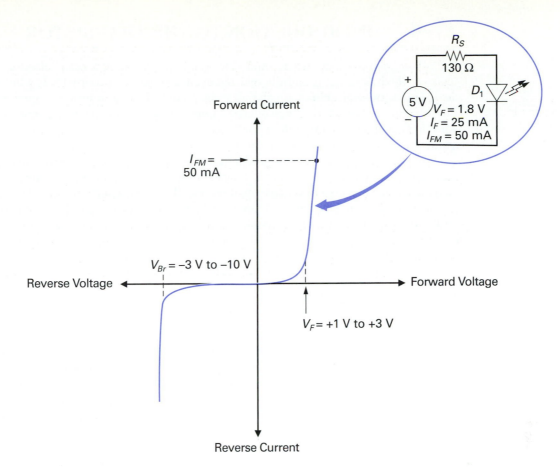

FIGURE 17-10 The LED Voltage–Current Characteristic Curve.

$$I_S = \frac{V_S - V_{LED}}{R_S}$$

$$= \frac{5\ V - 1.8\ V}{130\ \Omega} = \frac{3.2\ V}{130\ \Omega} = 24.6\ mA$$

Studying the reverse characteristics, you can see that LEDs have lower reverse breakdown voltage values than junction diodes (V_{Br} is typically $-3\ V$ to $-10\ V$), which means that even a low reverse voltage will cause the LED to break down and become damaged.

SELF-TEST EVALUATION POINT FOR SECTION 17-3

Now that you have completed this section, you should be able to:

- **Objective 8.** *Sketch the schematic symbol, and describe the different package types for the light-emitting diode or LED.*

- **Objective 9.** *Describe the forward and reverse voltage–current characteristics of a typical LED and the key parameters that govern its operation.*

- **Objective 10.** *Explain the basic operating principles of the LED.*

Use the following questions to test your understanding of Section 17-3.

1. What would be the typical voltage drop across a forward biased LED?

   a. 0.7 V

   b. 2 V

   c. 0.3 V

   d. 5.6 V

2. A _____ is normally always included to limit I_F to just below its maximum.

In 1948, a component known as a transistor sparked a whole new era in electronics, the effects of which have not been fully realized even to this day. A transistor is a three-element device made of semiconductor materials used to control electron flow, the amount of which can be controlled by varying the voltages applied to its three elements. Having the ability to control the amount of current through the transistor allows us to achieve two very important applications: switching and amplification.

In most cases it is easier to build a jigsaw puzzle when you can refer to the completed picture on the box. The same is true whenever you are trying to learn anything new, especially a science that contains many small pieces. These first-approximation descriptions are a means for you to quickly see the complete picture without having to wait until you connect all the pieces. Like the diode's first-approximation description, this general overview will cover the transistor's basic construction, schematic symbol, physical appearance, basic operation, and main applications.

17-4-1   *Transistor Types (NPN and PNP)*

Like the diode, a bipolar transistor is constructed from a semiconductor material. However, unlike the diode, which has two oppositely doped regions and one P-N junction, the transistor has three alternately doped semiconductor regions and two P-N junctions. These three alternately doped regions are arranged in one of two different ways, as shown in Figure 17-11.

With the **NPN transistor,** shown in Figure 17-11(a), a thin, lightly doped P-type region known as the **base** (symbolized *B*) is sandwiched between two N-type regions called the **emitter** (symbolized *E*) and the **collector** (symbolized *C*). Looking at the NPN transistor's schematic symbol in Figure 17-11(b), you can see that an arrow is used to indicate the emitter lead. As a memory aid for the NPN transistor's schematic symbol, you may want to remember that when the emitter arrow is "*n*ot *p*ointing i*n*" to the base, the transistor is an **NPN.** An easier method is to think of the arrow as a diode, with the tip of the arrow or cathode pointing to an N terminal and the back of the arrow or anode pointing to a P terminal, as seen in the inset in Figure 17-11(b).

The **PNP transistor** can be seen in Figure 17-11(c). With this transistor type, a thin, lightly doped N-type region (base) is placed between two P-type regions (emitter and collector). Figure 17-11(d) illustrates the PNP transistor's schematic symbol. Once again, if you think of the emitter arrow as a diode, as shown in the inset in Figure 17-11(d), the tip of the arrow or cathode is pointing to an N terminal, and the back of the arrow or anode is pointing to a P terminal.

17-4-2   *Transistor Construction and Packaging*

As in the diode, the three layers of an NPN or PNP transistor are not formed by joining three alternately doped regions. These three layers are formed by a "diffusion process" that first melts the base region into the collector region and then melts the emitter region into the base region. For example, with the NPN transistor shown in Figure 17-12(a), the construction process would begin by diffusing or melting a P-type base region into the N-type collector region. Once this P-type base region was formed, an N-type emitter region would be diffused or melted into the newly diffused P-type base region to form an NPN transistor. Keep in mind that manufacturers will generally construct thousands of these transistors simultaneously on a thin semiconductor wafer or disc, as shown in Figure 17-12(a). Once tested, these discs, which are about 3 in. in diameter, are cut to separate the individual transistors. Each transistor is placed in a package, as shown in Figure 17-12(b). The package protects the transistor from humidity and dust, provides a means for electrical connection between the three semiconductor regions and the three transistor terminals, and serves as a heat sink to conduct away any heat generated by the transistor.

NPN Transistor

A device in which a thin, lightly doped P-type region (base) is sandwiched between two N-type regions (emitter and collector).

Base

The region that lies between an emitter and a collector of a transistor and into which minority carriers are injected.

Emitter

A transistor region from which charge carriers are injected into the base.

Collector

A semiconductor region through which a flow of charge carriers leaves the base of the transistor.

PNP Transistor

A device in which a thin, lightly doped N-type region (base) is placed between two P-type regions (emitter and collector).

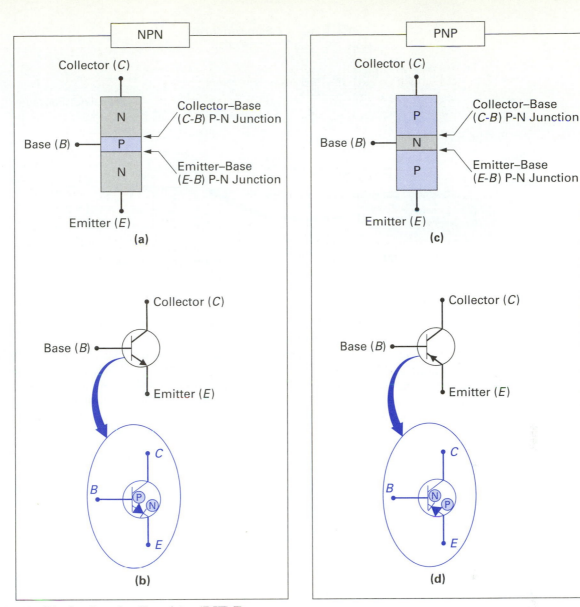

FIGURE 17-11 **Bipolar Junction Transistor (BJT) Types.**

Figure 17-13 illustrates some of the typical low-power and high-power transistor packages. Most low-power, small-signal transistors are hermetically sealed in a metal, plastic, or epoxy package. Four of the low-power packages shown in Figure 17-13(a) have their three leads protruding from the bottom of the package because these package types are usually inserted and soldered into holes in printed circuit boards (PCBs). The surface mount technology (SMT) low-power transistor package, on the other hand, has flat metal legs that mount directly onto the surface of the PCB. These transistor packages are generally used in high component density PCBs because they use less space than a "through-hole" package. To explain this in more detail, a through-hole transistor package needs a hole through the PCB and a connecting pad around the hole to make a connection to the circuit. With an SMT package, however, no holes are needed, only a small connecting pad. Without the need for holes, pads on printed circuit boards can be smaller and placed closer together, resulting in considerable space saving.

The high-power packages, shown in Figure 17-13(b), are designed to be mounted onto the equipment's metal frame or chassis so that the additional metal will act as a heat sink and conduct the heat away from the transistor. With these high-power transistor packages, two or three leads may protrude from the package. If only two leads are present, the

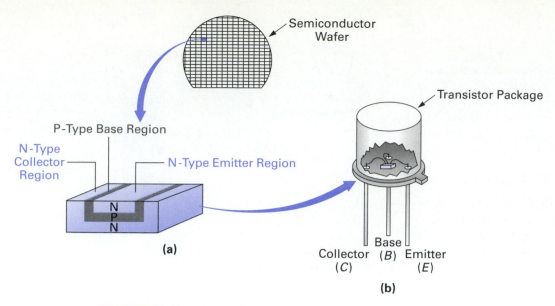

Semiconductor
Wafer

P-Type Base Region

N-Type
Collector
Region

N-Type Emitter Region

N
P
N

(a)

Transistor Package

Collector
(C)

Base
(B)

Emitter
(E)

(b)

FIGURE 17-12 **Bipolar Junction Transistor Construction and Packaging.**

metal case will serve as a collector connection, and the two pins will be the base and emitter.

Transistor package types are normally given a reference number. These designations begin with the letters "TO," which stands for transistor outline, and are followed by a number. Figure 17-13 includes some examples of TO reference designators.

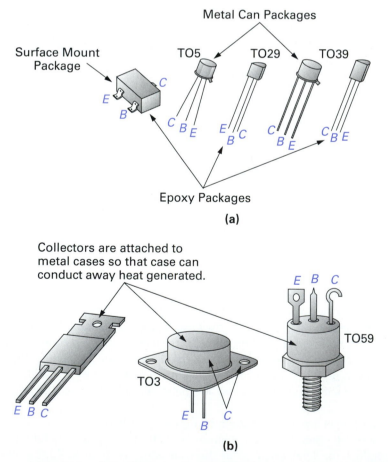

Metal Can Packages

Surface Mount
Package

TO5   TO29   TO39

E
B

C

C B E   E B C   C B E   C B E

Epoxy Packages

(a)

Collectors are attached to
metal cases so that case can
conduct away heat generated.

E B C

E B C

TO3

E   C
B

TO59

(b)

FIGURE 17-13 **Bipolar Junction Transistor Package Types. (a) Low Power. (b) High Power.**

17-4-3 *Transistor Operation*

Figure 17-14 shows an NPN bipolar transistor, and the inset shows how a transistor can be thought of as containing two diodes: a *base-to-collector diode* and a *base-to-emitter diode*. With an NPN transistor, both diodes will be back-to-back and "***n*ot *p*ointing *in***" (NPN) to the base, as shown in the inset in Figure 17-14. For a PNP transistor, the base–collector and base–emitter diodes will both be pointing into the base.

Transistors are basically controlled to operate as a switch, or they are controlled to operate as a variable resistor. Let us now examine each of these operating modes.

The Transistor's ON/OFF Switching Action

Figure 17-15 illustrates how the transistor can be made to operate as a switch. This ON/OFF switching action of the transistor is controlled by the transistor's base-to-emitter (B–E) diode. If the B–E diode of the transistor is forward biased, the transistor will turn ON; if the B–E diode of the transistor is reverse biased, the transistor will turn OFF.

To begin, let us see how the transistor can be switched ON. In Figure 17-15(a), the B–E diode of the transistor is forward biased (anode at base is +5 V, cathode at emitter is 0 V), and the transistor will turn ON. Its collector and emitter output terminals will be equivalent to a closed switch, as shown in Figure 17-15(b). This low resistance between the transistor's collector and emitter will cause a current (I), as shown in Figure 17-15(b). The output voltage in this condition will be 0 V because the entire +10 V supply voltage will be dropped across R. Another way to describe this would be to say that the low-resistance path between the transistor's emitter and collector connects the 0 V emitter potential through to the output.

Now let us see how the transistor can be switched OFF. In Figure 17-15(c), the transistor has 0 V being applied to its base input. In this condition, the B–E diode of the transistor is reverse biased (anode at base is 0 V, cathode at emitter is 0 V), and so the transistor will turn OFF, and its collector and emitter output terminals will be equivalent to an open switch, as shown in Figure 17-15(d). This high resistance between the transistor's collector and emitter will prevent any current and any voltage drop and will cause the full +10 V supply voltage to be applied to the output, as shown in Figure 17-15(d).

FIGURE 17-14 The Base–Collector and Base–Emitter Diodes within a Bipolar Transistor.

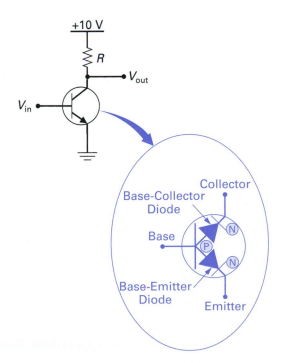

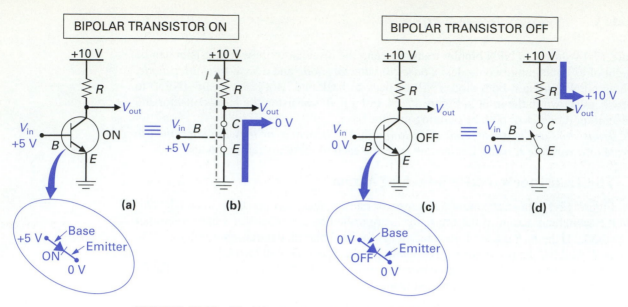

FIGURE 17-15 The Bipolar Transistor's ON/OFF Switching Action.

The Transistor's Variable-Resistor Action

In the previous section we saw how the transistor can be biased to operate in one of two states: ON or OFF. When operated in this two-state way, the transistor is being switched ON and OFF in almost the same way as a junction diode. The transistor, however, has another ability that the diode does not have—it can also function as a variable resistor, as shown in the equivalent circuit in Figure 17-16(a). In Figure 17-15 we saw how +5 V base input bias voltage would result in a low resistance between emitter and collector (closed switch) and how a 0 V base input bias would result in a high resistance between emitter and collector (open switch). The table in Figure 17-16(b) shows an example of the relationship

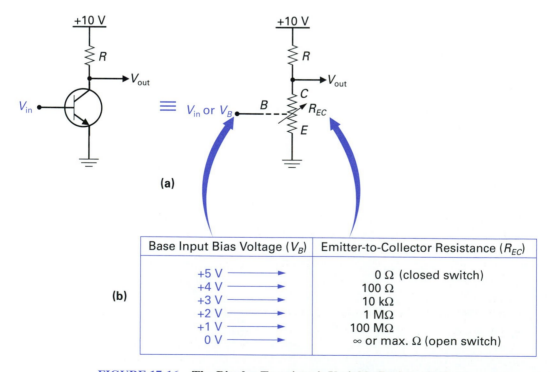

Base Input Bias Voltage (V_B)	Emitter-to-Collector Resistance (R_{EC})
+5 V ⟶	0 Ω (closed switch)
+4 V ⟶	100 Ω
+3 V ⟶	10 kΩ
+2 V ⟶	1 MΩ
+1 V ⟶	100 MΩ
0 V ⟶	∞ or max. Ω (open switch)

FIGURE 17-16 The Bipolar Transistor's Variable-Resistor Action.

between base input bias voltage (V_B) and emitter-to-collector resistance (R_{CE}). In this table, you can see that the transistor is going to be driven not only between the two extremes of fully ON and fully OFF. When the base input voltage is at some voltage level between +5 V and 0 V, the transistor is partially ON; therefore, the transistor's emitter-to-collector resistance is somewhere between 0 Ω and maximum Ω. For example, when $VB = +4$ V, the transistor is not fully ON, and its emitter-to-collector resistance will be slightly higher, at 100 Ω. If the base input bias voltage is further reduced to +3 V, for example, you can see in the table that the emitter-to-collector resistance will further increase to 10 kΩ. Further decreases in base input voltage ($V_B\downarrow$) will cause further increases in emitter-to-collector resistance ($R_{CE}\uparrow$) until $V_B = 0$ V and R_{CE} = maximum Ω.

As a matter of interest, the name transistor was derived from the fact that through base control different values of "resistance" can be "transferred" between the emitter and collector. This effect of "transferring resistance" is known as **transistance** and the component that functions in this manner is called the transistor.

Now that we have seen how the transistor can be made to operate as either a switch or a variable resistor, let us see how these characteristics can be made use of in circuit applications.

Transistance
The effect of transferring resistance.

17-4-4   *Transistor Applications*

The transistor's impact on electronics has been phenomenal. It initiated the multibillion dollar semiconductor industry and was the key element behind many other inventions such as integrated circuits (ICs), optoelectronic devices, and digital computer electronics. In all these applications, however, the transistor is basically made to operate in one of two ways: as a switch or as a variable resistor. Let us now briefly examine an example of each.

Digital Logic Gate Circuit

A digital logic gate circuit makes use of the transistor's ON/OFF switching action. Digital circuits are often referred to as "switching" or "two-state" circuits because their main control device (the transistor) is switched between the two states of ON and OFF. The transistor is at the very heart of all digital electronic circuits. For example, transistors are used to construct logic gate circuits, gates are used to construct flip-flop circuits, flip-flops are used to construct register and counter circuits, and these circuits are used to construct microprocessor, memory, and input/output circuits—the three basic blocks of a digital computer.

Figure 17-17(a) shows how the transistor can be used to construct a NOT gate or IN-VERTER gate. The basic NOT gate circuit is constructed using one NPN transistor and two resistors. This logic gate has only one input (A) and one output (Y), and its schematic symbol is shown in Figure 17-17(b). Figure 17-17(c) shows how this logic gate will react to the two different input possibilities. When the input is 0 V (logic 0), the transistor's base–emitter P-N diode will be reverse biased and so the transistor will turn OFF. Referring to the inset for this circuit condition in Figure 17-17(c), you can see that the OFF transistor is equivalent to an open switch between emitter and collector, and therefore the +5 V supply voltage will be connected to the output. In summary, a logic 0 input (0 V) will be converted to a logic 1 output (+5 V). On the other hand, when the input is +5 V (logic 1), the transistor's base–emitter P-N diode will be forward biased and so the transistor will turn ON. Referring to the inset for this circuit condition in Figure 17-17(c), you can see that the ON transistor is equivalent to a closed switch between emitter and collector, and therefore 0 V will be connected to the output. In summary, a logic 1 input (+5 V) will be converted to a logic 0 output (0 V).

Referring to the function table in Figure 17-17(c), you can see that the output logic level is "not" the same as the input logic level—hence the name NOT gate.

As an application, Figure 17-17(d) shows how a NOT or INV gate can be used to invert an input control signal. In this circuit, you can see that a normally closed push-button (NCPB) switch is used as a panic switch to activate a siren in a security system. Because the push button is normally closed, it will produce +5 V at A when it is not in alarm. If this voltage were connected directly to the siren, the siren would be activated incorrectly. Inclusion of the NOT

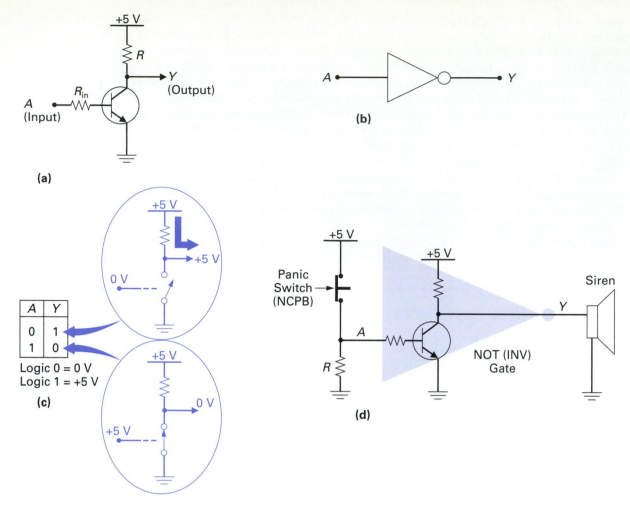

FIGURE 17-17 **A Transistor Being Used in a Digital Electronic Circuit. (a) Basic NOT or INVERTER Gate Circuit. (b) NOT Gate Schematic Symbol. (c) NOT Gate Function Table. (d) NOT Gate Security System Application.**

gate between the switch circuit and the siren inverts the normally HIGH output of the NCPB to a LOW, and the siren is not activated when not in alarm. When the panic switch is pressed, however, the NCPB contacts open, producing a LOW input voltage to the NOT gate. This LOW input is inverted to a HIGH output and activates the siren.

Analog Amplifier Circuit

When used as a variable resistor, the transistor is the controlling element in many analog or linear circuit applications such as amplifiers, oscillators, modulators, detectors, and regulators. The most important of these applications is **amplification,** which is the boosting in strength or increasing in amplitude of electronic signals.

Figure 17-18(a) shows a simplified transistor amplifier circuit, and Figure 17-18(b) shows the voltage waveforms present at different points in the circuit. As you can see, the transistor is labeled Q_1 because the letter Q is the standard letter designation used for transistors.

Before applying an ac sine-wave input signal, let us determine the dc voltage levels at the transistor's base and collector. The 6.3 kΩ/3.7 kΩ resistance ratio of the voltage divider R_1 and R_2 causes the +10 V supply voltage to be proportionally divided, producing +3.7 V dc across R_2. This +3.7 V dc is applied to the base of the transistor, causing the base–emitter junction of Q_1 to be forward biased and Q_1 to turn ON. With transistor Q_1 ON, a certain value of resistance will exist between the transistor's collector and emitter (R_{CE} or R_{EC}), and this resistance will form a voltage divider with R_E and R_C, as seen in the inset in Figure 17-18(a).

Amplification

Boosting in strength, or increasing amplitude, of electronic signals.

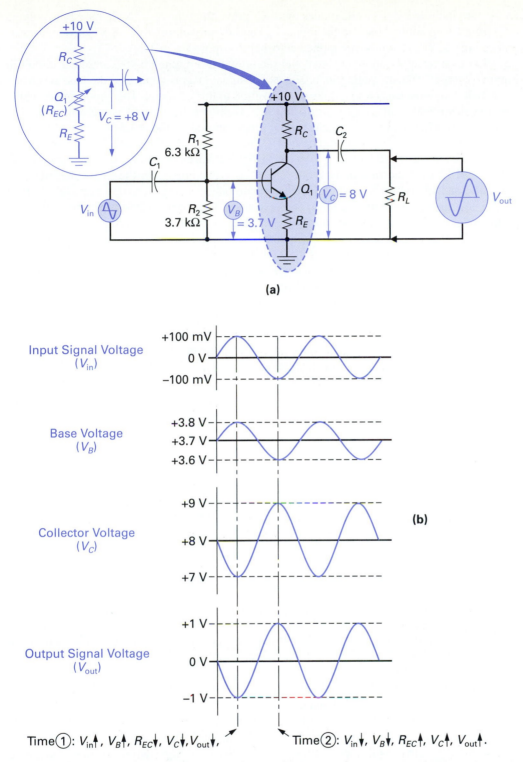

FIGURE 17-18 **A Transistor Being Used in an Analog Electronic Circuit. (a) Basic Amplifier Circuit. (b) Input/Output Voltage Waveforms.**

The dc voltage at the collector of Q_1 relative to ground (V_C) will be equal to the voltage developed across Q_1's collector–emitter resistance (R_{CE}) and R_E. In this example circuit, with no ac signal applied, a V_B of $+3.7$ V dc will cause R_{CE} and R_E to cumulatively develop $+8$ V at the collector of Q_1. The transistor has a base bias voltage (V_B) that is $+3.7$ V dc relative to ground and a collector voltage (V_C) that is $+8$ V dc relative to ground. Capacitors C_1

and C_2 are included to act as dc blocks, with C_1 preventing the +3.7 V dc base bias voltage (V_B) from being applied back to the input (V_{in}) and C_2 preventing the +8 V dc collector reference voltage (V_C) from being applied across the output (V_{RL} or V_{out}).

Let us now apply an input signal and see how it is amplified by the amplifier circuit in Figure 17-18(a). The alternating input sine-wave signal (V_{in}) is applied to the base of Q_1 via C_1, which, like most capacitors, offers no opposition to this ac signal. This input signal, which has a peak-to-peak voltage change of 200 mV, is shown in the first waveform in Figure 17-18(b). The alternating input signal will be superimposed on the +3.7 V dc base bias voltage and cause the +3.7 V dc at the base of Q_1 to increase by 100 mv (3.7 V + 100 mV = 3.8 V) and decrease by 100 mv (3.7 V − 100 mV = 3.6 V), as seen in the second waveform in Figure 17-18(b). An increase in the input signal ($V_{in} \uparrow$), and therefore the base voltage ($V_B \uparrow$), will cause an increase in the emitter diode's forward bias, causing Q_1 to turn more ON and the emitter-to-collector resistance of Q_1 to decrease ($R_{EC} \downarrow$). Because voltage drop is always proportional to resistance, a decrease in $R_{EC} \downarrow$ will cause a decrease in the voltage drop across R_{CE} and R_E ($V_C \downarrow$), and this decrease in V_C will be coupled to the output via C_2, causing a decrease in the output voltage developed across the load ($V_{out} \downarrow$).

Now let us examine what will happen when the sine-wave input signal decreases. A decrease in the input signal ($V_{in} \downarrow$), and therefore the base voltage ($V_B \downarrow$), will cause a decrease in the emitter diode's forward bias, causing Q_1 to turn less ON and the emitter-to-collector resistance of Q_1 to increase ($R_{EC} \uparrow$). Because voltage drop is always proportional to resistance, an increase in $R_{EC} \uparrow$ will cause an increase in the voltage drop across R_{EC} and R_E ($V_C \uparrow$). This increase in V_C will be coupled to the output via C_2, causing an increase in the output voltage developed across the load ($V_{out} \uparrow$).

Comparing the input signal voltage (V_{in}) with the output signal voltage (V_{out}) in Figure 17-18(b), you can see that a change in the input signal voltage produces a corresponding greater change in the output signal voltage. The ratio (comparison) of output signal voltage change to input signal voltage change is a measure of this circuit's **voltage gain (A_V)**. In this example, the **output signal voltage change (ΔV_{out})** is between +1 V and −1 V, and the **input signal voltage change (ΔV_{in})** is between +100 mV and −100 mV. The circuit's voltage gain between input and output will therefore be:

$$\text{Voltage gain } (A_V) = \frac{\text{output voltage change } (\Delta V_{out})}{\text{input voltage change } (\Delta V_{in})}$$

$$A_V = \frac{+1 \text{ to } -1 \text{ V}}{+100 \text{ mV to } -100 \text{ mV}} = \frac{2 \text{ V}}{200 \text{ mV}} = 10$$

A voltage gain of 10 means that the output voltage is 10 times larger than the input voltage. The transistor does not produce this gain magically within its NPN semiconductor structure. The gain or amplification is achieved by the input signal controlling the conduction of the transistor, which takes energy from the collector supply voltage and develops this energy across the load resistor. Amplification is achieved by having a small input voltage control a transistor and its large collector supply voltage, so that a small input voltage change results in a similar but larger output voltage change.

Comparing the input signal with the output signal at time 1 and time 2 in Figure 17-18(b), you can see that this circuit will invert the input signal voltage in the same way that the NOT gate inverts its input voltage (positive input voltage swing produces a negative output voltage swing, and vice versa). This inversion always occurs with this particular transistor circuit arrangement; however, it is not a problem, since the shape of the input signal is still preserved at the output (both input and output signals are sinusoidal).

A Switching Regulator Circuit

A zener diode can be used in a dc power supply to function as a voltage regulator. These regulator types maintain a constant output voltage because variations in input voltage or load current are dissipated as heat. These **series dissipative regulators** generally have a

Voltage Gain (A_V)

The ratio of the output signal voltage change to input signal voltage change.

Output Signal Voltage Change

Change in output signal voltage in response to a change in the input signal voltage.

Input Signal Voltage Change

The input voltage change that causes a corresponding change in the output voltage.

Series Dissipative Regulators

Voltage regulators that maintain a constant output voltage by causing variations in input voltage or load current to be dissipated as heat.

CHAPTER 17 / DIODES AND TRANSISTORS

low "conversion efficiency" of typically 60% to 70% and should be used only in low- to medium-load current applications.

Series switching regulators, on the other hand, have a conversion efficiency of typically 90%. To understand the operation of these regulator types, refer to the simplified circuit in Figure 17-19(a). To improve efficiency, a series-pass transistor (Q_1) is operated as a switch rather than as a variable resistor. This means that Q_1 is switched ON and OFF and therefore either switches the $+12$ V input at its collector through to its emitter or blocks the $+12$ V from passing through to the emitter. These $+12$ V pulses at the emitter of Q_1 charge capacitor C_1 to an average voltage (which in this example is $+5$ V), and this voltage is applied to the load (R_L). To explain this in more detail, when Q_1 is turned ON by a HIGH base voltage from the switching regulator IC, the unregulated $+12$ V at Q_1's collector is switched through to Q_1's emitter, where it reverse biases D_1 and is applied to the series-connected inductor L_1 and parallel capacitor C_1. Inductor L_1 and capacitor C_1 act as a low-pass filter because series-connected L_1 opposes the ON/OFF changes in current and passes a relatively constant current to the load; shunt- or parallel-connected C_1 opposes the ON/OFF changes in voltage and holds the output voltage relatively constant at $+5$ V. When Q_1 is turned OFF by a LOW base voltage from the switching regulator IC, the unregulated $+12$ V input is disconnected from the LC filter, the inductor's magnetic field will collapse and produce a current through the load, and the $+5$ V charge held by C_1 will still be applied across the load. Inductor L_1 therefore

Series Switching Regulators
A regulator circuit containing a power transistor in series with the load that is switched ON and OFF to regulate the dc output voltage delivered to the load.

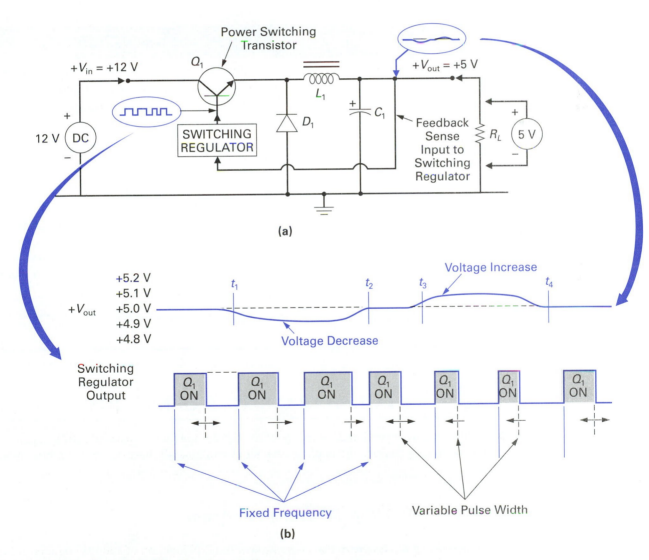

FIGURE 17-19 Basic Switching Regulator Action.

smoothes out the current changes, and capacitor C_1 smoothes out the voltage changes caused by the ON/OFF switching of transistor Q_1. The next, and most important, question is: How does this circuit regulate, or maintain constant, the output voltage? The answer is, through a closed-loop "sense and adjust" system controlled by a switching regulator IC. The switching regulator IC operates by comparing an internal fixed reference voltage with a sense input, which is taken from the +5 V output, as is shown in Figure 17-19(a). Referring to the waveforms shown in Figure 17-19(b), you can see that whenever the output voltage falls below +5 V (from time t_1 to t_2), the switching regulator responds by increasing the width of the positive output pulse applied to the base of Q_1. This increases the ON time of Q_1, which raises the average output voltage, bringing the output back up to +5 V. On the other hand, whenever the output voltage rises above +5 V (between time t_3 and t_4), the switching regulator responds by decreasing the width of the positive output pulse applied to the base of Q_1. This decreases the ON time of Q_1, which lowers the average output voltage, bringing the output back down to +5 V. The net result is that the output voltage will remain locked at +5 V despite variations in the input voltage and variations in the load.

A switching regulator, therefore, is a voltage regulator that chops up, or switches ON and OFF (at typically a 20 kHz rate), a dc input voltage to efficiently produce a regulated dc output voltage. A **switching power supply** uses switching regulators and is generally small in size and very efficient. The only disadvantage is that the circuitry is generally a little more complex and therefore a little more costly.

Switching Power Supply

A dc power supply that makes use of a series switching regulator controlled by a pulse-width modulator to regulate the output voltage.

SELF-TEST EVALUATION POINT FOR SECTION 17-4

Now that you have completed this section, you should be able to:

■ *Objective 11.* *Name the three terminals of the bipolar junction transistor.*

■ *Objective 12.* *Describe the difference between the construction and schematic symbol of the* NPN *and* PNP *bipolar transistor.*

■ *Objective 13.* *Identify the base, collector, and emitter terminals of typical low-power and high-power transistor packages.*

■ *Objective 14.* *Describe the two basic actions of a bipolar transistor:*
a. *ON/OFF switching action*
b. *Variable-resistor action*

■ *Objective 15.* *Define the terms* transistor *and* transistance.

■ *Objective 16.* *Explain how the transistor can be used in the following basic applications:*
a. *A digital (two-state) circuit, such as a logic gate*
b. *An analog (linear) circuit, such as an amplifier*

Use the following questions to test your understanding of Section 17-4.

1. What are the two basic types of bipolar transistor?
2. Name the three terminals of a bipolar transistor.
3. What are the two basic ways in which a transistor is made to operate?
4. Which of the modes of operation mentioned in question 3 is made use of in digital circuits and which is made use of in analog circuits?

17-5 DETAILED TRANSISTOR OPERATION

Now that we have a good understanding of the bipolar junction transistor's (BJT's) general characteristics, operation, and applications, let us examine all these aspects in a little more detail.

17-5-1 *Basic Bipolar Transistor Action*

When describing diodes previously, we saw how the P-N junction of a diode could be either forward or reverse biased to either permit or block the flow of current through the device.

The transistor must also be biased correctly; however, in this case, two P-N junctions rather than one must have the correct external supply voltages applied.

A Correctly Biased NPN Transistor Circuit

Figure 17-20(a) shows how an NPN transistor should be biased for normal operation. In this circuit, a +10 V supply voltage is connected to the transistor's collector (C) via a 1 kΩ collector resistor (R_C). The emitter (E) of the transistor is connected to ground via a 1.5 kΩ emitter resistor (R_E), and, as an example, an input voltage of +3.7 V is being applied to the base (B). The output voltage (V_{out}) is taken from the collector, and this collector voltage (V_C) will be equal to the voltage developed across the transistor's collector-to-emitter and the emitter resistor R_E.

As previously mentioned in the introduction to the transistor, the transistor can be thought of as containing two diodes, as shown in Figure 17-19(b). In normal operation, *the*

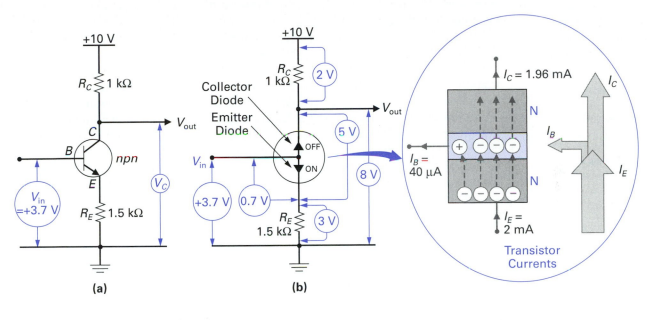

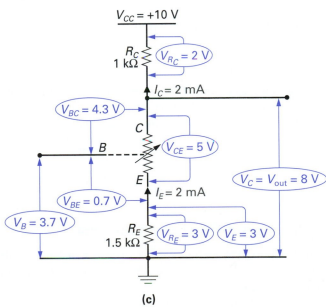

FIGURE 17-20 A Correctly Biased NPN Transistor Circuit.

transistor's emitter diode or junction is forward biased, and the transistor's collector diode or junction is reverse biased. To explain how these junctions are biased ON and OFF simultaneously, let us see how the input voltage of +3.7 V will affect this transistor circuit. An input voltage of +3.7 V is large enough to overcome the barrier voltage of the emitter diode (base–emitter junction), and so it will turn ON (base or anode is +, emitter or cathode is connected to ground or 0 V). Like any forward-biased silicon diode, the emitter diode will drop 0.7 V between base and emitter, and so the +3.7 V at the base will produce +3.0 V at the emitter. Knowing the voltage drop across the emitter resistor ($R_{RE} = 3$ V) and the resistance of the emitter resistor ($R_E = 1.5$ kΩ), we can calculate the value of current through the emitter resistor.

$$ I_{R_E} = \frac{V_{R_E}}{R_E} = \frac{3 \text{ V}}{1.5 \text{ k}\Omega} = 2 \text{ mA} $$

This emitter resistor current of 2 mA will leave ground, travel through R_E, and then enter the transistor's N-type emitter region. This current at the transistor's emitter terminal is called the **emitter current** (I_E). The forward-biased emitter diode will cause the steady stream of electrons entering the emitter to head toward the base region, as shown in the inset in Figure 17-20(b). The base is a very thin, lightly doped region with very few holes in relation to the number of electrons entering the transistor from the emitter. Consequently, only a few electrons combine with the holes in the base region and flow out of the base region. This relatively small current at the transistor's base terminal is called the **base current** (I_B). Because only a few electrons combine with holes in the base region, there is an accumulation of electrons in the base's P layer. These free electrons, feeling the attraction of the large positive collector supply voltage (+10 V), will travel through the N-type collector junction and out of the transistor to the positive external collector supply voltage. The current emerging out of the transistor's collector is called the **collector current** (I_C). Because both the collector current and base current are derived from the emitter current, we can state that:

Emitter Current (I_E)

The current at the transistor's emitter terminal.

Base Current (I_B)

The relatively small current at the transistor's base terminal.

Collector Current (I_C)

The current emerging out of the transistor's collector.

$$ I_E = I_B + I_C $$

In the example in the inset in Figure 17-20(b), you can see that this is true because

$$ I_E = I_B + I_C $$
$$ I_E = 40 \text{ μA} + 1.96 \text{ mA} = 2 \text{ mA} \ (40 \text{ μA} = 0.04 \text{ mA}) $$

Stated another way, we can say that the collector current is equal to the emitter current minus the current that is lost out of the base.

$$ I_C = I_E - I_B $$
$$ I_C = 2 \text{ mA} - 40 \text{ μA} = 1.96 \text{ mA} $$

Approximately 98% of the electrons entering the emitter of a transistor will arrive at the collector. Because of the very small percentage of current flowing out of the base (I_B equals about 2% of I_E), we can approximate and assume that I_C is equal to I_E.

$$ I_C = I_E $$
(I_C approximately equals I_E)

The Current-Controlled Transistor

In the previous section we discovered that because the collector and base currents (I_C and I_B) are derived from the emitter current (I_E), an increase in the emitter current ($I_E\uparrow$), for example, will cause a corresponding increase in collector and base current ($I_C\uparrow$, $I_B\uparrow$). Looking at this from a different angle, we see that an increase in the applied base voltage (base input increases to +3.8 V) will increase the forward bias applied to the emitter diode of the transistor, which will draw more electrons up from the emitter and cause an increase in I_E, I_B, and I_C. Similarly, a decrease in the applied base voltage (base input decreases to +3.6 V) will decrease the forward bias applied to the emitter diode of the transistor, which will decrease the number of electrons being drawn up from the emitter and cause a decrease in I_E, I_B, and I_C. The applied input base voltage will control the amount of base current, which will in turn control the amount of emitter and collector current and therefore the conduction of the transistor. This is why *the bipolar transistor is known as a current-controlled device.*

Continuing our calculations for the example circuit in Figure 17-20(b), let us apply this current relationship and assume that I_C is equal to I_E, which, as we previously calculated, is equal to 2 mA. Knowing the value of current for the collector resistor ($I_{R_C} = 2$ mA) and the resistance of the collector resistor ($R_C = 1$ kΩ), we can calculate the voltage drop across the collector resistor:

$$V_{R_C} = I_{R_C} \times R_C = 2\,\text{mA} \times 1\,\text{k}\Omega = 2\,\text{V}$$

With 2 V being dropped across R_C, the voltage at the transistor's collector (V_C) will be:

$$V_C = +10\,\text{V} - V_{R_C} = 10\,\text{V} - 2\,\text{V} = 8\,\text{V}$$

Because the voltage at the transistor's collector relative to ground is applied to the output, the output voltage will also be equal to 8 V:

$$V_C = V_{\text{out}} = 8\,\text{V}$$

At this stage we can determine a very important point about any correctly biased NPN transistor circuit. *A properly biased transistor will have a forward biased base–emitter junction (emitter diode is ON) and a reverse biased base–collector junction (collector diode is OFF).* We can confirm this with our example circuit in Figure 17-20(b), because we now know the voltages at each of the transistor's terminals.

Emitter diode (base–emitter junction) is forward biased (ON) because
Anode (base) is connected to +3.7 V (V_{in})
Cathode (emitter) is connected to 0 V via R_E

Collector diode (base–collector junction) is reverse biased (OFF) because
Anode (base) is connected to +3.7 V (V_{in})
Cathode (collector) is at +8 V (due to 2 V drop across R_C)

Keep in mind that even though the collector diode (base–collector junction) is reverse biased, current will still flow through the collector region. This is because most of the electrons traveling from emitter to base (through the forward-biased emitter diode) do not find many holes in the thin, lightly doped base region, and therefore the base current is always very small. Almost 98% of the electrons accumulating in the base region feel the strong attraction of the positive collector supply voltage and flow up into the collector region and then out of the collector as collector current.

With the example circuit in Figure 17-20(a) and (b), the emitter diode is ON and the collector diode is OFF, and the transistor is said to be operating in its normal, or *active, region.*

Operating a Transistor in the Active Region

A transistor is said to be in **active operation,** or in the **active region,** when its base–emitter junction is forward biased (emitter diode is ON), and the base–collector junction is reverse biased (collector diode is OFF). In this mode, the transistor is equivalent to a variable resistor between collector and emitter.

In Figure 17-20(c), the transistor circuit example has been redrawn with the transistor being shown as a variable resistor between collector and emitter and with all the calculated voltage and current values inserted. Before we go any further with this circuit, let us discuss some of the letter abbreviations used in transistor circuits. To begin, the term V_{CC} is

Active Operation or in the Active Region

A forward-biased base–emitter junction and reverse-biased base–collector junction. In this mode, the transistor is equivalent to a variable resistor between collector and emitter.

used to denote the "stable collector voltage," and this dc supply voltage is typically positive for an NPN transistor. Two Cs are used in this abbreviation $(+V_{CC})$ because V_C (V sub single C) is used to describe the voltage at the transistor's collector relative to ground. The double-letter subscripts such as V_{CC}, V_{EE}, or V_{BB} are used to denote a constant dc bias voltage for the collector (V_{CC}), emitter (V_{EE}), and base (V_{BB}). A single-letter subscript abbreviation such as V_C, V_E, or V_B is used to denote a transistor terminal voltage relative to ground. The other voltage abbreviations, V_{CE}, V_{BE}, and V_{CB}, are used for the voltage difference between two terminals of the transistor. For example, V_{CE} is used to denote the potential difference between the transistor's collector and emitter terminals. Finally, I_E, I_B, and I_C are, as previously stated, used to denote the transistor's emitter current (I_E), base current (I_B), and collector current (I_C).

Because the transistor's resistance between emitter and collector in Figure 17-20(c) is in series with R_C and R_E, we can calculate the voltage drop between collector and emitter (V_{CE}) because V_{RC} and V_{RE} are known.

$$V_{CE} = V_{CC} - (V_{R_E} + V_{R_C})$$
$$V_{CE} = 10 \text{ V} - (3 \text{ V} + 2 \text{ V}) = 10 \text{ V} - 5 \text{ V} = 5 \text{ V}$$

Now that we know the voltage drop between the transistor's collector and emitter (V_{CE}), we can calculate the transistor's equivalent resistance between collector and emitter (R_{CE}) because we know that the current through the transistor is 2 mA.

$$R_{CE} = \frac{V_{CE}}{I_C} = \frac{5 \text{ V}}{2 \text{ mA}} = 2.5 \text{ k}\Omega$$

Operating the Transistor in Cutoff and Saturation

Figure 17-21 shows the three basic ways in which a transistor can be operated. As we have already discovered, the bias voltages applied to a transistor control the transistor's operation by controlling the two P-N junctions (or diodes) in a bipolar transistor. For example, the center column reviews how a transistor will operate in the active region. As you can see, the previous circuit example with its values has been used. To summarize: *when a transistor is operated in the active region, its emitter diode is biased ON, its collector diode is biased OFF, and the transistor is equivalent to a variable resistor between the collector and the emitter.*

The left column in Figure 17-21 shows how the same transistor circuit can be driven into **cutoff**. A transistor is in cutoff when the bias voltage is reduced to the point at which it stops current in the transistor. In this example circuit, you can see that when the base input bias voltage (V_B) is reduced to 0 V, the transistor is cut off. *In cutoff, both the emitter and the collector diode of the transistor is biased OFF, the transistor is equivalent to an open switch between the collector and the emitter, and the transistor current is zero.*

The right column in Figure 17-21(c) shows how the same transistor circuit can be driven into **saturation**. A transistor is in saturation when the bias voltage is increased to a point at which any further increase in bias voltage will not cause any further increase in current through the transistor. In the equivalent circuit in Figure 17-21(c), you can see that when the base input bias voltage (V_B) is increased to $+6.7$ V, the emitter diode of the transistor will be heavily forward biased, and the emitter current will be large.

$$I_E = \frac{V_B - V_{BE}}{R_E} = \frac{6.7 \text{ V} - 0.7 \text{ V}}{1.5 \text{ k}\Omega} = 4.\text{mA}$$

Because I_B and I_C are both derived from I_E, an increase in I_E will cause a corresponding increase in both I_B and I_C. These high values of current through the transistor account for why a transistor operating in saturation is said to be equivalent to a closed switch (high conductance, low resistance). Although the transistor's resistance between the collector and emitter (R_{CE}) is assumed to be 0 Ω, there is still some small value of R_{CE}. Typically, a saturated transistor will have a 0.3 V drop between the collector and emitter ($V_{CE} = 0.3$ V), as shown in

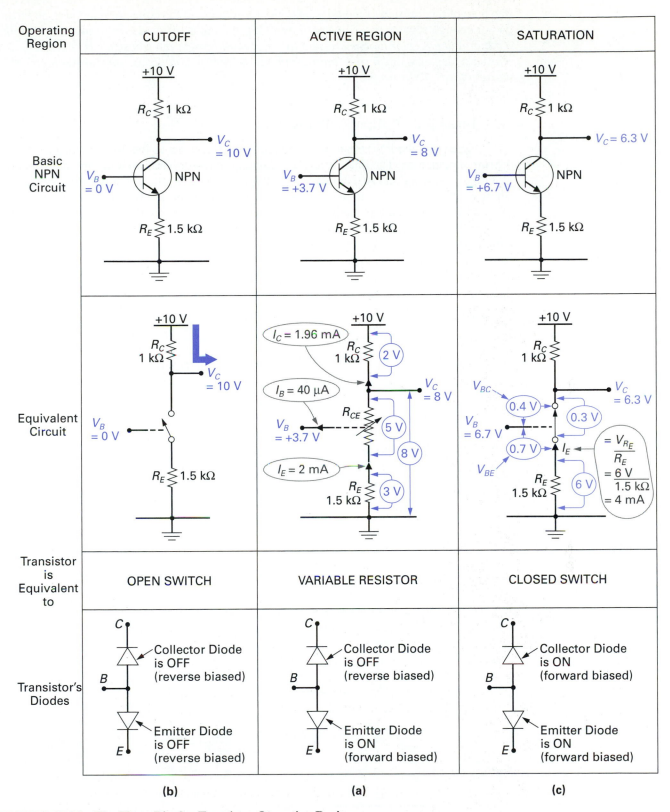

FIGURE 17-21 **The Three Bipolar Transistor Operating Regions.**

the equivalent circuit in Figure 17-21(c). If $V_{BE} = 0.7$ V and $V_{CE} = 0.3$ V, then the voltage drop across the collector diode of a saturated transistor (V_{BC}) will be 0.4 V. This means that the base of the transistor (anode) is now +0.4 V relative to the collector (cathode), and there is not enough reverse bias voltage to turn OFF the collector diode. *In saturation, both the emitter and collector diodes are said to be forward biased, the transistor is equivalent to a closed switch, and any further increase in bias voltage will not cause any further increase in current through the transistor.*

Biasing PNP Bipolar Transistors

Generally, the PNP transistor is not employed as often as the NPN transistor in most circuit applications. The only difference that occurs with PNP transistor circuits is that the polarity of V_{CC} and the base bias voltage (V_B) need to be reversed to a negative voltage, as shown in Figure 17-22. The PNP transistor has the same basic operating characteristics as the npn transistor, and all the previously discussed equations still apply. Referring to the inset in Figure 17-22, you will see that the −3.7 V base bias voltage will forward bias the emitter diode, and the −10 V V_{CC} will reverse bias the collector diode, so that the transistor is operating in the active region. Also, the electron transistor currents are in the opposite direction. This, however, makes no difference because the sum of the collector current entering the collector and base current entering the base is equal to the value of emitter current leaving the emitter, so $I_E = I_B + I_C$ still applies.

17-5-2 *Bipolar Transistor Circuit Configurations and Characteristics*

In the previous sections, we have seen how the bipolar junction transistor can be used in digital two-state switching circuits and analog or linear circuits such as the amplifier. In all these different circuit interconnections or **configurations,** the bipolar transistor was used as the main controlling element, with one of its three leads being used as a common reference and the other two leads being used as an input and an output. Although there are many thousands of different bipolar transistor circuit applications, all these circuits can be classified in one of three groups based on which of the transistor's leads is used as the **common** reference. These three different circuit configurations are shown in Figure 17-23. With the *common-emitter*

Configurations

Different circuit interconnections.

Common

Shared by two or more services, circuits, or devices. Although the term "common ground" is frequently used to describe two or more connections sharing a common ground, the term "common" alone does not indicate a ground connection, only a shared connection.

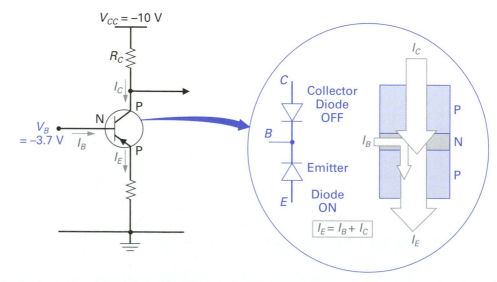

FIGURE 17-22 **A Correctly Biased PNP Transistor Circuit.**

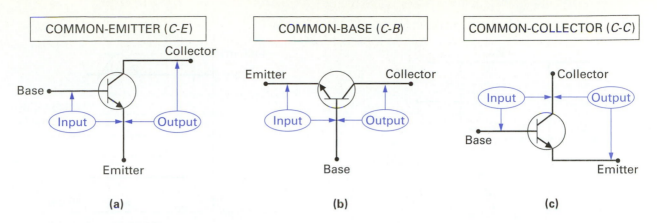

| COMMON-EMITTER (*C-E*) | COMMON-BASE (*C-B*) | COMMON-COLLECTOR (*C-C*) |

(a)   **(b)**   **(c)**

FIGURE 17-23   Bipolar Transistor Circuit Configurations.

(C-E) bipolar transistor circuit configuration, shown in Figure 17-23(a), the input signal is applied between the base and emitter, and the output signal appears between the transistor's collector and emitter. With this circuit arrangement, the input signal controls the transistor's base current, which in turn controls the transistor's output collector current, and the emitter lead is common to both the input and output. Similarly, with the *common-base (C-B)* circuit configuration shown in Figure 17-23(b), the input signal is applied between the transistor's emitter and base, the output signal is developed across the transistor's collector and base, and the base is common to both input and output. Finally, with the *common-collector (C-C)* circuit configuration, shown in Figure 17-23(c), the input is applied between the base and collector, the output is developed across the emitter and collector, and the collector is common to both the input and output.

To begin, we will discuss the common-emitter circuit configuration characteristics because this circuit arrangement has been used in all the circuit examples in this chapter.

Common-Emitter Circuits

With the **common-emitter circuit,** the transistor's emitter lead is common to both the input and output signals. In this circuit configuration, the base serves as the input lead, and the collector serves as the output lead. Figure 17-24 contains a basic common-emitter (*C-E*) circuit, its associated input/output voltage and current waveforms, characteristic curves, and table of typical characteristics. Using this illustration, we will examine the operation and characteristics of the *C-E* circuit configuration.

DC Current Gain   Referring to the *C-E* circuit in Figure 17-24(a), and its associated waveforms in Figure 17-24(b), let us now examine this circuit's basic operation.

Before applying the ac sine-wave input signal (V_{in}), let us assume that $V_{in} = 0$ V and examine the transistor's dc operating characteristics, or the "no input signal" condition. The voltage divider R_1 and R_2 will divide the V_{CC} supply voltage, producing a positive dc base bias voltage across R_2. This base bias voltage will be applied to the base of Q_1, and because it is generally greater than 0.7 V, it will forward bias the transistor's base–emitter junction, turning Q_1 ON (in the example in Figure 17-24, $V_B = 3$ V dc). Capacitor C_1 is included to act as a dc block, preventing the base bias voltage (V_B) from being applied back to the input (V_{in}). The value of dc base bias voltage will determine the value of base current (I_B) flowing out of the transistor's base, and this value of I_B will in turn determine the value of collector current (I_C) flowing out of the transistor's collector and through R_C. Because the transistor's output current (I_C) is so much larger than the transistor's very small input current (I_B), the circuit produces an increase in current, or a **current gain.** The current gain in a common-emitter configuration is called the transistor's **beta** (symbolized by β). A transistor's **dc beta** (β_{DC}) indicates a common-emitter transistor's "dc current

Common-Emitter *(C-E)* **Circuit**
Configuration in which the input signal is applied between the base and the emitter, and the output signal appears between the transistor's collector and emitter.

Current Gain
The increase in current produced by the transistor circuit.

Beta (β)
The transistor's current gain in a common-emitter configuration.

DC Beta (β_{DC})
The ratio of a transistor's dc output current to its input current.

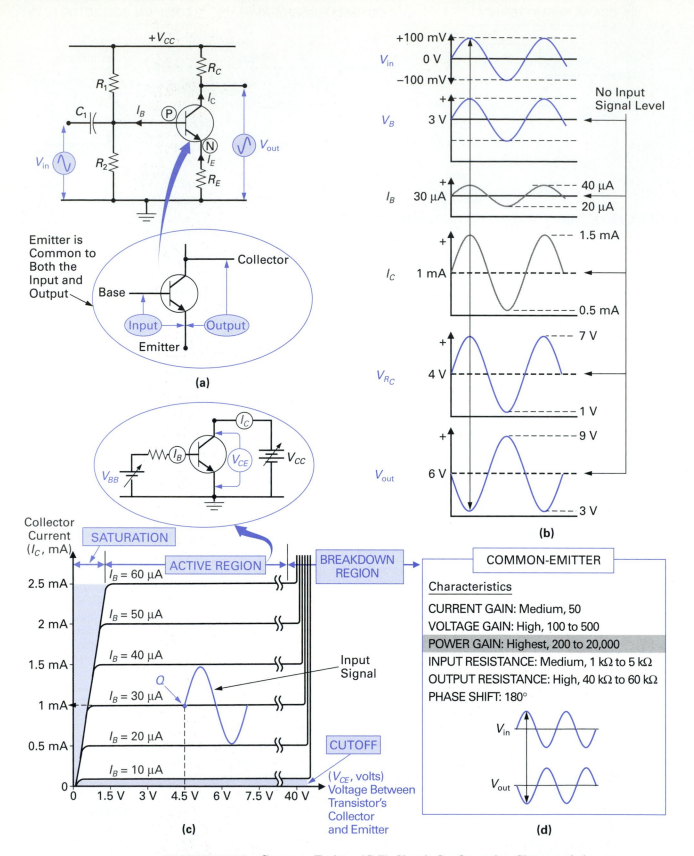

FIGURE 17-24 **Common-Emitter (*C-E*) Circuit Configuration Characteristics.**

gain," and it is the ratio of its output current (I_C) to its input current (I_B). This ratio can be expressed mathematically as:

$$\beta_{DC} = \frac{I_C}{I_B}$$

■ INTEGRATED MATH APPLICATION:

As can be seen in Figure 17-24(b), the no-input-signal level of $I_C = 1$ mA, and $I_B = 30$ μA. What is the transistor's dc beta?

■ *Solution:*

$$\begin{aligned}\beta_{DC} &= \frac{I_C}{I_B} \\ &= \frac{1 \text{ mA}}{30 \text{ μA}} \\ &= 33.3\end{aligned}$$

This value indicates that I_C is 33.3 times greater than I_B; therefore, the dc current gain between input and output is 33.3.

Another form of system analysis, known as "hybrid parameters," uses the term *hfe* instead of β_{DC} to indicate a transistor's dc current gain.

AC Current Gain   When amplifying an ac waveform, a transistor has applied to it both the dc voltage to make it operational and the ac signal voltage to be amplified that varies the base bias voltage and the base current. Referring to the first four waveforms in Figure 17-24(b), let us now examine what will happen when we apply a sine-wave input signal. As V_{in} increases above 0 V to a peak positive voltage ($V_{in}\uparrow$) it will cause the forward base bias voltage applied to the transistor to increase ($V_B\uparrow$) above its dc reference or no-input-signal level. This increase in V_B will increase the forward conduction of the transistor's emitter diode, resulting in an increase in the input base current ($I_B\uparrow$) and a corresponding larger increase in the output collector current ($I_C\uparrow$). The ratio of output collector current change (ΔI_C) to input base current change (ΔI_B) is the transistor's ac current gain or **ac beta** (β_{AC}). The formula for calculating a transistor's ac current gain is:

$$\beta_{AC} = \frac{\Delta I_C}{\Delta I_B}$$

AC Beta (β_{AC})
The ratio of a transistor's ac output current to input current.

■ INTEGRATED MATH APPLICATION:

Calculate the ac current gain of the example in Figure 17-24(b).

■ *Solution:*

$$\begin{aligned}\beta_{AC} &= \frac{\Delta I_C}{\Delta I_B} \\ &= \frac{1.5 \text{ to } 0.5 \text{ mA}}{40 \text{ to } 20 \text{ μA}} \\ &= \frac{1 \text{ mA}}{20 \text{ μA}} \\ &= 50\end{aligned}$$

This means that the alternating collector current at the output is 50 times greater than the alternating base current at the input.

Common-emitter transistors will typically have beta, or current gain values, of 50.

Voltage Gain The common-emitter circuit is used not only to increase the level of current between input and output but also to increase the amplitude of the input signal voltage or to produce a voltage gain. This action can be seen by examining the *C-E* circuit in Figure 17-24(a) and by following the changes in the associated waveforms in Figure 17-24(b). An input signal voltage increase from 0 V to a positive peak ($V_{in}\uparrow$) causes an increase in the dc base bias voltage ($V_B\uparrow$), causing the emitter diode of Q_1 to turn more ON and result in an increase in both I_B and I_C. Because the I_C flows through R_C, and because voltage drop is proportional to current, an increase in I_C will cause an increase in the voltage drop across R_C ($V_{R_C}\uparrow$). The output voltage is equal to the voltage developed across Q_1's collector-to-emitter resistance (R_{CE}) and R_E. Because R_C is in series with Q_1's collector-to-emitter resistance (R_{CE}) and R_E, an increase in $V_{R_C}\uparrow$ will cause a decrease in the voltage developed across R_{CE} and R_E, which is $V_{out}\downarrow$. This action is summarized with Kirchhoff's voltage law, which states that the sum of the voltages in a series circuit is equal to the voltage applied ($V_{R_C} + V_{out} = V_{CC}$). Using the example in Figure 17-24(b), you can see that when:

$V_{R_C}\uparrow$ to 7 V, $V_{out}\downarrow$ to 3 V. $\quad$ ($V_{R_C} + V_{out} = V_{CC}$, 7 V + 3 V = 10 V)

$V_{R_C}\downarrow$ to 1 V, $V_{out}\uparrow$ to 9 V. $\quad$ ($V_{R_C} + V_{out} = V_{CC}$, 1 V + 9 V = 10 V)

Although the input voltage (V_{in}) and output voltage (V_{out}) are out of phase with each other, you can see from the example values in Figure 17-24(b) that there is an increase in the signal voltage between input and output. This voltage gain between input and output is possible because the output current (I_C) is so much larger than the input current (I_B). The amount of voltage gain (which is symbolized by A_V) can be calculated by comparing the output voltage change (ΔV_{out}) with the input voltage change (ΔV_{in}).

$$A_V = \frac{\Delta V_{out}}{\Delta V_{in}}$$

■ INTEGRATED MATH APPLICATION:

Calculate the voltage gain of the circuit and its associated waveforms in Figure 17-24(a) and (b).

■ *Solution:*

$$A_V = \frac{\Delta V_{out}}{\Delta V_{in}} = \frac{+9 \text{ V to } +3 \text{ V}}{+100 \text{ mV to } -100 \text{ mV}} = \frac{6 \text{ V}}{200 \text{ mV}} = 30$$

This value indicates that the output ac signal voltage is 30 times larger than the input ac signal voltage.

Most common-emitter transistor circuits have high voltage gains between 100 to 500.

Power Gain As we have seen so far, the common-emitter circuit provides both current gain and voltage gain. Because power is equal to the product of current and voltage ($P = V \times I$), it is not surprising that the *C-E* circuit configuration also provides **power gain** (A_P). The power gain of a circuit can be calculated by dividing the output signal power (P_{out}) by the input signal power (P_{in}):

Power Gain (A_p)

The ratio of the output signal power to the input signal power.

$$A_P = \frac{P_{out}}{P_{in}}$$

To calculate the amount of input power (P_{in}) applied to the C-E circuit, we will have to multiply the change in input signal voltage (ΔV_{in}) by the accompanying change in input signal current (ΔI_{in} or ΔI_B):

$$P_{in} = \Delta V_{in} \times \Delta I_{in}$$

To calculate the amount of output power (P_{out}) delivered by the C-E circuit, we will have to multiply the change in output signal voltage (ΔV_{out}) produced by the change in output signal current (ΔI_{out} or ΔI_C).

$$P_{out} = \Delta V_{out} \times \Delta I_{out}$$

The power gain of a common-emitter circuit is therefore calculated with the formula:

$$A_P = \frac{P_{out}}{P_{in}} = \frac{\Delta V_{out} \times \Delta I_{out}}{\Delta V_{in} \times \Delta I_{in}}$$

■ **INTEGRATED MATH APPLICATION:**

Calculate the power gain of the example circuit in Figure 17-24.

■ *Solution:*

$$
\begin{aligned}
A_P &= \frac{P_{out}}{P_{in}} = \frac{\Delta V_{out} \times \Delta I_{out}}{\Delta V_{in} \times \Delta I_{in}} \\
&= \frac{(9 \text{ V to } 3 \text{ V}) \times (1.5 \text{ mA to } 0.5 \text{ mA})}{(+100 \text{ mV to } -100 \text{ mV}) \times (40 \text{ µA to } 20 \text{ µA})} \\
&= \frac{6 \text{ V} \times 1 \text{ mA}}{200 \text{ mV} \times 20 \text{ µA}} = \frac{6 \text{ mW}}{4 \text{ µW}} = 1500
\end{aligned}
$$

In this example, the common-emitter circuit has increased the input signal power from 4 µW to 6 mW—a power gain of 1500.

The power gain of the circuit in Figure 17-24 can also be calculated by multiplying the previously calculated C-E circuit voltage gain (A_V) by the previously calculated C-E circuit current gain (β_{AC}).

$$\text{Since } A_V = \frac{\Delta V_{out}}{\Delta V_{in}}, \text{ and } \beta_{AC} = \frac{\Delta I_{out} \text{ (or } \Delta I_C)}{\Delta I_{in} \text{ (or } \Delta I_B)}$$

$$A_P = V \times I = \frac{\Delta V_{out}}{\Delta V_{in}} \times \frac{\Delta I_{out} \text{ (or } \Delta I_C)}{\Delta I_{in} \text{ (or } \Delta I_B)} \text{ or } A_P = A_V \times \beta_{AC}$$

$$A_P = A_V \times \beta_{AC}$$

For the example circuit in Figure 17-24, this will be:

$$A_P = A_V \times \beta_{AC} = 30 \times 50 = 1500$$

indicating that the common-emitter circuit's output power in Figure 17-24 is 1500 times larger than the input power.

The power gain of common-emitter transistor circuits can be as high as 20,000, making this characteristic the circuit's key advantage.

Collector Characteristic Curves   One of the easiest ways to compare several variables is to combine all the values in a graph. Figure 17-24(c) shows a special graph, called the **collector characteristic curves,** for a typical common–emitter transistor circuit. The data for this graph are obtained by using the transistor test circuit, shown in the inset in Figure 17-24(c), which will apply different values of base bias voltage (V_{BB}) and collector bias voltage (V_{CC}) to an NPN transistor. The two ammeters and one voltmeter in this test circuit

Collector Characteristic Curves

Graph for a typical common-emitter transistor circuit.

are used to measure the circuit's I_B, I_C, and V_{CE} response to each different circuit condition. The values obtained from this test circuit are then used to plot the transistor's collector current (I_C in mA) on the vertical axis against the transistor's collector-emitter voltage (V_{CE}) drop on the horizontal axis for various values of base current (I_B in μA). This graph shows the relationship between a transistor's input base current, output collector current, and collector-to-emitter voltage drop.

Let us now examine the typical set of collector characteristic curves shown in Figure 17-24(c). When V_{CE} is increased from zero, by increasing V_{CC}, the collector current rises very rapidly, as indicated by the rapid vertical rise in any of the curves. When the collector diode of the transistor is reverse biased by the voltage V_{CE}, the collector current levels off. At this point, any one of the curves can be followed based on the amount of base current, which is determined by the value of base bias voltage (V_{BB}) applied. This flat part of the curve is known as the transistor's **active region.** The transistor is normally operated in this region, where it is equivalent to a variable resistor between the collector and emitter.

Active Region

Flat part of the collector characteristic curve. A transistor is normally operated in this region, where it is equivalent to a variable resistor between the collector and emitter.

As an example, let us use these curves in Figure 17-24(c) to calculate the value of output current (I_C) for a given value of input current (I_B) and collector-to-emitter voltage (V_{CE}). If V_{BB} is adjusted to produce a base current of 30 μA, and V_{CC} is adjusted until the voltage between the transistor's collector and emitter (V_{CE}) is 4.5 V, the output collector current (I_C) will be equal to approximately 1 mA. This is determined by first locating 4.5 V on the horizontal axis (V_{CE} = 4.5 V), following this point directly up to the I_B = 30 μA curve, and then moving directly to the left to determine the value of output current on the vertical axis (I_C = 1 mA). When these values of V_{CE} and I_B are present, the transistor is said to be operating at point "Q." This dc operating point is often referred to as a **quiescent operating point (Q point),** which means a dc steady-state or no-input-signal operating point. The Q point of a transistor is set by the circuit's dc bias components and supply voltages. For instance, in the circuit in Figure 17-24(a), R_1 and R_2 were used to set the dc base bias voltage (V_B, and therefore I_B), and the values of V_{CC}, R_C, and R_E were chosen to set the transistor's dc collector–emitter voltage (V_{CE}). At this dc operating point, we can calculate the transistor's dc current gain (β_{DC}), since both I_B and I_C are known:

Quiescent Operating Point (Q Point)

The voltage or current value that sets up the no-input-signal or operating point bias voltage.

$$\beta_{DC} = \frac{I_C}{I_B} = \frac{1 \text{ mA}}{30 \text{ μA}} = 33.3$$

If a sine-wave signal was applied to the circuit, as shown in the waveforms in Figure 17-24(b) and the characteristic curves in Figure 17-24(c), it would cause the transistor's input base current, and therefore output collector current, to alternate above and below the transistor's Q point (dc operating point). This ac input signal voltage will cause the input base current (I_B) to increase between 20 μA and 40 μA, and this input base current change will generate an output collector current change of 0.5 mA to 1.5 mA. The transistor's ac current gain (β_{AC}) will be:

$$\begin{aligned} \beta_{AC} &= \frac{\Delta I_C}{\Delta I_B} \\ &= \frac{1.5 \text{ to } 0.5 \text{ mA}}{40 \text{ to } 20 \text{ μA}} \\ &= \frac{1 \text{ mA}}{20 \text{ μA}} \\ &= 50 \end{aligned}$$

Returning to the collector characteristic curves in Figure 17-24(c), you can see that if the collector supply voltage (V_{CC}) is increased to an extreme, a point will be reached at which the V_{CE} voltage across the transistor will cause the transistor to break down, as indicated by the rapid rise in I_C. This section of the curve is called the **breakdown region** of the graph, and the damaging value of current through the transistor will generally burn out and destroy the device. As an example, for the 2N3904 bipolar transistor, breakdown will occur at a V_{CE} voltage of 40 V.

There are two shaded sections shown in the set of collector characteristic curves in Figure 17-24(c). These two shaded sections represent the other two operating regions of

Breakdown Region

The point at which the collector supply voltage will cause a damaging value of current through the transistor.

the transistor. To begin, let us examine the vertically shaded **saturation region.** If the base bias voltage (V_{BB}) is increased to a large positive value, the emitter diode of the transistor will turn ON heavily, I_B will be a large value, and the transistor will be operating in saturation. In this operating region, the transistor is equivalent to a closed switch between its collector and emitter (both the emitter and collector diode are forward biased), and therefore the voltage drop between collector and emitter will be almost zero (V_{CE} = typically 0.3 V, when the transistor is saturated), and I_C will be a large value that is limited only by the externally connected components. The horizontally shaded section represents the **cutoff region** of the transistor. If the base bias voltage (V_{BB}) is decreased to zero, the emitter diode of the transistor will turn OFF, I_B will be zero, and the transistor will be operating in cutoff. In this operating region, the transistor is equivalent to an open switch between its collector and emitter (both the emitter and collector diode are reverse biased), the voltage drop between collector and emitter will be equal to V_{CC}, and I_C will be zero.

A set of collector characteristic curves are therefore generally included in a manufacturer's device data sheet and can be used to determine the values of I_B, I_C, and V_{CE} at any operating point.

Input Resistance   The **input resistance (R_{in})** of a common–emitter transistor is the amount of opposition offered to an input signal by the input base–emitter junction (emitter diode). Because the base–emitter junction is normally forward biased when the transistor is operating in the active region, the opposition to input current is relatively small; however, the extremely small base region will support only a very small input base current. On average, if no additional components are connected in series with the transistor's base–emitter junction, the input resistance of a *C-E* transistor circuit is typically a medium value between 1 kΩ and 5 kΩ. This typical value is an average because the transistor's input resistance is a "dynamic or changing quantity" that will vary slightly as the input signal changes the conduction of the *C-E* transistor's emitter diode, and this changes I_B ($R \updownarrow = V/I \updownarrow$).

Because the transistor has a small value of input P-N junction capacitance and input terminal inductance, the opposition to the input signal is not only resistive but, to a small extent, reactive. For this reason, the total opposition offered by the transistor to an input signal is often referred to as the **input impedance (Z_{in})** because impedance is the total combined resistive and reactive input opposition.

Output Resistance   The **output resistance (R_{out})** of a common-emitter transistor is the amount of opposition offered to an output signal by the output base–collector junction (collector diode). This junction is normally reverse biased when the transistor is operating in the active region, and therefore the *C-E* transistor's output resistance is relatively high; however, because a unique action occurs within the transistor and allows current to flow through this reverse-biased junction (electron accumulation at the base and then conduction through collector diode due to attraction of $+V_{CC}$), the output current (I_C) is normally large, and so the output resistance is not an extremely large value. On average, if no load resistor is connected in series with the transistor's collector diode, the output resistance of a *C-E* transistor circuit is typically a high value between 40 kΩ and 60 kΩ.

Because the transistor has a small value of output P-N junction capacitance and output terminal inductance, the opposition to the output signal is not only resistive but also reactive. For this reason, the total opposition offered by the transistor to the output signal is often referred to as the **output impedance (Z_{out}).**

Common-Base Circuits

With the **common-base circuit,** the transistor's base lead is common to both the input and output signal. In this circuit configuration, the emitter serves as the input lead, and the collector serves as the output lead. Figure 17-25 contains a basic common-base (*C-B*) circuit, its associated input/output voltage and current waveforms, and table of typical characteristics.

Saturation Region

The point at which the collector supply voltage has the transistor operating at saturation.

Cutoff Region

The point at which the collector supply voltage has the transistor operating in cutoff.

Input Resistance (R_{in})

The amount of opposition offered to an input signal by the input base–emitter junction (emitter diode).

Input Impedance (Z_{in})

The total opposition offered by the transistor to an input signal.

Output Resistance (R_{out})

The amount of opposition offered to an output signal by the output base–collector junction (collector diode).

Output Impedance (Z_{in})

The total opposition offered by the transistor to the output signal.

Common-Base *(C-B)* Circuit

Configuration in which the input signal is applied between the transistor's emitter and base, and the output is developed across the transistor's collector and base.

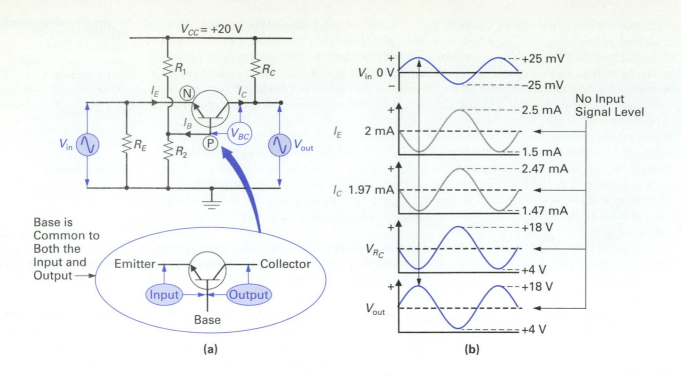

(a)

(b)

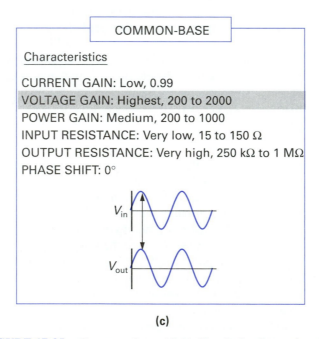

COMMON-BASE

Characteristics

CURRENT GAIN: Low, 0.99
VOLTAGE GAIN: Highest, 200 to 2000
POWER GAIN: Medium, 200 to 1000
INPUT RESISTANCE: Very low, 15 to 150 Ω
OUTPUT RESISTANCE: Very high, 250 kΩ to 1 MΩ
PHASE SHIFT: 0°

(c)

FIGURE 17-25 Common-Base *(C-B)* Circuit Configuration Characteristics.

Using this illustration, we will examine the operation and characteristics of the *C-B* circuit configuration.

DC Current Gain   Referring to the *C-B* circuit in Figure 17-25(a), and its associated waveforms in Figure 17-25(b), let us now examine this circuit's basic operation.

Before applying the ac sine-wave input signal (V_{in}), let us assume that $V_{in} = 0$ V and examine the transistor's dc operating characteristics, or the no-input-signal condition. The voltage divider R_1 and R_2 will divide the V_{CC} supply voltage, producing a positive dc base bias voltage across R_2. This base bias voltage will be applied to the base of Q_1. Because it is generally greater than 0.7 V, it will forward bias the transistor's base–emitter junction, turning Q_1 ON. Because the common-base circuit's input current is I_E and its output current is I_C, the current gain between input and output will be determined by the ratio of I_C to I_E. This ratio for calculating a C-B transistor's dc current gain is called the transistor's **dc alpha** (α_{DC}) and is equal to

DC Alpha (α_{DC}) Circuit
The ratio for calculating a C–B transistor's dc current gain.

$$\alpha_{DC} = \frac{I_C}{I_E}$$

The no-input-signal or steady-state dc levels of I_E and I_C are determined by the value of voltage developed across R_2, which is controlling the conduction of the transistor's forward biased base–emitter junction (emitter diode). Because the output current I_C is always slightly lower than the input current I_E (due to the small I_B current flow out of the base), the C-B transistor circuit does not increase current between input and output. In fact, there is a slight loss in current between input and output, which is why the C-B circuit configuration is said to have a current gain that is less than 1.

■ **INTEGRATED MATH APPLICATION:**

Calculate the dc alpha of the circuit in Figure 17-25(a) if $I_C = 1.97$ mA and $IE = 2$ mA.

■ *Solution:*

$$\alpha_{DC} = \frac{I_C}{I_E} = \frac{1.97 \text{ mA}}{2 \text{ mA}} = 0.985$$

This value of 0.985 indicates that I_C is 98.5% of I_E ($0.985 \times 100 = 98.5$).

As you can see from this example, the difference between I_C and I_E is generally so small that we always assume that the dc alpha is 1, which means that $I_C = I_E$.

AC Current Gain   When amplifying an ac waveform, a transistor has applied to it both the dc voltage to make it operational and the ac signal voltage to be amplified that varies the base–emitter bias and the input emitter current. Referring to the waveforms in Figure 17-25(b), let us now examine what will happen when we apply a sine-wave input signal. As mentioned previously, the positive voltage developed across R_2 will make the NPN transistor's base positive with respect to the emitter and forward bias the P-N base–emitter junction.

As the input voltage swings positive ($V_{in}\uparrow$) it will reduce the forward bias across the transistor's P-N base–emitter junction. For example, if $V_B = +5$ V and the transistor's n-type emitter is made positive, the P-N base–emitter diode will be turned more OFF. Turning the transistor's emitter diode less ON will cause a decrease in emitter current ($I_E\downarrow$), a decrease in collector current ($I_C\downarrow$), and a decrease in the voltage developed across R_C ($V_{R_C}\downarrow$). Because V_{R_C} and V_{out} are connected in series across V_{CC}, a decrease in $V_{R_C}\downarrow$ must be accompanied by an increase in $V_{out}\uparrow$. To explain this another way, the decrease in I_E and the subsequent decrease in both I_C and I_B means that the conduction of the transistor has decreased. This decrease in conduction means that the normally forward-biased base–emitter junction has turned less ON, and the normally reverse-biased base–collector junction has turned more OFF. Because the transistor's base-collector junction is in series with R_C and R_2 across V_{CC}, an increase in the transistor's base–collector resistance ($R_{BC}\uparrow$) will cause an increase in the voltage developed across the transistor's base–collector junction ($V_{BC}\uparrow$), which will cause an increase in $V_{out}\uparrow$.

Similarly, as the input voltage swings negative ($V_{in}\downarrow$), it will increase the forward bias across the transistor's P-N base–emitter junction. For example, if $V_B = -5$ V and the transistor's *n*-type emitter is made negative, the P-N base–emitter diode will be turned more ON. Turning the transistor's emitter diode more ON will cause an increase in emitter current ($I_E\uparrow$), an increase in collector current ($I_C\uparrow$), and an increase in the voltage developed across R_C ($V_{R_C}\uparrow$). Because V_{R_C} and V_{out} are connected in series across V_{CC}, an increase in $V_{R_C}\uparrow$ must be accompanied by a decrease in $V_{out}\downarrow$.

Now that we have seen how the input voltage causes a change in input current (I_E) and output current (I_C), let us examine the *C-B* circuit's ac current gain. The ratio of input emitter current change (ΔI_E) to output collector current change (ΔI_C) is the *C-B* transistor's ac current gain or **ac alpha (α_{AC})**. The formula for calculating a *C-B* transistor's ac current gain is:

AC Alpha (α_{AC}) Circuit
The ratio of input emitter current change to output collector current change.

$$\alpha_{AC} = \frac{\Delta I_C}{\Delta I_E}$$

■ **INTEGRATED MATH APPLICATION:**

Calculate the ac current gain of the example in Figure 17-25(b).

■ *Solution:*

$$\begin{aligned}\alpha_{AC} &= \frac{\Delta I_C}{\Delta I_E}\\ &= \frac{2.47 \text{ to } 1.47 \text{ mA}}{2.5 \text{ to } 1.5 \text{ mA}}\\ &= \frac{1 \text{ mA}}{1 \text{ mA}}\\ &= 1\end{aligned}$$

This means that the change in output collector current is equal to the change in input emitter current, and therefore the ac current gain is 1. (The output is 1 times larger than the input, $1 \text{ mA} \times 1 = 1 \text{ mA}$.)

Common-base transistors will typically have an ac alpha, or ac current gain, of 0.99.

Voltage Gain Although the common–base circuit does not achieve any current gain, it does make up for this disadvantage by achieving a very large voltage gain between input and output. Returning to the *C-B* circuit in Figure 17-25(a) and its waveforms in Figure 17-25(b), let us see how this very high voltage gain is obtained. Only a small input voltage (V_{in}) is needed to control the conduction of the transistor's emitter diode, and therefore the input emitter current (I_E) and output collector current (I_C). Even though I_C is slightly lower than I_E, it is still a relatively large value of current and will develop a large voltage change across R_C for a very small change in V_{in}. Because V_{R_C} and V_{out} are in series and connected across V_{CC}, a large change in voltage across R_C will cause a large change in the voltage developed across the transistor output (V_{BC}) and V_{out}. As before, the amount of voltage gain (which is symbolized by A_V) can be calculated by comparing the output voltage change (ΔV_{out}) with the input voltage change (ΔV_{in}):

$$A_V = \frac{\Delta V_{out}}{\Delta V_{in}}$$

Calculate the voltage gain of the circuit and its associated waveforms in Figure 17-25(a) and (b).

■ *Solution:*

$$A_V = \frac{\Delta V_{\text{out}}}{\Delta V_{\text{in}}} = \frac{+18 \text{ V to} +4 \text{ V}}{+25 \text{ mV to} -25 \text{ mV}} = \frac{14 \text{ V}}{50 \text{ mV}} = 280$$

This value indicates that the output ac signal voltage is 280 times larger than the ac input signal voltage.

Most common–base transistor circuits have very high voltage gains between 200 and 2000. Also, you can see by looking at Figure 17-25(b) that, unlike the *C-E* circuit, the common–base circuit has no phase shift between input and output (V_{out} is in phase with V_{in}).

Power Gain   Although the common-base circuit achieves no current gain, it does have a very high voltage gain and therefore can provide a medium amount of power gain ($P\uparrow = V\uparrow \times I$). The power gain of a circuit can be calculated by dividing the output signal power (P_{out}) by the input signal power (P_{in}).

$$A_P = \frac{P_{\text{out}}}{P_{\text{in}}}$$

To calculate the amount of input power (P_{in}) applied to the *C-B* circuit, we will have to multiply the change in input signal voltage (ΔV_{in}) by the accompanying change in input signal current (ΔI_{in} or ΔI_E).

$$P_{\text{in}} = \Delta V_{\text{in}} \times \Delta I_{\text{in}}$$
$$P_{\text{in}} = 50 \text{ mV} \times 1 \text{ mA} = 50 \text{ }\mu\text{W}$$

To calculate the amount of output power (P_{out}) delivered by the *C-B* circuit, we will have to multiply the change in output signal voltage (ΔV_{out}) produced by the change in output signal current (ΔI_{out} or ΔI_C):

$$P_{\text{out}} = \Delta V_{\text{out}} \times \Delta I_{\text{out}}$$
$$P_{\text{out}} = 14 \text{ V} \times 1 \text{ mA} \times 14 \text{ mW}$$

The power gain of a common–base circuit is calculated with the formula:

$$A_P = \frac{P_{\text{out}}}{P_{\text{in}}} = \frac{\Delta V_{\text{out}} \times \Delta I_{\text{out}}}{\Delta V_{\text{in}} \times \Delta I_{\text{in}}}$$

■ INTEGRATED MATH APPLICATION:

Calculate the power gain of the example circuit in Figure 17-25.

■ *Solution:*

$$A_P = \frac{P_{\text{out}}}{P_{\text{in}}} = \frac{\Delta V_{\text{out}} \times \Delta I_{\text{out}}}{\Delta V_{\text{in}} \times \Delta I_{\text{in}}}$$
$$= \frac{14 \text{ V} \times 1 \text{ mA}}{50 \text{ mV} \times 1 \text{ mA}} = \frac{14 \text{ mW}}{50 \text{ }\mu\text{W}} = 280$$

In this example, the common-base circuit has increased the input signal power from 50 μW to 14 mW—a power gain of 280.

The power gain of the circuit in Figure 17-25 can also be calculated by multiplying the previously calculated *C-B* circuit voltage gain (A_V) by the previously calculated *C-B* circuit current gain (α_{AC}).

$$A_P = A_V \times \alpha_{AC}$$

For the example circuit in Figure 17-25, this will be

$$A_P = A_V \times \alpha_{AC} = 280 \times 1 = 280$$

indicating that the common–base circuit's output power in Figure 17-25 is 280 times larger than the input power.

Typical common-base circuits will have power gains from 200 to 1000.

Input Resistance The input resistance (R_{in}) of a common–base transistor is the amount of opposition offered to an input signal by the input base–emitter junction (emitter diode). Because the base–emitter junction is normally forward biased, and the input emitter current (I_E) is relatively large, the input signal sees a very low input resistance. On average, if no additional components are connected in series with the transistor's base–emitter junction, the input resistance of a *C-B* transistor circuit is typically a low value between 15 Ω and 150 Ω. This typical value is an average because the transistor's input resistance is a dynamic or changing quantity that will vary slightly as the input signal changes the conduction of the *C-B* transistor's emitter diode, and this changes I_E ($R\updownarrow = V/I\updownarrow$).

Output Resistance The output resistance (R_{out}) of a common–base transistor is the amount of opposition offered to an output signal by the output base–collector junction (collector diode). This junction is normally reverse biased when the transistor is operating in the active region, and therefore the *C-B* transistor's output resistance is relatively high. On average, if no load resistor is connected in series with the transistor's collector diode, the output resistance of a *C-B* transistor circuit is typically a very high value between 250 kΩ and 1 MΩ.

Common-Collector Circuits

With the **common-collector circuit,** the transistor's collector lead is common to both the input and output signal. In this circuit configuration, therefore, the base serves as the input lead, and the emitter serves as the output lead. Figure 17-25 contains a basic common-collector (*C-C*) circuit, its associated input/output voltage and current waveforms, and table of typical characteristics. Using this illustration, we will examine the operation and characteristics of the *C-C* circuit configuration.

DC Current Gain Referring to the *C-C* circuit in Figure 17-26(a) and its associated waveforms in Figure 17-26(b), let us now examine this circuit's basic operation.

Before applying the ac sine-wave input signal (V_{in}), let us assume that $V_{in} = 0$ V and examine the transistor's dc operating characteristics, or the no-input-signal condition. The voltage divider R_1 and R_2 will divide the V_{CC} supply voltage, producing a positive dc base bias voltage across R_2. This base bias voltage will be applied to the base of Q_1. Because it is generally greater than 0.7 V, it will forward bias the transistor's base–emitter junction, turning Q_1 ON. Because the common-collector (*C-C*) circuit's input current I_B is much smaller than the output current I_E, the circuit provides a high current gain. In fact, the common-collector circuit provides a slightly higher gain than the *C-E* circuit because the common-collector's output current (I_E) is slightly higher than the *C-E*'s output current (I_C).

As in any other circuit configuration, the dc current gain is equal to the ratio of output current to input current. For the *C-C* circuit, this is equal to the ratio of I_E to I_B:

$$\text{DC current gain} = \frac{I_E}{I_B}$$

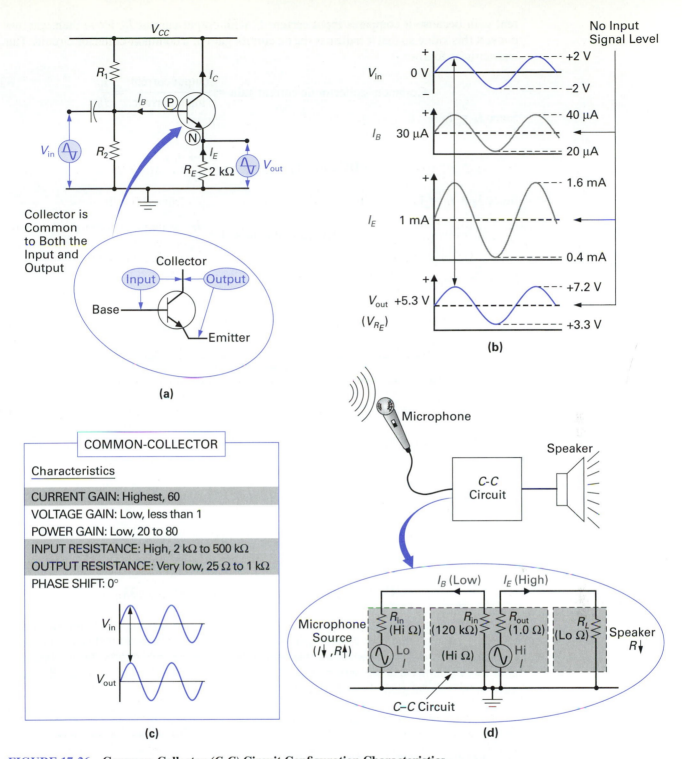

FIGURE 17-26 Common-Collector (*C-C*) Circuit Configuration Characteristics.

Transistor manufacturers do not generally provide specifications for all three circuit configurations. In most cases, because the common-emitter (*C-E*) circuit configuration is most frequently used, manufacturers will give the transistor's characteristics for only the *C-E* circuit configuration. In these instances, we will have to convert this *C-E* circuit data to equivalent specifications for other configurations. For example, in most data sheets the transistor's dc current gain will be listed as β_{DC}. As we know, dc beta is the measure of a *C-E* circuit's current gain because it compares input current I_B with output current I_C. How, then, can we convert this value so that it indicates the dc current gain of a common-collector circuit? The answer is as follows:

$$\text{Common-collector dc current gain} = \frac{\text{output current}}{\text{input current}} = \frac{I_E}{I_B}$$

Since $I_E = I_B + I_C$,

$$\text{DC current gain} = \frac{I_E}{I_B} = \frac{(I_B + I_C)}{I_B}$$

Since $I_B \div I_B = 1$,

$$\text{DC current gain} = 1 + \frac{I_C}{I_B}$$

Since $\frac{I_C}{I_B} = \beta_{DC}$,

$$\text{DC current gain} = 1 + \frac{I_C}{I_B} = 1 + \beta_{DC}$$

$$\boxed{\text{DC current gain} = 1 + \beta_{DC}}$$

■ **INTEGRATED MATH APPLICATION:**

Calculate the dc current gain of the *C-C* circuit in Figure 17-26 if the transistor's $\beta_{DC} = 32.33$.

■ *Solution:*

$$\text{DC current gain} = 1 + \frac{I_C}{I_B} = 1 + \frac{(I_E - I_B)}{I_B} = 1 + \frac{1\text{ mA} - 30\ \mu\text{A}}{30\ \mu\text{A}}$$

$$= 1 + \frac{970\ \mu\text{A}}{30\ \mu\text{A}} = 1 + 32.33 = 33.33$$

or,

$$\text{DC current gain} = 1 + \beta_{DC} = 1 + 32.33 = 33.33$$

As you can see in this example, the dc current gain of a common-collector circuit ($\beta_{DC} + 1$) is slightly higher than the dc current gain of a *C-E* circuit (β_{DC}). In most instances, the extra 1 makes so little difference when the transistor's dc current gain is a large value of about 30, as in this example, that we assume that the current gain of a common-collector circuit is equal to the current gain of a *C-E* circuit.

$$\boxed{\text{*C-C* dc current gain} \cong \text{*C-E* dc current gain } (\beta_{DC})}$$

AC Current Gain   When amplifying an ac waveform, a transistor has applied to it both the dc voltage to make it operational and the ac signal voltage that varies the base–emitter bias and the input base current. Referring to the waveforms in Figure 17-25(b), let us now examine what will happen when we apply a sine-wave input signal. As mentioned previously, the positive voltage developed across R_2 will make the NPN transistor's base positive with respect to the emitter, and therefore forward bias the P-N base–emitter junction.

As the input voltage swings positive ($V_{in}\uparrow$) it will add to the forward bias applied across the transistor's P-N base–emitter junction. This means that the transistor's emitter diode will turn more ON, cause an increase in the $I_B\uparrow$ and therefore a proportional but much larger increase in the output current $I_E\uparrow$.

Similarly, as the input voltage swings negative ($V_{in}\downarrow$) it will subtract from the forward bias applied across the transistor's P-N base–emitter junction. This means that the transistor's emitter diode will turn less ON, causing a decrease in the $I_B\downarrow$, and therefore a proportional but larger decrease in the output current $I_E\downarrow$.

The ac current gain of a common-collector transistor is calculated using the same formula as for dc current gain; however, with an ac current, we will compare the output current change (ΔI_E) with the input current change (ΔI_B).

$$\text{AC current gain} = \frac{\Delta I_E}{\Delta I_B}$$

■ **INTEGRATED MATH APPLICATION:**

Calculate the ac current gain of the circuit in Figure 17-26(a), using the values in Figure 17-26(b).

■ *Solution:*

$$\text{AC current gain} = \frac{\Delta I_E}{\Delta I_B} = \frac{1.6\,\text{mA} - 0.4\,\text{mA}}{40\,\mu\text{A} - 20\,\mu\text{A}} = 60$$

As with the common-collector's dc current gain, because there is so little difference between I_E and I_C, we can assume that the ac current gain is equivalent to β_{AC}.

$$C\text{-}C \text{ ac current gain} \cong C\text{-}E \text{ ac current gain } (\beta_{AC})$$

Common-collector transistor circuit configurations can have current gains as high as 60, indicating that I_E is 60 times larger than I_B.

Voltage Gain Although the common-collector circuit has a very high current gain rating, it cannot increase voltage between input and output. Returning to the *C-C* circuit in Figure 17-26(a) and its waveforms in Figure 17-26(b), let us see why this circuit has a very low voltage gain.

As the input voltage swings positive ($V_{in}\uparrow$) it will add to the forward bias applied across the transistor's P-N base–emitter junction ($V_{BE}\uparrow$). As the transistor's emitter diode turns more ON it will cause an increase in $I_B\uparrow$, a proportional but larger increase in $I_E\uparrow$, and therefore an increase in the voltage developed across RE (V_{R_E}, V_{out}, or $V_E\uparrow$). This increase in the voltage developed across R_E has a **degenerative effect** because an increase in the emitter voltage ($V_E\uparrow$) will counter the initial increase in base voltage ($V_B\uparrow$), and therefore the voltage difference between the transistor's base and emitter will remain almost constant (V_{BE} is almost constant). In other words, if the base goes positive and then the emitter goes positive, there is almost no increase in the potential difference between the base and the emitter and so the change in forward bias is almost zero. There is, in fact, a very small change in forward bias between base and emitter, and this will cause a small change in I_B and I_E, and therefore a small output voltage will be developed across R_E. Comparing the input and output voltage signals in Figure 17-25(b), you can see that both are about 4 V p-p, and both are in phase with each other. The common-collector circuit is often referred to as an **emitter-follower** or **voltage-follower** because the emitter output voltage seems to track or follow the phase and amplitude of the input voltage.

As with all other circuit configurations, the amount of voltage gain (A_V) can be calculated by comparing the output voltage change (ΔV_{out}) with the input voltage change (ΔV_{in}).

$$A_V = \frac{\Delta V_{out}}{\Delta V_{in}}$$

Degenerative Effect

An effect that causes a reduction in amplification due to negative feedback.

Emitter-Follower or Voltage-Follower

The common-collector circuit in which the emitter output voltage seems to track or follow the phase and amplitude of the input voltage.

Calculate the voltage gain of the circuit and its associated waveforms in Figure 17-26(a) and (b).

■ *Solution:*

$$A_V = \frac{\Delta V_{out}}{\Delta V_{in}} = \frac{+7.2 \text{ V to } +3.3 \text{ V}}{+2 \text{ V to } -2 \text{ V}} = \frac{3.9 \text{ V}}{4 \text{ V}} = 0.975$$

This value indicates that the output ac signal voltage is 0.975 or 97.5% of the ac input signal voltage ($0.975 \times 4 \text{ V} = 3.9 \text{ V}$).

Most common-collector transistor circuits have a voltage gain that is less than 1. However, in most circuit examples it is assumed that output voltage change equals input voltage change.

Power Gain   Although the common-collector circuit achieves no voltage gain, it does have a very high current gain and therefore can provide a small amount of power gain ($P\uparrow = V \times I\uparrow$). As before, the power gain of a circuit can be calculated by dividing the output signal power (P_{out}) by the input signal power (P_{in}):

$$A_P = \frac{P_{out}}{P_{in}}$$

To calculate the amount of input power (P_{in}) applied to the *C-C* circuit, we will have to multiply the change in input signal voltage (ΔV_{in}) by the accompanying change in input signal current (ΔI_{in} or ΔI_B):

$$P_{in} = \Delta V_{in} \times \Delta I_{in}$$
$$P_{in} = 4 \text{ V} \times 20 \text{ μA} = 80 \text{ μW}$$

To calculate the amount of output power (P_{out}) delivered by the *C-C* circuit, we will have to multiply the change in output signal voltage (ΔV_{out}) produced by the change in output signal current (ΔI_{out} or ΔI_E):

$$P_{out} = \Delta V_{out} \times \Delta I_{out}$$
$$P_{out} = 3.9 \text{ V} \times 1.2 \text{ mA} = 4.68 \text{ mW}$$

The power gain of a common-collector circuit is therefore calculated with the formula:

$$A_P = \frac{P_{out}}{P_{in}} = \frac{\Delta V_{out} \times \Delta I_{out}}{\Delta V_{in} \times \Delta I_{in}}$$

■ **INTEGRATED MATH APPLICATION:**

Calculate the power gain of the example circuit in Figure 17-26.

■ *Solution:*

$$A_P = \frac{P_{out}}{P_{in}} = \frac{\Delta V_{out} \times \Delta I_{out}}{\Delta V_{in} \times \Delta I_{in}}$$

$$= \frac{3.9 \text{ V} \times 1.2 \text{ mA}}{4 \text{ V} \times 20 \text{ μA}} = \frac{4.68 \text{ mW}}{80 \text{ μW}} = 58.5$$

In this example, the common-collector circuit has increased the input signal power from 80 μW to 4.68 mW—a power gain of 58.5.

The power gain of the circuit in Figure 17-26 can also be calculated by multiplying the previously calculated C-C circuit voltage gain (A_V) by the previously calculated C-C circuit current gain:

$$A_P = A_V \times \text{ac current gain}$$

For the example circuit in Figure 17-25, this will be

$$A_P = A_V \times \text{ac current gain} = 0.975 \times 60 = 58.5$$

indicating that the common-collector circuit's output power in Figure 17-26 is 58.5 times larger than the input power.

Typical common-collector circuits will have power gains from 20 to 80.

Input Resistance   An input signal voltage will see a very large input resistance when it is applied to a common-collector circuit. This is because the input signal sees the very large emitter-connected resistor ($R_E \uparrow\uparrow$) and, to a smaller extent, the resistance of the forward-biased base–emitter junction ($R_{\text{in}} \uparrow = V_{\text{in}}/I_B \downarrow$: R_{in} is large because I_{in} or I_B is small). Using these two elements, we can derive a formula for calculating the input resistance of a C-C transistor circuit:

$$R_{\text{in}} = R_E \times \text{ac current gain}$$

Since

$$\text{C-C ac current gain} \cong \text{C-E ac current gain } (\beta_{\text{AC}})$$

the input resistance can also be calculated with the formula:

$$R_{\text{in}} = R_E \times \beta_{\text{AC}}$$

■   **INTEGRATED MATH APPLICATION:**

Calculate the input resistance of the circuit in Figure 17-26, assuming $\beta_{\text{AC}} = 60$.

■   *Solution:*

$$R_{\text{in}} = R_E \times \beta_{\text{AC}} = 2 \text{ k}\Omega \times 60 = 120 \text{ k}\Omega$$

This means that an input voltage signal will see this C-C circuit as a resistance of 120 kΩ.

The input resistance of a C-C transistor circuit is typically a very large value between 2 kΩ and 500 kΩ.

Output Resistance   The output signal from a common-collector circuit sees a very low output resistance, as proved by this circuit's very high output current gain. Like this circuit's input resistance, the output resistance is largely dependent on the value of the emitter resistor R_E.

The output resistance of a typical C-C transistor circuit is a very low value between 25 Ω and 1 kΩ.

Impedance or Resistance Matching   Do not be misled into thinking that the very high input resistance and low output resistance of the common-collector transistor circuit are disadvantages. On the contrary, the very high input resistance and low output resistance of this configuration are made use of in many circuit applications, along with the C-C circuit's other advantage of high current gain.

To explain why a high input resistance and a low output resistance are good circuit characteristics, refer to the application circuit in Figure 17-26(d). In this example, a microphone is connected to the input of a C-C amplifier, and the output of this circuit is applied to a speaker. As we know, the sound-wave input to the microphone will physically move a magnet within the microphone, which will in turn interact to induce a signal voltage into a stationary coil. This voice signal voltage from the microphone, which is the source, is then applied across the input resistance of our example C-C circuit, which is the load.

In the inset in Figure 17-26(d), you can see that the microphone has been represented as a low-current ac source with a high internal resistance, and the input resistance of the C-C circuit is shown as a high value (in the previous example, 120 kΩ) resistor. Remembering our previous discussion on sources and loads, we know that a small load resistance will cause a large current to be drawn from the source, and this large current will drain or pull down the source voltage. Many signal sources, such as microphones, can generate only a small signal source voltage because they have a high internal resistance. If this small signal source voltage is applied across an amplifier with a small input resistance, a large current will be drawn from the source. This heavy load will pull the signal voltage down to such a small value that it will not be large enough to control the amplifier. A large amplifier input resistance ($R_{in}\uparrow$), on the other hand, will not load the source. Therefore, the input voltage applied to the amplifier will be large enough to control the amplifier circuit, to vary its transistor currents, and to achieve the gain between amplifier input and output. In summary, the high input resistance of the C-C circuit can be connected to a high-resistance source because it will not draw an excessive current and pull down the source voltage.

Referring again to the inset in Figure 17-26(d), you can see that at the output end, the C-C's output circuit has been represented as a high-current source with a low-value internal output resistor, and the speaker has been represented as a low-value resistance load. The low output resistance of the C-C circuit means that this circuit can deliver the high current output that is needed to drive the low-resistance load.

As you will see later in application circuits, most C-C circuits are used as a resistance or **impedance matching circuit** that can match, or isolate, a high-resistance (low-current) source, such as a microphone, to a low-resistance (high-current) load, such as a speaker. By acting as a **buffer current amplifier,** the C-C circuit can ensure that power is efficiently transferred from source to load.

17-5-3 *Bipolar Transistor Biasing Circuits*

As we discovered in the previous discussion on transistor circuit configurations, the ac operation of a transistor is determined by the dc bias level, or no-input-signal level. This steady-state or dc operating level is set by the value of the circuit's dc supply voltage (V_{CC}) and the value of the circuit's biasing resistors. This single supply voltage and the one or more biasing resistors set up the initial dc values of transistor current (I_B, I_E and I_C) and transistor voltage (V_{BE}, V_{CE} and V_{BC}).

In this section, we will examine some of the more commonly used methods for setting the initial dc operating point of a bipolar transistor circuit. As you encounter different circuit applications, you will see that many of these circuits include combinations of these basic biasing techniques and additional special-purpose components for specific functions. Because the common-emitter (C-E) circuit configuration is used more extensively than the C-B and the C-C, we will use this configuration in all the following basic biasing circuit examples.

Base Biasing

Figure 17-27(a) shows how a common-emitter transistor circuit could be base biased. With **base biasing,** the emitter diode of the transistor is forward biased by applying a positive base bias voltage ($+V_{BB}$) via a current-limiting resistor (R_B) to the base of Q_1. In Figure 17-27(b), the transistor circuit from Figure 17-27(a) has been redrawn so as to simplify the analysis of the circuit. The transistor is now represented as a diode between base and emitter (emitter diode), and the transistor's emitter to collector has been represented as a variable resistor. Assuming Q_1 is a silicon bipolar transistor, the forward-biased emitter diode will have a standard base–emitter voltage drop of 0.7 V (emitter diode drop = 0.7 V).

$$V_{BE} = 0.7 \text{ V}$$

Impedance Matching Circuit

A circuit that can match, or isolate, a high-resistance (low-current) source.

Buffer Current Amplifier

The *C-C* circuit that can ensure that power is efficiently transferred from source to load.

Base Biasing

A transistor biasing method in which the dc supply voltage is applied to the base of the transistor via a base-bias

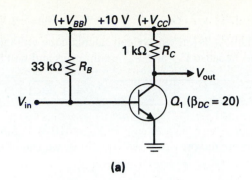

FIGURE 17-27 A Base-Biased Common Emitter Circuit. (a) Basic Circuit. (b) Simplified Equivalent Circuit.

(a)

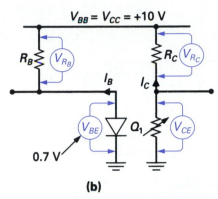

(b)

The base-bias resistor (R_B) and the transistor's emitter diode form a series circuit across V_{BB}, as seen in Figure 17-27(b). Therefore, the voltage drop across R_B (V_{RB}) will be equal to the difference between V_{BB} and V_{BE}:

$$V_{R_B} = V_{BB} - V_{BE}$$
$$= V_{BB} - 0.7 \text{ V}$$

$$V_{R_B} = V_{BB} - V_{BE} = 10 \text{ V} - 0.7 \text{ V} = 9.3 \text{ V}$$

Now that the resistance and voltage drop across R_B are known, we can calculate the current through R_B (I_{R_B}). Because a series circuit is involved, the current through R_B (I_{R_B}) will also be equal to the transistor base current I_B:

$$I_B = \frac{V_{R_B}}{R_B}$$

$$I_B = \frac{V_{R_B}}{R_B} = \frac{9.3 \text{ V}}{33 \text{ k}\Omega} = 282 \text{ }\mu\text{A}$$

Because the transistor's dc current gain (β_{DC}) is given in Figure 17-27(a), we can calculate I_C because β_{DC} tells us how much greater the output current I_C is compared with the input current I_B:

$$I_C = I_B \times \beta_{DC}$$

$$I_C = I_B \times \beta_{DC} = 282 \text{ }\mu\text{A} \times 20 = 5.6 \text{ mA}$$

Because the current through R_C is I_C, we can now calculate the voltage drop across R_C (V_{R_C}):

$$V_{R_C} = I_C \times R_C$$

$$V_{R_C} = I_C \times R_C = 5.6 \text{ mA} \times 1 \text{ k}\Omega = 5.6 \text{ V}$$

Now that V_{R_C} is known, we can calculate the voltage drop across the transistor's collector-to-emitter because V_{CE} and V_{R_C} are in series and will be equal to the applied voltage V_{CC}:

$$V_{CE} = V_{CC} - V_{R_C}$$

$$V_{CE} = V_{CC} - V_{R_C} = 10 \text{ V} - 5.6 \text{ V} = 4.4 \text{ V}$$

Combining the previous two equations, we can obtain the following V_{CE} formula:

$$V_{CE} = V_{CC} - V_{R_C}$$

Since

$$V_{R_C} = I_C \times R_C$$

$$V_{CE} = V_{CC} - (I_C \times R_C)$$

$$V_{CE} = V_{CC} - (I_C \times R_C) = 10 \text{ V} - (5.6 \text{ mA} \times 1 \text{ k}\Omega) = 4.4 \text{ V}$$

Using the preceding formulas, which are all basically Ohm's law, you can calculate the current and voltage values in a base-biased circuit.

DC Load Line In a transistor circuit, such as the example in Figure 17-27, V_{CC} and V_{R_C} are constants. On the other hand, the input current I_B and the output current I_C are variables. Using the example circuit in Figure 17-27, let us calculate what collector-to-emitter voltage drops (V_{CE}) will result for different values of I_C.

a. When Q_1 is OFF, $I_C = 0$ mA, and therefore V_{CE} equals:

$$V_{CE} = V_{CC} - (I_C \times R_C) = 10 \text{ V} - (0 \text{ mA} \times 1 \text{ k}\Omega) = 10 \text{ V} - 0 \text{ V} = 10 \text{ V}$$

This would make sense because Q_1 would be equivalent to an open switch between collector and emitter when it is OFF, and therefore the entire 10 V V_{CC} supply voltage would appear across the open. Figure 17-28 shows how this point would be plotted on a graph (point A).

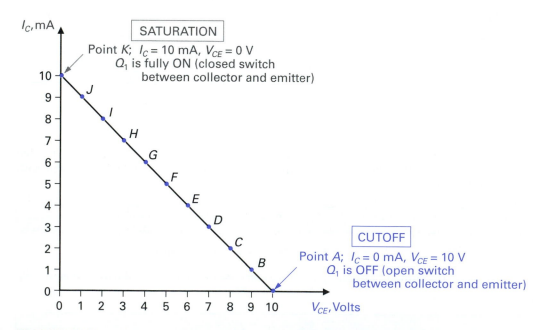

FIGURE 17-28 A Transistor DC Load Line with Cutoff and Saturation Points.

b. When $I_C = 1$ mA,

$$V_{CE} = 10 \text{ V} - (1 \text{ mA} \times 1 \text{ k}\Omega) = 10 \text{ V} - 1 \text{ V} = 9 \text{ V (point } B)$$

c. When $I_C = 2$ mA,

$$V_{CE} = 10 \text{ V} - (2 \text{ mA} \times 1 \text{ k}\Omega) = 10 \text{ V} - 2 \text{ V} = 8 \text{ V (point } C)$$

d. When $I_C = 3$ mA,

$$V_{CE} = 10 \text{ V} - (3 \text{ mA} \times 1 \text{ k}\Omega) = 10 \text{ V} - 3 \text{ V} = 7 \text{ V (point } D)$$

e. When $I_C = 4$ mA, $V_{CE} = 6$ V (point E)
f. When $I_C = 5$ mA, $V_{CE} = 5$ V (point F)
g. When $I_C = 6$ mA, $V_{CE} = 4$ V (point G)
h. When $I_C = 7$ mA, $V_{CE} = 3$ V (point H)
i. When $I_C = 8$ mA, $V_{CE} = 2$ V (point I)
j. When $I_C = 9$ mA, $V_{CE} = 1$ V (point J)
k. When $I_C = 10$ mA, the only resistance is that of R_C because Q_1 is fully ON and is equivalent to a closed switch between collector and emitter. It is not a surprise that the voltage drop across Q_1's collector-to-emitter is almost 0 V.

$$V_{CE} = 10 \text{ V} - (10 \text{ mA} \times 1 \text{ k}\Omega) = 10 \text{ V} - 10 \text{ V} = 0 \text{ V (point } K)$$

The line drawn in the graph in Figure 17-28 is called the **dc load line** because it is a line representing all the dc operating points of the transistor for a given load resistance. In this example, the transistor's load was the 1 kΩ collector-connected resistor R_C in Figure 17-27.

DC Load Line

A line representing all the dc operating points of the transistor for a given load resistance.

Cutoff and Saturation Points Let us now examine the two extreme points in a transistor's dc load line, which in the example in Figure 17-28 were points A and K. If a transistor's base input bias voltage is reduced to zero, its input current I_B will be zero, Q_1 will turn OFF and be equivalent to an open switch between the collector and emitter, the output current I_C will be 0 mA, and V_{CE} will be 10 V. This point in the transistor dc load line is called *cutoff* (point A in Figure 17-27) because the output collector current is reduced to zero, or cut off. In summary, at cutoff:

$$I_{C(\text{cutoff})} = 0 \text{ mA}$$
$$V_{CE(\text{cutoff})} = V_{CC}$$

In the example circuit in Figure 17-27 and its dc load line in Figure 17-28, with Q_1 cut OFF:

$$I_{C(\text{cutoff})} = 0 \text{ mA}, V_{CE(\text{cutoff})} = V_{CC} = 10 \text{ V}$$

If the base input bias voltage is increased to a large positive value, the transistor's collector diode (which is normally reverse biased) will be forward biased. In this condition, I_B will be at its maximum, Q_1 will be fully ON and equivalent to a closed switch between the collector and emitter, I_C will be at its maximum of 10 mA, and V_{CE} will be 0 V. This point in the transistor's dc load line is called *saturation* (point K in Figure 17-28) because, just as a point is reached at which a wet sponge is saturated and cannot hold any more water, the transistor at saturation cannot increase I_C beyond this point. In summary, at saturation:

$$I_{C(\text{sat.})} = \frac{V_{CC}}{R_C}$$
$$V_{CE(\text{sat.})} = 0 \text{ V}$$

In the example circuit in Figure 17-20 and its dc load line in Figure 17-21, with Q_1 saturated:

$$I_{C(\text{sat.})} = \frac{V_{CC}}{R_C} = \frac{10 \text{ V}}{1 \text{ k}\Omega} = 10 \text{ mA}$$
$$V_{CE(\text{sat.})} = 0 \text{ V}$$

Rearranging the formula $\beta_{DC} = I_C/I_B$, we can calculate the value of input base current that causes the output saturation current:

$$\beta_{DC} = \frac{I_C}{I_B}$$

therefore,

$$I_{B(\text{sat.})} = \frac{I_{C(\text{sat.})}}{\beta_{DC}}$$

In the example circuit in Figure 17-27 and its dc load line in Figure 17-28, the input current that will cause saturation will be

$$I_{B(\text{sat.})} = \frac{I_{C(\text{sat.})}}{\beta_{DC}} = \frac{10 \text{ mA}}{20} = 500 \text{ }\mu\text{A}$$

Figure 17-29 summarizes all our base-bias circuit calculations so far by including the dc load line from Figure 17-28 in a set of collector characteristic curves for the transistor circuit example in Figure 17-27. As you can see in the graph in Figure 17-29, at cutoff: $I_B = 0 \text{ }\mu\text{A}$, $I_C = 0 \text{ mA}$, and $V_{CE} = V_{CC}$, which is 10 V. On the other hand, at saturation: $I_B = 500 \text{ }\mu\text{A}$, $I_C = 10 \text{ mA}$, and $V_{CE} = 0 \text{ V}$.

Quiescent Point   Generally, the value of the base-bias resistor (R_B) is chosen so that the value of base current (IB) is near the middle of the dc load line. For example, if a base-bias resistance of 37.2 kΩ was used in the example circuit in Figure 17-27 ($R_B = 37.2$ kΩ), it would produce a base current of 250 mA ($I_B = 9.3$ V/37.2 k$\Omega = 250 \text{ }\mu$A). Referring to the dc load line in Figure 17-29, you can see that this value of base current is halfway between cutoff at 0 μA, and saturation at 500 μA. This point is called the *quiescent* (at rest) or *Q point* and is defined as *the dc bias point at which the circuit rests when no ac input signal is applied.* An ac input signal voltage will vary I_B above and below this Q point, resulting in a corresponding but larger change in I_C.

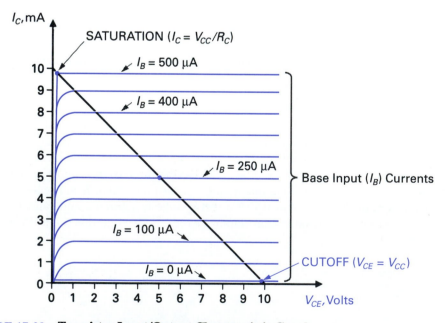

FIGURE 17-29   **Transistor Input/Output Characteristic Graph.**

Complete the following for the circuit shown in Figure 17-30.

   a. Calculate I_B.

   b. Calculate I_C.

   c. Calculate V_{CE}.

   d. Sketch the circuit's dc load line with saturation and cutoff points.

   e. Indicate where the Q point is on the circuit's dc load line.

■ *Solution:*

   a. Since $V_{BE} = 0.7$ V and $V_{BB} = 12$ V,

$$V_{R_B} = 12 \text{ V} - 0.7 \text{ V} = 11.3 \text{ V}$$

$$I_B = \frac{V_{R_B}}{R_B} = \frac{11.3 \text{ V}}{220 \text{ k}\Omega} = 51.4 \text{ μA}$$

   b. $\qquad\qquad I_C = I_B \times \beta_{DC} = 51.4 \text{ μA} \times 80 = 4.1 \text{ mA}$

   c. $\qquad\qquad V_{R_C} = I_C \times R_C = 4.1 \text{ mA} \times 1.2 \text{ k}\Omega = 4.92 \text{ V}$

$$V_{CE} = V_{CC} - V_{R_C} = 12 \text{ V} - 4.92 \text{ V} = 7.08 \text{ V}$$

   d. At cutoff, the transistor is OFF and therefore equivalent to an open switch between collector and emitter. The entire V_{CC} supply voltage will therefore be across $Q1$.

      At cutoff, $V_{CE} = V_{CC} = 12$ V (see cutoff in the dc load line in Figure 17-31).

      At saturation, the transistor is fully ON and therefore equivalent to a closed switch between the collector and emitter. The only resistance is that of R_C, and so:

      At saturation,

$$I_{C(\text{sat.})} = \frac{V_{CC}}{R_C} = \frac{12 \text{ V}}{1.2 \text{ k}\Omega} = 10 \text{ mA}$$

   (see saturation in the dc load line in Figure 17-30).

   e. The operating point or Q point of this circuit is set by the base-bias resistor R_B. This Q point will be at

$$I_C = 4.1 \text{ mA}$$

which produces a

$$V_{CE} = 7.08 \text{ V}$$

This quiescent (Q) point is also shown on Figure 17-31.

FIGURE 17-30 **Bipolar Transistor Example.**

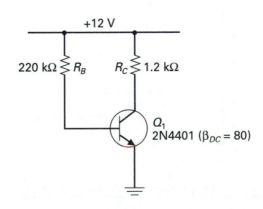

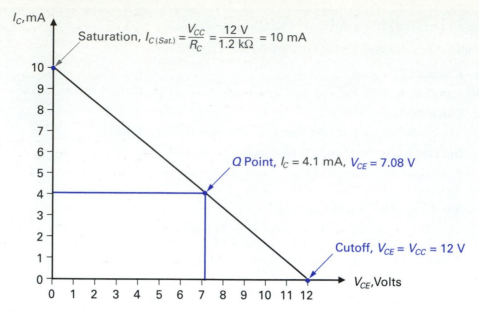

FIGURE 17-31 The DC Load Line for the Circuit in Figure 17-30.

INTEGRATED MATH APPLICATION:

Calculate the current through the lamp in Figure 17-32.

Solution:

$$V_{BE} = 0.7 \text{ V}, V_{\text{in}} = +5 \text{ V}$$

therefore,

$$\begin{aligned} V_{R_B} &= V_{\text{in}} - 0.7 \text{ V} \\ &= 5\text{V} - 0.7 \text{ V} = 4.3 \text{ V} \\ I_B &= \frac{V_{R_B}}{R_B} = \frac{4.3 \text{ V}}{10 \text{ k}\Omega} = 430 \text{ }\mu\text{A} \\ I_C &= I_B \times \beta_{DC} \\ &= 430 \text{ }\mu\text{A} \times 125 = 53.75 \text{ mA} \end{aligned}$$

An input of zero volts ($V_{\text{in}} = 0$ V) will turn OFF Q_1 and therefore lamp L_1. On the other hand, an input of +5 V will turn ON Q_1 and permit a collector current, and therefore lamp current, of 53.75 mA.

FIGURE 17-32 Two-State Lamp Circuit.

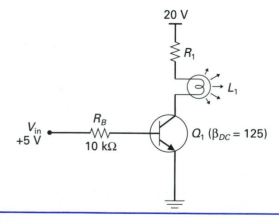

Base-Biasing Applications Base-bias circuits are used in switching circuit applications like the two-state ON/OFF lamp circuit discussed in the previous example. In these circuits, the bipolar transistor is equivalent to a switch and is controlled by a HIGH/LOW input voltage that drives the transistor between the two extremes of cutoff and saturation.

The advantage of this biasing technique is circuit simplicity because only one resistor is needed to set the base-bias voltage. The disadvantage of the base-biased circuit is that it cannot compensate for changes in its dc bias current due to changes in temperature. To explain this in more detail, a change in temperature will result in a change in the internal resistance of the transistor (all semiconductor devices have a negative temperature coefficient of resistance—temperature $\uparrow$ causes internal resistance $\downarrow$). This change in the transistor's internal resistance will change the transistor's dc bias currents (I_B and I_C), which will change or shift the transistor's dc operating point or Q point away from the desired midpoint.

Voltage-Divider Biasing

Figure 17-33(a) shows how a common-emitter transistor circuit could be **voltage-divider biased.** The name of this biasing method comes from the two-resistor series voltage divider (R_1 and R_2) connected to the transistor's base. In this most widely used biasing

FIGURE 17-33 A Voltage-Divider Biased Common–Emitter Circuit. (a) Basic Circuit. (b) Simplified Equivalent Circuit.

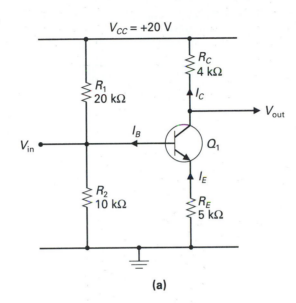

(a)

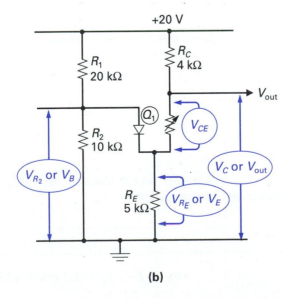

(b)

method, the emitter diode of Q_1 is forward biased by the voltage developed across R_2 (V_{R_2}), as seen in the simplified equivalent circuit in Figure 17-33(b). To calculate the voltage developed across R_2, and therefore the voltage applied to Q_1's base, we can use the voltage-divider formula:

$$V_{R_2} \text{ or } V_B = \frac{R_2}{R_1 + R_2} \times V_{CC}$$

$$V_{R_2} \text{ or } V_B = \frac{R_2}{R_1 + R_2} \times V_{CC} = \frac{10 \text{ k}\Omega}{20 \text{ k}\Omega + 10 \text{ k}\Omega} \times 20 \text{ V} = 0.333 \times 20 \text{ V} = 6.7 \text{ V}$$

Because the current through R_1 and R_2 (from ground to $+V_{CC}$) is generally more than 10 times greater than the base current of $Q1$ (I_B), it is normally assumed that IB will have no effect on the voltage-divider current through R_1 and R_2. The R_1 and R_2 voltage divider can be assumed to be independent of Q_1, and the preceding voltage-divider formula can be used to calculate V_{R_2} or V_B.

Because $V_B = 6.7$ V, the emitter diode of $Q1$ will be forward biased. Assuming a 0.7 V drop across the transistor's base–emitter junction $(V_{BE} = 0.7$ V), the voltage at the emitter terminal of $Q1$ (V_E) will be:

$$V_{R_E} \text{ or } V_E = V_B - 0.7 \text{ V}$$

$$V_{R_E} \text{ or } V_E = V_B - 0.7 \text{ V} = 6.7 \text{ V} - 0.7 \text{ V} = 6 \text{ V}$$

Now that the voltage drop across R_E (V_{R_E}) is known, along with its resistance, we can calculate the current through RE and the value of current being injected into the transistor's emitter:

$$I_{R_E} = I_E = \frac{V_{R_E}}{R_E}$$

$$I_{R_E} = I_E = \frac{V_{R_E}}{R_E} = \frac{6 \text{ V}}{5 \text{ k}\Omega} = 1.2 \text{ mA}$$

Because we know that a transistor collector current (I_C) is approximately equal to the emitter current (IE), we can state that:

$$I_E \cong I_C$$

$$I_E \cong I_C = 1.2 \text{ mA}$$

Now that I_C is known, we can calculate the voltage drop across R_C (V_{R_C}) because both its resistance and current are known:

$$V_{R_C} = I_C \times R_C$$

$$V_{R_C} = I_C \times R_C = 1.2 \text{ mA} \times 4 \text{ k}\Omega = 4.8 \text{ V}$$

The dc quiescent voltage at the collector of Q_1 with respect to ground (V_C), which is also V_{out}, will be equal to the dc supply voltage (V_{CC}) minus the voltage drop across RC.

$$V_C \text{ or } V_{out} = V_{CC} - V_{R_C}$$

$$V_C \text{ or } V_{out} = V_{CC} - V_{R_C} = 20\text{ V} - 4.8\text{ V} = 15.2\text{ V}$$

Because V_CC is connected across the series voltage divider formed by R_C, $Q1$'s collector-to-emitter resistance (R_{CE}), and RE, we can calculate V_{CE} if both V_{R_C} and V_E are known:

$$V_{CE} = V_{CC} - (V_{R_C} + V_E)$$

$$V_{CE} = V_{CC} - (V_{R_C} + VE) = 20\text{ V} - (4.8\text{ V} + 6\text{ V}) = 20\text{ V} - 10.8\text{ V} = 9.2\text{ V}$$

DC Load Line   Figure 17-34 shows the dc load line for the example circuit in Figure 17-33. Referring to the dc load line's two endpoints, let us examine this circuit's saturation and cutoff points.

When transistor Q_1 is fully ON or saturated, it will have approximately 0 Ω of resistance between its collector and emitter. As a result, R_C and R_E determine the value of I_C when Q_1 is saturated:

$$I_{C(sat.)} = \frac{V_{CC}}{R_C + R_E}$$

$$I_{C(sat.)} = \frac{V_{CC}}{R_C + R_E} = \frac{20\text{ V}}{4\text{ k}\Omega + 5\text{ k}\Omega} = \frac{20\text{ V}}{9\text{ k}\Omega} = 2.2\text{ mA}$$

As you can see in Figure 17-34, at saturation, I_C is maximum at 2.2 mA, and V_{CE} is 0 V because Q_1 is equivalent to a closed switch (0 Ω) between Q_1's collector and emitter.

$$V_{CE(sat.)} = 0\text{ V}$$

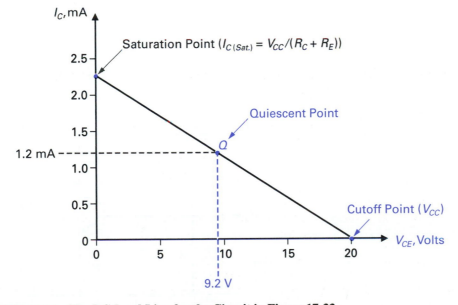

FIGURE 17-34   **The DC Load Line for the Circuit in Figure 17-33.**

At the other end of the dc load line in Figure 17-34, we can see how the transistor's characteristics are plotted when it is cut off. When Q_1 is cut OFF, it is equivalent to an open switch between collector and emitter. Therefore the enter V_{CC} supply voltage will appear across the series circuit open.

$$V_{CE(\text{cutoff})} = V_{CC}$$

$$V_{CE(\text{cutoff})} = V_{CC} = 20 \text{ V}$$

As you can see in Figure 17-34, when Q_1 is cut OFF, the entire V_{CC} supply voltage will appear across Q_1's collector-to-emitter terminals, and I_C will be blocked and equal to zero.

$$I_{C(\text{cutoff})} = 0 \text{ mA}$$

Generally, the values of the voltage-divider resistors R_1 and R_2 are chosen so that the value of base current (I_B) is near the middle of the dc load line. Referring to Figure 17-34, you can see that by plotting our previously calculated values of I_C (which at rest was 1.2 mA) and V_{CE} ($V_{CE} = 9.2$ V), we obtain a Q point that is near the middle of the dc load line.

■ INTEGRATED MATH APPLICATION:

Calculate the following for the circuit shown in Figure 17-35.

   a. V_B and V_E

   b. Determine whether C_E will have any effect on the dc operating voltages

   c. I_C

   d. V_C and V_{CE}

   e. Sketch the circuit's dc load line and include the saturation, cutoff, and Q points

FIGURE 17-35 A Common-Emitter Amplifier Circuit Example.

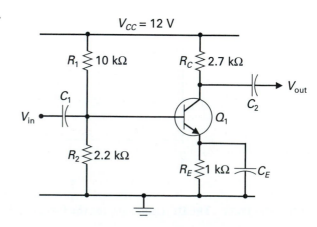

CHAPTER 17 / DIODES AND TRANSISTORS

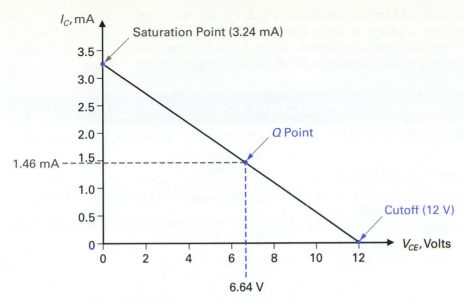

FIGURE 17-36 The DC Load Line for the Circuit in Figure 17-35.

■ *Solution:*

a. $V_B = \dfrac{R_2}{R_1 + R_2} \times V_{CC} = \dfrac{2.2\ k\Omega}{10\ k\Omega + 2.2\ k\Omega} \times 12\ V = 2.16\ V$

   $V_E = V_B - 0.7\ V = 2.16\ V - 0.7\ V = 1.46\ V$

b. Since all capacitors can be thought of as a dc block, C_E will have no effect on the circuit's dc operating voltages.

c. $I_E = \dfrac{V_E}{R_E} = \dfrac{1.46\ V}{1\ k\Omega} = 1.46\ mA$

   $I_C \cong I_E = 1.46\ mA$

d. $V_{R_C} = I_C \times R_C = 1.46\ mA \times 2.7\ k\Omega = 3.9\ V$

   V_{out} or $V_C = V_{CC\,-}\,V_{R_C} = 12\ V - 3.9\ V = 8.1\ V$

   $V_{CE} = V_{CC\,-}\,(V_{R_C} + V_E) = 12\ V - (3.9\ V + 1.46\ V) = 12\ V - 5.36\ V = 6.64\ V$

e. $I_{C(sat.)} = \dfrac{V_{CC}}{R_C + R_E} = \dfrac{12\ V}{2.7\ k\Omega + 1\ k\Omega} = \dfrac{12\ V}{3.7\ k\Omega} = 3.24\ mA$

   $V_{CE(cutoff)} = V_{CC} = 12\ V$

   Q point, $I_C = 1.46\ mA$ and $V_{CE} = 6.46\ V$
   (This information is plotted on the graph in Figure 17-36.)

Voltage-Divider Bias Applications Voltage-divider biased circuits are used in analog or linear circuit applications such as the amplifier circuit discussed in the previous example. In these circuits, the bipolar transistor is equivalent to a variable resistor and is controlled by an alternating input signal voltage.

Unlike the base-biased circuit, the voltage-divider biased circuit has very good temperature stability due to the emitter resistor R_E. To explain this in more detail, let us assume that there is an increase in the temperature surrounding a voltage-divider circuit, such as the example circuit in Figure 17-36. As temperature increases, it causes an increase in the transistor's internal currents ($I_B\uparrow, I_E\uparrow, I_C\uparrow$) because all semiconductor devices have a negative temperature coefficient of resistance (temperature $\uparrow, R\downarrow, I\uparrow$). An increase in $I_E\uparrow$ will cause an increase in the voltage drop across $R_E\uparrow$, which will decrease the voltage difference between the transistor's base and emitter ($V_{BE}\downarrow$). Decreasing the forward bias applied to the transis-

Emitter Feedback

The coupling from the emitter output to the base input in a transistor amplifier.

tor's emitter diode will decrease all the transistor's internal currents ($I_B\downarrow$, $I_E\downarrow$, $I_C\downarrow$) and return them to their original values. Therefore, a change in output current (I_C) due to temperature will effectively be fed back to the input and change the input current (I_B), which is why a circuit containing an emitter resistor is said to have **emitter feedback** for temperature stability.

SELF-TEST EVALUATION POINT FOR SECTION 17-5

Now that you have completed this section, you should be able to:

■ **Objective 17.** *Describe in more detail the bipolar junction transistor action.*

■ **Objective 18.** *Explain in more detail how the transistor can be used as an amplifier.*

■ **Objective 19.** *Identify and list the characteristics of the three transistor circuit combinations, namely:*

   a. *Common base*
   b. *Common emitter*
   c. *Common collector*

■ **Objective 20.** *Explain the meaning of the following:*

   a. *Transistor voltage and current abbreviations*
   b. *DC alpha*
   c. *DC beta*
   d. *Collector characteristic curve*
   e. *AC beta*
   f. *Input resistance or impedance*
   g. *Output resistance or impedance*

■ **Objective 21.** *Calculate a transistor circuit's:*

   a. *DC current gain*
   b. *AC current gain*
   c. *Voltage gain*
   d. *Power gain*

■ **Objective 22.** *Calculate the different values of circuit voltage and current for the following transistor biasing methods:*

   a. *Base biasing*
   b. *Voltage-divider biasing*

■ **Objective 23.** *Define the following terms:*

   a. *DC load line*
   b. *Cutoff point*
   c. *Saturation point*
   d. *Quiescent point*

Use the following questions to test your understanding of Section 17-5.

1. The bipolar transistor is a _____ (voltage/current) controlled device.

2. When a bipolar transistor is being operated in the active region, its emitter diode is _____ biased and its collector diode is _____ biased.

3. Which of the following is correct:

   a. $I_E = I_C - I_B$
   b. $I_C = I_E - I_B$
   c. $I_B = I_C - I_E$

4. When a transistor is in cutoff, it is equivalent to a/an _____ between its collector and emitter.

5. When a transistor is in saturation, it is equivalent to a/an _____ between its collector and emitter.

6. Which of the bipolar transistor circuit configurations has the best

   a. Voltage gain
   b. Current gain
   c. Power gain

7. Which biasing method makes use of two series-connected resistors across the V_{CC} supply voltage?

8. Which biasing technique has a single resistor connected in series with the base of the transistor?

REVIEW QUESTIONS

Multiple Choice Questions

1. Which of the following junction diodes are forward biased?

   a. Anode = +7 V, cathode = +10 V
   b. Anode = +5 V, cathode = +3 V
   c. Anode = +0.3 V, cathode = +5 V
   d. Anode = −9.6 V, cathode = −10 V

2. The junction diode _____ current when it is forward biased, and _____ current when it is reverse biased.

   a. Blocks, conducts
   b. Conducts, passes
   c. Blocks, prevents
   d. Conducts, blocks

3. Semiconductor devices need voltages of a certain amplitude and polarity to control their operation. These voltages are called:

   a. Barrier potentials
   b. Depletion voltages
   c. Knee voltages
   d. Bias voltages

4. When forward biased, a junction diode is equivalent to a/an _____ switch, whereas when it is reverse biased it is equivalent to a/an _____ switch.

   a. Open, closed
   b. Closed, closed
   c. Open, open
   d. Closed, open

5. The black band on a diode's package is always closest to the _____ .

   a. Anode
   b. Cathode
   c. *p*-type material
   d. Both (a) and (c) are true

6. When current dramatically increases, the voltage point on the diode's forward V–I characteristic curve is called the:

   a. Breakdown voltage   **c.** Barrier voltage
   b. Knee voltage   **d.** Both (b) and (c) are true

7. When current dramatically increases, the voltage point on the diode's reverse V–I characteristic curve is called the:

   a. Breakdown voltage   **c.** Barrier voltage
   b. Knee voltage   **d.** Both (b) and (c) are true

8. The _____ diode is designed to withstand high reverse currents that result when the diode is operated in the reverse breakdown region.

   a. Basic P-N junction   **c.** Zener
   b. Light emitting   **d.** Both (a) and (c) are true

9. When a zener diode's breakdown voltage is exceeded, the reverse current through the diode increases from a small leakage value to a high reverse current value.

   a. True   **b.** False

10. When operating in the reverse breakdown region, the _____ the zener will vary over a wide range, and the _____ the zener will vary by only a small amount.

    a. Voltage drop across, forward current through
    b. Forward current through, voltage drop across
    c. Reverse current through, forward current through
    d. Reverse current through, forward drop across

11. The zener diode's symbol is different from all other diode symbols due to its:

    a. Z-shaped cathode bar   **c.** Straight bar cathode
    b. Two exiting arrows   **d.** None of the above

12. The ability of a zener diode to maintain a relatively constant _____ regardless of variations in zener _____ is the key characteristic of a zener diode.

    a. Current, voltage   **c.** Current, impedance
    b. Impedance, voltage   **d.** Voltage, current

13. A 12 V ± 5% zener diode will have a voltage drop in the _____ range.

    a. 10.8 V to 13.2 V   **c.** 5.04 V to 18.96 V
    b. 11.4 V to 12.6 V   **d.** 11.88 V to 12.12 V

14. A 9.1 V zener diode has a power rating of 10 W. What is the diode's value of maximum zener current?

    a. 1.1 m   **b.** 91 mA   **c.** 1.1 A   **d.** None of the above

15. A/an _____ circuit maintains the output voltage of a voltage source constant despite variations in the input voltage and the load resistance.

    a. Encoder   **c.** Comparator
    b. Logic gate   **d.** Voltage regulator

16. The zener diode is able to maintain the voltage drop across its terminals constant by continually changing its _____ in response to a change in input voltage.

    a. Impedance   **c.** Power rating
    b. Voltage   **d.** Both (a) and (c) are true

17. In a voltage regulator circuit, the voltage developed across the zener diode remains constant, and therefore any changes in the input voltage must appear across the:

    a. Load   **c.** Source terminals
    b. Series resistor   **d.** Zener diode

18. The light-emitting diode is a semiconductor device that converts _____ energy into _____ energy.

    a. Chemical, electrical   **c.** Electrical, light
    b. Light, electrical   **d.** Heat, electrical

19. The _____ lead of an LED is distinguished from the other terminal by its longer lead, flattened lead, or its close proximity to the flat side of the case.

    a. Anode   **b.** Cathode

20. The forward voltage drop across a typical LED is usually:

    a. 0.7 V   **b.** 0.3 V   **c.** 5 V   **d.** 2.0 V

21. The bipolar junction transistor has three terminals called the:

    a. Drain, source, gate
    b. Anode, cathode, gate
    c. Main terminal 1, main terminal 2, gate
    d. Emitter, base, collector

22. The term *bipolar junction transistor* was given to the device because it has:

    a. Two p-n junctions
    b. Two magnetic poles
    c. One p region and one n region
    d. Two magnetic junctions

23. An NPN transistor is normally biased so that its base is _____.

    a. Positive   **b.** Negative

24. Which is considered the most common bipolar junction transistor configuration?

    a. Common-base   **c.** Common-emitter
    b. Common-collector   **d.** None of the above

25. A common-collector circuit is often called a/an _____.

    a. Base-follower   **c.** Collector-follower
    b. Emitter-follower   **d.** None of the above

26. With the NPN transistor schematic symbol, the emitter arrow will point _____ the base, whereas with the PNP transistor schematic symbol, the emitter arrow will point _____ the base.

    a. Toward, away from   **b.** Away from, toward

27. The transistor's ON/OFF switching action is made use of in _____ circuits.

    a. Analog   **c.** Linear
    b. Digital   **d.** Both (a) and (c) are true

28. The transistor's variable resistor action is made use of in _____ circuits.

    a. Analog   **c.** Linear
    b. Digital   **d.** Both (a) and (c) are true

29. Approximately 98% of the electrons entering the _____ of a bipolar transistor will arrive at the _____ , and the remainder will flow out of the _____ .

    a. Emitter, collector, base   c. Collector, emitter, base
    b. Base, collector, emitter   d. Emitter, base, collector

30. The common-base circuit configuration achieves the highest _____ gain, the common-emitter achieves the highest _____ gain, and the common-collector achieves the highest _____ gain.

    a. Voltage, current, power   c. Voltage, power, current
    b. Current, power, voltage   d. Power, voltage, current

31. Which of the following abbreviations is used to denote the voltage drop between a transistor's base and emitter?

    a. I_{BE} b. V_{CC} c. V_{CE} d. V_{BE}

32. Which of the following abbreviations is used to denote the voltage drop between a transistor's collector and emitter?

    a. V_C b. V_{CE} c. V_E d. V_{CC}

33. Consider the following for a base-biased bipolar transistor circuit: $R_B = 33$ kΩ, $R_C = 560$ Ω, Q_1 (β_{DC}) = 25, $V_{CC} = +10$ V. What is V_{BE}?

    a. 1.43 mV
    b. 25×33 kΩ
    c. 0.7 V
    d. Not enough information is given to calculate.

34. Which point on the dc load line results in an $I_C = V_{CC}/R_C$ and a $V_{CE} = 0$ V?

    a. Saturation point   c. Q point
    b. Cutoff point   d. None of the above

35. Which point on the dc load line results in a $V_{CE} = V_{CC}$, and an $I_C = 0$?

    a. Saturation point   c. Q point
    b. Cutoff point   d. None of the above

Communication Skill Questions

36. What is a junction diode? (17-1)

37. Sketch the schematic symbol for a diode and label its terminals. (17-1)

38. Describe the basic operation of a diode. (17-1)

39. Explain in detail the differences between a forward-biased diode and a reverse-biased diode. (17-1)

40. Referring to the junction diode's V–I characteristic curve, describe the: (17-1)

    a. Forward-bias curve   b. Reverse-bias curve

41. Sketch the zener diode schematic symbol. (17-2)

42. Define the following terms: (17-2)

    a. Reverse leakage current
    b. Zener breakdown voltage
    c. Zener knee current
    d. Zener current
    e. Zener voltage
    f. Zener impedance
    g. Maximum zener power dissipation
    h. Zener test current
    i. Maximum zener current

43. LED is an abbreviation for _____ . (17-3)

44. Sketch the schematic symbol used to represent an LED. (17-3)

45. What elements are contained in a typical LED package? (17-3)

Practice Problems

56. Which of the silicon diodes in Figure 17-37 are forward biased and which are reverse biased?

57. Calculate I_F for the circuits in Figure 17-38.

58. What would be the voltage drop across each of the resistors in Figure 17-38?

46. Briefly describe how the BJT is used in: (17-4-4)

    a. Digital circuit applications
    b. Analog circuit applications

47. Sketch and briefly describe the operation of: (17-4-4)

    a. A transistor logic gate   b. A transistor amplifier

48. Briefly describe how a transistor achieves a gain in voltage between input and output. (17-4-4)

49. What is the relationship between a bipolar transistor's emitter current, base current, and collector current? (17-5-1)

50. Why is the bipolar transistor known as a current-controlled device? (17-5-1)

51. In normal operation, which of the bipolar transistor's P-N junctions or diodes is forward biased, and which is reverse biased? (17-5-1)

52. What is the bipolar transistor equivalent to when it is operated in: (17-5-1)

    a. Saturation   b. The active region   c. Cutoff

53. Briefly describe the following terms: (17-5-2)

    a. DC beta   c. DC alpha
    b. AC beta   d. AC alpha

54. What are the collector characteristic curves? (17-5-2)

55. Why is the common-emitter circuit configuration the most widely used? (17-5-2)

59. In reference to the polarity of the applied voltage, which of the circuits in Figure 17-39 are correctly biased for normal zener operation?

60. In reference to the magnitude of the applied voltage, which of the circuits in Figure 17-39 are correctly biased for normal zener operation?

61. Calculate the value of circuit current for each of the zener diode circuits in Figure 17-39.

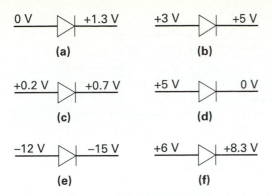

FIGURE 17-37 Biased Junction Diodes.

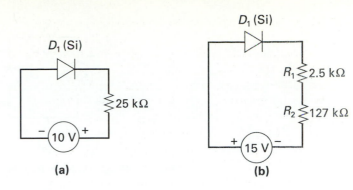

FIGURE 17-38 Forward Current Examples.

62. Would a 1 watt zener diode have a suitable power dissipation rating for the circuit in Figure 17-39(a)?

63. What wattage or maximum power dissipation rating would you choose for the zener diode in Figure 17-39(e)?

64. What would be the regulated output voltage and polarity at points X and Y for the circuits shown in Figure 17-40 (a) and (b)?

65. Which of the light-emitting diodes in Figure 17-41 are biased correctly?

66. Calculate the value of circuit current for each of the LEDs in Figure 17-41.

67. Would a maximum forward current rating of 18 mA be adequate for the LED in Figure 17-41(a)?

68. Would a maximum reverse voltage rating of 5 V be adequate for the LEDs in Figure 17-41(c)?

69. Identify the type and terminals of the transistors shown in Figure 17-42.

70. A bipolar transistor is correctly biased for operation in the active region when its emitter diode is forward biased and its collector diode is reverse biased. In Figure 17-43, which of the bipolar transistor circuits is correctly biased?

71. Calculate the value of the missing current in the following examples:

    a. $I_E = 25$ mA, $I_C = 24.6$ mA, $I_B = ?$
    b. $I_B = 600$ µA, $I_C = 14$ mA, $I_E = ?$
    c. $I_E = 4.1$ mA, $I_B = 56.7$ µA, $I_C = ?$

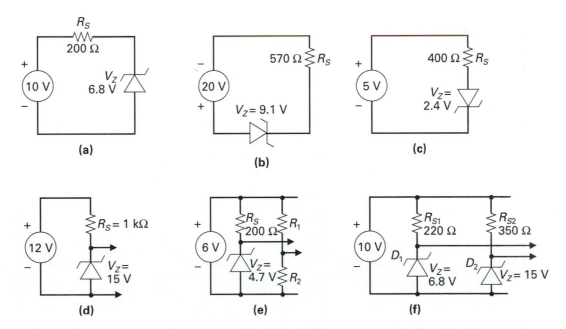

FIGURE 17-39 Biasing Voltage Polarity and Magnitude.

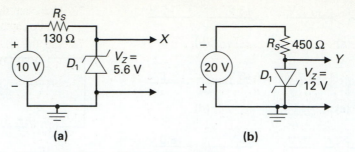

FIGURE 17-40   Voltage Regulator Circuits (Unloaded).

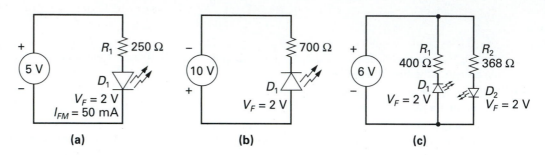

FIGURE 17-41   Biasing Light-Emitting Diodes.

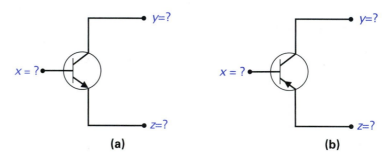

FIGURE 17-42   Identify the Transistor Type and Terminals.

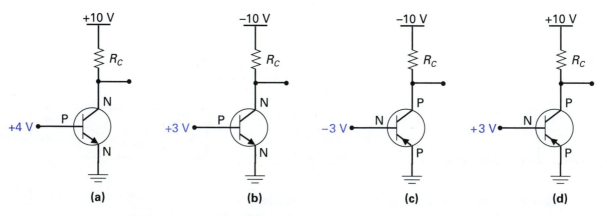

FIGURE 17-43   Identifying the Correctly Biased (Active Region) Bipolar Transistors.

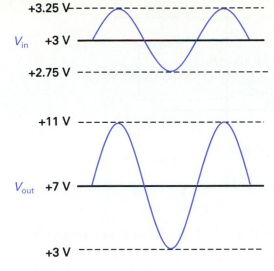

FIGURE 17-44 Transistor Amplifier
Input/Output Waveforms.

72. Calculate the voltage gain (A_V) of the transistor amplifier whose input/output waveforms are shown in Figure 17-44.

73. Identify the configuration of the actual bipolar transistor electronic system circuits shown in Figure 17-45.

74. Identify the bipolar transistor type and the biasing technique used in each of the circuits in Figure 17-45.

75. Calculate the following for the base biased transistor circuit shown in Figure 17-46:

    a. I_B **b.** I_C **c.** V_{CE}

76. Sketch the dc load line for the circuit in Figure 17-46, showing the saturation, cutoff, and Q points.

77. Calculate the following for the voltage-divider biased transistor circuit shown in Figure 17-47:

    a. V_B and V_E **b.** I_C **c.** V_C **d.** V_{CE}

78. Sketch the dc load line for the circuit in Figure 17-47, showing the saturation, cutoff, and Q points.

79. In Figure 17-47, does Kirchhoff's voltage law apply to the voltage divider made up of R_C, R_{CE}, and R_E?

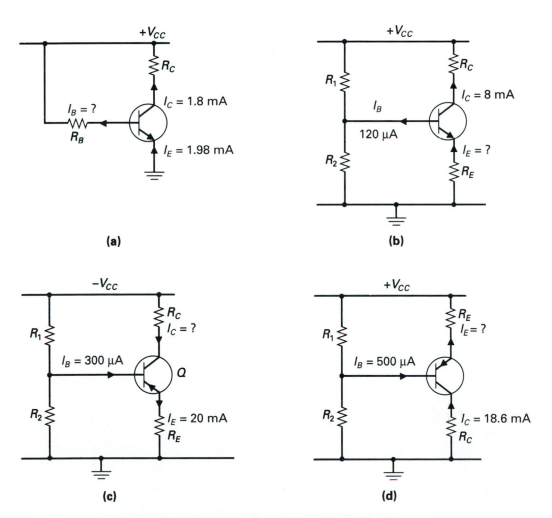

FIGURE 17-45 Identifying the Configuration of Actual BJT Circuits.

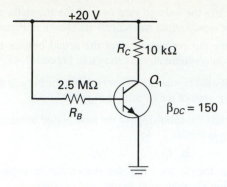

FIGURE 17-46 **Base-Biased Transistor Circuit.**

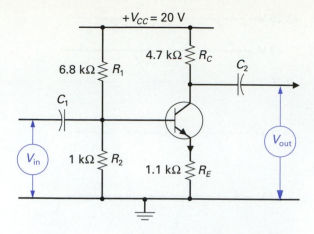

FIGURE 17-47 **Voltage-Divider Biased Transistor Circuit.**

Web Site Questions

Go to the web site http://www.prenhall.com/cook, select the textbook *Mathematics for Electronics and Comput-* *ers,* select this chapter, and then follow the instructions when answering the multiple-choice practice problems.

Analog to Digital

Moon Walk

In July of 1969 almost everyone throughout the world was caught at one stage or another gazing up to the stars and wondering. For U.S. astronaut Neil Armstrong, the age-old human dream of walking on the moon was close to becoming a reality. On earth, millions of people and thousands of newspapers and magazines waited anxiously to celebrate a successful moon landing. Then, the rather brief but eloquent message from Armstrong was transmitted 240,000 miles to Houston, Texas, where it was immediately retransmitted to a waiting world. This message was: "That's one small step for man, one giant leap for mankind." These words were received by many via television, but for magazines and newspapers this entire mission, including the speech from Armstrong, was converted into a special code made up of ON–OFF pulses that traveled from computer to computer. Every letter of every word was converted into a code that used the two symbols of the binary number system, 0 and 1. This code is still used extensively by all modern computers and is called *American Standard Code for Information Interchange,* or ASCII (pronounced "askey").

It is only fitting that these codes made up of zeros and ones conveyed the finale to this historic mission, since they played such an important role throughout. Commands encoded into zeros and ones were used to control almost everything, from triggering the take-off to keeping the spacecraft at the proper angle for reentry into the earth's atmosphere.

No matter what its size or application, a digital electronic computer is quite simply a system that manages the flow of information in the form of zeros and ones. Referring to the ASCII code table in Figure 18-2 on p. 594, see if you can decode the following famous Armstrong message that reads from left to right:

```
0100010  1010100  1101000  1100001  1110100  0100111  1110011  0100000  1101111  1101110
1100101  0100000  1110011  1101101  1100001  1101100  1101100  0100000  1110011  1110100
1100101  1110000  0100000  1100110  1101111  1110010  0100000  1100001  0100000  1101101
1100001  1101110  0101100  0100000  1101111  1101110  1100101  0100000  1100111  1101101
1100001  1101110  1110100  0100000  1101100  1100101  1100001  1110000  0100000  1100110
1101111  1110010  0100000  1101101  1100001  1101110  1101011  1101001  1101110  1100100
0101110  0100010  0100000  0100000  0101101  1001110  1100101  1101001  1101100  0100000
1000001  1110010  1101101  1110011  1110100  1110010  1101111  1101110  1100111  0101100
0100000  1000001  1110000  1101111  1101100  1101100  1101111  0100000  0110001  0110001
```

18

Outline and Objectives

VIGNETTE: MOON WALK

INTRODUCTION

18-1 ANALOG AND DIGITAL DATA AND DEVICES

Objective 1: Define the differences between analog and digital data and devices.

Objective 2: Explain why analog circuits are also referred to as linear circuits.

Objective 3: Describe the function of the digital ASCII code.

Objective 4: Explain why the 10-state digital system was replaced by the two-state digital system.

18-1-1 Analog Data and Devices

18-1-2 Digital Data and Devices

18-2 ANALOG AND DIGITAL SIGNAL CONVERSION

Objective 5: List the three basic blocks within a computer.

Objective 6: Describe how analog and digital devices are connected to a computer.

Objective 7: Define the function of the analog-to-digital converter (ADC) and the digital-to-analog converter (DAC).

18-2-1 The Computer Interface

MULTIPLE CHOICE QUESTIONS

COMMUNICATION SKILL QUESTIONS

WEB SITE QUESTIONS

Introduction

Since World War II, no branch of science has contributed more to the development of the modern world than electronics. It has stimulated dramatic advances in the fields of communication, computing, consumer products, industrial automation, test and measurement, and health care. It has now become the largest single industry in the world, exceeding the automobile and oil industries, with annual sales of electronic systems exceeding $2 trillion.

One of the most important trends in this huge industry has been a gradual shift from *analog electronics* to *digital electronics*. This movement began in the 1960s and is almost complete today. In fact, a recent statistic stated that, on average, 90% of the circuitry within electronic systems is now digital and only 10% is analog. This digitalization of the electronics industry is bringing sectors that were once separate closer together. For example, two of the largest sectors or branches of electronics are *computing* and *communications.* Being able to communicate with each other using the common language of digital has enabled computers and communications to interlink so that computers can now function within communication-based networks, and communications networks can now function through computer-based systems. Industry experts call this merging *convergence* and predict that digital electronics will continue to unite the industry and stimulate progress in practically every field of human endeavor.

Needless to say, this part of the book you are about to read involving digital electronic mathematics and concepts is an essential element in your study of electronics.

18-1 ANALOG AND DIGITAL DATA AND DEVICES

Before we begin, let's first review analog electronics and then compare it to digital, so that we can clearly see the differences between the two.

18-1-1 *Analog Data and Devices*

Figure 18-1(a) shows an electronic circuit designed to amplify speech information detected by a microphone. One of the easiest ways to represent data or information is to have a voltage change in direct proportion to the information it is representing. In the example in Figure 18-1(a), the *pitch and loudness* of the sound waves applied to the microphone should control the *frequency and amplitude* of the voltage signal from the microphone. The output voltage signal from the microphone is said to be an analog of the input speech signal. The word **analog** means "similar to," and in Figure 18-1(a) the electronic signal produced by the microphone is an analog (or similar) to the speech signal, since a change in speech "loudness or pitch" will cause a corresponding change in signal voltage "amplitude or frequency."

In Figure 18-1(b), a light detector or solar cell converts light energy into an electronic signal. This information signal represents the amount of light present, since changes in voltage amplitude result in a change in light-level intensity. Once again, the output electronic signal is an analog (or similar) to the sensed light level at the input.

Figure 18-1, therefore, indicates two analog circuits. The microphone in Figure 18-1(a) generates an **ac analog signal** that is amplified by an **ac amplifier circuit.** The microphone is considered an **analog device or component** and the amplifier an **analog circuit.** The light detector in Figure 18-1(b) would also be an analog component; however, in this example it generates a **dc analog signal** that is amplified by a **dc amplifier circuit.**

Analog

The representation of physical properties by a proportionally varying signal.

AC Analog Signal

An analog signal that alternates positive and negative.

AC Amplifier

An amplifier designed to increase the magnitude of an ac signal.

Analog Circuit

A circuit designed to manage analog signals.

Analog Device or Component

Advice or component that makes up a circuit designed to manage analog signals.

DC Analog Signal

An analog signal that is always either positive or negative.

DC Amplifier

An amplifier designed to increase the magnitude of a dc signal.

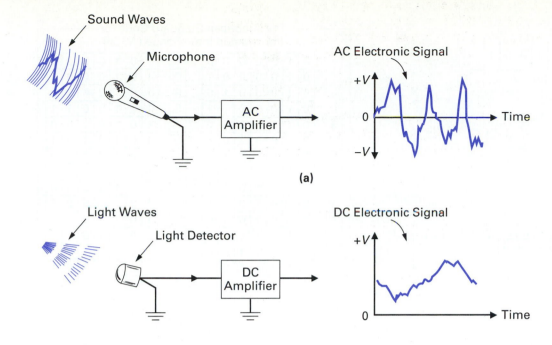

(a)

FIGURE 18-1 Analog Data and Devices.

Both of the information signals in Figure 18-1 vary smoothly and continuously, in accordance with the natural quantities they represent (sound and light). Analog circuits are often called **linear circuits,** since linear by definition is *the variation of an output in direct proportion to the input.* This linear circuit response is evident in Figure 18-1, where you can see that in both the dc and ac circuit, the output signal voltage varies in direct proportion to the sound or light signal input.

18-1-2 *Digital Data and Devices*

In **digital electronic circuits,** information is first converted into a coded group of pulses. To explain this concept, let's take a closer look at the example shown in Figure 18-2(a). This code consists of a series of HIGH and LOW voltages, in which the HIGH voltages are called "1s" (ones) and the LOW voltages are called "0s" (zeros). Figure 18-2(b) lists the ASCII code, which is one example of a digital code. Referring to Figure 18-2(a), you will notice that the "1101001" information or data stream code corresponds to lowercase *i* in the ASCII table shown highlighted in Figure 18-2(b). Computer keyboards are one of many devices that make use of the digital ASCII code. In Figure 18-2(c), you can see how the lowercase *i* ASCII code is generated whenever the *i* key is pressed, encoding the information *i* into a group of pulses (1101001).

The next question you may have is: Why do we go to all this trouble to encode our data or information into these two-state codes? The answer can best be explained by examining history. The early digital systems constructed in the 1950s made use of a decimal code that used 10 levels or voltages, with each of these voltages corresponding to one of the 10 digits in the decimal number system (0 = 0 V, 1 = 1 V, 2 = 2 V, 3 = 3 V, up to 9 = 9 V). The circuits that had to manage these decimal codes, however, were very complex since they had to generate one of 10 voltages and sense the difference among all 10 voltage levels. This complexity led to inaccuracy, since some circuits would periodically confuse one voltage level for a different voltage level. *The solution to the problem of circuit complexity and inaccuracy was solved by a adopting a two-state system instead of a 10-state system.* Using a two-state or two-digit system, you can generate codes for any number, letter, or symbol, as we have seen

Linear Circuit
A circuit in which the output varies in direct proportion to the input.

Digital Eletonic Circuit
A circuit designed to manage digital information signals.

HIGH (+ V)
LOW (0 V)
1 1 0 1 0 0 1
(a)

KEYBOARD (ASCII Code Generator) → COMPUTER

(c)

The American Standard Code for Information Interchange (ASCII)

Char	b	b	b	b	b	b	b		Char	b	b	b	b	b	b	b
SP	0	1	0	0	0	0	0		P	1	0	1	0	0	0	0
!	0	1	0	0	0	0	1		Q	1	0	1	0	0	0	1
"	0	1	0	0	0	1	0		R	1	0	1	0	0	1	0
#	0	1	0	0	0	1	1		S	1	0	1	0	0	1	1
$	0	1	0	0	1	0	0		T	1	0	1	0	1	0	0
%	0	1	0	0	1	0	1		U	1	0	1	0	1	0	1
&	0	1	0	0	1	1	0		V	1	0	1	0	1	1	0
'	0	1	0	0	1	1	1		W	1	0	1	0	1	1	1
(	0	1	0	1	0	0	0		X	1	0	1	1	0	0	0
)	0	1	0	1	0	0	1		Y	1	0	1	1	0	0	1
*	0	1	0	1	0	1	0		Z	1	0	1	1	0	1	0
+	0	1	0	1	0	1	1		[	1	0	1	1	0	1	1
,	0	1	0	1	1	0	0		/	1	0	1	1	1	0	0
−	0	1	0	1	1	0	1		]	1	0	1	1	1	0	1
.	0	1	0	1	1	1	0		^	1	0	1	1	1	1	0
/	0	1	0	1	1	1	1		_	1	0	1	1	1	1	1
0	0	1	1	0	0	0	0		'	1	1	0	0	0	0	0
1	0	1	1	0	0	0	1		a	1	1	0	0	0	0	1
2	0	1	1	0	0	1	0		b	1	1	0	0	0	1	0
3	0	1	1	0	0	1	1		c	1	1	0	0	0	1	1
4	0	1	1	0	1	0	0		d	1	1	0	0	1	0	0
5	0	1	1	0	1	0	1		e	1	1	0	0	1	0	1
6	0	1	1	0	1	1	0		f	1	1	0	0	1	1	0
7	0	1	1	0	1	1	1		g	1	1	0	0	1	1	1
8	0	1	1	1	0	0	0		h	1	1	0	1	0	0	0
9	0	1	1	1	0	0	1		**i**	**1**	**1**	**0**	**1**	**0**	**0**	**1**
:	0	1	1	1	0	1	0		j	1	1	0	1	0	1	0
;	0	1	1	1	0	1	1		k	1	1	0	1	0	1	1
<	0	1	1	1	1	0	0		l	1	1	0	1	1	0	0
=	0	1	1	1	1	0	1		m	1	1	0	1	1	0	1
>	0	1	1	1	1	1	0		n	1	1	0	1	1	1	0
?	0	1	1	1	1	1	1		o	1	1	0	1	1	1	1
@	1	0	0	0	0	0	0		p	1	1	1	0	0	0	0
A	1	0	0	0	0	0	1		q	1	1	1	0	0	0	1
B	1	0	0	0	0	1	0		r	1	1	1	0	0	1	0
C	1	0	0	0	0	1	1		s	1	1	1	0	0	1	1
D	1	0	0	0	1	0	0		t	1	1	1	0	1	0	0
E	1	0	0	0	1	0	1		u	1	1	1	0	1	0	1
F	1	0	0	0	1	1	0		v	1	1	1	0	1	1	0
G	1	0	0	0	1	1	1		w	1	1	1	0	1	1	1
H	1	0	0	1	0	0	0		x	1	1	1	1	0	0	0
I	1	0	0	1	0	0	1		y	1	1	1	1	0	0	1
J	1	0	0	1	0	1	0		z	1	1	1	1	0	1	0
K	1	0	0	1	0	1	1		(	1	1	1	1	0	1	1
L	1	0	0	1	1	0	0		¦	1	1	1	1	1	0	0
M	1	0	0	1	1	0	1		)	1	1	1	1	1	0	1
N	1	0	0	1	1	1	0		~	1	1	1	1	1	1	0
O	1	0	0	1	1	1	1		DEL	1	1	1	1	1	1	1

(b)

FIGURE 18-2 **Two-State Digital Information. (a) Information or Data Code for *i*. (b) ASCII Code. (c) Keyboard ASCII Generator.**

in the ASCII table in Figure 18-2. The electronic circuits that manage these two-state codes are less complex, since they have to generate and sense only either a HIGH or LOW voltage. In addition, two-state circuits are much more accurate, since there is little room for error between the two extremes of ON and OFF, or HIGH voltage and LOW voltage.

Abandoning the 10-state system and adopting the two-state system for the advantages of circuit simplicity and accuracy meant that we were no longer dealing with the decimal number system. As a result, having only two digits (0 and 1) means that we are now operating in the two-state number system, which is called **binary.** Figure 18-3(a) shows how the familiar decimal system has 10 digits or levels labeled 0 through 9, and Figure 18-3(b) shows how we could electronically represent each decimal digit. With the base 2, or binary

Binary

Having only two alternatives, two-state.

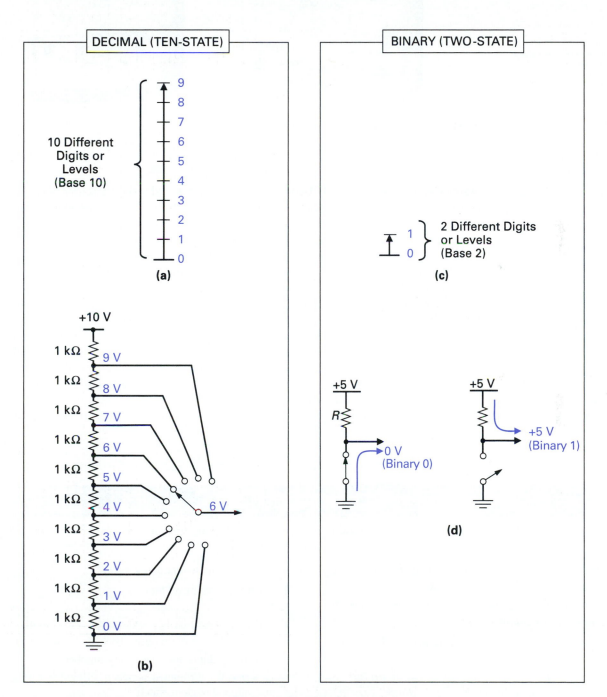

FIGURE 18-3   **Electronically Representing the Digits of a Number System.**

FIGURE 18-4 **Analog and Digital Readouts.**

Binary Digit

Abbreviated bit; it is either a 0 or 1 from the binary number system.

Bit

Binary digit.

Digital Signal

An electronic signal made up of binary digits.

Digitize

To convert an analog signal to a digital signal.

Analog Readout Multimeter

A multimeter that uses a movement on a calibrated scale to indicate a value.

Digital Readout Multimeter

A multimeter that uses digits to indicate a value.

number system, we have only two digits in the number scale, and therefore only the digits 0 (zero) and 1 (one) exist in binary, as shown in Figure 18-3(c). These two states are typically represented in an electronic circuit as two different values of voltage (binary 0 = LOW voltage; binary 1 = HIGH voltage), as shown in Figure 18-3(d).

Using combinations of **binary digits** (abbreviated as **bits**) we can represent information as a binary code. This code is called a **digital signal** because it is an *information signal* that makes use of *binary digits*. Today, almost all information, from your telephone conversations to the music on your compact discs, is **digitized** or converted to binary data form.

You will probably hear the terms *analog* and *digital* used to describe the difference between the two types of meters shown in Figure 18-4. On the **analog readout multimeter** shown in Figure 18-4(a), the amount of pointer deflection across the scale is an analog (or similar) to the magnitude of the electrical property being measured. On the other hand, with the **digital readout multimeter** shown in Figure 18-4(b), the magnitude of the electrical property being measured is displayed using digits, which in this case are decimal digits.

SELF-TEST EVALUATION POINT FOR SECTION 18-1

Now that you have completed this section, you should be able to:

■ *Objective 1.* *Define the differences between analog and digital data and devices.*

■ *Objective 2.* *Explain why analog circuits are also referred to as linear circuits.*

■ *Objective 3.* *Describe the function of the digital ASCII code.*

■ *Objective 4.* *Explain why the 10-state digital system was replaced by the two-state digital system.*

Use the following questions to test your understanding of Section 18-1:

1. What does the word *analog* mean?
2. What is an analog signal?
3. What is the ASCII code?
4. What were the two key reasons for adopting the two-state binary system in place of the 10-state decimal system?
5. How many digits does the binary number system have?
6. Describe the difference between an analog readout watch and a digital readout watch.

18-2 ANALOG AND DIGITAL SIGNAL CONVERSION

Whenever you have two different forms, like analog and digital, there is always a need to convert back and forth between the two. Computers are used at the heart of almost every electronic system today because of their ability to quickly process and store large amounts of data, make systems more versatile, and perform many functions. Many of the input signals applied to a computer for storage or processing are analog in nature, and therefore a data conversion circuit is needed to interface these analog inputs with a digital system. Similarly, a data conversion circuit may also be needed to interface the digital computer system with an analog output device.

Figure 18-5 shows a simplified block diagram of a computer. No matter what the application, a computer has three basic blocks: a microprocessor unit, a memory unit, and an input/output unit.

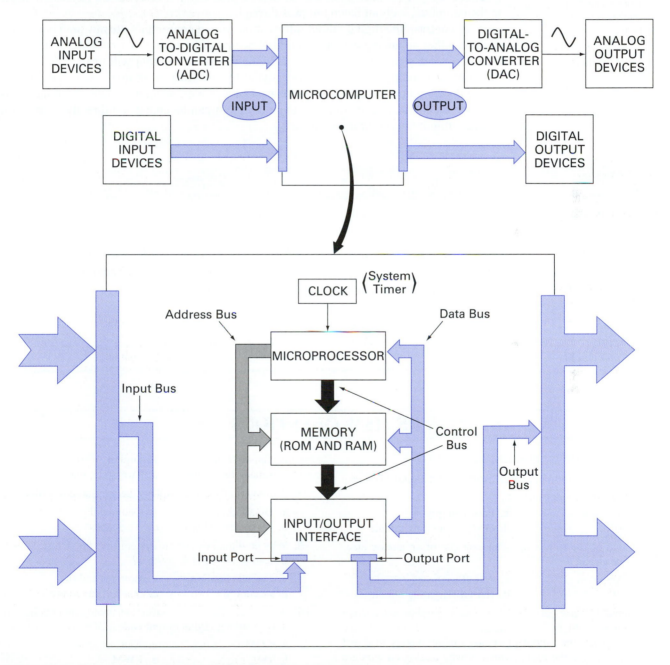

FIGURE 18-5 Connecting Analog and Digital Devices to a Computer.

Analog-to-Digital Converter

A circuit that converts analog input signals into equivalent digital output signals.

Input Port

An electronic doorway used for receiving information into a computer.

Output Port

An electronic doorway used for transmitting information out of a computer.

Digital-to-Analog Converter

A circuit that converts digital input signals into equivalent analog output signals.

Getting information into the computer is the job of input devices, and as you can see in Figure 18-5, there are two input paths into a computer: one path from analog input devices, and the other path from digital input devices. Information in the form of light, sound, heat, pressure, or any other real-world quantity is analog in nature and has an infinite number of input levels. Sensors or transducers of this type, such as photoelectric cells for light or microphones for sound, will generate analog signals. Since the computer operates on information in only digital or binary form, a translation or conversion is needed if the computer is going to be able to understand or interpret an analog signal. This signal processing is achieved by an **analog-to-digital converter (ADC),** which transforms the varying analog input voltage into equivalent digital codes that the computer can understand. Once the analog information has been encoded into digital form, the information can then enter the computer through an electronic doorway called the **input port.** Some input devices, such as a computer keyboard, automatically generate a digital code for each key that is pressed (ASCII codes). The codes generated by these digital input devices can be connected directly to the computer, without having to pass through a converter.

In contrast, the digital output information from a computer's **output port** can be used to drive either an analog device or a digital device. If an analog output is desired, as in the case of a voice signal for a speaker, the digital computer's output will have to be converted into a corresponding equivalent analog signal by a **digital-to-analog converter (DAC).** If only a digital output is desired, as in the case of a printer, which translates the binary codes into printed characters, the digital computer output can be connected directly to the output device without the need for a converter.

SELF-TEST EVALUATION POINT FOR SECTION 18-2

Now that you have completed this section, you should be able to:

■ *Objective 5.* *List the three basic blocks within a computer.*

■ *Objective 6.* *Describe how analog and digital devices are connected to a computer.*

■ *Objective 7.* *Define the function of the analog-to-digital converter (ADC) and the digital-to-analog converter (DAC).*

Use the following questions to test your understanding of Section 18-2:

1. What are the three basic blocks of a computer?
2. Give the full names of the following acronyms:
   a. ADC   b. DAC
3. An ADC will convert a/an _____ input into an equivalent _____ output.
4. A DAC will convert a/an _____ input into an equivalent _____ output.

REVIEW QUESTIONS

Multiple Choice Questions

1. Which of the following is an example of a digital electronic circuit?
   a. Amplifier
   b. MPU
   c. Power supply
   d. ON/OFF lighting circuit

2. Which of the following is an example of an analog electronic circuit?
   a. Amplifier
   b. CPU
   c. ASCII keyboard
   d. ON/OFF lighting circuit

3. Digital circuits based on the binary number system are often referred to as _____ circuits. On the other hand, the output of analog circuits varies in direct proportion to the input, which is why analog circuits are often called _____ circuits.

   a. Ten-state, linear
   b. Linear, two-state
   c. Digital, 10-state
   d. Two-state, linear

4. A digital multimeter displays the measured quantity using
   a. Binary digits
   b. A pointer and scale
   c. Decimal digits
   d. An abacus

5. Why do digital electronic systems use the two-state system instead of the 10-state system?
   a. Circuit simplicity
   b. Accuracy
   c. Both (a) and (b)
   d. None of the above

6. A _____ converts an analog input voltage into a proportional digital output code.
   a. ROM
   b. DAC
   c. ADC
   d. RWM

7. A _____ converts a digital input code into a proportional analog output voltage.

  a. ROM   **c.** ADC

  b. DAC   **d.** RWM

8. What are the three basic blocks of a computer?

  a. ADC, I/O, memory   **c.** MPU, I/O, memory

  b. DAC, ADC, I/O   **d.** ROM, MPU, I/O

Communication Skill Questions

9. Define the following: (18-1)

  a. Analog   **c.** Binary

  b. Convergence   **d.** Linear

10. Why do digital systems convert information into binary codes instead of decimal codes? (18-1-2)

11. Why are analog circuits also called linear circuits? (18-1-1)

12. Describe the differences between an analog and digital readout display. (18-1-1)

13. What are the three basic blocks of a computer? (18-2)

14. Give the full names of the following abbreviations and acronyms:

  a. Bit   **c.** ADC

  b. ASCII   **d.** DAC

15. What is the function of the ADC and the DAC? (18-2)

Web Site Questions

Go to the web site http://www.prenhall.com/cook, select the textbook *Mathematics for Electronics and Computers,* select this chapter, and then follow the instructions when answering the multiple-choice practice problems.

Number Systems and Codes

Leibniz's Language of Logic

Gottfried Wilhelm Leibniz was born in Leipzig, Germany, in 1646. His father was a professor of moral philosophy and spent much of his time discussing thoughts and ideas with his son. Tragically, his father died when Leibniz was only six, and from that time on, Leibniz spent hour upon hour in his late father's library reading all his books.

By the age of 12, Leibniz had taught himself history, Latin, and Greek. At 15 he entered the University of Leipzig, where he came across the works of scholars such as Johannes Kepler and Galileo. The new frontiers of science fascinated him, so he added mathematics to his curriculum.

In 1666, while finishing his university studies, the 20-year-old Leibniz wrote what he modestly called a schoolboy essay, titled "DeArte Combinatoria," which translated means "On the Art of Combination." In this work, he described how all thinking of any sort on any subject could be reduced to exact mathematical statements. Logic, or as he called it, the laws of thought, could be converted from the verbal realm, which is full of ambiguities, into precise mathematical statements. In order to achieve this, however, Leibniz stated that a "universal language" would be needed. Most of his professors found the paper either baffling or outrageous, and this caused Leibniz not to pursue the idea any further.

After graduating from the university, Leibniz was offered a professorship at the University of Nuremberg, which he turned down for a position as an international diplomat. This career proved not as glamorous as he imagined, since most of his time was spent in uncomfortable horse-drawn coaches traveling between European capitals.

In 1672 his duties took him to Paris, where he met Dutch mathematician and astronomer Christian Huygens. After seeing the hours that Huygens spent on endless computations, Leibniz set out to develop a mechanical calculator. A year later Leibniz unveiled the first machine that could add, subtract, multiply, and divide decimal numbers.

In 1676 Leibniz began to concentrate more on mathematics. It was at this time that he invented calculus, which was also independently discovered by Isaac Newton in England. Leibniz's focus, however, was on the binary number system, which preoccupied him for years. He worked tirelessly to document the long combinations of 1s and 0s that make up the modern binary number system, and perfected binary arithmetic. Ironically, for all his genius, Leibniz failed to make the connection between his 1666 essay and binary, which was the universal language of logic that he was seeking.

This discovery would remain hidden until another self-taught mathematician named George Boole uncovered it a century and a quarter after Leibniz's death in 1716.

Outline and Objectives

VIGNETTE: LEIBNIZ'S LANGUAGE OF LOGIC

INTRODUCTION

19-1   THE DECIMAL NUMBER SYSTEM

Objective 1: Explain how different number systems operate.

Objective 2: Describe the differences between the decimal and binary number systems.

Objective 3: Explain the positional weight and the reset and carry action for decimal.

19-1-1 Positional Weight

19-1-2 Reset and Carry

19-2   THE BINARY NUMBER SYSTEM

Objective 4: Explain the positional weight, the reset and carry action, and conversion for binary.

19-2-1 Positional Weight

19-2-2 Reset and Carry

19-2-3 Converting Binary Numbers to Decimal Numbers

19-2-4 Converting Decimal Numbers to Binary Numbers

19-2-5 Converting Information Signals

19-3   THE HEXADECIMAL NUMBER SYSTEM

Objective 5: Explain the positional weight, the reset and carry action, and conversion for hexadecimal.

19-3-1 Converting Hexadecimal Numbers to Decimal Numbers

19-3-2 Converting Decimal Numbers to Hexadecimal Numbers

19-3-3 Converting between Binary and Hexadecimal

19-4   THE OCTAL NUMBER SYSTEM

Objective 6: Explain the positional weight, the reset and carry action, and conversion for octal.

19-4-1 Converting Octal Numbers to Decimal Numbers

19-4-2 Converting Decimal Numbers to Octal Numbers

19-4-3 Converting between Binary and Octal

19-5   BINARY CODES

Objective 7: Describe the following binary codes: binary coded decimal (BCD), the excess-3 code, the gray code, and the American Standard Code for Information Interchange (ASCII).

19-5-1 The Binary Coded Decimal (BCD) Code

19-5-2 The Excess-3 Code

19-5-3 The Gray Code

19-5-4 The American Standard Code for Information Interchange (ASCII)

MULTIPLE CHOICE QUESTIONS

COMMUNICATION SKILL QUESTIONS

PRACTICE PROBLEMS

WEB SITE QUESTIONS

Introduction

One of the best ways to understand anything new is to compare it with something that you are already familiar with so that the differences are highlighted. In this chapter we will be examining the *binary number system,* which is the language used by digital computer electronic circuits. To best understand this new number system, we will compare it with the number system most familiar to you, the *decimal* (base 10) *number system.* Although this system is universally used to represent quantities, many people do not fully understand its weighted structure. This review of decimal has been included so that you can compare the base 10 number system with the system used internally by digital electronics, the binary (base 2) number system.

In addition to binary, two other number systems are widely used in conjunction with digital electronics: the octal (base 8) number system, and the hexadecimal (base 16) number system. These two number systems are used as a type of shorthand for large binary numbers and enable us to represent a large group of binary digits with only a few octal or hexadecimal digits.

The last section in this chapter covers many of the different binary codes, which use a combination of 1s and 0s to represent letters, numbers, symbols, and other information.

Keep in mind throughout this chapter that different number systems are just another way to count, and as with any new process, you will become more proficient and more at ease with practice as we proceed through the chapters that follow.

19-1 THE DECIMAL NUMBER SYSTEM

The decimal system of counting and keeping track of items was first created by Hindu mathematicians in India in A.D. 400. Since it involved the use of fingers and thumbs, it was natural that this system would have 10 digits. The system found its way to all the Arab countries by A.D. 800, where it was named the Arabic number system, and from there it was eventually adopted by nearly all the European countries by A.D. 1200, where it was called the **decimal number system.**

The key feature that distinguishes one number system from another is the number system's **base** or *radix.* This base indicates the number of digits that will be used. The decimal number system, for example, is a base 10 number system, which means that it uses 10 digits (0 through 9) to communicate information about an amount. A subscript is sometimes included after a number when different number systems are being used, to indicate the base of the number. For example, $12{,}567_{10}$ is a base 10 number, whereas 10110_2 is a base 2 number.

19-1-1 *Positional Weight*

The position of each digit of a decimal number determines the weight of that digit. A 1 by itself, for instance, is worth only 1, whereas a 1 to the left of three 0s makes the 1 worth 1000.

In decimal notation, each position to the left of the decimal point indicates an increased positive power of 10, as seen in Figure 19-1(a). The total quantity or amount of the number is therefore determined by the size and the weighted position of each digit. For ex-

Decimal Number System

A base 10 number system.

Base

With number systems, it describes the number of digits used.

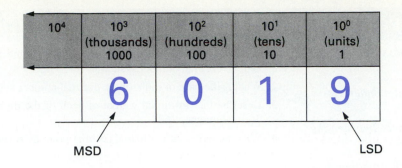

$$\text{Value of number} = (6 \times 10^3) + (0 \times 10^2) + (1 \times 10^1) + (9 \times 10^0)$$
$$= (6 \times 1000) + (0 \times 100) + (1 \times 10) + (9 \times 1)$$
$$= 6000 + 0 + 10 + 9 = 6019_{10}$$

(a)

(b)   **(c)**

FIGURE 19-1   **The Decimal (Base 10) Number System.**

ample, the value shown in Figure 19-1(a) has six thousands, zero hundreds, one ten, and nine ones, which combined makes a total of 6019_{10}.

In the decimal number system, the leftmost digit is called the **most significant digit (MSD),** and the rightmost digit is called the **least significant digit (LSD).** Applying these definitions to the example in Figure 19-1(a), we see that the 6 is the MSD, since its position carries the most weight, and the 9 is the LSD, since its position carries the least weight.

19-1-2   *Reset and Carry*

Before proceeding to the binary number system, let us review one other action that occurs when counting in decimal. This action, which is familiar to us all, is called **reset and carry.** Referring to Figure 19-1(b), you can see that a reset and carry operation occurs after a count of 9. The units column, which has reached its maximum count, resets to 0 and carries a 1 to the tens column, resulting in a final count of 10.

This reset and carry action will occur in any column that reaches its maximum count. For example, Figure 8-1(c) shows how two reset and carry operations will take place after a count of 99. The units column resets and carries a 1 to the tens column, which in turn resets and carries a 1 to the hundreds column, resulting in a final count of 100.

Most Significant Digit (MSD)

The leftmost, largest-weight digit in a number.

Least Significant Digit (LSD)

The rightmost, smallest-weight digit in a number.

Reset and Carry

An action that occurs when a column has reached its maximum count.

Now that you have completed this section, you should be able to:

■ **Objective 1.** *Explain how different number systems operate.*

■ **Objective 2.** *Describe the differences between the decimal and binary number systems.*

■ **Objective 3.** *Explain the positional weight and the reset and carry action for decimal.*

Use the following questions to test your understanding of Section 19-1:

1. What is the difference between the Arabic number system and the decimal number system?

2. What is the base or radix of the decimal number system?

3. Describe the positional weight of each of the digits in the decimal number 2639.

4. What action occurs when a decimal column advances beyond its maximum count?

19-2   THE BINARY NUMBER SYSTEM

Binary Point

A symbol used to separate the whole from the fraction in a binary number.

As in the decimal system, the value of a binary digit is determined by its position relative to the other digits. In the decimal system, each position to the left of the decimal point increases by a power of 10. Similarly, in the binary number system, since it is a base 2 (bi) number system, each place to the left of the **binary point** increases by a power of 2. Figure 19-2 shows how columns of the binary and decimal number systems have different weights. For example, with binary, the columns are weighted so that 2^0 is one, 2^1 is two, 2^2 is four, 2^3 is eight ($2 \times 2 \times 2 = 8$), and so on.

As we know, the base or radix of a number system also indicates the maximum number of digits used by the number system. The base 2 binary number system uses only the first two digits on the number scale, 0 and 1. The 0s and 1s in binary are called binary digits, or *bits,* for short.

19-2-1   *Positional Weight*

As in the decimal system, each column in binary carries its own weight, as seen in Figure 19-3(a). With the decimal number system, each position to the left increases 10 times. With binary, the weight of each column to the left increases 2 times. The first column, therefore, has a weight of 1, the second column has a weight of 2, the third column 4, the fifth 8, and so on. The value or quantity of a binary number is determined by the digit in each column

FIGURE 19-2   A Comparison of Decimal and Binary.

DECIMAL			BINARY			
10^2	10^1	10^0	2^3	2^2	2^1	2^0
100	10	1	8	4	2	1
		0				0
		1				1
		2			1	0
		3			1	1
		4		1	0	0
		5		1	0	1
		6		1	1	0
		7		1	1	1
		8	1	0	0	0
		9	1	0	0	1
	1	0	1	0	1	0
	1	1	1	0	1	1
	1	2	1	1	0	0

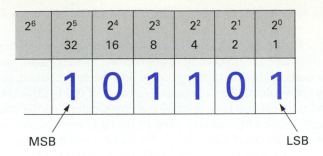

Value of number = (1×2^5) + (0×2^4) + (1×2^3) + (1×2^2) + (1×2^1) + (1×2^0)
= (1×32) + (0×16) + (1×8) + (1×4) + (0×2) + (1×1)
= 32 + 0 + 8 + 4 + 0 + 1
= 45_{10}

(a)

DECIMAL		BINARY			
10^1	10^0	2^3	2^2	2^1	2^0
10	1	8	4	2	1
	0				0
	1				1
	2			1	0
	3			1	1
	4		1	0	0
	5		1	0	1
	6		1	1	0
	7		1	1	1
	8	1	0	0	0
	9	1	0	0	1
1	0	1	0	1	0
1	1	1	0	1	1
1	2	1	1	0	0
1	3	1	1	0	1
1	4	1	1	1	0
1	5	1	1	1	1

(b)

FIGURE 19-3 The Binary (Base 2) Number System.

and the positional weight of the column. For example, in Figure 19-3(a), the binary number 101101_2 is equal in decimal to 45_{10}, since we have 1×32, 1×8, 1×4, and 1×1 (32 + 8 + 4 + 1 = 45). The 0s in this example are not multiplied by their weights (16 and 2), since they are "0" and of no value. The leftmost binary digit is called the most significant bit (MSB) since it carries the most weight, and the rightmost digit is called the least significant bit (LSB), since it carries the least weight. Applying these definitions to the example in Figure 19-3(a), we see that the 1 in the thirty-twos column is the MSB, and the 1 in the units column is the LSB.

19-2-2 *Reset and Carry*

The reset and carry action occurs in binary in exactly the same way as does in decimal; however, since binary has only two digits, a column will reach its maximum digit much sooner, and therefore the reset and carry action in the binary number system will occur much more frequently.

Referring to Figure 19-3(b), you can see that the binary counter begins with 0 and advances to 1. At this stage, the units column has reached its maximum, and therefore the next count forces the units column to reset and carry into the next column, producing a count of 0010_2 (2_{10}). The units column then advances to a count of 0011_2 (3_{10}). At this stage, both the units column and twos column have reached their maximums. As the count advances by 1 it will cause the units column to reset and carry a 1 into the twos column, which will also have to reset and carry a 1 into the fours column. This will result in a final count of 0100_2 (4_{10}). The count will then continue to 0101_2 (5_{10}), 0110_2 (6_{10}), 0111_2 (7_{10}), and then 1000_2 (8_{10}), which, as you can see in Figure 19-3(b), is a result of three reset and carries.

Comparing the binary reset and carry with the decimal reset and carry in Figure 19-3(b), you can see that since the binary number system uses only two digits, binary numbers quickly turn into multidigit figures. For example, a decimal eight (8) uses only one digit, whereas a binary eight (1000) uses four digits.

19-2-3   *Converting Binary Numbers to Decimal Numbers*

Binary numbers can easily be converted to their decimal equivalent by simply adding together all the column weights that contain a binary 1, as we did previously in Figure 19-3(a).

■ **INTEGRATED MATH APPLICATION:**

Convert the following binary numbers to their decimal equivalents.

   a. 1010

   b. 101101

■ *Solution:*

   a.

Binary column weights	32	16	8	4	2	1
Binary number			1	0	1	0

Decimal equivalent $= (1 \times 8) + (0 \times 4) + (1 \times 2) + (0 \times 1)$
$= 8 + 0 + 2 + 0 = 10_{10}$

   b.

Binary column weights	32	16	8	4	2	1
Binary number	1	0	1	1	0	1

Decimal equivalent $= (1 \times 32) + (0 \times 16) + (1 \times 8) + (1 \times 4)$
$+ (0 \times 2) + (1 \times 1)$
$= 32 + 0 + 8 + 4 + 0 + 1 = 45_{10}$

■ **INTEGRATED MATH APPLICATION:**

The LEDs in Figure 19-4 are being used as a 4-bit (4-binary-digit) display. When the LED is OFF, it indicates a binary 0, and when the LED is ON, it indicates a binary 1. Determine the decimal equivalent of the binary displays shown in Figure 19-4(a), (b), and (c).

■ *Solution:*

   a. $0101_2 = 5_{10}$

   b. $1110_2 = 14_{10}$

   c. $1001_2 = 9_{10}$

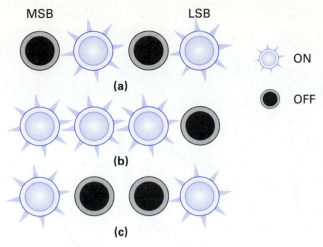

FIGURE 19-4 Using LEDs to Display Binary Numbers.

19-2-4 *Converting Decimal Numbers to Binary Numbers*

To convert a decimal number to its binary equivalent, continually subtract the largest possible power of two until the decimal number is reduced to zero, placing a binary 1 in columns that are used and a binary 0 in the columns that are not used. To see how simple this process is, refer to Figure 19-5, which shows how decimal 53 can be converted to its binary equivalent. As you can see in this example, the first largest power of two that can be subtracted from decimal 53 is 32, and therefore a 1 is placed in the thirty-twos column, and 21 remains. The largest power of two that can be subtracted from the remainder 21 is 16, and therefore a 1 is placed in the sixteens column, and 5 remains. The next largest power of two that can be subtracted from 5 is 4, and therefore a 1 is placed in the fours column, and 1 remains. The final 1 is placed in the units column, and therefore decimal 53 is represented in binary as 110101, which indicates that the value is the sum of 1×32, 1×16, 1×4, and 1×1.

■ **INTEGRATED MATH APPLICATION:**

Convert the following decimal numbers to their binary equivalents.

    a. 25

    b. 55

■ *Solution:*

    a. Binary column weights   32   16   8   4   2   1
        Binary number   1   1   0   0   1
        Decimal equivalent $25 - 16 = 9, 9 - 8 = 1, 1 - 1 = 0$

    b. Binary column weights   32   16   8   4   2   1
        Binary number   1   1   0   1   1   1
        Decimal equivalent $55 - 32 = 23, 23 - 16 = 7, 7 - 4 = 3,$
                                $3 - 2 = 1, 1 - 1 = 0$

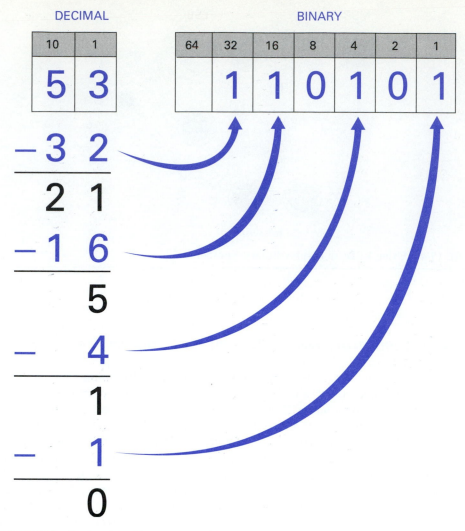

FIGURE 19-5 Decimal-to-Binary Conversion.

INTEGRATED MATH APPLICATION:

Figure 19-6 shows a strip of magnetic tape containing binary data. The shaded dots indicate magnetized dots (which are equivalent to binary 1s), and the unshaded circles on the tape represent unmagnetized points (which are equivalent to binary 0s). What binary values should be stored in rows 4 and 5 so that decimal 19_{10} and 33_{10} are recorded?

FIGURE 19-6 Binary Numbers on Magnetic Tape.

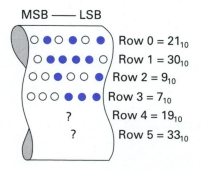

Solution:

a.

Binary column weights	32	16	8	4	2	1

Binary number   1   0   0   1   1

Decimal equivalent $19 - 16 = 3, 3 - 2 = 1, 1 - 1 = 0$

Stored magnetic data   ○ ● ○ ○ ● ●

b.

Binary column weights   32   16   8   4   2   1

Binary number   1   0   0   0   0   1

Decimal equivalent $33 - 32 = 1, 1 - 1 = 0$

Stored magnetic data   ● ○ ○ ○ ○ ●

19-2-5   *Converting Information Signals*

Figure 19-7(a) shows how analog data can be converted into digital data, and Figure 19-7 (b) shows how digital data can be converted into analog data. To give this process purpose, let us consider how the music information stored on a compact disc (CD) is first recorded and then played back. Like all sound, music is made up of waves of compressed air. When these waves strike the diaphragm of a microphone, an analog voltage signal is generated. In the past, this analog data was recorded on magnetic tapes or as a grooved track on a record. These data storage devices were susceptible to wear and tear, temperature, noise, and age. Using digital recording techniques, binary codes are stored on a compact disc to achieve near-perfect fidelity to live sound.

Figure 19-7(a) shows how an analog-to-digital converter (ADC) is used during the recording process to convert the analog music input into a series of digital output codes. These digital codes are used to control the light beam of a recording laser so that it will engrave the binary 0s and 1s onto a compact disc (CD) in the form of pits or spaces. The ADC is triggered into operation by a sampling pulse that causes it to measure the input voltage of the analog signal at that particular time and generate an equivalent digital output code. For example, on the active edge of sampling pulse 1, the analog input voltage is at 2 V, so the binary code 010 (decimal 2) is generated at the output of the ADC. On the active edge of sampling pulse 2, the analog voltage has risen to 4 V, so the binary code 100 (decimal 4) is generated at the output of the ADC. On the active edge of sampling pulses 3, 4, and 5, the binary codes 110 (decimal 6), 111 (decimal 7), and 110 (decimal 6) are generated at the output, representing the analog voltages 6 V, 7 V, and 6 V, respectively, and so on.

Figure 19-7(b) shows how a digital-to-analog converter (DAC) is used during the music playback process to convert the digital codes stored on a CD to an analog output signal. Another laser is used during playback to read the pits and spaces on the compact disc as 0s and 1s. These codes are then applied to the DAC, which converts the digital input codes into discrete voltages. The DAC is triggered into operation by a strobe pulse that causes it to convert the code currently being applied at the input into an equivalent output voltage.

For example, on the active edge of strobe pulse 1, the digital code 010 (decimal 2) is being applied to the DAC, so it will generate 2 V at its output. On the active edge of strobe pulse 2, the digital code 100 (decimal 4) is being applied to the DAC, so it will generate 4 V at its output. On the active edge of strobe pulse 3, the digital code 110 (decimal 6) is being applied to the DAC, so it will generate 6 V at its output. On the active edge of strobe pulse 4, the digital code 111 (decimal 7) is being applied to the DAC, so it will generate 7 V at its output. On the active edge of strobe pulse 5, the digital code 110 (decimal 6) is being applied to the DAC, so it will generate 6 V at its output, and so on. If the output of a DAC is then applied to a low-pass filter, the discrete voltage steps can be blended into a smooth wave that closely approximates the original analog wave, as shown by the dashed line in Figure 19-7(b).

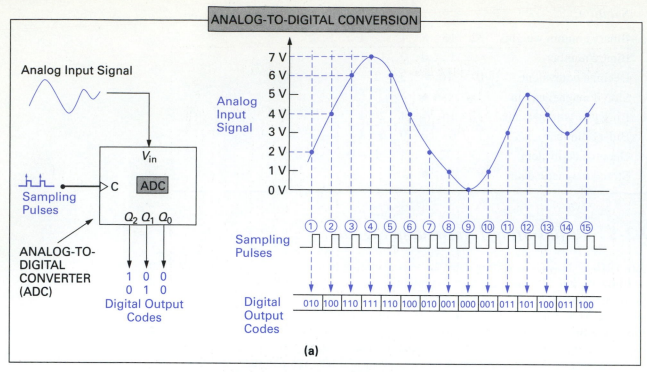

(a)

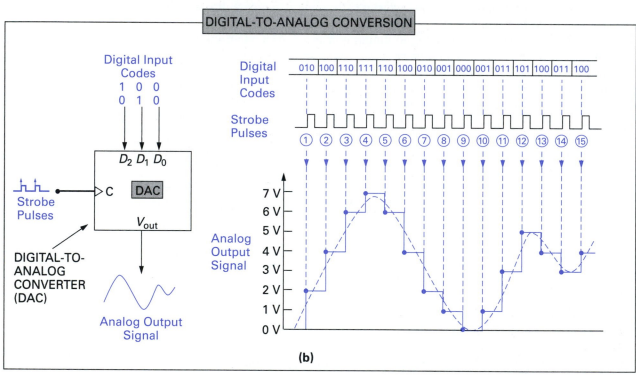

(b)

FIGURE 19-7   **Converting Information Signals.**

DECIMAL			HEXADECIMAL				BINARY					
10^2	10^1	10^0	16^3	16^2	16^1	16^0	2^5	2^4	2^3	2^2	2^1	2^0
100	10	1	4096	256	16	1	32	16	8	4	2	1
		0				0	0	0	0	0	0	0
		1				1	0	0	0	0	0	1
		2				2	0	0	0	0	1	0
		3				3	0	0	0	0	1	1
		4				4	0	0	0	1	0	0
		5				5	0	0	0	1	0	1
		6				6	0	0	0	1	1	0
		7				7	0	0	0	1	1	1
		8				8	0	0	1	0	0	0
		9				9	0	0	1	0	0	1
	1	0				A	0	0	1	0	1	0
	1	1				B	0	0	1	0	1	1
	1	2				C	0	0	1	1	0	0
	1	3				D	0	0	1	1	0	1
	1	4				E	0	0	1	1	1	0
	1	5				F	0	0	1	1	1	1
	1	6			1	0	0	1	0	0	0	0
	1	7			1	1	0	1	0	0	0	1
	1	8			1	2	0	1	0	0	1	0
	1	9			1	3	0	1	0	0	1	1
	2	0			1	4	0	1	0	1	0	0
	2	1			1	5	0	1	0	1	0	1
	2	2			1	6	0	1	0	1	1	0
	2	3			1	7	0	1	0	1	1	1
	2	4			1	8	0	1	1	0	0	0
	2	5			1	9	0	1	1	0	0	1
	2	6			1	A	0	1	1	0	1	0
	2	7			1	B	0	1	1	0	1	1
	2	8			1	C	0	1	1	1	0	0
	2	9			1	D	0	1	1	1	0	1
	3	0			1	E	0	1	1	1	1	0
	3	1			1	F	0	1	1	1	1	1
	3	2			2	0	1	0	0	0	0	0
	3	3			2	1	1	0	0	0	0	1
	3	4			2	2	1	0	0	0	1	0

(a)

A9	DE7	78	F9	10C
AA	DE8	79	FA	10D
AB	DE9	7A	FB	10E
AC	DEF	7B	FC	10F
AD	DF0	7C	FD	110
AE	DF1	7D	FE	111
AF	DF2	7E	FF	112
B0	DF3	7F	100	
B1		80	101	
		81		

(b)

FIGURE 19-8   Hexadecimal Reset and Carry (a) Number System Comparison (b) Hex Counting Examples.

Now that you have completed this section, you should be able to:

■ **Objective 4.** *Explain the positional weight, the reset and carry action, and conversion for binary.*

Use the following questions to test your understanding of Section 19-2:

1. What is the decimal equivalent of 11010_2?
2. Convert 23_{10} to its binary equivalent.
3. What are the full names for the acronyms LSB and MSB?
4. Convert 110_{10} to its binary equivalent.

19-3 THE HEXADECIMAL NUMBER SYSTEM

If digital electronic circuits operate using binary numbers and we operate using decimal numbers, why is there any need for us to have any other system? The hexadecimal, or hex, system is used as a sort of shorthand for large strings of binary numbers, as will be explained in this section. To begin, let us examine the basics of this number system and then look at its application.

Hexadecimal

A base 16 number system.

Hexadecimal means "sixteen," and this number system has 16 different digits, as shown in Figure 19-8(a), which shows a comparison between decimal, hexadecimal, and binary. Looking at the first 10 digits in the decimal and hexadecimal columns, you can see that there is no difference between the two columns; however, beyond 9, hexadecimal makes use of the letters A, B, C, D, E, and F. Digits 0 through 9 and letters A through F make up the 16 total digits of the hexadecimal number system. Comparing hexadecimal with decimal once again, you can see that $A_{16} = 10_{10}$, $B_{16} = 11_{10}$, $C_{16} = 12_{10}$, $D_{16} = 13_{10}$, $E_{16} = 14_{10}$ and $F_{16} = 15_{10}$. Having these extra digits means that a column reset and carry will not occur until the count has reached the last and largest digit, F. The hexadecimal column in Figure 19-8(a) shows how reset and carry will occur whenever a column reaches its maximum digit, F, and this action is further illustrated in the examples in Figure 19-8(b).

19-3-1 *Converting Hexadecimal Numbers to Decimal Numbers*

To find the decimal equivalent of a hexadecimal number, simply multiply each hexadecimal digit by its positional weight. For example, referring to Figure 19-9, you can see that the positional weight of each of the hexadecimal columns, as expected, is progressively 16 times larger as you move from right to left. The hexadecimal number 4C, therefore, indicates that the value is 4×16 ($4 \times 16 = 64$) and $C \times 1$. Since C is equal to 12 in decimal (12×1), the result of $12 + 64$ is 76, so hexadecimal 4C is equivalent to decimal 76.

■ **INTEGRATED MATH APPLICATION:**

Convert hexadecimal 8BF to its decimal equivalent.

■ *Solution:*

Hexadecimal column weights	4096	256	16	1
		8	B	F

Decimal equivalent
$$= (8 \times 256) + (B \times 16) + (F \times 1)$$
$$= 2048 + (11 \times 16) + (15 \times 1)$$
$$= 2048 + 176 + 15$$
$$= 2239$$

16^4	16^3	16^2	16^1	16^0
65,536	4096	256	16	1
			4	C

$$\text{Value of Number} = (4 \times 16) + (C \times 1)$$
$$= (4 \times 16) + (12 \times 1)$$
$$= \quad 64 \quad + \quad 12 \quad = 76_{10}$$

FIGURE 19-9 Positional Weight of the Hexadecimal Number System.

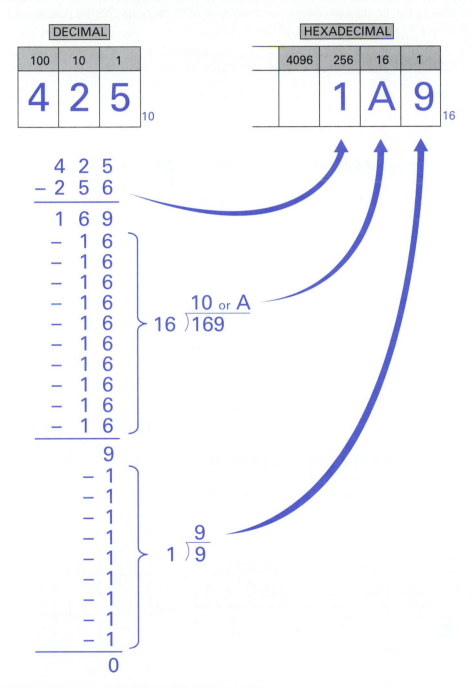

FIGURE 19-10 Decimal-to-Hexadecimal Conversion.

19-3-2 *Converting Decimal Numbers to Hexadecimal Numbers*

Decimal-to-hexadecimal conversion is achieved in the same way as decimal to binary. First, subtract the largest possible power of 16, and then keep subtracting the largest possible power of 16 from the remainder. Each time a subtraction takes place, add a 1 to the respective column until the decimal value has been reduced to zero. Figure 19-10 illustrates this procedure with an example showing how decimal 425 is converted to its hexadecimal equivalent. To begin, the largest possible power of sixteen (256) is subtracted once from 425 and therefore a 1 is placed in the 256s column, leaving a remainder of 169. The next largest power of 16 is 16, which can be subtracted 10 times from 169, and therefore the hexadecimal equivalent of 10, which is A, is placed in the sixteens column, leaving a remainder of 9. Since nine 1s can be subtracted from the remainder of 9, the units column is advanced nine times, giving us our final hexadecimal result, 1A9.

■ **INTEGRATED MATH APPLICATION:**

Convert decimal 4525 to its hexadecimal equivalent.

■ *Solution:*

Hexadecimal column weights	4096	256	16	1
	1	1	A	D

Decimal equivalent
$$= 4525 - 4096 = 429, 429 - 256 = 173,$$
$$173 - 16 - 16 - 16 - 16 - 16 - 16 -$$
$$16 - 16 - 16 - 16 = 13, 13 - 1 -$$
$$1 - 1 - 1 - 1 - 1 - 1 - 1 - 1 -$$
$$1 - 1 - 1 - 1 = 0$$

19-3-3 *Converting between Binary and Hexadecimal*

As mentioned at the beginning of this section, hexadecimal is used as a shorthand for representing large groups of binary digits. To illustrate this, Figure 19-11(a) shows how a 16-bit

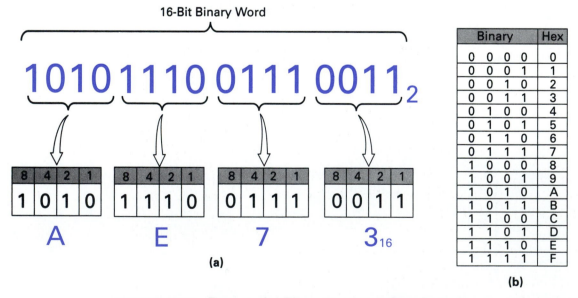

FIGURE 19-11 **Representing Binary Numbers in Hexadecimal.**

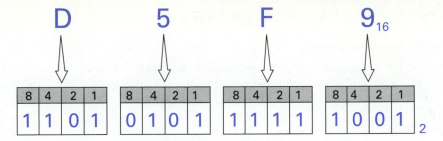

FIGURE 19-12 Hexadecimal-to-Binary Conversion.

binary number, which is more commonly called a 16-bit binary **word,** can be represented by 4 hexadecimal digits. To explain this, Figure 19-11(b) shows how a 4-bit binary word can have any value from 0_{10} (0000_2) to 15_{10} (1111_2), and since hexadecimal has the same number of digits (0 through F), we can use one hexadecimal digit to represent 4 binary bits. As you can see in Figure 19-11(a), it is much easier to work with a number like $AE73_{16}$ than with 1010111001110011_2.

To convert from hexadecimal to binary, we simply do the opposite, as shown in Figure 19-12. Since each hexadecimal digit represents 4 binary digits, a 4-digit hexadecimal number will convert to a 16-bit binary word.

> **Word**
>
> An ordered set of characters that is treated as a unit.

■ **INTEGRATED MATH APPLICATION:**

Convert hexadecimal 2BF9 to its binary equivalent.

■ *Solution:*

Hexadecimal number	2	B	F	9
Binary equivalent	0010	1011	1111	1001

CALCULATOR KEYS

Name: Entering and converting number bases

Function and Examples: Some calculators allow you to enter and convert number bases.

To enter a binary number, use the form:

0b*binaryNumber*   (for example: 0b11100110)

— Binary number with up to 32 digits
— Zero, not the letter *O*, and the letter *b*

To enter a hexadecimal number, use the form:

0h*hexadecimalNumber*   (for example: 0h89F2C)

— Hexadecimal number with up to 8 digits
— Zero, not the letter *O*, and letter *h*

If you enter without the 0b or 0h prefix, such as 11, it is always treated as a decimal number. If you omit the 0h prefix on a hexadecimal number containing A–F, all or part of the entry is treated as a variable.

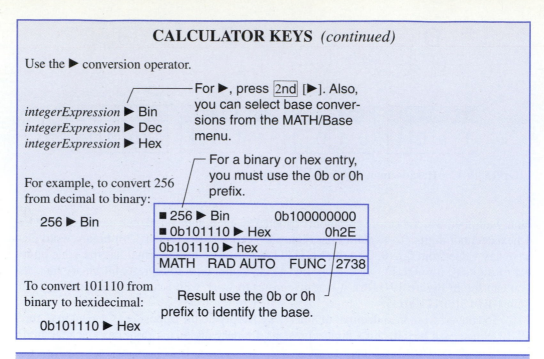

CALCULATOR KEYS *(continued)*

Use the ▶ conversion operator.

integerExpression ▶ Bin
integerExpression ▶ Dec
integerExpression ▶ Hex

For ▶, press 2nd [▶]. Also, you can select base conversions from the MATH/Base menu.

For example, to convert 256 from decimal to binary:

256 ▶ Bin

For a binary or hex entry, you must use the 0b or 0h prefix.

■ 256 ▶ Bin	0b100000000
■ 0b101110 ▶ Hex	0h2E
0b101110 ▶ hex	
MATH RAD AUTO FUNC	2738

To convert 101110 from binary to hexadecimal:

0b101110 ▶ Hex

Result use the 0b or 0h prefix to identify the base.

■ **INTEGRATED MATH APPLICATION:**

Convert binary 110011100001 to its hexadecimal equivalent.

■ *Solution:*

Binary number	1100	1110	0001
Hexadecimal equivalent	C	E	1

SELF-TEST EVALUATION POINT FOR SECTION 19-3

Now that you have completed this section, you should be able to:

■ *Objective 5.* *Explain the positional weight, the reset and carry action, and conversion for hexadecimal.*

Use the following questions to test your understanding of Section 19-3:

1. What is the base of each?
   a. Decimal number system
   b. Binary number system
   c. Hexadecimal number system
2. What are the decimal and hexadecimal equivalents of 10101_2?
3. Convert $1011\ 1111\ 0111\ 1010_2$ to its hexadecimal equivalent.
4. What are the binary and hexadecimal equivalents of 33_{10}?

19-4   THE OCTAL NUMBER SYSTEM

Although not as frequently used as hexadecimal, the *octal number system* is also used as a shorthand for large binary words. It is very much like hexadecimal in that it allows for easy conversion from binary to octal and from octal to binary. Before we look at its application, however, let us examine the basics of this number system.

Octal

A base 8 number system.

Octal means "eight," and the octal number system has eight digits: 0 through 7. Figure 19-13(a) shows a comparison among decimal, octal, binary, and hexadecimal. Having only eight possible digits means that a column reset and carry will occur when the count has reached the last and largest digit, 7. At this point, the first octal column will reset and carry, as shown in Figure 19-13(a). This action occurs when any of the octal columns reaches the maximum digit, 7, as illustrated in the additional counting examples shown in Figure 19-13(b).

DECIMAL				OCTAL				BINARY					HEXADECIMAL			
1000	100	10	1	512	64	8	1	16	8	4	2	1	4096	256	16	1
			0				0	0	0	0	0	0				0
			1				1	0	0	0	0	1				1
			2				2	0	0	0	1	0				2
			3				3	0	0	0	1	1				3
			4				4	0	0	1	0	0				4
			5				5	0	0	1	0	1				5
			6				6	0	0	1	1	0				6
			7				7	0	0	1	1	1				7
			8			1	0	0	1	0	0	0				8
			9			1	1	0	1	0	0	1				9
		1	0			1	2	0	1	0	1	0				A
		1	1			1	3	0	1	0	1	1				B
		1	2			1	4	0	1	1	0	0				C
		1	3			1	5	0	1	1	0	1				D
		1	4			1	6	0	1	1	1	0				E
		1	5			1	7	0	1	1	1	1				F
		1	6			2	0	1	0	0	0	0			1	0
		1	7			2	1	1	0	0	0	1			1	1
		1	8			2	2	1	0	0	1	0			1	2
		1	9			2	3	1	0	0	1	1			1	3
		2	0			2	4	1	0	1	0	0			1	4

(a)

44	75	4 7 4
45	76	4 7 5
46	77	4 7 6
47	1 0 0	4 7 7
50	1 0 1	5 0 0
51	1 0 2	5 0 1

(b)

FIGURE 19-13 **Octal Reset and Carry (a) Number System Comparison (b) Octal Counting Examples.**

19-4-1 *Converting Octal Numbers to Decimal Numbers*

To find the decimal equivalent of an octal number, simply multiply each octal digit by its positional weight. For example, referring to Figure 19-14, you can see that the positional weight of each of the octal columns is as expected, progressively eight times larger as you move from right to left. The octal number 126 therefore indicates that the value is the sum of 1×64, 2×8, and 6×1. The result of $64 + 16 + 6$ is decimal 86.

8^5	8^4	8^3	8^2	8^1	8^0
32,768	4096	512	64	8	1
			1	2	6

$$\text{Value of Number} = (1 \times 64) + (2 \times 8) + (6 \times 1)$$
$$= \quad 64 \quad + \quad 16 \quad + \quad 6 \quad = 86_{10}$$

FIGURE 19-14 **Positional Weight of the Octal Number System.**

■ **INTEGRATED MATH APPLICATION:**

Convert octal 2437 to its decimal equivalent.

■ *Solution:*

Octal column weights

512	64	8	1
2	4	3	9

Decimal equivalent
$$= (2 \times 512) + (4 \times 64) + (3 \times 8) + (7 \times 1)$$
$$= 1024 + 256 + 24 + 7$$
$$= 1311$$

19-4-2 *Converting Decimal Numbers to Octal Numbers*

Decimal-to-octal conversion is achieved in the same way as decimal to binary and decimal to hex. First, subtract the largest possible power of eight, and then keep subtracting the largest possible power of eight from the remainder. Each time a subtraction takes place, add 1 to the respective column until the decimal value has been reduced to zero. Figure 19-15 illustrates this procedure with an example showing how decimal 139 is converted to its octal equivalent. To begin, the largest possible power of eight (64) can be subtracted twice from

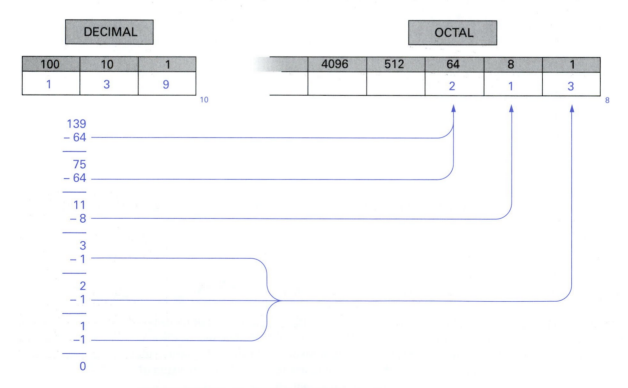

FIGURE 19-15 **Decimal-to-Octal Conversion.**

139, and therefore a 2 is placed in the 64 column. The next largest power of eight is 8 itself, and therefore the octal 8 column is advanced to 1, leaving a remainder of 3. Since 3 units can be subtracted from this final remainder of 3, a 3 is placed in the octal units column, giving us our final octal result, 213.

■ **INTEGRATED MATH APPLICATION:**

Convert decimal 3724 to its octal equivalent.

■ *Solution:*

Octal column weights

$$\frac{512 \quad 64 \quad 8 \quad 1}{7 \quad\;\; 2 \quad\;\; 1 \quad 4}$$

Decimal equivalent
$= 3724 - 512 = 3212, 3212 - 512 = 2700,$
$2700 - 512 = 2188, 2188 - 512 = 1676,$
$1676 - 512 = 1164, 1164 - 512 = 652,$
$652 - 512 = 140, 140 - 64 = 76, 76 - 64 = 12,$
$12 - 8 = 4, 4 - 1 = 3, 3 - 1 = 2, 2 - 1 = 1,$
$1 - 1 = 0$

19-4-3   *Converting between Binary and Octal*

Like hexadecimal, octal is used as a shorthand for representing large groups of binary digits. To illustrate this, Figure 19-16(a) shows how a 16-bit binary number, which is more commonly called a 16-bit binary word, can be represented by six octal digits. To explain this, Figure 19-16(b) shows how a 3-bit binary word can have any value from 0_{10} (000_2) to 7_{10} (111_2), and since octal has the same number of digits (0 through 7), we can use one octal digit to represent three binary bits. As you can see in the example in Figure 19-16(a), it is much easier to work with a number like 127163 than with 1010111001110011.

To convert from octal to binary, we simply do the opposite, as shown in the example in Figure 19-17. Since each octal digit represents three binary digits, a 4-digit octal number will convert to a 12-bit binary word.

■ **INTEGRATED MATH APPLICATION:**

Convert octal number 2635 to its binary equivalent.

■ *Solution:*

Octal Number	2	6	3	5
Binary Equivalent	010	110	011	101

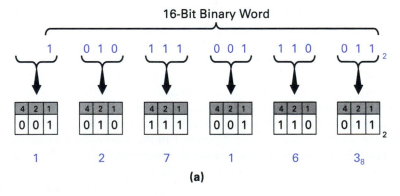

BINARY	OCTAL
0 0 0	0
0 0 1	1
0 1 0	2
0 1 1	3
1 0 0	4
1 0 1	5
1 1 0	6
1 1 1	7

(a)   (b)

FIGURE 19-16   Representing Binary Numbers in Octal.

FIGURE 19-17   **Octal-to-Binary**
Conversion.

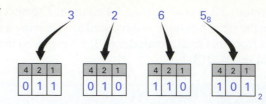

■   **INTEGRATED MATH APPLICATION:**

Convert binary 101111001000010110 to its octal equivalent.

■   *Solution:*

Binary Number	101	111	001	000	010	110
Octal Equivalent	5	7	1	0	2	6

SELF-TEST EVALUATION POINT FOR SECTION 19-4

Now that you have completed this section, you should be able to:

■   *Objective 6.   Explain the positional weight, the reset and carry action, and conversion for octal.*

Use the following questions to test your understanding of Section 19-4:

1. The octal number system has a base of _____.
2. Convert 154_{10} into octal.
3. A 12-bit binary word will convert into _____ octal digits.
4. What is the octal equivalent of 100010110111_2?

19-5   BINARY CODES

Pure Binary

Uncoded binary.

The process of converting a decimal number to its binary equivalent is called *binary coding.* The result of the conversion is a binary number or code that is called **pure binary.** There are, however, binary codes used in digital circuits other than pure binary, and in this section we will examine some of the most frequently used.

19-5-1   *The Binary Coded Decimal (BCD) Code*

Binary-Coded Decimal (BCD)

A code in which each decimal digit is represented by a group of 4 binary bits.

No matter how familiar you become with binary, it will always be less convenient to work with than the decimal number system. For example, it will always take a short time to convert 1111000_2 to 120_{10}. Designers realized this disadvantage early on and developed a binary code that had decimal characteristics and that was appropriately named **binary coded decimal (BCD).** Being a binary code, it has the advantages of a two-state system, and since it has a decimal format, it is also much easier for an operator to interface via a decimal keypad or decimal display with systems such as pocket calculators and wristwatches.

The BCD code expresses each decimal digit as a 4-bit word, as shown in Figure 19-18(a). In this example, decimal 1753 converts to a BCD code of 0001 0111 0101 0011, with the first 4-bit code ($0001_2 = 1_{10}$) representing the 1 in the thousands column, the second 4-bit code ($0111_2 = 7_{10}$) representing the 7 in the hundreds column, the third 4-bit code ($0101_2 = 5_{10}$) representing the 5 in the tens column, and the fourth 4-bit code ($0011_2 = 3_{10}$) representing the 3 in the units column. As can be seen in Figure 19-18(a), the subscript BCD is often used after a BCD code to distinguish it from a pure binary number ($1753_{10} = 0001$ $0111\ 0101\ 0011_{BCD}$).

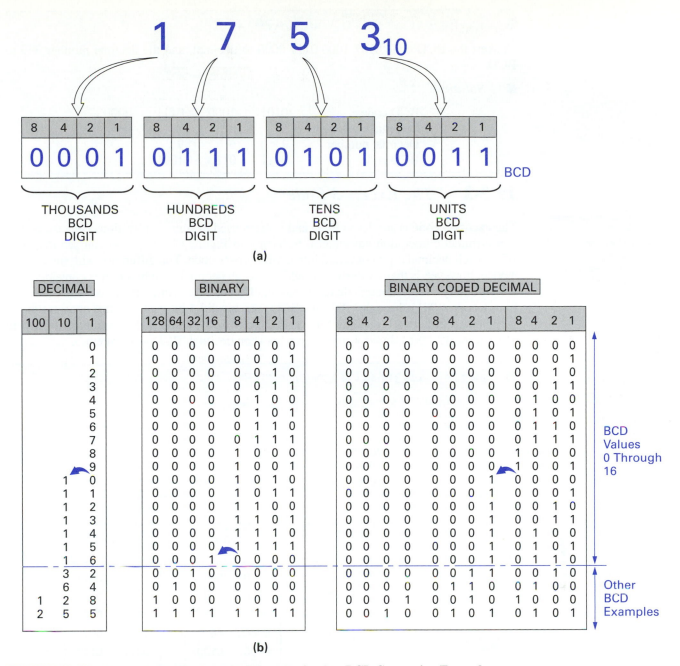

FIGURE 19-18 Binary Coded Decimal (BCD) (a) Decimal-to-BCD Conversion Example (b) A Comparison of Decimal, Binary, and BCD.

Figure 19-18(b) compares decimal, binary, and BCD. As you can see, the reset and carry action occurs in BCD at the same time it does in decimal. This is because BCD was designed to have only ten 4-bit binary codes, 0000, 0001, 0010, 0011, 0100, 0101, 0110, 0111, 1000, and 1001 (0 through 9), to make it easy to convert between this binary code and decimal. Binary codes 1010, 1011, 1100, 1101, 1110, and 1111 (A_{16} through F_{16}), are invalid codes that are not used in BCD because they are not used in decimal.

■ **INTEGRATED MATH APPLICATION:**

Convert the BCD code 0101 1000 0111 0000 to decimal, and the decimal number 369 to BCD.

■ *Solution:*

BCD code	0101	1000	0111	0000
Decimal equivalent	5	8	7	0
Decimal number	3	6	9	
BCD code	0011	0110	1001	

19-5-2 *The Excess-3 Code*

Excess-3 Code

A code in which the decimal digit *n* is represented by the 4-bit binary code *n* + 3.

The **excess-3 code** is similar to BCD and is often used in some applications because of certain arithmetic operation advantages. Referring to Figure 19-19(a), you can see that, like BCD, each decimal digit converted into a 4-bit binary code. The difference with the excess-3 code, however, is that a value of 3 is added to each decimal digit before it is converted into a 4-bit binary code. Although this code has an offset or excess of 3, it still uses only ten 4-bit binary codes (0011 through 1100) like BCD. The invalid 4-bit values in this code are therefore 0000, 0001, 0010 (decimal 0 through 2), and 1101, 1110, 1111 (decimal 13 through 15). Figure 19-19(b) shows a comparison between decimal, BCD, and excess-3.

■ **INTEGRATED MATH APPLICATION:**

Convert decimal 408 to its excess-3 code equivalent.

■ *Solution:*

Decimal number	4	0	8
+3 =	7	3	11
Excess-3 binary	0111	0011	1011

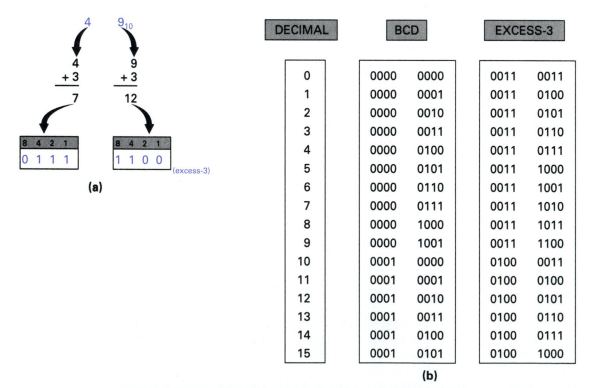

DECIMAL	BCD		EXCESS-3	
0	0000	0000	0011	0011
1	0000	0001	0011	0100
2	0000	0010	0011	0101
3	0000	0011	0011	0110
4	0000	0100	0011	0111
5	0000	0101	0011	1000
6	0000	0110	0011	1001
7	0000	0111	0011	1010
8	0000	1000	0011	1011
9	0000	1001	0011	1100
10	0001	0000	0100	0011
11	0001	0001	0100	0100
12	0001	0010	0100	0101
13	0001	0011	0100	0110
14	0001	0100	0100	0111
15	0001	0101	0100	1000

(b)

FIGURE 19-19 **The Excess-3 Code (a) Example (b) Code Comparison.**

19-5-3 *The Gray Code*

The **gray code,** shown in Figure 19-20, is a nonweighted binary code, which means that the position of each binary digit carries no specific weight or value. This code, named after its inventor, was developed so that only one of the binary digits changes as you step from one code group to the next code group in sequence. For example, in pure binary, a change from 3 (0011) to 4 (0100) results in a change in 3 bits, whereas in the gray code only 1 bit changes.

Gray Code
A nonweighted binary code in which sequential numbers are represented by codes that differ by one bit.

The minimum-change gray code is used to reduce errors in digital electronic circuitry. To explain this in more detail, when a binary count changes, it takes a very small amount of time for the bits to change from 0 to 1, or from 1 to 0. This transition time can produce an incorrect intermediate code. For example, if when changing from pure binary 0011 (3) to 0100 (4), the LSB were to switch slightly faster than the other bits, an incorrect code would be momentarily generated, as shown in the following example.

Binary	Decimal
0 0 1 ~~1~~	3
0 ~~0~~ ~~1~~ 0	Error
0 1 0 0	4

This momentary error could trigger an operation that should not occur, resulting in a system malfunction. Using the gray code eliminates these timing errors, since only 1 bit changes at a time. The disadvantage of the gray code is that the codes must be converted to pure binary if numbers need to be added, subtracted, or used in other computations.

19-5-4 *The American Standard Code for Information Interchange (ASCII)*

To this point we have discussed only how we can encode numbers into binary codes. Digital electronic computers must also be able to generate and recognize binary codes that represent

DECIMAL	BINARY	GRAY CODE
0	0000	0000
1	0001	0001
2	0010	0011
3	0011	0010
4	0100	0110
5	0101	0111
6	0110	0101
7	0111	0100
8	1000	1100
9	1001	1101
10	1010	1111
11	1011	1110
12	1100	1010
13	1101	1011
14	1110	1001
15	1111	1000

FIGURE 19-20 **The Gray Code.**

Alphanumeric Code

A code used to represent the letters of the alphabet and decimal numbers.

letters of the alphabet and symbols. The ASCII code is the most widely used **alphanumeric code** (alphabet and numeral code). As I write this book the ASCII codes for each of these letters is being generated by my computer keyboard, and the computer is decoding these ASCII codes and then displaying the alphanumeric equivalent (letter, number, or symbol) on my computer screen.

Figure 19-21 lists all the 7-bit ASCII codes and the full names for the abbreviations used. This diagram also shows how each of the 7-bit ASCII codes is made up of a 4-bit group that indicates the row of the table and a 3-bit group that indicates the column of the table. For example, the uppercase letter K is in column 100 (4_{10}) and in row 1011 (11_{10}) and is therefore represented by the ASCII code 1001011.

Column		0	1	2	3	4	5	6	7
Row Bits 4321 765 →		000	001	010	011	**100**	101	110	111
0	0000	NUL	DLE	SP	0	@	P	\	p
1	0001	SOH	DC1	!	1	A	Q	a	q
2	0010	STX	DC2	"	2	B	R	b	r
3	0011	ETX	DC3	#	3	C	S	c	s
4	0100	EOT	DC4	$	4	D	T	d	t
5	0101	ENQ	NAK	%	5	E	U	e	u
6	0110	ACK	SYN	&	6	F	V	f	v
7	0111	BEL	ETB	'	7	G	W	g	w
8	1000	BS	CAN	(	8	H	X	h	x
9	1001	HT	EM	)	9	I	Y	i	y
10	1010	LF	SUB	*	:	J	Z	j	z
11	**1011**	VT	ESC	+	;	**K**	[	k	{
12	1100	FF	FS	,	<	L	\	l	|
13	1101	CR	GS	-	=	M	]	m	}
14	1110	SO	RS	.	>	N	⌢	n	~
15	1111	SI	US	/	?	O	—	o	DEL

Example: Upper Case "K" = Column 4 (100), Row 11 (1011)

Bits	7	6	5		4	3	2	1
ASCII Code	1	0	0		1	0	1	1

NUL	Null	FF	Form Feed		CAN	Cancel
SOH	Start of Heading	CR	Carriage Return		EM	End of Medium
STX	Start of Text	SO	Shift Out		SUB	Substitute
ETX	End of Text	SI	Shift In		ESC	Escape
EOT	End of Transmission	SP	Space (blank)		FS	File Separator
ENQ	Enquiry	DLE	Data Link Escape		GS	Group Separator
ACK	Acknowledge	DC1	Device Control 1		RS	Record Separator
BEL	Bell (audible signal)	DC2	Device Control 2		US	Unit Separator
BS	Backspace	DC3	Device Control 3		DEL	Delete
HT	Horizontal Tabulation	DC4	Device Control 4			
	(punched card skip)	NAK	Negative Acknowledge			(ASCII
LF	Line Feed	SYN	Synchronous Idle			Abbreviations)
VT	Vertical Tabulation	ETB	End of Transmission Block			

FIGURE 19-21   **The ASCII Code.**

■ INTEGRATED MATH APPLICATION:

List the ASCII codes for the message "Digital."

■ *Solution:*

$$D = 1000100$$
$$i = 1101001$$
$$g = 1100111$$
$$i = 1101001$$
$$t = 1110100$$
$$a = 1100001$$
$$l = 1101100$$

■ INTEGRATED MATH APPLICATION:

Figure 19-22 illustrates a 7-bit register. A register is a circuit that is used to store or hold a group of binary bits. If a switch is closed, the associated "Q output" is grounded and 0 V (binary 0) is applied to the output. On the other hand, if a switch is open, the associated "Q output" is pulled HIGH due to the connection to +5 V via the 10 kΩ resistor, and therefore +5 V (binary 1) is applied to the output. What would be the pure binary value, hexadecimal value, and ASCII code stored in this register?

■ *Solution:*

	Q_6	Q_5	Q_4	Q_3	Q_2	Q_1	Q_0
Pure binary	1	0	0	0	0	1	1_2
Hexadecimal				43_{16}			
ASCII			C (uppercase C)				

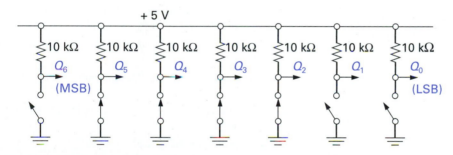

FIGURE 19-22   A 7-Bit Switch Register.

SELF-TEST EVALUATION POINT FOR SECTION 19-5

Now that you have completed this section, you should be able to:

■ *Objective 7.   Describe the following binary codes: binary coded decimal (BCD), the excess-3 code, the gray code, and the American Standard Code for Information Exchange (ASCII).*

Use the following questions to test your understanding of Section 19-5:

1. Convert 7629_{10} to BCD.

2. What is the decimal equivalent of the excess-3 code 0100 0100?

3. What is a minimum-change code?

4. What does the following string of ASCII codes mean?
   1000010   1100001   1100010   1100010   1100001
   1100111   1100101

REVIEW QUESTIONS

Multiple Choice Questions

1. Binary is a base _____ number system, decimal a base _____ number system, and hexadecimal a base _____ number system.

   a. 1, 2, 3   **c.** 10, 2, 3
   b. 2, 10, 16   **d.** 8, 4, 2

2. What is the binary equivalent of decimal 11?

   a. 1010   **c.** 1011
   b. 1100   **d.** 0111

3. Which of the following number systems is used as a shorthand for binary?

   a. Decimal   **c.** Binary
   b. Hexadecimal   **d.** Both a and b

4. What would be displayed on a hexadecimal counter if its count was advanced by 1 from $39FF_{16}$?

   a. 4000   **c.** 3900
   b. 3A00   **d.** 4A00

5. What is the base of 101?

   a. 10   **c.** 16
   b. 2   **d.** Could be any base

6. Which of the following is an invalid BCD code?

   a. 1010   **c.** 0111
   b. 1000   **d.** 0000

7. What is the decimal equivalent of 1001 0001 1000 0111_{BCD}?

   a. 9187   **c.** 8659
   b. 9A56   **d.** 254,345

8. What is the pure binary equivalent of 15_{16}?

   a. 0000 1111   **c.** 0101 0101
   b. 0001 0101   **d.** 0000 1101

9. Which of the following is an invalid excess-3 code?

   a. 1010   **c.** 0111
   b. 1000   **d.** 0000

10. Since the lower 4 bits of the ASCII code for the numbers 0 through 9 are the same as their pure binary equivalent, what is the 7-bit ASCII code for the number 6?

    a. 0111001   **c.** 0110011
    b. 0110111   **d.** 0110110

Communication Skill Questions

11. Give the full name of the following abbreviation and acronyms.

    a. Bit   **b.** BCD   **c.** ASCII

12. What is the Arabic number system?   (19-1)

13. Why does reset and carry occur after 9 in decimal and after 1 in binary?   (19-1)

14. Briefly describe how to: (19-2)

    a. convert a binary number to a decimal number.
    b. convert a decimal number to a binary number.

15. Why is the hexadecimal number system used in conjunction with digital circuits?   (19-3)

16. Briefly describe how to: (19-3)

    a. convert a hexadecimal number to a decimal number.
    b. convert a decimal number to a hexadecimal number.
    c. convert between binary and hexadecimal.

17. In what application is the octal number system used?   (19-4)

18. Briefly describe how to: (19-4)

    a. convert an octal number to a decimal number.
    b. convert a decimal number to an octal number.
    c. convert between binary and octal.

19. Why is the BCD code easier to decode than pure binary?   (19-5)

20. What is binary coded decimal and why is it needed?   (19-5)

21. Briefly describe how to convert a BCD code to its decimal equivalent.   (19-5)

22. What is the excess-3 code?   (19-5)

23. What is the advantage of the gray code?   (19-5)

24. What is the ASCII code and in what applications is it used?   (19-5)

Practice Problems

25. What is the pure binary output of the switch register shown in Figure 19-23 and what is its decimal equivalent?

26. Convert the following into their decimal equivalents:

    a. 110111_2 **c.** 10110_{10}
    b. $2F_{16}$

27. What would be the decimal equivalent of the LED display in Figure 19-24 if it were displaying each of the following:

    a. Pure binary   **c.** Excess-3
    b. BCD

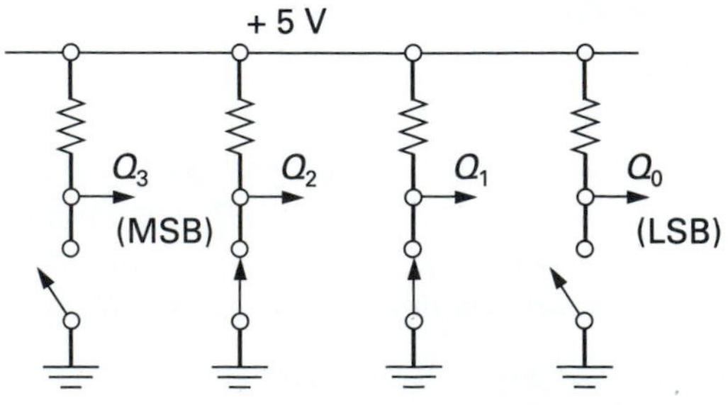

FIGURE 19-23   Four-Bit Switch Register.

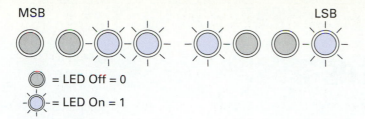

MSB   LSB

◯ = LED Off = 0

◯ = LED On = 1

FIGURE 19-24   **8-Bit LED Display.**

28. Convert the following binary numbers to octal and hexa-decimal:

   a. 111101101001   **c.** 111000

   b. 1011   **d.** 1111111

29. Convert the following decimal numbers to BCD:

   a. 2,365   **b.** 24

30. Give the ASCII codes for the following:

   a. ?   **c.** 6

   b. $

31. Identify the following codes.

DECIMAL	(A)		(B)	(C)		(D)
0	0000	0000	0000	0011	0011	011	0000
1	0000	0001	0001	0011	0100	011	0001
2	0000	0010	0011	0011	0101	011	0010
3	0000	0011	0010	0011	0110	011	0011
4	0000	0100	0110	0011	0111	011	0100
5	0000	0101	0111	0011	1000	011	0101

32. Convert the decimal number 23 into each of the following:

   a. ASCII   **d.** Hexadecimal

   b. Pure binary   **e.** Octal

   c. BCD   **f.** Excess-3

33. If a flashlight is used to transmit ASCII codes, with a dot (short flash) representing a 0 and a dash (long flash) representing a 1, what is the following message?

0000010 1010000 1110010 1100001 1100011 1110100 1101001
1100011 1100101 0100000 1001101 1100001 1101011 1100101
1110011 0100000 1010000 1100101 1110010 1100110 1100101
1100011 1110100 0000011

34. Convert the following:

   a. $1110110_2 = $ _____ 16

   b. $175_{10} = $ _____ 2

   c. $ABC_{16} = $ _____ 2

   d. $00110110_2 = $ _____ 8

Web Site Questions

Go to the web site http://prenhall.com/cook, select the textbook *Mathematics for Electronics and Computers*, select this chapter, and then follow the instructions when answering the multiple-choice practice problems.

Logic Gates

The Great Experimenter

Michael Faraday was born to James and Margaret Faraday on September 22, 1791. James Faraday, a blacksmith in failing health, had brought his family to London in hopes of finding work; however, his poor health meant that the Faradays were often penniless and hungry. At the age of 12, the young Faraday left school to work as an errand boy for a local bookseller and was later apprenticed to the bookbinder's trade. At 21, the gifted and engaging Faraday was at the right place at the right time, and with the right talents. He impressed the brilliant professor Humphry Davvy, who made him his research assistant at Count Rumford's Royal Institution. After only two years, Faraday was given a promotion and an apartment at the Royal Institution, and in 1821 he married. He lived at the Royal Institution for the rest of his active life, working in his laboratory, giving very dynamic lectures, and publishing over 450 scientific papers. Unlike many scientific papers, Faraday's papers never used a calculus formula to explain a concept or idea. Instead, Faraday would explain all his findings using logic and reasoning, so that a person trying to understand science did not need to be a scientist. It was this gift of being able to make even the most complex areas of science easily accessible to the student, coupled with his motivational teaching style, that made him so popular.

By 1855 he had written three volumes of papers on electromagnetism, the first dynamo, the first transformer, and the foundations of electrochemistry, a large amount on dielectrics, and even some papers on plasma. The unit of capacitance is the *farad,* in honor of his work in these areas of science.

Faraday began two series of lectures at the Royal Institution, which have continued to this day. Michael and Sarah Faraday were childless, but they both loved children, and in 1860 Faraday began a series of Christmas lectures expressly for children, the most popular of which was called "The Chemical History of a Candle." The other lecture series was the "Friday Evening Discourses," of which he himself delivered over a hundred. These very dynamic, enlightening, and entertaining lectures covered areas of science or technology for the layperson and were filled with demonstrations. On one evening in 1846, an extremely nervous and shy speaker ran off just moments before he was scheduled to give the Friday Evening Discourse. Faraday had to fill in, and to this day a tradition is still enforced whereby the lecturer for the Friday Evening Discourse is locked up for half an hour before the presentation with a single glass of whiskey.

Faraday was often referred to as "the great experimenter," and it was this consistent experimentation that led to many of his findings. He was fascinated by science and technology, and was always exploring new and sometimes dangerous horizons. In fact, in one of his reports he states, "I have escaped, not quite unhurt, from four explosions." When asked to comment on experimentation, his advice was to "Let your imagination go, guiding it by judgment and principle, but holding it in and directing it by experiment. Nothing is so good as an experiment which, while it sets an error right, gives you as a reward for your humility an absolute advance in knowledge."

Outline and Objectives

VIGNETTE: THE GREAT EXPERIMENTER

INTRODUCTION

20-1 HARDWARE FOR BINARY SYSTEMS

Objective 1: Describe the evolution of the two-state switch and how it is used to represent binary data.

Objective 2: Explain how the diode and transistor semiconductor switches can be used to construct digital electronic logic gates.

20-2 BASIC LOGIC GATES

Objective 3: Describe the discrete circuit construction, truth table, operation, and an application for the following basic logic-gate types.
   a. The OR Gate
   b. The AND Gate

20-2-1 The OR Gate

20-2-2 The AND Gate

20-3 INVERTING LOGIC GATES

Objective 4: Describe the discrete circuit construction, truth table, operation, and an application for the following inverting logic-gate types.
   a. The NOT Gate
   b. The NOR Gate
   c. The NAND Gate

20-3-1 The NOT Gate

20-3-2 The NOR Gate

20-3-3 The NAND Gate

20-4 EXCLUSIVE LOGIC GATES

Objective 5: Describe the discrete circuit construction, truth table, operation, and an application for the following exclusive logic-gate types.
   a. The XOR Gate
   b. The XNOR Gate

20-4-1 The XOR Gate

20-4-2 The XNOR Gate

20-5 IEEE/ANSI SYMBOLS FOR LOGIC GATES

Objective 6: Explain the major differences between the traditional and IEEE/ANSI logic-gate symbols.

MULTIPLE CHOICE QUESTIONS

COMMUNICATION SKILL QUESTIONS

PRACTICE PROBLEMS

WEB SITE QUESTIONS

Introduction

Within any digital electronic system you will find that diodes and transistors are used to construct **logic gate circuits.** Logic gates are in turn used to construct flip-flop circuits, and flip-flops are used to construct register, counter, and a variety of other circuits. These logic gates, therefore, are used as the basic building blocks for all digital circuits, and their purpose is to control the movement of binary data and binary instructions. Logic gates, and all other digital electronic circuits, are often referred to as **hardware** circuits. By definition, the hardware of a digital electronic system includes all the electronic, magnetic, and mechanical devices of a digital system. In contrast, the **software** of a digital electronic system includes the binary data (like pure binary and ASCII codes) and binary instructions that are processed by the digital electronic system hardware. To use an analogy, we could say that a compact disc player is hardware, and the music information stored on a compact disc and processed by the player is software. Just as a CD player is useless without CDs, a digital system's hardware is useless without software. In other words, the information on the CD determines what music is played, and similarly, the digital software determines the actions of the digital electronic hardware.

Every digital electronic circuit uses logic-gate circuits to manipulate the coded pulses of binary language. These logic-gate circuits are constructed using diodes and transistors, and they are the basic decision-making elements in all digital circuits.

20-1 HARDWARE FOR BINARY SYSTEMS

Digital circuits are often referred to as *switching circuits* because their control devices (diodes and transistors) are switched between the two extremes of ON and OFF. These digital circuits are also called **two-state circuits** because their control devices are driven into one of two states: either into the saturation state (fully ON), or cutoff state (fully OFF). These two modes of operation are used to represent the two binary digits of 1 and 0.

To develop a digital electronic system that could manipulate binary information, inventors needed a two-state electronic switch. Early machines used mechanical switches and electromechanical relays—like the examples in Figure 20-1(a) and (b)—to represent binary data by switching current ON or OFF. These mechanical devices were eventually replaced by the vacuum tube, shown in Figure 20-1(c), which, unlike switches and relays, had the advantage of no moving parts. The vacuum tube, however, was bulky, fragile, had to warm up, and consumed an enormous amount of power. Finally, compact and low-power digital electronic circuits and systems became a reality with the development of semiconductor diode and transistor switches, which are shown in Figures 20-1(d) and 20-1(e).

A logic gate accepts inputs in the form of HIGH or LOW voltages, judges these input combinations based on a predetermined set of rules, and then produces a single output in the form of a HIGH or LOW voltage. The term *logic* is used because the output is predictable or logical, and the term *gate* is used because only certain input combinations will "unlock the gate." For example, any HIGH input to an OR gate will unlock the gate and allow the HIGH at the input to pass through to the output.

SELF-TEST EVALUATION POINT FOR SECTION 20-1

Now that you have completed this section, you should be able to:

■ *Objective 1.* *Describe the evolution of the two-state switch and how it is used to represent binary data.*

■ *Objective 2.* *Explain how the diode and transistor semiconductor switches can be used to construct digital electronic gates.*

Use the following questions to test your understanding of Section 20-1:

1. The _____ includes the electronic, magnetic, and mechanical devices of a digital electronic system.

2. The _____ includes the binary data and binary instructions that are processed by the digital electronic system.

3. Which two-state switch was the first to have no moving parts?

4. Which two semiconductor switches are used to implement digital electronic circuits?

20-2   BASIC LOGIC GATES

20-2-1   *The OR Gate*

The **OR gate** can have two or more inputs but will always have a single output. Figure 20-2(a) shows a table listing all the input possibilities for this two-input OR gate. This table is often referred to as a **truth table** or **function table,** since it details the "truth" or the way in which

> **OR Gate**
>
> A logic gate that will give a HIGH output if either of its inputs are HIGH.
>
> **Truth Table or Function Table**
>
> A table used to show the action of a device as it reacts to all possible input combinations.

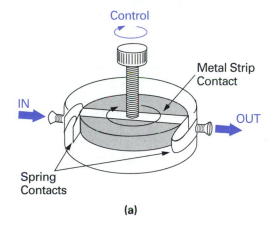

(a)

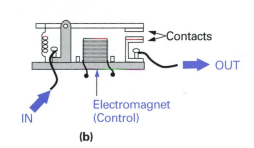

(b)

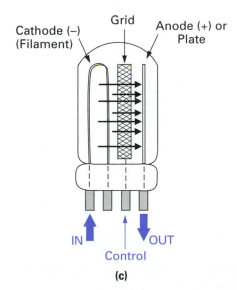

(c)

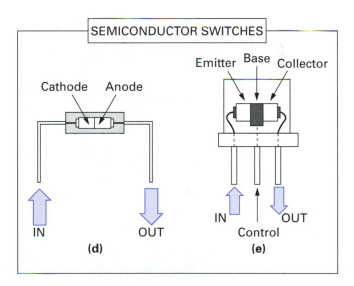

FIGURE 20-1   **The Evolution of the Switch for Digital Electronic Circuits (a) A Nineteenth-Century Mechanical Turn Switch (b) An Electromechanical Relay (c) A 1906 Vacuum Tube (d) A 1939 Semiconductor Diode (e) A 1948 Semiconductor Transistor.**

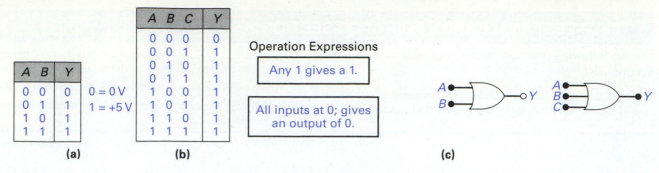

A	B	Y
0	0	0
0	1	1
1	0	1
1	1	1

$0 = 0\,V$
$1 = +5\,V$

(a)

A	B	C	Y
0	0	0	0
0	0	1	1
0	1	0	1
0	1	1	1
1	0	0	1
1	0	1	1
1	1	0	1
1	1	1	1

(b)

Operation Expressions

Any 1 gives a 1.

All inputs at 0; gives an output of 0.

(c)

FIGURE 20-2   The OR Gate (a) Two-Input Truth Table (b) Three-Input Truth Table (c) OR Gate Symbols and Operation Expressions.

this circuit will function. We could summarize the operation of this gate, therefore, by saying that *if either* A *OR* B *is HIGH, the output* Y *will be HIGH, and only when both inputs are LOW will the output be LOW.*

The truth table for a three-input OR gate is shown in Figure 20-2(b). Although another input has been added to this circuit, the action of the gate still remains the same, since any HIGH input will still produce a HIGH output, and only when all the inputs are LOW will the output be LOW.

Figure 20-2(c) shows the logic schematic symbols for a two- and three-input OR gate and the abbreviated operation expressions for the OR gate. Logic gates with a large number of inputs will have a large number of input combinations; however, the gate's operation will still be the same. To calculate the number of possible input combinations, you can use the following formula:

$$n = 2^x$$

where n = number of input combinations

2 is included because we are dealing with a base 2 number system

x = number of inputs

■   INTEGRATED MATH APPLICATION:

Construct a truth table for a four-input OR gate, and show the output logic level for every input combination.

■   *Solution:*

Figure 20-3(a) shows the truth table for a four-input OR gate. The number of possible input combinations will be

$$n = 2^x$$
$$n = 2^4 = 16$$

Calculator sequence: $\boxed{2}\ \boxed{y^x}\ \boxed{4}\ \boxed{=}$

With only two possible digits (0 and 1) and four inputs, we can have a maximum of 16 different input combinations. Whether the OR gate has 2, 3, 4, or 444 inputs, it will still operate in the same predictable or logical way: when all inputs are LOW, the output will be LOW, and when any input is HIGH, the output will be HIGH. This result is shown in the Y column in Figure 20-3(a).

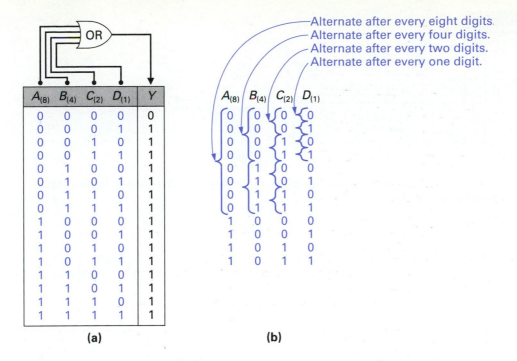

FIGURE 20-3 A Four-Input OR Gate.

An easy way to construct these truth tables is first to calculate the maximum number of input combinations and then start in the units (1s) column (D column) and move down, alternating the binary digits after every single digit (0101010101, and so on) up to the maximum count, as shown in Figure 20-3(b). Then go to the twos (2s) column (C column), and move down, alternating the binary digits after every 2 digits (001100110011, and so on) up to the maximum count. Then go to the fours (4s) column (B column), and move down, alternating the binary digits after every 4 digits (000011110000, and so on) up to the maximum count. Finally, go to the eights (8s) column (A column), and move down, alternating the binary digits after every 8 digits (0000000011111111, and so on) up to the maximum count.

These input combinations—or binary words—will start at the top of the truth table with binary 0 and then count up to a maximum value that is always 1 less than the maximum number of combinations. For example, with the four-input OR gate in Figure 20-3(a), the truth table begins with a count of 0000_2 (0_{10}) and then counts up to a maximum of 1111_2 (15_{10}). The maximum count within a truth table (1111_2 or 15_{10}) is always 1 less than the maximum number of combinations (16_{10}) because 0000_2 (0_{10}) is one of the input combinations (0000, 0001, 0010, 0011, 0100, 0101, 0110, 0111, 1000, 1001, 1010, 1011, 1100, 1101, 1110, 1111 is a total of 16 different combinations, with 15 being the maximum count). Stated in a formula:

$$\text{Count}_{max} = 2^x - 1$$

For example, in Figure 20-3: $\text{Count}_{max} = 2^4 - 1 = 16 - 1 = 15$

Calculator sequence: $\boxed{2}\ \boxed{y^x}\ \boxed{4}\ \boxed{-}\ \boxed{1}\ \boxed{=}$

20-2-2 *The AND Gate*

Like the OR gate, an **AND gate** can have two or more inputs but will always have a single output. Figure 20-4(a) shows the truth table for a two-input AND gate. We can summarize the operation of this gate by saying that *any LOW input will cause a LOW output, and only when both A AND B inputs are HIGH will the output Y be HIGH.*

AND Gate

A logic gate that will give a HIGH output only if all inputs are HIGH.

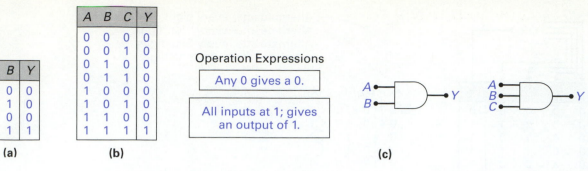

A	B	Y
0	0	0
0	1	0
1	0	0
1	1	1

(a)

A	B	C	Y
0	0	0	0
0	0	1	0
0	1	0	0
0	1	1	0
1	0	0	0
1	0	1	0
1	1	0	0
1	1	1	1

(b)

Operation Expressions

Any 0 gives a 0.

All inputs at 1; gives an output of 1.

(c)

FIGURE 20-4   The AND Gate (a) Two-Input Truth Table (b) Three-Input Truth Table (c) AND Gate Symbols.

Figure 20-4(b) shows the truth table for a three-input AND gate. Figure 20-4(c) shows the logic schematic symbols for a two- and three-input AND gate and the abbreviated operation expressions for the AND gate.

■   INTEGRATED MATH APPLICATION:

Develop a truth table for a five-input AND gate. Show the output logic level for every input combination, and indicate the range of values within the truth table.

■   *Solution:*

Figure 20-5 shows the truth table for a five-input AND gate. The number of possible input combinations will be

$$n = 2^x$$
$$n = 2^5 = 32$$

A five-input AND gate will still operate in the same predictable or logical way: when any input is LOW, the output will be LOW, and when all inputs are HIGH, the output will be HIGH. This result is shown in the Y column in Figure 20-5.

The range of values within the truth table will be

$$\text{Count}_{max} = 2^5 - 1 = 32 - 1 = 31$$

With the five-input OR gate in Figure 20-5, the truth table begins with a count of 00000_2 (0_{10}) and then counts up to a maximum of 11111_2 (31_{10}).

■   INTEGRATED MATH APPLICATION:

Referring to the circuit in Figure 20-6(a), describe what would be present at the output of the AND gate for each of the CONTROL switch positions.

■   *Solution:*

When the CONTROL switch is put in the ENABLE position, a HIGH is applied to the lower input of the AND gate, as shown in Figure 20-6(b). In this mode, the output Y will follow the square-wave input, since when the square-wave is HIGH, the AND gate inputs are both HIGH, and Y is HIGH. When the square-wave is LOW, the AND gate will have a LOW input, and Y will be LOW. The Y output, therefore, follows the square-wave input, and the AND gate is said to be equivalent to a closed switch, as shown in Figure 20-6(b). On the other hand, when the CONTROL switch is put in the DISABLE position, a LOW is applied to the lower input of the AND gate, as shown in Figure 20-6(c). In this mode, the output Y will always remain LOW, since any LOW input to an AND gate will always result in a LOW

FIGURE 20-5   A Five-Input AND Gate.

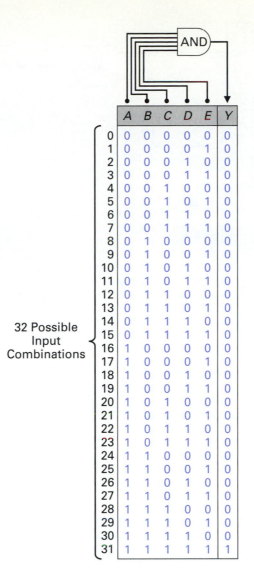

	A	B	C	D	E	Y
0	0	0	0	0	0	0
1	0	0	0	0	1	0
2	0	0	0	1	0	0
3	0	0	0	1	1	0
4	0	0	1	0	0	0
5	0	0	1	0	1	0
6	0	0	1	1	0	0
7	0	0	1	1	1	0
8	0	1	0	0	0	0
9	0	1	0	0	1	0
10	0	1	0	1	0	0
11	0	1	0	1	1	0
12	0	1	1	0	0	0
13	0	1	1	0	1	0
14	0	1	1	1	0	0
15	0	1	1	1	1	0
16	1	0	0	0	0	0
17	1	0	0	0	1	0
18	1	0	0	1	0	0
19	1	0	0	1	1	0
20	1	0	1	0	0	0
21	1	0	1	0	1	0
22	1	0	1	1	0	0
23	1	0	1	1	1	0
24	1	1	0	0	0	0
25	1	1	0	0	1	0
26	1	1	0	1	0	0
27	1	1	0	1	1	0
28	1	1	1	0	0	0
29	1	1	1	0	1	0
30	1	1	1	1	0	0
31	1	1	1	1	1	1

32 Possible Input Combinations

output. In this situation, the AND gate is said to be equivalent to an open switch, as shown in Figure 20-6(c).

When used in applications such as this, the AND gate is said to be acting as a **controlled switch.**

Controlled Switch
An electronically controlled switch.

SELF-TEST EVALUATION POINT FOR SECTION 20-2

Now that you have completed this section, you should be able to:

■ **Objective 3.**   *Describe the discrete circuit construction, truth table, operation, and an application for the following basic logic-gate types:*

   a. The OR Gate
   b. The AND Gate

Use the following questions to test your understanding of Section 20-2:

1. The two basic logic gates are the _____ gate and the _____ gate.

2. With an OR gate, any binary 1 input will give a binary _____ output.

3. Which logic gate can be used as a controlled switch?

4. Only when all inputs are LOW will the output of an _____ gate be LOW?

5. With an AND gate, any binary 0 input will give a binary _____ output.

6. Only when all inputs are HIGH will the output of an _____ gate be HIGH.

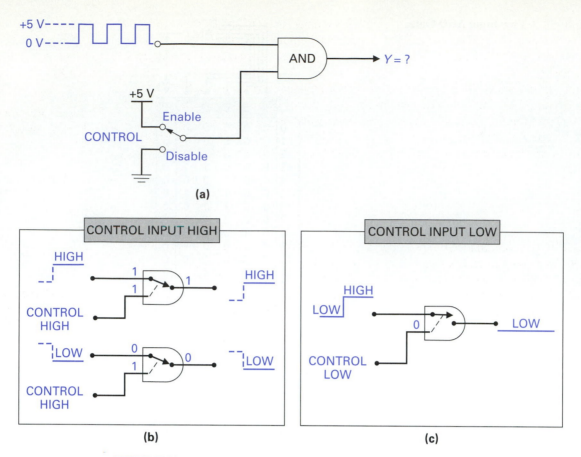

FIGURE 20-6 An Enable/Disable Control Circuit Using an AND Gate.

20-3 INVERTING LOGIC GATES

Although the basic OR and AND logic gates can be used to construct many digital circuits, in some circuit applications other logic operations are needed. For this reason, semiconductor manufacturers make other logic-gate types available. In this section, we will examine three inverting-type logic gates, called the NOT gate, the NOR gate, and the NAND gate. We will begin with the NOT or INVERTER gate.

20-3-1 *The NOT Gate*

The NOT, or logic INVERTER gate, is the simplest of all the logic gates because it has only one input and one output. Figure 20-7(a) shows the truth table for this gate. We can summarize

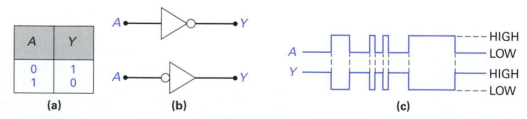

A	Y
0	1
1	0

(a) (b) (c)

FIGURE 20-7 The NOT Gate or Logic Inverter (a) Truth Table (b) Logic Symbols for a NOT Gate (c) Timing Diagram Showing Input-to-Output Inversion.

the operation of this gate by saying that *the output logic level is "NOT" the same as the input logic level* (a 1 input is NOT the same as the 0 output, or a 0 input is NOT the same as the 1 output), hence the name NOT gate. The output is, therefore, always the **complement,** or opposite, of the input.

Complement
Opposite.

Figure 20-7(b) shows the schematic symbols normally used to represent the NOT gate. The triangle is used to represent the logic circuit, and the **bubble,** or small circle either before or after the triangle, is used to indicate the complementary or inverting nature of this circuit. Figure 20-7(c) shows a typical example of a NOT gate's input/output waveforms. As you can see from these waveforms, the output waveform is an inverted version of the input waveform.

Bubble
A small circle symbol used to signify an invert function.

20-3-2 *The NOR Gate*

The name NOR is a combination of the two words NOT and OR and is used to describe a type of logic gate that contains an OR gate followed by a NOT gate, as shown in Figure 20-8(a).

Figure 20-8(b) shows the standard symbol used to represent the **NOR gate.** Studying the NOR gate symbol, you can see that the OR symbol is used to show how the two inputs are first ORed, and the bubble is used to indicate that the result from the OR operation is inverted before being applied to the output. As a result, the NOR gate's output is simply the opposite, or complement, of the OR gate, as can be seen in the truth table for a two-input NOR gate shown in Figure 20-8(c). If you compare the NOR gate truth table with the OR gate truth table, you can see that the only difference is that the output is inverted:

NOR Gate
A NOT-OR gate that will give a LOW output if any of its inputs are HIGH.

OR gate: When all inputs are 0, the output is **0,** and any 1 input gives a **1** output.

NOR gate: When all inputs are 0, the output is **1,** and any 1 input gives a **0** output.

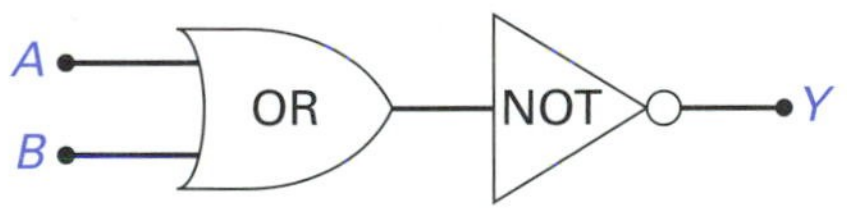

(a)

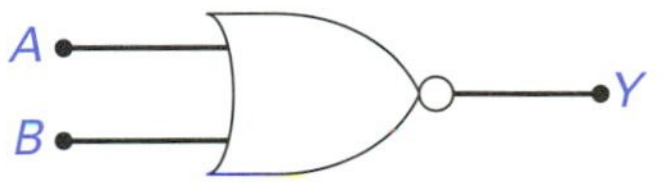

(b)

A	B	Y
0	0	1
0	1	0
1	0	0
1	1	0

(c)

Operation Expressions

Any 1 gives a 0.

All inputs at 0; gives an output of 1.

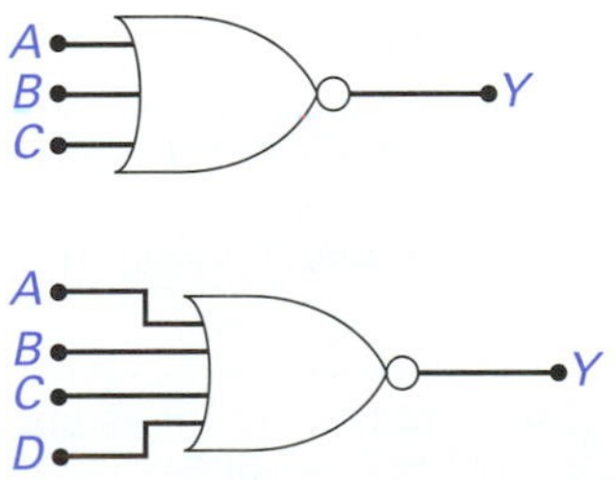

(d)

FIGURE 20-8 **The NOR Gate (a) Two-Input Circuit (b) Two-Input Logic Symbol (c) Two-Input Truth Table (d) NOR Gate Operation Expressions and Three- and Four-Input Logic Symbols.**

The operation expressions for the NOR gate are listed in Figure 20-8(d), along with the standard symbols for a three-input and a four-input NOR gate.

20-3-3 *The NAND Gate*

NAND Gate

A NOT-AND logic-gate circuit that will give a HIGH output if any of its inputs are low.

Like NOR, the name NAND is a combination of the two words NOT and AND, and it describes another type of logic gate that contains an AND gate followed by a NOT gate, as shown in Figure 20-9(a).

Figure 20-9(b) shows the standard symbol used to represent the **NAND gate.** The AND symbol is used to show how the two inputs are first "ANDed," and the triangle followed by the bubble is used to indicate that the result from the AND operation is inverted before being applied to the output. As a result, the NAND gate's output is simply the opposite, or complement, of the AND gate, as can be seen in the truth table for a two-input NAND gate shown in Figure 20-9(c). If you compare the NAND gate truth table with the AND gate truth table, you can see that the only difference is that the output is inverted. To compare:

AND Gate: Any 0 input gives a **0** output, and only when both inputs are 1 will the output be **1.**

NAND Gate: Any 0 input gives a **1** output, and only when both inputs are 1 will the output be **0.**

The operation expressions for the NAND gate are listed in Figure 20-9(d), along with the standard symbols for a three-input and four-input NAND gate.

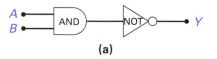

(a)

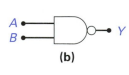

(b)

A	B	Y
0	0	1
0	1	1
1	0	1
1	1	0

(c)

Operation Expressions

Any 0 gives a 1.

All inputs at 1; gives an output of 0.

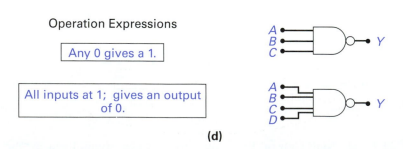

(d)

FIGURE 20-9   **The NAND Gate (a) Two-Input Circuit (b) Two-Input Logic Symbol (c) Two-Input Truth Table (d) NAND Gate Operation Expressions and Three- and Four-Input Logic Symbols.**

Now that you have completed this section, you should be able to:

■ *Objective 4.* *Describe the discrete circuit construction, truth table, operation, and an application for the following inverting logic-gate types:*

   a. The NOT Gate
   b. The NOR Gate
   c. The NAND Gate

Use the following questions to test your understanding of Section 20-3:

1. Which of the logic gates has only one input?
2. With a NOT gate any binary 1 input will give a binary _____ output.
3. Which logic gate is a combination of an AND gate followed by a logic INVERTER?
4. Only when all inputs are LOW, will the output of a(n) _____ gate be HIGH.
5. With a NAND gate, any binary 0 input will give a binary _____ output.
6. Only when all inputs are HIGH, will the output of a(n) _____ gate be LOW.

20-4   EXCLUSIVE LOGIC GATES

In this section, we will discuss the final two logic-gate types: the exclusive-OR (XOR) gate and exclusive-NOR (XNOR) gate. Although these two logic-gates are not used as frequently as the five basic OR, AND, NOT, NOR, and NAND gates, their function is ideal in some applications.

20-4-1   *The XOR Gate*

The **exclusive-OR (XOR) gate** logic symbol is shown in Figure 20-10(a), and its truth table is shown in Figure 20-10(b). As with the basic OR gate, the operation of the XOR is dependent

Exclusive-OR Gate
A logic-gate circuit that will give a HIGH output if any odd number of binary 1s are applied to the input.

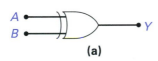

(a)

	A	B	Y
Even	0	0	0
Odd	0	1	1
Odd	1	0	1
Even	1	1	0

(b)

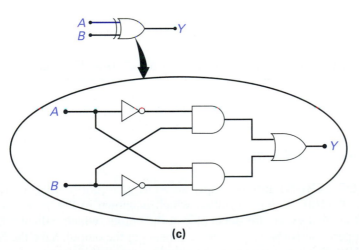

(c)

FIGURE 20-10   **The Exclusive-OR (XOR) Gate. (a) Two-Input Logic Symbol. (b) Two-Input Truth Table. (c) Basic Gate Circuit.**

on the HIGH or binary 1 inputs. With the basic OR, any 1 input will cause a 1 output. With the XOR, any odd number of binary 1s at the input will cause a binary 1 at the output. Looking at the truth table in Figure 20-10(b), you can see that when the binary inputs are 01 or 10, there is one binary 1 at the input (odd), and the output is 1. When the binary inputs are 00 or 11, however, there are two binary 0s or two binary 1s at the input (even), and the output is 0. To distinguish the XOR gate symbol from the basic OR gate symbol, you may have noticed that an additional curved line is included across the input.

Exclusive-OR logic gates are constructed by combining some of the previously discussed basic logic gates. Figure 20-10(c) shows how two NOT gates, two AND gates, and an OR gate can be connected to form an XOR gate. If inputs A and B are both HIGH or both LOW (even input), the AND gates will both end up with a LOW at one of their inputs, so the OR gate will have both of its inputs LOW, and therefore the final output will be LOW. On the other hand, if A is HIGH and B is LOW or if B is HIGH and A is LOW (odd input), one of the AND gates will have both of its inputs HIGH, giving a HIGH to the OR gate and, therefore, a HIGH to the final output. We can summarize the operation of this gate therefore by saying that *the output is HIGH only when there is an odd number of binary 1s at the input.*

■ INTEGRATED MATH APPLICATION:

List the outputs you would expect for all the possible inputs to a three-input XOR gate.

■ *Solution:*

$$n = 2^x$$
$$n = 2^3 = 8$$

A	B	C	Y	
0	0	0	0	Even
0	0	1	1	Odd
0	1	0	1	Odd
0	1	1	0	Even
1	0	0	1	Odd
1	0	1	0	Even
1	1	0	0	Even
1	1	1	1	Odd

The output is 1 only when there is an odd number of binary 1s at the input.

20-4-2   *The XNOR Gate*

Exclusive-NOR Gate

A NOT-exclusive OR gate that will give a HIGH output if any even number of binary 1s are applied to the input.

The **exclusive-NOR (XNOR) gate** is simply an XOR gate followed by a NOT gate, as shown in Figure 20-11(a). Its logic symbol, which is shown in Figure 20-11(b), is the same as the XOR symbol except for the bubble at the output, which indicates that the result of the XOR operation is inverted before it appears at the output. Like the NAND and NOR gates, the XNOR is not a new gate but simply a previously discussed gate with an inverted output.

CHAPTER 20 / LOGIC GATES

The truth table for the XNOR gate is shown in Figure 20-11(c). Comparing the XOR gate with the XNOR gate, you can see that the XOR gate's output is HIGH only when there is an odd number of binary 1s at the input, whereas with the XNOR gate, the output is HIGH only when there is an even number of binary 1s at the input. The operation expression for the XNOR gate is listed in Figure 20-11(d).

CALCULATOR KEYS

Name: Comparing or manipulating bits

Function: Some calculators let you compare or manipulate bits in a binary number. You can enter an integer in any number base. Your entries are converted to binary automatically for the operation, and results are displayed according to the Base mode.

Example:

OPERATOR WITH SYNTAX	DESCRIPTION
not *integer*	Returns the one's complement, in which each bit is flipped.
$(-)$ *integer*	Returns the two's complement, which is the one's complement +1.
integer1 **and** *integer2*	In a bit-by-bit **and** comparison, the result is 1 if both bits are 1; otherwise, the result is 0. The returned value represents the bit results.
integer1 **or** *integer2*	In a bit-by-bit **or** comparison, the result is 1 if either bit is 1; the result is 0 only if both bits are 0. The returned value represents the bit results.
integer1 **xor** *integer2*	In a bit-by-bit **xor** comparison, the result is 1 if either bit (but not both) is 1; the result is 0 if both bits are 0 or both bits are 1. The returned value represents the bit results.

Suppose you enter:

0h7AC36 and 0h3D5F

If Base mode = HEX:

■ 0h7AC36 and 0h3D5F	
	0h2C16
0h7ac36 and 0h3d5f	
MAIN RAD AUTO FUNC 1/30	

Internally, the hexadecimal integers are converted to a signed, 32-bit binary number.

If Base mode = BIN:

■ 0h7AC36 and 0h3D5F	
	0b0110000010110
0h7ac36 and 0h3d5f	
MAIN RAD AUTO FUNC 1/30	

Then corresponding bits are compared.

0h7AC36 = 0b00000000000001111010110000110110

   and   **and**

$$0h3D5F = \frac{0b00000000000000000011110101011111}{0b00000000000000000010110000010110} = 0h2C16$$

Leading zeros are not shown in the result.

The result is displayed according to the Base mode.

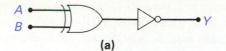

(a)

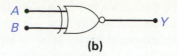

(b)

A	B	Number of 1s	Y
0	0	Even	1
0	1	Odd	0
1	0	Odd	0
1	1	Even	1

(c)

Operation Expression

Even number of 1s at the input
gives a 1 at the output.

(d)

FIGURE 20-11　The Exclusive-NOR (XNOR) Gate (a) Circuit (b) Logic Symbol (c) Truth Table (d) Operation Expression.

SELF-TEST EVALUATION POINT FOR SECTION 20-4

Now that you have completed this section, you should be able to:

■ *Objective 5.*　*Describe the discrete circuit construction, truth table, operation, and an application for the following exclusive logic-gate types.*
　　a.　The XOR Gate
　　b.　The XNOR Gate

Use the following questions to test your understanding of

Section 20-4:

1. Which type of logic gate will always give a HIGH output when any of its inputs are HIGH?
2. Only when an odd number of 1s is applied to the input, will the output of an _____ gate be HIGH.
3. Only when an even number of 1s is applied to the input, will the output of an _____ gate be HIGH.
4. Which type of logic gate will always give a LOW output when an odd number of 1s is applied at the input?

20-5　IEEE/ANSI SYMBOLS FOR LOGIC GATES

The logic symbols presented so far in this chapter have been used for many years in the digital electronics industry. In 1984 the *Institute of Electrical and Electronics Engineers (IEEE)* and the *American National Standards Institute (ANSI)* introduced a new standard for logic symbols, which is slowly being accepted by more and more electronics companies. The advantage of this new standard is that instead of using distinctive shapes to represent logic gates, it uses a special **dependency notation system.** Simply stated, the new standard uses a *notation* (or note) within a rectangular or square block to indicate how the output is *dependent* on the input. Figure 20-12 compares traditional logic symbols with the newer IEEE/ANSI logic symbols and describes the meaning of the dependency notations.

Dependency Notation System

A coding system used on schematic diagrams that uses notations to indicate how an output is dependent on inputs.

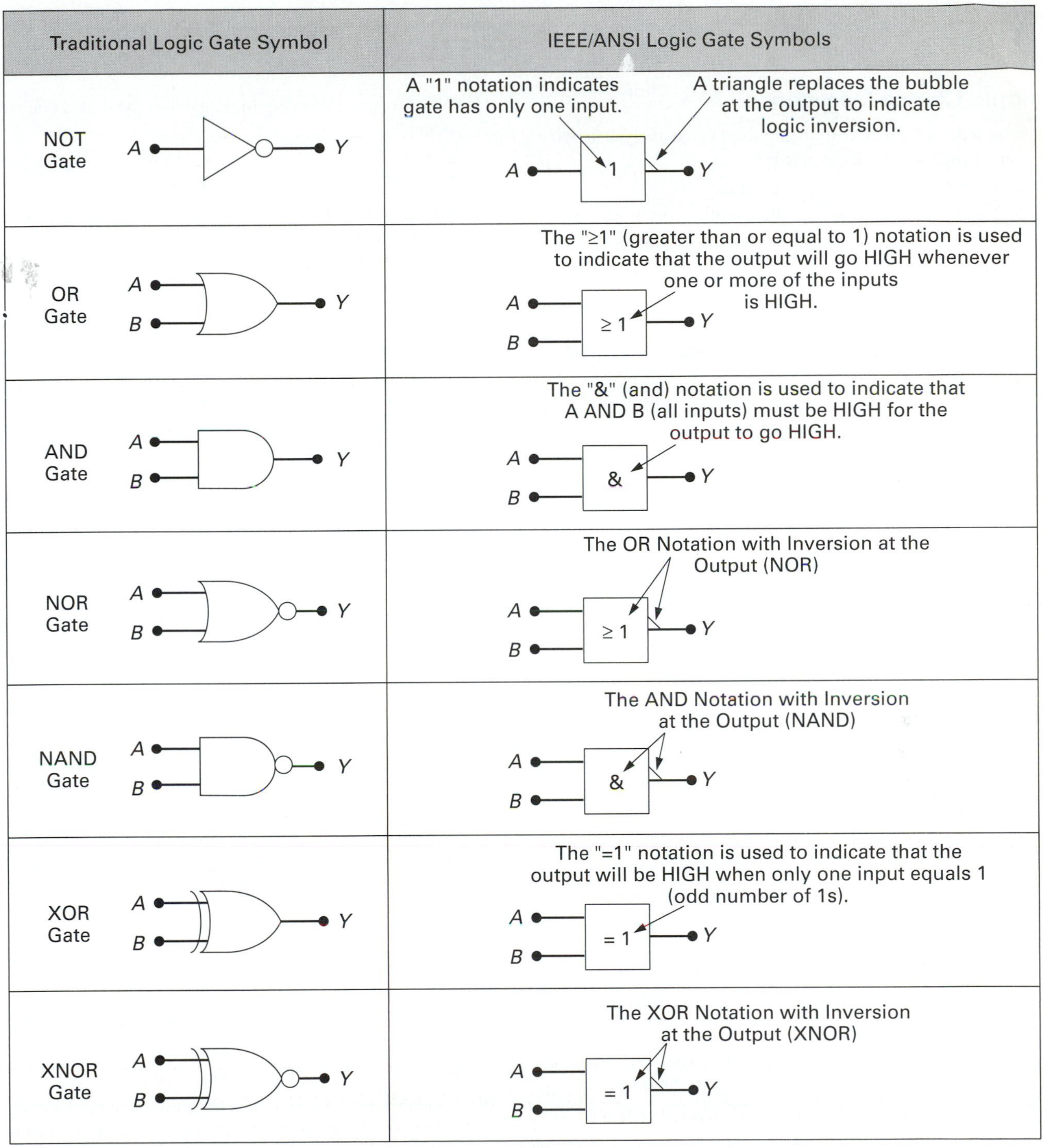

FIGURE 20-12   **Traditional and IEEE/ANSI Symbols for Logic Gates.**

Now that you have completed this section, you should be able to:

■ **Objective 6.** *Explain the major differences between the traditional and IEEE/ANSI logic-gate symbols.*

Use the following questions to test your understanding of Section 20-5:

1. Which type of logic-gate symbols make use of rectangular blocks?

2. What advantage does the IEEE/ANSI standard have over the traditional symbol standard?

REVIEW QUESTIONS

Multiple Choice Questions

1. Which device is currently being used extensively as a two-state switch in digital electronic circuits?
   - **a.** Transistor
   - **b.** Vacuum tube
   - **c.** Relay
   - **d.** Toggle switch

2. How many input combinations would a four-input logic gate have?
   - **a.** 8
   - **b.** 16
   - **c.** 32
   - **d.** 4

3. What would be the maximum count within the truth table for a four-input logic gate?
   - **a.** 7
   - **b.** 15
   - **c.** 17
   - **d.** 31

4. What would be the output from a three-input OR gate if its inputs were 101?
   - **a.** 1
   - **b.** 0
   - **c.** Unknown
   - **d.** None of the above

5. What would be the output from a three-input AND gate if its inputs were 101?
   - **a.** 1
   - **b.** 0
   - **c.** Unknown
   - **d.** None of the above

6. What would be the output from a three-input NAND gate if its inputs were 101?
   - **a.** 1
   - **b.** 0
   - **c.** Unknown
   - **d.** None of the above

7. What would be the output from a three-input NOR gate if its inputs were 101?
   - **a.** 1
   - **b.** 0
   - **c.** Unknown
   - **d.** None of the above

8. What would be the output from a three-input XOR gate if its inputs were 101?
   - **a.** 1
   - **b.** 0
   - **c.** Unknown
   - **d.** None of the above

9. What would be the output from a three-input XNOR gate if its inputs were 101?
   - **a.** 1
   - **b.** 0
   - **c.** Unknown
   - **d.** None of the above

10. What would be the output from a NOT gate if its input was 1?
    - **a.** 1
    - **b.** 0
    - **c.** Unknown
    - **d.** None of the above

11. Which of the following logic gates is ideal as a controlled switch?
    - **a.** OR
    - **b.** AND
    - **c.** NOR
    - **d.** NAND

12. Which of the following logic gates will always give a LOW output whenever a LOW input is applied?
    - **a.** OR
    - **b.** AND
    - **c.** NOR
    - **d.** NAND

13. Which of the following logic gates will always give a HIGH output whenever a LOW input is applied?
    - **a.** OR
    - **b.** AND
    - **c.** NOR
    - **d.** NAND

14. Which of the following logic gates will always give a HIGH output whenever a HIGH input is applied?
    - **a.** OR
    - **b.** AND
    - **c.** NOR
    - **d.** NAND

15. Which of the following logic gates will always give a LOW output whenever a HIGH input is applied?
    - **a.** OR
    - **b.** AND
    - **c.** NOR
    - **d.** NAND

16. An XOR gate will always give a HIGH output whenever an _____ number of 1s is applied to the input.
    - **a.** odd
    - **b.** even
    - **c.** unknown
    - **d.** none of the above

17. An XNOR gate will always give a HIGH output whenever an _____ number of 1s is applied to the input.
    - **a.** odd
    - **b.** even
    - **c.** unknown
    - **d.** none of the above

18. Which of the dependency notation logic symbols contains an "&" within the square block and no triangle at the output?
    - **a.** OR
    - **b.** AND
    - **c.** NOR
    - **d.** NAND

19. Which of the dependency notation logic symbols contains a "≥1" within the square block and no triangle at the output?
    - **a.** OR
    - **b.** AND
    - **c.** NOR
    - **d.** NAND

20. Which of the dependency notation logic symbols contains a "≥1" within the square block and a triangle at the output?
    - **a.** OR
    - **b.** AND
    - **c.** NOR
    - **d.** NAND

Communication Skill Questions

21. List some of the two-state devices that have been used in digital electronic switches. (20-1)
22. What were the reasons for using two-state switches instead of 10-state switches? (20-1)
23. Sketch and describe how the semiconductor diode can be used to construct a logic-gate circuit. (20-1)
24. What is the meaning of the term *logic gate?* (20-1)
25. Sketch and describe how the semiconductor transistor can be used to construct a logic-gate circuit. (20-1)
26. Sketch the traditional logic symbols used for the seven logic-gate types. (20-2, 20-3)
27. Describe the differences between the truth tables for the OR gate and the AND gate. (20-2)

28. Describe the basic differences between the following: (20-2 and 20-3)
    a. The OR and the NOR gate.
    b. The AND and the NAND gate.
29. Give the full names for the following abbreviations. (20-3, 20-4, 20-5)
    a. XOR   e. NAND
    b. &   f. XNOR
    c. = 1   g. NOR
    d. ≥1
30. Sketch the truth table and the IEEE/ANSI logic gate symbols for the OR, AND, NOT, NOR, NAND, XOR, and XNOR gates. (20-5)

Practice Problems

31. In the home security system shown in Figure 20-13(a), a two-input logic gate is needed to actuate an alarm (output Y) if either the window is opened (input A) or the door is opened (input B). Which decision-making logic gate should be used in this application?
32. In the office temperature-control system shown in Figure 20-13(b), a thermostat (input A) is used to turn a fan (output Y) ON and OFF. A thermostat-enable switch is connected to the other input of the logic gate (input B), and it will either enable thermostat control of the fan or disable thermostat control of the fan. Which decision-making logic gate should be used in this application?
33. Which logic gate would produce the output waveform shown in Figure 20-14?
34. Choose one of the six logic gates shown in Figure 20-15(a) and then sketch the output waveform that would result for the input waveforms shown in Figure 20-15(b).
35. Develop truth tables for each of the logic circuits shown in Figure 20-16.

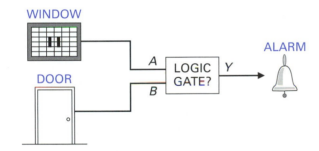

WINDOW	DOOR	ALARM
Closed (0)	Closed (0)	OFF (0)
Closed (0)	Open (1)	ON (1)
Open (1)	Closed (0)	ON (1)
Open (1)	Open (1)	ON (1)

(a)

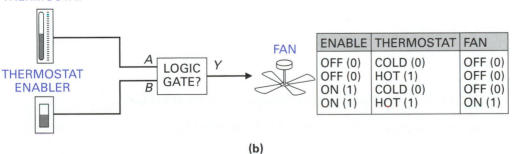

ENABLE	THERMOSTAT	FAN
OFF (0)	COLD (0)	OFF (0)
OFF (0)	HOT (1)	OFF (0)
ON (1)	COLD (0)	OFF (0)
ON (1)	HOT (1)	ON (1)

(b)

FIGURE 20-13 Decision-Making Logic Gates.

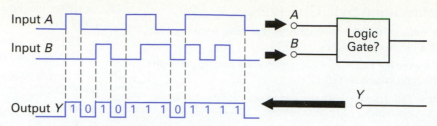

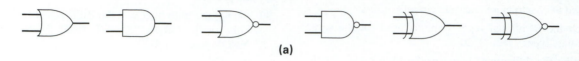

FIGURE 20-14 Timing Analysis of a Logic Gate.

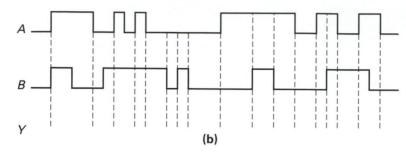

(a)

(b)

FIGURE 20-15 Timing Waveforms.

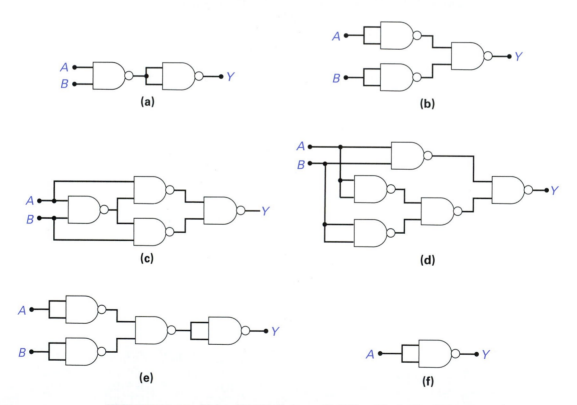

FIGURE 20-16 Using NAND Gates as Building Blocks for Other Gates.

Web Site Questions

Boolean Expressions and Algebra

From Folly to Foresight

George Boole was born in the industrial town of Lincoln in eastern England in 1815. His parents were poor trade people, and even though there was a school for boys in Lincoln, there is no record of his ever having attended. In those hard times, children of the working class had no hope of receiving any form of education, and their lives generally followed the familiar pattern of their parents'. George Boole, however, was to break the mold. He would rise up from these humble beginnings to become one of the most respected mathematicians of his day.

Boole's father had taught himself a small amount of mathematics, and since his son of six seemed to have a thirst for learning, he began to pass on all his knowledge. At 8 years old, George had surpassed his father's understanding and craved more. George quickly realized that his advancement was heavily dependent on his understanding Latin. Luckily, a family friend who owned a local bookshop knew enough about the basics of Latin to get him on his way, and once he had taught Boole all he knew, Boole continued with the books at his disposal. By the age of 12 he had conquered Latin, and by 14 he had added Greek, French, German, and Italian to his repertoire.

At the age of 16, however, poverty stood in his way. Since his parents could no longer support him, he was forced to take a job as a poorly paid teaching assistant. After studying the entire school system, he left 4 year laters and opened his own school in which he taught all subjects. It was in this role that he discovered that his mathematics was weak, so he began studying the mathematical journals at the local library in an attempt to stay ahead of his students. He quickly discovered that he had a talent for mathematics. As well as mastering all the present-day ideas, he began to develop some of his own, which were later accepted for publication. After a stream of articles, he became so highly regarded that he was asked to join the mathematics faculty at Queens College in 1849.

After accepting the position, Boole concentrated more on his ideas, one of which was to develop a system of symbolic logic. In this system he created a form of algebra that had its own set of symbols and rules. Using this system, Boole could encode any statement that had to be proven (a proposition) into his symbolic language and then manipulate it to determine whether it was true or false. Boole's algebra had three basic operations that are often called logic functions—AND, OR, and NOT. Using these three operations, Boole could perform such operations as add, subtract, multiply, divide, and compare. These logic functions were binary in nature, and therefore dealt with only two entities—TRUE or FALSE, YES or NO, OPEN or CLOSED, ZERO or ONE, and so on. Boole's theory was that if all logical arguments could be reduced to one of two basic levels, the questionable middle ground would be removed, making it easier to arrive at a valid conclusion.

At the time, Boole's system, which was later called Boolean algebra, was either ignored or criticized by colleagues, who called it a folly with no practical purpose. Almost a century later, however, scientists would combine George Boole's Boolean algebra with binary numbers and make possible the digital electronic computer.

Outline and Objectives

VIGNETTE: FROM FOLLY TO FORESIGHT

INTRODUCTION

21-1 BOOLEAN EXPRESSIONS FOR LOGIC GATES

Objective 1: Give the Boolean expressions for each of the seven basic logic gates.

Objective 2: Describe how to write an equivalent Boolean expression for a logic circuit and how to sketch an equivalent logic circuit for a Boolean expression.

21-1-1 The NOT Expression

21-1-2 The OR Expression

21-1-3 The AND Expression

21-1-4 The NOR Expression

21-1-5 The NAND Expression

21-1-6 The XOR Expression

21-1-7 The XNOR Expression

21-2 BOOLEAN ALGEBRA LAWS AND RULES

Objective 3: List and explain the laws and rules that apply to Boolean algebra.

21-2-1 The Commutative Law

21-2-2 The Associative Law

21-2-3 The Distributive Law

21-2-4 Boolean Algebra Rules
OR Gate Rules
AND Gate Rules
Double-Inversion Rule
DeMorgan's Theorems
The Duality Theorem

21-3 FROM TRUTH TABLE TO GATE CIRCUIT

Objective 4: Describe the logic circuit design process from truth table to gate circuit.

Objective 5: Explain the difference between a sum-of-products equation and a product-of-sums equation.

21-4 GATE CIRCUIT SIMPLIFICATION

Objective 6: Describe how Boolean algebra and Karnaugh maps can be used to simplify logic circuits.

21-4-1 Boolean Algebra Simplification

21-4-2 Karnaugh Map Simplification

MULTIPLE CHOICE QUESTIONS

COMMUNICATION SKILLS QUESTIONS

PRACTICE PROBLEMS

WEB SITE QUESTIONS

In 1854 George Boole invented a symbolic logic that linked mathematics and logic. Boole's logical algebra, which today is known as **Boolean algebra,** states that each variable (input or output) can assume one of two values or states—true or false.

Boolean algebra had no practical application until Claude Shannon used it in 1938 to analyze telephone switching circuits in his MIT thesis. In his paper he described how the two variables of the Boolean algebra (true and false) could be used to represent the two states of the switching relay (open and closed).

Today the mechanical relays have been replaced by semiconductor switches; however, Boolean algebra is still used to express both simple and complex two-state logic functions in a convenient mathematical format. These mathematical expressions of logic functions make it easier for technicians to analyze digital circuits, and are a primary design tool for engineers. Boolean algebra allows circuits to be made simpler, less expensive, and more efficient.

In this chapter we will discuss the basics of Boolean algebra and then see how it can be applied to logic-gate circuit simplification.

Boolean Algebra

A form of algebra invented by George Boole that deals with classes, propositions, and ON/OFF circuit elements associated with operators such as NOT, OR, NOR, AND, NAND, XOR, and XNOR.

21-1   BOOLEAN EXPRESSIONS FOR LOGIC GATES

The operation of each of the seven basic logic gates can be described with a Boolean expression. To explain this in more detail, let us first consider the most basic of all the logic gates, the NOT gate.

21-1-1   *The NOT Expression*

In Figure 21-1 you can see the previously discussed NOT gate, and because of the inversion that occurs between input and output, Y is always equal to the opposite, or complement, of input A.

$$Y = \text{NOT } A$$

Therefore, if the input is 0, the output will be 1.

$$Y = \text{NOT } 0 = 1$$

If, on the other hand, the A input is 1, the output will be 0.

$$Y = \text{NOT } 1 = 0$$

In Boolean algebra, this inversion of the A input is indicated with a bar over the letter A as follows:

$$Y = \overline{A} \text{ (pronounced "Y equals not A")}$$

Boolean Expression
$Y = \overline{A}$

FIGURE 21-1   The Boolean Expression for a NOT Gate.

CHAPTER 21 / BOOLEAN EXPRESSIONS AND ALGEBRA

Using this Boolean expression, we can easily calculate the output Y for either of the two input A conditions. For example, if $A = 0$,

$$Y = \overline{A} = \overline{0} = 1$$

If, on the other hand, the A input is 1,

$$Y = \overline{A} = \overline{1} = 0$$

21-1-2   *The OR Expression*

The operation of the OR gate can also be described with a Boolean expression, as seen in Figure 21-2(a). In Boolean algebra, a + sign is used to indicate the OR operation.

> $Y = A + B$ (pronounced "Y equals A or B")

The expression $Y = A + B$ (Boolean equation) is the same as $Y = A$ OR B (word equation). If you find yourself wanting to say "plus" instead of "OR," don't worry; it will require some practice to get used to using this traditional symbol in a new application. Just as the word "wind" can be used in two different ways ("a gusty wind" or "to wind a clock"), then so can the + symbol be used to mean either addition or the OR operation. With practice you will get more comfortable with this symbol's dual role.

If OR gates have more than two inputs, as seen in Figure 21-2(b) or (c), the OR expression is simply extended, as follows.

For a three-input OR gate: $Y = A + B + C$

For a four-input OR gate: $Y = A + B + C + D$

The truth table for a two-input OR gate has been repeated in Figure 21-2(d). Using the Boolean expression for this gate, we can determine the output Y by substituting all the possible 1 and 0 input combinations for the input variables A and B. Studying this truth table you can see that the output of the OR gate is a 1 when either A OR B is 1. George Boole described the OR function as *Boolean addition,* which is why the "plus" (addition) symbol was used. Boolean addition, or *logical addition,* is different from normal addition, and the OR gate is said to produce at its output the *logical sum* of the inputs.

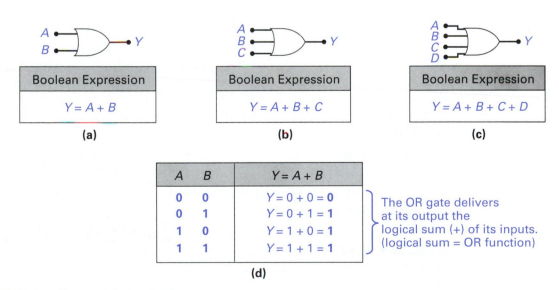

FIGURE 21-2   The Boolean Expression for an OR Gate.

A	B	Y = A • B
0	**0**	Y = 0 • 0 = **0**
0	**1**	Y = 0 • 1 = **0**
1	**0**	Y = 1 • 0 = **0**
1	**1**	Y = 1 • 1 = **1**

The AND gate delivers at its output the logical product (•) of its inputs. (logical product = AND function)

(d)

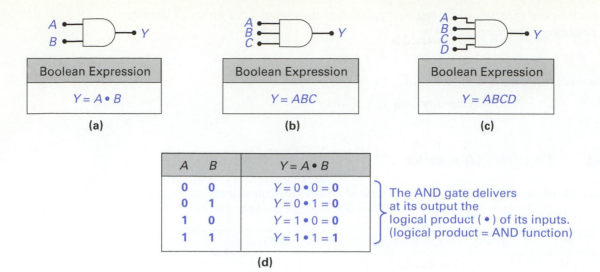

FIGURE 21-3 The Boolean Expression for an AND Gate.

21-1-3 *The AND Expression*

In Boolean algebra, the AND operation is indicated with a ·, and therefore the Boolean expression for the AND gate shown in Figure 21-3(a) would be

$Y = A \cdot B$ (pronounced "Y equals A and B")

This · is also used to indicate multiplication, and therefore like the + symbol, the · symbol has two meanings. In multiplication, the · is often dropped; however, the meaning is still known to be multiplication. For example:

$$V = I \times R \text{ is the same as } V = I \cdot R \text{ or } V = IR$$

In Boolean algebra the same situation occurs, in that

$$Y = A \cdot B \text{ is the same as } Y = AB$$

If AND gates have more than two inputs, as seen in Figure 21-3(b) and (c), the AND expression is simply extended.

The truth table for a two-input AND gate is shown in Figure 21-3(d). Using the Boolean expression, we can calculate the output Y by substituting all the possible 1 and 0 input combinations for the input variables A and B. Looking at the truth table, you can see that the output of the AND gate is a 1 only when both A and B are 1. George Boole described the AND function as *Boolean multiplication* or *logical multiplication,* which is why the period (multiplication) symbol was used. Boolean multiplication is different from standard multiplication as you have seen, and the AND gate is said to produce at its output the *logical product* of its inputs.

■ INTEGRATED MATH APPLICATION:

What would be the Boolean expression for the circuit shown in Figure 21-4, and what would be the output if $A = 0$ and $B = 1$?

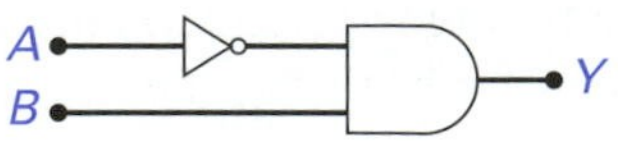

FIGURE 21-4 Single Bubble Input AND Gate.

■ *Solution:*

The *A* input in Figure 21-4 is first inverted before it is ANDed with the *B* input. The Boolean equation therefore, would be

$$Y = \overline{A} \cdot B$$

Using this expression, we can substitute the *A* and *B* for the 0 and 1 inputs, and then determine the output *Y.*

$$Y = \overline{A} \cdot B = \overline{0} \cdot 1 = 1 \cdot 1 = 1$$
(*Y* equals NOT *A* and *B* = NOT 0 AND 1 = 1 AND 1 = 1)

■ **INTEGRATED MATH APPLICATION:**

Give the Boolean expression and truth table for the logic circuit shown in Figure 21-5(a).

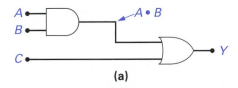

(a)

A	B	C	$Y = AB + C$
0	0	0	$Y = (0 \bullet 0) + 0 = 0 + 0 = \mathbf{0}$
0	0	1	$Y = (0 \bullet 0) + 1 = 0 + 1 = \mathbf{1}$
0	1	0	$Y = (0 \bullet 1) + 0 = 0 + 0 = \mathbf{0}$
0	1	1	$Y = (0 \bullet 1) + 1 = 0 + 1 = \mathbf{1}$
1	0	0	$Y = (1 \bullet 0) + 0 = 0 + 0 = \mathbf{0}$
1	0	1	$Y = (1 \bullet 0) + 1 = 0 + 1 = \mathbf{1}$
1	1	0	$Y = (1 \bullet 1) + 0 = 1 + 0 = \mathbf{1}$
1	1	1	$Y = (1 \bullet 1) + 1 = 1 + 1 = \mathbf{1}$

(b)

FIGURE 21-5   (*A* AND *B*) or (*C*) **Logic Circuit.**

■ *Solution:*

First, there are three inputs or variables to this circuit: *A, B,* and *C.* Inputs *A* and *B* are ANDed together ($A \cdot B$) and the result is then ORed with input *C* ($+ C$). The resulting equation is

$$Y = (A \cdot B) + C \quad \text{or} \quad Y = AB + C$$

As you can see in the truth table in Figure 21-5(b), the *Y* output is HIGH when *A* AND *B* are HIGH, OR if *C* is HIGH.

In the two previous examples, we have gone from a logic circuit to a Boolean equation. Now let us reverse the procedure and see how easy it is to go from a Boolean equation to a logic circuit.

Sketch the logic-gate circuit for the following Boolean equation.

$$Y = AB + CD$$

■ *Solution:*

Studying the equation, we can see that there are four variables: A, B, C, and D. Input A is ANDed with input B, giving $A \cdot B$ or AB, as seen in Figure 21-6(a). Input C is ANDed with input D, giving CD, as seen in Figure 21-6(b). The ANDed AB output and the ANDed CD output are then ORed to produce a final output Y, as seen in Figure 21-6(c).

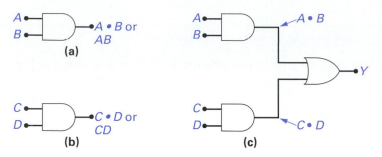

FIGURE 21-6   (*A* AND *B*) or (*C* AND *D*) Logic Circuit.

$$Y = \overline{A + B} \text{ (pronounced "Y equals not A or B")}$$

21-1-4   *The NOR Expression*

The Boolean expression for the two-input NOR gate shown in Figure 21-7(a) is

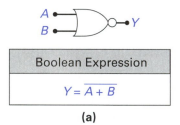

Boolean Expression
$Y = \overline{A + B}$

(a)

A	B	$Y = \overline{A + B}$
0	0	$Y = \overline{0 + 0} = \overline{0} = 1$
0	1	$Y = \overline{0 + 1} = \overline{1} = 0$
1	0	$Y = \overline{1 + 0} = \overline{1} = 0$
1	1	$Y = \overline{1 + 1} = \overline{1} = 0$

(b)

FIGURE 21-7   The Boolean Expression for a NOR Gate.

To explain why this expression describes a NOR gate, let us examine each part in detail. In this equation the two input variables A and B are first ORed (indicated by the $A + B$ part of the equation), and then the result is complemented, or inverted (indicated by the bar over the whole OR output expression).

The truth table in Figure 21-7(b) tests the Boolean expression for a NOR gate for all possible input combinations.

While studying Boolean algebra Augustus DeMorgan discovered two important theorems. The first theorem stated that a bubbled input AND gate was logically equivalent to a NOR gate. These two logic gates are shown in Figure 21-8(a) and (b), respectively, with their associated truth tables. Comparing the two truth tables you can see how any combination of inputs to either gate circuit will result in the same output, making the two circuits interchangeable.

■ INTEGRATED MATH APPLICATION:

Give the two Boolean equations for the two logic-gate circuits shown in Figure 21-8(a) and (b).

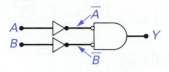

A	B	Y
0	0	1
0	1	0
1	0	0
1	1	0

$Y = \overline{A} \bullet \overline{B}$

(a)

A	B	Y
0	0	1
0	1	0
1	0	0
1	1	0

$Y = \overline{A + B}$

(b)

FIGURE 21-8   DeMorgan's First Theorem.

■ *Solution:*

With the bubbled input AND gate shown in Figure 21-8(a), you can see that input *A* is first inverted by a NOT gate, giving $\overline{A}$. Input *B* is also complemented by a NOT gate, giving $\overline{B}$. These two complemented inputs are then ANDed, giving the following final function.

$$Y = \overline{A} \cdot \overline{B}$$

With the NOR gate shown in Figure 21-8(b), inputs *A* and *B* are first ORed and then the result is complemented, giving

$$Y = \overline{A + B}$$

DeMorgan's first theorem, therefore, is as follows:

DeMorgan's first theorem: $\overline{A} \cdot \overline{B} = \overline{A + B}$

Studying the differences between the left equation and the right equation, you can see that there are actually two basic changes: the line is either broken or solid, and the logical sign is different. Therefore, if we wanted to convert the left equation into the right equation, all we would have to do is follow this simple procedure:

$$Y = \overline{A} \cdot \overline{B} \rightarrow Y = \overline{A + B}$$

Mend
the
line, and
change
the sign.

To convert the right equation into the left equation, all we would have to do is follow this simple procedure:

Break
the
line, and
change
the sign.

$$Y = \overline{A} \cdot \overline{B} \leftarrow Y = \overline{A + B}$$

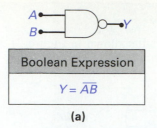

A	B	$Y = \overline{A \cdot B}$
0	0	$Y = \overline{0 \cdot 0} = \overline{0} = 1$
0	1	$Y = \overline{0 \cdot 1} = \overline{0} = 1$
1	0	$Y = \overline{1 \cdot 0} = \overline{0} = 1$
1	1	$Y = \overline{1 \cdot 1} = \overline{1} = 0$

(a)　　　　　　　　　　　　　　　　(b)

FIGURE 21-9　The Boolean Expression for a NAND Gate.

This ability to interchange an AND function with an OR function and to interchange a bubbled input with a bubbled output will come in handy when we are trying to simplify logic circuits, as you will see in this chapter. For now, remember the rules on how to "De-Morganize" an equation.

> *Mend the line, and change the sign.*
> *Break the line, and change the sign.*

21-1-5　*The NAND Expression*

The Boolean expression for the two-input NAND gate shown in Figure 21-9(a) is

> $Y = \overline{A \cdot B}$ or $Y = \overline{AB}$ (pronounced "Y equals not A and B")

This expression states that the two inputs A and B are first ANDed (indicated by the $A \cdot B$ part of the equation), and then the result is complemented (indicated by the bar over the AND expression).

　　The truth table shown in Figure 21-9(b) tests this Boolean equation for all possible input combinations.

■　**INTEGRATED MATH APPLICATION:**

DeMorganize the following equation, and then sketch the logically equivalent circuits with their truth tables and Boolean expressions.

$$Y = \overline{A} + \overline{B}$$

■　*Solution:*

To apply DeMorgan's theorem to the equation $Y = \overline{A} + \overline{B}$, simply apply the rule "mend the line, and change the sign," as follows:

$$\overline{A} + \overline{B} = \overline{A + B}$$

The circuits for these two equations can be seen in Figure 21-10(a) and (b), and if you study the truth tables you can see that the outputs of both logic gates are identical for all input combinations.

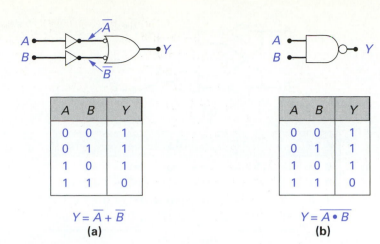

A	B	Y
0	0	1
0	1	1
1	0	1
1	1	0

$Y = \overline{A} + \overline{B}$

(a)

A	B	Y
0	0	1
0	1	1
1	0	1
1	1	0

$Y = \overline{A \cdot B}$

(b)

FIGURE 21-10   DeMorgan's Second Theorem.

This observation—that a bubbled input OR gate is interchangeable with a bubbled output AND gate, or NAND gate—was first made by DeMorgan and is referred to as DeMorgan's second theorem.

> DeMorgan's second theorem: $\overline{A} + \overline{B} = \overline{A \cdot B}$

21-1-6   *The XOR Expression*

In Boolean algebra, the $\oplus$ symbol is used to describe the exclusive-OR action. This means that the Boolean expression for the two-input XOR gate shown in Figure 21-11(a) is

> $Y = A \oplus B$ (pronounced "Y equals A exclusive-OR B")

The truth table in Figure 21-11(b) shows how this Boolean expression is applied to all possible input combinations.

■ **INTEGRATED MATH APPLICATION:**

Give the Boolean expression and truth table for the circuit shown in Figure 21-12(a).

■ *Solution:*

Looking at the circuit in Figure 21-12(a), you can see that the upper AND gate has an inverted A input ($\overline{A}$) and a B input, giving a result of $\overline{A} \cdot B$. This $\overline{A} \cdot B$ output is then ORed with

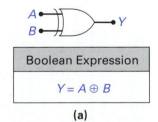

Boolean Expression
$Y = A \oplus B$

(a)

A	B	$Y = A \oplus B$
0	0	$Y = 0 \oplus 0 = 0$
0	1	$Y = 0 \oplus 1 = 1$
1	0	$Y = 1 \oplus 0 = 1$
1	1	$Y = 1 \oplus 1 = 0$

(b)

FIGURE 21-11   The Boolean Expression for an XOR Gate.

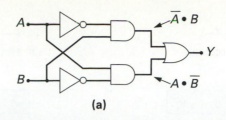

(a)

A B	$Y = \overline{A}B + A\overline{B}$
0 0	$Y = \overline{(0 \bullet 0)} + (0 \bullet \overline{0}) = (1 \bullet 0) + (0 \bullet 1) = 0 + 0 = \mathbf{0}$
0 1	$Y = \overline{(0 \bullet 1)} + (0 \bullet \overline{1}) = (1 \bullet 1) + (0 \bullet 0) = 1 + 0 = \mathbf{1}$
1 0	$Y = \overline{(1 \bullet 0)} + (1 \bullet \overline{0}) = (0 \bullet 0) + (1 \bullet 1) = 0 + 1 = \mathbf{1}$
1 1	$Y = \overline{(1 \bullet 1)} + (1 \bullet \overline{1}) = (0 \bullet 1) + (1 \bullet 0) = 0 + 0 = \mathbf{0}$

(b)

FIGURE 21-12   **(NOT *A* AND *B*) or (*A* AND NOT *B*).**

the input from the lower AND gate, which has an inverted *B* input and *A* input giving a result of $A \cdot \overline{B}$. The final equation therefore is

$$Y = (\overline{A} \cdot B) + (A \cdot \overline{B}) \quad \text{or} \quad Y = \overline{A}\,B + A\,\overline{B}$$

The truth table for this logic circuit that combines five logic gates is shown in Figure 21-12(b). Looking at the input combinations and the output, you can see that the output is 1 only when an odd number of 1s appears at the input. This circuit is therefore acting as an XOR gate, and so

$$\overline{A}\,B + A\,\overline{B} = A \oplus B$$

21-1-7   *The XNOR Expression*

The Boolean expression for the two-input XNOR gate shown in Figure 21-13(a) will be

$$Y = \overline{A \oplus B} \text{ (pronounced "Y equals not A exclusive-or B")}$$

From this equation, you can see that the two inputs *A* and *B* are first XORed and then the result is complemented.

The truth table in Figure 21-13(b) tests this Boolean expression for all possible input combinations.

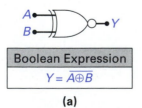

Boolean Expression
$Y = \overline{A \oplus B}$

(a)

A B	$Y = \overline{A \oplus B}$
0 0	$Y = \overline{0 \oplus 0} = \overline{0} = \mathbf{1}$
0 1	$Y = \overline{0 \oplus 1} = \overline{1} = \mathbf{0}$
1 0	$Y = \overline{1 \oplus 0} = \overline{1} = \mathbf{0}$
1 1	$Y = \overline{1 \oplus 1} = \overline{0} = \mathbf{1}$

(b)

FIGURE 21-13   **The Boolean Expression for an XNOR Gate.**

CHAPTER 21 / BOOLEAN EXPRESSIONS AND ALGEBRA

Give the Boolean expression and truth table for the circuit shown in Figure 21-14(a).

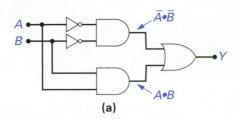

(a)

A	B	$Y = \overline{A}\overline{B} + AB$
0	0	$Y = (\overline{0} \bullet \overline{0}) + (0 \bullet 0) = (1 \bullet 1) + (0 \bullet 0) = 1 + 0 = \mathbf{1}$
0	1	$Y = (\overline{0} \bullet \overline{1}) + (0 \bullet 1) = (1 \bullet 0) + (0 \bullet 1) = 0 + 0 = \mathbf{0}$
1	0	$Y = (\overline{1} \bullet \overline{0}) + (1 \bullet 0) = (0 \bullet 1) + (1 \bullet 0) = 0 + 0 = \mathbf{0}$
1	1	$Y = (\overline{1} \bullet \overline{1}) + (1 \bullet 1) = (0 \bullet 0) + (1 \bullet 1) = 0 + 1 = \mathbf{1}$

(b)

FIGURE 21-14   **(NOT A AND NOT B) or (A AND B).**

■ *Solution:*

Studying the circuit in Figure 21-14(a), you can see that the upper AND gate has an inverted A input and an inverted B input, giving $\overline{A} \cdot \overline{B}$ at its output. The lower AND gate simply ANDs the A and B input, giving $A \cdot B$ at its output. The two AND gate outputs are then ORed, so the final equation for this circuit will be

$$Y = \overline{A} \cdot \overline{B} + A \cdot B \quad \text{or} \quad Y = \overline{A}\,\overline{B} + A\,B$$

The truth table for this circuit is shown in Figure 21-14(b). Comparing this truth table with an XNOR gate's truth table, you can see that the two are equivalent, since a 1 output is present only when an even number of 1s are applied to the input; therefore,

$$\overline{A}\,\overline{B} + A\,B = \overline{A \oplus B}$$

SELF-TEST EVALUATION POINT FOR SECTION 21-1

Now that you have completed this section, you should be able to:

■ *Objective 1.* *Give the Boolean expressions for each of the seven basic logic gates.*

■ *Objective 2.* *Describe how to write an equivalent Boolean expression for a logic circuit and how to sketch an equivalent logic circuit for a Boolean expression.*

Use the following questions to test your understanding of Section 21-1:

1. Which logic gate would be needed to perform each of the following Boolean expressions?
   a. $A\,B$ **c.** $\overline{N + M}$ **e.** $\overline{S + T}$
   b. $A \oplus B$ **d.** $X \cdot Y$ **f.** $\overline{G H}$

2. State DeMorgan's first theorem using Boolean expressions.

3. Apply DeMorgan's theorem to the following:
   a. $\overline{A + B}$ **b.** $Y = \overline{(A + B)} \cdot C$

4. Sketch the logic gates or circuits for the following expressions.
   a. ABC **c.** $\overline{A + B}$ **e.** $A \cdot (L \oplus D)$
   b. $\overline{L \oplus D}$ **d.** $(\overline{A} \cdot \overline{B}) + \overline{C}$ **f.** $\overline{(A + B) \cdot C}$

In this section we will discuss some of the rules and laws that apply to Boolean algebra. Many of these rules and laws are the same as ordinary algebra and, as you will see in this section, are quite obvious.

21-2-1 *The Commutative Law*

Commutative

Combining elements in such a manner that the result is independent of the order in which the elements are taken.

The word **commutative** is defined as "combining elements in such a manner that the result is independent of the order in which the elements are taken." This means that the order in which the inputs to a logic gate are ORed or ANDed, for example, is not important since the result will be the same, as shown in Figure 21-15.

In Figure 21-15(a), you can see that ORing A and B will achieve the same result as reversing the order of the inputs and ORing B and A. As stated in the previous section, the OR function is described in Boolean algebra as logical addition, so the *commutative law of addition* can be algebraically written as

Commutative law of addition: $A + B = B + A$

Figure 21-15(b) shows how this law relates to an AND gate, which in Boolean algebra is described as logical multiplication. The *commutative law of multiplication* can be algebraically written as

Commutative law of multiplication: $A \cdot B = B \cdot A$

21-2-2 *The Associative Law*

Associative

Combining elements such that when the order of the elements is preserved, the result is independent of the grouping.

The word **associative** is defined as "combining elements such that when the order of the elements is preserved, the result is independent of the grouping." To explain this in simple terms, Figure 21-16 shows that how you group the inputs in an ORing process or ANDing process has no effect on the output.

In Figure 21-16(a), you can see that ORing B and C and then ORing the result with A will achieve the same result as ORing A and B and then ORing the result with C. This *associative law of addition* can be algebraically written as

Associative law of addition: $A + (B + C) = (A + B) + C$

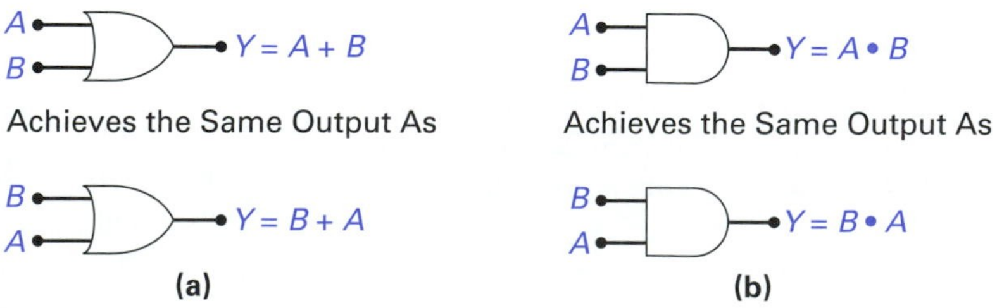

A •———⟩ •—— $Y = A + B$
B •———

Achieves the Same Output As

B •———⟩ •—— $Y = B + A$
A •———

(a)

A •———⟩ •—— $Y = A \cdot B$
B •———

Achieves the Same Output As

B •———⟩ •—— $Y = B \cdot A$
A •———

(b)

FIGURE 21-15 **The Commutative Law (a) Logical Addition (b) Logical Multiplication.**

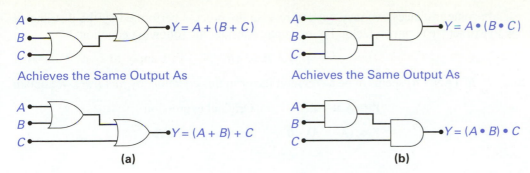

Achieves the Same Output As   Achieves the Same Output As

(a)   (b)

FIGURE 21-16   The Associative Law (a) Logical Addition (b) Logical Multiplication.

Figure 21-16(b) shows how this law relates to an AND gate, which in Boolean algebra is described as logical multiplication. The *associative law of multiplication* can be algebraically written as

$$\text{Associative law of multiplication: } A \cdot (B \cdot C) = (A \cdot B) \cdot C$$

21-2-3   *The Distributive Law*

By definition, the word **distributive** means "producing the same element when operating on a whole, as when operating on each part, and collecting the results." It can be algebraically stated as

Distributive

Producing the same element when operating on a whole as when operating on each part, and collecting the results.

$$\text{Distributive law: } A \cdot (B + C) = (A \cdot B) + (A \cdot C)$$

Figure 21-17 illustrates this law by showing that ORing two or more inputs and then ANDing the result, as shown in Figure 21-17(a), achieves the same output as ANDing the single variable (A) with each of the other inputs (B and C) and then ORing the results, as shown in Figure 21-17(b). To help reinforce this algebraic law, let us use an example involving actual values for A, B, and C.

■   **INTEGRATED MATH APPLICATION:**

Prove the distributive law by inserting the values $A = 2$, $B = 3$ and $C = 4$.

■   *Solution:*

$$A \cdot (B + C) = (A \cdot B) + (A \cdot C)$$
$$2 \times (3 + 4) = (2 \times 3) + (2 \times 4)$$
$$2 \times 7 = 6 + 8$$
$$14 = 14$$

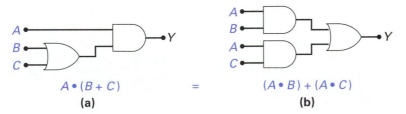

$A \bullet (B + C)$ = $(A \bullet B) + (A \bullet C)$

(a)   (b)

FIGURE 21-17   The Distributive Law.

In some instances, we may wish to reverse the process performed by the distributive law to extract the common factor. For example, consider the following equation:

$$Y = \overline{A}\,B + AB$$

Since "· B" (AND B) seems to be a common factor in this equation, by *factoring,* we obtain

$$Y = \overline{A}\,B + AB \qquad \text{Original expression}$$
$$Y = (\overline{A} + A) \cdot B \qquad \text{Factoring AND } B$$

21-2-4 *Boolean Algebra Rules*

Now that we have covered the three laws relating to Boolean algebra, let us now concentrate on these Boolean algebra rules, beginning with those relating to OR operations.

OR Gate Rules

The first rule can be seen in Figure 21-18(a). This shows what happens when one input to an OR gate is always 0 and the other input (A) is a variable. If $A = 0$, the output equals 0, and if $A = 1$, the output equals 1. Therefore, a variable input ORed with 0 will always equal the variable input. Stated algebraically:

$$A + 0 = A$$

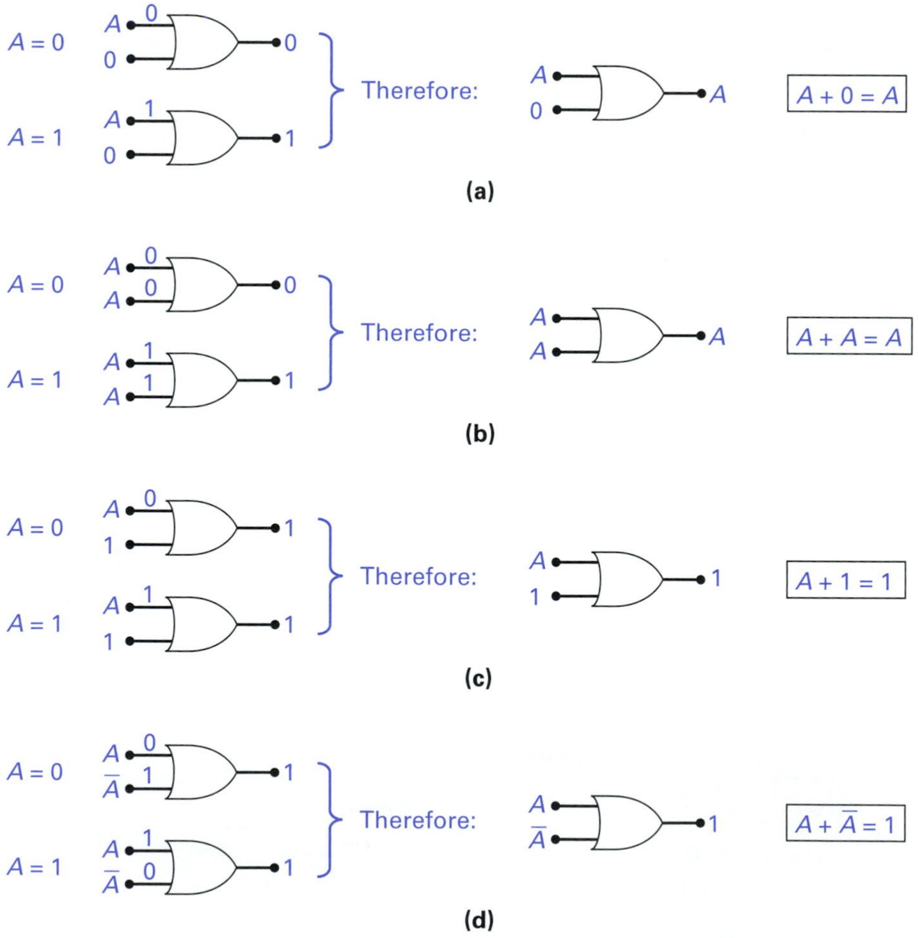

(a)

(b)

(c)

(d)

FIGURE 21-18 **Boolean Rules for OR Gates.**

CHAPTER 21 / BOOLEAN EXPRESSIONS AND ALGEBRA

Another OR gate Boolean rule is shown in Figure 21-18(b), and it states that

$$A + A = A$$

As you can see in Figure 21-18(b), when a variable is ORed with itself, the output will always equal the logic level of the variable input.

Figure 21-18(c) shows the next OR gate Boolean rule. This rule states that any 1 input to an OR gate will result in a 1 output, regardless of the other input. Stated algebraically:

$$A + 1 = 1$$

The final Boolean rule for OR gates is shown in Figure 21-18(d). This rule states that when any variable (A) is ORed with its complement ($\overline{A}$) the result will be a 1.

$$A + \overline{A} = 1$$

AND Gate Rules

Like the OR gate, the AND gate has four Boolean rules. The first is illustrated in Figure 21-19(a) and states that if a variable input A is ANDed with a 0, the output will be 0 regardless of the other input.

$$A \cdot 0 = 0$$

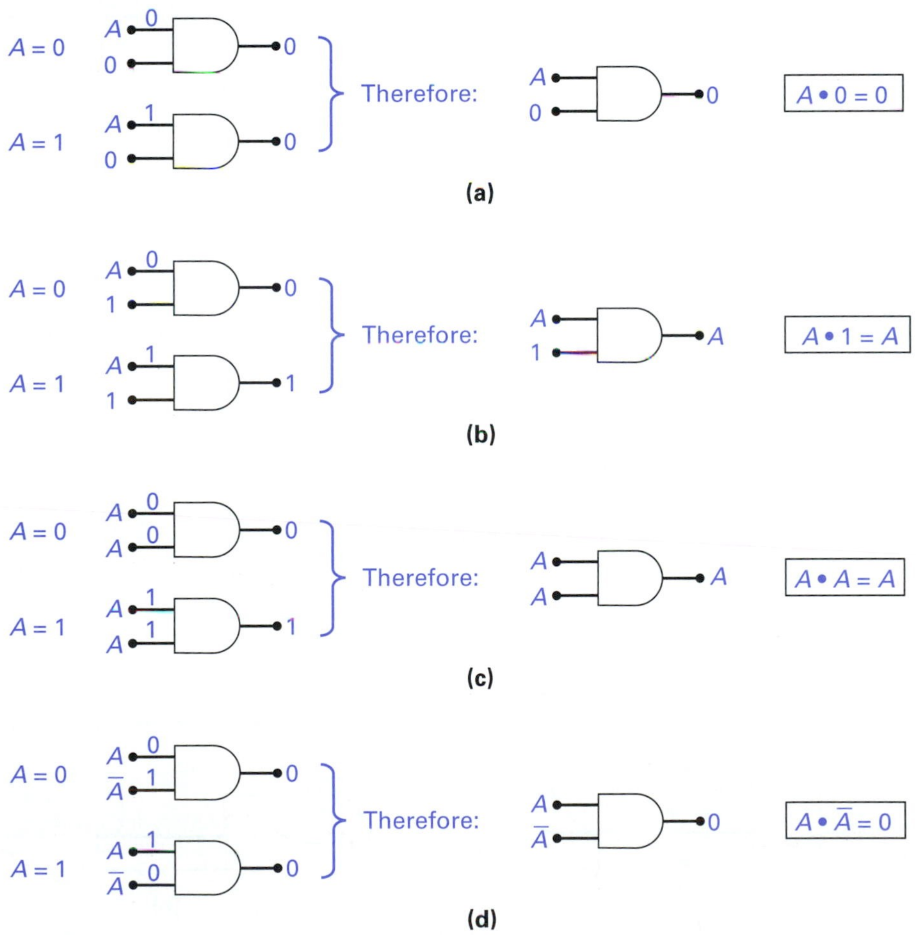

(a)

(b)

(c)

(d)

FIGURE 21-19 **Boolean Rules for AND Gates.**

The second Boolean rule for AND gates is shown in Figure 21-19(b). In this illustration you can see that if a variable is ANDed with a 1, the output will equal the variable.

$$A \cdot 1 = A$$

Another AND gate Boolean rule is shown in Figure 21-19(c). In this instance, any variable that is ANDed with itself will always give an output that is equal to the variable.

$$A \cdot A = A$$

The last of the AND gate Boolean rules is shown in Figure 21-19(d). In this illustration you can see that if a variable (A) is ANDed with its complement ($\overline{A}$), the output will always equal 0.

$$A \cdot \overline{A} = 0$$

Double-Inversion Rule

The *double-inversion rule* is illustrated in Figure 21-20(a) and states that if a variable is inverted twice, then the variable will be back to its original state. To state this algebraically, we use the double bar, as follows:

$$\overline{\overline{A}} = A$$

In Figure 21-20(b), you can see that if the NOR gate were replaced with an OR gate, the INVERTER would not be needed, since the double inversion returns the logic level to its original state.

DeMorgan's Theorems

DeMorgan's first and second theorems were discussed earlier in the previous section and are repeated here, since they also apply as Boolean rules.

DeMorgan's first theorem: $\overline{A} \cdot \overline{B} = \overline{A + B}$

DeMorgan's second theorem: $\overline{A} + \overline{B} = \overline{A \cdot B}$

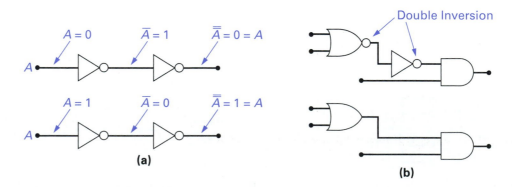

FIGURE 21-20   Double-Inversion Rule.

Apply DeMorgan's theorem and the double-inversion rule to find equivalent logic circuits of the following:

$$\text{a. } \overline{\overline{A} + B} \qquad \text{b. } \overline{A \cdot \overline{B}}$$

■ *Solution:*

With both equations we simply "break the line and change the sign" and then cancel any double inversions to find the equivalent logic circuits, as follows:

$$\text{a. } \overline{\overline{A} + B} = \overline{\overline{A}} \cdot \overline{B} = A \cdot \overline{B}$$

In the first example we found that a NOR gate with a bubbled *A* input is equivalent to an AND gate with a bubbled *B* input.

$$\text{b. } \overline{A \cdot \overline{B}} = \overline{A} + \overline{\overline{B}} = \overline{A} + B$$

In the second example we found that a NAND gate with a bubbled *B* input is equivalent to an OR gate with a bubbled *A* input.

The Duality Theorem

The *duality theorem* is very useful, since it allows us to change Boolean equations into their *dual* and to produce new Boolean equations. The two steps to follow for making the change are very simple.

a. First, change each OR symbol to an AND symbol and each AND symbol to an OR symbol.

b. Second, change each 0 to a 1 and each 1 to a 0.

To see if this works, let us apply the duality theorem to a simple example.

■ INTEGRATED MATH APPLICATION:

Apply the duality theorem to $A + 1 = 1$ and then state whether the resulting rule is true.

■ *Solution:*

Changing the OR symbol to an AND symbol and changing the 1s to 0s, we arrive at the following:

Original rule: $A + 1 = 1$
Dual rule: $A \cdot 0 = 0$

OR symbol is changed to AND symbol. ——————— 1s are changed to 0s.

The dual Boolean rule states that any variable ANDed with a 0 will always yield a 0 output, which is true.

■ INTEGRATED MATH APPLICATION:

Determine the dual Boolean equation for the following:

$$A \cdot (B + C) = (A \cdot B) + (A \cdot C)$$

Once you have determined the new Boolean equation, sketch circuits and give the truth tables for both the left and right parts of the equation.

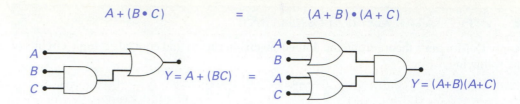

$$A + (B \cdot C) \qquad = \qquad (A + B) \bullet (A + C)$$

$Y = A + (BC)$ $=$ $Y = (A+B)(A+C)$

A	B	C	$Y = A + (B \bullet C)$
0	0	0	$Y = 0 + (0 \bullet 0) = 0$
0	0	1	$Y = 0 + (0 \bullet 1) = 0$
0	1	0	$Y = 0 + (1 \bullet 0) = 0$
0	1	1	$Y = 0 + (1 \bullet 1) = 1$
1	0	0	$Y = 1 + (0 \bullet 0) = 1$
1	0	1	$Y = 1 + (0 \bullet 1) = 1$
1	1	0	$Y = 1 + (1 \bullet 0) = 1$
1	1	1	$Y = 1 + (1 \bullet 1) = 1$

$=$

A	B	C	$Y = (A + B) \bullet (A + C)$
0	0	0	$Y = (0 + 0) \bullet (0 + 0) = 0$
0	0	1	$Y = (0 + 0) \bullet (0 + 1) = 0$
0	1	0	$Y = (0 + 1) \bullet (0 + 0) = 0$
0	1	1	$Y = (0 + 1) \bullet (0 + 1) = 1$
1	0	0	$Y = (1 + 0) \bullet (1 + 0) = 1$
1	0	1	$Y = (1 + 0) \bullet (1 + 1) = 1$
1	1	0	$Y = (1 + 1) \bullet (1 + 0) = 1$
1	1	1	$Y = (1 + 1) \bullet (1 + 1) = 1$

FIGURE 21-21 *A* **OR** (*B* **AND** *C*) **Equals** (*A* **OR** *B*) **AND** (*A* **OR** *C*).

■ *Solution:*

Applying the duality theorem to $A \cdot (B + C) = (A \cdot B) + (A \cdot C)$ means that we will have to change all the AND symbols to OR symbols, and vice versa.

Original equation: $A \cdot (B + C) = (A \cdot B) + (A \cdot C)$
Dual equation: $A + (B \cdot C) = (A + B) \cdot (A + C)$

Figure 21-21 shows the dual equation and its circuits and truth tables. As you can see by comparing the truth tables, the left and right parts of the equation are equivalent.

As a summary, here is a list of the Boolean laws and rules described in this section, with their dual equations.

		ORIGINAL EQUATION	DUAL EQUATION
1.	Commutative law	$A + B = B + A$	$A \cdot B = B \cdot A$
2.	Associative law	$A \cdot (B \cdot C) = (A \cdot B) \cdot C$	$A + (B + C) = (A + B) + C$
3.	Distributive law	$A + (B \cdot C) = (A + B) \cdot (A + C)$	$A \cdot (B + C) = (A \cdot B) + (A \cdot C)$
4.	OR-AND rules	$A + 0 = A$	$A \cdot 1 = A$
		$A + A = A$	$A \cdot A = A$
		$A + 1 = 1$	$A \cdot 0 = 0$
		$A + \overline{A} = 1$	$A \cdot \overline{A} = 0$
5.	DeMorgan's laws	$\overline{A} \cdot \overline{B} = \overline{A + B}$	$\overline{A} + \overline{B} = \overline{A \cdot B}$

SELF-TEST EVALUATION POINT FOR SECTION 21-2

Now that you have completed this section, you should be able to:

■ *Objective 3.* *List and explain the laws and rules that apply to Boolean algebra.*

Use the following questions to test your understanding of Section 21-2:

1. Give the answers to the following Boolean algebra rules.
   a. $A + 1 = ?$ **c.** $A \cdot 0 = ?$
   b. $A + \overline{A} = ?$ **d.** $A \cdot A = ?$
2. What is the dual equation of $A + (B \cdot C) = (A + B) \cdot (A + C)$?
3. Apply the associative law to the equation $A \cdot (B \cdot C)$.
4. Apply the distributive law to the equation $(A \cdot B) + (A \cdot C) + (A \cdot D)$.

Now that you understand the Boolean rules and laws for logic gates, let us put this knowledge to some practical use. If you need a logic-gate circuit to perform a certain function, the best way to begin is with a truth table that details what input combinations should drive the output HIGH and what input combinations should drive the output LOW. As an example, let us assume that we need a logic circuit that will follow the truth table shown in Figure 21-22(a). This means that the output (Y) should be 1 only when

$$A = 0, B = 0, C = 1$$
$$A = 0, B = 1, C = 1$$
$$A = 1, B = 0, C = 1$$
$$A = 1, B = 1, C = 1$$

In the truth table in Figure 21-22(a), you can see the **fundamental products** listed for each of these HIGH outputs. A fundamental product is a Boolean expression that describes what the inputs will need to be in order to generate a HIGH output. For example, the first fundamental product ($Y = \overline{A} \cdot \overline{B} \cdot C$) states that A must be 0, B must be 0, and C must be 1 for the output Y to be 1 ($Y = \overline{A} \cdot \overline{B} \cdot C = \overline{0} \cdot \overline{0} \cdot 1 = 1 \cdot 1 \cdot 1 = 1$). The second fundamental product ($Y = \overline{A} \cdot B \cdot C$) states that A must be 0, B must be 1, and C must be 1 for the output Y to be 1 ($Y = \overline{A} \cdot B \cdot C = \overline{0} \cdot 1 \cdot 1 = 1 \cdot 1 \cdot 1 = 1$). The third fundamental product ($Y = A \cdot \overline{B} \cdot C$) states that A must be 1, B must be 0, and C must be 1 for Y to be 1 ($Y = A \cdot \overline{B} \cdot C = 1 \cdot \overline{0} \cdot 1 = 1 \cdot 1 \cdot 1 = 1$). The fourth and final fundamental product ($Y = A \cdot B \cdot C$) states that A must be 1, B must be 1, and C must be 1 for Y to be 1 ($Y = A \cdot B \cdot C = 1 \cdot 1 \cdot 1 = 1$). The fundamental products for each of the HIGH outputs are listed in the truth table in Figure 21-22(a). By ORing all these fundamental products, we can derive a Boolean equation as follows:

$$Y = \overline{A}\,\overline{B}\,C + \overline{A}\,B\,C + A\,\overline{B}\,C + A\,B\,C$$

From this Boolean equation we can create a logic network that is the circuit equivalent of the truth table. To complete this step, we will need to study in detail the different parts of the Boolean equation. First, the Boolean equation states that the outputs of four three-input AND gates are connected to the input of a four-input OR gate. The first AND gate is connected to inputs $\overline{A}\ \overline{B}$ and C, the second to inputs $\overline{A}\ B$ and C, the third to inputs $A\ \overline{B}$ and C, and finally the fourth to inputs $A\ B$ and C. The resulting logic circuit equivalent for the truth table in Figure 21-22(a) is shown in Figure 21-22(b).

The Boolean equation for the logic circuit in Figure 21-22 is in the **sum-of-products (SOP) form.** To explain what this term means, we must recall once again that in Boolean

Fundamental Products

The truth table input combinations that are essential, since they produce a HIGH output.

Sum-of-Products (SOP) Form

A Boolean expression that describes the ORing of two or more AND functions.

A	B	C	Y	Fundamental Product
0	0	0	0	
0	0	1	1	$\overline{A}\overline{B}C$
0	1	0	0	
0	1	1	1	$\overline{A}BC$
1	0	0	0	
1	0	1	1	$A\overline{B}C$
1	1	0	0	
1	1	1	1	ABC

(a)

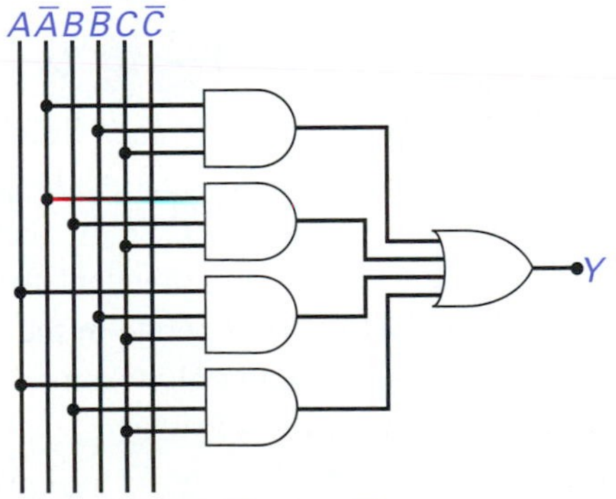

$A\overline{A}B\overline{B}C\overline{C}$

$Y = \overline{A}\overline{B}C + \overline{A}BC + A\overline{B}C + ABC$

(b)

FIGURE 21-22   Sum-of-Products (SOP) Form.

algebra, an AND gate's output is the *logical product* of its inputs. On the other hand, the OR gate's output is the *logical sum* of its inputs. A sum-of-products expression, therefore, describes the ORing together of (sum of) two or more AND functions (products). Here are some examples of sum-of-product expressions:

$$Y = AB + \overline{AB}$$
$$Y = A\overline{B} + C\overline{D}E$$
$$Y = \overline{A}BCD + AB\overline{CD} + A\overline{B}C\overline{D}$$

Product-of-Sums (POS) Form

A Boolean expression that describes the ANDing of two or more OR functions.

The other basic form for Boolean expressions is the **product-of-sums (POS) form.** This form describes the ANDing of (product of) two or more OR functions (sums). Here are some examples of product-of-sums expressions:

$$Y = (A + B + \overline{C}) \cdot (C + D)$$
$$Y = (\overline{A} + \overline{B}) \cdot (A + B)$$
$$Y = (A + B + C + \overline{D}) \cdot (\overline{A} + B + \overline{C} + D) \cdot (A + \overline{B} + \overline{C} + \overline{D})$$

■ **INTEGRATED MATH APPLICATION:**

Is the circuit shown in Figure 21-23 an example of sum-of-products or product-of-sums, and what is this circuit's Boolean expression?

■ *Solution:*

The circuit in Figure 21-23 is a product-of-sums equivalent circuit, and its Boolean expression is

$$Y = (A + B) \cdot (C + D)$$

FIGURE 21-23
SOP or POS?

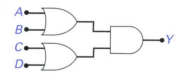

■ **INTEGRATED MATH APPLICATION:**

Determine the logic circuit needed for the truth table shown in Figure 21-24(a).

■ *Solution:*

The first step is to write the fundamental product for each HIGH output in the truth table, as seen in Figure 21-24(a). By ORing (+) these fundamental products, we can obtain the following sum-of-products Boolean equation:

$$Y = \overline{ABCD} + \overline{ABC}\,\overline{D} + A\overline{B}\,\overline{C}D + AB\overline{CD}$$

The next step is to develop the circuit equivalent of this equation. The equation states that the outputs of four four-input AND gates are connected to a four-input OR gate. The first AND gate has inputs $\overline{A}\,\overline{B}\,\overline{C}\,D$, the second has $\overline{A}B\overline{C}\,\overline{D}$, the third has $A\overline{B}\,\overline{C}\,D$, and the fourth has $AB\overline{C}D$. The circuit equivalent for this Boolean expression can be seen in Figure 21-24(b). This circuit will operate in the manner detailed by the truth table in Figure 21-24(a) and generate a 1 output only when

$$A = 0, B = 0, C = 0, D = 1$$
$$A = 0, B = 1, C = 0, D = 0$$
$$A = 1, B = 0, C = 0, D = 1$$
$$A = 1, B = 1, C = 0, D = 1$$

In all other instances, the output of this circuit will be 0.

CHAPTER 21 / BOOLEAN EXPRESSIONS AND ALGEBRA

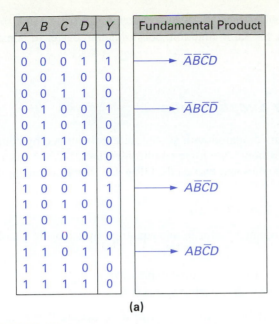

A	B	C	D	Y	Fundamental Product
0	0	0	0	0	
0	0	0	1	1	$\overline{A}\,\overline{B}\,\overline{C}D$
0	0	1	0	0	
0	0	1	1	0	
0	1	0	0	1	$\overline{A}B\overline{C}\,\overline{D}$
0	1	0	1	0	
0	1	1	0	0	
0	1	1	1	0	
1	0	0	0	0	
1	0	0	1	1	$A\overline{B}\,\overline{C}D$
1	0	1	0	0	
1	0	1	1	0	
1	1	0	0	0	
1	1	0	1	1	$AB\overline{C}D$
1	1	1	0	0	
1	1	1	1	0	

(a)

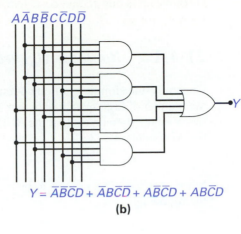

$$Y = \overline{A}\,\overline{B}\,\overline{C}D + \overline{A}B\overline{C}\,\overline{D} + A\overline{B}\,\overline{C}D + AB\overline{C}D$$

(b)

FIGURE 21-24 Sum-of-Products Example.

■ **INTEGRATED MATH APPLICATION:**

Apply the distributive law to the following Boolean expression in order to convert it into a sum-of-products form.

$$Y = A\overline{B} + C(D\overline{E} + F\overline{G})$$

■ *Solution:*

$Y = A\overline{B} + C(D\overline{E} + F\overline{G})$	Original expression
$Y = A\overline{B} + CD\overline{E} + CF\overline{G}$	Distributive law applied

---

SELF-TEST EVALUATION POINT FOR SECTION 21-3

Now that you have completed this section, you should be able to:

■ *Objective 4.* Describe the logic circuit design process from truth table to gate circuit.

■ *Objective 5.* Explain the difference between a sum-of-products equation and a product-of-sums equation.

Use the following questions to test your understanding of Section 21-3:

1. A two-input AND gate will produce at its output the logical _____ of A, and B, while an OR gate will produce at its output the logical _____ of A and B.

2. What are the fundamental products of each of the following input words?
   a. A, B, C, D = 1011   **b.** A, B, C, D = 0110

3. The Boolean expression $Y = AB + \overline{A}B$ is an example of a(n) _____ (SOP/POS) equation.

4. Describe the steps involved to develop a logic circuit from a truth table.

21-4 GATE CIRCUIT SIMPLIFICATION

In the last section we saw how we could convert a desired set of conditions in a truth table to a sum-of-products equation, and then to an equivalent logic circuit. In this section we will see how we can simplify the Boolean equation using Boolean laws and rules and therefore simplify the final logic circuit. As to why we would want to simplify a logic circuit, the answer is: A simplified circuit performs the same function, but is smaller, easier to construct, consumes less power, and costs less.

21-4-1 *Boolean Algebra Simplification*

Generally, the simplicity of one circuit compared with another circuit is judged by counting the number of logic-gate inputs. For example, the logic circuit in Figure 21-25(a) has a total of six inputs (two on each of the AND gates and two on the OR gate). The Boolean equation for this circuit is

$$Y = \overline{A}B + AB$$

Since "· *B*" (AND B) seems to be a common factor in this equation, by factoring we obtain the following:

$$Y = \overline{A}B + AB \quad \text{Original expression}$$
$$Y = (\overline{A} + A) \cdot B \quad \text{Factoring AND } B$$

This equivalent logic circuit is shown in Figure 21-25(b), and if you count the number of logic-gate inputs in this circuit (two for the OR and two for the AND, a total of four), you can see that the circuit has been simplified from a six-input circuit to a four-input circuit.

This, however, is not the end of our simplification process, since another rule can be applied to further simplify the circuit. Remembering that any variable ORed with its complement results in a 1, we can replace $\overline{A} + A$ with a 1, as follows:

$$Y = (\overline{A} + A) \cdot B$$
$$Y = (1) \cdot B$$
$$Y = 1 \cdot B$$

The resulting equation $Y = 1 \cdot B$ can be further simplified, since any variable ANDed with a 1 will always cause the output to equal the variable. Therefore, $Y = 1 \cdot B$ can be replaced with *B*, as follows:

$$Y = 1 \cdot B$$
$$Y = B$$

This means that the entire circuit in Figure 21-25(a) can be replaced with a single wire connected from the *B* input to the output, as seen in Figure 21-25(c). This fact is confirmed in the truth table in Figure 21-25(d), which shows that output *Y* exactly follows the input *B*.

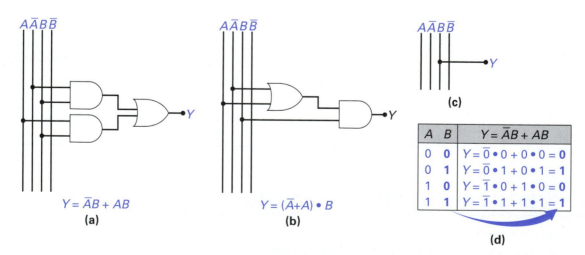

A	B	$Y = \overline{A}B + AB$
0	0	$Y = \overline{0} \bullet 0 + 0 \bullet 0 = 0$
0	1	$Y = \overline{0} \bullet 1 + 0 \bullet 1 = 1$
1	0	$Y = \overline{1} \bullet 0 + 1 \bullet 0 = 0$
1	1	$Y = \overline{1} \bullet 1 + 1 \bullet 1 = 1$

FIGURE 21-25 **Simplifying Gate Circuits.**

■ INTEGRATED MATH APPLICATION:

Simplify the Boolean equation shown in Figure 21-26(a).

■ *Solution:*

The Boolean equation for this circuit is

$$Y = A\bar{B}\,\bar{C} + \bar{A}B\bar{C} + A\bar{B}C + \bar{A}\,\bar{B}\,\bar{C}$$

The first step in the simplification process is to try and spot common factors in each of the ANDed terms, and then to rearrange these terms as shown in Figure 21-26(b). The next step is to factor out the common expressions, which are $A\bar{B}$ and $\bar{A}\,\bar{C}$, resulting in

$$
\begin{aligned}
Y &= A\bar{B}\,\bar{C} + \bar{A}B\bar{C} + A\bar{B}C + \bar{A}\,\bar{B}\,\bar{C} && \text{Original equation} \\
Y &= A\bar{B}\,\bar{C} + A\bar{B}C + \bar{A}B\bar{C} + \bar{A}\,\bar{B}\,\bar{C} && \text{Rearranged equation} \\
Y &= A\bar{B}(\bar{C} + C) + \bar{A}\,\bar{C}(B + \bar{B}) && \text{Factored equation}
\end{aligned}
$$

Studying the last equation you can see that we can apply the OR rule $(\bar{A} + A) = 1$ and simplify the terms within the parentheses to obtain the following:

$$
\begin{aligned}
Y &= A\bar{B}(\bar{C} + C) + \bar{A}\,\bar{C}(B + \bar{B}) \\
Y &= A\bar{B} \cdot (1) + \bar{A}\,\bar{C} \cdot (1)
\end{aligned}
$$

Now, we can apply the $A \cdot 1 = A$ rule to further simplify the terms, since $A\bar{B} \cdot 1 = A\bar{B}$ and $\bar{A}\,\bar{C} \cdot 1 = \bar{A}\,\bar{C}$; therefore:

$$Y = A\bar{B} + \bar{A}\,\bar{C}$$

This equivalent logic circuit is illustrated in Figure 21-26(c) and has only 6 logic gate inputs compared with the original circuit shown in Figure 21-26(a), which has 16 logic gate inputs. The truth table in Figure 21-26(d) compares the original equation with the simplified equation and shows how the same result is obtained.

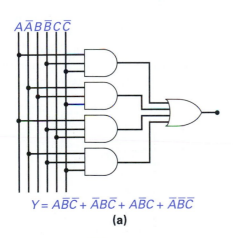

$$Y = A\bar{B}\bar{C} + \bar{A}B\bar{C} + A\bar{B}C + \bar{A}\bar{B}\bar{C}$$
(a)

$$Y = \overbrace{A\bar{B}\bar{C}} + \overbrace{\bar{A}B\bar{C}} + \overbrace{A\bar{B}C} + \overbrace{\bar{A}\bar{B}\bar{C}}$$

$$Y = A\bar{B}\bar{C} + A\bar{B}C + \bar{A}B\bar{C} + \bar{A}\bar{B}\bar{C}$$
(b)

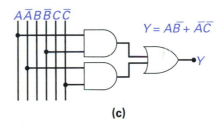

(c)

A	B	C	$Y = A\bar{B}\bar{C} + \bar{A}B\bar{C} + A\bar{B}C + \bar{A}\bar{B}\bar{C}$	$Y = A\bar{B} + \bar{A}\bar{C}$
0	0	0	$Y = 0\bar{0}\bar{0} + \bar{0}0\bar{0} + 0\bar{0}0 + \bar{0}\bar{0}\bar{0} = 1$	$Y = 0\bar{0} + \bar{0}\bar{0} = 1$
0	0	1	$Y = 0\bar{0}\bar{1} + \bar{0}0\bar{1} + 0\bar{0}1 + \bar{0}\bar{0}\bar{1} = 0$	$Y = 0\bar{0} + \bar{0}\bar{1} = 0$
0	1	0	$Y = 0\bar{1}\bar{0} + \bar{0}1\bar{0} + 0\bar{1}0 + \bar{0}\bar{1}\bar{0} = 1$	$Y = 0\bar{1} + \bar{0}\bar{0} = 1$
0	1	1	$Y = 0\bar{1}\bar{1} + \bar{0}1\bar{1} + 0\bar{1}1 + \bar{0}\bar{1}\bar{1} = 0$	$Y = 0\bar{1} + \bar{0}\bar{1} = 0$
1	0	0	$Y = 1\bar{0}\bar{0} + \bar{1}0\bar{0} + 1\bar{0}0 + \bar{1}\bar{0}\bar{0} = 1$	$Y = 1\bar{0} + \bar{1}\bar{0} = 1$
1	0	1	$Y = 1\bar{0}\bar{1} + \bar{1}0\bar{1} + 1\bar{0}1 + \bar{1}\bar{0}\bar{1} = 1$	$Y = 1\bar{0} + \bar{1}\bar{1} = 1$
1	1	0	$Y = 1\bar{1}\bar{0} + \bar{1}1\bar{0} + 1\bar{1}0 + \bar{1}\bar{1}\bar{0} = 0$	$Y = 1\bar{1} + \bar{1}\bar{0} = 0$
1	1	1	$Y = 1\bar{1}\bar{1} + \bar{1}1\bar{1} + 1\bar{1}1 + \bar{1}\bar{1}\bar{1} = 0$	$Y = 1\bar{1} + \bar{1}\bar{1} = 0$

(d)

FIGURE 21-26 **Boolean Equation Simplification.**

FIGURE 21-27   The Karnaugh Map—A Rearranged Truth Table.

A	B	Y
0	0	$\overline{A}\overline{B}$
0	1	$\overline{A}B$
1	0	$A\overline{B}$
1	1	AB

	$\overline{B}$	B
$\overline{A}$	$\overline{A}\overline{B}$	$\overline{A}B$
A	$A\overline{B}$	AB

21-4-2   *Karnaugh Map Simplification*

In most instances, engineers and technicians will simplify logic circuits using a **Karnaugh map,** or *K-map*. The Karnaugh map, named after its inventor, is quite simply a rearranged truth table, as shown in Figure 21-27. With this map the essential gating requirements can be more easily recognized and reduced to their simplest form.

The total number of boxes or cells in a K-map depends on the number of input variables. For example, in Figure 21-27, only two inputs (*A* and *B*) and their complements ($\overline{A}$ and $\overline{B}$) are present, and therefore the K-map contains (like a two-variable truth table) only four combinations (00, 01, 10, and 11).

Each cell of this two-variable K-map represents one of the four input combinations. In practice, the input labels are placed outside the cells, as shown in Figure 21-28(a), and apply to either a column or row of cells. For example, the row label $\overline{A}$ applies to the two upper cells, and the row label *A* applies to the two lower cells. Running along the top of the K-map, the label $\overline{B}$ applies to the two left cells, and the label *B* applies to the two right cells. As an example, the upper right cell represents the input combination $\overline{A}\, B$. Figure 21-28(b) and (c) shows the formats for a three-variable ($2^3 = 8$ cells) and four-variable ($2^4 = 16$ cells) K-map.

Now that we have an understanding of the K-map, let us see how it can be used to simplify a logic circuit. As an example, imagine that we need to create an equivalent logic circuit for the truth table given in Figure 21-29(a). The first step is to develop a sum-of-products Boolean expression. This is achieved by writing the fundamental product for each HIGH output in the truth table and then ORing all the fundamental products, as shown in Figure 21-29(b). The equivalent logic circuit for this equation is shown in Figure 21-29(c). The next step is to plot this Boolean expression on a two-variable K-map, as seen in Figure 21-29(d). When plotting a sum-of-products expression on a K-map, remember that each cell

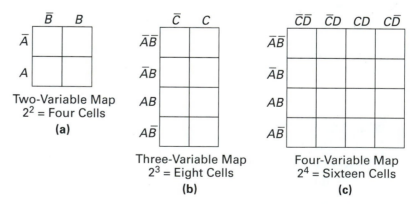

FIGURE 21-28   Karnaugh Map (a) Two-Variable (b) Three-Variable (c) Four-Variable.

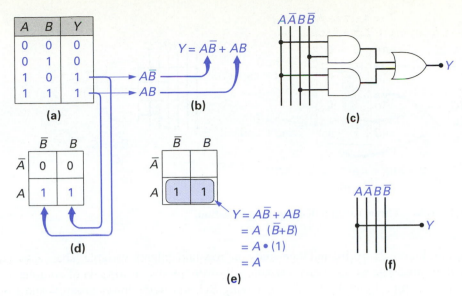

FIGURE 21-29   Karnaugh Map Simplification.

corresponds to each of the input combinations in the truth table. A HIGH output in the truth table should appear as a 1 in its equivalent cell in the K-map, and a LOW output in the truth table should appear as a 0 in its equivalent cell. A 1, therefore, will appear in the lower left cell (corresponding to $A\overline{B}$) and in the lower right cell (corresponding to AB). The other input combinations ($\overline{A}\,\overline{B}$ and $\overline{A}\,B$) both yield a 0 output, and therefore a 0 should be placed in these two upper cells.

Reducing Boolean equations is largely achieved by applying the rule of complements, which states that $A + \overline{A} = 1$. Now that the SOP equation has been plotted on the K-map shown in Figure 21-29(d), the next step is to group terms and then factor out the common variables. If you study the K-map in Figure 21-29(d), you will see that adjacent cells differ by only one input variable. This means that if you move either horizontally or vertically from one cell to an adjacent cell, only one variable will change. Grouping adjacent cells containing a 1, as seen in Figure 21-29(e), allows cells to be compared and simplified (using the rule of complements) to create one-product terms. In this example, cells $A\overline{B}$ and AB contain B and $\overline{B}$, so these opposites or complements cancel, leaving A, as follows:

$$Y = A\overline{B} + AB \qquad \text{Grouped pair}$$
$$Y = A \cdot (\overline{B} + B) \qquad \text{Factoring } A \text{ AND}$$
$$Y = A \cdot 1$$
$$Y = A$$

This procedure can be confirmed by studying the original truth table in Figure 21-29(a) in which you can see that output Y exactly follows input A. The equivalent circuit is shown in Figure 21-29(f).

■ INTEGRATED MATH APPLICATION:

Determine the simplest logic circuit for the truth table shown in Figure 21-30(a). Illustrate each step in the process.

■ *Solution:*

Since this example has three input variables, the first step is to draw a three-variable K-map, as shown in Figure 21-30(b). The next step is to look for the HIGH outputs in the truth table in Figure 21-30(a) and to plot these 1s in their equivalent cells in the K-map, as shown in Figure 21-30(b). After inserting 0s in the remaining cells, group the 1s into pairs,

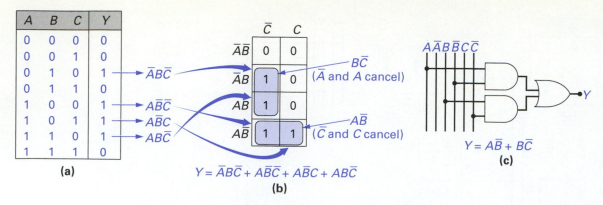

A	B	C	Y
0	0	0	0
0	0	1	0
0	1	0	1
0	1	1	0
1	0	0	1
1	0	1	1
1	1	0	1
1	1	1	0

(a)

$Y = \overline{A}B\overline{C} + A\overline{B}\,\overline{C} + A\overline{B}C + AB\overline{C}$

(b)

$Y = A\overline{B} + B\overline{C}$

(c)

FIGURE 21-30 Three-Variable K-Map Simplification.

as shown in Figure 21-30(b), and then study the row and column variable labels associated with that grouped pair to see which variable will drop out due to the rule of complements. In the upper group, $\overline{A}$ and A will cancel, leaving $B\overline{C}$, and in the lower group $\overline{C}$ and C will cancel, leaving $A\overline{B}$. These reduced products will form an equivalent Boolean equation and logic circuit, as seen in Figure 21-30(c). In this example, the original 16-input equation in Figure 21-30(b) has been simplified to the equivalent 6-input logic gate circuit shown in Figure 21-30(c).

The 1s in a K-map can be grouped in pairs (groups of two), quads (groups of four), octets (groups of eight), and all higher powers of 2. Figure 21-31 shows some examples of grouping and how the K-map has been used to reduce large Boolean equations. Notice that the larger groups will yield smaller terms and therefore gate circuits with fewer inputs. For this reason, you should begin by looking for the largest possible groups, and then step down in group size if none are found (for example, begin looking for octets, then quads, and finally pairs). Also notice that you can capture 1s on either side of a map by wrapping a group around behind the map.

■ **INTEGRATED MATH APPLICATION:**

What would be the sum-of-products Boolean expression for the truth table in Figure 21-32(a)? After determining the Boolean expression, plot it on a K-map to see if it can be simplified.

■ *Solution:*

The first step in developing the sum-of-products Boolean expression is to write down the fundamental products for each HIGH output in the truth table, as seen in Figure 21-32(a). From this, we can derive an SOP equation:

$$Y = \overline{A}\,\overline{B}\,\overline{C}\,D + \overline{A}\,\overline{B}CD + \overline{A}BC\overline{D} + \overline{A}BCD + A\overline{B}CD + ABCD$$

The next step is to draw a four-variable K-map, as seen in Figure 21-32(b), and then plot the HIGH outputs of the truth table in their equivalent cells in the K-map. Looking at Figure 21-32(b), you can see that the 1s can be grouped into two quads. With the square-shaped quad, row labels B and $\overline{B}$ will cancel, and column labels C and $\overline{C}$ will cancel, leaving $\overline{A}D$. With the rectangular-shaped quad, all its row labels will cancel, leaving CD. The simplified Boolean expression, therefore, is

$$Y = \overline{A}D + CD$$

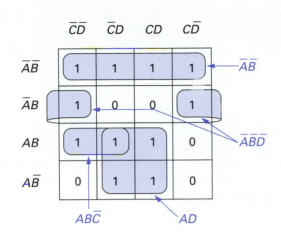
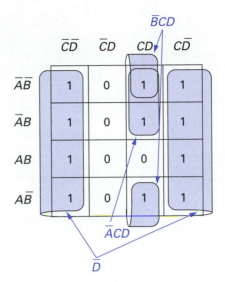

Before : $Y = \overline{A}\overline{B}\overline{C}\overline{D} + \overline{A}\overline{B}\overline{C}D + \overline{A}\overline{B}CD + \overline{A}\overline{B}C\overline{D} + \overline{A}B\overline{C}\overline{D} + \overline{A}B\overline{C}D + AB\overline{C}\overline{D} + AB\overline{C}D + ABCD + A\overline{B}\overline{C}\overline{D} + A\overline{B}CD$

After : $Y = AB\overline{C} + AD + \overline{A}B\overline{D} + \overline{A}\overline{B}$

(a)

Before : $Y = \overline{A}\overline{B}\overline{C}\overline{D} + \overline{A}\overline{B}CD + \overline{A}\overline{B}C\overline{D} + \overline{A}B\overline{C}\overline{D} + \overline{A}BCD + \overline{A}BC\overline{D} + AB\overline{C}\overline{D} + ABC\overline{D} + A\overline{B}\overline{C}\overline{D} + A\overline{B}CD + A\overline{B}C\overline{D}$

After : $Y = \overline{A}CD + \overline{B}CD + \overline{D}$

(b)

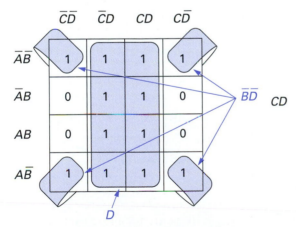

Before : $Y = \overline{A}\overline{B}\overline{C}\overline{D} + \overline{A}\overline{B}\overline{C}D + \overline{A}\overline{B}CD + \overline{A}\overline{B}C\overline{D} + \overline{A}B\overline{C}D + \overline{A}BCD + AB\overline{C}D + ABCD + A\overline{B}\overline{C}\overline{D} + A\overline{B}\overline{C}D + A\overline{B}CD + A\overline{B}C\overline{D}$

After : $Y = \overline{B}\overline{D} + D$

(c)

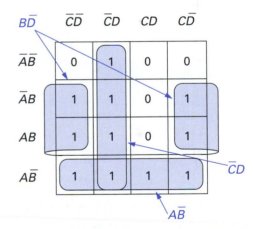

Before : $Y = \overline{A}\overline{B}\overline{C}D + \overline{A}B\overline{C}\overline{D} + \overline{A}B\overline{C}D + \overline{A}BC\overline{D} + AB\overline{C}\overline{D} + AB\overline{C}D + ABC\overline{D} + A\overline{B}\overline{C}\overline{D} + A\overline{B}\overline{C}D + A\overline{B}CD + A\overline{B}C\overline{D}$

After : $Y = A\overline{B} + B\overline{D} + \overline{C}D$

(d)

FIGURE 21-31 K-Map Grouping Examples.

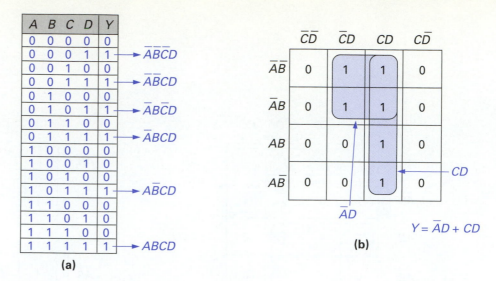

A	B	C	D	Y	
0	0	0	0	0	
0	0	0	1	1	→ $\overline{A}\overline{B}\overline{C}D$
0	0	1	0	0	
0	0	1	1	1	→ $\overline{A}\overline{B}CD$
0	1	0	0	0	
0	1	0	1	1	→ $\overline{A}B\overline{C}D$
0	1	1	0	0	
0	1	1	1	1	→ $\overline{A}BCD$
1	0	0	0	0	
1	0	0	1	0	
1	0	1	0	0	
1	0	1	1	1	→ $A\overline{B}CD$
1	1	0	0	0	
1	1	0	1	0	
1	1	1	0	0	
1	1	1	1	1	→ $ABCD$

(a)

$Y = \overline{A}D + CD$

(b)

FIGURE 21-32   **Example of Four-Variable K-Map.**

Referring to the truth table in Figure 21-33(a), you can see that input words 0000 through 1001 have either a 0 or 1 at the output Y. Input words 1010 through 1111, on the other hand, have an "X" written in the output column to indicate that the output Y for these input combinations is unimportant (output can be either a 0 or 1). These Xs in the output are appropriately called *don't-care conditions,* and can be treated as either a 0 or 1. In K-maps, these Xs can be grouped with 1s to create larger groups, and therefore a simpler logic circuit, as we will see in the next example.

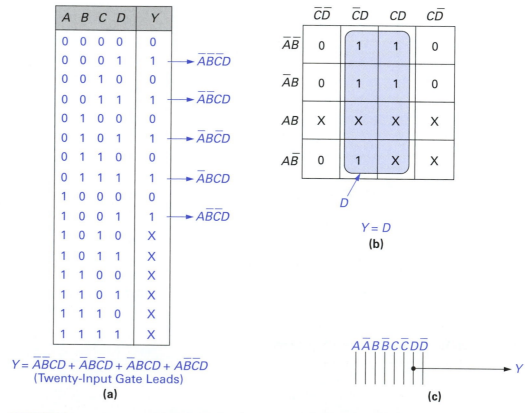

A	B	C	D	Y	
0	0	0	0	0	
0	0	0	1	1	→ $\overline{A}\overline{B}\overline{C}D$
0	0	1	0	0	
0	0	1	1	1	→ $\overline{A}\overline{B}CD$
0	1	0	0	0	
0	1	0	1	1	→ $\overline{A}B\overline{C}D$
0	1	1	0	0	
0	1	1	1	1	→ $\overline{A}BCD$
1	0	0	0	0	
1	0	0	1	1	→ $A\overline{B}\overline{C}D$
1	0	1	0	X	
1	0	1	1	X	
1	1	0	0	X	
1	1	0	1	X	
1	1	1	0	X	
1	1	1	1	X	

$Y = \overline{A}\overline{B}\overline{C}D + \overline{A}\overline{B}CD + \overline{A}B\overline{C}D + \overline{A}BCD$
(Twenty-Input Gate Leads)

(a)

$Y = D$

(b)

$A\overline{A}B\overline{B}C\overline{C}D\overline{D}$

Y

(c)

FIGURE 21-33   **Four-Variable K-Map Example with Don't-Care Conditions.**

CHAPTER 21 / BOOLEAN EXPRESSIONS AND ALGEBRA

■ INTEGRATED MATH APPLICATION:

Sketch an equivalent logic circuit for the truth table given in Figure 21-33(a) using the fewest number of inputs.

■ Solution:

Figure 21-33(b) shows a four-variable K-map with the 0s, 1s, and Xs (don't-care conditions) inserted into their appropriate cells. Since a larger group will always yield a smaller term, visualize the Xs as 1s, and then try for as big a group as possible. In this example, the Xs have allowed us to group an octet, resulting in the single product D. As a result, the 20-input Boolean equation in Figure 21-33(a) can be replaced with a single line connected to D, as shown in Figure 21-33(c).

SELF-TEST EVALUATION POINT FOR SECTION 21-4

Now that you have completed this section, you should be able to:

■ **Objective 6.** *Describe how Boolean algebra and Karnaugh maps can be used to simplify logic circuits.*

Use the following questions to test your understanding of Section 21-4.

1. A two-variable K-map will have _____ cells.
2. Simplify the equation $Y = \overline{A} B + A B$.
3. When grouping 1s in a K-map, the _____ (smaller/larger) the groups, the smaller the terms.
4. Can the don't-care conditions in a K-map be grouped with 0s or with 1s, or does it matter?

REVIEW QUESTIONS

Multiple Choice Questions

1. In Boolean, the OR function is described as logical _____ and the AND function is described as logical _____.

   a. Addition, division
   b. Multiplication, addition
   c. Addition, multiplication
   d. Multiplication, subtraction

2. The OR gate is said to produce at its output the logical _____ of its inputs, while the AND gate is said to produce at its output the logical _____ of its inputs.

   a. Sum, product
   b. Product, quotient
   c. Sum, subtrahend
   d. Product, sum

3. If inputs A and B are ANDed together and then the result is ORed with C, the Boolean expression equation describing this operation would be _____.

   a. $Y = AB + BC$
   b. $Y = (A + B)C$
   c. $Y = AB + C$
   d. $Y = AB + AC$

4. The Boolean expression $Y = (A \cdot B) + (A \cdot C)$ describes a logic circuit that has _____.

   a. Six inputs
   b. Two OR and one AND gate
   c. Two AND and one OR gate
   d. Four inputs

5. The Boolean expression $Y = \overline{A + B}$ describes a two-input _____.

   a. NAND gate
   b. NOR gate
   c. AND gate
   d. OR gate

6. DeMorgan's first theorem states that _____.

   a. $\overline{A \cdot B} = \overline{A} + \overline{B}$
   b. $\overline{A + B} = \overline{A} + B$
   c. $\overline{A} + \overline{B} = \overline{A} \cdot \overline{B}$
   d. $\overline{A \cdot B} = A \cdot \overline{B}$

7. DeMorgan's second theorem states that _____.

   a. $\overline{A \cdot B} = \overline{A} + \overline{B}$
   b. $\overline{A + B} = \overline{A} + B$
   c. $\overline{A} + \overline{B} = \overline{A} \cdot \overline{B}$
   d. $\overline{A \cdot B} = A \cdot \overline{B}$

8. In Boolean algebra which symbol is used to describe the exclusive OR action?

   a. $+$
   b. $\cdot$
   c. $\times$
   d. $\oplus$

9. Which of the following algebraically states the commutative law?

   a. $A + B = B + A$
   b. $A \cdot (B \cdot C) = (A \cdot B) \cdot C$
   c. $A + (B \cdot C) = (A + B) \cdot (A + C)$
   d. None of the above

10. $A + (B \cdot C) = (A + B) \cdot (A + C)$ is the algebraic definition of the _____.

   a. Associative law
   b. Distributive law
   c. Commutative law
   d. Complements law

11. The four Boolean rules for an OR gate are $A + 0 = ?$, $A + A = ?$, $A + 1 = ?$, $A + \overline{A} = ?$.

    a. $A, 0, 1, 1$ **c.** $A, A, 1, 1$
    b. $A, A, 0, 0$ **d.** $A, A, 0, 1$

12. The four Boolean rules for an AND gate are $A \cdot 1 = ?$, $A \cdot A = ?$, $A \cdot 0 = ?$, $A \cdot \overline{A} = ?$.

    a. $A, 0, 1, 1$ **c.** $A, A, 1, 1$
    b. $A, A, 0, 0$ **d.** $A, A, 0, 1$

13. If you applied DeMorgan's theorem and the double-inversion rule to the equation $Y = \overline{A} + BC$, what would be the result?

    a. $A\overline{B} + C$ **c.** $\overline{A}\,B + C$
    b. $A + \overline{BC}$ **d.** $AB + \overline{C}$

14. What would be the dual equation of $A + (B \cdot C) = (A + B) \cdot (A + C)$?

    a. $A \cdot (B + C) = (A \cdot B) \cdot (A \cdot C)$
    b. $\overline{A} + B = AB + \overline{C}$
    c. $A \cdot (B \cdot C) = (A \cdot B) + (A \cdot C)$
    d. $A + (B \cdot C) = (A \cdot B) \cdot (A + C)$

15. $Y = AB + BC + AC$ is an example of a _____.

    a. Product-of-sums expression
    b. Sum-of-products expression
    c. Quotient-of-sums expression
    d. Sum-of-quotients expression

16. What would be the fundamental product of the input word $ABCD = 1101$?

    a. $ABCD$ **c.** $AB\overline{C}D$
    b. $\overline{ABCD}$ **d.** $AB\overline{C}\,\overline{D}$

17. How many cells would a three-variable Karnaugh map have?

    a. 4   **c.** 8
    b. 6   **d.** 16

18. What would be the Boolean expression for a four-variable logic circuit that NANDs inputs A and B and also NANDs inputs C and D, and then NANDs the results?

    a. $\overline{\overline{AB} + \overline{CD}}$ **c.** $\overline{(\overline{AB}) + (\overline{CD})}$
    b. $(\overline{AB}) \cdot (\overline{CD})$ **d.** $\overline{(A + B) \cdot (C + D)}$

19. If you applied DeMorgan's theorem and the double-inversion rule to the answer in question 18, you would obtain which of the following equations?

    a. $AB + CD$ **c.** $(A + B) \cdot (C + D)$
    b. $A + B + C + D$ **d.** $(A + B) + (CD)$

20. In reference to questions 18 and 19, you can see that the DeMorgan equivalent of a NAND-NAND circuit is a(n) _____.

    a. OR-AND   **c.** AND-OR
    b. AND-NOR   **d.** NOR-AND

Communication Skill Questions

21. List the Boolean expression for the seven basic logic gates. (21-1)

22. Describe briefly, with an example, the process of converting a desired set of output conditions into an equivalent logic circuit. (21-2)

23. What is the difference between an SOP and a POS equation? (21-3)

24. What is a Karnaugh map? (21-4-2)

25. Briefly describe how K-maps can be used to simplify a logic circuit. (21-4-2)

26. How is the simplicity of a logic circuit generally judged? (21-4)

27. Why is it necessary to simplify logic circuits? (21-4)

28. What is meant by the terms *logical product* and *logical sum*? (21-1)

29. How can DeMorgan's theorems be stated using Boolean algebra? (21-2)

30. Briefly describe why larger groups of 1s on a K-map yield smaller terms, and why smaller terms are desired. (21-4-2)

Practice Problems

31. Determine the Boolean expressions for the logic circuits shown in Figure 21-34.

32. Which of the expressions in question 31 would be considered POS, and which SOP?

33. Sketch the equivalent logic circuit for the following Boolean equations.

    a. $Y = (AB) + (\overline{AB}) + (\overline{A + B}) + (A \oplus B)$
    b. $Y = (\overline{A} + B) \oplus ABC$

34. Determine the Boolean expression for the Y_2 output in Table 21-1.

35. Use a K-map to simplify the Y_2 expression from question 34.

36. How much was the Y_2 logic circuit simplified in question 35?

37. Referring to Table 21-1, (a) determine the Boolean expression for the Y_1 output. (b) Simplify the expression using a K-map. (c) Describe how much the logic circuit was simplified.

38. Determine the Boolean expression for the Y_3 output in Table 21-1, simplify the expression using a K-map, and then describe how much the logic circuit was simplified.

39. Sketch the simplified logic circuits for the Y_1, Y_2, and Y_3 outputs listed in Table 21-1.

40. If the don't-care conditions in the Y_3 output in Table 21-1 were 0s, what would the simplified expression be? Is the expression more or less simplified with don't-care conditions?

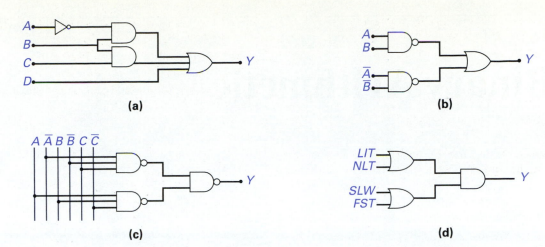

FIGURE 21-34 From Logic Circuit to Boolean Equation.

TABLE 21-1 The Y_1, Y_2, and Y_3 Truth Table

A	B	C	D	Y_1	Y_2	Y_3
0	0	0	0	1	1	1
0	0	0	1	1	1	0
0	0	1	0	1	0	1
0	0	1	1	1	0	0
0	1	0	0	0	0	1
0	1	0	1	0	0	1
0	1	1	0	0	1	1
0	1	1	1	0	1	1
1	0	0	0	0	1	0
1	0	0	1	1	1	0
1	0	1	0	0	0	0
1	0	1	1	1	0	0
1	1	0	0	1	0	1
1	1	0	1	1	0	x
1	1	1	0	1	0	x
1	1	1	1	1	0	x

Web Site Questions

Go to the web site http://www.prenhall.com/cook, select the textbook *Mathematics for Electronics and Computers,* select this chapter, and then follow the instructions when answering the multiple-choice practice problems.

Binary Arithmetic

Woolen Mill Makes Minis

At the age of 24 in 1950, Kenneth Olsen went to work at MIT's Digital Computer Laboratory as a research assistant. For the next seven years, Olsen worked on the SAGE (Semi Automatic Ground Environment) project, which was a new computer system designed to store the constantly changing data obtained from radars tracking long-range enemy bombers capable of penetrating American air space. IBM won the contract to build the SAGE network and Olsen traveled to the plant in New York to supervise production. After completing the SAGE project, Olsen was ready to move on to bigger and better things; however, he didn't want to go back to MIT and was not happy with the rigid environment at a large corporation like IBM.

In September 1957, Olsen and fellow MIT colleague Harlan Anderson leased an old brick woolen mill in Maynard, Massachusetts, and after some cleaning and painting they began work in their new company, which they named Digital Equipment Corporation or DEC. After three years, the company came out with their first computer, the PDP 1 (programmed data processor), and its ease of use, low cost, and small size made it an almost instant success. In 1965, the desktop PDP 8 was launched, and since in that era miniskirts were popular, it was probably inevitable that the small machine was nicknamed a minicomputer. In 1969, the PDP 11 was unveiled, and like its predecessors, this number-crunching data processor went on to be accepted in many applications, such as tracking the millions of telephone calls received on the 911 emergency line, directing welding machines in automobile plants, recording experimental results in laboratories, and processing all types of data in offices, banks, and department stores. Just 20 years after introducing low-cost computing in the 1960s, DEC was second only to IBM as a manufacturer of computers in the United States.

22

Outline and Objectives

VIGNETTE: WOOLEN MILL MAKES MINIS

INTRODUCTION

22-1 BINARY ARITHMETIC

Objective 1. Describe how to complete the four basic arithmetic operations—binary addition, binary subtraction, binary multiplication, and binary division.

22-1-1 Binary Addition

22-1-2 Binary Subtraction

22-1-3 Binary Multiplication

22-1-4 Binary Division

22-2 REPRESENTING POSITIVE AND NEGATIVE NUMBERS

Objective 2. Explain how the sign-magnitude, one's complement, and two's complement number systems are used to represent positive and negative numbers.

22-2-1 The Sign and Magnitude Number System

22-2-2 The One's Complement (1's Complement) Number System

22-2-3 The Two's Complement (2's Complement) Number System

22-3 TWO'S COMPLEMENT ARITHMETIC

Objective 3. Describe how an adder circuit can add, subtract, multiply, and divide two's complement numbers.

22-3-1 Adding Positive Numbers

22-3-2 Adding Positive and Negative Numbers

22-3-3 Adding Negative Numbers

22-3-4 Subtraction Through Addition

22-3-5 Multiplication Through Repeated Addition

22-3-6 Division Through Repeated Subtraction

22-4 REPRESENTING LARGE AND SMALL NUMBERS

Objective 4. Describe how the floating-point number system can be used to represent a large range of values.

MULTIPLE CHOICE QUESTIONS

COMMUNICATION SKILLS QUESTIONS

PRACTICE PROBLEMS

WEB SITE QUESTIONS

Introduction

At the heart of every digital electronic computer system is a circuit called a microprocessor unit (MPU). To perform its function, the MPU contains several internal circuits, one of which is called the arithmetic–logic unit, or ALU. This unit is the number-crunching circuit of every digital system and, as its name implies, it can be controlled to perform either arithmetic operations (such as addition and subtraction) or logic operations (such as AND and OR).

In order to understand the operation of an ALU, we must begin by discussing how binary numbers can represent any value, whether it is large, small, positive, or negative. We must also discuss how these numbers can be manipulated to perform arithmetic operations such as addition, subtraction, multiplication, and division.

22-1 BINARY ARITHMETIC

The addition, subtraction, multiplication, and division of binary numbers is performed in exactly the same way as the addition, subtraction, multiplication, and division of decimal numbers. The only difference is that the decimal number system has 10 digits, whereas the binary number system has 2.

As we discuss each of these arithmetic operations, we will use decimal arithmetic as a guide to binary arithmetic so that we can compare the known to the unknown.

22-1-1 *Binary Addition*

Figure 22-1(a) illustrates how the decimal values 29,164 and 63,729 are added together. First the units 4 and 9 are added, producing a sum of 13, or 3 carry 1. The carry is then added to the 6 and 2 in the tens column, producing a sum of 9 with no carry. This process is then repeated for each column, moving right to left, until the total sum of the addend, augend, and their carries has been obtained.

In the addition of two decimal values, many combinations are available, since decimal has 10 different digits. With binary, however, only 2 digits are available, and therefore only four basic combinations are possible. These four rules of binary addition are shown in Figure 22-1(b). To help explain these rules, let us apply them to the example in Figure 22-1(c), in which we will add 11010 (decimal 26) to 11100 (decimal 28). As in decimal addition, we will begin at the right in the ones column and proceed to the left. In the ones column $0 + 0 = 0$ with a carry of 0 (Rule 1). In the twos column $1 + 0$ and the 0 carry from the ones column $= 1$ with no carry (Rule 2). In the fours column we have $0 + 1$ plus a carry of 0 from the twos column. This results in a fours column total of 1, with a carry of 0 (Rule 2). In the eights column we have $1 + 1$ plus a 0 carry from the fours column, producing a total of 10_2 ($1 + 1 =$ decimal 2, which is 10 in binary). In the eights column this 10 (decimal 2) is written down as a total of 0 with a carry of 1 (Rule 3). In the sixteens column, we have $1 + 1$, plus a carry of 1 from the eights column, producing a total of 11_2 ($1 + 1 + 1 =$ decimal 3, which is 11 in binary). In the sixteens column, this 11 (decimal 3) is written down as a total of 1 with a carry of 1 (Rule 4). Since there is only a carry of 1 in the thirty-twos column, the total in this column will be 1.

Converting the binary addend and augend to their decimal equivalents as seen in Figure 22-1(c), we see that 11010 (decimal 26) plus 11100 (decimal 28) will yield a sum or total of 110110 (decimal 54).

$$(10{,}000)\ (1{,}000)\ (100)\ (10)\ (1)$$

Carry	0	1	0	0	1
Addend:		2	9, 1	6	4
Augend:	+	6	3, 7	2	9
Sum:		9	2, 8	9	3

$4 + 9 = 13$ or 3, 1 carry

(a)

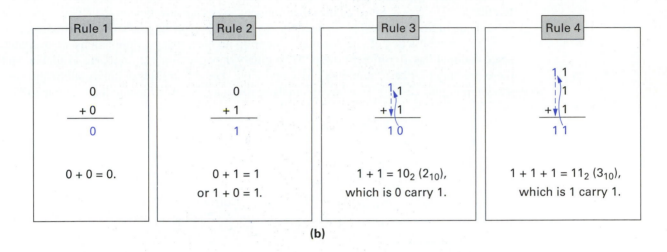

Rule 1
$$\begin{array}{r} 0 \\ +\,0 \\ \hline 0 \end{array}$$
$0 + 0 = 0$.

Rule 2
$$\begin{array}{r} 0 \\ +\,1 \\ \hline 1 \end{array}$$
$0 + 1 = 1$ or $1 + 0 = 1$.

Rule 3
$$\begin{array}{r} 1 \\ +\,1 \\ \hline 1\,0 \end{array}$$
$1 + 1 = 10_2\ (2_{10})$, which is 0 carry 1.

Rule 4
$$\begin{array}{r} 1 \\ +\,1 \\ \hline 1\,1 \end{array}$$
$1 + 1 + 1 = 11_2\ (3_{10})$, which is 1 carry 1.

(b)

$$(32)\ (16)\ (8)\ (4)\ (2)\ (1) \qquad (10)\ (1)$$

		(32)	(16)	(8)	(4)	(2)	(1)			(10)	(1)
Carry:		1	1	0	0	0				1	
Addend:			1	1	0	1	0_2	=		2	6_{10}
Augend:	+		1	1	1	0	0_2		+	2	8_{10}
Sum:		1	1	0	1	1	0_2			5	4_{10}

(c)

FIGURE 22-1   Binary Addition.

■ **INTEGRATED MATH APPLICATION:**

Find the sum of the following binary numbers.

   a. $1011 + 1101 = ?$ b. $1000110 + 1100111 = ?$

■ *Solution:*

a. *Carry:*   1 1 1 1

Augend:		1	0	1	1_2	(11_{10})
Addend:	+	1	1	0	1_2	(13_{10})
Sum:		1 1	0	0	0_2	(24_{10})

b. *Carry:*   1 0 0 0 1 1 0

Augend:		1	0	0	0	1	1	0_2	(70_{10})
Addend:	+	1	1	0	0	1	1	1_2	(103_{10})
Sum:	1	0	1	0	1	1	0	1_2	(173_{10})

22-1-2 *Binary Subtraction*

Figure 22-2(a) reviews the decimal subtraction procedure by subtracting 4615 from 7003. Starting in the units column, you can see that since we cannot take 5 away from 3, we will have to go to a higher-order minuend unit and borrow. Since the minuend contains no tens or hundreds, we will have to go to the thousands column. From this point a chain of borrowing occurs as a thousand is borrowed and placed in the hundreds column (leaving 6 thousands), one of the hundreds is borrowed and placed in the tens column (leaving 9 hundreds), and one of the tens is borrowed and placed in the units column (leaving 9 tens). After borrowing 10, the minuend units digit has a value of 13, and if we now perform the subtraction, the result or difference of $13 - 5 = 8$. Since all the other minuend digits are now greater than their respective subtrahend digits, there will be no need for any further borrowing to obtain the difference.

In the subtraction of one decimal value from another, many combinations are available since decimal has 10 different digits. With binary, however, only 2 digits are available, and therefore only four basic combinations are possible. These four rules of binary subtraction are shown in Figure 22-2(b). To help explain these rules, let us apply them to the example in Fig-

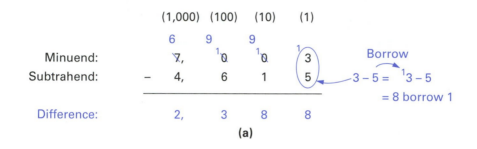

(a)

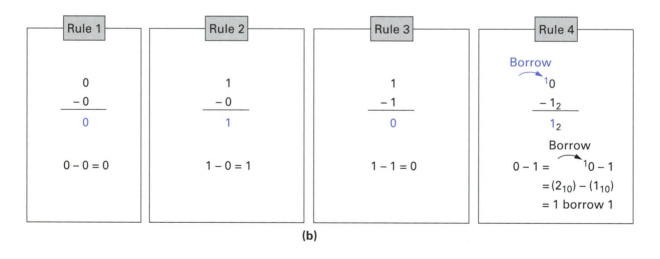

(b)

(c)

FIGURE 22-2   Binary Subtraction.

ure 22-2(c) in which we will subtract 11100 (decimal 28) from 1010110 (decimal 86). As in decimal subtraction, we will begin at the right in the ones column and proceed to the left. In the ones column $0 - 0 = 0$ (Rule 1). In the twos column $1 - 0 = 1$ (Rule 2), and in the fours column $1 - 1 = 0$ (Rule 3). In the eights column we cannot subtract 1 from nothing or 0, so therefore we must borrow 1 from the sixteens column, making the minuend in the eights column 10 (decimal 2). The difference can now be calculated in the eights column since $10 - 1 = 1$ (decimal $2 - 1 = 1$, Rule 4). Due to the previous borrow, the minuend in the sixteens column is now 0 and therefore we need to borrow a 1 from the thirty-twos column. Since the thirty-twos column is also 0, we will need to borrow from the sixty-fours column. Borrowing 1 from the sixty-fours column will leave a minuend of 0 and make the minuend in the thirty-twos column equal to 10 (decimal 2). Borrowing 1 from the 10 (borrowing decimal 1 from 2) in the thirty-twos column will leave a minuend of 1 and make the minuend in the sixteens column equal to 10 (decimal 2). We can now subtract the subtrahend of 1 from the minuend of 10 (decimal 2) in the sixteens column to obtain a difference of 1. Due to the previous borrow, the minuend in the thirty-twos column is equal to 1, and $1 - 0 = 1$ (Rule 2). Finally, since the minuend and subtrahend are both 0 in the sixty-fours column, the subtraction is complete.

Converting the binary minuend and subtrahend to their decimal equivalents, as seen in Figure 22-2(c), we see that 1010110 (decimal 86) minus 11100 (decimal 28) will result in a difference of 111010 (decimal 58).

■ INTEGRATED MATH APPLICATION:

Find the difference between the following binary numbers.

a. $100111 - 11101 = ?$ b. $101110111 - 1011100 = ?$

■ *Solution:*

a.

Minuend:		$\cancel{1}^{0}$	$\cancel{0}^{1}$	0^{1}	1	1	1_2	(39_{10})
Subtrahend:	$-$		1	1	1	0	1_2	(29_{10})
Difference:		0	0	1	0	1	0_2	(10_{10})

b.

Minuend:		1	0	1	$\cancel{1}^{0}$	$\cancel{1}^{0\ 1}$	0^{1}	1	1	1_2	(375_{10})
Subtrahend:	$-$			1	0	1	1	1	0	0_2	(92_{10})
Difference:		1	0	0	0	1	1	0	1	1_2	(283_{10})

22-1-3   *Binary Multiplication*

Figure 22-3(a) reviews the decimal multiplication procedure by multiplying 463 by 23. To perform this operation, we begin by multiplying each digit of the multiplicand by the units digit of the multiplier to obtain the first partial product. Second, we multiply each digit of the multiplicand by the tens digit of the multiplier to obtain the second partial product. Finally, we add all the partial products to obtain the final product.

In the multiplication of one decimal value by another, many combinations are available, since decimal has 10 different digits. With binary multiplication, however, only 2 digits are available, and therefore only four basic combinations are possible. These four rules of binary multiplication are shown in Figure 22-3(b). To help explain these rules, let us apply them to the example in Figure 22-3(c) in which we will multiply 1011 (decimal 11) by 101 (decimal 5). As in decimal multiplication, we begin by multiplying each bit of the multiplicand by the ones column multiplier bit to obtain the first partial product. Second, we multiply each bit of the multiplicand by the twos column multiplier bit to obtain the second partial product. Remember that the LSB of the partial product must always be directly under its respective multiplier bit, and therefore the LSB of the second partial product should be under the twos column multiplier bit. In the next step, we add together the first and second partial products to obtain a sum of the partial products. With decimal multiplication we usually calculate all the partial products and then add them together to obtain the final product.

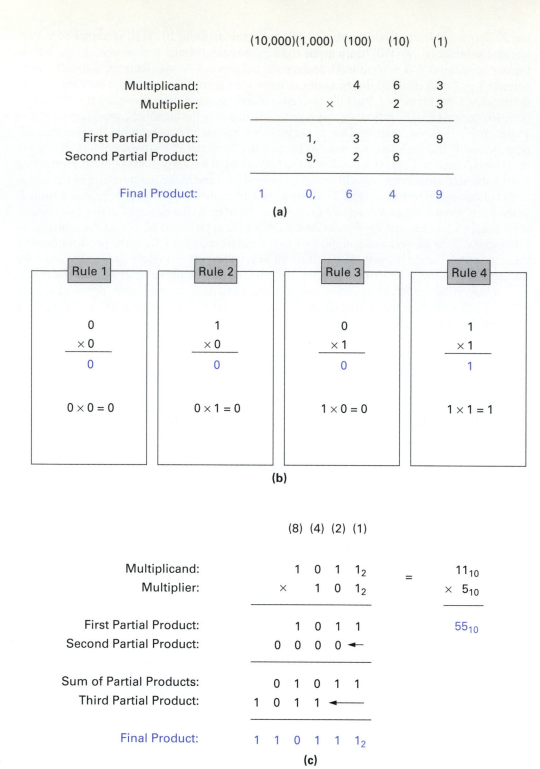

FIGURE 22-3   Binary Multiplication.

This procedure can also be used with binary multiplication; however, it is better to add only two partial products at a time, since the many carries in binary become hard to keep track of and easily lead to errors. In the next step, we multiply each bit of the multiplicand by the fours column multiplier bit to obtain the third partial product. Remember that the LSB of this third partial product must be placed directly under its respective multiplier bit. The final step is to add the third partial product to the previous partial sum to obtain the final product or result.

If you study the partial products obtained in this example, you will notice that they were either exactly equal to the multiplicand (when the multiplier was 1) or all 0s (when the multiplier was 0). You can use this shortcut in future examples; however, be sure always to place the LSB of the partial products directly below their respective multiplier bits.

■ **INTEGRATED MATH APPLICATION:**

Multiply 1101_2 by 1011_2.

■ *Solution:*

Multiplicand:					1	1	0	1_2		13_{10}
Multiplier:				$\times$	1	0	1	1_2		$\times 11_{10}$
First Partial Product:					1	1	0	1		143_{10}
Second Partial Product:				1	1	0	1	$\leftarrow$		
Sum of Partial Product:			1	0	0	1	1	1		
Third Partial Product:			0	0	0	0	$\leftarrow$	$\leftarrow$		
Sum of Partial Product:			1	0	0	1	1	1		
Fourth Partial Product:		1	1	0	1	$\leftarrow$	$\leftarrow$	$\leftarrow$		
Final Product:	1	0	0	0	1	1	1	1_2		

22-1-4 *Binary Division*

Figure 22-4(a) reviews the decimal division procedure by dividing 830 by 23. We begin this long-division procedure by determining how many times the divisor (23) can be subtracted from the first digit of the dividend (8). Since the dividend is smaller, the quotient is 0. Next,

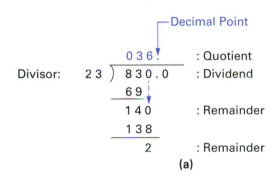

(a)

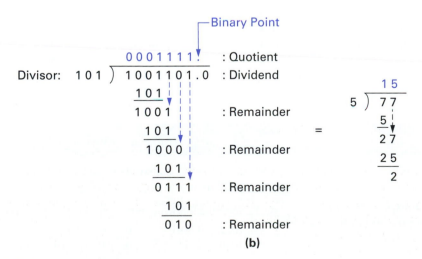

(b)

FIGURE 22-4 **Binary Division.**

we see how many times the divisor (23) can be subtracted from the first two digits of the dividend (83). Since 23 can be subtracted three times from 83, a quotient of 3 results, and the product of 3×23, or 69, is subtracted from the first two digits of the dividend, giving a remainder of 14. Then the next digit of the dividend (0) is brought down to the remainder (14), and we see how many times the divisor (23) can be subtracted from the new remainder (140). In this example, 23 can be subtracted six times from 140, a quotient of 6 results, and the product of 6×23, or 138, is subtracted from the remainder, giving a new remainder of 2. Therefore, $830 \div 23 = 36$, remainder 2.

Figure 22-4(b) illustrates an example of binary division, which is generally much simpler than decimal division. To perform this operation, we must first determine how many times the divisor (101_2) can be subtracted from the first bit of the dividend (1_2). Since the dividend is smaller than the divisor, a quotient of 0_2 is placed above the first digit of the dividend. Next, we see how many times the divisor (101_2) can be subtracted from the first two bits of the dividend (10_2), and this again results in a quotient of 0_2. Similarly, the divisor (101_2) cannot be subtracted from the first three bits of the dividend, so the quotient is once again 0_2. Finally, we find a dividend (1001_2) that is greater than the divisor (101_2), and therefore a 1_2 is placed in the quotient, and the divisor is subtracted from the first four bits of the dividend, resulting in a remainder of 100_2. Then the next bit of the dividend (1_2) is brought down to the remainder (100_2), and we see how many times the divisor (101_2) can be subtracted from the new remainder (1001_2). Since 101_2 is smaller than 1001_2, a quotient of 1_2 results. The divisor is subtracted from the remainder, resulting in a new remainder of 100_2. Then, the next bit of the dividend (0_2) is brought down to the remainder (100_2), and we see how many times the divisor (101_2) can be subtracted from the new remainder (1000_2). Since 101_2 is smaller than 1000_2, a quotient of 1 results. The divisor is subtracted from the remainder, resulting in a new remainder of 11_2. Then, the next bit of the dividend (1_2) is brought down to the remainder (11_2), and we see how many times the divisor (101_2) can be subtracted from the new remainder (111_2). Since 101_2 is smaller than 111_2, a quotient of 1 results. The divisor is subtracted from the remainder, resulting in a final remainder of 10. Therefore, $1001101_2 \div 101_2 = 1111_2$, remainder 10 ($77 \div 5 = 15$, remainder 2).

■ INTEGRATED MATH APPLICATION:

Divide 1110101_2 by 110_2.

■ *Solution:*

```
                  0 0 1 0 0 1 1₂        Quotient            019₁₀
Divisor:   110)1 1 1 0 1 0 1₂          Dividend       6₁₀)117₁₀
              1 1 0 ↓ ↓ ↓ ↓
              0 0 1 0 1 0 ↓            Remainder
                  1 1 0 ↓                                    6↓
                  1 0 0 1              Remainder             57
                    1 1 0                                    54
                  0 0 1 1₂            Remainder              3₁₀
```

Now that you have completed this section, you should be able to:

■ *Objective 1.* *Describe how to complete the four basic arithmetic operations—binary addition, binary subtraction, binary multiplication, and binary division.*

Use the following questions to test your understanding of Section 22-1:

1. Perform the following binary arithmetic operations.
   a. $1011 + 11101 = ?$ **c.** $101 \times 10 = ?$
   b. $1010 - 1011 = ?$ **d.** $10111 \div 10 = ?$

22-2 REPRESENTING POSITIVE AND NEGATIVE NUMBERS

Many digital systems, such as calculators and computers, are used to perform mathematical functions on a wide range of numbers. Some of these numbers are positive and negative signed numbers; therefore, a binary code is needed to represent these numbers.

22-2-1 *The Sign and Magnitude Number System*

The **sign and magnitude number system** is a binary code system used to represent positive and negative numbers. A sign–magnitude number contains a *sign bit* (0 for positive, 1 for negative) followed by the *magnitude bits.*

As an example, Figure 22-5(a) shows some 4-bit sign–magnitude positive numbers. All these sign–magnitude binary numbers—0001, 0010, 0011, and 0100—have an MSB, or sign bit, of 0, and therefore they are all positive numbers. The remaining 3 bits in these 4-bit words indicate the magnitude of the number based on the standard 421 binary column weight ($001_2 = 1_{10}$, $010_2 = 2_{10}$, $011_2 = 3_{10}$, $100_2 = 4_{10}$).

As another example, Figure 22-5(b) shows some 4-bit sign–magnitude negative numbers. All of these sign–magnitude binary numbers—1100, 1101, 1110, and 1111—have an MSB, or sign bit, of 1, and therefore they are all negative numbers. The remaining 3 bits in these 4-bit words indicate the magnitude of the number based on the standard 421 binary column weight ($100_2 = 4_{10}$, $101_2 = 5_{10}$, $110_2 = 6_{10}$, $111_2 = 7_{10}$).

If you need to represent larger decimal numbers, you simply use more bits, as shown in Figure 22-5(c) and (d). The principle still remains the same in that the MSB indicates the sign of the number, and the remaining bits indicate the magnitude of the number.

The sign–magnitude number system is an ideal example of how binary numbers could be coded to represent positive and negative numbers. This code system, however, requires complex digital hardware for addition and subtraction, and is therefore seldom used.

> **Sign and Magnitude Number System**
>
> A binary code used to represent positive and negative numbers in which the MSB signifies sign and the following bits indicate magnitude.

Sign Bit	Magnitude Bits
0 = +	
1 = −	= Binary Equivalent

4-Bit Sign-Magnitude Numbers

0 0 0 1$_2$	=	+ 1$_{10}$
0 0 1 0$_2$	=	+ 2$_{10}$
0 0 1 1$_2$	=	+ 3$_{10}$
0 1 0 0$_2$	=	+ 4$_{10}$

Sign Bit   Magnitude Bits

(a)

1 1 0 0$_2$	=	− 4$_{10}$
1 1 0 1$_2$	=	− 5$_{10}$
1 1 1 0$_2$	=	− 6$_{10}$
1 1 1 1$_2$	=	− 7$_{10}$

Sign Bit   Magnitude Bits

(b)

8-Bit Sign-Magnitude Numbers

0 0 0 0 1 1 1 1$_2$	=	+ 15$_{10}$
0 1 0 1 0 0 0 0$_2$	=	+ 80$_{10}$
1 0 0 0 1 0 1 0$_2$	=	− 10$_{10}$
1 1 1 1 1 1 1 1$_2$	=	− 127$_{10}$

(c)

16-Bit Sign-Magnitude Numbers

0 0 0 0 0 0 0 0 0 0 0 0 0 0 0 1$_2$	=	+ 1$_{10}$
0 0 0 0 0 0 0 0 1 0 1 0 0 1 0 0$_2$	=	+ 164$_{10}$
1 0 0 0 0 0 0 0 0 0 1 1 0 0 1 0$_2$	=	− 50$_{10}$
1 0 0 0 1 1 1 1 0 0 0 0 0 0 0 0$_2$	=	− 3840$_{10}$

(d)

FIGURE 22-5 The Sign–Magnitude Number System.

What are the decimal equivalents of the following sign–magnitude number system codes?

   a. 0111   b. 1010   c. 0001 1111   d. 1000 0001

■ *Solution:*

   a. $+7$ b. -2 c. $+31$ d. -1

What are the 8-bit sign–magnitude number system codes for the following decimal values?

   a. $+8$ b. -65 c. $+127$ d. -127

■ *Solution:*

   a. 0000 1000   b. 1100 0001   c. 0111 1111   d. 1111 1111

22-2-2   *The One's Complement (1's Complement) Number System*

One's Complement Number System

A binary code used to represent positive and negative numbers, in which negative values are determined by inverting all the binary digits of an equivalent positive value.

The **one's complement number system** became popular in early digital systems. It also used an MSB sign bit that was either a 0 (indicating a positive number) or a 1 (indicating a negative number).

    Figure 22-6(a) shows some examples of 8-bit one's complement positive numbers. As you can see, one's complement positive numbers are no different from sign–magnitude positive numbers. The only difference between the one's complement number system and the sign–magnitude number system is the way in which negative numbers are represented. To understand the code used to represent negative numbers, we must first understand the meaning of the term *one's complement*. To one's complement a binary number means to invert or change all the 1s in the binary word to 0s, and all the 0s in the binary word to 1s. Performing the one's complement operation on a positive number will change the number from a one's complement positive number to a one's complement negative number. For example,

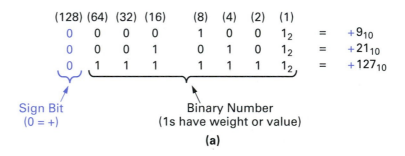

(a)

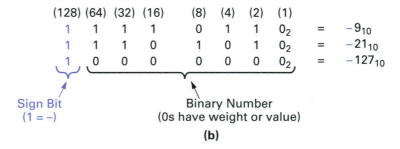

(b)

FIGURE 22-6 The One's Complement Number System.

by one's complementing the number 00000101, or +5, we will get 11111010, which is −5. In Figure 22-6(b), you can see the negative one's complement codes of the positive one's complement codes shown in Figure 22-6(a).

The next question you may have is, How can we determine the decimal value of a negative one's complement code? The answer is to do the complete opposite, or complement, of the positive one's complement code. This means count the 1s in a positive one's complement code, and count the 0s in a negative one's complement code. For example, with a positive one's complement code (MSB = 0), the value is determined by simply adding together all the column weights that contain a 1 (00000110 has a 1 in the fours column and twos column, so this number is 4 + 2, or +6). On the other hand, for a negative one's complement code (MSB = 1), the value is determined by doing the opposite and adding together all the column weights that contain a 0 (11110101 has a 0 in the eights column and twos column, so this number is 8 + 2, or −10).

■ **INTEGRATED MATH APPLICATION:**

What are the decimal equivalents of the following one's complement number system codes?

    a. 0010 0001   b. 1101 1101

■ *Solution:*

    a. MSB = 0 (number is positive, so add together all the column weights that contain a 1). The number 0010 0001 has a 1 in the thirty-twos column and a one in the units column, so 0010 0001 = +33.

    b. MSB = 1 (number is negative, so add together all the column weights that contain a 0). The number 1101 1101 has a 0 in the thirty-twos column and a 0 in the twos column, so 1101 1101 = −34.

Since the term *one's complement* can be used to describe both an operation (change all of the 1s to 0s, and all the 0s to 1s), and a positive and negative number system, some confusion can occur. To avoid this, remember that if you are asked to one's complement a number, it means to change all the 1s in the binary word to 0s and all 0s to 1s. If, on the other hand, you are asked to find the decimal equivalent of a one's complement number, you should decode the sign bit and binary value to determine the decimal equivalent value.

■ **INTEGRATED MATH APPLICATION:**

One's complement the following numbers.

    a. 0101 1111   b. 1110 0101

■ *Solution:*

To one's complement a number means to change all the 1s to 0s and all the 0s to 1s.

    a. 1010 0000   b. 0001 1010

■ **INTEGRATED MATH APPLICATION:**

Find the decimal equivalent of the following one's complement numbers.

    a. 0101 1111   b. 1110 0101

■ *Solution:*

    a. MSB = 0 (number is positive, so add together all the column weights that contain a 1). The number 0101 1111 = +95.

    b. MSB = 1 (number is negative, so add together all the column weights that contain a 0). The number 1110 0101 = −26.

22-2-3　The Two's Complement (2's Complement) Number System

The **two's complement number system** is used almost exclusively in digital systems to represent positive and negative numbers.

Positive two's complement numbers are no different from positive sign–magnitude and one's complement numbers in that an MSB sign bit of 0 indicates a positive number, and the remaining bits of the word indicate the value of the number. Figure 22-7(a) shows two examples of positive two's complement numbers.

Negative two's complement numbers are determined by two's complementing a positive two's complement number code. To two's complement a number first one's complement it (change all the 1s to 0s and 0s to 1s), and then add 1. For example, if we were to two's complement the number 0000 0101 (+5), we would obtain the two's complement number code for −5. The procedure to follow to two's complement 0000 0101 (+5) would be as follows:

0000	0101	Original number (+5)
1111	1010	One's complement (inverted)
+	1	Plus 1
1111	1011	Two's complement code for −5

The new two's complement code obtained after this two's complement operation is the negative equivalent of the original number. In our example, therefore:

Original number:　　　　　　　　　　　　0000 0101 = +5
After two's complement operation:　　　1111 1011 = −5

Any negative equivalent two's complement code therefore can be obtained by simply two's complementing the same positive two's complement number. For example, to find the negative two's complement code for −125, we would perform the following operation.

Two's	0111 1101	Original number (+125)
complement ⟶	1000 0010	One's complement (inverted)
	+ 1	Plus 1
operation	1000 0011	Two's complement code for −125

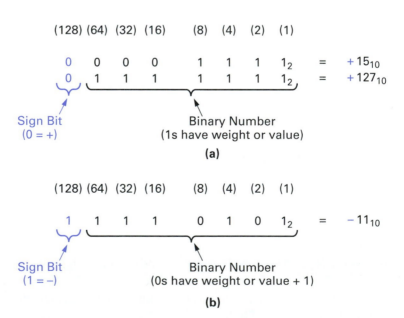

(a)

(b)

FIGURE 22-7　**The Two's Complement Number System.**

CHAPTER 22 / BINARY ARITHMETIC

■ **INTEGRATED MATH APPLICATION:**

Find the negative two's complement codes for

a. −26   b. −67

■ *Solution:*

a. 0001 1010   Two's complement code for +26
   1110 0101   One's complement (inverted)
   +   1   Plus 1
   1110 0110   Two's complement code for −26

b. 0100 0011   Two's complement code for +67
   1011 1100   One's complement (inverted)
   +   1   Plus 1
   1011 1101   Two's complement code for −67

Performing the two's complement operation on a number twice will simply take the number back to its original value and polarity. For example, let's take the two's complement code for +6 (0000 0110) and two's complement it twice.

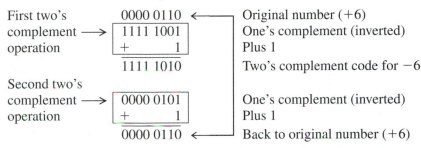

First two's   0000 0110 ←   Original number (+6)
complement →   1111 1001   One's complement (inverted)
operation   +   1   Plus 1
                     1111 1010   Two's complement code for −6

Second two's
complement →   0000 0101   One's complement (inverted)
operation   +   1   Plus 1
                     0000 0110 ←   Back to original number (+6)

The decimal equivalent of a positive two's complement number (MSB = 0) is easy to determine, since the binary value following the sign bit applies the same as any binary number, as shown in Figure 22-7(a). On the other hand, the decimal equivalent of a negative two's complement number (MSB = 1) is not determined in the same way. To determine the value of a negative two's complement number, refer to the remaining bits following the sign bit, add the weight of all columns containing a 0, and then add 1 to the result. For example, the negative two's complement number in Figure 22-7(b) has 0s in the eights and twos columns, and therefore its value will equal the sum of 8 and 2, plus the additional 1 (due to two's complement). As a result:

$$\overset{(8)\,(4)\,(2)\,(1)}{1\ 1\ 1\ 1\ 0\ 1\ 0\ 1} = (8 + 2) + 1 = -11$$

■ **INTEGRATED MATH APPLICATION:**

Determine the decimal equivalent of the following negative two's complement numbers.

a. 1110 0001   c. 1000 0000
b. 1011 1011   d. 1111 1111

■ *Solution:*

$$\overset{(\pm)\ (64)\ (32)\ (16)\ (8)\ (4)\ (2)\ (1)}{\text{a. }\ 1\quad 1\quad 1\quad 0\quad 0\quad 0\quad 0\quad 1_2} = (16 + 8 + 4 + 2) + 1 = -31_{10}$$

$$\overset{(\pm)\ (64)\ (32)\ (16)\ (8)\ (4)\ (2)\ (1)}{\text{b. }\ 1\quad 0\quad 1\quad 1\quad 1\quad 0\quad 1\quad 1_2} = (64 + 4) + 1 = -69_{10}$$

$$\overset{(\pm)\ (64)\ (32)\ (16)\ (8)\ (4)\ (2)\ (1)}{\text{c. }\ 1\quad 0\quad 0\quad 0\quad 0\quad 0\quad 0\quad 0_2} = (64 + 32 + 16 + 8 + 4 + 2 + 1) + 1 = -128_{10}$$

$$\overset{(\pm)\ (64)\ (32)\ (16)\ (8)\ (4)\ (2)\ (1)}{\text{d. }\ 1\quad 1\quad 1\quad 1\quad 1\quad 1\quad 1\quad 1_2} = (0) + 1 = -1_{10}$$

4-Bit Word	Standard Binary Value	2's Complement Value
0 0 0 0	0	0
0 0 0 1	1	+1
0 0 1 0	2	+2
0 0 1 1	3	+3 8 Positive
0 1 0 0	4	+4 Numbers
0 1 0 1	5	+5
0 1 1 0	6	+6
0 1 1 1	7	+7
1 0 0 0	8	−8
1 0 0 1	9	−7
1 0 1 0	10	−6
1 0 1 1	11	−5 8 Negative
1 1 0 0	12	−4 Numbers
1 1 0 1	13	−3
1 1 1 0	14	−2
1 1 1 1	15	−1

$2^4 =$ 16 Words

FIGURE 22-8   **Four-Bit Two's Complement Codes.**

As with the sign–magnitude and one's complement number systems, if you need to represent larger decimal numbers with a two's complement code, you will have to use a word containing more bits. For example, a total of 16 different combinations is available for a 4-bit word ($2^4 = 16$), as shown in Figure 22-8. With standard binary, these 16 different combinations give a count from 0 to 15, as shown in the center column in Figure 22-8. As for two's complement, we split these 16 different codes in half, and end up with 8 positive two's complement codes and 8 negative two's complement codes, as shown in the right column in Figure 22-8. Since decimal zero uses one of the positive two's complement codes, the 16 different combinations give a range of -8_{10} to $+7_{10}$ (the maximum positive value, $+7$, is always one less than the maximum negative value, -8, since decimal zero uses one of the positive codes).

An 8-bit word will give 256 different codes ($2^8 = 256$), as shown in Figure 22-9. Dividing this total in half will give 128 positive two's complement codes and 128 negative two's complement codes. Since decimal zero uses one of the positive two's complement codes, the 256 different combinations give a range of -128_{10} to $+127_{10}$.

■ **INTEGRATED MATH APPLICATION:**

What would be the two's complement range for a 16-bit word?

■ *Solution:*

$2^{16} = 65,536$; half of $65,536 = 32,768$; therefore, the range will be $-32,768$ to $+32,767$.

SELF-TEST EVALUATION POINT FOR SECTION 22-2

Now that you have completed this section, you should be able to:

■ *Objective 2.*   *Explain how the sign–magnitude, one's complement, and two's complement number systems are used to represent positive and negative numbers.*

Use the following questions to test your understanding of Section 22-2:

1. Two's complement the following binary numbers.
   a. 1011   **b.** 0110 1011   **c.** 0011   **d.** 1000 0111

2. What is the decimal equivalent of the following two's complement numbers?
   a. 1011   **b.** 0110   **c.** 1111 1011   **d.** 0000 0110

8-Bit Word	Standard Binary Value	2's Complement Value
0 0 0 0 0 0 0 0	0	0
0 0 0 0 0 0 0 1	1	+1
0 0 0 0 0 0 1 0	2	+2
0 0 0 0 0 0 1 1	3	+3
0 0 0 0 0 1 0 0	4	+4
0 0 0 0 0 1 0 1	5	+5
0 0 0 0 0 1 1 0	6	+6
0 0 0 0 0 1 1 1	7	+7
.	.	.
0 1 1 1 1 0 1 1	123	+123
0 1 1 1 1 1 0 0	124	+124
0 1 1 1 1 1 0 1	125	+125
0 1 1 1 1 1 1 0	126	+126
0 1 1 1 1 1 1 1	127	+127
1 0 0 0 0 0 0 0	128	−128
1 0 0 0 0 0 0 1	129	−127
1 0 0 0 0 0 1 0	130	−126
1 0 0 0 0 0 1 1	131	−125
1 0 0 0 0 1 0 0	132	−124
1 0 0 0 0 1 0 1	133	−123
.	.	.
1 1 1 1 1 0 1 1	251	−5
1 1 1 1 1 1 0 0	252	−4
1 1 1 1 1 1 0 1	253	−3
1 1 1 1 1 1 1 0	254	−2
1 1 1 1 1 1 1 1	255	−1

$2^8 =$ 256 Words

128 Positive Numbers

128 Negative Numbers

FIGURE 22-9 Eight-Bit Two's Complement Codes.

22-3 TWO'S COMPLEMENT ARITHMETIC

To perform its function, a computer's MPU contains several internal circuits, and one of these circuits is called the **arithmetic–logic unit,** or **ALU.** This unit is the number-crunching circuit of every digital system, and as its name implies, it can be controlled to perform either arithmetic operations (such as addition or subtraction) or logic operations (such as AND and OR). The arithmetic–logic unit has two parallel inputs, one parallel output, and a set of function select control lines, as shown in Figure 22-10.

The ALU has to be able to perform the four basic arithmetic operations, namely, addition, subtraction, multiplication, and division. One would expect, therefore, that an ALU would contain a separate circuit for each of these operations. This is not so, since all four operations can be performed using a binary-adder circuit. To explain this point, because the binary-adder circuit's function is to perform addition, and since multiplication is simply repeated addition, the binary-adder circuit can also be used to perform any multiplication operations. For example, 3×4 means that the number 4 is added three times ($4 + 4 + 4 = 12$, therefore $3 \times 4 = 12$). But what about subtraction and division? The answer to this problem is the two's complement number system. As you will see in this section, by representing all values as two's complement numbers, we can perform a subtraction operation by

Arithmetic–Logic Unit (ALU)

The section of a computer that performs all arithmetic and logic operations.

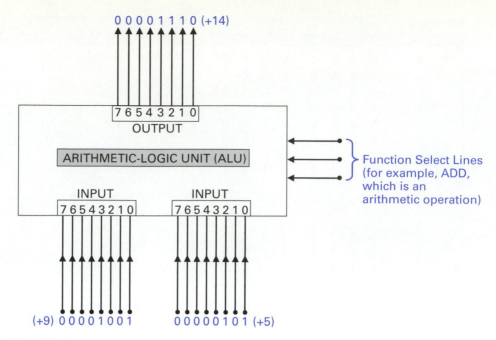

FIGURE 22-10   **Block Diagram of the Arithmetic-Logic Unit (ALU).**

slightly modifying the second number and then adding the two numbers together. This means that the binary-adder circuit can also be used to perform subtraction operations, and since division is simply repeated subtraction, the subtraction operation of the binary-adder circuit can also be used to perform any division operation. For example, $12 \div 4$ means that you take the number 12 and see how many times you can subtract 4 from it ($12 - 4 - 4 - 4 = 0$; therefore, there are three 4s in 12, $12 \div 4 = 3$).

By using two's complement numbers, therefore, we can use a binary-adder circuit within an ALU to add, subtract, multiply, and divide. Having one circuit able to perform all four mathematical operations means that the circuit will be small in size and fast to operate, will consume very little power, and will be cheap to manufacture. In this section we will see how we can perform all four mathematical operations by always adding two's complement numbers.

22-3-1   *Adding Positive Numbers*

To begin, let us prove that the addition of two positive two's-complement numbers will render the correct result.

■   **INTEGRATED MATH APPLICATION:**

Add +9 (0000 1001) and +5 (0000 0101).

■   *Solution:*

$$
\begin{array}{r}
0000\ 1001 \\
+\ 0000\ 0101 \\
\hline
\end{array}
$$

MSB of 0 = positive number $\longrightarrow$ 0000 1110

When both numbers to be added are positive two's complement numbers (both have an MSB of 0), it makes sense that the sum will have an MSB of 0, indicating a positive result. This is always true, unless the range of the two's complement word is exceeded. For example, the maximum positive number that can be represented by an 8-bit word is +127 (0111 1111). If the sum of the two positive two's complement numbers to be added exceeds the upper positive range limit, there will be a *two's complement overflow* into the sign bit. For instance:

$$
\begin{array}{c}
(\pm)\ (64)\ (32)\ (16)\ (8)\ (4)\ (2)\ (1) \\
\ \ \ \ 0\ \ \ 1\ \ \ \ 1\ \ \ \ 0\ \ \ \ 0\ \ 1\ \ 0\ \ 0 \qquad (+100) \\
+\ \ 0\ \ \ 0\ \ \ \ 0\ \ \ \ 1\ \ \ \ 1\ \ 1\ \ 1\ \ 0 \qquad +\ \ (+30) \\
\hline
\text{MSB of 1 = negative number} \longrightarrow\ \ 1\ \ \ 0\ \ \ \ 0\ \ \ \ 0\ \ \ \ 0\ \ 0\ \ 1\ \ 0 \qquad (-126)
\end{array}
$$

If two positive numbers are applied to an arithmetic circuit and it is told to add the two, the result should also be a positive number. Most arithmetic circuits are able to detect this problem if it occurs by simply monitoring the MSB of the two input words and the MSB of the output word, as shown in Figure 22-11. If the two input sign bits are the same, but the output sign bit is different, then a two's complement overflow has occurred.

22-3-2 *Adding Positive and Negative Numbers*

By using the two's complement number system, we can also add numbers with unlike signs and obtain the correct result. As an example, let us add +9 (0000 1001) and −5 (1111 1011).

$$
\begin{array}{c}
0000\ 1001 \qquad\qquad (+9) \\
+\ 1111\ 1011 \qquad\qquad +\ (-5) \\
\hline
\text{Ignore the final carry} \longrightarrow 1 \quad 0000\ 0100 \qquad\qquad (+4)
\end{array}
$$

A final carry will always be generated whenever the sum is a positive number; however, if this carry is ignored because it is beyond the 8-bit two's complement word, the answer will be correct.

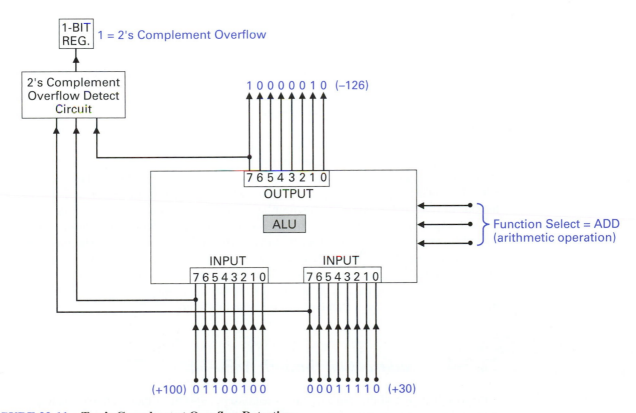

FIGURE 22-11 Two's Complement Overflow Detection.

Add +10 (0000 1010) and −12 (1111 0100).

■ *Solution:*

$$
\begin{array}{ll}
\text{0000 1010} & (+10) \\
\underline{+\ \text{1111 0100}} & \underline{+\ (-12)} \\
\text{1111 1110} & (-2)
\end{array}
$$

22-3-3 *Adding Negative Numbers*

As a final addition test for the two's complement number system, let us add two negative numbers to see if we can obtain the correct negative sum.

■ **INTEGRATED MATH APPLICATION:**

Add −5 (1111 1011) and −4 (1111 1100).

■ *Solution:*

$$
\begin{array}{ll}
\phantom{\text{1}\ }\text{1111 1011} & (-5) \\
\underline{+\ \text{1111 1100}} & \underline{+\ (-4)} \\
\text{Ignore the final carry} \longrightarrow 1\ \ \text{1111 0111} & (-9)
\end{array}
$$

The sum in this example is the correct two's complement code for −9.

Two's complement overflow can also occur in this condition if the sum of the two negative numbers exceeds the negative range of the word. For example, with an 8-bit two's complement word, the largest negative number is −128, and if we add −100 (1001 1100) and −30 (1110 0010), we will get an error, as follows:

$$
\begin{array}{ll}
\phantom{\text{1}\ }\text{1001 1100} & (-100) \\
\underline{+\ \text{1110 0010}} & \underline{+\ (-30)} \\
\text{Ignore the final carry} \longrightarrow 1\ \ \text{0111 1110} & (+126)
\end{array}
$$

Once again, the two's complement overflow detect circuit, shown in Figure 22-11, will indicate such an error, since two negative numbers were applied to the input and a positive number appeared at the output.

In all the addition examples mentioned so far (adding positive numbers, adding positive and negative numbers, and adding negative numbers) you can see that an ALU added the bits of the input bytes without any regard for whether they were signed or unsigned numbers. This brings up a very important point, which is often misunderstood. The ALU adds the bits of the input words using the four basic rules of binary addition, and it does not know if these input and output bit patterns are coded two's complement words or standard unsigned binary words. It is therefore up to us to maintain a consistency, which means that if we are dealing with coded two's complement words at the input of an ALU, we must interpret the output of an ALU as a two-coded two's complement word. On the other hand, if we are dealing with standard unsigned binary words at the input of an ALU, we must interpret the output of an ALU as a standard unsigned binary word.

22-3-4 *Subtraction through Addition*

The key advantage of the two's complement number system is that it enables us to perform a subtraction operation using the adder circuit within the ALU. To explain how this is done, let us take an example and assume that we want to subtract 3 (0000 0011) from 8 (0000 1000), as follows:

Minuend:	0000 1000	(+8)
Subtrahend:	+ 0000 0011	− (+3)
Difference:	0000 0101	+5

If we subtract the subtrahend of +3 from the minuend of +8, we will obtain the correct result of +5 (0000 0101), however, we must find the difference to this problem by using addition and not subtraction. The solution to this is achieved by *two's complementing the subtrahend, and then adding the result to the minuend.* The following display shows how this is done.

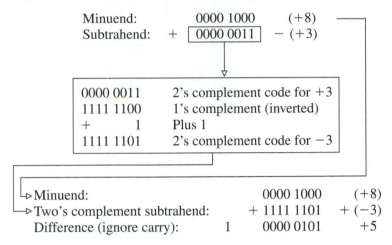

As you can see from this operation, adding the −3 to +8 achieves exactly the same result as subtracting +3 from +8. We can therefore perform a subtraction by two's complementing the subtrahend of a subtraction problem, and then adding the result to the minuend.

Digital electronic calculators and computer systems perform the previously described steps automatically. That is, they first instruct the ALU to one's complement the subtrahend input (function select control inputs = invert, which is a logic operation), then add 1 to the result (function select control inputs = +1, which is an arithmetic operation), and finally add the result of the two previous operations (which is the two's complemented subtrahend) to the minuend (function select = ADD, arithmetic operation). Although this step-by-step procedure or **algorithm** seems complicated, it is a lot simpler than having to include a separate digital electronic circuit to perform all subtraction operations. Having one circuit able to perform both addition and subtraction means that the circuit will be small in size and fast to operate, will consume very little power, and will be cheap to manufacture.

Flowcharts will be discussed in more detail in the next chapter, but for now Figure 22-12 shows a **flowchart** summarizing the subtraction through addition procedure. Flowcharts provide a graphic way of describing a step-by-step procedure (algorithm) or a program of instructions.

Algorithm
A set of rules for the solution of a problem.

Flowchart
A graphic representation of a step-by-step procedure or program of instructions.

22-3-5   *Multiplication through Repeated Addition*

Most digital electronic calculators and computers do not include a logic circuit that can multiply, since the adder circuit can also be used to perform this arithmetic operation. For example, to multiply 6 × 3, you simply add 6 three times.

Multiplicand:	6		6
Multiplier:	× 3	=	+ 6
Product:	18		+ 6
			18

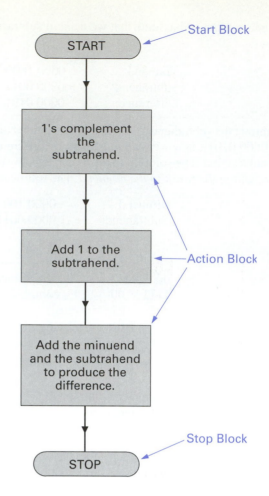

FIGURE 22-12 **Flowchart for Subtraction through Addition.**

As with subtraction through addition, an algorithm must be followed when a digital system needs to perform a multiplication operation. Figure 22-13 shows the flowchart for multiplication through repeated addition. Let us apply this procedure to our previous example of 6 × 3 and see how the multiplicand 6 is added three times to obtain a final product of 18. The steps are as follows:

Start: multiplicand = 6, multiplier = 3, product = 0
After first pass: multiplicand = 6, multiplier = 2, product = 6
After second pass: multiplicand = 6, multiplier = 1, product = 12
After third pass: multiplicand = 6, multiplier = 0, product = 18
Stop.

In this example we cycled through the flowchart three times (until the multiplier was 0) and then finally obtained a product of 18.

Although this algorithm seems complicated, it is a lot simpler than having to include a separate digital circuit for multiplication. Having one circuit able to perform addition, subtraction, and multiplication means that the circuit will be small in size and fast to operate, will consume very little power, and will be cheap to manufacture.

22-3-6 *Division through Repeated Subtraction*

Up to this point we have seen how we can use the two's complement number system and the adder circuit within an ALU for addition, subtraction, and multiplication. Just as multiplication can be achieved through repeated addition, division can be achieved through repeated

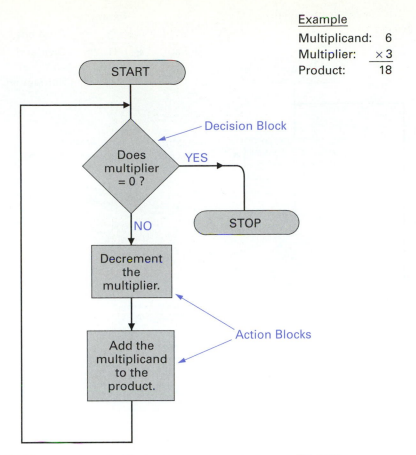

Decision Block

Does multiplier = 0 ?

YES

NO

STOP

Decrement the multiplier.

Action Blocks

Add the multiplicand to the product.

FIGURE 22-13 **Flowchart for Multiplication through Repeated Addition.**

subtraction. For example, to divide 31 by 9 we simply see how many times 9 can be subtracted from 31, as follows:

$$
\begin{array}{r}
3 \\
9\overline{)31} \\
27 \\
4
\end{array}
\quad
\begin{array}{l}
\text{Quotient} \\
\text{Dividend} \\
\\
\text{Remainder}
\end{array}
\qquad = \qquad
\begin{array}{r}
31 \\
-9 \\
22 \\
-9 \\
13 \\
-9 \\
4
\end{array}
\quad
\begin{array}{l}
\\
\text{First subtraction} \\
\\
\text{Second subtraction} \\
\\
\text{Third subtraction} \\
\\
\end{array}
$$

Since we were able to subtract 9 from 31 three times with a remainder of 4, then $31 \div 9 = 3$, remainder 4.

As with multiplication, an algorithm must be followed when a digital system needs to perform a division. Figure 22-14 shows the flowchart for division through repeated subtraction. Let us apply this procedure to our previous example of $31 \div 3$ and see how the quotient is incremented to 3, leaving a remainder of 4. The steps would be as follows:

Start: dividend = 31, divisor = 9, quotient = 0, remainder = 31
After first pass: dividend = 22, divisor = 9, quotient = 1, remainder = 22
After second pass: dividend = 13, divisor = 9, quotient = 2, remainder = 13
After third pass: dividend = 4, divisor = 9, quotient = 3, remainder = 4
Stop.

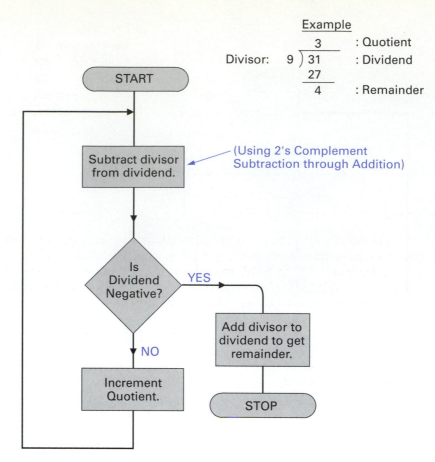

FIGURE 22-14   **Flowchart for Division through Repeated Subtraction.**

When this program of instructions cycles or loops back for the third pass, the divisor of 9 is subtracted from the remaining dividend of 4, resulting in −5. When the question, Is the dividend negative? is asked, the answer is yes, so the branch on the right is followed. This branch of the flowchart will add the divisor of 9 to the dividend of −5 to restore it to the correct remainder of 4 before the program stops. To summarize, these are the steps followed by this program:

$$
\begin{array}{rl}
(+31) & \\
-\,(+\,9) & \\
\hline
(+22) & \text{Quotient} = 1 \\
-\,(+\,9) & \\
\hline
(+13) & \text{Quotient} = 2 \\
-\,(+\,9) & \\
\hline
(+\,4) & \text{Quotient} = 3 \\
-\,(+\,9) & \\
\hline
(-\,5) & \\
-\,(+\,9) & \text{Add divisor to dividend} \\
\hline
(+\,4) & \text{Remainder} = 4
\end{array}
$$

When a digital system is instructed to perform a division, it will use the ALU for all arithmetic and logic operations. Since division is achieved through repeated subtraction, and subtraction is achieved through addition, the ALU's adder circuit can also be used for division.

Although this algorithm seems complicated, it is a lot simpler than having to include a separate digital circuit for division. As previously mentioned, having one circuit able to perform addition, subtraction, multiplication, and division means that the circuit will be small in size and fast to operate, will consume very little power, and will be cheap to manufacture.

Now that you have completed this section, you should be able to:

■ **Objective 3.** *Describe how an adder circuit can add, subtract, multiply, and divide two's complement numbers.*

Use the following questions to test your understanding of Section 22-3.

1. The two's complement number system makes it possible for us to perform addition, multiplication and division all with a/an _____ circuit.

22-4   REPRESENTING LARGE AND SMALL NUMBERS

Although two's complementing enables us to represent positive and negative numbers, the range is limited to the number of bits in the word. For example, the range of an 8-bit two's complement word is from -128 to $+127$, and the range of a 16-bit two's complement word is from $-32,768$ to $+32,767$. To extend this range, we can keep increasing the number of bits; however, these large words become difficult to work with. Another problem arises when we wish to represent fractional numbers such as 1.57.

These problems are overcome by using the **floating-point number system,** which uses scientific notation to represent a wider range of values. Figure 22-15 explains the 2-byte (16-bit) floating-point number system. The first byte is called the **mantissa,** and it is always an unsigned binary number that is interpreted as a value between 0 and 1. The second byte is called the **exponent,** and it is always a two's complement number and therefore represents a range from 10^{-128} to 10^{+127}. As an example, let us see how the floating point number system can be used to represent a large-value number such as 27,000,000,000,000 and a small-value number such as 0.0000000015.

Floating-Point Number System
A number system in which a value is represented by a mantissa and an exponent.

Mantissa
The fractional part of a logarithm.

Exponent
A small number placed to the right and above a value to indicate the number of times that symbol is a factor.

■ **INTEGRATED MATH APPLICATION:**

Give the 2-byte floating-point binary values for 27,000,000,000,000.

■ *Solution:*

The value 27,000,000,000,000 is written in scientific notation as 0.27×10^{14} and, when coded as a floating-point number, appears as follows:

$$27,000,000,000,000$$
$$= \quad 0.27 \times 10^{14}$$
$$= \boxed{0001\ 1011} \qquad \boxed{0000\ 1110}$$

Mantissa (unsigned binary) = 27, which is interpreted as 0.27.

Exponent (two's complement = $+14$, which is interpreted as 10^{14}.

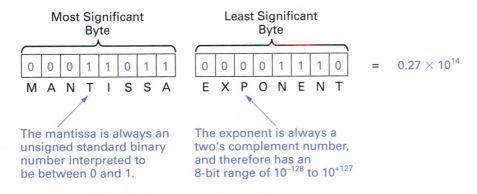

The mantissa is always an unsigned standard binary number interpreted to be between 0 and 1.

The exponent is always a two's complement number, and therefore has an 8-bit range of 10^{-128} to 10^{+127}

FIGURE 22-15 **The Floating-Point Number System.**

In this example, the mantissa byte (standard binary number) is equal to 27 (0001 1011) and is therefore interpreted as 0.27, and the exponent byte (two's complement number) is equal to +14 (0000 1110) and is therefore interpreted as 10^{14}.

■ **INTEGRATED MATH APPLICATION:**

Give the 2-byte floating-point binary values for 0.0000000015.

■ *Solution:*

The value 0.0000000015 is written in scientific notation as $0.15 \times 3~10^{-28}$ and, when coded as a floating-point number, appears as follows:

$$
\begin{aligned}
&\quad\quad\quad\quad\quad 0.0000000015 \\
=&\quad\quad\quad\quad\quad 0.15 \times 10^{-8} \\
=&\quad \boxed{0000~1111} \quad\quad \boxed{1111~1000}
\end{aligned}
$$

Mantissa (unsigned binary) = 15, which is interpreted as 0.15.

Exponent (two's complement = −8, which is interpreted as 10^{-8}.

In this example, the mantissa byte (standard binary number) is equal to 15 (0000 1111) and is therefore interpreted as 0.15, and the exponent byte (two's complement number) is equal to −8 (1111 1000) and is therefore interpreted as 10^{-8}.

As with all binary number coding methods, we must know the number representation system being used if we are to accurately decipher the bit pattern. For instance, the 16-bit word in Figure 22-15 would be very different if it were interpreted as an unsigned binary number, a two's complement number, and as a floating-point number.

SELF-TEST EVALUATION POINT FOR SECTION 22-4

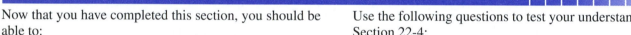

Now that you have completed this section, you should be able to:

■ *Objective 4.* *Describe how the floating-point number system can be used to represent a large range of values.*

Use the following questions to test your understanding of Section 22-4:

1. Give the 2-byte floating-point binary values for each.
   a. 65,000,000
   b. 0.18×10^{-12}

REVIEW QUESTIONS

Multiple Choice Questions

1. The sum of 101_2 and 011_2 is
   a. 0111_2.
   b. 8_{10}.
   c. 1001_2.
   d. 6_{10}.

2. Which digital logic unit is used to perform arithmetic operations and logic functions?
   a. MPU
   b. ALU
   c. CPU
   d. ICU

3. With binary addition, $1 + 1 = ?$
   a. 10_2
   b. 0 carry 1
   c. 2_{10}
   d. All the above

4. With binary subtraction, $0 - 1 = ?$
   a. 10_2
   b. 1 borrow 0
   c. 1 borrow 1
   d. All the above

5. What would be the result of $1011101_2 \times 01_2$?
   a. 1011101_2
   b. 1110101_2
   c. 000111_2
   d. 1011100_2

6. The four rules of binary multiplication are $0 \times 0 = $ _____, $1 \times 0 = $ _____, $0 \times 1 = $ _____, and $1 \times 1 = $ _____.
   a. 0, 1, 0, 1
   b. 0, 0, 1, 1
   c. 0, 0, 0, 0
   d. 0, 0, 0, 1

7. What would be the result of $10012 \div 0011_2$?

   a. 3_{10} **c.** 0011_2
   b. 3, remainder zero   **d.** All the above

8. The one's complement of $0101\ 0001_2$ would be _____.

   a. $0101\ 0010_2$ **c.** $1010\ 1111_2$
   b. $1010\ 1110_2$ **d.** All the above

9. What would be the decimal equivalent of 1001_2 if it were a two's complement number?

   a. $+9$ **c.** $+7$
   b. -9 **d.** -7

Communication Skill Questions

11. List the rules for each of the following: (22-1)

   a. Binary addition   **c.** Binary multiplication
   b. Binary subtraction   **d.** Binary division

12. Using an example, briefly describe how to perform the following operations. (22-1)

   a. Binary addition   **c.** Binary multiplication
   b. Binary subtraction   **d.** Binary division

13. Briefly describe how the following number systems are used to represent positive and negative binary numbers. (22-2)

   a. Sign–magnitude   **c.** Two's complement
   b. One's complement

14. Why is the two's complement number system most frequently used to represent positive and negative numbers? (22-2-3)

Practice Problems

19. Add the following 8-bit numbers.

   a. 0110 1000 and 0001 1111
   b. 0111 1111 and 0100 0011
   c. 0001 1010 and 0011 1010

20. Subtract the following 8-bit numbers.

   a. 0001 1010 from 1010 1010
   b. 0001 1111 from 0011 0000
   c. 0010 0000 from 1000 0000

21. Multiply the following binary numbers.

   a. 0101 1011 by 011   **c.** 1110 by 010
   b. 0001 1111 by 1011

22. Divide the following binary numbers.

   a. 0110 into 0111 0110   **c.** 1001 into 1011 1110
   b. 101 into 1010 1010

23. Convert each of the following decimal numbers to an 8-bit sign–magnitude number.

   a. $+34$ **c.** $+83$
   b. -83

10. Which number system makes it possible for us to add, subtract, multiply, and divide binary numbers all with an adder circuit?

   a. Sign–magnitude   **c.** Two's complement
   b. One's complement   **d.** Floating-point

15. How would you determine the maximum positive and maximum negative value of a 10-bit two's complement number? (22-2-3)

16. Describe how to two's complement a binary number. (22-2-3)

17. Using the two's complement number system, briefly describe how to perform each of the following operations. (22-3)

   a. Add two positive numbers
   b. Add two negative numbers
   c. Add a positive and negative number
   d. Subtract through addition
   e. Multiply through repeated addition
   f. Divide through repeated subtraction

18. What is the floating-point number system? (22-4)

24. One's complement the following numbers.

   a. 0110 1001   **c.** 1111 0000
   b. 1010 1010

25. Convert all the following two's complement numbers to their decimal equivalents.

   a. 0001 0000   **c.** 1111 0101
   b. 1011 1110

26. Convert each of the following decimal numbers to an 8-bit two's complement number.

   a. $+119$ **c.** -34
   b. -119

27. Two's complement the following numbers.

   a. 0101 1010   **c.** 1111 1101
   b. 1000 0001

28. What would be the two's complement range of a 6-bit binary number?

Web Site Questions

Go to the web site http://www.prenhall.com/cook, select the textbook *Mathematics for Electronics and Computers,* select this chapter, and then follow the instructions when answering the multiple-choice practice problems.

Introduction to Computer Programming

Making an Impact

John Von Neuman, a mathematics professor at the Institute of Advanced Studies, delighted in amazing his students by performing complex computations in his head faster than they could with pencil, paper, and reference books. He possessed a photographic memory, and at his frequently held lavish parties in his home in Princeton, New Jersey, he gladly occupied center stage to recall from memory entire pages of books read years previously, the lineage of European royal families, and a store of controversial limericks. His memory, however, failed him in his search for basic items in a house he had lived in for 17 years. On many occasions when traveling, he would become so completely absorbed in mathematics that he would have to call his office to find out where he was going and why.

Born in Hungary, he was quick to demonstrate his genius. At the age of six he would joke with his father in classical Greek. At the age of eight, he had mastered calculus, and in his mid-20s he was teaching and making distinct contributions to the science of quantum mechanics, which is the cornerstone of nuclear physics.

Next to fine clothes, expensive restaurants, and automobiles, which he had to replace annually due to smash-ups, was his love of his work. His interest in computers began when he became involved in the top-secret Manhattan Project at Los Alamos, New Mexico, where he proved mathematically the implosive method of detonating an atom bomb. Working with the then-available computers, he became aware that they could become much more than a high-speed calculator. He believed that they could be an all-purpose scientific research tool, and he published these ideas in a paper. This was the first document to outline the logical organization of the electronic digital computer and was widely circulated to all scientists throughout the world. In fact, even to this day, scientists still refer to computers as "Von Neuman machines."

Von Neuman collaborated on a number of computers of advanced design for military applications, such as the development of the hydrogen bomb and ballistic missiles.

In 1957, at the age of 54, he lay in a hospital dying of bone cancer. Under the stress of excruciating pain, his brilliant mind began to break down. Since Von Neuman had been privy to so much highly classified information, the Pentagon had him surrounded with only medical orderlies specially cleared for security for fear he might, in pain or sleep, give out military secrets.

Outline and Objectives

VIGNETTE: MAKING AN IMPACT

INTRODUCTION

Objective 1: Define the term *microprocessor.*

23-1 COMPUTER HARDWARE

Objective 2: Describe the difference between a hard-wired digital system and a microprocessor-based digital system.

Objective 3: List and describe the functions of the three basic blocks that make up a microcomputer.

Objective 4: Define the difference between computer hardware and software.

23-2 SOFTWARE

Objective 5: Describe how flowcharts can help a programmer write programs.

Objective 6: Describe the meaning, structure, and differences among programs that are written in machine language, assembly language, and a high-level language.

23-2-1 The Microcomputer as a Logic Device

23-2-2 Flowcharts

23-2-3 Programming Languages

23-2-4 A Programming Example

MULTIPLE CHOICE QUESTIONS

COMMUNICATION SKILLS QUESTIONS

PRACTICE PROBLEMS

WEB SITE QUESTIONS

Note: Material in this chapter relating to the SAM microcomputer (5036A) courtesy of Hewlett-Packard Company.

Introduction

The microprocessor is a large, complex integrated circuit containing all the computation and control circuitry for a small computer. The introduction of the microprocessor caused a dramatic change in the design of digital electronic systems. Before microprocessors, random or hard-wired digital circuits were designed using individual logic blocks, such as gates, flip-flops, registers, and counters. These building blocks were interconnected to achieve the desired end, as required by the application. Using random logic, each application required a unique design, and there was little similarity between one system and another. This approach was very similar to analog circuit design, in that the structure of the circuit was governed by the function that needed to be performed. Once constructed, the function of the circuit was difficult to change.

The microprocessor, on the other hand, provides a general-purpose control system that can be adapted to a wide variety of applications with only slight circuit modification. The individuality of a microprocessor-based system is provided by a list of instructions (called the program) that controls the system's operation. A microprocessor-based system therefore has two main elements—the actual components, or hardware, and the programs, or software.

23-1   COMPUTER HARDWARE

The earliest electronic computers were constructed using thousands of vacuum tubes. These machines were extremely large and unreliable and were mostly a laboratory curiosity. Figure 23-1 compares the electronic numeric integrator and computer, or ENIAC, which was unveiled in 1946, with today's personal computer. The ENIAC weighed 38 tons, measured 18 feet wide and 88 feet long, and used 17,486 vacuum tubes. These vacuum tubes produced a great deal of heat and developed frequent faults requiring constant maintenance. Today's personal computer, on the other hand, makes use of semiconductor integrated circuits and is

FIGURE 23-1   Microcomputer Systems—Past and Present. (Photo (b) Courtesy of Apple Computer, Inc.)

far more powerful, versatile, portable, and reliable than the ENIAC. Another advantage is cost: In 1946 the ENIAC calculator cost $400,000 to produce, whereas a present-day high-end personal computer can be purchased for about $1,000.

Figure 23-2 shows the block diagram of a basic microcomputer system. The microprocessor (also called central processor unit, or CPU) is the "brains" of the system. It contains all the logic circuitry needed to recognize and execute the program of instructions stored in memory. The input port connects the processor to the keyboard so that data can be read from this input device. The output port connects the processor to the display so that we can write data to this output device. Combined in this way, the microprocessor unit, memory unit, and input/output unit form a **microcomputer.**

The blocks within the microcomputer are interconnected by three buses. The microprocessor uses the address bus to select which memory location, input port, or output port it wishes to put information into or take information out of. Once the microprocessor has selected the location using the address bus, data or information is transferred via the data bus. In most cases, data will travel either from the processor to memory, from memory to the processor, from the input port to the processor, or from the processor to the output port. The control bus is a group of control signal lines that are used by the processor to coordinate the transfer of data within the microcomputer.

A list of instructions is needed to direct a microcomputer system so that it will perform a desired task. For example, if we wanted the system in Figure 23-2 to display the number of any key pressed on the keyboard, the program would be as follows:

1. Read the data from the keyboard.
2. Write the data to the display.
3. Repeat (go to step 1).

For a microprocessor to perform a task from a list of instructions, the instructions must first be converted into codes that the microprocessor can understand. These instruction codes are stored, or programmed, into the system's memory. When the program is run, the microprocessor begins by reading the first coded instruction from memory, decoding its meaning, and then performing the indicated operation. The processor then reads the instruction from the next location in memory, decodes the meaning of this next instruction, and performs the indicated operation. This process is then repeated, one memory location after another, until all the instructions within the program have been fetched, decoded and then executed.

Microcomputer

Complete system, including CPU, memory, and I/O interfaces.

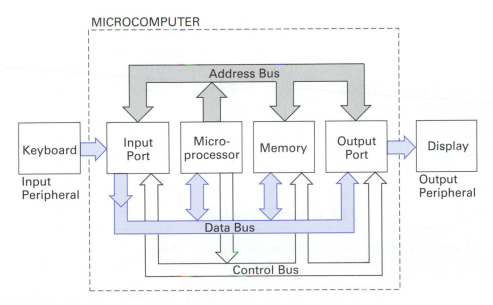

FIGURE 23-2 **Basic Microcomputer System.**

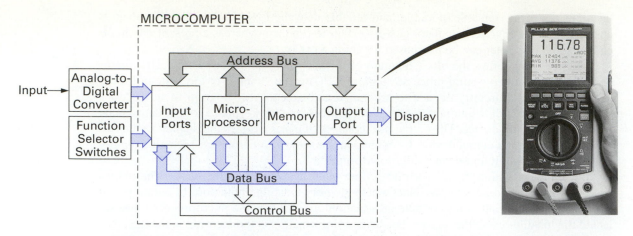

MICROCOMPUTER

Address Bus

Input → Analog-to-Digital Converter → Input Ports → Micro-processor → Memory → Output Port → Display

Function Selector Switches

Data Bus

Control Bus

FIGURE 23-3   A Microprocessor-Based Digital Voltmeter. (Photo courtesy of Fluke Corporation. Reproduced with permission.)

The input/output devices connected to the input and output ports (the keyboard and display, for example) are called the **peripherals.** Peripherals are the system's interface with the user, or the system's interface with other equipment such as printers or data storage devices. As an example, Figure 23-3 shows how a microcomputer could be made to operate as a microprocessor-based digital voltmeter. Its input peripherals are an analog-to-digital converter (ADC), and a range and function switches. The output peripheral is a digital display.

As you come across different microcomputer applications, you will see that the basic microcomputer system is always the same. The only differences between one system and another are in the program and peripherals. Whether the system is a personal computer or a digital voltmeter, the only differences are in the program of instructions that controls the microcomputer and the peripherals connected to the microcomputer.

SELF-TEST EVALUATION POINT FOR SECTION 23-1

Now that you have completed this section, you should be able to:

■ *Objective 1. Define the term* microprocessor.

■ *Objective 2. Describe the difference between a hard-wired digital system and a microprocessor-based digital system.*

■ *Objective 3. List and describe the function of the three basic blocks that make up a microcomputer.*

■ *Objective 4. Define the difference between computer hardware and software.*

Use the following questions to test your understanding of Section 23-1:

1. What three blocks are interconnected to form a microcomputer, and what is the function of each block?

2. What three buses are needed to interconnect the three blocks within a microcomputer, and what is the function of each bus?

3. What is the difference between hardware and software?

23-2   COMPUTER SOFTWARE

Microcomputer programs are first written in a way that is convenient for the person writing the program, or *programmer.* Once written, the program must then be converted and stored as a code that can be understood by the microprocessor. When writing a computer program, the programmer must tell the computer what to do down to the most minute detail. Computers can act with tremendous precision and speed for long periods of time, but they must be told exactly what to do. A computer can respond to a change in conditions, but only if it contains a program that says "if this condition occurs, do this."

23-2-1 *The Microcomputer as a Logic Device*

To understand how a program controls a microcomputer, let us see how we could make a microcomputer function as a simple two-input AND gate. As can be seen in Figure 23-4, the two AND gate inputs will be applied to the input port, and the single AND gate output will appear at the output port. To make the microcomputer operate as an AND gate, a program of instructions will first have to be loaded into the system's memory. This list of instructions will be as follows:

1. Read the input port.
2. Go to step 5 if all inputs are HIGH; otherwise, continue.
3. Set output LOW.
4. Go to step 1.
5. Set output HIGH.
6. Go to step 1.

Studying the steps in this program, you can see that first the input port is read. Then, the inputs are examined to see if they are all HIGH, since that is the function of an AND gate. If the inputs are all HIGH, the output is set HIGH. If the inputs are not all HIGH, the output is set LOW. Once the procedure has been completed, the program jumps back to step 1 and repeats indefinitely, so that the output will continuously follow any changes at the input.

You may be wondering why we would use this complex system to perform a simple AND gate function. If this were the only function that we wanted the microcomputer to perform, then it would definitely be easier to use a simple AND logic gate; however, the microcomputer provides tremendous flexibility. It allows the function of the gate to be arbitrarily redefined just by changing the program. You could easily add more inputs, and the gate function could be extremely complex. For example, we could have the microcomputer function as an eight-input electronic lock. The output of this lock would go HIGH only if the inputs were turned ON in a specific order. Using traditional logic, this would require a complex circuit; however, this function could be easily implemented using the same microprocessor system used for the simple AND gate. Of course, a new program would be needed, more complex than the simple AND gate program, but the hardware would not change. Additionally, the combination of this electronic lock could easily be changed by simply modifying the program.

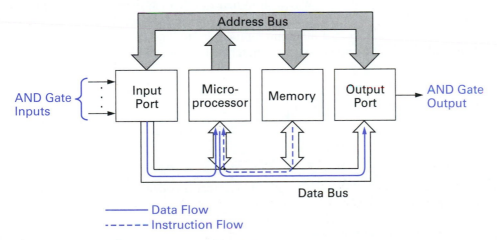

FIGURE 23-4   A Microprocessor-Based AND Gate.

23-2-2   *Flowcharts*

To develop a program of instructions to solve a problem, a programmer would generally follow three steps:

1. Define the problem.
2. Flowchart the solution.
3. Write the program.

To understand these steps, let us begin by examining the elements of the flowchart in more detail.

Flowcharts provide a graphic way of describing the operation of a program. They are composed of different types of blocks interconnected with lines, as shown in Figure 23-5. A rectangular block describes each action that the program takes. A diamond-shaped block is used for each decision, such as testing the value of a variable. An oval block is used at the beginning of the flowchart with the name of the program placed inside it. This oval block can also be used to mark the end of the flowchart.

As an example, Figure 23-6 shows a flowchart for the AND gate program discussed previously. For each line of the program there is a block, except for the two "go to" instructions. These are represented as lines indicating how the program will flow from one block to another. The flowchart contains the same information as the program list but in a graphic form. When you first set out to write a program, a flowchart is a good way to organize your thoughts and document what the program must do. By going through the flowchart manually, you can check the logic of the flowchart and then write the actual program from it. Flowcharts are also useful for understanding a program that has been written in the past.

23-2-3   *Programming Languages*

Writing programs in English is convenient, since it is the language most people understand. Unfortunately, English is meaningless to a microprocessor. The language understood by the microprocessor is called **machine language,** or machine code. Because microprocessors deal directly with digital signals, machine language instructions are binary codes, such as 00111100 and 11100111.

Machine language is not easy for people to use, since 00111100 has no obvious meaning. It can be made easier to work with by using the hexadecimal representation of 0011 1100, which is 3C; however, this still does not provide the user with any clue as to the meaning of this instruction. To counteract this problem, microprocessor manufacturers replace each instruction code with a short name called a **mnemonic,** or memory aid. As an example, Intel's 8085 microprocessor uses the mnemonic "INR A" for the code 3C, since this code instructs the 8085 microprocessor to "increment its internal A register." The mnemonics are much easier to remember than the machine codes. By assigning

Flowchart or Flow Diagram

Graphical representation of program logic. Flowcharts enable the designer to visualize the procedure necessary for each item in the program. A complete flowchart leads directly to the final code.

Machine Language

Binary language (often represented in hex) that is directly understood by the processor. All other programming languages must be translated into binary code before they can be entered into the processor.

Mnemonic Code

To assist the human memory, the binary numbered codes are assigned groups of letters (or mnemonic symbols) that suggest the definition of the instruction.

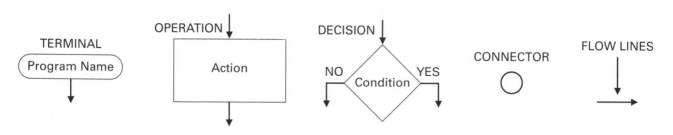

FIGURE 23-5   Flowchart Symbols.

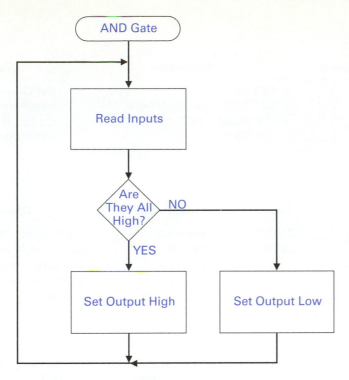

FIGURE 23-6 **AND Gate Flowchart.**

a mnemonic to each instruction code, you can write programs using the mnemonics instead of machine codes. Once the program is written, the mnemonics can easily be converted to their equivalent machine codes. Programs written using mnemonics are called **assembly language** programs.

The machine language is generally determined by the design of the microprocessor chip and cannot be modified. The assembly language mnemonics are made up by the microprocessor's manufacturer as a convenience for programmers and are not set by the microprocessor's design. For example, you could write INC A instead of INR A, as long as both were translated to the machine code 3C. A microprocessor is designed to recognize a specific list or group of codes called the **instruction set,** and each microprocessor type has its own set of instructions.

Although assembly language is a vast improvement over machine language, it is still difficult to use for complex programs. To make programming easier, **high-level languages** have been developed. These are similar to English and are generally independent of any particular microprocessor. For example, a typical instruction might be "LET COUNT = 10" or "PRINT COUNT." These instructions give a more complicated command than those that the microprocessor can understand. Therefore, microcomputers using high-level languages also contain long, complex programs (permanently stored in their memory) that translate the high-level language program into a machine language program. A single high-level instruction may translate into dozens of machine language instructions. Such translator programs are called **compilers.**

23-2-4   *A Programming Example*

To show the difference between machine language, assembly language, and a high-level language, let us use a simple programming example. Figure 23-7(a) shows the flowchart for a program that counts to 10. There are no input or output operations in this program, since

Assembly Language

A program written as a series of statements using mnemonic symbols that suggest the definition of the instruction. It is then translated into machine language by an assembler program.

Instruction Set

Total group of instructions that can be executed by a given microprocessor.

High-level Language

A language closer to the needs of the problem to be handled than to the language of the machine on which it is to be implemented.

Compiler

Translation program that converts high-level instructions into a set of binary instructions (machine code) for execution.

Line No.	Instruction	Description
1	LET COUNT = 0	Set Count to 0
2	LET COUNT = COUNT + 1	Increment Count
3	IF COUNT = 10 THEN 1	Go to 1 if Count = 10
4	GO TO 2	Otherwise go to 2

(b)

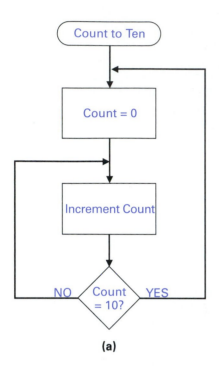

(a)

Label	Instruction	Comments
START:	MVI A,0	;Set A register to 0
LOOP:	INR A	;Increment A register
	CPI 10	;Compare A register to 10
	JZ START	;Go to beginning if A = 10
	JMP LOOP	;Repeat

(c)

Memory Address (Hex)	Memory Contents (Hex)	(Binary)
07F0	3E	00111110
07F1	00	00000000
07F2	3C	00111100
07F3	FE	11111110
07F4	0A	00001010
07F5	CA	11001010
07F6	F0	11110000
07F7	07	00000111
07F8	C3	11000011
07F9	F2	11110010
07FA	07	00000111

(d)

BASIC Language		8085 Assembly Language			8085 Machine Language		
Line No.	Instruction	Label	Instruction		Address	Contents	
1	LET COUNT = 0	START:	MVI	A,0	07F0	3E	*Opcode*
					07F1	00	*Data*
2	LET COUNT = COUNT + 1	LOOP:	INR	A	07F2	3C	*Opcode*
3	IF COUNT = 10 THEN 1		CPI	10_{10}	07F3	FE	*Opcode*
					07F4	0A	*Data*
			JZ	START	07F5	CA	*Opcode*
					07F6	F0	*Address*
					07F7	07	
4	GO TO 2		JMP	LOOP	07F8	C3	*Opcode*
					07F9	F2	*Address*
					07FA	07	

(e)

FIGURE 23-7   A Count-to-10 Programming Example. (a) Flowchart. (b) High-Level Language Program. (c) Assembly Language Program. (d) Machine Language Program. (e) Summary of Program Languages.

all we will be doing is having the contents of a designated memory location count from zero to 10, and then repeat.

The translation from the flowchart to a high-level language is fairly simple, as seen in Figure 23-7(b). The high-level language used in this example is called **BASIC,** which stands for Beginner's All-purpose Symbolic Instruction Code and has the advantage of being simple and similar to English. Following the program listing shown in Figure 23-7(b), you can see that the first two lines of the program correspond exactly to the first two action blocks of the count-to-10 flowchart. In the first line, the memory location called COUNT is set to 0. In the second line, LET COUNT = COUNT + 1 is simply a way of saying "increment the count." Lines 3 and 4 perform the function of the decision block in the count-to-10 flow chart. Line 3 specifies that if COUNT = 10, then the next instruction executed should be line 1. If the count is not equal to 10, the COUNT = 10 instruction has no effect, and the program continues with line 4, which says "go to line 2." To test whether this BASIC program will count to 10 and then repeat, try following the program step by step to see if it works, starting with a value of zero.

Assembly language is not one specific language but a class of languages. Each microprocessor has its own machine language and therefore its own assembly language, which is defined by the manufacturer. Figure 23-7(c) shows the assembly language listing for the count-to-10 program. This program is certainly more cryptic than the BASIC language program, but it performs exactly the same function.

The three columns in Figure 23-7(c) are for labels, instructions, and comments. The label provides the same function as the line number in the BASIC program. Instead of numbering every line, you simply make up a name (called a label) for each line to which you need to refer. A colon (:) is used to identify the label. A line needs a label only if there is another instruction in the program that refers to that line. The comments are an aid to understanding the program, and a semicolon (;) is used to identify the beginning of a comment. High-level language programs do not always need many comments because the instructions themselves are more descriptive. For assembly language programs, however, comments are an invaluable aid for people other than the programmer, or when the programmer returns to a program after some time.

The first instruction is MVI A, 0, which means "move immediately to the accumulator the data zero." The **accumulator** is also called the A register and is a storage location inside the microprocessor. This assembly language instruction is equivalent to the BASIC instruction LET COUNT = 0, except instead of making up a name for the variable (COUNT), we used a preassigned name (A) for a register inside the microprocessor. This MVI A, 0 instruction will therefore load the data 0 into the microprocessor's internal A register. The next instruction, INR A, means "increment the value in the accumulator," and this assembly language instruction is equivalent to the BASIC instruction LET COUNT = COUNT + 1. The next three instructions implement the decision function. The instruction CPI 10 means "compare the value in the accumulator with the value 10." The result of this comparison will determine whether a special flip-flop within the microprocessor called the *zero flag* is SET or RESET. If the value in the accumulator is not equal to 10, the zero flag is RESET LOW, whereas if the value in the accumulator is equal to 10, the zero flag is SET HIGH. The next instruction, JZ START, means "jump if the zero flag is SET to the line with the label START." This instruction tests the zero flag, and if it is SET (accumulator = 10), it will cause the program to jump to the line with the label START. Together, these two instructions (CPI 10 and JZ START) perform the function of the BASIC statement IF COUNT = 0 THEN 1. The last instruction, JMP LOOP, means "jump to the line with the label LOOP." This instruction simply causes the program to jump to the line with the label LOOP and is equivalent to the BASIC statement GO TO 2.

Figure 23-7(d) shows the machine language listing for the count-to-10 program. Although this language looks the most alien to us, its sequence of 1s and 0s is the only language that the microprocessor understands. In this example, each memory location holds eight bits of data. To program the microcomputer, we will have to store these 8-bit codes in the microcomputer's memory. Each instruction begins with an **opcode,** or operation code, that specifies the operation to be performed, and since all 8085 opcodes are eight bits, each

BASIC

An easy-to-learn and easy-to-use language available on most microcomputer systems.

Accumulator

One or more registers associated with the arithmetic and logic unit (ALU), which temporarily store sums and other arithmetical and logical results of the ALU.

Opcode

The first part of a machine language instruction that specifies the operation to be performed. The other parts specify the data, address, or port. For the 8085, the first byte of each instruction is the opcode.

opcode will occupy one memory location. An 8085 opcode may be followed by one byte of data, two bytes of data, or no bytes, depending upon the instruction type used.

Stepping through the machine language program shown in Figure 23-7(d), you can see that the first byte ($3E_{16}$) at address $07F0_{16}$ is the opcode for the instruction MVI A. The MVI A instruction is made up of two bytes. The first byte is the opcode specifying that you want to move some data into the accumulator, and the second byte stored in the very next memory location (address $07F1_{16}$) contains the data 00_{16} to be stored in the accumulator. The third memory location (address 07F2) contains the opcode for the second instruction, INR A. This opcode (3C) tells the microprocessor to increment the accumulator. The INR A instruction has no additional data, and therefore the instruction occupies only one memory location. The next code, FE, is the opcode for the compare instruction, CPI. Like the MVI A, 0 instruction, the memory location following the opcode contains the data required by the instruction. Since the machine language program is shown in hexadecimal notation, the data 10 (decimal) appears as 0A (hex). This instruction compares the accumulator with the value 10 and sets the zero flag if they are equal, as described earlier. The next instruction, JZ, has the opcode CA and appears at address 07F5. This opcode tells the microprocessor to jump if the zero flag is set. The next two memory locations contain the address to jump to. Because addresses in an 8085 system are 16 bits long, it takes two memory locations (8 bits each) to store an address. The two parts of the address are stored in an order that is the reverse of what you might expect. The least significant half of the address is stored first and then the most significant half of the address is stored next. The address 07F0 is therefore stored as F0 07. The assembly language instruction JZ START means that the processor should jump to the instruction labeled START. The machine code must therefore use the actual address that corresponds to the label START, which in this case is address 07F0. The last instruction, JMP LOOP, is coded in the same way as the previous jump instruction. The only difference is that this jump instruction is independent of any flag condition, and therefore no flags will be tested to determine whether this jump should occur. When the program reaches the C3 jump opcode, it will always jump to the address specified in the two bytes following the opcode, which in this example is address 07F2. The machine language program, therefore, contains a series of bytes, some of which are opcodes, some are data, and some are addresses.

You must know the size and format of an instruction if you want to be able to interpret the operation of the program correctly. To compare the differences and show the equivalents, Figure 23-7(e) combines the high-level language, assembly language, and machine language listings for the count-to-10 program.

SELF-TEST EVALUATION POINT FOR SECTION 23-2

Now that you have completed this section, you should be able to:

■ **Objective 5.** *Describe how flowcharts can help a programmer write programs.*

■ **Objective 6.** *Describe the meaning, structure, and differences among programs that are written in machine language, assembly language, and a high-level language.*

Use the following questions to test your understanding of Section 23-2:

1. Briefly describe the three following languages:
   a. machine language
   b. assembly language
   c. high-level language

2. What is a flowchart?

3. Define the following terms:
   a. instruction set
   b. peripheral
   c. programmer
   d. mnemonic
   e. compiler
   f. accumulator
   g. flag register
   h. opcode

CHAPTER 23 / INTRODUCTION TO COMPUTER PROGRAMMING

Multiple Choice Questions

1. Microprocessor-based systems are more flexible than hard-wired logic designs because _____.

   a. They are faster
   b. They use LSI devices
   c. Their operation is controlled by software
   d. The hardware is specialized

2. The peripherals of a microcomputer system are _____.

   a. The memory devices
   b. The microprocessor
   c. The software
   d. The I/O devices

3. The personality of a microprocessor-based system is determined primarily by _____.

   a. The microprocessor used
   b. The program and peripherals
   c. The number of data bus lines
   d. The type of memory ICs used

4. The language that the microprocessor understands directly is called _____.

   a. Assembly language   c. English language
   b. High-level language   d. Machine language

5. Programs written using mnemonics are called _____ programs.

   a. Assembly language   c. English language
   b. High-level language   d. Machine language

6. A program designed to translate a high-level language into machine language is called _____.

   a. An instruction set   c. An opcode
   b. A compiler   d. A mnemonic

7. Which unit within the microprocessor performs arithmetic and logic operations?

   a. Accumulator   c. ALU
   b. I/O   d. Memory

8. "INRA" is an example of _____.

   a. Assembly language   c. Machine language
   b. BASIC   d. C language

9. "LET COUNT = 0" is an example of _____.

   a. Assembly language   c. Machine language
   b. BASIC   d. C language

10. An _____ is the first part of a machine language instruction.

    a. Accumulator   c. Assembler
    b. ALU   d. Opcode

Communication Skill Questions

11. What is a microprocessor?   (Introduction)

12. What three blocks form a microcomputer?   (23-1)

13. What is the difference between programming a microcomputer and running a program in a microcomputer?   (23-1)

14. What is the difference between software and hardware? (23-1)

15. What are peripherals?   (23-1)

16. What is a flowchart?   (23-2-2)

17. Briefly describe each of the following programming languages: (23-2-3)

    a. machine language   c. high-level language
    b. assembly language

18. Define the following terms: (23-2)

    a. opcode   f. mnemonic
    b. instruction   g. program
    c. flag   h. peripheral
    d. accumulator   i. flowchart
    e. BASIC

19. List and describe the steps a programmer would follow to develop a program.   (23-2-4)

20. Why, do you think, are flowcharts useful?   (23-2-2)

Practice Problems

21. Develop a flowchart to control two traffic signals (A and B) at an intersection with the following specifications and sequence:

    Light A: Green for 3 minutes
              Yellow for 1 minute
              Red for as long as light B is green and yellow

    Light B: Green for 2 minutes
              Yellow for 1 minute
              Red for as long as light A is green and yellow

    Repeat cycle.

22. Modify the flowchart in question 21 so that light B will go into its cycle only if a car is detected by the B road sensors.

23. Develop a flowchart that monitors the number of stadium tickets sold and when that number equals 10,000, activates the "Stadium Full" sign.

24. Referring to the flowchart in Figure 23-8, step through the instructions to see if the example value will be converted from BCD (3 digits) to binary (8-bit).

25. Referring to the flowchart in Figure 23-9, step through the instructions to see if a value will be converted from binary (8-bit) to BCD (3 digit).

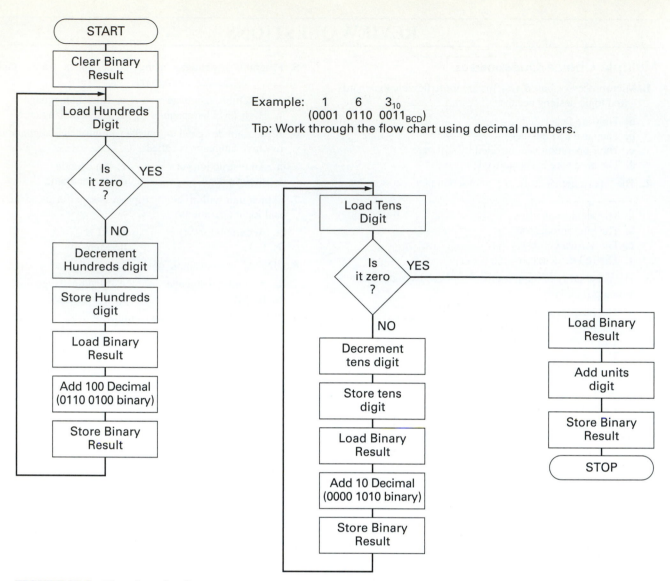

Example: 1 6 3$_{10}$
(0001 0110 0011$_{BCD}$)
Tip: Work through the flow chart using decimal numbers.

FIGURE 23-8 **Flowchart for Converting BCD to Binary.**

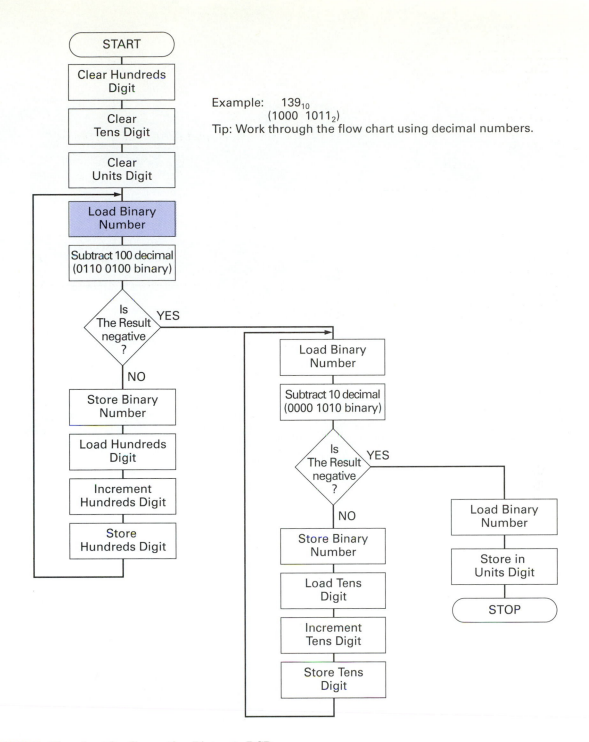

FIGURE 23-9 Flowchart for Converting Binary to BCD.

Web Site Questions

Go to the web site http://prenhall.com/cook, select the textbook *Mathematics for Electronics and Computers,* select this chapter, and follow the instructions when answering the multiple-choice practice problems.

Answers to Self-Test Evaluation Points

STEP 1-1

1. (a) $\dfrac{6}{8}$ **(b)** $\dfrac{3}{6}$ **(c)** $\dfrac{5}{16}$

2. (a) $2\dfrac{1}{2}$ **(b)** $4\dfrac{1}{4}$ **(c)** $5\dfrac{3}{4}$ **(d)** $6\dfrac{7}{8}$ **(e)** $8\dfrac{3}{8}$

STEP 1-2

1. $\dfrac{4}{6} + \dfrac{1}{6} = \dfrac{4+1}{6} = \dfrac{5}{6}$

2. $\dfrac{3}{64} + \dfrac{2}{64} + \dfrac{25}{64} = \dfrac{3+2+25}{64} = \dfrac{30}{64} \div \dfrac{2}{2} = \dfrac{15}{32}$

3. $\dfrac{3}{9} + \dfrac{6}{18} = \dfrac{6+6}{18} = \dfrac{12}{18} \div \dfrac{6}{6} = \dfrac{2}{3}$

4. $\dfrac{1}{3} + \dfrac{1}{4} + \dfrac{3}{15} = \dfrac{?}{?}$

$$
\begin{array}{c|ccc}
2 & 3 & 4 & 15 \\
2 & 3 & 2 & 15 \\
3 & 3 & 1 & 15 \\
5 & 1 & 1 & 5 \\
  & 1 & 1 & 1
\end{array}
$$
$\longrightarrow 2 \times 2 \times 3 \times 5 = 60$

$\dfrac{1}{3} + \dfrac{1}{4} + \dfrac{3}{15} = \dfrac{20+15+12}{60} = \dfrac{47}{60}$

5. $\dfrac{3}{4} + \dfrac{4}{5} = \dfrac{15+16}{20} = \dfrac{31}{20} = 1\dfrac{11}{20}$

$$
\begin{array}{c|cc}
2 & 4 & 5 \\
2 & 2 & 5 \\
5 & 1 & 5 \\
  & 1 & 1
\end{array}
$$
$\longrightarrow 2 \times 2 \times 5 = 20$

6. $\dfrac{9}{12} + \dfrac{4}{24} + \dfrac{3}{4} = \dfrac{18+4+18}{24} = \dfrac{40}{24} = 1\dfrac{16}{24} \div \dfrac{8}{8} = 1\dfrac{2}{3}$

7. $\dfrac{7}{9} + \dfrac{5}{9} + \dfrac{4}{18} = \dfrac{14+10+4}{18} = \dfrac{28}{18} = 1\dfrac{10}{18} \div \dfrac{2}{2} = 1\dfrac{5}{9}$

8. $\dfrac{15}{6} = 2\dfrac{3}{6} \div \dfrac{3}{3} = 2\dfrac{1}{2}$

9. $\dfrac{4}{24} = \dfrac{1}{6}$

10. $\dfrac{25}{100} \div \dfrac{25}{25} = \dfrac{1}{4}$

STEP 1-3

1. $\dfrac{4}{32} - \dfrac{2}{32} = \dfrac{4-2}{32} = \dfrac{2}{32} \div \dfrac{2}{2} = \dfrac{1}{16}$

2. $\dfrac{4}{5} - \dfrac{3}{10} = \dfrac{8-3}{10} = \dfrac{5}{10} \div \dfrac{5}{5} = \dfrac{1}{2}$

3. $\dfrac{6}{12} - \dfrac{4}{16} = \dfrac{24-12}{48} = \dfrac{12}{48} = \dfrac{1}{4}$

$$
\begin{array}{c|cc}
2 & 12 & 16 \\
2 & 6 & 8 \\
2 & 3 & 4 \\
2 & 3 & 2 \\
3 & 3 & 1 \\
  & 1 & 1
\end{array}
$$
$\longrightarrow 2 \times 2 \times 2 \times 2 \times 3 = 48$

4. $15\dfrac{1}{3} - 2\dfrac{4}{12} = 13\dfrac{4-4}{12} = 13\dfrac{0}{12} = 13$

5. $5\dfrac{15}{36} - 2\dfrac{31}{36} = 3\dfrac{15-31}{36} \leftarrow$ need to borrow

$= 4\left(\dfrac{36}{36}\right) + \dfrac{15}{36} - 2\dfrac{31}{36}$

$= 4\dfrac{51}{36} - 2\dfrac{31}{36} = 2\dfrac{51-31}{36} = 2\dfrac{20}{36} = 2\dfrac{5}{9}$

STEP 1-4

1. $\dfrac{1}{3} \times \dfrac{2}{3} = \dfrac{1 \times 2}{3 \times 3} = \dfrac{2}{9}$

2. $\dfrac{4}{8} \times \dfrac{2}{9} = \dfrac{8}{72} \div \dfrac{8}{8} = \dfrac{1}{9}$

3. $\dfrac{4}{17} \times \dfrac{3}{34} = \dfrac{12}{578} = \dfrac{6}{289}$

4. $\dfrac{1}{16} \times 5\dfrac{3}{7} = \dfrac{1}{16} \times \dfrac{38}{7} = \dfrac{38}{112} = \dfrac{19}{56}$

5. $2\dfrac{1}{7} \times 7\dfrac{2}{3} = \dfrac{15}{7} \times \dfrac{23}{3} = \dfrac{345}{21} = 16\dfrac{9}{21} = 16\dfrac{3}{7}$

STEP 1-5

1. $\dfrac{2}{3} \div \dfrac{1}{12} = \dfrac{2}{3} \times \dfrac{12}{1} = \dfrac{24}{3} = 8$

2. $3\dfrac{1}{4} \div \dfrac{3}{4} = \dfrac{13}{4} \div \dfrac{3}{4} = \dfrac{13}{4} \times \dfrac{4}{3} = \dfrac{52}{12} = 4\dfrac{4}{12} = 4\dfrac{1}{3}$

3. $8\dfrac{16}{20} \div 4\dfrac{4}{20} = \dfrac{176}{20} \div \dfrac{84}{20} = \dfrac{176}{20} \times \dfrac{20}{84}$

$= \dfrac{3520}{1680} = 2\dfrac{160}{1680} \div \dfrac{10}{10} = 2\dfrac{16}{168} \div \dfrac{4}{4} = 2\dfrac{4}{42} \div \dfrac{2}{2}$

$= 2\dfrac{2}{21}$

4. $4\dfrac{3}{8} \div 2\dfrac{5}{16} = \dfrac{35}{8} \div \dfrac{37}{16} = \dfrac{35}{8} \times \dfrac{16}{37}$

$= \dfrac{560}{296} = 1\dfrac{264}{296} \div \dfrac{4}{4} = 1\dfrac{66}{74} \div \dfrac{2}{2} = 1\dfrac{33}{37}$

STEP 1-6

1. $\dfrac{16}{\cancel{27}} \times \dfrac{\cancel{27}}{240} = \dfrac{\cancel{16}}{1} \times \dfrac{1}{\cancel{240}} = \dfrac{1}{1} \times \dfrac{1}{15} = \dfrac{1}{15}$

2. $\dfrac{\cancel{8}}{17} \times \dfrac{3}{\cancel{4}} = \dfrac{2}{17} \times \dfrac{3}{1} = \dfrac{6}{17}$

3. $\dfrac{15}{32} \div \dfrac{30}{4} = \dfrac{\cancel{15}}{32} \times \dfrac{4}{\cancel{30}} = \dfrac{1}{\cancel{32}} \times \dfrac{\cancel{4}}{2}$

$= \dfrac{1}{8} \times \dfrac{1}{2} = \dfrac{1}{16}$

4. $6\dfrac{2}{3} \times 2\dfrac{7}{10} = \dfrac{\cancel{20}}{3} \times \dfrac{27}{\cancel{10}} = \dfrac{2}{\cancel{3}} \times \dfrac{\cancel{27}}{1}$

$= \dfrac{2}{1} \times \dfrac{9}{1} = \dfrac{18}{1} = 18$

5. $6\dfrac{1}{2} \div \dfrac{9}{32} = \dfrac{13}{\cancel{2}} \times \dfrac{\cancel{32}}{9} = \dfrac{13}{1} \times \dfrac{16}{9} = \dfrac{208}{9} = 23\dfrac{1}{9}$

STEP 2-1

1. 10

2.
$$
\begin{array}{rcr}
3 \times 10{,}000 & = & 30{,}000 \\
5 \times 1{,}000 & = & 5{,}000 \\
7 \times 100 & = & 700 \\
2 \times 10 & = & 20 \\
9 \times 1 & = & 9 \\
\hline
& & 35{,}729
\end{array}
$$

3. See Section 2-1-3.

4.
499,997	
499,998	1
499,999	2
500,000	3
500,001	4
500,002	5

The units, tens, hundreds, thousands, and ten-thousands columns will reset and carry when the count is advanced by 3. The final count will be 500,002.

STEP 2-2

1.
$$
\begin{array}{rcr}
1 \times 100 & = & 100 \\
7 \times 10 & = & 70 \\
8 \times 1 & = & 8 \\
6 \times 0.1 & = & 0.6 \\
4 \times 0.01 & = & 0.04 \\
9 \times 0.001 & = & 0.009 \\
\hline
& & 178.649
\end{array}
$$

2. (a) $1\dfrac{5}{2} = 1 + (5 \div 2) = 1 + 2.5 = 3.5$

(b) $192\dfrac{3}{4} = 192 + (3 \div 4) = 192 + 0.75 = 192.75$

(c) $67\dfrac{6}{9} = 67 + (6 \div 9) = 67 + 0.667 = 67.667$

3. (a) $\dfrac{1}{2500} = 0.0004$

(b) $\dfrac{1}{0.25} = 4$

(c) $\dfrac{1}{0.000025} = 40{,}000$

4. (a) $7.25 = 7\dfrac{4}{16} \left(\dfrac{1}{16} = 0.0625 \right)$

(b) $156.90625 = 156\dfrac{29}{32} \left(\dfrac{1}{32} = 0.03125 \right)$

STEP 2-3

1. Addition, subtraction

2. (a) Subtraction

(b) Addition

(c) Division

(d) Multiplication

3. (a)
$$
\begin{array}{r}
26.443 \\
197.1 \\
2.1103 \\
+\ 0.004 \\
\hline
225.6573
\end{array}
$$

(b)
$$
\begin{array}{r}
19637.224 \\
-\ 866.43 \\
\hline
18770.794
\end{array}
$$

(c)
$$
\begin{array}{r}
894.357 \\
\times\ 8.6 \\
\hline
5366142 \\
7154856 \\
\hline
7691.4702
\end{array}
$$

(d)
$$
0.015\,\overline{)1.3397}^{\,89.313333}
$$

or $15\overline{)1339.7}$

4. $\dfrac{176 \div 8}{8 \div 8} = \dfrac{22}{1}$ or 22:1 or 22.0 to 1 (ratio of twenty-two to one)

5. (a) 86.44

(b) 12,263,415.01

(c) 0.18

6. 86.43760   (a) 7

               (b) 6

12,263,415.00510   (a) 13

                     (b) 12

0.176600   (a) 6

              (b) 4

7. (a) 0.075

(b) 220

(c) 0.235

(d) 19.2

8. (a) 35

(b) 4004

STEP 3-1

1. (a) $+3 - (-4) = +7$

(b) $12 \times (+4) = +48$

(c) $-5 \div (-7) = 0.7142$

(d) $-0.63 \times (+6.4) = -4.032$

2. (a) $+6 + (-7)$

(b) $-0.75 \times (18)$ or -0.75×18

3. (a) $\boxed{7}\ \boxed{.}\ \boxed{5}\ \boxed{+/-}\ \boxed{+}\ \boxed{4}\ \boxed{.}\ \boxed{6}\ \boxed{+/-}\ \boxed{=}$

(b) $\boxed{5}\ \boxed{\times}\ \boxed{2}\ \boxed{+/-}\ \boxed{=}$

(c) $\boxed{3}\ \boxed{1}\ \boxed{6}\ \boxed{.}\ \boxed{6}\ \boxed{2}\ \boxed{9}\ \boxed{\div}\ \boxed{1}\ \boxed{.}\ \boxed{4}\ \boxed{4}\ \boxed{+/-}\ \boxed{=}$

STEP 3-2

1. $+15 + (+4) = +19$

2. $-3 + (+45) = +42$

3. $+114 + (-111) = +3$

4. $-357 + (-74) = -431$

5. $17 + (15) = +32$

6. $-8 + (-177) = -185$

7. $+4600 + (-3400) = +1200$

8. $-6.25 + (+0.34) = -5.91$

STEP 3-3

1. $+18 - (+7) = +11$

2. $-3.4 - (-5.7) = +2.3$

3. $19{,}665 - (-5{,}031) = +24{,}696$

4. $-8 - (+5) = -13$

5. $467 - 223 = +244$

6. $-331 - (-2.6) = -328.4$

7. $8 - (+25) = -17$

8. $-0.64 - (-0.04) = -0.6$

STEP 3-4

1. $4 \times (+3) = +12$

2. $+17 \times (-2) = -34$

3. $-8 \times (+16) = -128$

4. $-8 \times (-5) = +40$

5. $+12.6 \times (+15) = +189$
6. $-3.3 \times (+1.4) = -4.62$
7. $+0.3 \times (-4) = -1.2$
8. $-4.6 \times (-3.3) = +15.18$

STEP 3-5
1. $+16.7 \div (+2.3) = 16.7 \div 2.3 = 7.26$
2. $+18 \div (+6) = 3$
3. $-6 \div (+2) = -3$
4. $+18 \div (-4) = -4.5$
5. $+2 \div (-8) = -0.25$
6. $-8 \div (+5) = -1.6$
7. $-15 \div (-5) = 3$
8. $0.664 \div (-0.2) = -3.32$

STEP 3-6
1. $+6 + (+3) + (-7) + (-5) = -3$
2. $+9 - (+2) - (-13) - (-4) = +24$
3. $-6 \times (-4) \times (-5) = -120$
4. $+8 \div (+2) \div (-5) = -0.8$
5. $-4 \div (-2) \times (+8) = +16$
6. $-9 + (+5) - (-7) = +3$

STEP 4-1
1. (a) $16^4 = 16 \times 16 \times 16 \times 16 = 65,536$
(b) $32^3 = 32 \times 32 \times 32 = 32,768$
(c) $112^2 = 112 \times 112 = 12,544$
(d) $15^6 = 15 \times 15 \times 15 \times 15 \times 15 \times 15 = 11,390,625$
(e) $2^3 = 2 \times 2 \times 2 = 8$
(f) $3^{12} = 3 \times 3 \times 3 \times 3 \times 3 \times 3 \times 3 \times 3 \times 3 \times 3 \times 3 \times 3$
$\quad = 531,441$

2. (a) $\sqrt[2]{144} = 12$
(b) $\sqrt[3]{3375} = 15$
(c) $\sqrt[2]{20} = 4.47$
(d) $\sqrt[3]{9} = 2.08$

3. (a) $(9^2 + 14^2)^2 - \sqrt[3]{3 \times 7}$
$\quad = (81 + 196)^2 - \sqrt[3]{21}$
$\quad = 277^2 - 2.76$
$\quad = 76,729 - 2.76 = 76,726.24$

(b) $\sqrt{3^2 \div 2^2} + \dfrac{151 - 9^2}{3.5^2}$

$\quad = \sqrt{9 \div 4} + \dfrac{151 - 81}{12.25}$

$\quad = \sqrt{2.25} + \dfrac{70}{12.25}$

$\quad = 1.5 + 5.71 = 7.21$

STEP 4-2
1. (a) 10^2
(b) 10^0
(c) 10^1
(d) 10^6
(e) 10^{-3}
(f) 10^{-6}

2. (a) $6.3 \times 10^3 = 6.300. = 6300.0$ or 6300
(b) $114,000 \times 10^{-3} = 114.000 = 114.0$ or 114
(c) $7,114,632 \times 10^{-6} = 7.114632 = 7.114632$
(d) $6624 \times 10^6 = 6624.000000. = 6,624,000,000.0$

3. (a) $\sqrt{3 \times 10^6} = \sqrt{3,000,000} = 1732.05$
(b) $(2.6 \times 10^{-6}) - (9.7 \times 10^{-9}) = 0.0000025$ or 2.5×10^{-6}
(c) $\dfrac{(4.7 \times 10^3)^2}{3.6 \times 10^6} = (4.7 \times 10^3)^2 \div (3.6 \times 10^6) = 6.14$

4. (a) $47,000 = 47000. = 47 \times 10^3$
(b) $0.00000025 = 0.000000250. = 250 \times 10^{-9}$
(c) $250,000,000 = 250.000\,000. = 250 \times 10^6$
(d) $0.0042 = 0.004.2 = 4.2 \times 10^{-3}$

STEP 4-3
1. (a) 10^3
(b) 10^{-2}
(c) 10^{-3}
(d) 10^6
(e) 10^{-6}

2. (a) cm $\times 0.4 =$ inches, 15 cm $\times 0.4 = 6$ inches
(b) kg $\times 2.2 =$ pounds, 23 kg $\times 2.2 = 50.6$ pounds
(c) liters $\times 0.26 =$ gallons, 37 L $\times 0.26 = 9.62$ gallons
(d) $\left(\dfrac{9}{5} \times {}^\circ\text{C}\right) + 32 = \left(\dfrac{9}{5} \times 23\right) + 32 = 73.4{}^\circ\text{F}$

3. (a) Miles/hour $\times 1.6 =$ kilometers/hour,
$\quad$ 55 mph $\times 1.6 = 88$ km/hour
(b) gallons $\times 3.8 =$ liters, 16 gallons $\times 3.8 = 60.8$ liters
(c) square yards $\times 0.8 =$ square meters, 3 yd$^2 \times 0.8 = 2.4$ m^2
(d) $\dfrac{5}{9} \times ({}^\circ\text{F} - 32) = \dfrac{5}{9} \times (92 - 32) = 33.33{}^\circ\text{C}$

4. (a) meter (m) $\qquad$ **(f)** ampere (A)
(b) gram (g) $\qquad$ **(g)** volt (V)
(c) degree Celsius (°C) $\qquad$ **(h)** liter (L)
(d) second (s) $\qquad$ **(i)** joule (J)
(e) watt (W) $\qquad$ **(j)** ohm (Ω)

5. (a) 25,000 volts $= 25.000. \times 10^3$ volts
$\quad = 25$ kilovolts
(b) 0.014 watts $= 0.014. = 14 \times 10^{-3}$ watts
$\quad = 14$ milliwatts
(c) 0.000016 microfarad $= 0.000016 \times 10^{-6}$ farad
$\quad = 0.000.016 \times 10^{-9}$ farad
$\quad = 0.016 \times 10^{-9}$ farad
$\quad = 0.016$ nanofarad

STEP 5-1
1. Yes
2. No
3. $\dfrac{144}{12} \times \square = \dfrac{36}{6} \times 2 \times \square = 60$
$\quad 12 \times \square = 6 \times 2 \times \square = 60$
$\quad 12 \times 5 = 12 \times 5 = 60$
$\quad\quad \square = 5$

4. $\dfrac{(8 - 4) + 26}{5} = \dfrac{81 - 75}{2}$

$\quad \dfrac{4 + 26}{5} = \dfrac{6}{2}$

$\quad \dfrac{30}{5} = \dfrac{6}{2}$

$\quad\quad 6 = 3 \qquad$ (equation is not equal)

5. Yes

STEP 5-2
1. (a) $x + x = 2x$
(b) $x \times x = x^2$
(c) $7a + 4a = 11a$
(d) $2x - x = 1x$ or x
(e) $\dfrac{x}{x} = 1$
(f) $x - x = 0$

2. (a) $\quad\quad x + 14 = 30$
$\quad x + 14 - 14 = 30 - 14 \qquad$ (-14 from both sides)
$\quad\quad\quad\quad x = 30 - 14$
$\quad\quad\quad\quad x = 16$

(b) $8 \times x = \dfrac{80 - 40}{10} \times 12$

$$8 \times x = 4 \times 12$$
$$8 \times x = 48$$
$$\dfrac{8 \times x}{8} = \dfrac{48}{8} \qquad (\div\, 8)$$
$$x = \dfrac{48}{8}$$
$$x = 6$$

(c) $\quad y - 4 = 8$
$$y - 4 + 4 = 8 + 4 \qquad (+\,4)$$
$$y = 8 + 4$$
$$y = 12$$

(d) $\quad (x \times 3) - 2 = \dfrac{26}{2}$
$$(x \times 3) - 2 + 2 = \dfrac{26}{2} + 2 \qquad (+\,2)$$
$$x \times 3 = \dfrac{26}{2} + 2$$
$$x \times 3 = 15$$
$$\dfrac{x \times 3}{3} = \dfrac{15}{3} \qquad (\div\, 3)$$
$$x = \dfrac{15}{3}$$
$$x = 5$$

(e) $\quad x^2 + 5 = 14$
$$x^2 + 5 - 5 = 14 - 5 \qquad (-\,5)$$
$$x^2 = 14 - 5$$
$$\sqrt{x^2} = \sqrt{14 - 5} \qquad (\sqrt{\ })$$
$$x = \sqrt{14 - 5}$$
$$x = \sqrt{9}$$
$$x = 3$$

(f) $\quad 2(3 + 4x) = 2(x + 13)$
$$6 + 8x = 2x + 26 \qquad \text{(remove parentheses)}$$
$$6 + 8x - 2x = 2x + 26 - 2x \qquad (-\,2x)$$
$$6 + (8x - 2x) = 26$$
$$6 + 6x = 26$$
$$6 + 6x - 6 = 26 - 6 \qquad (-\,6)$$
$$6x = 26 - 6$$
$$6x = 20$$
$$\dfrac{6 \times x}{6} = \dfrac{20}{6} \qquad (\div\, 6)$$
$$x = \dfrac{20}{6}$$
$$x = 3.3333$$

3. (a) $x + y = z,\ y = ?$
$$x + y - x = z - x \qquad (-\,x)$$
$$y = z - x$$

(b) $Q = C \times V,\ C = ?$
$$\dfrac{Q}{V} = \dfrac{C \times V}{V} \qquad (\div\, V)$$
$$\dfrac{Q}{V} = C$$
$$C = \dfrac{Q}{V}$$

(c) $\quad X_L = 2 \times \pi \times f \times L,\ L = ?$
$$\dfrac{X_L}{2 \times \pi \times f} = \dfrac{2 \times \pi \times f \times L}{2 \times \pi \times f} \qquad (\div\, 2 \times \pi \times f)$$

$$\dfrac{X_L}{2 \times \pi \times f} = L$$
$$L = \dfrac{X_L}{2 \times \pi \times f}$$

(d) $V = I \times R,\ R = ?$
$$\dfrac{V}{I} = \dfrac{I \times R}{I} \qquad (\div\, I)$$
$$\dfrac{V}{I} = R$$
$$R = \dfrac{V}{I}$$

4. (a) $\quad I^2 = 9$
$$\sqrt{I^2} = \sqrt{9} \qquad (\sqrt{\ })$$
$$I = \sqrt{9}$$
$$I = 3$$

(b) $\quad \sqrt{Z} = 8$
$$\sqrt{Z}^2 = 8^2$$
$$Z = 8^2$$
$$Z = 64$$

STEP 5-3

1. $x = y \times z$ and $a = x \times y,\ y = 14,\ z = 5,\ a = ?$
$$a = x \times y$$
$$a = y \times z \times y \qquad \text{(substitute } y \times z \text{ for } x)$$
$$a = y^2 \times z \qquad (y \times y = y^2)$$
$$a = 14^2 \times 5$$
$$a = 980$$

2. (a) For $I = \sqrt{\dfrac{P}{R}}$,
$$P = V \times I \text{ to } P = I^2 \times R$$
$$\dfrac{P}{R} = \dfrac{I^2 \times R}{R} \qquad (\div\, R)$$
$$\dfrac{P}{R} = I^2$$
$$\dfrac{P}{R} = \sqrt{I^2} \qquad (\sqrt{\ })$$
$$\sqrt{\dfrac{P}{R}} = I$$
$$I = \sqrt{\dfrac{P}{R}}$$

(b) For $I = \dfrac{P}{V}$,
$$P = V \times I$$
$$\dfrac{P}{V} = \dfrac{V \times I}{V} \qquad (\div\, V)$$
$$\dfrac{P}{V} = I$$
$$I = \dfrac{P}{V}$$

(c) For $V = \sqrt{P \times R}$
$$P = V \times I \text{ to } P = \dfrac{V^2}{R}$$
$$P \times R = \dfrac{V^2 \times R}{R} \qquad (\times\, R)$$
$$P \times R = V^2$$
$$\sqrt{P \times R} = \sqrt{V^2} \qquad (\sqrt{\ })$$
$$\sqrt{P \times R} = V$$
$$V = \sqrt{P \times R}$$

(d) For $R = \dfrac{V^2}{P}$,

$$P = V \times I \text{ to } P = \frac{V^2}{R}$$

$$P \times R = \frac{V^2}{\cancel{R}} \times \cancel{R} \qquad (\times R)$$

$$P \times R = V^2$$

$$\frac{\cancel{P} \times R}{\cancel{P}} = \frac{V^2}{P}$$

$$R = \frac{V^2}{P}$$

3. What percentage of 12 gives 2.5?

$$x \times 12 = 2.5$$

$$\frac{x \times \cancel{12}}{\cancel{12}} = \frac{2.5}{12} \div 12$$

$$x = \frac{2.5}{12}$$

$$x = 0.2083 \quad (0.2083 \times 12 = 2.5)$$

$$x\% = 20.83\% \quad (20.83\% \times 12 = 2.5)$$

4. 60 is 22% of what number?

$$60 = 22\% \times x$$

$$60 = 0.22 \times x$$

$$\frac{60}{0.22} = \frac{\cancel{0.22} \times x}{\cancel{0.22}} \qquad (\div 0.22)$$

$$\frac{60}{0.22} = x$$

$$x = 272.73 \quad (60 = 0.22 \times 272.73$$

$$\text{or } 60 = 22\% \times 272.73)$$

STEP 5-4

1. Parentheses, Exponents, Multiplication, Division, Addition, Subtraction.

2. Monomial: $3x$
Binomial: $3x + 4y$
Trinomial: $3x + 4y + 5z$
Polynomial: $2a + 2b + 2c + 2d + 2e$

3. **(a)** $a \div 1 = a$
(b) $a + a = 2a$
(c) $a - a = 0$
(d) $a \times a = a^2$
(e) $3x + b - 2x = x + b$

4. **(a)** $16 + a = 5 \times a$
$a = 4$
(b) $\dfrac{42}{b} = 3 \times 7$
$b = 2$

STEP 6-1

1. **(a)** $40°$ **(d)** $90°$ **(g)** $165°$ **(j)** $165°$
(b) $60°$ **(e)** $115°$ **(h)** $180°$ **(k)** $65°$
(c) $80°$ **(f)** $135°$ **(i)** $15°$ **(l)** $90°$

2. **(a)** Acute **(d)** Right **(g)** Obtuse **(j)** Obtuse
(b) Acute **(e)** Obtuse **(h)** Obtuse **(k)** Acute
(c) Acute **(f)** Obtuse **(i)** Acute **(l)** Right

3. (See Figure 6-1)

STEP 6-2

1. (Section 6-2 Introduction)

2. **(a)** Figure 6-2 **(e)** Figure 6-6 **(i)** Figure 6-12
(b) Figure 6-3 **(f)** Figure 6-7 **(j)** Figure 6-13
(c) Figure 6-4 **(g)** Figure 6-10
(d) Figure 6-5 **(h)** Figure 6-11

3. **(a)** $A = \sqrt{C^2 - B^2}$
$= \sqrt{3^2 - 2^2}$
$= \sqrt{9 - 4}$
$= \sqrt{5}$
$= 2.24$ feet
(b) $B = \sqrt{C^2 - A^2}$
$= \sqrt{160^2 - 80^2}$
$= \sqrt{25{,}600 - 6400}$
$= \sqrt{19{,}200}$
$= 138.56$ km
(c) $C = \sqrt{A^2 + B^2}$
$= \sqrt{112^2 + 25^2}$
$= \sqrt{12{,}544 + 625}$
$= \sqrt{13{,}169}$
$= 114.76$ mm

4. **(a)** $x = \sqrt{A^2 + B^2}$
$= \sqrt{40^2 + 30^2}$
$= \sqrt{1600 + 900}$
$= \sqrt{2500}$
$= 50$ volts
(b) $x = \sqrt{A^2 + B^2}$
$= \sqrt{75^2 + 26^2}$
$= \sqrt{5625 + 676}$
$= \sqrt{6301}$
$= 79.38$ watts
(c) $x = \sqrt{A^2 + B^2}$
$= \sqrt{93^2 + 36^2}$
$= \sqrt{8649 + 1296}$
$= \sqrt{9945}$
$= 99.72$ mm
$y = \sqrt{A^2 + B^2}$
$= \sqrt{48^2 + 96^2}$
$= \sqrt{2304 + 9216}$
$= \sqrt{11{,}520}$
$= 107.33$ mm

STEP 6-3

1. **(a)** $O = 35$ mm $\qquad \theta = 36°$
$H = \quad ?$

$$SOH \quad \text{or} \quad \sin \theta = \frac{O}{H}$$

$$\sin 36° = \frac{35 \text{ mm}}{H}$$

$$0.59 = \frac{35}{H}$$

$$0.59 \times H = \frac{35}{\cancel{H}} \times \cancel{H} \qquad (\times H)$$

$$\frac{0.59 \times H}{0.59} = \frac{35}{0.59} \qquad (\div 0.59)$$

$$H = \frac{35}{0.59} = 59.32 \text{ mm}$$

(b) $\widehat{H} = 160$ km $\qquad \theta = 38°$
$\widehat{A} = ?$

$\qquad CAH \qquad$ or $\qquad \cos\theta = \dfrac{A}{H}$

$\qquad\qquad\qquad\qquad \cos 38° = \dfrac{A}{160 \text{ km}}$

$\qquad\qquad\qquad\qquad 0.79 = \dfrac{A}{160}$

$\qquad\qquad 0.79 \times 160 = \dfrac{A}{\cancel{160}} \times \cancel{160} \qquad (\times 160)$

$\qquad\qquad 0.79 \times 160 = A$

$\qquad\qquad\qquad\qquad\qquad A = 126.4$ km

(c) $\widehat{O} = 163$ cm $\qquad \theta = 72°$
$\widehat{A} = ?$

$\qquad TOA \qquad$ or $\qquad \tan\theta = \dfrac{O}{A}$

$\qquad\qquad\qquad\qquad \tan 72° = \dfrac{163 \text{ cm}}{A}$

$\qquad\qquad\qquad\qquad 3.08 = \dfrac{163}{A}$

$\qquad\qquad 3.08 \times A = \dfrac{163 \times \cancel{A}}{\cancel{A}} \qquad (\times A)$

$\qquad\qquad \dfrac{3.08 \times A}{3.08} = \dfrac{163}{3.08} \qquad (\div 3.08)$

$\qquad\qquad\qquad\qquad A = \dfrac{163}{3.08} = 52.92$ cm

2. (a) $\widehat{H} = 120$ miles $\qquad \theta = ?$
$\widehat{O} = 38$ miles

$\qquad SOH \qquad$ or $\qquad \sin\theta = \dfrac{O}{H}$

$\qquad\qquad\qquad\qquad \sin\theta = \dfrac{38 \text{ miles}}{120 \text{ miles}}$

$\qquad\qquad\qquad\qquad \sin\theta = 0.32$

$\qquad \cancel{\text{invsin}} \times \cancel{\sin}\theta = \text{invsin } 0.32 \qquad (\times \text{ invsin})$

$\qquad\qquad\qquad\qquad \theta = \text{invsin } 0.32$

$\qquad\qquad\qquad\qquad \theta = 18.66°$

(b) $\widehat{H} = 25$ feet $\qquad \theta = ?$
$\widehat{A} = 17$ feet

$\qquad CAH \qquad$ or $\qquad \cos\theta = \dfrac{A}{H}$

$\qquad\qquad\qquad\qquad \cos\theta = \dfrac{17 \text{ feet}}{25 \text{ feet}}$

$\qquad\qquad\qquad\qquad \cos\theta = 0.68$

$\qquad \cancel{\text{invcos}} \times \cancel{\cos}\theta = \text{invcos } 0.68$

$\qquad\qquad\qquad\qquad \theta = \text{invcos } 0.68$

$\qquad\qquad\qquad\qquad \theta = 47.16°$

(c) $\widehat{A} = 69$ cm $\qquad \theta = ?$
$\widehat{O} = 51$ cm

$\qquad TOA \qquad$ or $\qquad \tan\theta = \dfrac{O}{A}$

$\qquad\qquad\qquad\qquad \tan\theta = \dfrac{51 \text{ cm}}{69 \text{ cm}}$

$\qquad\qquad\qquad\qquad \tan\theta = 0.74$

$\qquad \cancel{\text{invtan}} \times \cancel{\tan}\theta = \text{invtan } 0.74$

$\qquad\qquad\qquad\qquad \theta = \text{invtan } 0.74$

$\qquad\qquad\qquad\qquad \theta = 36.5°$

STEP 6-4

1. (Section 6-4 Introduction)
2. (Section 6-4-1)
3. (a) Figure 6-22 **(c)** Figure 6-24 **(e)** Figure 6-26
(b) Figure 6-23 **(d)** Figure 6-25
4. A frustum is the base section of a solid pyramid or cone.

STEP 7-1

1. Exponent
2. (c) 4
3. (a) 1 **(c)** 1.4 **(e)** 4.02 **(g)** -0.456
(b) 2 **(d)** 2.176 **(f)** 3.57 **(h)** 0.243
4. (a) 56.23 **(c)** 100,000 **(e)** 316,227.76 **(g)** 1.26
(b) 223.87 **(d)** 10 **(f)** 1.78 **(h)** 1^{10}

STEP 7-2

1.

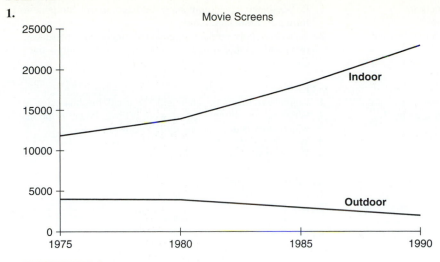

FIGURE 7-2-1

2.

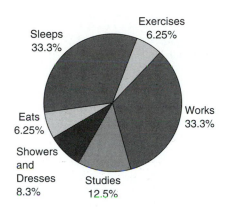

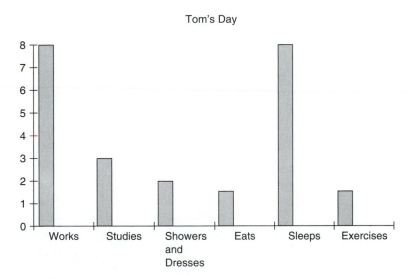

FIGURE 7-2-2

3. Graphing calculators have all the functions of scientific calculators, and in addition, they can graph functions.
4. (a) Y-Editor: Key is used to enter equation.
   (b) Table: Shows every possible result for all possible input values.
   (c) Graph: Displays the plotted data in graphical form.

STEP 8-1
1. Amp
2. Current = Q/t (number of coulombs divided by time in seconds).
3. The ammeter

STEP 8-2
1. Volts
2. 3000 kV
3. The voltmeter

STEP 8-3
1. False
2. Copper
3. 28.6 millisiemens

STEP 8-4
1. True
2. Mica
3. The voltage needed to cause current to flow through a material
4. Small

STEP 9-1
1. A circuit is said to have a resistance of 1 ohm when 1 volt produces a current of 1 ampere.
2. $I = V/R = 24$ V/6 $\Omega = 4$ A.
3. A memory aid to help remember Ohm's law:

4. Current is proportional to voltage and inversely proportional to resistance.
5. $V = I \times R = 25$ mA $\times 1$ k$\Omega = (25 \times 10^{-3}) \times (1 \times 10^{3}) = 25$ V.
6. $R = V/I = 12$ V/100 μA $= 12$ V/100 $\times 10^{-6}$ A $= 120$ kΩ.

STEP 9-2
1. A fixed-valve resistor's resistance cannot be changed, whereas a variable-valve resistor's resistance can be adjusted.
2. Linear means that the resistance changes in direct proportion to the amount of change of the input, while a tapered potentiometer varies nonuniformly.
3. True
4. Photoresistor
5. The resistance of a coil of wire within a lightbulb.

STEP 9-3
1. Ohmmeter
2. See Figure 9-14.

STEP 9-4
1. General purpose are $\pm5\%$ or greater; precision are $\pm2\%$ or less.
2. $222 \times 0.1, \pm0.25\% = 22.2$ Ω, $\pm0.25\%$
3. Yellow, violet, red, silver
4. True (General purpose, $\pm20\%$ tolerance)

STEP 9-5
1. It is designed to accommodate and interconnect components to form experimental circuits.
2. A power supply voltage is generally connected across these two points so that voltages can be tapped off to supply the circuit.

STEP 9-6
1. Light, heat, magnetic, chemical, electrical, mechanical
2. When energy is transformed, work is done. Power is the rate at which work is done or energy is transformed.
3. $W = Q \times V, P = V \times I$
4. When 1 kW of power is used in 1 hour

STEP 10-1 and 10-2
1. A circuit in which current has only one path
2. 8 A

STEP 10-3
1. $R_T = R_1 + R_2 + R_3 + \cdots$
2. $R_T = R_1 + R_2 + R_3 = 2$ k$\Omega + 3$ k$\Omega + 4700$ $\Omega = 9.7$ kΩ

STEP 10-4
1. True
2. True
3. $R_T = R_1 + R_2 = 6 + 12 = 18$ Ω; $I_T = V_S/R_T = 18/18 = 1$ A
   $V_{R_1} = 1$ A $\times 6$ $\Omega = 6$ V
   $V_{R_2} = 1$ A $\times 12$ $\Omega = 12$ V
4. $V_X = (R_X/R_T) = V_S$
5. Potentiometer
6. No

STEP 10-5
1. $P = I \times V$ or $P = V^2/R$ or $P = I^2 \times R$
2. $P = V^2/R = 12^2/12 = 144/12 = 12$ W
3. Wirewound, at least 12 W, ideally a 15 W
4. $P_T = P_1 + P_2 = 25$ W $+ 3800$ mW $= 28.8$ W

STEP 11-1 and 11-2
1. When two or more components are connected across the same voltage source so that current can branch out over two or more paths
2. False
3. $V_{R_1} = V_S = 12$ V
4. No

STEP 11-3
1. The sum of all currents entering a junction is equal to sum of all currents leaving that junction.
2. $I_2 = I_T - I_1 = 4$ A $- 2.7$ A $= 1.3$ A
3. $I_X = (R_T/R_X) \times I_T$
4. $I_T = V_T/R_T = 12$ V/1 k$\Omega = 12$ mA; $I_1 = 1$ kΩ/2 k$\Omega \times 12$ mA $= 6$ mA

STEP 11-4
1. $R_T = \dfrac{R_1 \times R_2}{R_1 + R_2}$
2. $R_T = \dfrac{1}{(1/R_1) + (1/R_2) + (1/R_3)} + \cdots$
3. $R_T = \dfrac{\text{common value of resistors } (R)}{\text{number of parallel resistors } (n)}$
4. $R_T = \dfrac{1}{(1/2.7\text{ k}\Omega) + (1/24\text{ k}\Omega) + (1/1\text{ M}\Omega)} = 2.421$ kΩ

STEP 11-5
1. True
2. $P_1 = I_1 \times V = 2 \text{ mA} \times 24 \text{ V} = 48 \text{ mW}$
3. $P_T = P_1 + P_2 = 22 \text{ mW} + 6400 \text{ } \mu\text{W} = 28.4 \text{ mW}$
4. Yes

STEP 12-1 and 12-2
1. By tracing current to see if it has one path (series connection) or more than one path (parallel connection)
2. $R_{1,2} = R_1 + R_2 = 12 \text{ k}\Omega + 12 \text{ k}\Omega = 24 \text{ k}\Omega$

$$R_{1,2,3} = \frac{R_{1,2} \times R_3}{R_{1,2} + R_3} = \frac{24 \text{ k}\Omega \times 6 \text{ k}\Omega}{24 \text{ k}\Omega + 6 \text{ k}\Omega}$$
$$= \frac{144 \text{ k}\Omega^2}{30 \text{ k}\Omega} = 4.8 \text{ k}\Omega$$

3. STEP A: Find equivalent resistances of series-connected resistors. STEP B: Find equivalent resistances of parallel-connected combinations. STEP C: Find equivalent resistances of remaining series-connected resistances.
4. $R_{1,2} = R_1 + R_2 = 470 + 330 = 800 \text{ } \Omega$

$$R_{1,2,3} = \frac{R_{1,2} \times R_3}{R_{1,2} + R_3} = \frac{800 \times 270}{800 + 270} = \frac{216 \text{ k}\Omega^2}{1.07 \text{ k}\Omega} = 201.9 \text{ } \Omega$$

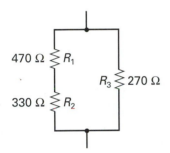

FIGURE 12-1,4

STEP 12-3
1. STEP 1: Find total resistance (STEPs A, B and C). STEP 2: Find total current. STEP 3: Find voltage drop with $I_T \times R_X$.
2. The voltage drops previously calculated would not change since the ratio of the series resistor to the series equivalent resistors remains the same, and therefore the voltage division will remain the same.

STEP 12-4, 12-5, and 12-6
1. Find total resistance; find total current; find the voltage across each series and parallel combination resistors; find the current through each branch of parallel resistors; find the total and individual power dissipated.
2. This is a do-it-yourself question; each answer will vary.

STEP 12-7
1. When a load resistance changes the circuit and lowers output voltage

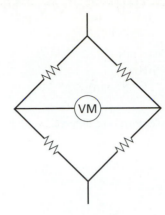

FIGURE 12-7,2

2. Used to check for an unknown resistor's resistance
3. R
4. Current divider for a digital-to-analog converter

STEP 12-8
1. Small, large
2. Voltage source
3. Voltage, resistor
4. Current

STEP 13-1
1. a. Alternating current
   b. Direct current
2. DC; current only flows in one direction
3. AC
4. DC
5. Power transfer, information transfer
6. DC flows in one direction, whereas ac first flows in one direction and then in the opposite direction.

STEP 13-2-1
1. AC generators can be larger, less complex, and cheaper to run; transformers can be used with ac to step up/down, so low-current power lines can be used; easy to change ac to dc, but hard the other way around.
2. False
3. $P = I^2 \times R$
4. A device that can step up or down ac voltages
5. 120 V ac
6. AC, DC

STEP 13-2-2 and 13-2-3
1. The property of a signal or message that conveys something meaningful to the recipient; the transfer of information between two points
2. Sound, electromagnetic, electrical
3. a. Sound wave
   b. Electromagnetic wave
   c. Electrical wave
4. 1133 feet per second; 186,000 miles per second
5. Sound, electromagnetic, electrical
6. a. Microphone   c. Antenna
   b. Speaker   d. Human ear
7. Electronic
8. Electrical

STEP 13-3

1.

(a)   (b)   (c)   (d)   (e)

2. Time, frequency

3. (Sound wave); $\lambda(mm) = \dfrac{344.4 \text{ m/s}}{f(Hz)}$;

   (Electromagnetic wave): $\lambda(m) = \dfrac{3 \times 10^8 \text{ m/s}}{f(Hz)}$;

   different because sound waves travel at a different speed than do electromagnetic waves.

4. Odd

STEP 13-4

1. Oscilloscope
2. Cathode ray tube (CRT)
3. $t = 80 \ \mu s$
   $f = 12.5$ kHz
4. $V_p = 4$ V
   $V_{p\text{-}p} = 8$ V
5. Waveforms can be compared.

STEP 13-5

1. The logarithmic unit for the difference in power at the output of an amplifier compared with the input.
2. 6 dB
3. 3 dB

STEP 14-1

1. The farad
2. Capacitance, C (farads) = charge, Q (coulombs)/voltage, V (volts)
3. $30,000 \times 10^{-6} = 0.03 \times 10^0 = 0.03$ F
4. $C = Q/V = 17.5/9 = 1.94$ F

STEP 14-2

1. Plate area, distance between plates, type of dielectric used
2. $C = \dfrac{(8.85 \times 10^{-12}) \times K \times A}{d}$
3. Double
4. Double

STEP 14-3

1. $C_T = \dfrac{1}{(1/C_1) + (1/C_2) + (1/C_3)}$
   $= \dfrac{1}{(1/2 \ \mu F) + (1/3 \ \mu F) + (1/5 \ \mu F)}$
   $= 0.968 \ \mu F$ or 1 μF
2. $C_T = C_1 + C_2 + C_3 = 7$ pF $+ 2$ pF $+ 14$ pF $= 23$ pF
3. $V_{CX} = (C_T/C_X) \times V_T$
4. True

STEP 14-4

1. The time it takes a capacitor to charge to 63.2%
2. 63.2%
3. 36.8%
4. False

STEP 14-5

1. Opposition to current flow without the dissipation of energy
2. $X_C = 1/2\pi fC$

3. When frequency or capacitance goes up, there is more charge and discharge current; so X_C is lower.
4. $X_C = 1/2\pi fC = 1/2\pi \times 4$ kHz $\times 4 \ \mu F = 9.95 \ \Omega$

STEP 14-6

1. Current leads voltage by some phase angle less than 90%.
2. An arrangement of vectors to illustrate the magnitude and phase relationships between two or more quantities of the same frequency
3. Z = total opposition to current flow; $Z = \sqrt{R^2 + X_C^2}$
4. a. 0°
   b. 90°
   c. Between 0 and 90°

STEP 14-7

1. Resistor current is in phase with voltage; capacitor current is 90° out of phase (leading) with voltage.
2. No
3. a. $I_T = \sqrt{I_R^2 + I_C^2}$;   b. $Z = V_S/I_T$;
   also $Z = (R \times X_C)/\sqrt{R^2 + X_C^2}$
4. Lead

STEP 15-1

1. When the current-carrying coil of a conductor induces a voltage within itself
2. The induced voltage, which opposes the applied emf; $V_{ind} = L \times (\Delta i/\Delta t)$
3. $V_{ind} = L \times (\Delta i/\Delta t) = 2$ mH $\times 4$ kA/s $= 8$ V

STEP 15-2

1. Different applications: An electromagnet is used to generate a magnetic field; an inductor is used to oppose any changes of circuit current.
2. True

STEP 15-3

1. Number of turns, area of coil, length of coil, core material used

STEP 15-4

1. False
2. a. $L_T = L_1 + L_2 + L_3 + \cdots$
   b. $L_T = 1/((1/L_1) + (1/L_2) + (1/L_3) + \cdots)$
3. a. $L_T = L_1 + L_2 = 4$ mH $+ 2$ mH $= 6$ mH
   b. $L_T = \dfrac{L_1 \times L_2}{L_1 + L_2}$ (using product over sum)
      $= \dfrac{4 \text{ mH} \times 2 \text{ mH}}{4 \text{ mH} + 2 \text{ mH}} = \dfrac{8 \text{ mH}^2}{6 \text{ mH}} = 1.33$ mH

STEP 15-5

1. Current in an inductive circuit builds up in the same way that voltage does in a capacitive circuit, but the capacitive time constant is proportional to resistance, where the inductive time constant is inversely proportional to resistance.
2. True
3. False
4. It will continuously oppose the alternating current.

STEP 15-6

1. The opposition to current flow offered by an inductor without the dissipation of energy: $X_L = 2 \times \pi \times f \times L$
2. The higher the frequency (and therefore the faster the change in current) or the larger the inductance of the inductor, the larger the magnetic field created, the larger the counter emf will be to oppose applied emf; so the inductive reactance (opposition) will be large also.

3. Inductive reactance is an opposition, so it can be used in Ohm's law in place of resistance.
4. False

STEP 15-7
1. False
2. $V_S = \sqrt{V_R^2 + V_L^2} = \sqrt{4^2 + 2^2} = \sqrt{16 + 4} = \sqrt{20} = 4.47$ V
3. The total opposition to current flow offered by a circuit with both resistance and reactance: $Z = \sqrt{R^2 + X_L^2}$
4. $+45°$
5. Power consumption above the zero line, caused by positive current and voltage or negative current and voltage
6. Quality factor of an inductor that is the ratio of the energy stored in the coil by its inductance to the energy dissipated in the coil by the resistance: $Q = X_L/R$
7. True power is energy dissipated and lost by resistance; reactive power is energy consumed and then returned by a reactive device.
8. PF = true power (P_R)/apparent power (P_A) or PF = R/Z or PF = $\cos \theta$

STEP 15-8
1. False
2. $I_T = \sqrt{I_R^2 + I_L^2}$

STEP 15-9
1. Mutual inductance is the process by which an inductor induces a voltage in another inductor, whereas self-inductance is the process by which a coil induces a voltage within itself.
2. False

STEP 15-10
1. True
2. True
3. Primary and secondary

STEP 15-11
1. True
2. False

STEP 15-12
1. Turns ratio = N_s/N_p = 1608/402 = 4; step up
2. $V_s = N_s/N_p \times V_p$
3. False
4. $I_s = N_p/N_s \times I_p$
5. Turns ratio = $\sqrt{Z_L/Z_S} = \sqrt{75\ \Omega/25\ \Omega} = \sqrt{3} = 1.732$
6. $V_s = (N_s/N_p) \times V_p = (200/112) \times 115 = 205.4$ V

STEP 15-13
1. 10 kVA is the apparent power rating, 200 V the maximum primary voltage, 100 V the maximum secondary voltage, at 60 cycles per second (Hz)
2. $R_L = V_s/I_s$ = 1 kV/8 A = 125 Ω; 125 > 100, so the transformer will overheat and possibly burn out

STEP 16-1
1. Calculate the inductive and capacitive reactance (X_L and X_C), the circuit impedance (Z), the circuit current (I), the component voltage drops (V_R, V_L, and V_C), and the power distribution and power factor (PF).
2. **a.** $Z = \sqrt{R^2 + (X_L \sim X_C)^2}$
   b. $I = V_s/Z$
   c. Apparent power = $V_s \times I$ (volt-amperes)
   d. $V_S = \sqrt{V_R^2 + (V_L \sim V_C)^2}$

e. True power = $I^2 \times R$(watts)
f. $V_R = I \times R$
g. $V_L = I \times X_L$
h. $V_C = I \times X_C$
i. $\theta = \arctan \dfrac{V_L \sim V_C}{V_R}$
j. PF = $\cos \theta$

STEP 16-2
1. **a.** $I_R = V/R$
   b. $I_T = \sqrt{I_R^2 + I_X^2}$
   c. $I_C = V/X_C$
   d. $I_L = V/X_L$
2. **a.** $P_R = I^2 \times R$
   b. $P_X = I^2 \times X_L$
   c. $P_A = V_S \times I_T$
   d. PF = $\cos \theta$

STEP 16-3-1
1. A circuit condition that occurs when the inductive reactance (X_L) and the capacitive reactance (X_C) have been balanced
2. A series RLC circuit that at resonance X_L equals X_C, so V_L and V_C will cancel, and $Z = R$
3. Voltage across L and C will measure 0; impedance only equals R; voltage drops across inductor or capacitor can be higher than source voltage
4. Q factor indicates the quality of the series resonant circuit, or is the ratio of the reactance to the resistance
5. Group or band of frequencies that causes the larger current flow
6. BW = f_0/Q = 12 kHz/1000 = 12 Hz

STEP 16-3-2
1. In a series resonant RLC circuit, source current is maximum; in a parallel resonant circuit, source current is minimum at resonance
2. Oscillating effect with continual energy transfer between capacitor and inductor
3. $Q = X_L/R$ = 50 Ω/25 Ω = 2
4. Yes
5. Ability of a tuned circuit to respond to a desired frequency and ignore all others

STEP 16-4
1. **a.** High pass
   b. Low pass
   c. Band stop
   d. Band pass
2. Television, radio, and other communications equipment

STEP 16-5
1. Real, imaginary
2. Magnitude = $\sqrt{5^2 + 6^2}$ = 7.81
   Angle = arctan (6/5) = 50.2°
   Rectangular number 65 + j6 = polar number 7.81 ∠50.2°
3. Real number = 33 cos 25° = 29.9
   Imaginary number = 33 sin 25° = 13.9
   Polar number 33 ∠25° = rectangular number 29.9 + j13.9
4. Combination of a real and imaginary number

STEP 17-1
1. 0.7 V
2. 0.7 V
3. One

STEP 17-2
1. True
2. Reverse
3. The cathode bar is shaped like a "z."

STEP 17-3
1. (a)
2. Series current limiting resistor

STEP 17-4
1. NPN and PNP
2. Emitter, base, and collector
3. As a switch, and as a variable-resistor
4. The two-state switching action is used in digital circuits, while the variable-resistor action is used in analog circuits.

STEP 17-5
1. Current
2. Forward, reverse
3. (b)
4. Open switch
5. Closed switch
6. a. Common-base
   b. Common-collector
   c. Common-emitter
7. Voltage-divider bias
8. Base biasing

STEP 18-1
1. Section 18-1-1
2. Section 18-1-1
3. Section 18-1-2
4. Circuit Simplicity and Accuracy
5. 2
6. Section 18-1-2

STEP 18-2
1. CPU, Memory, Input-Output
2. Analog to Digital Converter
3. Analog, Digital
4. Digital, Analog

STEP 19-1
1. No difference
2. Base 10
3. $2 = 1000, 6 = 100, 3 = 10, 9 = 1$
4. Reset and carry

STEP 19-2
1. 26_{10}
2. 10111_2
3. LSB = least significant bit, MSB = most significant bit
4. 1101110_2

STEP 19-3
1. (a) Base 10   (b) Base 2   (c) Base 16
2. 21_{10} 15_{16}
3. BF7A
4. Binary equivalent = 100001
   Hexadecimal equivalent = 21

STEP 19-4
1. 8
2. 232_8
3. 4
4. 4267_8

STEP 19-5
1. $0111\ 0110\ 0010\ 1001_{BCD}$
2. 11_{10}

STEP 19-5 (continued)
3. Only one digit changes as you step to the next code group.
4. B a b b a g e

STEP 20-1
1. Hardware
2. Software
3. Vacuum tube
4. Diode, transistor

STEP 20-2
1. OR, AND
2. Binary 1
3. AND gate
4. OR
5. Binary 0
6. AND

STEP 20-3
1. NOT gate
2. Binary 0
3. NAND
4. NOR
5. Binary 1
6. NAND

STEP 20-4
1. OR
2. XOR
3. XNOR
4. XNOR

STEP 20-5
1. IEEE/ANSI
2. Instead of using distinctive shapes to represent logic gates, it uses a special dependency notation system to indicate how the output is dependent on the input.

STEP 21-1
1. a. AND   c. OR   e. NOR
   b. XOR   d. NAND   f. NAND
2. $\overline{A \cdot B} = \overline{A} + \overline{B}$
3. a. $\overline{AB}$ b. $(\overline{A} \cdot \overline{B}) \cdot C$
4. a.

   b.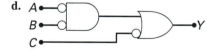

   c.

   d.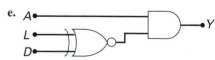

   e.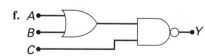

   f.

732   APPENDIX A/ANSWERS TO SELF-TEST EVALUATION POINTS

STEP 21-2
1. **a.** 1   **b.** 1   **c.** 0   **d.** A
2. $A(B + C) = (AB) + (AC)$
3. $(AB)C$ or $(AC)B$
4. $A(B + C + D)$

STEP 21-3
1. Product, sum
2. **a.** $A\overline{B}CD$ **b.** $\overline{A}BC\overline{D}$
3. SOP
4. Fundamental products are ORed and then the circuit is developed.

STEP 21-4
1. 4
2. $y = B(A + \overline{A}) = B(1) = B$
3. Larger
4. With 1s for larger grouping

STEP 22-1
1. **(a)** $1011 + 11101 = 101000$ **(c)** $101 \times 10 = 1010$
   (b) $1010 - 1011 = 11110001$ **(d)** $10111 \div 10 = 1011$

STEP 22-2
1. **(a)** 0101   **(b)** 1001 0101   **(c)** 1101   **(d)** 0111 1001
2. **(a)** -5 **(b)** $+6$ **(c)** -5 **(d)** $+6$

STEP 22-3
1. Adder circuit

STEP 22-4
1. **(a)** 0100 0001 0000 1000
   (b) 0001 0010 1111 0100

STEP 23-1
1. The three blocks are the microprocessor unit (MPU), the memory unit, and the input/output unit. Functions are:

   Microprocessor—logic circuitry recognizes and executes the program stored in memory

   Memory—contains the program of instruction that is fetched and executed by the MPU

   Input/output—allows data to be read in from or sent out to peripheral devices

2. The three buses are address, data, and control. Functions are:

   Address—used by the MPU to select memory locations or input/output ports

   Data—used to transfer data or information between devices

   Control—carries signals used by the processor to coordinate the transfer of data within the microcomputer

3. Hardware—the units and components of the system

   Software—written programs stored as code that are understood by the MPU

STEP 23-2
1. **a.** Machine language—the digital code understood by the MPU

   b. Assembly language—programs written using mnemonics, which are easier to remember

   c. High-level language—an independent language designed to make programming easier that later is compiled into machine language for the MPU

2. A flowchart is a graphic form that describes the operation of a program using oval, rectangular, and diamond-shaped blocks.

3. **a.** Instruction set—contains a list of the machine codes that are recognized by a specific type of microprocessor

   b. Peripherals—input and output devices connected to microprocessor system ports

   c. Programmer—a person who writes a program in any language that will be understood by the MPU

   d. Mnemonic—a short name, or memory aid, used in assembly language programming

   e. Compiler— a program that translates high-level language into machine code

   f. Accumulator—an MPU internal register (register A), used in conjunction with the ALU; a storage location within an MPU chip

   g. Flag register—an internal register within an MPU chip that is a collection of flip-flops used to indicate the results of certain instructions

   h. Op-code—the operation code or "do" portion of an instruction

Answers to Odd-Numbered Problems

Chapter 1
1. c
3. b
5. d
7. c
9. c

(The answers to Communication Skill Questions 11 through 19 can be found in the indicated sections that follow the questions.)

21. **(a)** $3\frac{3}{4}$ **(b)** $2\frac{1}{3}$

23. **(a)** $\dfrac{5}{8} + \dfrac{2}{8} = \dfrac{5+2}{8} = \dfrac{7}{8}$

(b) $\dfrac{47}{76} + \dfrac{15}{76} + \dfrac{1}{76} = \dfrac{63}{76}$

25. **(a)** $\dfrac{5}{2} = ?$, 2 into 5 = 2 with 1 remaining

$\dfrac{5}{2} = 2\dfrac{1}{2}$

(b) $\dfrac{17}{4} = 4\dfrac{1}{4}$

(c) $\dfrac{25}{16} = 1\dfrac{9}{16}$

(d) $\dfrac{37}{3} = 12\dfrac{1}{3}$

27. **(a)** $\dfrac{4 \div 4}{16 \div 4} = \dfrac{1}{4}$

(b) $\dfrac{16 \div 2}{18 \div 2} = \dfrac{8}{9}$

(c) $\dfrac{74 \div 2}{128 \div 2} = \dfrac{37}{64}$

(d) $\dfrac{28}{45} = \dfrac{28}{45}$

29. **(a)** $\dfrac{1}{9} \times 4 = \dfrac{1 \times 4}{9 \times 1} = \dfrac{4}{9}$

(b) $\dfrac{3}{6} \times \dfrac{4}{5} = \dfrac{3 \times 4}{6 \times 5} = \dfrac{12 \div 6}{30 \div 6} = \dfrac{2}{5}$

(c) $\dfrac{1}{3} \times 4\dfrac{1}{2} = \dfrac{1}{3} \times \dfrac{9}{2} = \dfrac{9}{6} = 1\dfrac{3}{6} = 1\dfrac{1}{2}$

(d) $2\dfrac{3}{4} \times 4\dfrac{4}{11} = \dfrac{\cancel{11}}{\cancel{4}} \times \dfrac{\cancel{48}}{\cancel{11}} = \dfrac{1}{1} \times \dfrac{12}{1} = 1 \times 12 = 12$

Chapter 2
1. b
3. a
5. b
7. a
9. c

(The answers to Communication Skill Questions 11 through 19 can be found in the indicated sections that follow the questions.)

21. $9 \times 10{,}000 = 90{,}000$
$6 \times\ \ 1{,}000 = \ \ 6{,}000$
$2 \times\ \ \ \ \ 100 = \ \ \ \ \ 200$
$3 \times\ \ \ \ \ \ \ 10 = \ \ \ \ \ \ \ 30$
$7 \times\ \ \ \ \ \ \ \ \ 1 = \underline{\ \ \ \ \ \ \ \ \ 7}$
$ 96{,}237$

23. **(a)** Units and tens
(b) Units
(c) Units
(d) Units, tens, hundreds, and thousands

25. **(a)** 2.3
(b) 0.507
(c) 9030.04

27. **(a)** $\dfrac{16}{32} \div \dfrac{2}{2} = \dfrac{1}{2} = 1 \div 2 = 0.5$

(b) $3\dfrac{8}{9} = 3 + (8 \div 9) = 3 + 0.889 = 3.889$

(c) $4\dfrac{9}{8} = 4 + \dfrac{8}{8} + \dfrac{1}{8} = 4 + 1 + \dfrac{1}{8} = 5\dfrac{1}{8} = 5 + (1 \div 8)$
$= 5 + 0.125 = 5.125$

(d) $195\dfrac{7}{3} = 195 + \dfrac{3}{3} + \dfrac{3}{3} + \dfrac{1}{3} = 197\dfrac{1}{3} = 197 + (1 \div 3)$
$= 197 + 0.333 = 197.333$

29. **(a)** $0.777 = \dfrac{7}{9}$

(b) $0.6149069 = \dfrac{99}{161}$

(c) $43.125 = 43\dfrac{1}{8}$

31. **(a)** $\dfrac{20 \text{ ft}}{5 \text{ ft}}$ ← Reduce to lowest terms

$\dfrac{20 \div 5}{5 \div 5} = \dfrac{4}{1}$

The ratio of 20 ft to 5 ft is 4 to 1 (4:1).
(b) Both quantities must be alike, so we must first convert minutes to seconds.

$2\dfrac{1}{2} \text{ min} = 2\dfrac{1}{2} \times 60 \text{ s} = 150 \text{ s}$

$\dfrac{150 \text{ s} \div}{30 \text{ s} \div} = \dfrac{5}{1}$

The ratio of $2\dfrac{1}{2}$ min to 30 s is 5 to 1 (5:1).

33. 10.9
35. **(a)** 6
(b) 17.25 or $17\dfrac{1}{4}$ s
(c) 116 V

(d) 505.75, or $505\frac{3}{4}\,\Omega$

(e) 155 m

Chapter 3
1. c   **7.** b
3. a   **9.** a
5. a

(The answers to Communication Skill Questions 11 through 19 can be found in the indicated sections that follow the questions.)

21. **(a)** $+8 - (-5)$
    (b) -0.6×13
    (c) $-22.3 \div (-17)$
    (d) $(+4) \div (-9)$
23. **(a)** $+6 - (+8) = -2$
    (b) $+9 - (-6) = +15$
    (c) $-75 - (+62) = -137$
    (d) $-39 - (-112) = +73$
25. **(a)** $+19 \div (+3) = 6.33$
    (b) $+36 \div (-3) = -12$
    (c) $-80 \div (+5) = -16$
    (d) $-44 \div (-2) = 22$
27. $+15 \div (+5) \times (-3.5) = -10.5$
29. $-6 \div (-4) \div (-3) \times (-15) = 7.5$

Chapter 4
1. c   **7.** a
3. a   **9.** b
5. d

(The answers to Communication Skill Questions 11 through 19 can be found in the indicated sections that follow the questions.)

21. **(a)** $9^2 = 9 \times 9 = 81$
    (b) $6^2 = 6 \times 6 = 36$
    (c) $2^2 = 2 \times 2 = 4$
    (d) $0^2 = 0 \times 0 = 0$
    (e) $1^2 = 1 \times 1 = 1$
    (f) $12^2 = 12 \times 12 = 144$
23. **(a)** $9^3 = 9 \times 9 \times 9 = 729$
    (b) $10^4 = 10 \times 10 \times 10 \times 10 = 10,000$
    (c) $4^6 = 4 \times 4 \times 4 \times 4 \times 4 \times 4 = 4096$
    (d) $2.5^3 = 2.5 \times 2.5 \times 2.5 = 15.625$
25. **(a)** $\dfrac{1}{100} = 10^{-2}$
    (b) $1,000,000,000 = 10^9$
    (c) $\dfrac{1}{1000} = 10^{-3}$
    (d) $1000 = 10^3$

27.

	Scientific Notation Base of 1 to 10, plus power of ten	Engineering Notation Base greater than 1, plus $\times$ 3 power of ten
(a) 475	$= 4.75 \times 10^2$	475.0
(b) 8200	$= 8.2 \times 10^3$	8.2×10^3
(c) 0.07	$= 7 \times 10^{-2}$	70×10^{-3}
(d) 0.00045	$= 4.5 \times 10^{-4}$	450×10^{-6}

29. **(a)** $\text{km} \times 0.6 = \text{mi}, \ 100 \text{ km} \times 0.6 = 60 \text{ mi}$
    (b) $\text{m}^2 \times 1.2 = \text{yd}^2, \ 29 \text{ m}^2 \times 1.2 = 34.8 \text{ yd}^2$
    (c) $\text{kg} \times 2.2 = \text{lb}, \ 67 \text{ kg} \times 2.2 = 147.4 \text{ lb}$
    (d) $\text{L} \times 2.1 = \text{pt}, \ 2 \text{ L} \times 2.1 = 4.2 \text{ pt}$

31. **(a)** $\dfrac{5}{9} \times (°\text{F} - 32) = °\text{C}$

$$= \frac{5}{9} \times (32°\text{F} - 32) = 0.5555 \times 0 = 0°\text{C}$$

$$32°\text{F} = 0°\text{C}$$

(b) $\dfrac{5}{9} \times (72°\text{F} - 32)$

$$= \frac{5}{9} \times 40 = 22.22°\text{C}$$

33. **(a)** meter
    (b) square meter
    (c) gram
    (d) cubic meter
    (e) liter
    (f) degrees Celsius
35. **(a)** $8000 \text{ ms} = \underline{\hspace{1cm}} \mu\text{s}$
    $= 8000 \times 10^{-3} = \underline{\hspace{1cm}} 10^{-6}$
    $= 8000.000. \times 10^{-6} = 8,000,000 \times 10^{-6} \text{ s}$
    $= 8000 \text{ ms} = 8,000,000 \ \mu\text{s}$
    (b) $0.02 \text{ MV} = \underline{\hspace{1cm}} \text{kV}$
    $0.02 \times 10^6 \text{ V} = \underline{\hspace{1cm}} \times 10^3 \text{ V}$
    $0.020. \times 10^3 \text{ V} = 20 \times 10^3 \text{ V}$
    $0.02 \text{ MV} = 20 \text{ kV}$
    (c) $10 \text{ km} = \underline{\hspace{1cm}}$
    $10 \times 10^3 \text{ m} = \underline{\hspace{1cm}} \times 100 \text{ m}$
    $10.000. \times 100 \text{ m} = 10,000 \text{ m}$
    $10 \text{ km} = 10,000 \text{ m}$
    (d) $250 \text{ mm} = \underline{\hspace{1cm}} \text{cm}$
    $250 \times 10^{-3} \text{ m} = \underline{\hspace{1cm}} \times 10^{-2} \text{ m}$
    $25.0. \times 10^{-3} \text{ m} = 25 \times 10^{-2} \text{ m}$
    $250 \text{ mm} = 25 \text{ cm}$

Chapter 5
1. a   **7.** c
3. d   **9.** b
5. c

(The answers to Communication Skill Questions 11 through 25 can be found in the indicated sections that follow the questions.)

27. **(a)**
$$4x = 11$$
$$4 \times x = 11$$
$$\frac{\cancel{4} \times x}{\cancel{4}} = \frac{11}{4} \qquad (\div 4)$$
$$x = \frac{11}{4}$$
$$x = 2.75$$

(b) $6a + 4a = 70$
$$10a = 70$$
$$10 \times a = 70$$
$$\frac{\cancel{1}0 \times a}{\cancel{1}0} = \frac{70}{10} \qquad (\div 10)$$
$$a = \frac{70}{10}$$
$$a = 7$$

(c) $5b - 4b = \dfrac{7.5}{1.25}$
$$1b = \frac{7.5}{1.25} \qquad (1b = 1 \times b = b)$$
$$b = \frac{7.5}{1.25}$$
$$b = 6$$

(d) $\dfrac{2z \times 3z}{4.5} = 2z$

$$\frac{6z^2}{4.5} = 2z$$

$$\begin{bmatrix} 2z \times 3z = 2 \times z \times 3 \times z \\ = (2 \times 3) \times (z \times z) \\ = 6 \times z^2 \\ = 6z^2 \end{bmatrix}$$

$$\frac{6z^2}{4.5} \times 4.5 = 2z \times 4.5 \qquad (\times 4.5)$$

$$6z^2 = 2z \times 4.5$$

$$\frac{\overset{3}{6} \times \cancel{z} \times z}{\cancel{2} \times \cancel{z}} = \frac{(2 \times z) \times 4.5}{2 \times z} \qquad (\div 2z)$$

$$3z = 4.5$$

$$\frac{\cancel{3} \times z}{\cancel{3}} = \frac{4.5}{3} \qquad (\div 3)$$

$$z = \frac{4.5}{3}$$

$$z = 1.5$$

29. (a) Power (P) = voltage $(V) \times$ current (I)
1500 watts = 120 volts $\times$?
Transpose formula to solve for I.
$$P = V \times I$$
$$\frac{P}{V} = \frac{\cancel{V} \times I}{\cancel{V}} \qquad (\div V)$$
$$\frac{P}{V} = I$$
$$I = \frac{P}{V}$$

Current (I) in amperes $= \dfrac{\text{power } (P) \text{ in watts}}{\text{voltage } (V) \text{ in volts}}$

$$I = \frac{1500 \text{ W}}{120 \text{ V}}$$

Current $(I) = 12.5$ amperes

(b) Voltage (V) in volts = current (I) in amperes $\times$ resistance (R) in ohms
120 volts = 12.5 amperes $\times$?
Transpose formula to solve for R.
$$V = I \times R$$
$$\frac{V}{I} = \frac{\cancel{I} \times R}{\cancel{I}}$$
$$\frac{V}{I} = R$$
$$R = \frac{V}{I}$$

Resistance (R) in ohms $= \dfrac{\text{voltage } (V) \text{ in volts}}{\text{current } (I) \text{ in amperes}}$

$$R = \frac{120 \text{ V}}{12.5 \text{ A}}$$

Resistance $(R) = 9.6$ ohms

Chapter 6
1. b   9. c
3. c   11. d
5. d   13. a
7. b   15. d

(The answers to Communication Skill Questions 17 through 25 can be found in the indicated sections that follow the questions.)

27. (a) $C = \sqrt{A^2 + B^2}$
$$= \sqrt{20^2 + 53^2}$$
$$= \sqrt{400 + 2809}$$
$$= \sqrt{3209}$$
$$= 56.65 \text{ mi}$$
(b) $C = \sqrt{2^2 + 3^2}$
$$= 3.6 \text{ km}$$

(c) $C = \sqrt{4^2 + 3^2}$
$$= 5 \text{ inches}$$
(d) $C = \sqrt{12^2 + 12^2}$
$$= 16.97 \text{ mm}$$

29. (a) $\sin 0° = 0$ **(i)** $\cos 60° = 0.5$
(b) $\sin 30° = 0.5$ **(j)** $\cos 90° = 0$
(c) $\sin 45° = 0.707$ **(k)** $\tan 0° = 0$
(d) $\sin 60° = 0.866$ **(l)** $\tan 30° = 0.577$
(e) $\sin 90° = 1.0$ **(m)** $\tan 45° = 1.0$
(f) $\cos 0° = 1.0$ **(n)** $\tan 60° = 1.73$
(g) $\cos 30° = 0.866$ **(o)** $\tan 90° = \infty$ (infinity)
(h) $\cos 45° = 0.707$

31. (a)

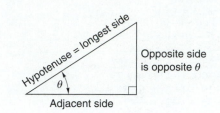

Known values	Unknown values
$H = 27$ miles	$O = ?$
$\theta = 37°$	$A = ?$

We must calculate the length of the opposite and adjacent sides. To achieve this we can use either H and θ to calculate O (SOH), or H and θ to calculate A (CAH).

$$\sin \theta = \frac{O}{H}$$
$$\sin 37° = \frac{O}{27}$$
$$0.6 = \frac{O}{27}$$
$$0.6 \times 27 = \frac{O}{\cancel{27}} \times \cancel{27}$$
$$O = 0.6 \times 27$$
$$\text{Opposite} = 16.2 \text{ mi}$$

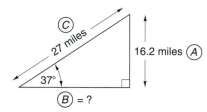

The next step is to use the Pythagorean theorem to calculate the length of the unknown side now that the length of two sides are known.

$$B = \sqrt{C^2 - A^2}$$
$$= \sqrt{27^2 - 16.2^2}$$
$$= \sqrt{729 - 262.44}$$
$$= \sqrt{466.56}$$
$$\text{Adjacent or } B = 21.6 \text{ mi}$$

(b)

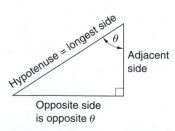

Known values	Unknown values
$O = 29$ cm	$H = ?$
$\theta = 21°$	$A = ?$

We can use either O and θ to calculate H (*SOH*), or O and θ to calculate A (*TOA*).

$$\tan \theta = \frac{O}{A}$$

$$\tan 21° = \frac{29 \text{ cm}}{A}$$

$$0.384 = \frac{29}{A}$$

$$0.384 \times A = \frac{29}{A} \times A \, (\times A)$$

$$\frac{0.384 \times A}{0.384} = \frac{29}{0.384} \, (\div 0.384)$$

$$A = \frac{29}{0.384}$$

$$A = 75.5 \text{ cm}$$

Now that O and A are known, we can calculate H.

$$C \text{ (or } H) = \sqrt{A^2 + B^2}$$
$$= \sqrt{75.5^2 - 29^2}$$
$$= \sqrt{5700.25 + 841}$$
$$= \sqrt{6541.25}$$
$$= 80.88 \text{ cm}$$

(c)

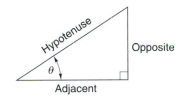

Known values	Unknown values
$H = 34$ volts	$\theta = ?$
$A = 18.5$ volts	$O = ?$

Because both H and A are known, we can use cosine to calculate θ (*CAH*)

$$\cos \theta = \frac{A}{H}$$

$$\cos \theta = \frac{18.5 \text{ V}}{34 \text{ V}}$$

$$\cos \theta = 0.544$$

$$\text{invcos} \times \cos \theta = \text{invcos } 0.544 \quad (\times \text{invcos})$$

$$\theta = \text{invcos } 0.544$$

$$\theta = 57°$$

To calculate the length of the unknown side, we can use the Pythagorean theorem.

$$B = \sqrt{C^2 - A^2}$$
$$= \sqrt{34^2 - 18.5^2}$$
$$= \sqrt{813.75}$$
$$= 28.5 \text{ V}$$

33. $S = \pi d^2 = \pi \times 2^2 = 12.57 \text{ in}^2$, $V = \frac{1}{6} \pi \times d^3 = 4.189 \text{ in}^3$

35. $L = \frac{1}{2} ps$

$$= \frac{1}{2} \times 24 \times 8 = 96 \text{ in}^2$$

Chapter 7
1. b
3. c
5. a
7. b
9. c

(The answers to Communication Skill Questions 11 through 19 can be found in the indicated sections that follow the questions.)

21. (a) 1   (g) -0.3
     (b) 4   (h) -0.125
     (c) 6.09   (i) -0.9
     (d) 1.7   (j) -0.02
     (e) 0.176   (k) 1.088
     (f) -0.6 (l) 1.88
23. *Step 1:* Find the log of each number.
     Step 2: Change each number to base 10.
     Step 3: Add the base 10 exponents.
     Step 4: Convert the log answer back to a number using antilog.
25. *Step 1:* Find the log of the dividend and divisor.
     Step 2: Change each number to base 10.
     Step 3: Subtract the log of the divisor from the log of the dividend.
     Step 4: Convert the log answer back to a number using antilog.
27. (a) Line graph
     (b) Bar graph
29. (a) Children are able to access educational material.
     (b) 1. Children are able to access educational material.
         2. Video conferencing
         3. Accessing information on hobbies or interests
         4. Accessing information for work
         5. Sending notes to friends and family

Chapter 8
1. a
3. a
5. d
7. b

(The answers to Communication Skill Questions 9 through 13 can be found in the sections indicated that follow the questions.)

15. $n = Q \times (6.24 \times 10^{18}) = 6.5 \times (6.24 \times 10^{18}) = 4.056 \times 10^{19}$ electrons
17. a. 14 mA   c. 776 nA
     b. 1.374 kA   d. 910 μA
19. $0.945/1 = 0.945$
21. 10 V/750 kV/cm = 0.0000133 cm or 0.000133 mm
23. $35 \times (2000 \text{ kV/cm} \div 10) = 7000 \text{ kV}$

Chapter 9
1. a	11. b
3. a	13. b; $P = I^2R$
5. d	$= (20 \text{ mA})^2 \times 2 \text{ k}\Omega$
7. a	$= 0.8$ W
9. a	15. c
6. b	17. c
7. a	19. a

(The answers to Communication Skill Questions 21 through 35 can be found in the sections indicated that follow the questions.)

37. $V = I \times R = 8 \text{ mA} \times 16 \text{ k}\Omega = 128 \text{ V}$

39. $P = I \times V, I = P/V$

  a. $300/120 = 2.5 \text{ A}$ **c.** $60/120 = 0.5 \text{ A}$

  b. $100/120 = 833 \text{ mA}$ **d.** $25/120 = 208.3 \text{ mA}$

41. $I = V/R = 12 \text{ V}/12 \text{ mA} = 1 \text{ k}\Omega$

43. $R = V/I = 120 \text{ V}/500 \text{ mA} = 240 \ \Omega$

    $P = I \times V = 500 \text{ mA} \times 120 \text{ V} = 60 \text{ W}$

45. **a.** 1 kW **c.** 1.25 MW

     b. 345 mW **d.** $1250 \ \mu\text{W}$

47. **a.** $7500 = 7.5 \text{ kW}$

       Energy Consumed $= 7.5 \text{ kW} \times 1 \text{ hr} = 7.5 \text{ kWh}$

     b. $25 \text{ W} = 0.025 \text{ kW}$

       Energy Consumed $= 0.025 \text{ kW} \times 6 \text{ hrs} = 0.15 \text{ kWh}$

     c. $127{,}000 \text{ W} = 127 \text{ kW}$

       Energy Consumed $= 127 \text{ kW} \times 0.5 \text{ hr} = 63.5 \text{ kWh}$

49. $5.6 \text{ k}\Omega \times 0.1 = 560 \ \Omega, 5.6 \text{ k}\Omega \pm 560 \text{ k}\Omega = 5.04 \text{ k}\Omega$ to $6.16 \text{ k}\Omega$

Chapter 10

1. d   **9.** a

3. c   **11.** c

5. a   **13.** d

7. d   **15.** d

(The answers to Communication Skill Questions 16 through 25 can be found in the indicated sections that follow the questions.)

27. $I = \dfrac{V_S}{R_T}, R_T = R_1 + R_2 = 40 + 35 = 75 \ \Omega$

    $I = 24/75 = 320 \text{ mA}; 150 \ \Omega$ (double $75 \ \Omega$) needed to halve current

29. $40 \ \Omega, 20 \ \Omega, 60 \ \Omega.$ (Any values can be used, as long as the ratio remains the same.)

31. $I_{R_1} = I_T = 6.5 \text{ mA}$

33. $P_T = P_1 + P_2 + P_3 = 120 + 60 + 200 = 380 \text{ W}; I_T = P_T/V_S$

    $= 380/120 = 3.17 \text{ A}$

    $V_1 = P_1/I_T = 120 \text{ W}/3.17 \text{ A} = 38 \text{ V}; V_2 = P_2/I_T =$

    $60 \text{ W}/3.17 \text{ A} = 18.9 \text{ V}$

    $V_3 = P_3/I_T = 200 \text{ W}/3.17 \text{ A} = 63.1 \text{ V}$

35. **a.** $R_T = R_1 + R_2 + R_3 = 22 \text{ k}\Omega + 3.7 \text{ k}\Omega + 18 \text{ k}\Omega = 43.7 \text{ k}\Omega$

       $I = V/R = 12 \text{ V}/43.7 \text{ k}\Omega = 274.6 \ \mu\text{A}$

     b. $R_T = V/I = 12 \text{ V}/10 \text{ mA} = 1.2 \text{ k}\Omega$

       $P_T = V \times I = 12 \text{ V} \times 10 \text{ mA} = 120 \text{ mW}$

     c. $R_T = R_1 + R_2 + R_3 + R_4 = 5 + 10 + 6 + 4 = 25 \ \Omega; V_S =$

       $I \times R_T = 100 \text{ mA} \times 25 \ \Omega = 2.5 \text{ V}$

       $V_{R_1} = I \times R_1 = 100 \text{ mA} \times 5 \ \Omega = 500 \text{ mV}, V_{R_2} = I \times R_2 =$

       $100 \text{ mA} \times 10 \ \Omega = 1 \text{ V}, V_{R_3} = I \times R_3 = 100 \text{ mA}$

       $\times 6 \ \Omega = 600 \text{ mV}; V_{R_4} = I \times R_4 = 100 \text{ mA} \times 4 \ \Omega = 400 \text{ mV}$

       $P_1 = I \times V_1 = 100 \text{ mA} \times 600 \text{ mV} = 60 \text{ mW},$

       $P_2 = I \times V_2 = 100 \text{ mA} \times 1 \text{ V} = 100 \text{ mW}, P_3 = I \times V_3 =$

       $100 \text{ mA} \times 600 \text{ mV} = 60 \text{ mW}, P_4 = I \times V_4 =$

       $100 \text{ mA} \times 400 \text{ mV} = 40 \text{ mW}$

     d. $P_T = P_1 + P_2 + P_3 + P_4 = 12 \text{ mW} + 7 \text{ mW} + 16 \text{ mW} +$

       $3 \text{ mW} = 38 \text{ mW}$

       $I = P_T/V_S = 38 \text{ mW}/12.5 \text{ V} = 3.04 \text{ mA}$

       $R_1 = P_1/I^2 = 12 \text{ mW}/(3.04 \text{ mA})^2 = 1.3 \text{ k}\Omega$

       $R_2 = P_2/I^2 = 7 \text{ mW}/(3.04 \text{ mA})^2 = 757.4 \ \Omega$

       $R_3 = P_3/I^2 = 16 \text{ mW}/(3.04 \text{ mA})^2 = 1.73 \text{ k}\Omega:$

       $R_4 = P_4/I^2 = 3 \text{ mW}/(3.04 \text{ mA})^2 = 324.6 \ \Omega$

Chapter 11

1. b

3. d

5. c

7. b

9. a

(The answers to Communication Skill Questions 11 through 20 can be found in the sections indicated that follow the questions.)

21. $R_T = R/\text{no. of } R\text{'s} = 30 \text{ k}\Omega/4 = 7.5 \text{ k}\Omega$

23. $R_T = R/\text{no. of } Rs = 25 \ \Omega/3 = 8.33 \ \Omega, I_T = V_S/R_T = 10/8.33$

    $= 1.2 \text{ A}$

    $I_1 = I_2 = I_3 = R_T/R_x \times I_T = 8.33/25 \times 1.2 = I_T/\text{no. of } Rs =$

    $1.2 \text{ A}/3 = 400 \text{ mA}$

25. $I_T = V_S/R_T = 14/700 = 20 \text{ mA}; I_X = I_T/\text{no. of } Rs = 20 \text{ mA}/3$

    $= 6.67 \text{ mA}$

27. **a.** $R_T = \dfrac{R_1 \times R_2}{R_1 + R_2} = \dfrac{33 \times 22}{33 + 22} = \dfrac{726 \text{ k}\Omega^2}{55 \text{ k}\Omega} = 13.2 \text{ k}\Omega$

     b. $I_T = V_S/R_T = 20 \text{ V}/13.2 \text{ k}\Omega = 1.5 \text{ mA}$

     c. $I_1 = R_T/R_1 \times I_T = (13.2 \text{ k}\Omega/33 \text{ k}\Omega) \times 1.5 \text{ mA} = 600 \ \mu\text{A}$

       $I_2 = R_T/R_2 \times I_T = (13.2 \text{ k}\Omega/22 \text{ k}\Omega) \times 1.5 \text{ mA} = 900 \ \mu\text{A}$

     d. $P_T = I_T \times V_S = 1.5 \text{ mA} \times 20 \text{ V} = 30 \text{ mW}$

     e. $P_1 = I_1 \times V_1, (V_1 = V_S = 20 \text{ V}), 600 \ \mu\text{A} \times 20 \text{ V} = 12 \text{ mW}$

       $P_2 = I_2 \times V_2, (V_2 = V_S = 20 \text{ V}), 900 \ \mu\text{A} \times 20 \text{ V} = 18 \text{ mW}$

29. **a.** $R_T = \dfrac{R_1 \times R_2}{R_1 + R_2} = \dfrac{22 \text{ k}\Omega \times 33 \text{ k}\Omega}{22 \text{ k}\Omega + 33 \text{ k}\Omega} = \dfrac{726 \text{ k}\Omega^2}{55 \text{ k}\Omega} = 13.2 \text{ k}\Omega$

       $I_T = \dfrac{V_S}{R_T} = \dfrac{10 \text{ V}}{13.2 \text{ k}\Omega} = 757.6 \ \mu\text{A}$

       $I_1 = \dfrac{R_T}{R_1} \times I_T = \dfrac{13.2 \text{ k}\Omega}{22 \text{ k}\Omega} \times 757.6 \ \mu\text{A} = 454.56 \ \mu\text{A}$

       $I_2 = \dfrac{R_T}{R_2} \times I_T = \dfrac{13.2 \text{ k}\Omega}{33 \text{ k}\Omega} \times 757.6 \ \mu\text{A} = 303.04 \ \mu\text{A}$

     b. $R_T = \dfrac{1}{(1/R_1) + (1/R_2) + (1/R_3)}$

       $= \dfrac{1}{(1/220 \ \Omega) + (1/330 \ \Omega) + (1/470 \ \Omega)} = 103 \ \Omega$

       $I_T = V_S/R_T = 10/103 = 97 \text{ mA}, I_1 = R_T/R_1 \times I_T =$

       $103/220 \times 97 \text{ mA} = 45.4 \text{ mA}$

       $I_2 = (R_T/R_2) \times I_T = (103/330) \times 97 \text{ mA} = 30.3 \text{ mA}$

       $I_3 = (R_T/R_3) \times I_T = (103/470) \times 97 \text{ mA} = 21.3 \text{ mA}$

31. **a.** $G_T = \dfrac{1}{R_1} + \dfrac{1}{R_2} + \dfrac{1}{R_3} = \dfrac{1}{5} + \dfrac{1}{5} + \dfrac{1}{5} = 0.6 \text{ S},$

       $R_T = \dfrac{1}{G} = \dfrac{1}{0.6} = 1.67 \ \Omega$

     b. $G_T = \dfrac{1}{R_1} + \dfrac{1}{R_2} = \dfrac{1}{200} + \dfrac{1}{200} = 10 \text{ mS},$

       $R_T = \dfrac{1}{G} = \dfrac{1}{10 \text{ mS}} = 100 \ \Omega$

     c. $G_T = \dfrac{1}{R_1} + \dfrac{1}{R_2} + \dfrac{1}{R_3} = \dfrac{1}{1 \text{ M}\Omega} + \dfrac{1}{500 \text{ M}\Omega} + \dfrac{1}{3.3 \text{ M}\Omega}$

       $= 1.305 \ \mu\text{S}, R_T = \dfrac{1}{G} = \dfrac{1}{1.305 \ \mu\text{S}} = 766.3 \text{ k}\Omega$

     d. $G_T = \dfrac{1}{R_1} + \dfrac{1}{R_2} + \dfrac{1}{R_3} = \dfrac{1}{5} + \dfrac{1}{3} + \dfrac{1}{2} = 1.033 \text{ S},$

       $R_T = \dfrac{1}{G} = \dfrac{1}{1.033} = 967.7 \text{ m}\Omega$

33. a. $R_T = \dfrac{R_1 \times R_2}{R_1 + R_2} = \dfrac{15 \times 7}{15 + 7} = \dfrac{105}{22} = 4.77 \ \Omega$

b. $R_T = \dfrac{1}{(1/R_1) + (1/R_2) + (1/R_3)}$

$= \dfrac{1}{(1/26 \ \Omega) + (1/15 \ \Omega) + (1/30 \ \Omega)} = 7.22 \ \Omega$

c. $R_T = \dfrac{R_1 \times R_2}{R_1 + R_2} = \dfrac{5.6 \ k\Omega \times 2.2 \ k\Omega}{5.6 \ k\Omega + 2.2 \ k\Omega} = \dfrac{12.32 \ M\Omega^2}{7.8 \ k\Omega}$

$= 1.58 \ k\Omega$

d. $R_T = \dfrac{1}{(1/R_1) + (1/R_2) + (1/R_3) + (1/R_4) + (1/R_5)} =$

$\dfrac{1}{(1/1 \ M\Omega) + (1/3 \ M\Omega) + (1/4.7 \ M\Omega) + (1/10 \ M\Omega) + (1/33 \ M\Omega)}$

$= 596.5 \ k\Omega$

35. a. $I_2 = I_T - I_1 - I_3 = 6 \ mA - 2 \ mA - 3.7 \ mA = 300 \ \mu A$

b. $I_T = I_1 + I_2 + I_3 = 6 \ A + 4 \ A + 3 \ A = 13 \ A$

c. $R_T = \dfrac{R_1 \times R_2}{R_1 + R_2} = \dfrac{5.6 \ M \times 3.3 \ M}{5.6 \ M + 3.3 \ M} = \dfrac{18.48 \ (M\Omega)^2}{8.9 \ M\Omega}$

$= 2.08 \ M\Omega$

$V_S = I_T \times R_T = 100 \ mA \times 2.08 \ M\Omega = 208 \ kV$

$I_1 = \dfrac{R_T}{R_1} \times I_T = \dfrac{2.08 \ M\Omega}{5.6 \ M\Omega} \times 100 \ mA = 37 \ mA$

$I_2 = \dfrac{R_T}{R_2} \times I_T = \dfrac{2.08 \ m\Omega}{3.3 \ m\Omega} \times 100 \ mA = 63 \ mA$

d. $I_1 = V_{R_1}/R_1 = 2 \ V/200 \ k\Omega = 10 \ \mu A, \ I_2 = I_T - I_1 = 100 \ mA - 10 \ \mu A = 99.99 \ mA$

$R_2 = \dfrac{V_{R_2}}{I_2} = \dfrac{2}{99.99 \ mA} = 20.002 \ \Omega,$

$P_T = I_T \times V_S = 100 \ mA \times 2 \ V = 200 \ mW$

Chapter 12

1. c	**11.** b
3. b	**13.** a
5. c	**15.** a
7. a	**17.** c
9. c	**19.** c

(The answers to Communication Skill Questions 21 through 32 can be found in the sections indicated that follow the questions.)

33. a. $R_{1,2} = R_1 + R_2 = 2.5 \ k\Omega + 10 \ k\Omega = 12.5 \ k\Omega;$
$R_{3,4} = R_3 + R_4 = 7.5 \ k\Omega + 2.5 \ k\Omega = 10 \ k\Omega$

$R_{3,4,5} = \dfrac{R_{3,4} \times R_5}{R_{3,4} + R_5} = \dfrac{10 \ k\Omega \times 2.5 \ M\Omega}{10 \ k\Omega + 2.5 \ M\Omega} = \dfrac{25 \ G\Omega^2}{2.51 \ M\Omega}$

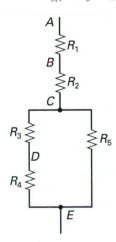

FIGURE 12-33

$= 9.96 \ k\Omega$

$R_{1,2,3,4,5} = R_{1,2} + R_{3,4,5} = 12.5 \ k\Omega + 9.96 \ k\Omega = 22.46 \ k\Omega$

b. $I_T = V_S/R_T = 100 \ V/22.46 \ k\Omega = 4.45 \ mA$

c. $V_{R_1} = I_T \times R_1 = 4.45 \ mA \times 2.5 \ k\Omega = 11.125 \ V$

$V_{R_2} = I_T \times R_2 = 4.45 \ mA \times 10 \ k\Omega = 44.5 \ V$
$V_{R_{3,4,5}} = I_T \times R_{3,4,5} = 4.45 \ mA \times 9.96 \ k\Omega = 44.3 \ V$

d. $I_{R_1} = I_T = 4.45 \ mA, \ I_{R_2} = I_T = 4.45 \ mA$

$I_{R_3} = I_{R_4} = V_{R_{3,4,5}}/R_{3,4} = \dfrac{44.3 \ V}{10 \ k\Omega} = 4.43 \ mA$

$I_{R_5} = V_{R_{3,4,5}}/R_5 = 44.3 \ V/2.5 \ M\Omega = 17.73 \ \mu A$

e. $P_T = I_T \times V_S = 4.45 \ mA \times 100 \ V = 445 \ mW$
$P_{R_1} = I_{R_1} \times V_{R_1} = 4.45 \ mA \times 11.125 \ V = 49.5 \ mW$
$P_{R_2} = I_{R_2} \times V_{R_2} = 4.45 \ mA \times 44.5 \ V = 198.025 \ mW$
$P_{R_3} = I_3^2 \times R_3 = (4.43 \ mA)^2 \times 7.5 \ k\Omega = 147.2 \ mW$
$P_{R_4} = I_4^2 \times R_4 = (4.43 \ mA)^2 \times 2.5 \ k\Omega = 49.1 \ mW$
$P_{R_5} = V_{R_{3,4,5}} \times I_{R_5} = 44.3 \ V \times 17.73 \ \mu A = 785 \ \mu W$

35. $R_{3,4} = \dfrac{R_3 \times R_4}{R_3 + R_4} = \dfrac{200 \times 300}{200 + 300} = \dfrac{60 \ k\Omega}{500} = 120 \ \Omega$

$R_{2,3,4} = R_2 + R_{3,4} = 100 + 120 = 220 \ \Omega$

$R_{1,2,3,4} = \dfrac{R_1 \times R_{2,3,4}}{R_1 + R_{2,3,4}} = \dfrac{100 \times 220}{100 + 220} = \dfrac{22 \ k\Omega}{320} = 68.75 \ \Omega = R_T$

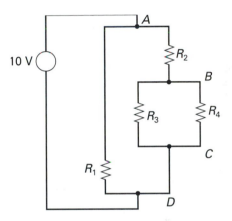

FIGURE 12-35

$I_T = V_S/R_T = 10/68.75 = 145.45 \ mA$

$V_{R_2} = I_{R_2} \times R_2, \ I_{R_2} = I_{R_{2,3,4}} = \dfrac{R_T}{R_{2,3,4}} \times I_T$

$= \dfrac{68.76}{220} \times 145.45 \ mA = 45.45 \ mA = I_2$

$V_{R_2} = 45.45 \ mA \times 100 \ \Omega = 4.545 \ V, \ V_{R_{3,4}} = I_{R_{3,4}} \times R_{3,4}$

$= 45.45 \ mA \times 120 \ \Omega = 5.45 \ V$
$V_{R_1} = V_S = 10 \ V, \ I_{R_1} = V_S/R_1 = 10 \ V/100 \ \Omega = 100 \ mA$
$I_{R_2} = 45.45 \ mA, \ I_{R_3} = (R_{3,4}/R_3) \times I_{R_{3,4}}$
$= (120/200) \times 45.45 \ mA = 27.27 \ mA$
$I_{R_4} = I_{R_{3,4}} - I_{R_3} = 45.45 \ mA - 27.27 \ mA = 18.18 \ mA$
$P_T = V_S \times I_T = 10 \ V \times 145.45 \ mA = 1.4545 \ W$
$P_{R_1} = I_{R_1} \times V_{R_1} = 100 \ mA \times 10 \ V = 1 \ W$
$P_{R_2} = I_{R_2} \times V_{R_2} = 45.45 \ mA \times 4.545 \ V = 207 \ mW$
$P_{R_3} = I_{R_3} \times V_{R_3} = 27.27 \ mA \times 5.45 \ V = 149 \ mW$
$P_{R_4} = I_{R_4} \times V_{R_4} = 18.18 \ mA \times 5.45 \ V = 99 \ mW$

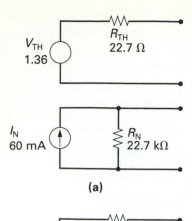

(a)

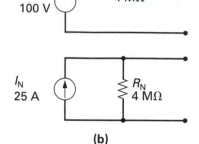

(b)

FIGURE 12-41

$V_A = V_S = 10\text{ V}, V_B = V_A - V_{R_2} = 10 - 4.545 = 5.455\text{ V}$

$V_C = V_D = 0\text{ V}$

37. **a.** $I_{RL} = \dfrac{V_S}{R_T} = \dfrac{V_S}{R_L + R_{\text{int}}} = \dfrac{25\text{ V}}{25\ \Omega + 2.5\text{ M}\Omega} = 9.9999\ \mu\text{A}$

b. $I_{RL} = \dfrac{V_S}{R_T} = \dfrac{V_S}{R_L + R_{\text{int}}} = \dfrac{25\text{ V}}{25\text{ k}\Omega + 2.5\text{ M}\Omega} = 9.99\ \mu\text{A}$

c. $I_{RL} = \dfrac{V_S}{R_T} = \dfrac{V_S}{R_L + R_{\text{int}}} = \dfrac{25\text{ V}}{25\text{ M}\Omega + 2.5\text{ M}\Omega} = 5\ \mu\text{A}$

39. **a.** $I = V_S/R_{\text{int}} = 10/15 = 667\text{ mA}$

b. $I = V_S/R_{\text{int}} = 36/18 = 2\text{ A}$

c. $I = V_S/R_{\text{int}} = 110/7 = 15.7\text{ A}$

41. **a.** $V_{\text{TH}} = (R_3/R_T) \times V_S, R_T = R_1 + R_2 + R_3 = 100 + 150 + 25$
$= 275\ \Omega$

$V_{\text{TH}} = \dfrac{25}{275} \times 15\text{ V} = 1.36\text{ V}, R_{\text{TH}} = \dfrac{R_3 \times R_{1,2}}{R_3 + R_{1,2}},$

$R_{1,2} = R_1 + R_2 = 100 + 150 = 250\ \Omega$

$R_{\text{TH}} = \dfrac{25 \times 250}{25 + 250} = \dfrac{6250}{275} = 22.7\ \Omega$

$I_N = V_S/(R_1 + R_2) = 15/250 = 60\text{ mA}$
$R_N = R_{\text{TH}} = 22.7\ \Omega$

b. $V_{\text{TH}} = 100\text{ V}, R_{\text{TH}} = 4\text{ M}\Omega$
$I_N = V_S/R_3 = 100\text{ V}/4\text{ M}\Omega = 25\ \mu\text{A}$
$R_N = R_{\text{TH}} = 4\text{ M}\Omega$

Chapter 13

1. c	**15.** b
3. d	**17.** c
5. b	**19.** d
7. b	**21.** a
9. a	**23.** c
11. d	**25.** a
13. b	

(The answers to Communication Skill Questions 26 through 35 can be found in the sections indicated that follow the questions.)

37. Frequency = 1/time:
    a. 1/16 ms = 62.5 Hz;   **d.** 1/0.05 s = 20 Hz;
    b. 1/1 s = 1 Hz;   **e.** 1/200 μs = 5 kHz;
    c. 1/15 μs = 66.67 kHz;   **f.** 1/350 ms = 2.86 Hz

39. **a.** Peak = 1.414 × rms = 1.414 × 40 mA = 56.56 mA
    b. Peak to peak = 2 × peak = 2 × 56.56 mA = 113.12 mA
    c. Average = 0.637 × peak = 0.637 × 56.56 mA = 36 mA

41. Duty cycle % = $(P_w/t) \times 100, t = 1/f = 1/10\text{ kHz} = 100\ \mu\text{s}$;
duty cycle = (10 μs/100 μs) × 100 = 10%

43. I_{avg} = baseline + (duty cycle × I_p), duty cycle = P_w/t × 100,
$t = 1/f = 1/10\text{ kHz} = 100\ \mu\text{s}$, duty cycle = 10 μs/100 μs ×
100 = 10%, I_{avg} = 0 + (0.1 × 15 A) = 1.5 A

45. **a.** Third harmonic = 3 × fundamental = 3 × 1 kHz = 3 kHz
    b. Second harmonic = 2 × fundamental = 2 × 1 kHz
      = 2 kHz
    c. Seventh harmonic = 7 × fundamental = 7 × 1 kHz
      = 7 kHz

47. 3.5 cm × 10 V/cm = 35 V pk-pk

49. $A_{\text{P(dB)}} = 10 \times \log_{10}\left(\dfrac{25\text{ W}}{23.7\text{ W}}\right)$

$A_{\text{P(dB)}} = 0.232$

Chapter 14

1. d	**11.** b
3. d	**13.** d
5. a	**15.** b
7. b	**17.** d
9. b	**19.** a

(The answers to Communication Skill Questions 21 through 29 can be found in the indicated sections that follow the questions.)

31. Field strength = V/d = 6 V/32 μm = 187.5 kV/m

33. $C = \dfrac{(8.85 \times 10^{-12}) \times k \times A}{d} =$

$\dfrac{(8.85 \times 10^{-12}) \times 2.5 \times 0.008\text{ m}^2}{0.00095\text{ m}} = 186.3\text{ pF}$

35. $X_C = \dfrac{1}{2\pi FC}$:   **a.** $\dfrac{1}{2\pi(1\text{ kHz})\,2\ \mu\text{F}} = 79.6\ \Omega$;

    b. $\dfrac{1}{2\pi(100\text{ Hz})0.01\ \mu\text{F}} = 159.2\text{ k}\Omega$;

    c. $\dfrac{1}{2\pi(17.3\text{ MHz})\,47\ \mu\text{F}}$

37. **a.** $Z = \sqrt{R^2 + X_C^2}, X_C = \dfrac{1}{2\pi fC}$

$= \dfrac{1}{2\pi(20\text{ kHz})\,3.7\ \mu\text{F}} = 2.15\ \Omega,$

$Z = \sqrt{2.7\text{ M}^2 + (2.15)^2} = 2.7\text{ M}\Omega$

    b. $Z = \sqrt{R^2 + X_C^2},$

$X_C = \dfrac{1}{2\pi fc} = \dfrac{1}{2\pi(3\text{ MHz})0.005\ \mu\text{F}} = 10.61\ \Omega,$

    c. $Z = \sqrt{350^2 + 10.61^2} = 350.16\ \Omega$
$Z = \sqrt{R^2 + X_C^2} = \sqrt{8.6\text{ k}^2 + 2.4^2} = 8.6\text{ k}\Omega$

    d. $Z = \sqrt{R^2 + X_C^2} = \sqrt{4.7\text{ k}^2 + 2\text{ k}^2} = 5.1\text{ k}\Omega$

39. **a.** $X_{C_T} = X_{C_1} + X_{C_2} + X_{C_3} = 200\ \Omega + 300\ \Omega +$
$400\ \Omega = 900\ \Omega$

b. $X_{C_T} = \dfrac{X_{C_1} \times X_{C_2}}{X_{C_1} + X_{C_2}} = \dfrac{3.3 \text{ k}\Omega \times 2.7 \text{ k}\Omega}{3.3 \text{ k}\Omega + 2.7 \text{ k}\Omega} = \dfrac{8.91 \text{ M}\Omega^2}{6 \text{ k}\Omega}$

$= 1.485 \text{ k}\Omega$

41. $F = \dfrac{1}{2\pi X_C C} = \dfrac{1}{2\pi(2000 \ \Omega)4.7 \ \mu\text{F}} = 16.93 \text{ Hz}$

43.

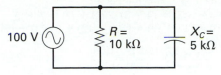

FIGURE 14-43(a)

b. $I_R = V/R = \dfrac{100 \text{ V}}{10 \text{ k}\Omega} = 10 \text{ mA}$

$I_C = V/X_C = 100 \text{ V}/5 \text{ k}\Omega = 20 \text{ mA}$

$I_T = \sqrt{I_R^2 + I_C^2} = \sqrt{(10 \text{ mA})^2 + (20 \text{ mA})^2}$

$= 22.36 \text{ mA}$

$Z = V/I_T = 100 \text{ V}/22.36 \text{ mA} = 4.5 \text{ k}\Omega$

$V_R = V_C = V_S = 100 \text{ V}$

$\theta = \arctan(R/X_C) = \arctan(10 \text{ k}\Omega/5 \text{ k}\Omega) = 63.4°$

45. a. $Z = \sqrt{R^2 + X_C^2} = \sqrt{30^2 + 25^2} = 39.05 \ \Omega$

b. $X_C = \dfrac{1}{2\pi f c} = \dfrac{1}{2\pi(100 \text{ kHz})50 \text{ pF}} = 31.8 \text{ k}\Omega$

$Z = \sqrt{(12\text{k}\Omega)^2 + (31.8 \text{ k}\Omega)^2} = 33.94 \text{ k}\Omega$

c. $C_T = \dfrac{C_1 \times C_2}{C_1 + C_2} = \dfrac{100 \ \mu\text{F} \times 330 \ \mu\text{F}}{100 \ \mu\text{F} + 330 \ \mu\text{F}} = 76.7 \ \mu\text{F},$

$R_T = R_1 + R_2 = 5000 + 4700 = 9.7 \text{ k}\Omega$

$X_C = \dfrac{1}{2 \ \mu f C} = \dfrac{1}{2\pi(10 \text{ kHz})76.7 \ \mu\text{F}} = 207.5 \text{ m}\Omega$

$Z = \sqrt{R^2 + X_C^2} = \sqrt{(9.7 \text{ k}\Omega)^2 + (207.5 \text{ m}\Omega)^2} = 9.7 \text{ k}\Omega$

d. $C_T = C_1 + C_2 = 0.002 \ \mu\text{F} + 0.005 \ \mu\text{F} = 0.007 \ \mu\text{F}$

$X_C = \dfrac{1}{2\pi f c} = \dfrac{1}{2\pi(1 \text{ MHz})0.007 \ \mu\text{F}} = 22.7 \ \Omega$

$Z = \sqrt{R^2 + X_C^2} = \sqrt{(2.5 \text{ k}\Omega)^2 + (22.7 \ \Omega)^2} = 2.5 \text{ k}\Omega$

47. Lead, 90

49. a. $X_C = \dfrac{1}{2\pi f c} = \dfrac{1}{2\pi(45 \text{ kHz}) 2 \ \mu\text{F}} = 1.77 \ \Omega$

$Z = \sqrt{R^2 + X_C^2} = \sqrt{330^2 + 1.77^2} = 330 \ \Omega$

$I = V/Z = 10 \text{ V}/330 \ \Omega = 30.3 \text{ mA}, I_R = I = 30.3 \text{ mA}$

$\theta = \arctan(X_C/R) = \arctan 1.77/330 = 0.3°$

$V_R = I \times R = 30.3 \text{ mA} \times 330 = 9.999 \text{ V}$

$V_C = I \times X_C = 30.3 \text{ mA} \times 1.77 \ \Omega = 53.631 \text{ mV}$

b. $V_R = V_C = V_S = 18 \text{ V}$

$I_R = V/R = \dfrac{18 \text{ V}}{1.2 \text{ k}\Omega} = 15 \text{ mA}$

$I_C = V/X_C = 18 \text{ V}/2.7 \text{ k}\Omega = 6.67 \text{ mA}$

$I_T = \sqrt{I_R^2 + I_C^2} = \sqrt{(15 \text{ mA})^2 + (6.67 \text{ mA})^2} = 16.4 \text{ mA}$

$Z = V/I_T = \dfrac{18 \text{ V}}{16.4 \text{ mA}} = 1.1 \text{ k}\Omega$

$\theta = \arctan(R/X_C) = \arctan \dfrac{1.2 \text{ k}\Omega}{2.7 \text{ k}\Omega} = 23.96°$

Chapter 15

1. c	**11.** d
3. d	**13.** a
5. b	**15.** a
7. d	**17.** b
9. e	**19.** a

(The answers to Communication Skill Questions 21 through 30 can be found in the sections indicated that follow the questions.)

31. a. $0.037 \text{ H} = 37 \times 10^{-3} \text{ H} = 37 \text{ mH}$

b. $1760 \ \mu\text{H} = 1760 \times 10^{-6} \text{ H} = 1.76 \times 10^{-3} \text{ H} = 1.76 \text{ mH}$

c. $862 \text{ mH} = 862 \times 10^{-3} \text{ H} = 0.862 \times 10^0 \text{ H} = 0.862 \text{ H}$

d. $0.256 \text{ mH} = 0.256 \times 10^{-3} \text{ H} = 256 \times 10^{-6} \text{ H} = 256 \ \mu\text{H}$

33. a. $V_{induced} = L \times (\Delta i/\Delta t) = 2 \ \mu\text{H} \times 120 \text{ mA/ms} = 240 \ \mu\text{V}$

b. $V_{induced} = L \times (\Delta i/\Delta t) = 463 \ \mu\text{H} \times 62 \ \mu\text{A}/\mu\text{s} = 28.706 \text{ V}$

c. $V_{induced} = L \times (\Delta i/\Delta t) = 25 \text{ mH} \times 4 \text{ A/s} = 100 \text{ mV}$

35. a. $L_T = \dfrac{L_1 \times L_2}{L_1 + L_2} = \dfrac{12 \text{ mH} \times 8 \text{ mH}}{12 \text{ mH} + 8 \text{ mH}} = \dfrac{96 \ \mu\text{H}^2}{20 \text{ mH}} = 4.8 \text{ mH}$

b. $L_T = \dfrac{1}{(1/L_1) + (1/L_2) + (1/L_3)}$

$= \dfrac{1}{(1/75 \ \mu\text{H}) + (1/34 \ \mu\text{H}) + (1/27 \ \mu\text{H})} = 12.53 \ \mu\text{H}$

37. a. $V_S = \sqrt{V_R^2 + V_L^2} = \sqrt{12^2 + 6^2} = \sqrt{144 + 36}$

$= \sqrt{180} = 13.4 \text{ V}$

b. $I = V_S/Z = 13.4 \text{ V}/14 \text{ k}\Omega = 957.1 \ \mu\text{A}$

c. $\angle = \arctan V_L/V_R = \arctan 2 = 63.4°$

d. $Q = V_L/V_R = 12/6 = 2.$

e. $PF = \cos \theta = 0.448.$

39. $f = X_L/2\pi L = 27 \text{ k}\Omega/2\pi330 \ \mu\text{H} = 13.02 \text{ MHz}$

41. $\tau = \dfrac{L}{R} = \dfrac{400 \text{ mH}}{2 \text{ k}\Omega} = 200 \ \mu\text{s}$

V_L will start at 12 V and then exponentially drop to 0 V

Time	Factor	V_S	V_L
0	1.0	12	12
1 TC	0.365	12	4.416
2 TC	0.135	12	1.62
3 TC	0.05	12	0.6
4 TC	0.018	12	0.216
5 TC	0.007	12	0.084

43. a. $R_T = R_1 + R_2 = 250 + 700 = 950 \ \Omega$

b. $L_T = L_1 + L_2 = 800 \ \mu\text{H} + 1200 \ \mu\text{H} = 2 \text{ mH}$

c. $X_L = 2\pi f L = 2\pi(350 \text{ Hz}) 2 \text{ mH} = 4.4 \ \Omega$

d. $Z = \dfrac{R \times X_L}{\sqrt{R^2 + X_L^2}} = \dfrac{950 \times 4.4}{\sqrt{950^2 + 4.4^2}} = 4.39$

e. $V_{R_T} = V_{L_T} = V_S = 20 \text{ V}$

f. $I_{R_T} = \dfrac{V_S}{R_T} = \dfrac{20 \text{ V}}{950 \ \Omega} = 21 \text{ mA},$

$I_L = \dfrac{V_S}{X_{L_T}} = \dfrac{20 \text{ V}}{4.4 \ \Omega} = 4.5 \text{ A}$

g. $I_T = \sqrt{I_R^2 + I_L^2} = \sqrt{(21 \text{ mA})^2 + (4.5 \text{ A})^2} = 4.5 \text{ A}$

h. $\theta = \arctan(R/X_L) = \arctan 950/4.4 = 89.7°$

i. $P_R = I^2 \times R = (21 \text{ mA})^2 \times 950 \ \Omega = 418.95 \text{ mW}$

$P_X = I^2 \times X_L = (4.5 \text{ A})^2 \times 4.4 \ \Omega = 89.1 \text{ VAR}$

$P_A = \sqrt{P_R^2 + P_X^2} = \sqrt{(418.95 \text{ mW})^2 + (89.1 \text{ VA})^2}$

$= 89.1 \text{ VA}$

j. $PF = P_R/P_A = 418.95 \text{ mW}/89.1 \text{ W} = 0.0047$

45. a. $V_s = \dfrac{N_s}{N_p} \times V_p = 24/12 \times 100 \text{ V} = 200 \text{ V}$

b. $V_s = \dfrac{N_s}{N_p} \times V_p = 250/3 \times 100 \text{ V} = 8.33 \text{ kV}$

c. $V_s = \dfrac{N_s}{N_p} \times V_p = 5/24 \times 100 \text{ V} = 20.83 \text{ V}$

d. $V_s = \dfrac{N_s}{N_p} \times V_p = 120/240 \times 100 \text{ V} = 50 \text{ V}$

47. a. Turns ratio $= N_s/N_p = V_s/V_p = 240\text{ V}/120\text{ V} = 2{:}1$ step up
b. Turns ratio $= N_s/N_p = V_s/V_p = 720\text{ V}/240\text{ V} = 3{:}1$ step up
c. Turns ratio $= N_s/N_p = V_s/V_p = 14\text{ V}/30\text{ V} = 0.467$ (2.14:1) step down
d. Turns ratio $= N_s/N_p = V_s/V_p = 6\text{ V}/24\text{ V} = 0.25$ (4:1) step down

49. $I_s =$ apparent power$/V_s$, $V_s =$ ratio $\times V_p = 1/2 \times 120\text{ V} = 60\text{ V}$
$I_s = 2\text{ kVA}/60\text{ V} = 33.3\text{ A}$
$R_L = V_s/I_s = 60\text{ V}/33.3\text{ A} = 1.8\ \Omega$

Chapter 16

1. b	**9.** c
3. b	**11.** a
5. d	**13.** a
7. c	**15.** d

(The answers to Communication Skill Questions 17 through 29 can be found in the sections indicated that follow the questions.)

31. a. $X_C = \dfrac{1}{2\pi fC} = \dfrac{1}{2\pi(60)0.02\mu F} = 132.6\text{ k}\Omega$

b. $X_C = \dfrac{1}{2\pi fC} = \dfrac{1}{2\pi(60)18\mu F} = 147.4\ \Omega$

c. $X_C = \dfrac{1}{2\pi fC} = \dfrac{1}{2\pi(60)360\text{ pF}} = 7.37\text{ M}\Omega$

d. $X_C = \dfrac{1}{2\pi fC} = \dfrac{1}{2\pi(60)2700\text{ nF}} = 982.4\ \Omega$

e. $X_L = 2\pi fL = 2\pi(60)4\text{ mH} = 1.5\ \Omega$
f. $X_L = 2\pi fL = 2\pi(60)8.18\text{ H} = 3.08\text{ k}\Omega$
g. $X_L = 2\pi fL = 2\pi(60)150\text{ mH} = 56.5\ \Omega$
h. $X_L = 2\pi fL = 2\pi(60)2\text{ H} = 753.98\ \Omega$

33. a. $X_L = 2\pi fL = 2\pi(60\text{ Hz})150\text{ mH} = 56.5\ \Omega$

b. $X_C = \dfrac{1}{2\pi fc} = \dfrac{1}{2\pi(60\text{ Hz})20\ \mu F} = 132.6\ \Omega$

c. $I_R = V/R = 120\text{ V}/270\ \Omega = 444.4\text{ mA}$
d. $I_L = V/X_L = 120\text{ V}/56.5\ \Omega = 2.12\text{ A}$
e. $I_C = V/X_C = 120\text{ V}/132.6\ \Omega = 905\text{ mA}$
f. $I_T = \sqrt{I_R^2 + I_X^2} = \sqrt{(444.4\text{ mA})^2 + (1.215\text{ A})^2} = 1.29\text{ A}$
g. $Z = V/I_T = 120\text{ V}/1.29\text{ A} = 93.02\ \Omega$

h. Resonant frequency $= \dfrac{1}{2\pi\sqrt{LC}}$

$= \dfrac{1}{2\pi\sqrt{150\text{ mH} \times 20\ \mu F}}$

$= 91.89\text{ Hz}$

i. $X_L = 2\pi fL$
$= 6.28 \times 91.89\text{ Hz} \times 150\text{ mH}$
$= 86.54\ \Omega$

$Q = \dfrac{X_L}{R} = \dfrac{86.54\ \Omega}{270\ \Omega} = 0.5769$

j. BW $= \dfrac{f_0}{Q} = \dfrac{91.89\text{ Hz}}{0.5769} = 159.28\text{ Hz}$

35. Using a source voltage of 1 volt:
$Z = V/I_T, I_T = \sqrt{I_R^2 + I_X^2}, I_R = V/R = 1/750 = 1.33\text{ mA}$
$I_L = V/X_L = 1/25 = 40\text{ mA}, I_C = V/X_C = 1/160 = 6.25\text{ mA}$
$I_X = I_L - I_C = 40\text{ mA} - 6.25\text{ mA} = 33.75\text{ mA}$
$I_T = \sqrt{(1.33\text{ mA})^2 + (33.75\text{ mA})^2} = 33.78\text{ mA}$
$Z = 1\text{ V}/33.78\text{ mA} = 29.6\ \Omega$

37. a. Real number $= 25\cos 37° = 19.97$; imaginary number $= 25\sin 37° = 15$, $19.97 + j15$
b. Real number $= 19\cos(-20°) = 17.9$; imaginary number $= 19\sin(-20°) = -6.5$, $17.9 - j6.5$
c. Real number $= 114\cos(-114°) = -46.4$; imaginary number $= 114\sin(-114°) = -104.1$, $-46.4 - j104.1$*
d. Real number $= 59\cos 99° = +9.2$; imaginary number $= 59\sin 99° = 58.3$, $+9.2 + j58.3$

39. a. $(4 + j3) + (3 + j2) = (4 + 3) + (j3 + j2) = 7 + j5$
b. $(100 - j50) + (12 + j9) = (100 + 12) + (-j50 + j9)$
$= 112 - j41$

41. a. $Z_T = 73 - j23$, $\sqrt{73^2 + 23^2} = 76.5$, $\angle = \arctan(23/73)$
$= -17.5°$, $76.5 \angle -17.5°$; $Z_T = 76.5\ \Omega$ at $-17.5°$ phase angle
b. $Z_T = 40 + j15$, $\sqrt{40^2 + 15^2} = 42.7$, $\angle = \arctan(15/40)$
$= 20.6°$, $42.7 \angle 20.6°$; $Z_T = 42.7\ \Omega$ at $20.6°$ phase angle
c. $Z_T = 8\text{ k}\Omega - j3\text{ k}\Omega + j20\text{ k}\Omega = 8\text{ k}\Omega + j17\text{ k}\Omega$,
$\sqrt{(8\text{ k}\Omega)^2 + (17\text{ k}\Omega)^2} = 18.8\text{ k}\Omega$, $\angle =$
$\arctan(17\text{ k}\Omega/8\text{ k}\Omega) = 64.8°$, $18.8\text{ k}\Omega \angle 64.8°$; $Z_T = 18.8\text{ k}\Omega$ at $64.8°$ phase angle

43. a. $Z_T = 47 - j40 + j30 = 47 - j10$, $\sqrt{47^2 + 10^2} = 48.05$,
$\angle = \arctan(-10/47) = -12°$, $Z_T = 48.05 \angle = -12°$

b. $I = \dfrac{V_s}{Z_T} = \dfrac{20 \angle 0°}{48.05 \angle -12°} = \dfrac{20}{48.05} \angle 0 - (-12)°$
$= 416.2\text{ mA} \angle 12°$

c. $V_R = I \times R = 416.2 \angle 12°\text{ mA} \times 47 \angle 0°\ \Omega = 416.2\text{ mA} \times 47\ \Omega \angle (12° + 0°) = 19.56 \angle 12°\text{V}$
$V_C = I \times X_C = 416.2 \angle 12°\text{ mA} \times 40 \angle -90°\ \Omega = 416.2\text{ mA} \times 40\ \Omega \angle (12° - 90°) = 16.65 \angle -78°\text{ V}$
$V_L = I \times X_L = 416.2 \angle 12°\text{ mA} \times 30 \angle 90°\ \Omega = 416.2\text{ mA} \times 30\ \Omega \angle (12° + 90°) = 12.49 \angle 102°\text{ V}$

d. V_C lags I_T by $90°$, V_L leads I by $90°$, V_R is in phase with I.

45. a. $Z_1 = 0 + j27 - j17 = j10$, $\sqrt{0^2 + 10^2} = 10$, $\angle = 90°$,
$Z_1 = 10 \angle 90°$; $Z_2 = 37 - j20 = 42.06$, $\angle -28.39°$;
Z combined = Product/Sum; Product $= 420.59 \angle 61.61°$;
Sum $= 37 - j10 = 38.33 \angle -15.12°$; Z combined $= 10.97 \angle 76.73°$
b. $I_1 = 3.72 \angle -34.48°\text{ A}$; $I_2 = 884 \angle 83.91°\text{ mA}$
c. $I_T = 3.39 \angle -21.21°\text{ A}$
d. $Z_T = 29.52 \angle 21.21°\ \Omega$

47. $V_R = 19.56 \angle 12° = 19.56\cos 12° + j19.56\sin 12°$
$= 19.13 + j4.07$
$V_C = 16.65 \angle -78° = 16.65\cos(-78°) + j16.65\sin(-78°)$
$= 3.46 - j16.29$
$V_L = 12.49 \angle 102° = 12.49\cos 102° + j12.49\sin 102°$
$= -2.6 + j12.22$

49. Easier with use of complex numbers

Chapter 17

1. b	**19.** a
3. d	**21.** d
5. b	**23.** a
7. a	**25.** b
9. a	**27.** b
11. d	**29.** a
13. b	**31.** d
15. d	**33.** c
17. b	

(The answers to Communication Skill Questions 35 through 55 can be found in the sections indicated that follow the questions.)

*These examples were for practice purposes only; real numbers are always positive when they represent impedances.

57. a. Since D_1 is reverse biased, I_F will be 0.

b. $I = (V_S - V_{diode})/R$
$I = (1.5\,V - 0.7\,V)/(2.5\,k\Omega + 127\,k\Omega)$
$I = 14.3\,V/129.5\,k\Omega$
$I = 110.4\,\mu A$

59. a. Polarity correct ($+ \rightarrow$ cathode, $- \rightarrow$ anode)

b. Polarity incorrect ($+ \rightarrow$ anode, $- \rightarrow$ cathode)

c. Polarity incorrect

d. Polarity correct

e. Polarity correct

f. Polarity for both D_1 and D_2 are correct

61. a. $I_S = (V_{in} - V_Z)/R_S = (10\,V - 6.8\,V)/200\,\Omega = 16\,mA$

b. Since the zener diode is forward biased, we will assume a 0.7 V forward voltage drop. Therefore
$I_S = (V_{in} - V_Z)/R_S = (20\,V - 0.7\,V)/570\,\Omega = 33.9\,mA$

c. Since the zener diode is forward biased, we will assume a 0.7 V forward voltage drop. Therefore
$I_S = (V_{in} - V_Z)/R_S = (5\,V - 0.7\,V)/400\,\Omega = 10.75\,mA$

d. Since the input coltage is not large enough to send the zener into its reverse zener breakdown region, the circuit current will be equal to that of the reverse leakage current, which is almost zero.

e. $I_S = (V_{in} - V_Z)/R_S = (6\,V - 4.7\,V)/200\,\Omega = 6.5\,mA$

f. For zener D_1, $I_S = (V_{in} - V_Z)/R_S = (10\,V - 6.8\,V)/220\,\Omega$
$= 14.5\,mA$

For zener D_2 since the input voltage is not large enough to send the zener into its reverse zener breakdown region, the circuit current will be equal to that of the reverse leakage current, which is almost zero.

63. $P_D = I_{ZM} \times V_Z = 6.5\,mA \times 4.7\,V = 30.55\,mW$
A 50 mW zener diode would be adequate in this application.

65. a. Forward biased ($+ \rightarrow$ anode, $- \rightarrow$ cathode)

b. Forward biased

c. D_1 is reverse biased, D_2 is forward biased

67. Yes

69. a. NPN, $x =$ base, $y =$ collector, $z =$ emitter.

b. PNP, $x =$ base, $y =$ collector, $z =$ emitter.

71. a. $I_E = 25\,mA$, $I_C = 24.6\,mA$, $I_B = I_E - I_C = 0.4\,mA$ or $400\,\mu A$

b. $I_B = 600\,\mu A$, $I_C = 14\,mA$, $I_E = I_B + I_C = 14.6\,mA$

c. $I_E = 4.1\,mA$, $I_B = 56.7\,\mu A$, $I_C = I_E - I_B = 4.04\,mA$

73. a. Common-collector circuit, since the input is applied to the base and the output is taken from the emitter (collector is common to both input and output).

b. Common-emitter circuit, since the input is applied to the base and the output is taken from the collector (emitter is common to both input and output).

c. Common-emitter circuit, since the input is applied to the base and the output is taken from the collector (emitter is common to both input and output).

d. Common-emitter circuit, since the input is applied to the base and the output is taken from the collector (emitter is common to both input and output).

75. a. $V_{R_B} = 20\,V - 0.7\,V = 19.3\,V$
$I_B = V_{R_B}/R_B = 19.3\,V/2.5\,M\Omega = 7.72\,\mu A$

b. $I_C = I_B \times B_{DC} = 7.72\,\mu A \times 150 = 1.16\,mA$

c. $V_{R_C} = I_C \times R_C = 1.16\,mA \times 10\,k\Omega = 11.6\,V$
$V_{CE} = V_{CC} - V_{R_C} = 20\,V - 11.6\,V = 8.4\,V$

77. a. $V_B = R_2/(R_1 + R_2) \times V_{CC} = 1\,k\Omega/(6.8\,k\Omega + 1\,k\Omega) \times 20\,V$
$= 2.56\,V$
$V_E = V_B - 0.7\,V = 2.56\,V - 0.7\,V = 1.86\,V$

b. $I_E = V_E/R_E = 1.86\,V/1.1\,k\Omega = 1.7\,mA$
$I_E = I_C = 1.7\,mA$

c. $V_{R_C} = I_C \times R_C = 1.7\,mA \times 4.7\,k\Omega = 8\,V$
V_C or $V_{out} = V_{CC} - V_{R_C} = 20\,V - 8\,V = 12\,V$

d. $V_{CE} = V_{CC} - (V_{R_C} - V_E) = 20\,V - (8\,V + 1.86\,V)$
$= 10.14\,V$

79. Yes, since; $V_{R_C} + V_{CE} + V_{R_E} = V_{CC}$
$8\,V + 10.14\,V + 1.86\,V = 20\,V$

Chapter 18

1. b
3. d
5. c
7. b

(The answers to Communication Skill Questions 9 through 15 can be found in the indicated sections that follow the questions.)

Chapter 19

1. b
3. c
5. d
7. a
9. d

(The answers to Communication Skill Questions 11 through 23 can be found in the indicated sections that follow the questions.)

25. 1001, decimal 9
27. Decimal equivalents are:
  a. 57   **b.** 39   **c.** 06
29. BCD equivalents are:
  a. 0010 0011 0110 0101   **b.** 0010 0100
31. Codes in order are: BCD, gray, excess-3, and ASCII.
33. Stx practice makes perfect etx.

Chapter 20

1. a	**11.** b
3. b	**13.** b
5. b	**15.** c
7. b	**17.** b
9. b	**19.** a

(The answers to Communication Skill Questions 21 through 29 can be found in the indicated sections that follow the questions.)

31. OR gate
33. OR
35. Truth table outputs (Y) should be:

a. 0 (AND)	**c.** 0 (XOR)	**e.** 1 (NOR)
0	1	0
0	1	0
1	0	0
b. 0 (OR)	**d.** 1 (XNOR)	**f.** 0 (NOT)
1	0	1
1	0	
1	1	

Chapter 21

1. c	**11.** c
3. c	**13.** a
5. b	**15.** b
7. c	**17.** c
9. a	**19.** c

(The answers to Communication Skill Questions 21 through 29 can be found in the indicated sections that follow the questions.)

31. a. $Y = (\overline{A} \cdot B) + (B \cdot C) + D$

b. $Y = \overline{(A \cdot B)} + (\overline{A} \cdot \overline{B})$

c. $Y = \overline{(A \cdot B \cdot C)} + (A \cdot B \cdot \overline{C})$

d. $Y = (LIT + NLT) \cdot (SLW + FST)$

33. a.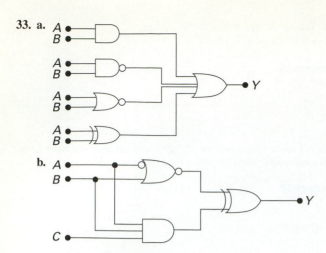

b.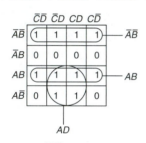

35.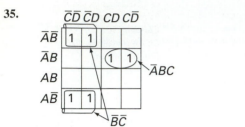

$$Y_2 = \overline{BC} + \overline{A}BC$$

37. $Y_1 = \overline{A}\overline{B}\overline{C}\overline{D} + \overline{A}\overline{B}C\overline{D} + \overline{A}\overline{B}C\overline{D} + \overline{A}\overline{B}CD + A\overline{B}CD$
$\quad\quad A\overline{B}CD + AB\overline{C}\overline{D} + AB\overline{C}D + ABC\overline{D} + ABCD$

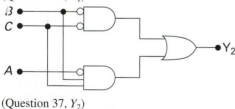

$$Y_1 = \overline{AB} + AB + AD$$
(50 Inputs to 9 Inputs)

39. (Question 35, Y_1)

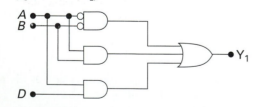

(Question 37, Y_2)

(Question 38, Y_3)

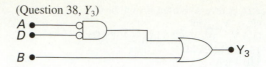

Chapter 22
 1. b
 3. d
 5. a
 7. d
 9. d

(The answers to Communication Skill Questions 11 through 17 can be found in the indicated sections that follow the questions.)

19. a. 1000 0111	**b.** 1100 0010	**c.** 0101 0100
21. a. 0001 0001 0001	**b.** 0001 0101 0101	**c.** 0001 1100

23. $+34 = 0010\ 0010$
$\quad\ -83 = 1010\ 1101$
$\quad\ +83 = 0101\ 0011$

25. a. $+16$	**b.** -66	**c.** -11
27. a. 1010 0110	**b.** 0111 1111	**c.** 0000 0011

Chapter 23
 1. c
 3. b
 5. a
 7. c
 9. b

(The answers to Communication Skill Questions 11 through 19 can be found in the indicated sections that follow the questions.)

21.
23.
25. Choose value and then follow flow chart.

Index

Calculator Key Functions

Decimal-Point key, 31
Reciprocal key, 36
Pi key, 36
Conversion function, 38
Add key, 40
Equals key, 40
Subtract key, 40
Multiply key, 41
Divide key, 41
Round key, 51
Percent key, 52
Change-sign key, 62
Order of evaluation key, 73
Square key, 80
Square root key, 81
Xth root of the Y key, 82
Exponent key, 87
Normal, scientific, engineering
    modes, 89
Convert key, 97
Factor function, 111
Expand function, 112
Solve function key, 123
Sine key, 153
Inverse sine, 153
Cosine key, 155
Inverse cosine, 155
Tangent key, 156
Inverse tangent, 156
Common logarithm, 171
Common antilogarithm, 173
Complex numbers, 519
Converting number bases, 615
Comparing or manipulating bits,
    641

AC meter, 378
AC waves, 352
Accumulator, 715
Active region, 549
Acute angle, 140
Addend, 42
Adding fractions, 7
Adding mixed numbers, 12
Adding polynomials, 132
Adding positive and negative
    numbers, 63
Addition, 41
Adjacent, 151
Algebra, 102
Algorithm, 699
Alphanumeric code, 624
Alphanumeric coding, 226
Alternating current, 344

Ammeter, 195
Amp, 193
Ampere, 193
Amplification, 542
Amplifier, 542
Amplitude, 353
Analog, 592
Analog-to-digital converter, 598
AND expression, 652
AND gate rules, 663
AND gate, 633
Antilogarithm, 173
Apex, 161
Apparent power, 427
Arabic number system, 28
Arc sine, 154
Area, 93
Arithmetic-logic unit, 695
Armstrong, Neil 590
ASCII, 623
Assembly language, 713
Associative law, 660
Associative property of addition,
    132
Associative property of
    multiplication, 132
Augend, 42
Average value, 356
Average, 367
Averages, 53

Babbage, Charles 100
Band-pass filter, 506
Band-stop filter, 507
Bandwidth, 497
Bar graph, 177
Base biasing, 570
Base, 28, 536, 602
BASIC, 715
Basics of algebra, 102
Battery, 199
Bell, Alexander Graham, 385
Bias voltage, 529
Binary addition, 682
Binary arithmetic, 682
Binary coded decimal, 620
Binary division, 687
Binary multiplication, 685
Binary subtraction, 684
Binary, 595, 604
Binomial, 132
Bit, 596
Bleeder current, 316
Bolometer, 218
Boole, George, 648
Boolean algebra, 650
Boot, Henry, 361

Borrowing, 16
Branch current, 280
Breakdown region, 558
Breakdown voltage, 204, 531
Bubble, 637

Calculating in decimal, 39
Calibration, 381
Canceling fractions, 20
Capacitance formula, 404
Capacitance, 400
Capacitive reactance, 415
Capacitive time constant, 410
Capacitor, 400
Capacitors in parallel, 405
Capacitors in series, 406
Carry, 41
Cold resistance, 222
Celsius, 94
Christie, S. H., 300
Circle formula example, 124
Circle graph, 178
Circle, 149
Coding resistors, 223
Coefficients, 132
Collector characteristic curves, 557
Collector, 536
Color code, 224
Common base, 559
Common collector, 564
Common denominator, 8
Common emitter, 553
Common logarithm, 170
Common, 552
Communication, 349
Commutative law, 660
Commutative property of addition,
    132
Commutative property of
    multiplication, 132
Compiler, 713
Complement, 637
Complex numbers, 509
Complex plane, 509
Conductance, 203
Conductor, 201
Cone, 165
Controlled switch, 635
Conventional current, 195
Converting decimal fractions to
    written fractions, 37
Converting to powers of ten, 86
Converting written fractions to
    decimal fractions, 33
Core material, 445
Cosine, 154
Coulomb of charge, 192

Coulomb, Charles, 191
Counter emf, 442
Counting in decimal, 28
Cray, Seymour, 208
Cube root, 81
Current divider, 283
Current in parallel, 280
Current in series circuit, 244
Current in series parallel, 309
Current source, 327
Current, 190
Cutoff frequency, 497
Cutoff, 550
Cylinder, 163

Decibel, 385
Decimal fraction, 30
Decimal number system, 602
Decimal numbers, 28
Decimal point, 30
Degenerative effect, 567
Degree, 151
Degrees Celsius, 94
Degrees Fahrenheit, 94
De Morgan's first theorem, 655
De Morgan's second theorem, 657
Denominator, 4
Dependency notation, 643
Descartes, René, 76
Dielectric constant, 404
Dielectric strength, 204
Difference, 43
Digital circuit, 593
Digital-to-analog converter, 598
Diode, 526
Direct current, 344
Dissipation, 213
Distributive law, 661
Distributive property of
    multiplication, 132
Dividend, 46
Dividing fractions, 20
Dividing mixed fractions, 20
Dividing polynomials, 132
Dividing positive and negative
    numbers, 70
Division, 46
Divisor, 46
Double inversion rule, 664
Double negative, 67
Drawing an angle, 139
du Fay, Charles, 203
Duality theorem, 665
Duty cycle, 366

Edison, Thomas, 438
Effective value, 355

Electric current, 191
Electrical conversions, 95
Electrical equipment, 352
Electrical prefixes, 95
Electrical units, 95
Electrical wave, 349
Electromagnetic wave, 349, 360
Electromotive force, 196
Electron flow, 195
Electron, 190
Electronic equipment, 351
Emitter follower, 567
Emitter, 536
Encoder circuit, 527
Energy, 228
Engineering notation, 88
ENIAC, 708
Equality on both sides of the equal
    sign, 102
Equation, 102
Equilateral triangle, 148
Equivalent resistance, 246
Excess-3 code, 622
Exponent, 78, 703
Expressing positive and negative
    numbers, 61

Factor, 9
Factoring, 110
Fahrenheit, 94
Farad, 400
Faraday, Daniel, 628
Fermat, Pierre, 136
Filament resistor, 220
Five-step series parallel analysis, 311
Fixed voltage divider, 256
Fixed-value resistor, 213
Floating point number system, 89,
    703
Flow chart, 699, 712
Flywheel action, 500
Formula, 102
Fraction bar, 4
Fractions, 4
Franklin, Benjamin, 198
Frequency response curve, 497
Frequency, 358
Frequency-domain analysis, 368
Frustum of cone, 165
Frustum of pyramid, 164
Full time, 367
Fundamental frequency, 369
Fundamental products, 667

Galvani, Luigi, 195
Gate simplification, 669
Gauss, Carl Friedrich, 3
General-purpose resistor, 225
Generator, 347
Geometry, 138
Graph, 175
Gray code, 623
Gray, Stephen, 203

Half-power point, 497
Hardware, 630, 708
Harmonic frequency, 369

Henry, Joseph, 442
Hertz, 358
Hexadecimal, 609
High level language, 713
High-pass filter, 505
Hopper, Grace Murray, 240
Hot resistance, 222
Hypotenuse, 151

Ideal voltage source, 255
Imaginary number, 509
Imaginary power, 427
Impedance ratio, 474
Impedance, 421
Improper fractions, 11
Incandescent lamp, 222
Increment, 29
Inductance formula, 446
Inductance, 442
Inductive reactance, 453
Inductive time constant, 449
Inductor, 443
Inductors in parallel, 447
Inductors in series, 447
Information transfer, 349
Input port, 598
Instruction set, 713
Insulator, 204
Inverse sine, 154
Isosceles triangle, 149

J operator, 509
Joule, James, 228
Junction diode, 526

Karnaugh mapping, 672
Kilowatt-hour, 234
Kinetic energy, 228
Kirchhoff, Gustav, 252
Kirchhoff's current law, 280
Kirchhoff's voltage law, 252

Laser, 342
Lateral surface area, 161
Least significant digit, 603
Leibniz, Wilhelm, 600
Length, 92
Light emitting diode, 534
Line graph, 175
Linear circuit, 593
Linear response, 181
Linear, 215, 373
Load line, 573
Loading, 255, 316
Logarithm, 170
Logarithmic response, 179
Logic gate circuits, 630
Logic gate, 541
Lowest common denominator, 9
Low-pass filter, 505

Machine language, 712
Maiman, Theodore, 342
Mantissa, 703
Marconi, Guglielmo, 351
Maximum power transfer theorem,
    474

Maximum power transfer, 268
Measuring an angle, 138
Measuring power, 233
Measuring resistance, 222
Measuring voltage, 201
Metric prefixes, 90
Metric system, 90
Metric unit of area, 93
Metric unit of length, 92
Metric unit of temperature, 94
Metric unit of volume, 94
Metric unit of weight, 94
Metric units, 90
Microcomputer, 709
Minuend, 43
Mixed number, 11
Mnemonic, 712
Monomial, 132
Most significant digit, 603
Multiplicand, 44
Multiplication, 44
Multiplier, 44
Multiplying fractions, 17
Multiplying mixed fractions, 18
Multiplying polynomials, 132
Multiplying positive and negative
    numbers, 68
Mutual inductance, 465

NAND expression, 656
NAND gate, 638
Napier, Charles, 168
Natural logarithm, 170
Negative charge, 190
Negative ion, 190
Newton, Isaac, 26
Non-linear, 215
NOR expression, 654
NOR gate, 637
Norton's theorem, 335
NOT expression, 650
NOT gate, 636
Numerator, 4

Obtuse angle, 140
Octal, 616
Ohm, Georg, 210
Ohm's law example, 118
Ohm's law, 210
Ohmmeter, 222
Olsen, Kenneth, 680
One's compliment, 690
Opcode, 715
Opposite, 151
Optical resistor, 219
OR expression, 651
OR gate rules, 662
OR gate, 631
Order of operations, 72
Oscilloscope, 378
Output port, 598

Packard, David, 274
Parallel circuit, 276
Parallel RC circuit, 430
Parallel resonance, 499
Parallel RL circuit, 463

Parallel RLC circuits, 488
Parallelogram, 140
Pascal, Blaise, 58
Peak inverse voltage, 531
Peak value, 353
Peak-to-peak value, 354
Percentages, 52
Period, 358
Peripheral, 710
Phase shift, 363
Phase, 362
Phasor, 146
Photo-resistor, 219
Pi, 36
Pie chart, 178
Plain figure, 140
Plotting a graph, 179
Polar coordinates, 510
Polygon, 140
Polygons, 150
Polynomial, 132
Positional weight, 29, 31
Positive and negative numbers,
    60
Positive charge, 190
Positive ion, 190
Potential difference, 196
Potential energy, 228
Potentiometer, 215
Power factor, 428
Power formula example, 122, 128
Power in parallel, 292
Power in series circuit, 265
Power in series parallel, 310
Power transfer, 347
Power, 230
Powers and roots in combination,
    82
Powers of ten in combination,
    87
Powers of ten, 85
Precision resistor, 225
Prime number, 10
Prism, 162
Product, 44
Product-of-sums form, 668
Programmer, 710
Programming language, 712
Proportional, 116
Proto-board, 227
Protractor, 138
Pulse repetition frequency, 370
Pulse repetition time, 370
Pulse wave, 370
Pulse width, 371
Pure binary, 620
Pyramid, 164
Pythagorean theorem, 143

Quadrilateral, 140
Quality factor, 459
Quiescent point, 558
Quotient, 46

Radix, 28, 602
Raising a base number to a higher
    power, 78

Ratios, 47
Reactive power, 427
Real numbers, 509
Reciprocals, 35
Rectangle, 141
Rectangular coordinates, 510
Rectangular wave, 370
Rectifier, 348
Reducing fractions, 13
Remainder, 46
Removing parenthesis, 110
Representing fractions, 4
Reset and carry, 29, 603
Resistance in parallel, 286
Resistance in series circuit, 246
Resistance in series parallel, 305
Resistance power, 426
Resistance, 210
Resistive temperature detector, 219
Resistor, 212
Resonance, 491
Resultant vector, 147
Reverse leakage current, 532
Rhombus, 142
Right-angle triangle, 140
Rise time, 367
RLC circuits, 482
RMS, 355
Root of a number, 80
Rounding off, 49
R-2R ladder, 321
Rules of algebra, 131

Saturation, 550
Sawtooth wave, 375
Scalene triangle, 148
Scientific notation, 88
Selectivity, 503

Self inductance, 440
Series circuit, 242
Series *RC* circuit, 417
Series resonance, 492
Series *RL* circuit, 454
Series *RLC* circuits, 482
Series-parallel circuit, 302
Shockley, William, 480
Sides, 138
Siemens, Werner, 300
Significant places, 51
Sign-magnitude system, 689
Sine of theta, 152
Sine wave, 158, 353
Sine, 152
Software, 630, 710
Solid figure, 161
Square and square root, 114
Square of a number, 79
Square wave, 365
Square, 141
Static, 193
Statistics, 54
Steinmetz, Charles, 188
Story problems, 108
Subscript, 28
Substitution, 124
Subtracting fractions, 15
Subtracting mixed fractions, 15
Subtracting polynomials, 132
Subtracting positive and negative numbers, 65
Subtraction, 43
Subtrahend, 43
Sum, 42
Sum-of-products form, 667
Superposition theorem, 328
Switching regulator, 544
Tangent, 156

Tank circuit, 500
Tapered, 215
Temperature, 94
Terms ratio, 469
Tesla, Nikola, 524
Theorems for DC, 325
Thermal resistor, 216
Thermistor, 219
Thermometry, 218
Theta, 151
Thévenin's theorem, 331
Thin-film detector, 219
Time constant, 411
Time-domain analysis, 368
Tolerance, 214
Total surface area, 161
Transducer, 229, 349
Transformer loading, 468
Transformer ratings, 476
Transformer ratios, 469
Transformer, 347, 467
Transistance, 541
Transistors, 536
Transposing formulas, 115
Transposition, 104
Trapezoid, 143
Treating both sides equally, 103
Triangular wave, 372
Triggering, 381
Trigonometry, 151
Trinomial, 132
True power, 426
Truth table, 631
Tuned circuit, 503
Turing, Alan, 398
Two's complement arithmetic, 695
Two's complement system, 692

Two-state circuits, 630

Variable voltage divider, 261
Variable-value resistor, 215
Varian, Russel and Sigurd, 361
Vector addition, 147
Vector diagram, 147, 419
Vector, 146, 353
Vertex, 138
Vinculum, 4
Volta, Alessandro, 195
Voltage divider biasing, 577
Voltage gain, 544
Voltage in parallel, 278
Voltage in series circuit, 249
Voltage in series parallel, 307
Voltage source, 325
Voltage, 196
Voltmeter, 201
Volume, 94, 161
Von Neuman, John, 706

Watt, James, 230
Wattage rating, 213
Wavelength, 360
Weight, 94
Wheatstone bridge, 317
Wheatstone, Charles, 300
Wireless communication, 349
Word, 614
Work, 228

XNOR expression, 658
XNOR gate, 640
XOR expression, 657
XOR gate, 639

Zener diode, 531
Zener voltage, 532

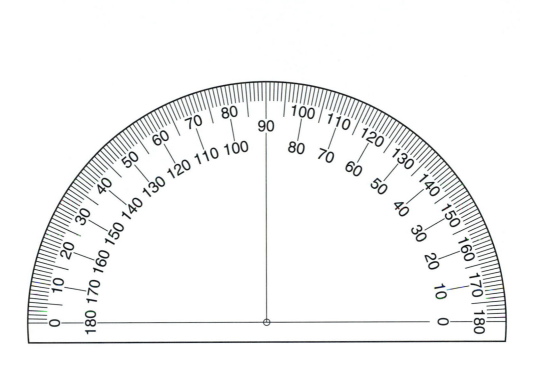